Geophysical Monograph Series

Including
IUGG Volumes
Maurice Ewing Volumes
Mineral Physics Volumes

Geophysical Monograph Series

166 **Back-Arc Spreading Systems: Geological, Biological, Chemical, and Physical Interactions** David M. Christie, Charles Fisher, Sang-Mook Lee, and Sharon Givens (Eds.)

167 **Recurrent Magnetic Storms: Corotating Solar Wind Streams** Bruce Tsurutani, Robert McPherron, Walter Gonzalez, Gang Lu, José H. A. Sobral, and Natchimuthukonar Gopalswamy (Eds.)

168 **Earth's Deep Water Cycle** Steven D. Jacobsen and Suzan van der Lee (Eds.)

169 **Magnetospheric ULF Waves: Synthesis and New Directions** Kazue Takahashi, Peter J. Chi, Richard E. Denton, and Robert L. Lysak (Eds.)

170 **Earthquakes: Radiated Energy and the Physics of Faulting** Rachel Abercrombie, Art McGarr, Hiroo Kanamori, and Giulio Di Toro (Eds.)

171 **Subsurface Hydrology: Data Integration for Properties and Processes** David W. Hyndman, Frederick D. Day-Lewis, and Kamini Singha (Eds.)

172 **Volcanism and Subduction: The Kamchatka Region** John Eichelberger, Evgenii Gordeev, Minoru Kasahara, Pavel Izbekov, and Johnathan Lees (Eds.)

173 **Ocean Circulation: Mechanisms and Impacts—Past and Future Changes of Meridional Overturning** Andreas Schmittner, John C. H. Chiang, and Sidney R. Hemming (Eds.)

174 **Post-Perovskite: The Last Mantle Phase Transition** Kei Hirose, John Brodholt, Thorne Lay, and David Yuen (Eds.)

175 **A Continental Plate Boundary: Tectonics at South Island, New Zealand** David Okaya, Tim Stem, and Fred Davey (Eds.)

176 **Exploring Venus as a Terrestrial Planet** Larry W. Esposito, Ellen R. Stofan, and Thomas E. Cravens (Eds.)

177 **Ocean Modeling in an Eddying Regime** Matthew Hecht and Hiroyasu Hasumi (Eds.)

178 **Magma to Microbe: Modeling Hydrothermal Processes at Oceanic Spreading Centers** Robert P. Lowell, Jeffrey S. Seewald, Anna Metaxas, and Michael R. Perfit (Eds.)

179 **Active Tectonics and Seismic Potential of Alaska** Jeffrey T. Freymueller, Peter J. Haeussler, Robert L. Wesson, and Göran Ekström (Eds.)

180 **Arctic Sea Ice Decline: Observations, Projections, Mechanisms, and Implications** Eric T. DeWeaver, Cecilia M. Bitz, and L.-Bruno Tremblay (Eds.)

181 **Midlatitude Ionospheric Dynamics and Disturbances** Paul M. Kintner, Jr., Anthea J. Coster, Tim Fuller-Rowell, Anthony J. Mannucci, Michael Mendillo, and Roderick Heelis (Eds.)

182 **The Stromboli Volcano: An Integrated Study of the 2002–2003 Eruption** Sonia Calvari, Salvatore Inguaggiato, Giuseppe Puglisi, Maurizio Ripepe, and Mauro Rosi (Eds.)

183 **Carbon Sequestration and Its Role in the Global Carbon Cycle** Brian J. McPherson and Eric T. Sundquist (Eds.)

184 **Carbon Cycling in Northern Peatlands** Andrew J. Baird, Lisa R. Belyea, Xavier Comas, A. S. Reeve, and Lee D. Slater (Eds.)

185 **Indian Ocean Biogeochemical Processes and Ecological Variability** Jerry D. Wiggert, Raleigh R. Hood, S. Wajih A. Naqvi, Kenneth H. Brink, and Sharon L. Smith (Eds.)

186 **Amazonia and Global Change** Michael Keller, Mercedes Bustamante, John Gash, and Pedro Silva Dias (Eds.)

187 **Surface Ocean–Lower Atmosphere Processes** Corinne Le Quèrè and Eric S. Saltzman (Eds.)

188 **Diversity of Hydrothermal Systems on Slow Spreading Ocean Ridges** Peter A. Rona, Colin W. Devey, Jérôme Dyment, and Bramley J. Murton (Eds.)

189 **Climate Dynamics: Why Does Climate Vary?** De-Zheng Sun and Frank Bryan (Eds.)

190 **The Stratosphere: Dynamics, Transport, and Chemistry** L. M. Polvani, A. H. Sobel, and D. W. Waugh (Eds.)

191 **Rainfall: State of the Science** Firat Y. Testik and Mekonnen Gebremichael (Eds.)

192 **Antarctic Subglacial Aquatic Environments** Martin J. Siegert, Mahlon C. Kennicut II, and Robert A. Bindschadler

193 **Abrupt Climate Change: Mechanisms, Patterns, and Impacts** Harunur Rashid, Leonid Polyak, and Ellen Mosley-Thompson (Eds.)

194 **Stream Restoration in Dynamic Fluvial Systems: Scientific Approaches, Analyses, and Tools** Andrew Simon, Sean J. Bennett, and Janine M. Castro (Eds.)

195 **Monitoring and Modeling the *Deepwater Horizon* Oil Spill: A Record-Breaking Enterprise** Yonggang Liu, Amy MacFadyen, Zhen-Gang Ji, and Robert H. Weisberg (Eds.)

196 **Extreme Events and Natural Hazards: The Complexity Perspective** A. Surjalal Sharma, Armin Bunde, Vijay P. Dimri, and Daniel N. Baker (Eds.)

197 **Auroral Phenomenology and Magnetospheric Processes: Earth and Other Planets** Andreas Keiling, Eric Donovan, Fran Bagenal, and Tomas Karlsson (Eds.)

198 **Climates, Landscapes, and Civilizations** Liviu Giosan, Dorian Q. Fuller, Kathleen Nicoll, Rowan K. Flad, and Peter D. Clift (Eds.)

199 **Dynamics of the Earth's Radiation Belts and Inner Magnetosphere** Danny Summers, Ian R. Mann, Daniel N. Baker, and Michael Schulz (Eds.)

200 **Langrangian Modeling of the Atmosphere** John Lin, Dominik Brunner, Christopher Gerbig, Andreas Stohl, Ashok Luhar, Peter Webley

Geophysical Monograph 201

Modeling the Ionosphere-Thermosphere System

Joseph Huba
Robert Schunk
George Khazanov
Editors

American Geophysical Union
Washington, DC

Published under the aegis of the AGU Books Board

Kenneth R. Minschwaner, Chair; Gray E. Bebout, Kenneth H. Brink, Jiasong Fang, Ralf R. Haese, Yonggang Liu, W. Berry Lyons, Laurent Montési, Nancy N. Rabalais, Todd C. Rasmussen, A. Surjalal Sharma, David E. Siskind, Rigobert Tibi, and Peter E. van Keken, members.

Library of Congress Cataloging-in-Publication Data

Modeling the ionosphere-thermosphere system / Joseph Huba, Robert Schunk, George Khazanov, editors.
  pages cm.—(Geophysical monograph, ISSN 0065-8448 ; 201)
  Includes bibliographical references and index.
  ISBN 978-0-87590-491-7

 1. Ionosphere–Mathematical models. 2. Thermosphere–Mathematical models. 3. Atmospheric thermodynamics. 4. Dynamic meteorology. I. Huba, J. D. (Joseph D.), 1950– editor of compilation. II. Schunk, R. W. (Robert W.), editor of compilation. III. Khazanov, G. V. (Georgii Vladimirovich), editor of compilation.

QC881.2.I6M62 2014

551.51–dc23

2013050020

  ISBN: 978-0-87590-490-0
  ISSN: 0065-8448

Cover Image: Ash plume from Augustine Volcano on 30 January 2006 during its eruptive stage. Photograph of the plume at 13:09 AKST (22:09 UTC). Photograph credit: Game McGimsey. Image courtesy of Alaska Volcano Observatory/United States Geological Survey. (inset) PUFF volcanic ash Lagrangian Dispersion Particle Model (LDPM) at 22:09 UTC with ash particles indicated by altitude above sea level. Graph courtesy of Peter Webley, Geophysical Institute, University of Alaska Fairbanks.

Copyright 2013 by the American Geophysical Union
2000 Florida Avenue, N.W.
Washington, DC 20009

Figures, tables and short excerpts may be reprinted in scientific books and journals if the source is properly cited.

Authorization to photocopy items for internal or personal use, or the internal or personal use of specific clients, is granted by the American Geophysical Union for libraries and other users registered with the Copyright Clearance Center (CCC). This consent does not extend to other kinds of copying, such as copying for creating new collective works or for resale. The reproduction of multiple copies and the use of full articles or the use of extracts, including figures and tables, for commercial purposes requires permission from the American Geophysical Union. Geopress is an imprint of the American Geophysical Union.

Printed in the United States of America.

CONTENTS

Preface
Joseph D. Huba, Robert W. Schunk, and George V. Khanzanov vii

Introduction
Joseph D. Huba, Robert W. Schunk, and George V. Khanzanov 1

Section I: Physical Processes

Ionosphere-Thermosphere Physics: Current Status and Problems
R. W. Schunk 3

Physical Characteristics and Modeling of Earth's Thermosphere
Tim Fuller-Rowell 13

Solar Cycle Changes in the Photochemistry of the Ionosphere and Thermosphere
P. G. Richards 29

Energetics and Composition in the Thermosphere
A. G. Burns, W. Wang, S. C. Solomon, and L. Qian 39

Section II: Numerical Methods

Numerical Methods in Modeling the Ionosphere
J. D. Huba and G. Joyce 49

Ionospheric Electrodynamics Modeling
A. D. Richmond and A. Maute. 57

Section III: IT Models

The NCAR TIE-GCM: A Community Model of the Coupled Thermosphere/Ionosphere System
*Liying Qian, Alan G. Burns, Barbara A. Emery, Benjamin Foster, Gang Lu, Astrid Maute,
Arthur D. Richmond, Raymond G. Roble, Stanley C. Solomon, and Wenbin Wang* 73

The Global Ionosphere-Thermosphere Model and the Nonhydrostatics Processes
Yue Deng and Aaron J. Ridley 85

Traveling Atmospheric Disturbance and Gravity Wave Coupling in the Thermosphere
L. C. Gardner and R. W. Schunk 101

Air Force Low-Latitude Ionospheric Model in Support of the C/NOFS Mission
*Yi-Jiun Su, John M. Retterer, Ronald G. Caton, Russell A. Stoneback, Robert F. Pfaff,
Patrick A. Roddy, and Keith M. Groves* 107

Long-Term Simulations of the Ionosphere Using SAMI3
S. E. Mcdonald, J. L. Lean, J. D. Huba, G. Joyce, J. T. Emmert, and D. P. Drob 119

Section IV: Validation of IT Models

Comparative Studies of Theoretical Models in the Equatorial Ionosphere
Tzu-Wei Fang, David Anderson, Tim Fuller-Rowell, Rashid Akmaev, Mihail Codrescu, George Millward, Jan Sojka, Ludger Scherliess, Vince Eccles, John Retterer, Joe Huba, Glenn Joyce, Art Richmond, Astrid Maute, Geoff Crowley, Aaron Ridley, and Geeta Vichare 133

Systematic Evaluation of Ionosphere/Thermosphere (IT) Models: CEDAR Electrodynamics Thermosphere Ionosphere (ETI) Challenge (2009–2010)
J. S. Shim, M. Kuznetsova, L. Rastätter, D. Bilitza, M. Butala, M. Codrescu, B. A. Emery, B. Foster, T. J. Fuller-Rowell, J. Huba, A. J. Mannucci, X. Pi, A. Ridley, L. Scherliess, R. W. Schunk, J. J. Sojka, P. Stephens, D. C. Thompson, D. Weimer, L. Zhu, D. Anderson, J. L. Chau, and E. Sutton 145

Section V: IT Coupling: Above and Below

Aspect of Coupling Processes in the Ionosphere and Thermosphere
R. A. Heelis 161

Use of NOGAPS-ALPHA as a Bottom Boundary for the NCAR/TIEGCM
David E. Siskind and Douglas P. Drob 171

WACCM-X Simulation of Tidal and Planetary Wave Variability in the Upper Atmosphere
H.-L. Liu 181

Inductive-Dynamic Coupling of the Ionosphere With the Thermosphere and the Magnetosphere
P. Song and V. M. Vasyliūnas 201

Section VI: Equatorial Ionospheric Processes

Ionospheric Irregularities: Frontiers
D. L. Hysell, H. C. Aveiro, and J. L. Chau 217

Three-Dimensional Numerical Simulations of Equatorial Spread F: Results and Diagnostics in the Peruvian Sector
H. C. Aveiro and D. L. Hysell 241

Density and Temperature Structure of Equatorial Spread F Plumes
J. Krall and J. D. Huba 251

Low-Latitude Ionosphere and Thermosphere: Decadal Observations From the CHAMP Mission
Claudia Stolle and Huixin Liu 259

Section VII: Data Assimilation

Upper Atmosphere Data Assimilation With an Ensemble Kalman Filter
Tomoko Matsuo 273

Scientific Investigation Using IDA4D and EMPIRE
G. S. Bust and S. Datta-Barua 283

Section VIII: Applications

Customers and Requirements for Ionosphere Products and Services
Rodney Viereck, Joseph Kunches, Mihail Codrescu, and Robert Steenburgh 299

Model-Based Inversion of Auroral Processes
Joshua Semeter and Matthew Zettergren 309

AGU Category Index 323

Index 325

PREFACE

The importance of large-scale numerical models to understand the complex dynamics of the ionosphere/thermosphere (IT) system has been recognized for over three decades. Many ionosphere and thermosphere models have been developed, both as separate and coupled models; they have been used to investigate IT dynamics and compare model results to observational data. However, until a few years ago, there have been very few (if any) conference sessions or workshops devoted solely to the development and understanding of computational IT models.

To address this problem, a session on ionosphere/thermosphere modeling was organized by myself, Aaron Ridley, and Bob Schunk at the 2009 NSF CEDAR workshop held in Santa Fe, New Mexico. The session description was as follows:

> The workshop will focus on IT modeling of the low solar activity (solar minimum or quiet) time, low- to mid-latitude ionosphere. It is hoped that a description of each model will be presented, highlighting (1) basic equations actually solved, (2) numerical techniques, (3) strong and weak points (both physics and numerics), i.e., the good, the bad, and the ugly, and (4) simulation results from a specified day. Results from the different studies can be compared and an ensemble average could be presented and compared to data. Finally, issues that need to be resolved to improve models could be addressed.

The session was extremely successful (i.e., well attended with ample discussion; perhaps, in part, because of a favorable time slot early in the week). Given the enthusiasm for the topic, Bob Schunk suggested we hold a Chapman Conference on IT modeling. I agreed to look into the matter and subsequently submitted a proposal to AGU requesting a Chapman Conference with myself, Bob Schunk, and Aaron Ridley as the conveners. The proposal was accepted, and we held a Chapman Conference on "Modeling the Ionosphere/Thermosphere System" in Charleston, South Carolina, on 9–12 May 2011.

This monograph is an outgrowth of the conference and represents a compilation of different aspects of modeling the IT system. The papers include tutorials on basic ionosphere/thermosphere physics, descriptions of numerical methods and models, and applications to important ionospheric phenomena (e.g., onset and evolution of irregularities) and space weather (e.g., data assimilation). As such, this book serves to provide a basic introduction to IT modeling and to make the IT community aware of the strengths, as well as limitations, of current modeling capabilities and the need for future development.

J.D. Huba
Naval Research Laboratory

Introduction

The science focus of the monograph is the physics of the coupled ionosphere/thermosphere (IT) system. This system is controlled largely by local ion-neutral processes, but there can be strong forcings from below (e.g., tides, gravity waves, and upper atmosphere winds) and above (e.g., solar EUV, high-latitude heating from precipitating electrons, and region 1 and 2 current systems) that impact its behavior. Thus, it is not an isolated system but can be thought of as a transition layer between the Earth's atmosphere and space; viewed from this vantage point, it is clear that it plays a vital role in forecasting space weather.

Given the complexity of the IT system, large-scale computational models of the ionosphere and thermosphere are required to provide a basic understanding of the key physical processes that govern the system, as well as to provide a quantitative description of its behavior that can be compared to observational data. Such models have been developed and are being used extensively to understand and model the IT system, as well as to aid in the development of space weather operational systems. The objective of the monograph is to provide the IT community with the following: (1) a basic description of IT models including the equations that are solved and the numerical methods and algorithms used, (2) examples of applications to the IT system with comparisons to data, (3) assessment of strengths and weaknesses of the models, (4) test simulations that elucidate those strengths and weaknesses, and (5) identification of future efforts to improve the IT modeling capability.

The monograph is divided into the following sections: (1) Physical Processes and Numerical Methods, (2) Ionosphere/Thermosphere Models, (3) Response From Forcings Below and Above, (4) Ionospheric Irregularities, (5) Data Assimilation Models, (6) Metrics and Validation, and (7) Space Weather and the Future. Each section contains papers that describe the current state of research in these areas, as well as providing insight into future development of models to improve our understanding of the ionosphere/thermosphere system.

J. D. Huba, Naval Research Laboratory, Plasma Physics Division, Code 6790, Washington, DC 20375-5320, USA. (huba@ppd.nrl.navy.mil)

G. V. Khazanov, NASA/GSFC, code 673, 8800 Greenbelt Rd, Greenbelt, MD 20771, USA.

R. W. Schunk, Utah State University, Center for Atmospheric and Space Sciences, Utah State University, 4405 Old Main Hill, Logan, UT 84322-4405, USA.

Ionosphere-Thermosphere Physics: Current Status and Problems

R. W. Schunk

Center for Atmospheric and Space Sciences, Utah State University, Logan, Utah, USA

The ionosphere-thermosphere (I-T) system is a highly dynamic, nonlinear, and complex medium that varies with altitude, latitude, longitude, time, season, solar cycle, and geomagnetic activity. Despite its complex nature, significant progress has been made during the last three decades in modeling the global I-T system. The climatology of the system has been clearly established, and the global I-T models have been able to reproduce the major I-T features. However, the global I-T models have been less successful in modeling weather features, and even with regard to climatology, there has been limited quantitative success when comparing global I-T models with measurements. The problem with the global models is that they are usually based on simple mathematical formulations, the model resolutions are coarse, the models contain uncertain parameters, the coupling between the I-T models is incomplete, and there is missing physics in all of the global models. Here the focus is on providing examples of the missing physics and how it affects the ionosphere and/or thermosphere.

1. INTRODUCTION

The ionosphere-thermosphere (I-T) system is a highly dynamic and complex medium that varies significantly, and this variation is particularly strong during geomagnetic storms and substorms. The complex nature of the I-T system results primarily from the fact that it is an open and externally driven system. It is subjected to solar UV/EUV radiation that varies continuously, and it exchanges mass, momentum, and energy with the lower atmosphere and magnetosphere. At high-latitudes, plasma convection, particle precipitation, and Joule heating are the main sources of momentum and energy for the I-T system, and all of the global I-T models include these processes. However, if these drivers are not properly/rigorously described, then the I-T model simulations can display significant errors. Despite this problem, the physics underlying the I-T climatology has been clearly established, and the global I-T models have been able to reproduce the major I-T features. However, the I-T models have been less successful in modeling weather features, especially when attempting long-term forecasts.

In addition to the need to properly describe the drivers of the I-T system, there are other problems connected with the global I-T models that need to be addressed if more reliable specifications and forecasts are desired. Some of the problems are that the coupled global models are usually based on relatively simple mathematical formulations, the spatial and temporal resolutions are coarse, many of the parameters in the models are uncertain, the coupling between the models is incomplete, and there is missing physics in all of the global models.

If the ionosphere simulated by a global I-T model is not correct, then the resulting thermosphere will be wrong and vice versa. This problem can be illustrated with the aid of the National Center for Atmospheric Research (NCAR) thermosphere-ionosphere nested grid (TING) model and the USU GAIM-GM data assimilation model [*Jee et al.*, 2007, 2008]. The NCAR TING model was run in its "standard" coupled mode for the period 1–4 April 2004, which contained both quiet and disturbed periods. The Utah State University Global Assimilation of Ionospheric Measurements - Gauss Markov (USU GAIM-GM) data assimilation model was run

Modeling the Ionosphere-Thermosphere System
Geophysical Monograph Series 201
© 2013. American Geophysical Union. All Rights Reserved.
10.1029/2012GM001351

for the same period using slant TEC from ground receivers, bottomside N_e profiles from ionosondes, and in situ N_e densities along DMSP satellite orbits. As expected, the TING and GAIM-GM ionospheres were significantly different, particularly during disturbed times. Since the GAIM-GM model results were consistent with the available measurements, its reconstructed ionosphere is expected to be more realistic than that obtained from the coupled I-T TING ionosphere. To get a feel for what a different, and more realistic, ionosphere would do to the TING thermosphere, the TING model was rerun with the GAIM-GM ionosphere supplied to it at each TING time step in order to see the effect on the thermosphere of using a different ionosphere. There were large neutral wind, temperature, and composition differences when the GAIM-GM ionosphere was used in place of the self-consistent TING ionosphere, with T_n increases as large as 40% (409 K).

In the GAIM-TING study described above, the main problem with the "standard" TING simulation was probably related to the use of empirical plasma convection and particle convection models for the high-latitude drivers. Empirical models are not capable of describing high-latitude weather features, and the uncertainty in the high-latitude drivers can produce the largest errors in global I-T simulations, particularly during geomagnetic storms.

In addition, concerted efforts have been made to compare global, physics-based I-T models. In the Equatorial PRIMO (Problems Related to Ionospheric Models and Observations) study [*Fang et al.*, 2013], 12 models were compared for the same geophysical conditions in order to see how well the models reproduced the equatorial ionization anomaly (EIA). N_mF_2 versus latitude was compared at selected times. Typically, the spread in model results was more than a factor of 2 and was as large as a factor of 5. In general, the performance of the coupled models was worse than the stand-alone models; the coupled models had difficulty in describing the latitudinal variation of the EIA. There was also a coupling energetics and dynamics of atmospheric regions challenge for a systematic quantitative comparison of physics-based I-T models with observations; eight global models were evaluated. Nine events (two strong and four moderate storms, three quiet periods), three parameters (N_mF_2, h_mF_2, vertical drifts), and all latitudes were considered [*Shim et al.*, 2011]. As expected, no model ranked the best when all events, parameters, and latitudes were taken together. The physics-based I-T models frequently displayed significant differences from each other and from the data.

As noted above, there are many reasons why the global physics-based I-T models have problems. Here the focus will be on "some" of the physics that is missing in the global physics-based models, and therefore, this study is by no means complete. Other issues, such as instabilities, turbulence, uncertain parameters, numerical techniques, etc., will be addressed in other papers in this monograph. Ten topics relevant to missing physics have been selected as examples. Some of the topics are primarily important in local regions and, therefore, are relevant to local weather, whereas others are important for global I-T weather simulations. Subsections 2.6, 2.7, and 2.8 provide examples of the problems that arise when relatively simple mathematical formulations are used in global modeling.

2. MISSING PHYSICS IN GLOBAL PHYSICS-BASED IONOSPHERE-THERMOSPHERE MODELS

In what follows, some of the physical processes that are not included in most of the global physics-based I-T models will be highlighted. The physics may not be included in the I-T models for several reasons: there are insufficient data to warrant its inclusion, it is only applicable in a local geographical domain, the global model resolution is too coarse to incorporate the physics, it is too difficult to include it, an entirely new I-T model needs to be developed to include the physics, etc. With this in mind, it should be noted that a major advance in I-T modeling has been made during the last three decades, and the next major advance will come when the missing physics is included in the global physics-based I-T models.

2.1. Polar Wind and Auroral Ion Outflow

Figure 1 shows processes that affect the polar wind and auroral ion outflow at high latitudes. Most of these processes have been included in recent ionosphere-polar wind simulations [*Barakat and Schunk*, 2006], but the continual loss of plasma due to the polar wind and energetic ion outflow is not taken into account in the global I-T models. Typically, the upper boundary condition adapted in these global models allows the plasma to flow upward, when the electron and ion temperatures increase, and then downward, when the temperatures decrease, so that there is no net loss of plasma. However, the continual loss of plasma due to the polar wind and auroral ion outflow is significant and should have an appreciable effect on the I-T system. The H^+ outflow varies from about 1 to 5×10^8 cm^{-2} s^{-1}, and the O^+ outflow can be as large as $1-2 \times 10^9$ cm^{-2} s^{-1} in the auroral oval and during geomagnetic storms [cf. *Schunk*, 2007]. Unfortunately, the outflow is not uniform; there are propagating and stationary polar wind jets, polar wind tongues that extend across the polar cap, pulsating geomagnetic storms, flickering aurora, auroral arcs, etc. The nonuniform and continuous plasma outflow needs to be taken into account in the global I-T models if more reliable model predictions are desired.

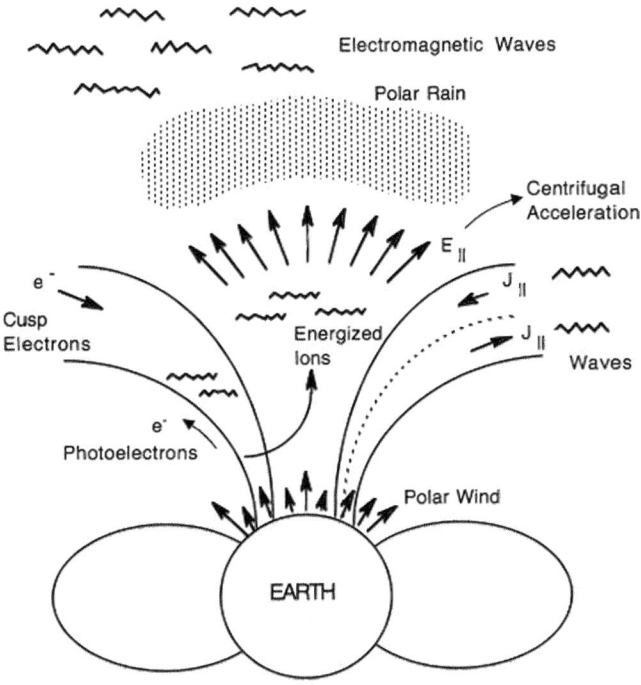

Figure 1. A schematic diagram showing the processes that affect the polar wind and energetic ion outflow from the ionosphere at high latitudes. From *Schunk and Sojka* [1997].

2.2. Downward Electron Heat Flow in the Polar Cap

As the polar wind plasma flows up and out of the topside ionosphere, it also interacts with the overlying polar rain. The energy gained by the polar wind electrons from their interaction with the hot polar rain electrons is subsequently conducted down into the underlying ionosphere, which acts to increase ionospheric electron temperatures [*Schunk et al.*, 1986]. The elevated electron temperatures then affect the ion temperatures and densities. Deductions based on model-measurement comparisons indicate that the downward electron heat flux varies from 0.5 to 1.5×10^{10} eV cm^{-2} s^{-1} over a range of solar cycle, seasonal, and geomagnetic activity conditions [*Bekerat et al.*, 2007]. Recently, time-dependent ionosphere model [*Schunk*, 1988; *Sojka*, 1989] simulations have been conducted of the effect that downward electron heat flows have on the high-latitude ionosphere [*David et al.*, 2011]. Three topside electron heat flux values were adopted in three separate simulations (0.0, 0.5, and 1.5×10^{10} eV cm^{-2} s^{-1}). Relative to the no heat flux case, the largest downward electron heat flow produced N_mF_2 changes of up to a factor of 10 in some regions of the polar cap. This effect is not included in most of current global I-T models.

2.3. Thermoelectric Heat Flow in Return Current Regions

In the ionosphere, the flow of heat is usually described by thermal conduction. In this case, $\mathbf{q} = -\lambda \nabla T$, where $\mathbf{q}$ is the heat flow vector, λ is the thermal conductivity, and T is the temperature. However, an electron heat flow can occur in response to both an electron temperature gradient (thermal conduction) and an electron current (thermoelectric heat flow). Therefore, in auroral return current regions, the electron heat flow along geomagnetic field lines is given by $\mathbf{q} = -\lambda \nabla T - \beta \mathbf{J}$, where β is the thermoelectric coefficient, and $\mathbf{J}$ is the field-aligned ionospheric return current. *Schunk et al.* [1987] studied the effect of ionospheric return currents on auroral electron temperatures for different seasonal, solar cycle, and upper boundary conditions. They found that thermoelectric heat flow is important for current densities greater than 10^{-5} A m^{-2} and that thermoelectric heat flow corresponds to an upward transport of electron energy. The upward transport of energy can result in electron temperatures that decrease with altitude, as shown in Figure 2. It is apparent that thermoelectric heat flow can be significant, but it is not included in the existing global I-T models.

2.4. Ion Temperature Anisotropy

When the convection electric field in the ionosphere is greater than about 50 mV m^{-1}, two processes occur. First, there is a rapid conversion of O$^+$ into NO$^+$, with the result that NO$^+$ becomes an important ion in the F region [*Schunk et al.*, 1975]. This rapid conversion is a consequence of the energy dependence of the O$^+$ + N$_2$ chemical reaction, and this process is included in all (or nearly all) of the global I-T models. However, in addition to the conversion of O$^+$ into NO$^+$, the ion temperature becomes anisotropic with the perpendicular temperature ($T_{i\perp}$) greater than the parallel temperature ($T_{i\parallel}$). Therefore, $T_{i\parallel}$ should be used in the ion momentum equation along the magnetic field, not T_i. Since T_i is greater than $T_{i\parallel}$, the use of T_i in the ion momentum equation results in an overestimation of the plasma density scale height above the F region peak (Figure 3). For a 100 mV m^{-1} electric field, the electron density at 600 km can be more than a factor of 2 too large if T_i is used in the momentum equation instead of $T_{i\parallel}$. This ion temperature anisotropy is probably not taken into account in most of the global I-T models.

2.5. Subauroral Red (SAR) Arcs

SAR arcs correspond to 6300 Å emission that is confined to a narrow latitudinal region just equatorward of the auroral oval [cf. *Schunk and Nagy*, 2009]. The emission occurs during elevated magnetic activity and can be seen in both

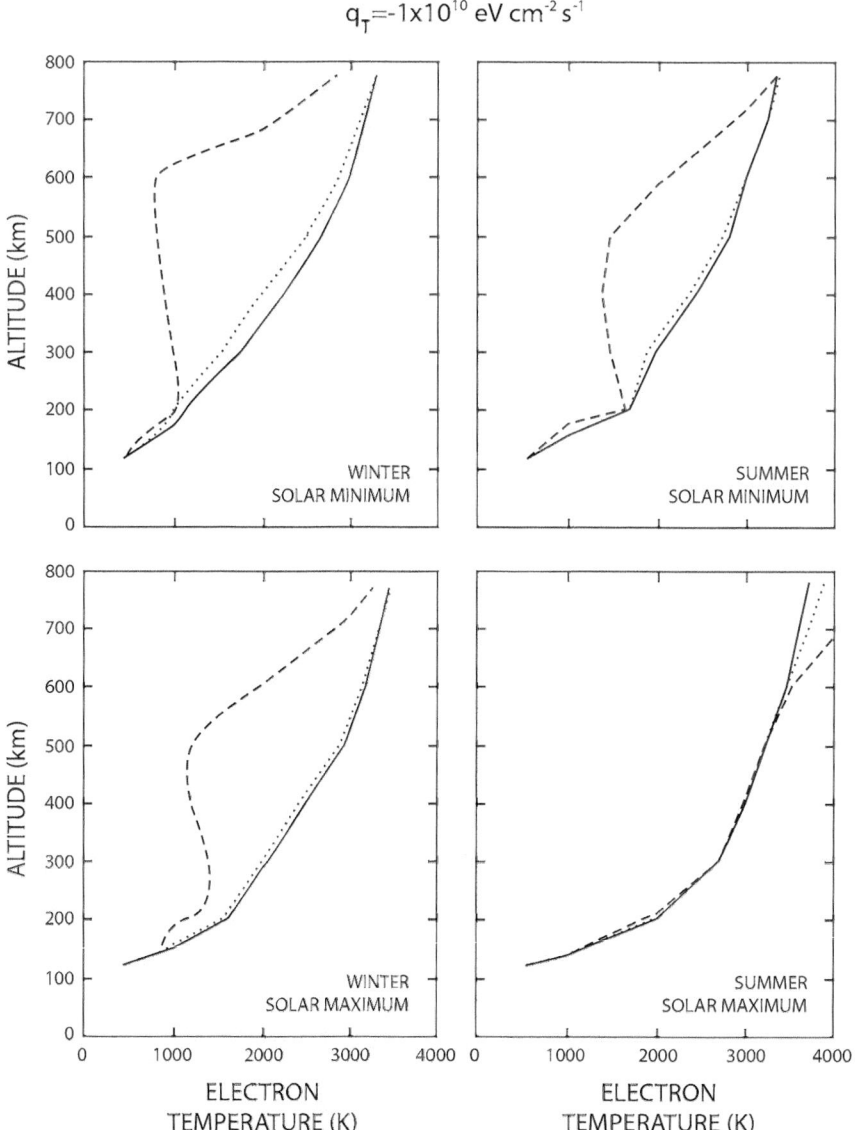

Figure 2. Electron temperature profiles for three values of the field-aligned auroral return current for winter and summer conditions at both solar minimum and maximum. The field-aligned current values are 0 (solid curves), -1×10^{-5} (dotted curves), and -5×10^{-5} (dashed curves) A m^{-2}. An upper boundary (800 km) heat flux of -1×10^{10} eV cm^{-2} s^{-1} was used for these simulations to account for the interaction of the ionospheric electrons with the hot polar rain electrons. From *Schunk et al.* [1987].

hemispheres and at all longitudes. The peak emission rate typically is localized in the 350–400 km altitude range. The emission originates from the interaction of the ring current with plasma on outer plasmaspheric flux tubes. Through Coulomb collisions and wave-particle interactions, energy is transferred from the ring current to the thermal electrons, and then, the energy is conducted down into the ionosphere. The elevated electron temperature is then capable of exciting the oxygen red line.

SAR arcs are useful for illustrating an important process that is not included in all of the global coupled I-T models. This process involves N_2 vibrational excitation. In addition to exciting the oxygen red line, elevated electron temperatures can increase the population of vibrationally excited N_2, which then acts to increase the rate of the $O^+ + N_2 \Rightarrow NO^+ + N$ reaction. The net result can be a rapid conversion of O^+ into NO^+. Figure 4 shows the possible effect of vibrationally excited N_2 on the N_e profile via the associated O^+ to NO^+ conversion

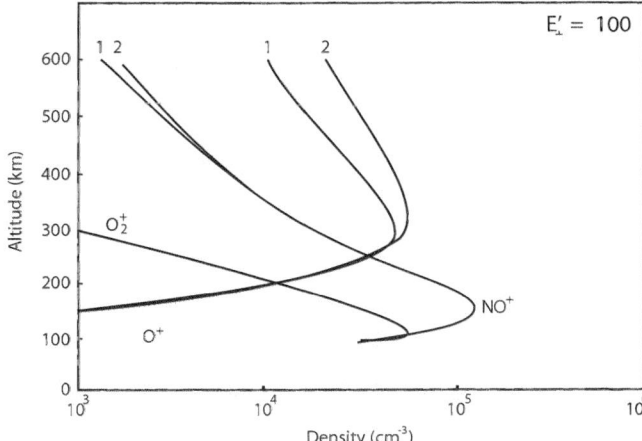

Figure 3. Ion density profiles calculated for a daytime high-latitude ionosphere subjected to a 100 mV m^{-1} electric field. The curves labeled 2 were calculated with T_i, and the curves labeled 1 were calculated with $T_{i\parallel}$. From *Schunk et al.* [1975].

process. The top panel shows the adopted SAR arc T_e profile, and the bottom panel shows the calculated N_e. Note that N_2 vibrational excitation can have a dramatic effect on the shape of the N_e profile. Since excited N_2 molecules are prevalent in and around the auroral oval, these molecules need to be taken into account in the global coupled I-T models.

2.6. Collisionless Plasma Flow

The current global ionosphere and ionosphere-plasmasphere models [*Bailey and Sellek*, 1990; *Millward et al.*, 1996; *Richards and Torr*, 1996; *Schunk et al.*, 2004] are based on relatively simple mathematical formulations. Specifically, the adopted continuity, momentum, and energy equations are simplified by ignoring nonlinear and/or complicated terms. It is also assumed that the plasma is collision dominated, which means that the momentum equation reduces to a diffusion equation (see section 2.7 for further details).

With regard to the energy equation, either an empirical model is adopted for the plasma temperatures or collision-dominated energy and heat flow equations are solved. With the collision-dominated transport formulation, the temperatures are isotropic, and the heat flow is simply given by the collision-dominated expression $\mathbf{q} = -\lambda \nabla T$. However, above about 3000 km, the plasma becomes collisionless in the polar wind, along SAR arc and plasmapause field lines, and in the plasmasphere after geomagnetic storms [*Demars and Schunk*, 1987a, 1987b]. When the plasma becomes collisionless, the use of isotropic temperatures and collision-dominated thermal conductivities is not valid. In a collisionless plasma, there are different species temperatures parallel and perpendicular to the magnetic field, and there are separate heat flow vectors for the transport of parallel and perpendicular energies. Hence, a *rigorous formulation* of the plasma flow requires a kinetic, semikinetic, generalized transport, or macroscopic particle-in-cell approach, all of which are difficult to implement for a global coupled I-T-P model. An example of collisionless heat flow is shown in Figure 5, where the heat flow vectors (parallel to **B**) for parallel and perpendicular energies are plotted versus altitude for SAR arc conditions [*Demars and Schunk*, 1986]. The simulation was from the

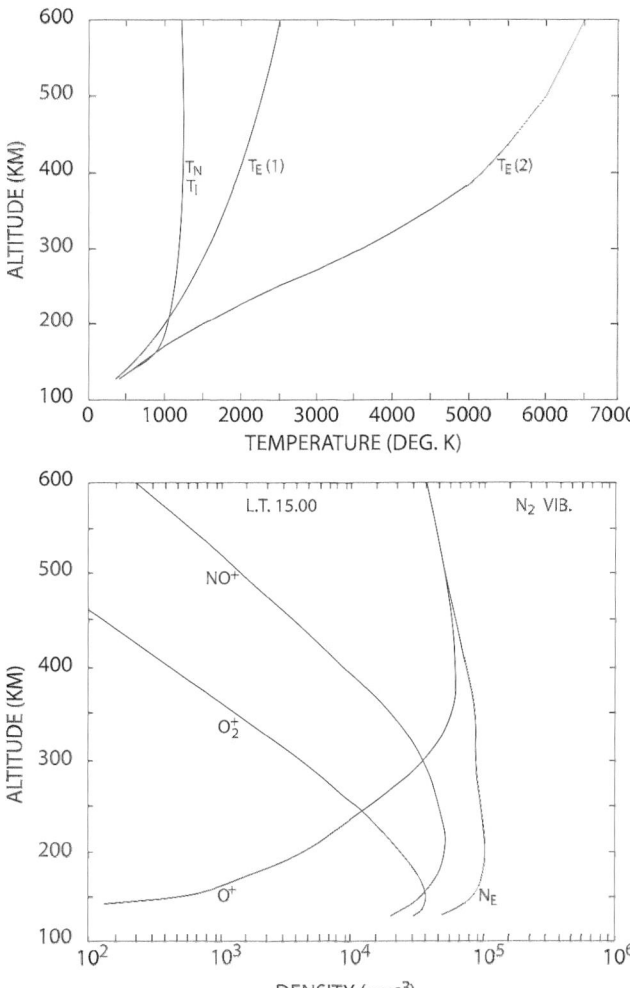

Figure 4. (top) Altitude profiles of the adopted electron, ion, and neutral temperatures used in subauroral red (SAR) arc calculations; $T_e(1)$ and $T_e(2)$ are the electron temperatures outside and inside the SAR arc, respectively. (bottom) Calculated ion and electron density profiles in a SAR arc including the effect of N_2 vibrational excitation and the associated increase in the $O^+ + N_2 \Rightarrow NO^+ + N$ reaction. From *Raitt et al.* [1976].

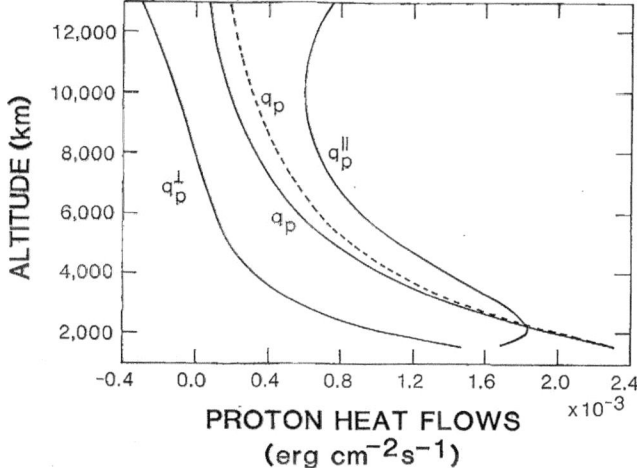

Figure 5. Proton heat flows along **B** for the transport of parallel energy ($q_p^{\parallel}$), perpendicular energy ($q_p^{\perp}$), and total energy (q_p) along a SAR arc field line, where $q_p = (q_p^{\parallel} + 2q_p^{\perp})/2$. Solid curves correspond to the solution of the 16-moment bi-Maxwellian transport equations. The dashed curve is not relevant to the discussion in the paper. From *Demars and Schunk* [1986].

solution of the 16-moment bi-Maxwellian transport equations. Note that with the more rigorous mathematical formulation, the density, drift velocity, and temperature solutions are significantly different from those obtained from the simplified diffusion and heat conduction equations commonly used in global coupled I-T-P models [see *Demars and Schunk*, 1986, 1987a, 1987b].

2.7. Ionosphere-Plasmasphere Coupling

As noted above, the four well-known physics-based global models of the coupled ionosphere-plasmasphere are based on a relatively simple diffusion formulation, which means that the nonlinear inertial term in the momentum equation ($\partial \mathbf{u}/\partial t + \mathbf{u}\cdot\nabla\mathbf{u}$) is not included [*Bailey and Sellek*, 1990; *Richards and Torr*, 1996; *Millward et al.*, 1996; *Schunk et al.*, 2004; *Scherliess et al.*, 2004]. The neglect of this term is useful for numerical reasons, but there are two negative consequences. Specifically, wave phenomena are not included, and the model cannot rigorously describe supersonic flow. The latter restriction is serious because after a geomagnetic storm, the upflow from the ionosphere that refills the depleted plasmasphere is supersonic. The neglect of the nonlinear inertial term, which acts to slow the upflow, not only means that the altitude profiles are wrong but that the refilling rate is too fast.

Another simplification is that none of the global I-P models couple to the ring current via wave-particle interactions, which means that the models do not properly describe the electron and ion thermal structure in the plasmasphere. Typically, the temperatures in the outer plasmasphere obtained from the global I-P models are too low (~4000–5000 K), whereas measurements indicate they are typically 8000–10,000 K [*Titheridge*, 1998]. To circumvent this problem, *Schunk et al.* [2004] adopted the empirical plasmasphere temperature model developed by *Titheridge* [1998]. Although this temperature model is based on an extensive satellite database, it is simplified in that it is a static empirical model.

Typically, the transport equations adopted to describe plasmasphere refilling determine the physics that is obtained. As noted above, a global I-P model that is based on a momentum equation that includes the nonlinear inertial term produces a different solution than that obtained from the four global I-P models that ignore this term (diffusion approximation), especially after geomagnetic storms. However, more advanced mathematical formulations can still lead to other completely different solutions [*Rasmussen and Schunk*, 1988]. In this latter study, the plasmasphere refilling was simulated with both a single-stream and a two-stream H^+ model. Specifically, the authors solved the H^+ continuity and momentum equations along a closed geomagnetic flux tube for a depleted plasmasphere. The momentum equation included the nonlinear inertial term so that wave phenomena and supersonic plasma flows could be properly modeled. In one simulation, a single H^+ stream was assumed, and in the second simulation, two independent H^+ streams were assumed (one from the Northern Hemisphere and one from the Southern Hemisphere). Figure 6 compares the plasmasphere refilling for the two cases. In both cases, the upflow is supersonic. For the single-stream simulation, there is only one H^+ velocity at each location along the flux tube, and when the counterstreaming H^+ flows from the conjugate hemispheres meet, a zero velocity results, and a pair of shocks is automatically triggered. The shocks then propagate toward lower altitudes, creating high-density plasma between the shock pair. In this case, the plasmasphere fills from the top down. On the other hand, for the case when the refilling is modeled with separate northern and southern H^+ streams, the counterstreaming supersonic flows penetrate each other, and shocks do not form. In this case, the plasmasphere fills from the bottom up. Hence, totally different results are obtained depending on how the plasmasphere refilling is modeled, with more rigorous mathematical formulations yielding more reliable solutions.

2.8. Plasma and Neutral Density Structures

Troposphere weather features can take on global characteristics, but most of the weather features are more localized, including hurricanes, tornados, snowstorms, fog banks,

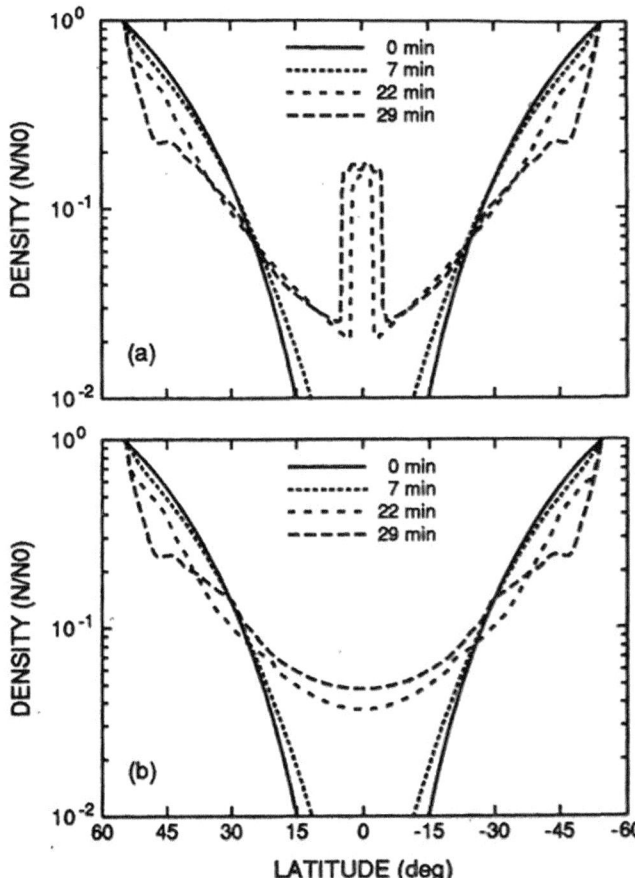

Figure 6. Electron density as a function of dipole latitude for the (a) single-stream H^+ model and for the (b) two-stream H^+ model. The multiple curves show the temporal evolution of the electron density as the flux tube fills, with the 0 min curves corresponding to the start of the simulations. From *Rasmussen and Schunk* [1988].

torrential rains, and sleet/hail. It is these more localized weather features that have the greater impact on human activities. Likewise, in addition to global characteristics, the weather features in the I-T-E system also displays more localized features, including mesoscale (50–1000 km) density structures. At high latitudes, the mesoscale plasma structures include tongues of ionization, aurora and boundary blobs, theta aurora, subauroral ion drift events, propagating plasma patches, and sun-aligned polar cap arcs. At middle and low latitudes, the structures include storm-enhanced density ridges and equatorial plasma bubbles. Examples of neutral density/wind structures include the cusp neutral fountain, propagating atmospheric holes, and supersonic neutral wind gusts. These localized features can have a significant impact on global-scale flows, and the resolution in the global I-T models needs to be fine enough to automatically include mesoscale I-T structures.

An important point to note about I-T structures is that if one observes a plasma structure, there is an associated neutral structure, and vice versa. For example, Figures 7a and 7b show I-T simulation results for the effect on the thermosphere of a series of propagating plasma patches [*Ma and Schunk*, 2001]. In this simulation, the width (200 km), length (1000 km), spacing (200 km), direction of propagation (antisunward), and density factor (10 above the background) of the cigar-shaped plasma patches were determined from measurements [*Fukui et al.*, 1994]. In the simulation, a diurnally reproducible, global I-T was first calculated, and then, at 02:65 UT ($t = 0$), the series of propagating plasma patches was introduced in the southern polar region. The patches were imposed, one at a time, near the cusp at a half-hour interval, and then, they propagated across the polar cap at the prevailing convection speed, yielding a 200 km separation between the plasma patches. The simulation results are for moderate solar activity ($F10.7 = 150$), a two-cell convection pattern with a 100 kV cross-tail magnetospheric potential, a Gaussian-shaped N_e profile in the horizontal direction, and a peak-to-background N_e ratio of 10.

In Figure 7a, the snapshot is for $t = 3$ h, which is 3 h after the first plasma patch was introduced. Figure 7a shows the N_e distribution at 300 km, the neutral density perturbation ($\Delta\rho$) at 300 km, and $\Delta\rho$ via a 2-D (altitude and latitude) day-night cut across the polar cap through the center of the series of plasma patches. In general, propagating plasma patches act as a snowplow, creating a hole in the thermosphere in and behind the individual plasma patches and neutral density enhancements in front of the patches. For plasma patches that have a factor of 10 density enhancement above the background plasma density, there is a 30%–35% neutral density perturbation due to the propagating plasma patches. The neutral disturbance moves along with the propagating plasma patches and is characterized by an increased wind speed ($\Delta u > 100$ m s^{-1}), a temperature enhancement ($\Delta T \approx$ 100–300 K), neutral gas upwelling, and O/N$_2$ composition changes. The propagating plasma patches also excite waves in the thermosphere that propagate away from the neutral disturbance, as shown in Figure 7a, which is for $t = 3$ h, and in Figure 7b, which is for $t = 4.41$ h.

2.9. Neutral Rain on the Thermosphere

As the polar wind and auroral H^+ and O^+ ions flow upward, they can undergo charge exchange reactions with the background thermal and energetic neutrals, thereby creating upflowing H_s and O_s stream neutrals [*Gardner and Schunk*, 2004, 2005]. The streaming neutrals are superthermal because at creation, they have the same velocity and energy as their parent ions. Upflowing H_s and O_s stream neutrals are

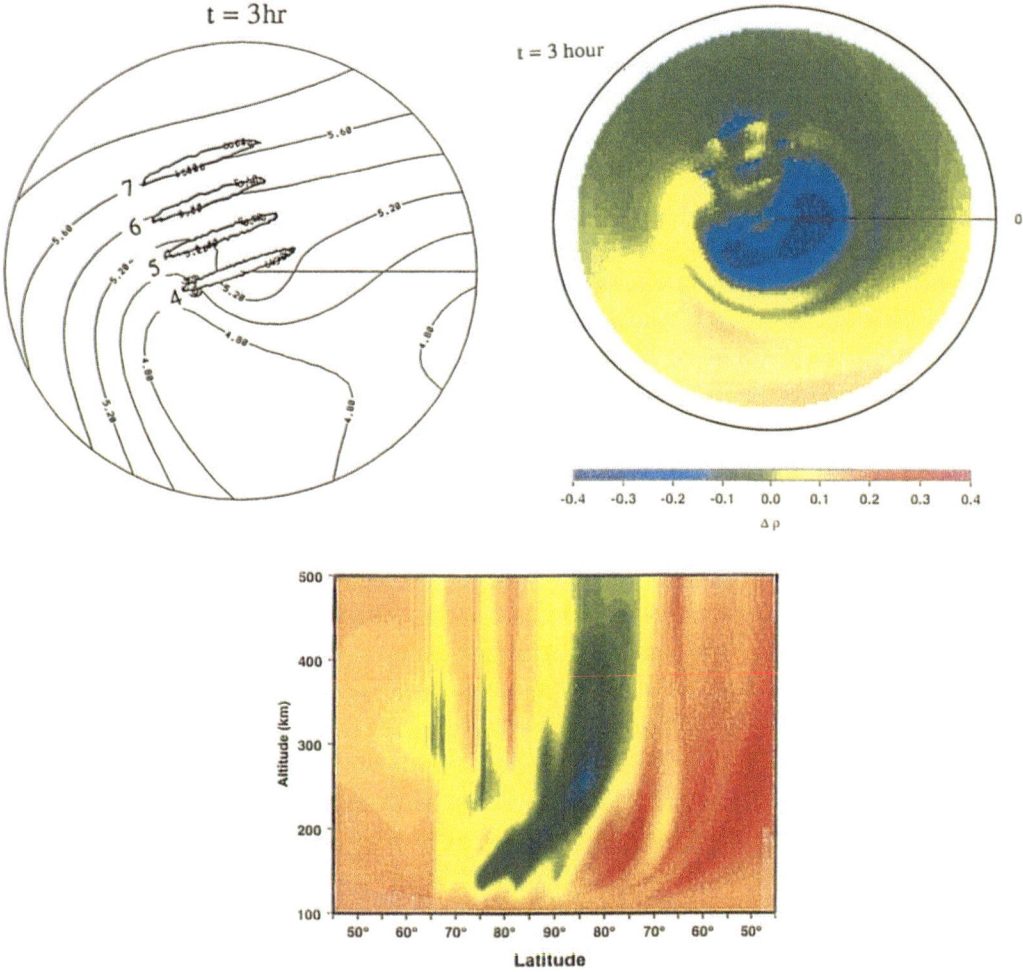

Figure 7a. Effect of multiple propagating plasma patches on the thermosphere at $t = 3$ h. (left top) The N_e distribution at 300 km in units of $\log_{10} N_e$ (cm^{-3}), (right top) the neutral density perturbation at 300 km, and (bottom) the neutral density perturbation versus altitude and latitude across the polar cap. From *Ma and Schunk* [2001].

also created on closed magnetic flux tubes, particularly after the tubes are depleted by geomagnetic storms. At all latitudes, the H_s stream neutrals have sufficient energy to escape, but most of the O_s stream neutrals do not have enough energy to escape and then rain down on the I-T system. The neutral rain provides a global energy source for the thermosphere as the downstreaming neutrals collide with the background thermal neutrals. This effect is not included in the current global I-T models, but the exact effect of this process has not been fully elucidated.

2.10. Lower Atmosphere Wave Effects

At low altitudes, the I-T system is continually subjected to planetary, tidal, and gravity waves that propagate upward from the lower atmosphere. These waves then affect the neutral winds and electrodynamics in the E region. They also interact with the waves generated internally in the thermosphere. The upward propagating planetary, tidal, and large-scale gravity waves can be described by models that extend from the Earth's surface to the upper thermosphere. However, high-resolution I-T models are needed to properly account for small-scale gravity waves and for the consequent wave-wave coupling in the thermosphere. This work has only recently begun (see other papers in this monograph).

3. SUMMARY

Community-wide initiatives have shown that the output of global physics-based I-T models can be significantly different

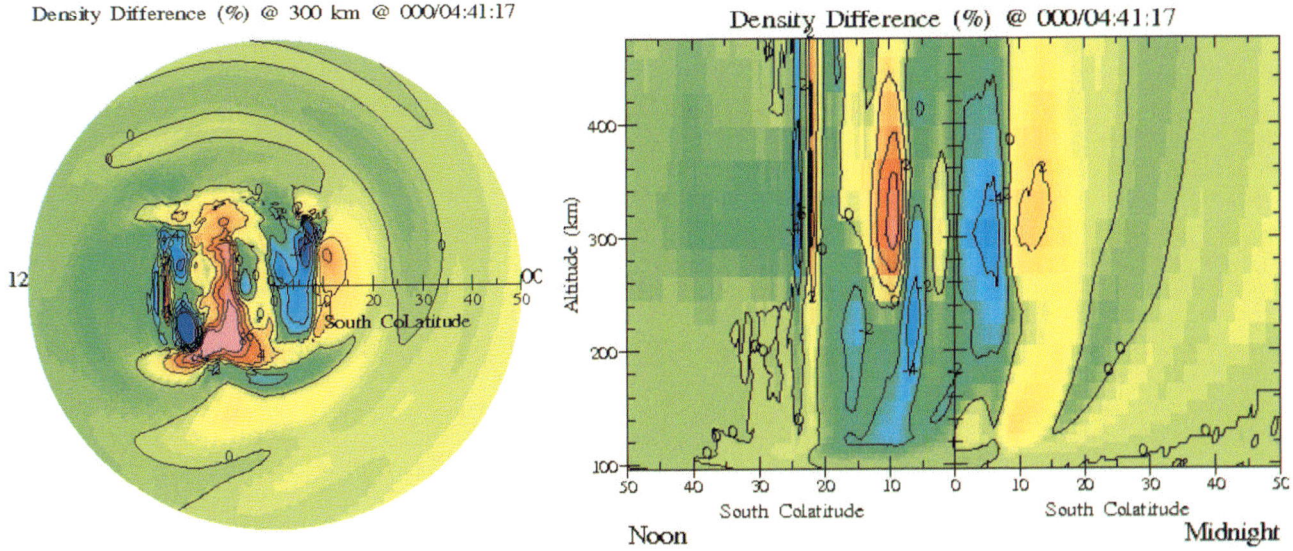

Figure 7b. Snapshot at $t = 4.41$ h of the neutral density perturbation due to multiple propagating plasma patches. The neutral perturbation is shown (left) at 300 km and (right) as a function of altitude and latitude across the polar cap. From *Schunk et al.* [2008].

from each other and from the measurements when they are compared for the same geophysical case. As noted above, there are numerous problems with the existing I-T models, but in this paper, the focus was on some of the physics that is missing from the global I-T models. At high latitudes, the models should be expanded to include polar wind and auroral ion outflow, the downward electron heat at the upper boundary in the polar cap, thermoelectric heat flow in the return current regions, and the ion temperature anisotropy due to strong electric fields. Near the plasmapause and in SAR arcs, the electron density effects associated with N_2 vibrational excitation need to be taken into account. In I-T models that couple to the plasmasphere, the plasma becomes collisionless above about 3000 km, and the heat flow is not governed by the standard thermal conductivity. In addition, during plasmasphere refilling after geomagnetic storms, a multistream formulation is needed for each ion species. In general, the resolution in the global I-T models needs to be improved so that mesoscale plasma and neutral density structures can be taken into account in a self-consistent way. The improved resolution is also needed to take account of the upward propagating gravity waves from the lower atmosphere. Another process that should be included is the neutral rain on the thermosphere.

Ten topics relevant to missing physics in I-T models were selected as examples. However, some of the topics are primarily important in local regions and, therefore, are relevant to local weather, whereas others are important for global I-T weather simulations. The missing physics relevant to local weather simulations includes thermoelectric heat flow in return current regions (section 2.3), ion temperature anisotropies in regions with large electric fields (section 2.4), SAR arcs (section 2.5), and plasma and neutral density structures (section 2.8). The missing physics that should have a significant effect on global I-T weather simulations includes polar and auroral ion outflow (section 2.1), downward electron heat flow in the polar cap (section 2.2), ionosphere-plasmasphere coupling (section 2.7), neutral rain on the thermosphere (section 2.9), and lower atmosphere wave effects (section 2.10). However, the cumulative effect of plasma and neutral density structures (section 2.8) may also affect the global mean circulation and temperature of the thermosphere [*Smith*, 2000].

In general, the global I-T and I-T-P models that are based on diffusion and hydrodynamic (fluid) formulations are not valid when the plasma and neutral gas become collisionless, which occurs above about 3000 km for the ionosphere-polar wind and ionosphere-plasmasphere models, and above about 500 km for the thermosphere models. As a consequence, the densities, drift velocities, temperatures, and heat flows, as well as their variations with altitude, are wrong if a diffusion or hydrodynamic formulation is used in the collisionless region.

The focus of this work was on "some" of the physics that is missing in the global physics-based models, and therefore, the examples of missing physics that were presented here do not constitute a complete list. Other issues, such as instabilities, turbulence, uncertain parameters, numerical techniques, etc., are addressed in other papers in this monograph.

REFERENCES

Bailey, G. J., and R. Sellek (1990), A mathematical model of the Earth's plasmasphere and its application in a study of He+ at L=3.0, *Ann. Geophys.*, *8*, 171–190.

Barakat, A. R., and R. W. Schunk (2006), A three-dimensional model of the generalized polar wind, *J. Geophys. Res.*, *111*, A12314, doi:10.1029/2006JA011662.

Bekerat, H., R. W. Schunk, and L. Scherliess (2007), Estimation of the high-latitude topside electron heat flux using DMSP plasma density measurements, *J. Atmos. Sol. Terr. Phys.*, *69*, 1029–1048.

David, M., R. W. Schunk, and J. J. Sojka (2011), The effect of downward electron heat flow and electron cooling processes in the high-latitude ionosphere, *J. Atmos. Sol. Terr. Phys.*, *73*(16), 2399–2409, doi:10.1016/j.jastp.2011.08.009.

Demars, H. G., and R. W. Schunk (1986), Solutions to bi-Maxwellian transport equations for SAR-arc conditions, *Planet. Space Sci.*, *34*, 1335–1348.

Demars, H. G., and R. W. Schunk (1987a), Temperature anisotropies in the terrestrial ionosphere and plasmasphere, *Rev. Geophys.*, *25*, 1659–1679.

Demars, H. G., and R. W. Schunk (1987b), Comparison of solutions to bi-Maxwellian and Maxwellian transport equations for subsonic flow, *J. Geophys. Res.*, *92*, 5969–5990.

Fang, T.-W., et al. (2013), Equatorial-PRIMO (Problems Related to Ionospheric Models and Observations), in *Modeling the Ionosphere/Thermosphere System*, Geophys. Monogr. Ser., doi:10.1029/2012GM001280, this volume.

Fukui, K., J. Buchau, and C. Valladares (1994), Convection of polar cap patches observed at Qaanaaq, Greenland during the winter of 1989–1990, *Radio Sci.*, *29*(1), 231–248.

Gardner, L. C., and R. W. Schunk (2004), The neutral polar wind, *J. Geophys. Res.*, *109*, A05301, doi:10.1029/2003JA010291.

Gardner, L. C., and R. W. Schunk (2005), Global neutral polar wind model, *J. Geophys. Res.*, *110*, A10302, doi:10.1029/2005JA011029.

Jee, G., A. G. Burns, W. Wang, S. C. Solomon, R. W. Schunk, L. Scherliess, D. C. Thompson, J. J. Sojka, and L. Zhu (2007), Duration of an ionospheric data assimilation initialization of a coupled thermosphere-ionosphere model, *Space Weather*, *5*, S01004, doi:10.1029/2006SW000250.

Jee, G., A. G. Burns, W. Wang, S. C. Solomon, R. W. Schunk, L. Scherliess, D. C. Thompson, J. J. Sojka, and L. Zhu (2008), Driving the TING model with GAIM electron densities: Ionospheric effects on the thermosphere, *J. Geophys. Res.*, *113*, A03305, doi:10.1029/2007JA012580.

Ma, T.-Z., and R. W. Schunk (2001), The effects of multiple propagating plasma patches on the polar thermosphere, *J. Atmos. Sol. Terr. Phys.*, *63*, 355–366.

Millward, G. W., R. J. Moffett, S. Quegan, and T. J. Fuller-Rowell (1996), A coupled thermosphere-ionosphere-plasmasphere model (CTIP), in *STEP Handbook of Ionospheric Models*, edited by R. W. Schunk, pp. 239–280, Utah State Univ., Logan.

Raitt, W. J., R. W. Schunk, and P. M. Banks (1976), Ionospheric composition in SAR-arcs, *Planet. Space Sci.*, *24*, 105–114.

Rasmussen, C. E., and R. W. Schunk (1988), Multi-stream hydrodynamic modeling of interhemispheric plasma flow, *J. Geophys. Res.*, *93*, 14,557–14,565.

Richards, P. G., and D. G. Torr (1996), The field line interhemispheric plasma model, in *STEP Handbook of Ionospheric Models*, edited by R. W. Schunk, pp. 239–280, Utah State Univ., Logan.

Scherliess, L., R. W. Schunk, J. J. Sojka, and D. C. Thompson (2004), Development of a physics-based reduced state Kalman filter for the ionosphere, *Radio Sci.*, *39*, RS1S04, doi:10.1029/2002RS002797.

Schunk, R. W. (1988), A mathematical model of the middle and high latitude ionosphere, *Pure Appl. Geophys.*, *127*, 255–303.

Schunk, R. W. (2007), Time-dependent simulations of the global polar wind, *J. Atmos. Sol. Terr. Phys.*, *69*, 2028–2047.

Schunk, R. W., and A. F. Nagy (2009), *Ionospheres*, 2nd ed., Cambridge Univ. Press, Cambridge, U. K.

Schunk, R. W., and J. J. Sojka (1997), Global ionosphere-polar wind system during changing magnetic activity, *J. Geophys. Res.*, *102*, 11,625–11,651.

Schunk, R. W., P. M. Banks, and W. J. Raitt (1975), Effect of electric fields on the daytime high-latitude E- and F-regions, *J. Geophys. Res.*, *80*, 3121–3130.

Schunk, R. W., J. J. Sojka, and M. D. Bowline (1986), Theoretical study of the electron temperature in the high latitude ionosphere for solar maximum and winter condition, *J. Geophys. Res.*, *91*, 12,041–12,054.

Schunk, R. W., J. J. Sojka, and M. D. Bowline (1987), Theoretical study of the effect of ionospheric return currents on the electron temperature, *J. Geophys. Res.*, *92*, 6013–6022.

Schunk, R. W., et al. (2004), Global Assimilation of Ionospheric Measurements (GAIM), *Radio Sci.*, *39*, RS1S02, doi:10.1029/2002RS002794.

Schunk, R. W., L. Gardner, L. Scherliess, D. C. Thompson, and J. J. Sojka (2008), Effect of lower atmospheric waves on the ionosphere and thermosphere, paper presented at the 2008 Ionospheric Effects Symposium, JMG Assoc., Natl. Tech. Inf. Serv., Springfield, Va.

Shim, J. S., et al. (2011), CEDAR electrodynamics thermosphere ionosphere (ETI) challenge for systematic assessment of ionosphere/thermosphere models: NmF2, hmF2, and vertical drift using ground-based observations, *Space Weather*, *9*, S12003, doi:10.1029/2011SW000727.

Smith, R. W. (2000), The global-scale effect of small-scale thermospheric disturbances, *J. Atmos. Sol. Terr. Phys.*, *62*, 1623–1628.

Sojka, J. J. (1989), Global scale, physical models of the F region ionosphere, *Rev. Geophys.*, *27*, 371–403.

Titheridge, J. E. (1998), Temperatures in the upper ionosphere and plasmasphere, *J. Geophys. Res.*, *103*, 2261–2277.

R. W. Schunk, Center for Atmospheric and Space Sciences, Utah State University, Logan, UT 84322-4405, USA. (robert.schunk@usu.edu)

Physical Characteristics and Modeling of Earth's Thermosphere

Tim Fuller-Rowell

CIRES University of Colorado, Boulder, Colorado, USA

NOAA Space Weather Prediction Center, Boulder, Colorado, USA

Earth's neutral thermosphere extends from roughly 100 to 600 km altitude and comprises more than 99% of the medium. The lower limit of the thermosphere is the mesopause, where the average temperature profile starts to increase sharply, heated by absorption of solar ultraviolet (UV) radiation. The thermosphere is collision dominated enabling the gas laws and fluid equations to be used. The upper limit of the thermosphere, the exobase, is the level were fluid properties break down. Hydrostatic balance and pressure coordinates are often invoked. In addition to advection, pressure gradients, Coriolis, and diffusion, the presence of plasma and the magnetic field give rise to ion drag and Joule heating. In the lower 10 or 20 km, turbulence mixes momentum, potential temperature, and the main O_2 and N_2 species. At higher altitudes, molecular diffusion allows species to separate out depending on their molecular mass, with lighter species above and heavier species below. Solar radiation also dissociates molecular oxygen; the lighter atomic oxygen becomes the third major species, which dominates in the upper thermosphere. Hydrostatic balance can support large vertical winds and the propagation of large-scale gravity waves. Thermal expansion does not change the relative contribution of species on pressure levels, but the global circulation transports species vertically and horizontally, enhancing heavier species in regions of upwelling and lighter species in regions of downwelling. At high latitudes, ion drag drives high-velocity, nonlinear neutral wind vortices and can stimulate inertial resonances. Thermospheric winds, temperature, and composition have a strong impact on the ionosphere.

1. INTRODUCTION

Earth's upper atmosphere is a gravitationally bound weakly ionized fluid extending from roughly 100 to 600 km altitude. More than 99% of the gas is "neutral," or not ionized, so understanding and modeling the neutral component, the thermosphere, is crucial for an adequate representation and understanding of the ionized component. The ions make up less than 1% of the total mass. This chapter describes the basic physical processes in the thermosphere, or Earth's neutral upper atmosphere, which need to be captured in a physics-based model. The thermosphere is the medium from which the ionosphere is created, and the neutral dynamics and composition is an important driver of the ionized component. The ionosphere, in turn, has important impacts on the neutral medium through ion drag and Joule heating.

The fluid properties of the neutral gas result from the frequent collisions between the atoms and molecules. Rather than having to accommodate the random nature of the forces exerted on individual gas particles, the principles of kinetic theory can be invoked so that the medium can be described by the bulk properties of the fluid, such as pressure, density, temperature, and velocity. The fluid properties also enable the use of the Navier-Stokes momentum equations. The upper extent of the atmosphere is usually defined as the altitude at which the fluid approximation is no longer valid, referred to as the exobase. Below the exobase, the distance traveled or time taken between collisions is short compared to the scale sizes of interest in the dynamics and energetics of the fluid. For the Earth, this altitude is usually around 600 km, but can be higher since it depends on the gas kinetic temperature, and hence the degree of thermal expansion of the medium. It is more appropriate to state the vertical extent of the atmosphere in terms of pressure level. Most physical models of Earth's upper atmosphere are limited to a top pressure level of about 10^{-7} Pa for this reason. At lower pressures, or greater heights, the mean free path or the distance between collisions exceeds tens of kilometers, and the medium acts as free particles rather than a fluid.

2. THE GAS LAW AND HYDROSTATIC BALANCE

The frequent collisions of a gas close to thermal equilibrium enable the Maxwellian energy distribution of the individual particles to be replaced by the basic fluid properties of pressure, p, temperature, T, number density, n, and mass density, ρ, that are related by the perfect gas law:

$$p = nkT \quad (1)$$

or

$$p = \rho R \frac{T}{M} \quad (2)$$

where k and R are the Boltzmann and gas constants, respectively, and M is the molecular mass in atomic units.

The gas under the influence of the planet's gravitational force gives rise to the concept of hydrostatic balance, which states that the change in pressure with height, ∂p, is closely balanced by the weight of the fluid, $nmg\partial h$, under the action of the planet's gravitational field. The concept is expressed mathematically as

$$\partial p = -nmg\partial h \quad (3)$$

or

$$\frac{\partial p}{\partial h} = -\rho g \quad (4)$$

where m is the mean molecular mass in kilograms, h is the height, and g is the planet's gravitational acceleration. These basic equations describe the exponential decrease in gas density with altitude and introduce the concept of scale height, $H = RT/Mg$, which represents the altitude through which the gas density will decrease by a factor of $1/e$. Most of the physical processes controlling the global thermosphere dynamics, energy budget, and composition can be described assuming the atmosphere is in "quasi-hydrostatic balance," which assumes vertical acceleration is small compared with gravity.

The assumption of hydrostatic balance implies that a vertical column of gas responds to a heat source instantaneously. In the real atmosphere, the information about heating at a given altitude or pressure level is transferred to other regions by acoustic gravity waves, which have speeds of hundreds of meters per second, corresponding to a time scale for adjustment of typically 5 to 10 min. This time scale is similar to the buoyancy or Brunt-Väisälä period. Treatment or understanding of dynamical time scales shorter than this period, or spatial scales less than ~50 km, has to explicitly include acoustic waves. So the physics and model described here is limited to the larger temporal and spatial scales. This chapter makes the implicit assumption of hydrostatic equilibrium and, consequently, will not cover acoustic waves and their potential impact on the thermosphere. The impact of non-hydrostatic processes during impulsive energy injection on short time scales will be addressed by *Deng and Ridley* [this volume]. During these times, vertical acceleration can be a significant fraction (20%) of the gravitational acceleration.

The assumption of quasi-hydrostatic balance enables pressure to be used as the vertical coordinate in neutral atmosphere models and enables the concept of a reduced height z^*, where $p = p_0 e^{-z^*}$, such that

$$\frac{\partial h}{\partial z^*} = H \quad (5)$$

and the height of the pressure surfaces can be evaluated by integrating from the lower boundary, h_0,

$$h = h_0 + \int H \partial z^* \quad (6)$$

and

$$\partial z^* = -\frac{\partial h}{H} = \frac{\partial p}{p} = \frac{\partial n}{n} + \frac{\partial H}{H}. \quad (7)$$

In these types of models, the horizontal pressure gradient is replaced by the horizontal gradient in the height of the pressure levels. The pressure coordinate system has several benefits. It reduces the dimensionality of the problem; simplifies the continuity equation (see below); simplifies solar and auroral absorption, since pressure surfaces are levels of constant optical depth; changes of temperature in a column of gas do not change the relative contribution of neutral species on pressure levels; and vertical winds can be separated into physically meaningful processes (see below).

The equilibrium condition implied by hydrostatic balance, however, does not exclude the possibility of vertical winds. The assumption simply demands that the rate of heating is such that the atmosphere adjusts at a comparable rate. The term "quasi-hydrostatic balance" is the more correct expression in the case of accommodating vertical winds in the system. One component of the vertical wind is then defined as the rate of change in the height of a pressure surface in the column of gas $(\partial h/\partial t)_p$ and is termed the "barometric wind." Vertical winds in Earth's upper atmosphere of the order of 100 m s^{-1} can be accommodated within the quasi-hydrostatic assumption. The assumption of hydrostatic balance has enabled the wide use of pressure as the vertical coordinate in atmospheric models. In fact, only recently have Earth upper-atmosphere models begun to relax this assumption and explicitly include a realistic adjustment process by acoustic waves [e.g., *Ridley et al.*, 2006; *Deng et al.*, 2008; *Deng and Ridley*, this volume]. Such models are able to examine the physical response at small-scale sizes and on short time scales, albeit at the expense of increased computation.

3. CONTINUITY EQUATION

The continuity equation is also one of the most widely used and universal fluid concepts. In the pressure coordinate system, the continuity equation can be expressed as

$$\frac{\partial \omega}{\partial p} = -\nabla_p \cdot \mathbf{V} \qquad (8)$$

where ω is the vertical wind in the pressure coordinate system, dp/dt, and the right-hand side represents the horizontal divergence of neutral wind, $\mathbf{V}$, on a pressure surface p. The fluid, therefore, appears incompressible in this natural coordinate system, where horizontal divergence or convergence must be balanced by a vertical flow. This, in fact, describes the second component of vertical winds, the so-called "divergence wind." A local heat source will cause the local column of gas to thermally expand (the barometric wind), while the horizontal pressure gradients so induced will drive a divergent wind that must be balanced by a vertical flow across the pressure surfaces. The total vertical wind, V_z, in this system can be expressed as the sum of the barometric and divergence wind, $-\omega/\rho g$, components thus

$$V_z = \left(\frac{\partial h}{\partial t}\right)_p - \frac{\omega}{\rho g}. \qquad (9)$$

The vertical wind becomes a diagnostic, rather than prognostic equation, and represents the reduced dimensionality of the problem.

4. LAGRANGIAN VERSUS EULERIAN FRAMES OF REFERENCE

Atmospheric models typically solve the equations in an Eulerian coordinate system fixed with respect to the Earth, usually spherical polar coordinates in radius, r, latitude, θ, and longitude, ϕ. The rate of change of a state parameter at a fixed point in the spherical polar coordinate space (r,θ,ϕ) is represented by the partial derivative $\partial/\partial t$. The partial and total derivatives are connected by the advection terms:

$$\frac{d}{dt} = \frac{\partial}{\partial t} + \frac{V_\theta}{r}\frac{\partial}{\partial \theta} + \frac{V_\phi}{r\sin\theta}\frac{\partial}{\partial \phi} + \omega\frac{\partial}{\partial p}. \qquad (10)$$

Physically, the advection terms represent the transport of the fluid properties, such as momentum or temperature, across the fixed grid. If the planet was not rotating, the inertial motion of a parcel of gas would follow a great circle trajectory. In a spherical polar coordinate system, the great circle trajectory requires two extra terms in the momentum equation, which are shown later.

5. HORIZONTALLY STRATIFIED FLUID AND OTHER COMMON ASSUMPTION

One common assumption when addressing large-scale dynamics in atmospheric fluids is that the horizontal scale size is significantly greater than the vertical scale. This assumption implies that horizontal motions are constrained to follow the curvature of the planet. Another common assumption is that the vertical wind is significantly less than the horizontal wind, so the advection cross terms in the spherical polar coordinate system, $V_\theta V_z/r$ and $V_\phi V_z/r$, can be neglected. The shallow atmosphere approximation also enables several other assumptions: that gravity is independent of altitude and that the following terms can be neglected: vertical component and second-order terms of the Coriolis force, Earth's nonsphericity, centrifugal terms, and the increase in geocentric distance.

6. CORIOLIS EFFECT

Planetary rotation gives rise to the Coriolis force. The Coriolis effect is an apparent deflection of moving objects from a straight path when they are viewed from a rotating frame of reference. The apparent force is a consequence of the inertia of the fluid being constrained to move on a horizontal curved surface. For planets with a rotational direction, the same as that of Earth's, a prograde rotation, the Coriolis effect away from the geographic equator causes a parcel of fluid moving with respect to the planetary rotation, to be directed toward the right in the Northern Hemisphere and to the left in the south. The basic force can be expressed as

$$-2\boldsymbol{\Omega} \times \mathbf{V} \tag{11}$$

where $\boldsymbol{\Omega}$ is the planet's angular velocity.

7. VISCOUS DRAG

Viscosity is the process that tends to smooth out gradients in the fluid and can be caused by molecular or turbulent diffusion. Strictly speaking, both processes smear out second-order gradients, $\partial^2/\partial z^2$. However, the imposed boundary condition at the top of the atmosphere that assumes that there is no mass flux through the upper boundary sets vertical gradients to zero and has the effect that viscosity acts to smear out first-order gradients, $\partial/\partial z$. In the upper thermosphere where diffusion is rapid, vertical viscosity is very effective in smoothing out the neutral wind and temperature profiles. Horizontal viscosity tends to have a weaker effect on the dynamics in global models due to the fact that horizontal spacing is large (200 to 500 km) compared to the vertical spacing (5 to 20 km). Note that horizontal viscosity is sometimes included in global models to assist with numerical stability.

8. ION DRAG

The basic processes driving the dynamics in an un-ionized fluid are advection, pressure gradients, Coriolis, and viscosity. In Earth's weakly ionized upper atmosphere, however, the presence of the intrinsic internal magnetic field gives rise to an additional force in the atmosphere known as ion drag. In the absence of electric fields and collisions with neutral particles, the ions are strongly constrained by the magnetic field. Plasma can flow freely parallel to the magnetic field direction, but flow perpendicular to the field is restricted.

In the upper thermosphere, the collision between the ions and neutral particles are relatively infrequent but are sufficient to cause a drag on the neutral flow. The basic force can be expressed as

$$-\nu_{ni}(\mathbf{V} - \mathbf{U}) \tag{12}$$

where the force is proportional to the difference in the neutral velocity, $\mathbf{V}$, and the ion velocity, $\mathbf{U}$, scaled by the neutral ion collision frequency, ν_{ni}. The simplicity of this formulism, however, hides a great deal of complexity in both the ion motion and the drag force as a function of altitude through the atmospheric domain. By adopting some reasonable assumptions, the ion drag force can also be written as $\mathbf{J} \times \mathbf{B}/\rho$ for a current, $\mathbf{J}$, in the presence of a magnetic field, $\mathbf{B}$.

9. MOMENTUM EQUATION

The momentum equations for the meridional and zonal direction for a unit mass of gas in the spherical polar coordinate system can therefore be stated thus

$$\begin{aligned}\frac{\partial}{\partial t}V_\theta = &\underbrace{-\frac{V_\theta}{r}\frac{\partial}{\partial \theta}V_\theta - \frac{V_\phi}{r\sin\theta}\frac{\partial}{\partial \phi}V_\theta - \omega\frac{\partial}{\partial p}V_\theta}_{\text{horizontal and vertical advection}} \underbrace{- \frac{g}{r}\frac{\partial}{\partial \theta}h}_{\text{pressure}} \\ &\underbrace{+\left(2\Omega + \frac{V_\phi}{r\sin\theta}\right)V_\phi\cos\theta}_{\text{Coriolis \& "curvature"}} + \underbrace{g\frac{\partial}{\partial p}\left[(\mu_m + \mu_T)\frac{p}{H}\frac{\partial}{\partial p}V_\theta\right]}_{\text{vertical viscosity}} \\ &\underbrace{-\nu_{ni}(V_\theta - U_\theta)}_{\text{ion drag}}\end{aligned} \tag{13}$$

and

$$\begin{aligned}\frac{\partial}{\partial t}V_\phi = &-\frac{V_\theta}{r}\frac{\partial}{\partial \theta}V_\phi - \frac{V_\phi}{r\sin\theta}\frac{\partial}{\partial \phi}V_\phi - \omega\frac{\partial}{\partial p}V_\phi - \frac{g}{r\sin\theta}\frac{\partial}{\partial \phi}h \\ &-\left(2\Omega + \frac{V_\phi}{r\sin\theta}\right)V_\theta\cos\theta + g\frac{\partial}{\partial p}\left[(\mu_m + \mu_T)\frac{p}{H}\frac{\partial}{\partial p}V_\phi\right] \\ &-\nu_{ni}(V_\phi - U_\phi)\end{aligned} \tag{14}$$

where V_θ and V_ϕ are the meridional and zonal neutral winds, and U_θ and U_ϕ are the meridional and zonal ion drifts, and the terms on the right are horizontal and vertical advection, horizontal pressure gradient, Coriolis and "curvature," vertical viscosity, and ion drag, respectively. The μ_m and μ_T expressions in the viscous drag term represent the molecular and turbulent viscous coefficients, respectively. Note that the terms mentioned previously that accommodate the inertial motion in the spherical polar coordinate system are included with the Coriolis term. These additional expressions are sometimes referred to as "curvature" terms. The momentum equations combined with the hydrostatic and continuity equations can be used to solve for the horizontal neutral wind across the globe. The vertical wind, ω, in the pressure coordinates is obtained by integrating the continuity equation from the top of the atmosphere down, where ω is set to zero at the top of the atmosphere, which assumes that there is no mass flux through the upper boundary.

10. ENERGY EQUATIONS

In a similar way, the energy equation can be stated thus

$$\frac{\partial}{\partial t}\varepsilon = \underbrace{-\frac{V_\theta}{r}\frac{\partial}{\partial \theta}(\varepsilon+gh) - \frac{V_\phi}{r\sin\theta}\frac{\partial}{\partial \phi}(\varepsilon+gh) - \omega\frac{\partial}{\partial p}(gh)}_{\text{advection and adiabatic processes}} + Q_{euv}$$

$$\underbrace{+Q_{ir} + Q_{vis}}_{\text{sinks}} + \underbrace{g\frac{\partial}{\partial p}\left[(\kappa_m + \kappa_T)\frac{p}{H}\frac{\partial}{\partial p}T\right]}_{\text{vertical heat conduction}}$$

$$\underbrace{-g\frac{\partial}{\partial p}\frac{g\kappa_T}{c_p}}_{\text{kinetic energy dissipation}} - \underbrace{\frac{J_\theta E_\theta + J_\phi E_\phi}{\rho}}_{\text{Joule heating}} \quad (15)$$

where ε represents the sum of the specific enthalpy, or internal energy of the gas, $c_p T$, and the kinetic energy of a unit mass of gas, $(V_\theta^2 + V_\phi^2)/2$.

The terms on the right are horizontal and vertical advection of the sum of internal, kinetic, and potential energy; sources and sinks of energy such as solar EUV heating, IR cooling, and viscous heating; vertical heat conduction; and the sum of Joule heating and kinetic energy dissipation from ion drag, resulting from a current, **J**, and electric field, **E**. A parcel of gas that is displaced vertically by a turbulent eddy will undergo adiabatic heating or cooling, depending on whether the parcel has been displaced upward to a lower-pressure level where it will expand and cool or downward to a region of higher pressure where the gas will compress and heat. In the absence of local heat sources, the equilibrium vertical temperature profile under the action of turbulence is the adiabatic lapse rate, g/c_p. The extra conduction term, involving K_T, represents this process. In the region of molecular diffusion, individual atoms and molecules exchange locations, rather than parcels of gas, so the equilibrium temperature profile migrates toward isothermal.

Adiabatic processes cause changes in the temperature of a fluid in response to compression or expansion of a gas parcel. Solar heating initially imparts energy and provides the heat source for the upper atmosphere. Local or regional heating imposes horizontal pressure gradients that set the atmosphere in motion horizontally. The continuity of the global circulation is closed by upwelling in the region of divergence and downwelling in the region of convergence. Upwelling of a parcel of gas to a region of reduced pressure causes a parcel of gas to expand and adiabatically cool; downwelling transports parcels of air to regions of higher pressure causing compression and adiabatic heating. In the pressure coordinate system, the gas appears mathematically incompressible (see the continuity equation (8) above), so the physical concept of expansion and compression and adiabatic heating and cooling is represented by the term $\omega\partial gh/\partial p$ in equation (15).

As well as momentum transfer involved in the neutral/plasma interactions, there is also an energy exchange via the frictional dissipation from the movement of the ions through the neutrals or the neutrals through the ions. This frictional dissipation from the perspective of the neutral gas is known as Joule heating. This effect can also be described as the dissipation of a current flowing through the resistive medium of the neutral gas. The last term in equation (15), $(\mathbf{J}\cdot\mathbf{E})/\rho$, is the sum of Joule heating, $\mathbf{J}\cdot(\mathbf{E} + \mathbf{V} \times \mathbf{B})/\rho$, and kinetic energy dissipation from ion drag, $\mathbf{V}\cdot(\mathbf{J} \times \mathbf{B})/\rho$.

11. NEUTRAL COMPOSITION

Using a combination of the generalized diffusion equation [*Chapman and Cowling*, 1970] and the continuity equations, the change in composition of the three major thermospheric species (O, O_2, and N_2) can be evaluated self-consistently with the wind and temperature fields. The major species are atomic oxygen, molecular oxygen, and molecular nitrogen. Allowance is made for mutual molecular diffusion of the three species, horizontal and vertical advection, turbulent mixing vertically and horizontally, and production and loss mechanisms.

The continuity equation for mass mixing ratio, $\psi_i = n_i m_i/\rho$, of species, i, is given by

$$\underbrace{\frac{\partial \psi_i}{\partial t}}_{\substack{\text{rate of change}\\\text{of mass mixing}\\\text{ratio of species } i}} = \underbrace{\frac{1}{\rho}m_i S_i}_{\substack{\text{sources}\\\text{and sinks}}} - \underbrace{\frac{V_\theta}{r}\frac{\partial}{\partial \theta}\psi_i - \frac{V_\phi}{r\sin\theta}\frac{\partial}{\partial \phi}\psi_i}_{\text{horizontal advection}}$$

$$- \underbrace{\omega\frac{\partial}{\partial p}\psi_i}_{\text{vertical advection}} - \underbrace{\frac{1}{\rho}\nabla\cdot(n_i m_i \mathbf{C}_i)}_{\text{molecular diffusion}}$$

$$+ \underbrace{\frac{1}{\rho}\frac{\partial}{\partial p}\left(K_T\frac{\partial}{\partial p}m\psi_i\right)}_{\text{eddy diffusion}} \quad (16)$$

where m_i is its molecular mass, S_i represents sources and sinks of the species, n_i, its number density, and C_i are the relative diffusion velocities. The terms on the right-hand side of the continuity equation for the species are, in their respective order, chemical sources and sinks of the species, horizontal meridional and zonal advection, vertical advection, mutual molecular diffusion between species, and eddy diffusion.

The molecular diffusion velocities are evaluated from the general diffusion equation for a multispecies gas given by

$$\frac{1}{n}\sum_{j\neq i}\left(\frac{\psi_i}{m_j D_{ij}}n_j m_j C_j - \frac{\psi_j}{m_j D_{ij}}n_i m_i C_i\right) = \nabla\psi_i + \frac{\psi_i}{m}\nabla m + \left(1-\frac{m_i}{m}\right)\frac{\psi_i}{mp}\nabla p \quad (17)$$

where D_{ij} is the mutual diffusion coefficient between species i and j, and m is the mean molecular mass of the gas. For Earth's thermosphere, consisting of O, O_2, and N_2, equation (17) is a system of three coupled equations for the three major constituents, $i = 1, 2$, and 3.

In the lower atmosphere of planets, the fluid is very well mixed, and the mean molecular mass of the fluid is constant. For instance, the relative proportion of molecular nitrogen and molecular oxygen in the atmosphere is remarkably consistent from the surface to about 110 km altitude. The altitude where mixing gives way to molecular diffusion is known as the homopause or turbopause. Below this level, the air is constantly being mixed by turbulent wave eddies. In large-scale or global physics-based models, this process is usually parameterized as an eddy diffusion coefficient, K_T, (see equation (16)). Above about 110 km, turbulent mixing gives way to molecular mixing processes, and each species begins to be distributed vertically under its own pressure scale height or hydrostatic balance (see equation (4)). A heavy species, such as carbon dioxide, will decrease in concentration with height more rapidly than a light species such as atomic oxygen. Each species will have its own characteristic scale height, $RT/m_i g$, which is the vertical distance a species will decrease in number density by a fraction of $1/e$. Thus, above the homopause, the mean mass of the fluid will change with altitude, as well as other gas parameters such as the specific heat, c_p.

The vertical distribution of species is therefore affected by the balance between turbulent mixing and diffusive separation. Traditionally, the point of transition where both processes contribute equally has been termed the turbopause, which is typically assigned to an altitude of about 110 km for Earth's atmosphere. This altitude, of course, can vary with location and season depending on the strength of the gravity waves and other sources from the lower atmosphere responsible for the mixing and the likelihood the waves will break. The process of gravity wave breaking is a complex field and will not be discussed further.

The global seasonal/latitudinal structure of composition is also affected by a physical process called wind-induced diffusion [*Mayr et al.*, 1978]. This is somewhat of a misnomer because the process is actually advection or the simple transport of species. During solstice, the warmer summer hemisphere and colder winter hemisphere introduces a pressure gradient force in the direction from the summer to winter hemisphere. In the absence of drag, zonal winds would develop such that the Coriolis force from the zonal winds would balance the meridional pressure gradient, a condition known as geostrophic balance. In reality, the zonal winds experience drag from collisions with ions or from viscosity so that this pure geostrophic balance rarely occurs. The imbalance result is an interhemispheric circulation from summer to winter. Closure of this circulation drives an upwelling of material across pressure surfaces in the summer hemisphere and a downwelling in the winter hemisphere. The upwelling causes the heavier molecular-rich gas, which had diffusively separated out at lower altitudes, to be transported upward to increase the mean molecular mass in summer. In winter, the downwelling reduces the mean mass.

The large-scale global seasonal circulation is analogous to a huge interhemispheric mixing cell, or "thermospheric spoon," a global equivalent of the small-scale turbulent mixing cells in the lower atmosphere [*Fuller-Rowell*, 1998]. The implication is that the upper atmosphere is better mixed at solstice. Through the Earth year, there will be peaks in mixing in June and December. At equinox, the weaker global circulation allows the atmosphere to separate out more by molecular diffusion. The difference is subtle, but there are several consequences of this semiannual variation.

First, the globally averaged mean mass will vary semiannually. Second, since the scale height, or thickness of the upper atmosphere, depends on the mean molecular mass, the atmosphere will be more compressed at solstice and more expanded at equinox. This semiannual breathing of the upper atmosphere introduces a semiannual variation in neutral density. Third, since the production and loss rate of the ionosphere is dependent on neutral composition, a semiannual variation in plasma density is introduced. The "thermospheric spoon" does not appear to generate all of the observed magnitude of the semiannual variation. Additional processes from lower atmosphere mixing, and changes in eddy diffusion, are also operating [*Qian et al.*, 2009].

The relatively simple pattern of summer-to-winter circulation is augmented by the addition of the high-latitude magnetospheric, or "geomagnetic," sources. These heat sources tend to reinforce the solar radiation-driven equatorward flow in the summer hemisphere and compete with the poleward flow in winter. During geomagnetic storms, the high-latitude source can dominate the solar-driven circulation.

12. GLOBAL WIND, TEMPERATURE, DENSITY, AND COMPOSITION STRUCTURE

The basic momentum and energy equations and physical processes are captured in the thermospheric components of the Coupled Thermosphere Ionosphere Model (CTIM) and Coupled Thermosphere Ionosphere Plasmasphere electrodynamics (CTIPe) models [*Fuller-Rowell and Rees*, 1980, 1983; *Fuller-Rowell et al.*, 1987, 1996a] and the Thermosphere General Circulation Model (TGCM), Thermosphere Ionosphere General Circulation Model (TIGCM), and Thermosphere Ionosphere Electrodynamics General Circulation Model (TIEGCM) suite of NCAR models [*Dickinson et al.*, 1981, 1984; *Roble et al.*, 1982, 1987, 1988]. An example of the global distribution of temperature and horizontal winds, at a fixed pressure level near 300 km altitude is shown in the top panels of Figure 1. The conditions are December solstice at fairly high solar activity and moderate geomagnetic activity. On the top left are contours of temperature with winds as vectors, and on the top right are contours of the meridional wind. December solstice conditions render the Southern Summer Hemisphere hotter, and Earth's rotation introduces a significant diurnal or day/night temperature and meridional wind difference. The winds at high latitude are the largest because a modest magnetospheric convection electric field has been imposed to drive the ions, which subsequently accelerate the neutrals through ion drag (see term in equations (13) and (14)). The solar-driven winds tend to blow away from the high-pressure area under the subsolar point to the low-pressure area at the antipodal solar point on the nightside. Thus, daytime meridional winds tend to be poleward, while nighttime meridional winds tend to be equatorward and larger partly because of the lower ion drag on the nightside.

The left lower panel of Figure 1 shows the neutral density at a fixed height of 300 km. The same seasonal latitude structure is apparent with higher densities in the summer hemisphere and with a large diurnal variation at mid and low latitudes. The lowest density and temperature is near the antipodal solar point on the nightside. The right lower panel of Figure 1 shows the ratio of the height-integrated atomic oxygen to molecular nitrogen O/N_2. The upwelling in the Southern Summer Hemisphere raises the concentration of the heavier molecular species giving the smaller O/N_2 and larger mean mass compared with the downwelling and increases in O/N_2 or decreases in the mean mass in the winter hemisphere. Since, as described above, the heating and cooling itself does not directly change the O/N_2 ratio, the composition distribution is somewhat different from the temperature and density. The main feature is summer-to-winter circulation creating the seasonal/latitude structure. The peak in the winter hemisphere is displaced to mid latitudes by the high-latitude Joule heating. A secondary feature is the weaker diurnal variation. The mixing of neutral composition by the global circulation impacts the ionosphere. Ion loss rates are faster in the molecular-rich atmosphere so the neutral composition structure gives rise to a seasonal ionospheric anomaly, where dayside winter plasma densities are greater than in winter.

The equations of momentum and energy described above can also support the propagation of large-scale gravity waves. Figure 2 shows a snapshot of the change in neutral wind at mid and low latitudes at 250 km at the December solstice 3 h into a numerical simulation of a step-function increase in high-latitude forcing in the auroral oval (65° to 75° geomagnetic latitude). The wind response is shown within 50° latitude of the geographic equator, to allow for a scale that clearly shows the mid- and low-latitude dynamic response. Whereas at auroral latitudes, the peak neutral winds would be several hundred m s^{-1}, at mid and low latitudes, the wind surges are typically 100 to 200 m s^{-1} above the background circulation. At this time, 3 h into the simulation, the disturbance winds have reached the equator and are beginning to penetrate the other hemisphere and interact with the opposing wavefront from the other pole. The arrival of the wavefront at the geographic equator within 3 h indicates a propagation speed of about 700 m s^{-1}, in this case. This speed is consistent with observations of traveling ionosphere disturbances (TIDs). A vertical cut through the thermosphere would reveal a tilted wave front with the wave propagating more slowly at the lower altitudes [*Richmond and Matsushita*, 1975].

Observations and model simulations reveal a "sloshing" of winds between hemispheres in response to the high-latitude heating during a storm. The net integrated wind effect is for an increase in the global circulation from pole to equator in both hemispheres [*Roble*, 1977; *Forbes*, 2007]. The change in circulation transports all neutral parameters including temperature, density, and species composition. The neutral composition changes and their impact on the ionosphere are dealt with below.

13. THERMAL EXPANSION

There are several misconceptions regarding temperature changes and changes in neutral composition. Heating the atmosphere locally and the resultant thermal expansion does not change the ratio of neutral species (e.g., O/N_2) on pressure surfaces, but it does change on height levels. Since the pressure levels are levels of constant optical depth, ionization rates from solar photons or auroral particles do not change as a result of the heating because the photons or particles have to penetrate the same amount of atmosphere. So to first order, neither the ion production nor loss rates change on the pressure surface. "Real" changes in neutral composition on a pressure surface are caused by the upwelling through the pressure surfaces by the global circulation.

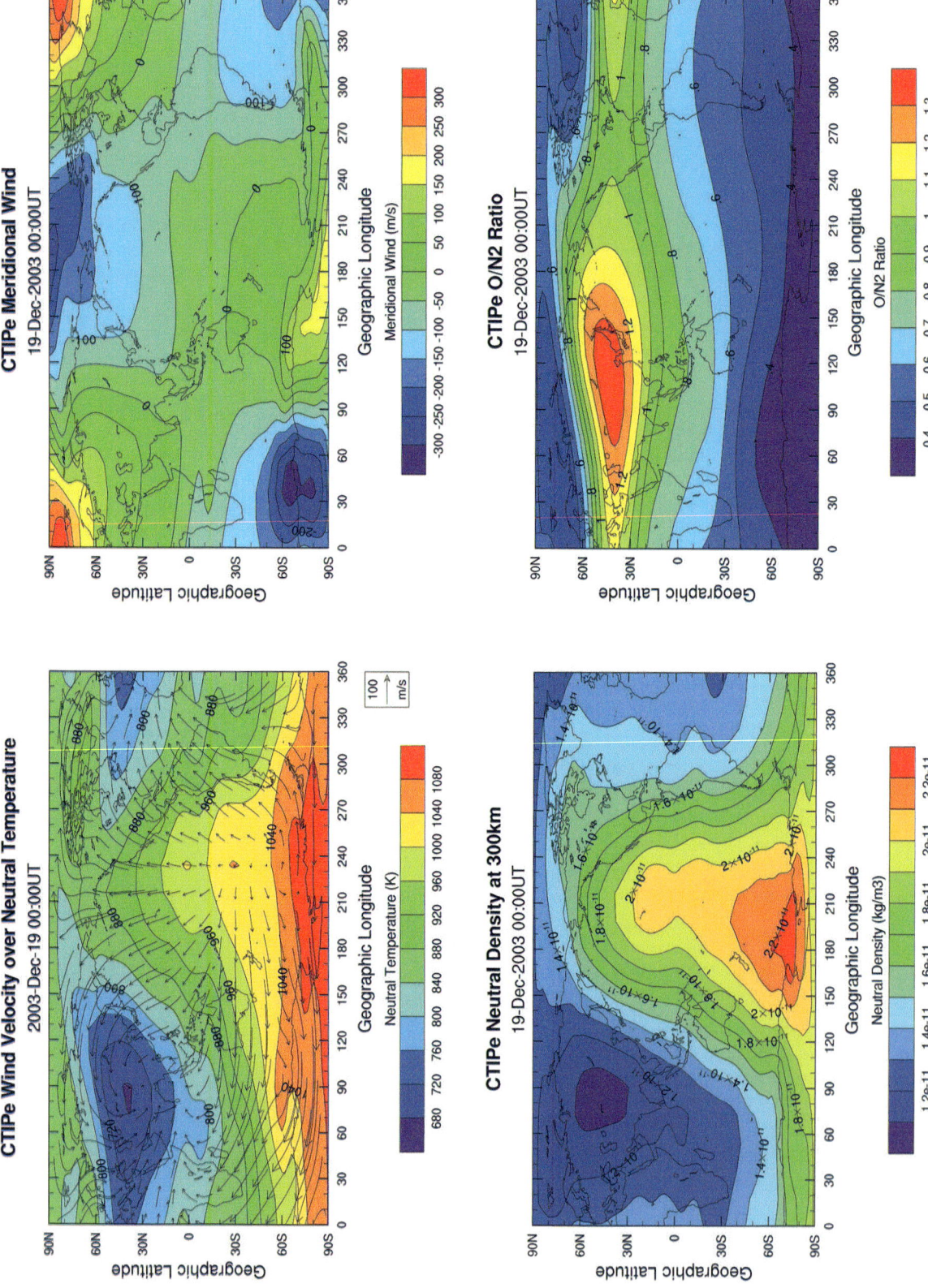

Figure 1. Typical global distribution of some of the major model parameters at the December solstice at moderate solar and geomagnetic activity. (left top) The contours of neutral temperature and total wind vector and (right top) the contours of meridional wind on a fixed pressure level in the upper thermosphere close to 300 km. (left bottom) Neutral density at a fixed altitude of 300 km and (right bottom) the height-integrated O/N$_2$ ratio (figure courtesy of *Mariangel Fedrizzi*, 2012).

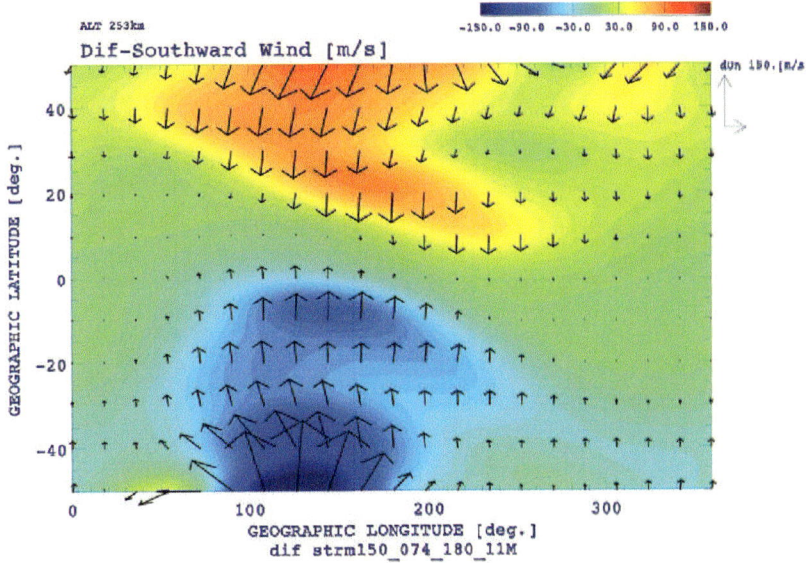

Figure 2. Illustration of the change in the neutral wind at midlatitude and low latitudes at 250 km altitude shortly after a sudden increase in the high-latitude magnetospheric energy input. The region within 50° of the geographic equator is shown at 15 UT, 3 h after the increase in the high-latitude magnetospheric forcing, equivalent to a K_p ~7.

This requires differential heat in one region, the generation of horizontal pressure gradient (or in our case, gradients in the height of the pressure surfaces), divergence in horizontal winds, and hence upwelling (see the continuity equation in equation (8)).

Figure 3 illustrates the consequence of heating and thermal expansion. The physics of this process has been described in detail by *Rishbeth and Garriott* [1969] and *Garriott and Rishbeth* [1963]. If the temperature, T, in a parcel of gas of thickness δh increases to θT, the parcel expands to $\theta \delta h$. The scale-height increases from H to θH, and the number density decreases from n to n/θ. Each species responds to the temperature change in the same way so the ratio of the species remains the same.

14. NEUTRAL COMPOSITION BULGE

During elevated geomagnetic activity, the increase in Joule heating induces additional upwelling of neutral composition at high latitudes and changes in the global circulation [*Prölss*, 2008]. *Fuller-Rowell et al.* [1994, 1996b] referred to the storm time increases as neutral composition "bulges." Figure 4 shows the development and evolution of a composition bulge from a simulation of a 12-h increase in Joule heating during northern summer solstice. Figure 4 shows the increase in mean molecular mass, compared to a background simulation without a storm, on a fixed pressure level in the upper thermosphere around 300 km altitude from 10° latitude to the pole. The generic storm started at 12 UT, so the first two panels show the development of the composition bulges at storm times of 6 and 12 h. The mean mass has increased by 2 to 3 atomic mass units at the end of the storm, at storm time 12 h.

In the subsequent panels, the evolution and recovery of the bulge is followed for another day and a half, at storm times of 18, 24, 36, and 48 h. Once the bulge is created, the storm time winds that created the bulge begin to die down, and Earth's rotation will carry the feature through LT. The magnitude of the "bulge" will gradually decay away as molecular diffusion returns the global distribution to quiet time levels. The subsequent evolution of the shape of the feature as it slowly recovers is controlled by two other distinct processes.

First, the prevailing daytime solar-driven background wind field tends to transport and spread the bulge equatorward in the summer hemisphere, shown in Figure 4. Comparing the zero contour line (i.e., no change in mean mass) in panel (b) at the end of the storm, and in (e) 24 h later, the area of the bulge has more than doubled. The response in the other seasons will be different. In the winter hemisphere, the prevailing daytime solar-driven winds will be toward the polar region, which will constrain the bulge and restrict the transport to midlatitude and low latitudes.

Second, the shape of the bulge and the position of the zero contour boundary are controlled by the diurnal variation of the meridional winds. On the dayside, the meridional wind is poleward, which tends to push the bulge to higher latitudes in

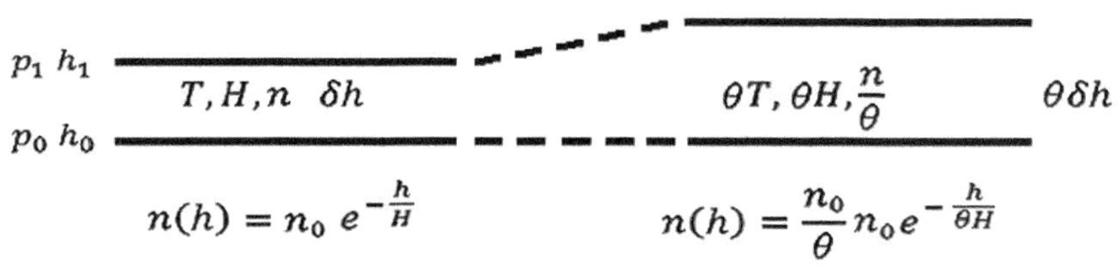

$$p_1 \; h_1 \; \rule{3cm}{0.4pt}$$
$$T, H, n \quad \delta h \qquad\qquad\qquad \theta T, \theta H, \frac{n}{\theta} \qquad\qquad \theta \delta h$$
$$p_0 \; h_0 \; \rule{3cm}{0.4pt}$$

$$n(h) = n_0 \, e^{-\frac{h}{H}} \qquad\qquad n(h) = \frac{n_0}{\theta} n_0 e^{-\frac{h}{\theta H}}$$

Figure 3. Illustration of the response of the thermosphere to heating and thermal expansion (figure courtesy of *Karen O'Loughlin*, 2012).

this LT sector. On the nightside, the winds tend to be more equatorward and larger, which tends to speed up the transport to low latitudes. This effect can be seen by comparing panels (d), (e), and (f). The 90°E longitude sector is at 6 LT in panel (d), and the zero mean mass contour has reached the perimeter latitude of the plot. In panel (e), 12 h later, this sector is now at 18 LT and has been moving through the dayside so has been under the influence of the dayside poleward winds. The zero contour line has moved back poleward a little in response, in spite of the general trend of the bulge to expand equatorward. In panel (f), again another 12 h later, this sector is back at 6 LT and has been moving through the nightside equatorward winds. During this time, as the magnitude of the bulge continues to decay, the zero contour line has again moved equatorward. Note also that by comparing panels (d) and (f) at the same UT, the bulge appears to rotate slower than the Earth, due to the prevailing westward zonal winds in the summer hemisphere.

Figure 4 shows the importance of the neutral winds in controlling both the initial development of the neutral composition structure and also the subsequent evolution and transport of the storm time changes. This dynamics and composition transport can explain the seasonal/diurnal dependence of the mid-latitude ionospheric response [*Rodger et al.*, 1989; *Prölss*, 2008]. The storm time dynamics will also drive the disturbance dynamo electric field [*Blanc and Richmond*, 1980], and the storm dynamics in Figure 2 can also explain the rapid onset of the storm time dynamo processes [*Fuller-Rowell et al.*, 2008].

15. WINDS AND NONLINEARITIES AT HIGH LATITUDE

The basic equations of motion include nonlinearities of the system, which are embedded in the total derivative, dV/dt, or advective transport. At low velocities, the effect of transport by the wind field is relatively minor. The tenuous gas of the upper atmosphere, however, can support wind speeds far exceeding those typical of the lower atmosphere. The degree of nonlinearity of a dynamic system is often defined by the Rossby number, which is the ratio of the acceleration under curvature, V^2/R, and the Coriolis force, fV, where R is the scale length or the radius of curvature of a dynamical system, and f is the Coriolis parameter. The magnitude of the Coriolis parameter, $2\Omega\sin\phi$, depends on the angular velocity of the planet, Ω, and latitude, ϕ.

A Rossby number significantly greater than unity denotes a highly nonlinear system, where transport or advection by the wind field is important. For example, a class 5 hurricane in Earth's troposphere with winds in excess of 70 m s^{-1} and scale sizes of 100 to 200 km is a highly nonlinear system with a Rossby number greater than 5. On the other hand, the hurricane-like vortex associated with the Great Red Spot in Jupiter's lower atmosphere, with wind speeds in excess of 120 m s^{-1}, faster planetary rotation, and a scale size of 10,000 km, has a significantly smaller Rossby number, although nonlinearities are still significant. In comparison, typical upper atmosphere winds on the Earth are 150 m s^{-1} but can exceed 1000 m s^{-1} under the strong ion drag forcing imposed from the magnetosphere at high latitudes. The scale sizes tend to be larger so that Rossby numbers at high latitudes can be similar to Earth's hurricanes.

At high latitudes, the ions respond directly to the imposition of strong magnetospheric electric fields that cause ion drifts of up to a few thousand meters per second. Although collisions between ions and the neutral gas are relatively infrequent above ~160 km, they are sufficient to accelerate the thermosphere to many hundreds of meters per second over periods of tens of minutes or more. The left panel of Figure 5 shows the typical pattern of neutral winds at 200 km altitude poleward of 40° geographic latitude in response to a strong two-cell magnetospheric convection pattern during active geomagnetic activity (K_p ~7). Winds driven by solar heating alone would typically be antisunward (day to night) and reach ~150 m s^{-1} at 300 km in the polar region. With the imposition of magnetospheric convection, there is a sufficient momentum source to accelerate the medium to about 600 m s^{-1}. The first reaction of the neutrals is to follow the ion convection, but other inertial and viscous forces act to introduce asymmetries in the circulation. The clockwise

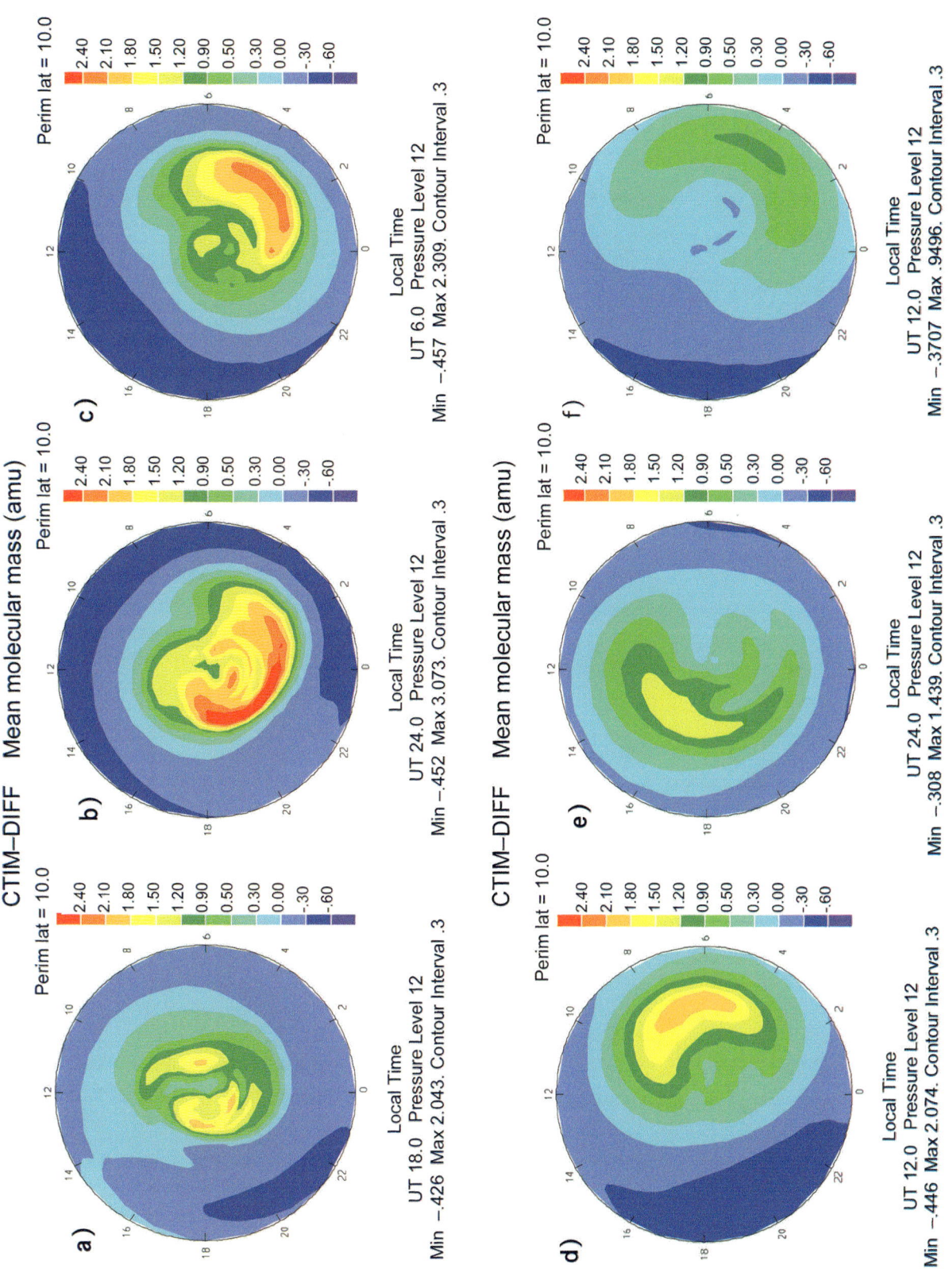

Figure 4. Illustration of the response and evolution of a composition "bulge" in the summer hemisphere upper thermosphere in response to a 12 h geomagnetic storm from 10° latitude to the pole. The response is followed through 36 h of recovery. The figure shows the increase in mean molecular mass in atomic mass units, compared to a background simulation without a storm, on a fixed pressure level in the upper thermosphere around 300 km altitude, from 10° latitude to the pole.

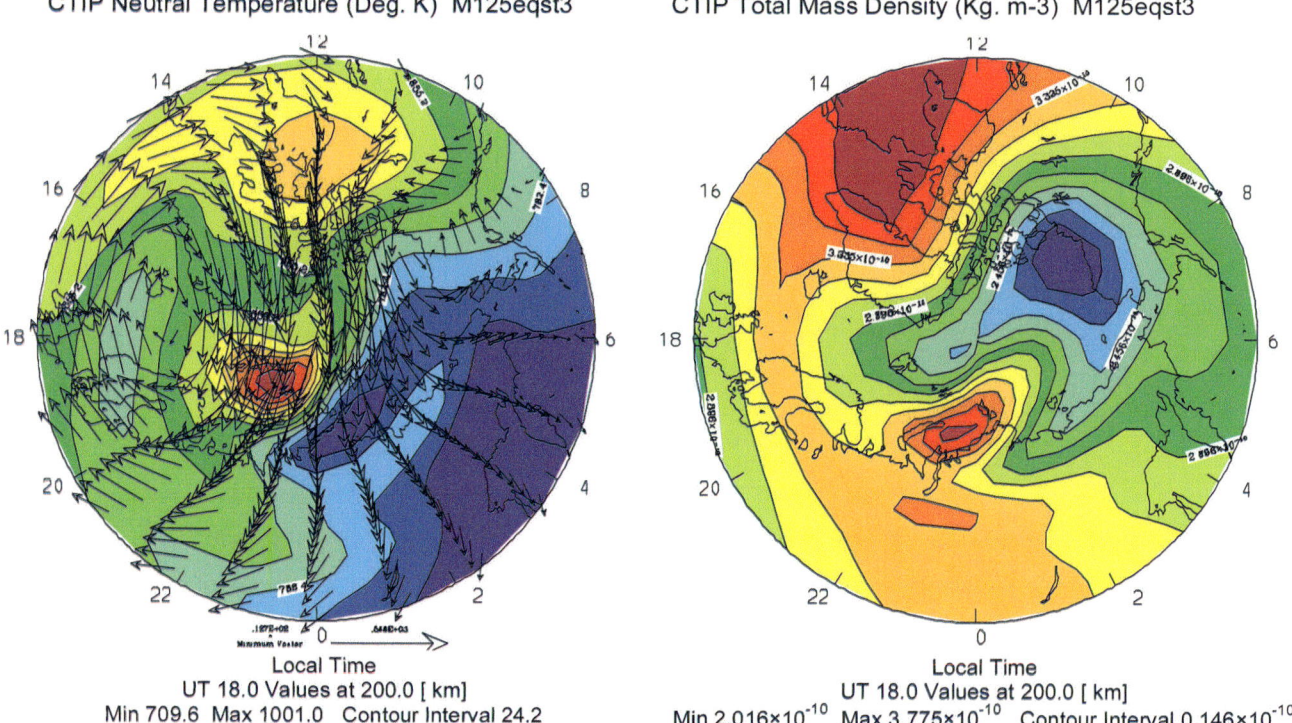

Figure 5. Illustration of the response to elevated geomagnetic activity (K_p ~7) of (left) neutral wind and temperature and (right) neutral density. The region poleward of 50° latitude is shown at 200 km altitude.

vortex excited in the dusk sector is particularly strong owing to an inertial resonance between the ion and neutral convection. The natural motion of the neutral gas is to form a clockwise vortex, owing to the action of the Coriolis force, so there is a natural tendency for this cell to develop and resonate [*Fuller-Rowell et al.*, 1984; *Fuller-Rowell*, 1985].

In contrast, the dawn cell is less well formed. This anticlockwise cell does not have a resonance with the ion convection since momentum is continually being transported out of the cell by the tendency for the vortex to diverge. The neutral temperature structure at this altitude and level of activity, shown in the color contours, is controlled by Joule heating, transport of the gas by the elevated circulation, and adiabatic processes. The warmer dayside gas tends to advect poleward by the neutral circulation, in much the same way as the tongue of ionization is transported over the pole by the magnetospheric convection electric field.

The direction of the forces perpendicular to the motion, in vortices forced by ion drag in a clockwise (dusk sector) or counterclockwise (dawn sector) direction in the Northern Hemisphere polar regions, is shown in Figure 6. The frame of these forces is the "natural" vortex coordinate as described by *Holton* [1972], traditionally used in the lower atmosphere.

In this frame, Coriolis acts to the right tending to converge a clockwise vortex and diverge a counterclockwise vortex. The centrifugal force, V^2/r, is always diverge, and the pressure gradient depends upon whether it is low or high pressure in the core. Note that in this frame, for a radius of curvature of about 2000 km, Coriolis and curvature can balance for a clockwise vortex of about 300 m s^{-1}, which is typical for the wind speed at high latitudes. This is the natural inertial motion of the gas, or inertial "resonance," and is the reason the dusk sector vortex is much more likely to respond to ion drag. In the more normal Earth frame coordinate system, this resonance is just referring to the fact that at these velocities (~300 m s^{-1}), the Coriolis effect will transport material in a clockwise vortex with a radius about 2000 km in the high northern latitudes.

If the wind continues to increase, as in Figure 5, the vortex becomes divergent as the curvature, or the centrifugal effect becomes greater than the Coriolis effect causing a low-pressure core to form. The dawn cell would also tend to have a low-pressure core because it tends to be naturally divergent, and the gas tends to be thrown out of the vortex. The impact on neutral density of these various vortex strengths and directions has been analyzed by *Crowley et al.* [1996] and *Schoendorf et al.* [1996].

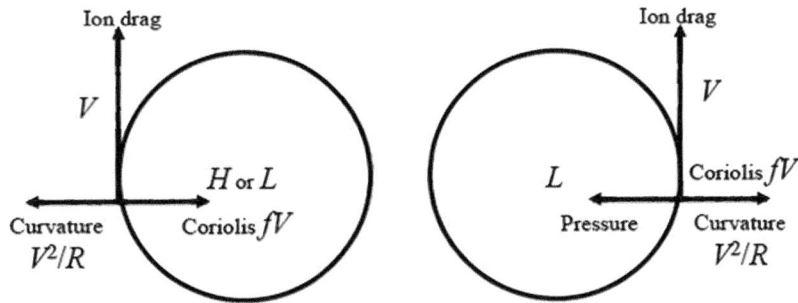

Figure 6. Illustration of the direction of forces for vortices driven by ion drag in the (left) clockwise and (right) counterclockwise direction. The forces are shown in a "natural" vortex coordinate system as suggested by *Holton* [1972].

On the right of Figure 5, the neutral density pattern is shown at high latitudes at 200 km as a consequence of the vortices. Initially, the antisunward flow from the dayside over the pole produces a tongue of dense air over the polar region. The dynamically driven vortices subsequently impose additional structure. On the dawnside, the counterclockwise divergent vortex imposes the low-density region or density "hole" [*Crowley et al.*, 1996]. On the duskside, for modest geomagnetic activity with winds less than ~300 m s^{-1}, there would be no matching hole. As the winds pick up, however, as in Figure 5, another dusk sector density hole starts to form as the vortex becomes more divergent and splits the tongue in two.

16. CONCLUSIONS

Our level of understanding of Earth's upper atmosphere is mature, and many of the controlling physical processes are understood at a reasonable level of sophistication. Large-scale thermospheric models, such as CTIPe and TIEGCM, have contributed significantly to this understanding and do well in capturing the global dynamics, energy budget, and composition, and their response to solar and geomagnetic activity. The basic structure of these models is a natural adaptation of the so-called "primitive equation" models of the troposphere. The application to the thermosphere differs in that neutral composition is no longer well mixed, and each species separates out vertically according to its mass. The gas constant and specific heats thus need to be height dependent. This becomes all the more important in the mid and upper thermosphere since lighter species, such as atomic oxygen, can be created by dissociation of the molecular species and become the dominant species.

Another difference from the troposphere is that ion-neutral interactions need to be included because the medium is weakly ionized. This results in the addition of an ion drag term in the momentum equation and a Joule heating term in the energy equation. The ion drag from magnetosphere-imposed electric fields causes wind speeds of up to ~1000 m s^{-1} at high latitudes making nonlinear advection effects more important in the thermosphere, and inertial resonances can form. The ionosphere clearly has an important impact on the thermosphere.

The thermosphere also has a strong impact on the ionosphere because it is the medium from which the ionosphere is produced by solar and auroral ionization and destroyed by recombination chemistry. As the thermosphere expands and contracts, the height at which the plasma is produced rises and falls. Existing plasma is also transported vertically with the neutral medium as it expands and contracts at mid and high latitudes. In addition, horizontal neutral winds can push plasma parallel to the magnetic field, raising and lowering the peak height of the ionosphere at mid latitudes. At low latitudes, the plasma vertical motion is constrained by the magnetic field, but horizontal winds can transport plasma along field lines.

The coupling with the ionosphere renders the system complex, so that it is not always possible to understand the relative contribution of various physical processes. It is challenging to separate the physics processes resulting from geomagnetic storms and solar flares so as to bring our understanding to a quantitative level where a specification can lead to an accurate forecast of the system. Combining data with complex physical models, and the adoption of data assimilation techniques, puts a much more severe constraint on our perceived understanding, and it offers one of the challenges for the future.

The CTIPe and TIEGCM hydrostatic thermospheres can support large vertical winds and are valid for scale sizes of 50 km or more and time scales longer than about ~5 min, which is the hydrostatic adjustment time scale. Nonhydrostatic processes that are important on shorter spatial and temporal scales are described by *Deng and Ridley* [this volume]. These new types of models are much more computationally intensive and have additional numerical and accuracy challenges because the vertical momentum equation must handle the sometimes subtle departures between two large forces: gravity and pressure. Detailed comparisons between hydrostatic and nonhydrostatic models will lead to a better understanding of the importance of acoustic waves. This transition from hydrostatic to nonhydrostatic models is proceeding for the troposphere as well.

Another exciting area is the development of whole atmosphere models [*Akmaev*, 2011; *Fuller-Rowell et al.*, 2011; *Liu and Roble*, 2002; *Liu et al.*, 2010] and the effort to quantify the impact of the lower atmosphere dynamics on the upper atmosphere. The energy of the dense lower atmosphere far exceeds that of the upper atmosphere. Even a very small fraction of this huge source propagating upward has a dramatic influence on the upper atmosphere system. The energy and momentum can be carried upward by any number of physical processes, such as gravity waves, tides, and planetary waves. Our understanding of the nature of this coupling is at a rudimentary level and is a significant challenge for the future. The combined effects of the apparently chaotic lower atmosphere and the strong external forcing of the upper atmosphere make the prospect of forecasting the system extremely challenging.

REFERENCES

Akmaev, R. A. (2011), Whole atmosphere modeling: Connecting terrestrial and space weather, *Rev. Geophys.*, *49*, RG4004, doi:10.1029/2011RG000364.

Blanc, M., and A. D. Richmond (1980), The ionospheric disturbance dynamo, *J. Geophys. Res.*, *85*(A4), 1669–1686, doi:10.1029/JA085iA04p01669.

Chapman, S., and T. G. Cowling (1970), *The Mathematical Theory of Non-Uniform Gases: An Account of the Kinetic Theory of Viscosity, Thermal Conduction and Diffusion in Gases*, 3rd ed., 422 pp., Cambridge Univ. Press, Cambridge, U. K.

Crowley, G., J. Schoendorf, R. G. Roble, and F. A. Marcos (1996), Cellular structures in the high-latitude thermosphere, *J. Geophys. Res.*, *101*(A1), 211–223, doi:10.1029/95JA02584.

Deng, Y., and A. J. Ridley (2012), The global ionosphere-thermosphere model and the non-hydrostatic processes, in *Modeling the Ionosphere-Thermosphere System*, *Geophys. Monogr. Ser.*, doi:10.1029/2012GM001296, this volume.

Deng, Y., A. D. Richmond, A. J. Ridley, and H.-L. Liu (2008), Assessment of the non-hydrostatic effect on the upper atmosphere using a general circulation model (GCM), *Geophys. Res. Lett.*, *35*, L01104, doi:10.1029/2007GL032182.

Dickinson, R. E., E. C. Ridley, and R. G. Roble (1981), A three-dimensional general circulation model of the thermosphere, *J. Geophys. Res.*, *86*(A3), 1499–1512, doi:10.1029/JA086iA03p01499.

Dickinson, R. E., E. C. Ridley, and R. G. Roble (1984), Thermospheric general circulation with coupled dynamics and composition, *J. Atmos. Sci.*, *41*(2), 205–219, doi:10.1175/1520-0469(1984)041<0205:TGCWCD>2.0.CO;2.

Forbes, J. M. (2007), Dynamics of the thermosphere, *J. Meteorol. Soc. Jpn.*, *85B*, 193–213.

Fuller-Rowell, T. J. (1985), A two-dimensional, high-resolution, nested-grid model of the thermosphere: 2. Response of the thermosphere to narrow and broad electrodynamic features, *J. Geophys. Res.*, *90*(A7), 6567–6586, doi:10.1029/JA090iA07p06567.

Fuller-Rowell, T. J. (1998), The "thermospheric spoon": A mechanism for the semiannual density variation, *J. Geophys. Res.*, *103*(A3), 3951–3956, doi:10.1029/97JA03335.

Fuller-Rowell, T. J., and D. Rees (1980), A three-dimensional time-dependent, global model of the thermosphere, *J. Atmos. Sci.*, *37*(11), 2545–2567, doi:10.1175/1520-0469(1980)037<2545:ATDTDG>2.0.CO;2.

Fuller-Rowell, T. J., and D. Rees (1983), Derivation of a conservation equation for mean molecular weight for a two-constituent gas within a three-dimensional, time-dependent model of the thermosphere, *Planet. Space Sci.*, *31*(10), 1209–1222, doi:10.1016/0032-0633(83)90112-5.

Fuller-Rowell, T. J., D. Rees, S. Quegan, G. J. Bailey, and R. J. Moffett (1984), The effect of realistic conductivities on the high-latitude neutral thermospheric circulation, *Planet. Space Sci.*, *32*(4), 469–480, doi:10.1016/0032-0633(84)90126-0.

Fuller-Rowell, T. J., D. Rees, S. Quegan, R. J. Moffett, and G. J. Bailey (1987), Interactions between neutral thermospheric composition and the polar ionosphere using a coupled ionosphere-thermosphere model, *J. Geophys. Res.*, *92*(A7), 7744–7748, doi:10.1029/JA092iA07p07744.

Fuller-Rowell, T. J., M. V. Codrescu, R. J. Moffett, and S. Quegan (1994), Response of the thermosphere and ionosphere to geomagnetic storms, *J. Geophys. Res.*, *99*(A3), 3893–3914, doi:10.1029/93JA02015.

Fuller-Rowell, T. J., D. Rees, S. Quegan, R. J. Moffett, M. V. Codrescu, and G. H. Millward (1996a), A coupled thermosphere-ionosphere model (CTIM), in *STEP Handbook of Ionospheric Models*, edited by R. W. Schunk, pp. 217–238, Utah State Univ., Logan.

Fuller-Rowell, T. J., M. V. Codrescu, H. Rishbeth, R. J. Moffett, and S. Quegan (1996b), On the seasonal response of the thermosphere and ionosphere to geomagnetic storms, *J. Geophys. Res.*, *101*(A2), 2343–2353, doi:10.1029/95JA01614.

Fuller-Rowell, T. J., A. D. Richmond, and N. Maruyama (2008), Global modeling of storm-time thermospheric dynamics and electrodynamics, in *Midlatitude Ionospheric Dynamics and Disturbances*, *Geophys. Monogr. Ser.*, vol. 181, edited by P. M. Kintner Jr. et al., pp. 187–200, AGU, Washington, D. C., doi:10.1029/181GM18.

Fuller-Rowell, T., H. Wang, R. Akmaev, F. Wu, T.-W. Fang, M. Iredell, and A. Richmond (2011), Forecasting the dynamic and electrodynamic response to the January 2009 sudden stratospheric warming, *Geophys. Res. Lett.*, *38*, L13102, doi:10.1029/2011GL047732.

Garriott, O. K., and H. Rishbeth (1963), Effects of temperature changes on the electron density profile in the F_2 layer, *Planet. Space Sci.*, *11*(6), 587–590, doi:10.1016/0032-0633(63)90165-X.

Holton, J. R. (1972), *An Introduction to Dynamic Meteorology*, *Int. Geophys. Ser.*, vol. 48, Academic Press, San Diego, Calif.

Liu, H.-L., and R. G. Roble (2002), A study of a self-generated stratospheric sudden warming and its mesospheric–lower thermospheric impacts using the coupled TIME-GCM/CCM3, *J. Geophys. Res.*, *107*(D23), 4695, doi:10.1029/2001JD001533.

Liu, H.-L., et al. (2010), Thermosphere extension of the Whole Atmosphere Community Climate Model, *J. Geophys. Res.*, *115*, A12302, doi:10.1029/2010JA015586.

Mayr, H. G., I. Harris, and N. W. Spencer (1978), Some properties of upper atmosphere dynamics, *Rev. Geophys.*, *16*(4), 539–565, doi:10.1029/RG016i004p00539.

Prölss, G. W. (2008), Ionospheric storms at mid-latitude: A short review, in *Midlatitude Ionospheric Dynamics and Disturbances*, *Geophys. Monogr. Ser.*, vol. 181, edited by P. M. Kintner Jr. et al., pp. 9–24, AGU, Washington, D. C., doi:10.1029/181GM03.

Qian, L., S. C. Solomon, and T. J. Kane (2009), Seasonal variation of thermospheric density and composition, *J. Geophys. Res.*, *114*, A01312, doi:10.1029/2008JA013643.

Richmond, A. D., and S. Matsushita (1975), Thermospheric response to a magnetic substorm, *J. Geophys. Res.*, *80*(19), 2839–2850, doi:10.1029/JA080i019p02839.

Ridley, A. J., Y. Deng, and G. Tóth (2006), The global ionosphere–thermosphere model, *J. Atmos. Sol. Terr. Phys.*, *68*(8), 839–864, doi:10.1016/j.jastp.2006.01.008.

Rishbeth, H., and O. K. Garriott (1969), *Introduction to Ionospheric Physics*, *Int. Geophys. Ser.*, vol. 14, 331 pp., Academic Press, New York.

Rodger, A. S., G. L. Wrenn, and H. Rishbeth (1989), Geomagnetic storms in the Antarctic *F*-region. II. Physical interpretation, *J. Atmos. Terr. Phys.*, *5*(11–12), 851–866, doi:10.1016/0021-9169(89)90002-0.

Roble, R. G. (1977), *The Upper Atmosphere and Magnetosphere*, Natl. Acad. of Sci., Washington, D. C.

Roble, R. G., R. E. Dickinson, and E. C. Ridley (1982), Global circulation and temperature structure of thermosphere with high-latitude plasma convection, *J. Geophys. Res.*, *87*(A3), 1599–1614, doi:10.1029/JA087iA03p01599.

Roble, R. G., E. C. Ridley, and R. E. Dickinson (1987), On the global mean structure of the thermosphere, *J. Geophys. Res.*, *92*(A8), 8745–8758, doi:10.1029/JA092iA08p08745.

Roble, R. G., E. C. Ridley, A. D. Richmond, and R. E. Dickinson (1988), A coupled thermosphere/ionosphere general circulation model, *Geophys. Res. Lett.*, *15*(12), 1325–1328, doi:10.1029/GL015i012p01325.

Schoendorf, J., G. Crowley, and R. G. Roble (1996), Neutral density cells in the high latitude thermosphere—1. Solar maximum cell morphology and data analysis, *J. Atmos. Terr. Phys.*, *58*(15), 1751–1768, doi:10.1016/0021-9169(95)00165-4.

T. Fuller-Rowell, CIRES, University of Colorado and NOAA Space Weather Prediction Center, Boulder, CO 80305, USA. (Tim.Fuller-Rowell@noaa.gov)

Solar Cycle Changes in the Photochemistry of the Ionosphere and Thermosphere

P. G. Richards

School of Physics, Astronomy and Computational Sciences, George Mason University, Fairfax, Virginia, USA

This paper uses a comprehensive ionosphere model to examine the solar cycle variation of the chemistry of O^+, N_2^+, NO^+, O_2^+, and N^+ between 100 and 500 km altitude in the Earth's ionosphere at equinox. It is found that solar cycle increases in the atomic ion densities are substantial, but the changes in the molecular ion densities are relatively small. This difference is primarily because increases in the molecular ion production rates are counterbalanced by the increased losses due to greatly increased electron recombination rates above 200 km. Also, in the F_1 region of the ionosphere between 120 and 200 km, potential solar cycle increases in ion production rates due to increasing solar EUV intensity are largely offset by increased attenuation from the increased O and N_2 column densities. The solar cycle variation of the molecular ion densities is largest in the E region due to the large solar cycle increases in the soft X-ray irradiances that create a large increase in photoelectron flux. In general, the sources and sinks of the ions scale with solar activity as would be expected from the changes in solar EUV intensity with the notable exceptions of reactions involving electrons and the O^+ reaction with N_2, which increases much more than the increase in solar EUV intensity due to the increased vibrational excitation of N_2.

1. INTRODUCTION

In a recent paper, *Richards* [2011] reexamined the chemistry of the major ions by comparing a local equilibrium photochemical model with measurements by the AE-C satellite between 140 and 400 km altitude. This study took into account advances in laboratory measurements of a number of important reaction rates and improvement in our understanding of the solar EUV irradiances and photoelectron fluxes. It showed that the photochemistry model reproduced the average measured densities of N_2^+, NO^+, O_2^+, and N^+ very well for the solar minimum conditions that prevailed in 1974. Unfortunately, there are no similar altitude profiles of ion density measurements for solar maximum conditions to compare model and data.

This paper uses the field line interhemispheric plasma (FLIP) model to compare the photochemistry of O^+, N_2^+, NO^+, O_2^+, and N^+ in the Earth's ionosphere for solar minimum and solar maximum [*Richards et al.*, 2010]. *Richards* [2011] did not examine the chemistry of the $O^+(^4S)$ ion because it is controlled by diffusion and cannot be represented by a simple local equilibrium photochemical model. The FLIP model incorporates diffusion in addition to the full ion photochemistry. This paper also presents the results between 100 and 140 km that were not included in the previous paper.

Figure 1 based on the work of *Richards* [2011] schematically summarizes the major processes in the ion chemistry of the Earth's ionosphere. The circles represent the major neutral species, the boxes represent the ions, and the arrows with labels show the various reactions and ionization processes. The arrows are labeled with the reacting species. The important reaction rate coefficients were discussed by *Richards* [2011] and are listed in Table 1. The major sources of ionization are photoionization and photoelectron impact ionization of O, O_2, and N_2. Photoionization is labeled $h\nu$ and

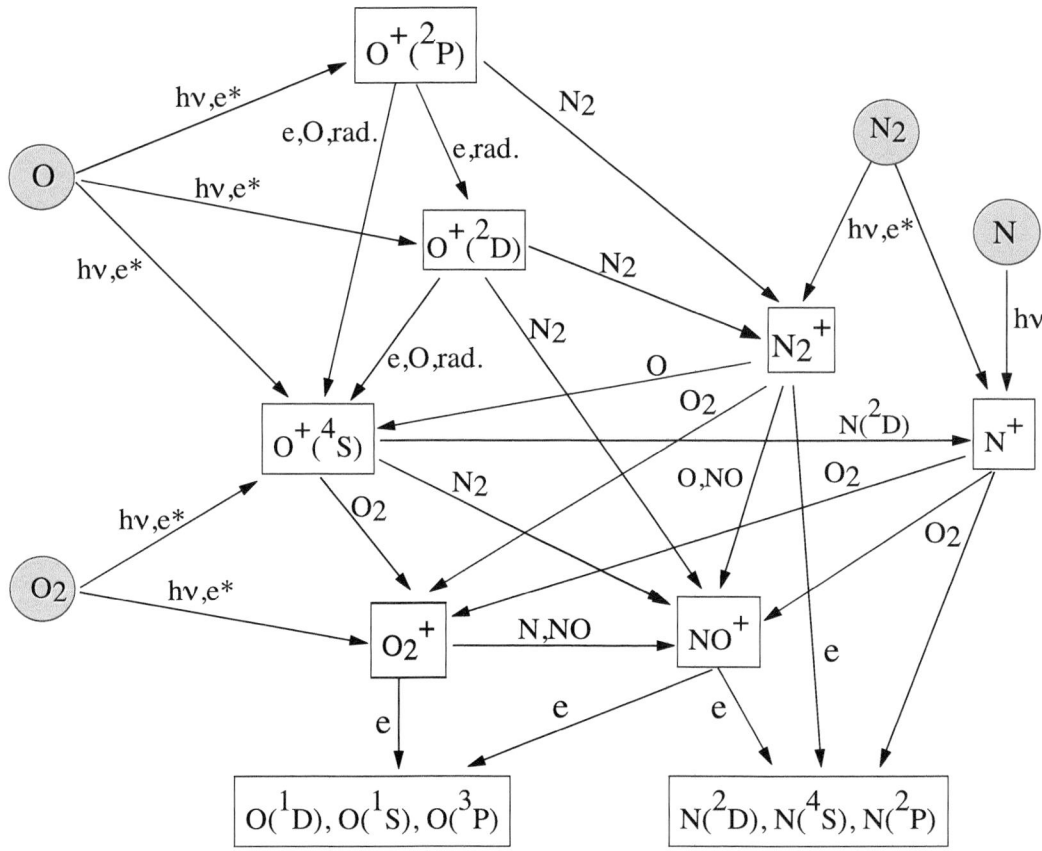

Figure 1. Schematic representation of the major processes in the ion chemistry of the Earth's ionosphere based on the work of *Richards* [2011]. The circles represent the major neutral species, the boxes represent the important ions, and the arrows with labels show the various reactions and ionization processes. Photoelectron impact ionization is labeled e*, photoionization is labeled $h\nu$, and spontaneous radiation is labeled rad. The chemical reactions are labeled with the reacting species.

photoelectron impact ionization is labeled e*. Photoelectron impact ionization constitutes 25%–30% of the total source of ionization above 200 km but becomes the dominant source below 150 km.

2. RESULTS

This section presents a comparison of the ion photochemistry at noon for 6 September 2005 (solar minimum) and 17 March 1990 (solar maximum) for the location of the Millstone Hill radar (43°N, 288°E). These dates were chosen because they were periods of quiet solar and magnetic conditions with good data availability.

The FLIP model calculations in this paper were performed with the high-resolution version of the solar EUV irradiance model for aeronomic calculations (HEUVAC) model [*Richards et al.*, 2006]. This is a high-resolution version of the solar EUV irradiance model for aeronomic calculations (EUVAC) [*Richards et al.*, 1994]. The magnitudes of the model irradiances are the same above 50 Å, but HEUVAC includes irradiances below 50 Å that are important in the E region ionosphere. The HEUVAC and EUVAC models agree with recent irradiance measurements within measurement uncertainties [*Solomon and Qian*, 2005, *Woods et al.*, 2008]. The FLIP model uses the HEUVAC irradiances to calculate the photoelectron fluxes. These photoelectron fluxes agree well with measurements by the FAST satellite [*Richards and Peterson*, 2008; *Peterson et al.*, 2009]. The ionospheric photoelectron flux is determined by solar EUV irradiances with wavelengths below 400 Å.

Figure 2 shows the neutral densities for noon on 5 September 2005 and 17 March 1990 (broken lines) from the National Research Laboratory Mass Spectrometer, Incoherent Scatter Radar Extended thermosphere model [*Picone et al.*, 2002]. These dates are for different equinoxes, but they are representative of the model solar cycle variation for either the equinox. The behavior of the three major gases is distinct. The N_2 density does not change much below 175 km but

Table 1. Ion Reaction Rate Coefficients

Reaction	Rate Coefficient (cm^3 s^{-1})	Reference
$N^+ + NO \to N_2^+ + O$	$8.33 \times 10^{-11} (300/T_i)^{0.24}$	Fahey et al. [1981]
$N^+ + NO \to NO^+ + N(^4S)$	$4.72 \times 10^{-10} (300/T_i)^{0.24}$	Anicich [1993]
$N^+ + O \to O^+ + N$	2.2×10^{-12}	Constantinides et al. [1979], Bates [1989]
$N^+ + O_2 \to NO^+ + O(^1D)$	36%	Dotan et al. [1997], O'Keefe et al. [1986]
$N^+ + O_2 \to NO^+ + O(^3P)$	9%	Dotan et al. [1997], O'Keefe et al. [1986]
$N^+ + O_2 \to O^+(^4S) + NO$	5%	Dotan et al. [1997], O'Keefe et al. [1986]
$N^+ + O_2 \to O_2^+ + N(^2D)$	15%	Dotan et al. [1997], O'Keefe et al. [1986]
$N^+ + O_2 \to O_2^+ + N(^4S)$	35%	Dotan et al. [1997], O'Keefe et al. [1986]
$N_2^+ + e \to N + N$	$2.2 \times 10^{-7} (300/T_e)^{0.39}$	Zipf [1980]
$N_2^+ + N(^4S) \to N_2 + N^+$	1.0×10^{-11}	Ferguson [1973]
$N_2^+ + NO \to N_2 + NO^+$	3.6×10^{-10}	Scott et al. [1999]
$N_2^+ + O \to NO^+ + N$	$1.33 \times 10^{-10} (300/T_i)^{0.44}$ $(T_i \leq 1500)$	Scott et al. [1999]
$N_2^+ + O \to O^+ + N_2$	$7.0 \times 10^{-12} (300/T_i)^{0.21}$ $(T_i \leq 1500)$	McFarland et al. [1974]
$N_2^+ + O_2 \to O_2^+ + N_2$	$5.1 \times 10^{-11} (300/T_i)^{1.16}$ $(T_i \leq 1000)$	Scott et al. [1999]
	$1.26 \times 10^{-11} (T_i/1000)^{0.67}$ $(T_i > 1000)$	
$NO^+ + e \to N(^2D) + O$	$3.4 \times 10^{-7} (300/T_e)^{0.85}$	Vejby-Christensen et al. [1998]
$NO^+ + e \to N(^4S) + O$	$0.6 \times 10^{-7} (300/T_e)^{0.85}$	Vejby-Christensen et al. [1998]
$O^+ + H \to O + H^+$	6.4×10^{-10}	Anicich [1993]
$O^+ + N(^2D) \to N^+ + O$	1.3×10^{-10}	Constantinides et al. [1979], Bates et al. [1989]
$O^+ + N_2 \to NO^+ + N$	$1.2 \times 10^{-12} (300/T_i)^{0.45}$ $(T_i \leq 1000)$	Hierl et al. [1997]
	$7.0 \times 10^{-13} (T_i/1000)^{2.12}$ $(T_i > 1000)$	
$O^+ + NO \to NO^+ + O$	$7.0 \times 10^{-13} (300/T_i)^{-0.87}$	Dotan and Viggiano [1999]
$O^+ + O_2 \to O_2^+ + O$	$1.6 \times 10^{-11} (300/T_i)^{0.52}$ $(T_i \leq 900)$	Hierl et al. [1997]
	$9.0 \times 10^{-12} (T_i/900)^{0.92}$ $(T_i > 900)$	
$O_2^+ + e \to O + O$	$1.95 \times 10^{-7} (300/T_e)^{0.70}$ $(T_i \leq 1200)$	Mehr and Biondi [1969]
	$7.39 \times 10^{-8} (1200/T_e)^{0.56}$ $(T_i > 1200)$	
$O_2^+ + N(^2D) \to NO^+ + O$	1.3×10^{-10}	Goldan et al. [1966]
$O_2^+ + N(^2D) \to N^+ + O$	8.65×10^{-10}	O'Keefe et al. [1986]
$O_2^+ + N(^2P) \to N + O_2^+$	2.2×10^{-11}	Zipf et al. [1980]
$O_2^+ + N(^4S) \to NO^+ + O$	1.0×10^{-10}	Scott et al. [1999]
$O_2^+ + NO \to NO^+ + O_2$	4.5×10^{-10}	Midey and Viggiano [1999]
$O^+(^2D) + e \to O^+(^4S) + e$	$6.03 \times 10^{-8} (300/T_e)^{0.5}$	McLaughlin and Bell [1998]
$O^+(^2D) + N \to N^+ + O$	1.5×10^{-10}	Dalgarno [1979]
$O^+(^2D) + N_2 \to N_2^+ + O$	$1.5 \times 10^{-10} (300/T_i)^{-0.55}$	Li et al. [1997], see text
$O^+(^2D) + N_2 \to NO^+ + N$	2.5×10^{-11}	Li et al. [1997]
$O^+(^2D) + NO \to NO^+ + O$	1.2×10^{-9}	Glosik et al. [1978]
$O^+(^2D) + O \to O^+(^4S) + O$	1.0×10^{-11}	Torr and Torr [1979]
$O^+(^2D) + O_2 \to O_2^+ + O$	7.0×10^{-10}	Johnsen and Biondi [1980]
$O^+(^2P) + e \to O^+(^2D) + e$	$1.84 \times 10^{-7} (300/T_e)^{0.5}$	McLaughlin and Bell [1998]
$O^+(^2P) + e \to O^+(^4S) + e$	$3.03 \times 10^{-8} (300/T_e)^{0.5}$	McLaughlin and Bell [1998]
$O^+(^2P) + N_2 \to N_2^+ + O$	$2.0 \times 10^{-10} (300/T_i)^{-0.55}$	Li et al. [1997], Chang et al. [1993]
$O^+(^2P) + O \to O^+ + O$	4.0×10^{-10}	Chang et al. [1993]
$O^+(^2P) + O_2 \to O^+(^4S) + O_2$	1.3×10^{-10}	Glosik et al. [1978]
$O^+(^2P) + O_2 \to O_2^+ + O$	1.3×10^{-10}	Glosik et al. [1978]

then increases above that altitude due to the higher neutral temperature that increases the scale height. The behavior of O and O_2 does not follow the same straightforward scale height response to the higher temperatures. The O_2 density shows little change, while O increases at all altitudes. The likely explanation is that both are affected by dynamics and the increased photodissociation of O_2 with increased solar activity. The scale height effect is also seen in the O density above 200 km. As the major gas throughout most of the atmosphere, the N_2 density is not greatly affected by either dynamics or chemistry. Also, when N_2 is dissociated, it mostly recombines locally, whereas when O_2 is dissociated,

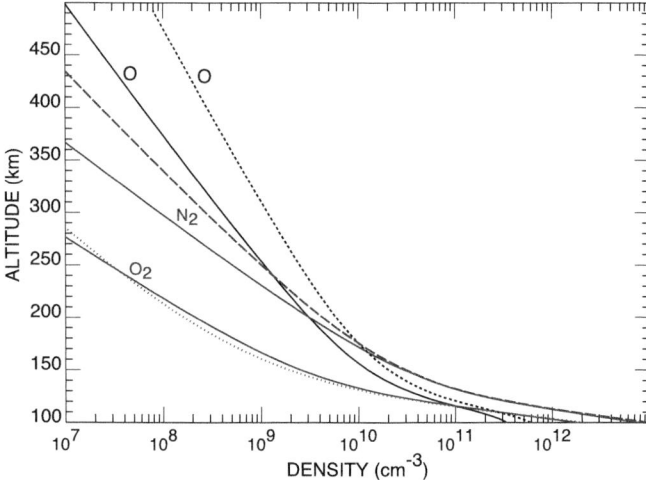

Figure 2. Model neutral densities for 5 September 2005 (full lines) and 17 March 1990 (broken lines).

it must diffuse below 100 km, where it is able to recombine via three-body reactions.

Figure 3 shows the FLIP model ion and electron densities for 5 September 2005 and 17 March 1990. The electron density equals the total ion density. The solid circles are the measured electron densities. Both days are magnetically quiet, and the model reproduces the measured density very well except in the E region on 6 September 2005 near 108 km. The model does not show an E region peak, and it underestimates the measured density by about 40%. The discrepancy is also seen in the data on the four following days. The 40% difference represents about a factor of 2 in either the ion sources or sinks, or a combination of both. This is a long-standing and well-known problem in modeling the E region electron density [*Solomon*, 2006]. The FLIP model does generally produce reasonable agreement with measured E region peak densities at solar maximum.

In the region above 200 km where O^+ is dominant, the electron density increases by almost an order of magnitude between solar minimum and maximum. An interesting aspect of the ion densities in Figure 3 is that the relative increase in the peak electron density is greater than the increase in the solar irradiance intensity because of the increase in the atomic to molecular neutral gas density ratio. The model electron density increase would be even greater if not for the greater effect of vibrational excitation of N_2 on the $O^+(^4S) + N_2$ loss rate at solar maximum.

Another interesting aspect is that the solar cycle changes in the molecular ion densities are much smaller than the changes in the O^+ and N^+ densities even though the production rate changes are comparable. The basic reason for this difference is that, for molecular ions, the enhanced production rates are counteracted by the greatly increased recombination rate as a result of the higher electron density. The molecular ion loss rate is proportional to the electron density, while the atomic ion loss rates are proportional to the neutral densities. The increases in the E region molecular ion densities are larger than in the F region primarily because the increases in the solar soft X-ray irradiances are larger than the increases at longer wavelengths that create the ionization above 150 km.

There are seasonal as well as solar cycle variations in the neutral densities. In general, for all ion species, reactions with N_2 will be more important in summer than in winter above 200 km because of the smaller O to N_2 density ratio. The change in neutral density ratio along with increased vibration excitation of N_2 are the major causes of the well-known seasonal anomaly in the peak electron density where the summer density is lower than the winter density at midlatitudes.

2.1. $O^+(^4S)$ Sources and Sinks

The O^+ ion occupies a pivotal place in the chemical scheme because it is produced in large quantities above 150 km, and it acts as a substantial source for all the molecular ions. The ionization of atomic oxygen by photons and photoelectrons primarily creates O^+ in the ground state $O^+(^4S)$ and the two metastable states $O^+(^2D)$ and $O^+(^2P)$. The metastable species undergo chemical reactions because they correspond to forbidden electronic transitions and have long radiative lifetimes. Ionization can also produce the two minor states, $O^+(^4P^*)$ and $O^+(^2P^*)$, where the * indicates that the electron was ejected from the inner

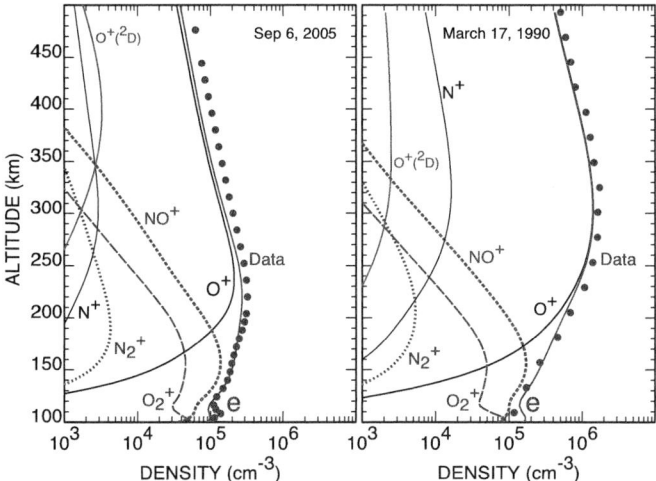

Figure 3. Modeled ion densities (lines) for 6 September 2005 and 17 March 1990. The symbols are the measured ion densities.

subshell 2S. These upper ionization states do not undergo reactions because $O^+(^4P^*)$ decays promptly to $O^+(^4S)$, while $O^+(^2P^*)$ decays promptly to either $O^+(^2D)$ and $O^+(^2P)$.

Figure 4 shows the FLIP model sources and sinks for $O^+(^4S)$ for noon on 6 September 2005 and 17 March 1990 for Millstone Hill. Photoionization ($h\nu$) is the most important source above 150 km, but photoelectron impact ionization of O (e*) becomes the dominant source in the E region. Photoelectron production rate increases by an order of magnitude in the E region at solar maximum due to the large solar cycle increase in the soft X-ray irradiance. Photoionization of O_2 is also important in the E region. However, O^+ production is not a major source of ions in the E region due to the relatively low O and O_2 densities.

The solar cycle increase in the photoionization rate is about a factor of 3 near its peak around 175 km mostly because of about a factor of 2 increase in O density rather than the increase in solar EUV irradiance. The relatively modest increase in production rate below 200 km is because the increase in solar irradiance is somewhat offset by the increased attenuation due to the larger O and N_2 column densities. The solar cycle difference in production rate increases with altitude to around an order of magnitude at 400 km because of the increase in the solar EUV irradiance and the atomic oxygen density. Attenuation of the EUV solar irradiance is small above 250 km. Electron quenching of $O^+(^2D)$ becomes a more important source of $O^+(^4S)$ with increasing solar activity as the high electron densities makes it more competitive with the charge exchange reaction with N_2.

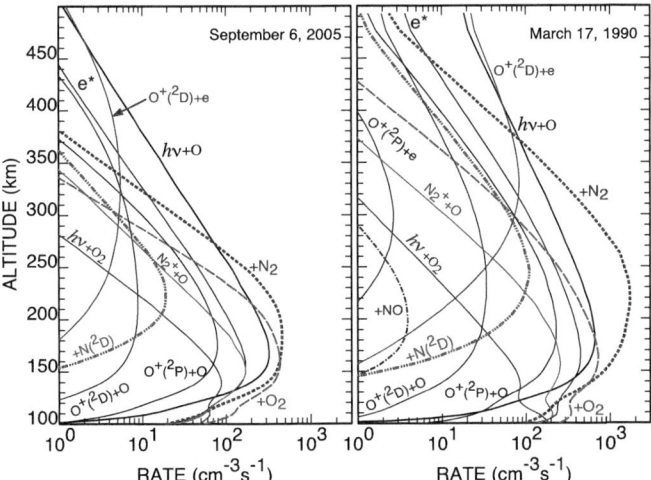

Figure 4. $O^+(^4S)$ production (solid lines) and loss (broken lines) rates for 6 September 2005 and 17 March 1990. The sources and sinks are labeled with the reacting species. Photoionization is labeled $h\nu$ and photoelectron impact ionization is labeled e*.

The main sinks of O^+ come from the reactions with O_2 and N_2. The reaction with $N(^2D)$ is a minor sink for O^+ but an important source for N^+. Figure 4 shows that the reaction with N_2 becomes increasingly important with increasing solar activity. There are two reasons for this. First, there is a large increase in N_2 density driven by the larger scale height, while the change in O_2 density is small. Second, vibrational excitation of N_2 increases with altitude, and the reaction rate is increased by a factor of 2 to 3 near 300 km.

The main ion reaction rate coefficients are given in Table 1. The largest $O^+(^4S)$ loss rates are $O^+(^4S) + N_2 \rightarrow NO^+ + N$ and $O^+(^4S) + O_2 \rightarrow O_2^+ + O$. The most recent laboratory measurements for these reactions rates agree well with a number of other measurements below 1000 K but are much larger at higher temperatures due to vibrational excitation of N_2 as was noted by *Hierl et al.* [1997]. Although N_2 is vibrationally excited in the thermosphere, the distribution is unlikely to be the same as in the laboratory because of the many chemical sources in the ionosphere.

So the laboratory rates are not appropriate for the ionosphere whenever the ion temperature is much larger than 1000 K. Vibrational excitation is important because it can lower the electron density by accelerating the $O^+ + N_2$ reaction rate. The FLIP model solves for vibrationally excited nitrogen, and the O^+ loss rate can increase by more than a factor of 2 at solar maximum [*Richards et al.*, 2010]. It is a small effect at solar minimum decreasing the model peak electron density by about 15%. For the present study, the temperature variation of the $O^+(^4S) + N_2 \rightarrow NO^+ + N$ reaction rate is taken from the work of *St.-Maurice and Torr* [1978], and the rate is normalized to the laboratory measurements at 900 K. The temperature variation of the $O^+(^4S) + O_2 \rightarrow O_2^+ + O$ reaction rate is taken from the work of *McFarland et al.* [1974]. At solar minimum, the FLIP model N_2 and O_2 loss rates have similar magnitudes, but the N_2 loss rate is dominant at solar maximum.

The vibrational state of O_2 is unknown in the ionosphere, but it is unlikely to play an important role in the O^+ chemistry because the loss rate to N_2 is much greater at solar maximum when vibrational excitation is most important. For the $O^+(^4S) + O_2 \rightarrow O_2^+ + O$ reaction rate, we have adopted the *Hierl et al.* [1997] measurements as parameterized by *Fox and Sung* [2001] below 900 K, but we have reduced the rate of increase above 900 K as given in Table 1.

2.2. $O^+(^2D)$ Sources and Sinks

Figure 5 shows the model $O^+(^2D)$ production (solid lines) and loss rates (broken lines) for noon on 6 September 2005 and 17 March 1990 for Millstone Hill. Photoionization ($h\nu$) is the largest source above 150 km, while photoelectron

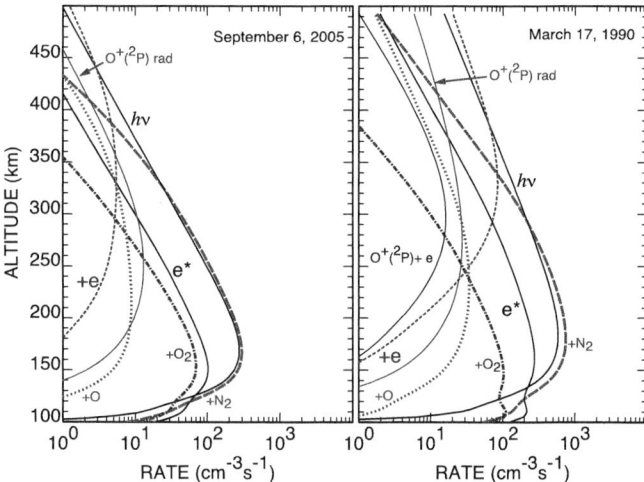

Figure 5. $O^+(^2D)$ production (solid lines) and loss (broken lines) rates for 6 September 2005 and 17 March 1990. The sources and sinks are labeled with the reacting species. Photoionization is labeled $h\nu$, and photoelectron impact ionization is labeled e*.

impact is the largest source in the E region. The relative changes in ionization rates with solar cycle are the same as those of the $O^+(^4S)$. Radiative relaxation of $O^+(^2P)$ is a small source of $O^+(^2D)$. The electron quenching of $O^+(^2P)$ is a small source of $O^+(^2D)$ at solar maximum but is negligible at solar minimum because of the low electron density. The $O^+(^2D) + N_2$ reaction is by far the largest sink between 150 and 300 km, but electron quenching becomes important at high altitudes, and the charge exchange reaction with O_2 becomes important in the E region. Quenching by atomic oxygen is a minor sink at all times.

2.3. $O^+(^2P)$ Sources and Sinks

Although $O^+(^2P)$ is a significant source of ions, its density is too low to appear in Figure 3 because its loss rates are relatively large. Figure 6 shows the model $O^+(^2P)$ production (solid lines) and loss rates (broken lines) for noon on 6 September 2005 and 17 March 1990 for Millstone Hill. There is a striking similarity to the chemistry of $O^+(^2D)$. The sources and sinks are almost the same, but the relative importance of the sinks is different. As in the case of $O^+(^2D)$, photoionization ($h\nu$) is the largest source above 150 km, while photoelectron impact is the largest source in the E region. The solar cycle changes in ionization rates are the same as those of the $O^+(^4S)$. Electron quenching and the reaction with N_2 are not as important for $O^+(^2P)$, and this allows radiation and O quenching to be more important. Unlike in the case of $O^+(^2D)$, charge exchange with O_2 does not become important in the E region.

2.4. N_2^+ Sources and Sinks

Although N_2^+ is produced in large quantities and is the dominant source of ionization throughout much of the ionosphere, it is a minor species because it reacts very quickly with electrons and atomic oxygen. It is the major source of NO^+ below 200 km and a key player in the ion and energy channels at all altitudes.

The photochemistry of N_2^+ has been controversial since early theoretical studies gave densities that were a factor of two larger than those observed by the AE-C satellite [*Abdou et al.*, 1984]. In a recent study, *Richards* [2011] showed that the discrepancy between the measured and modeled N_2^+ densities can be largely resolved by using recent laboratory measurements for the reaction rate coefficients.

Figure 7 shows the model N_2^+ production (solid lines) and loss rates (broken lines) for noon on 6 September 2005 and 17 March 1990 for Millstone Hill. Photoionization ($h\nu$) is the largest source around 200 km, while the $O^+(^2D) + N_2$ reaction is the largest source above 250 km. The photoelectron source (e*) is about half the photoionization rate above 200 km but becomes the dominant source below 150 km. There is very little solar cycle increase in the N_2^+ photoionization rate below 200 km because there is very little change in N_2 density there, and the increase in solar irradiance is offset by the increased attenuation due to the larger O and N_2 column densities.

The $N_2^+ + O \rightarrow NO^+ + N$ reaction is the dominant sink below 250 km, but electron recombination is the largest sink

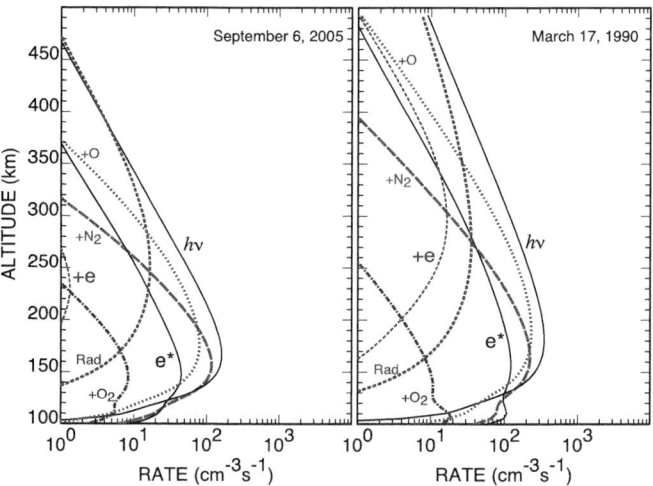

Figure 6. $O^+(^2P)$ production (solid lines) and loss (broken lines) rates for 6 September 2005 and 17 March 1990. The sources and sinks are labeled with the reacting species. Photoionization is labeled $h\nu$, and photoelectron impact ionization is labeled e*.

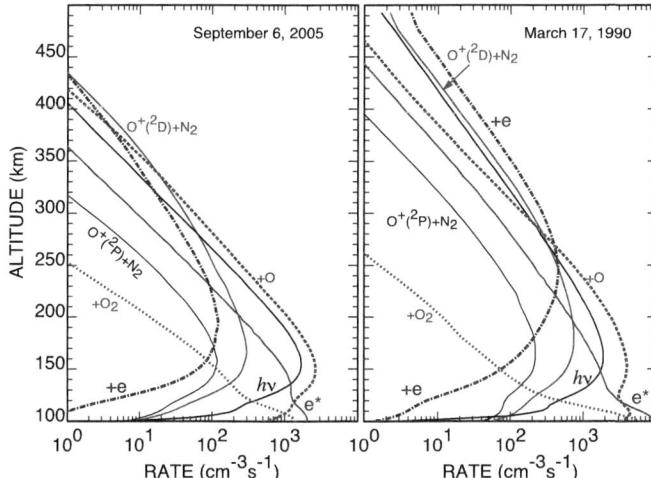

Figure 7. N_2^+ production (solid lines) and loss (broken lines) rates for 6 September 2005 and 17 March 1990. The sources and sinks are labeled with the reacting species. Photoionization is labeled $h\nu$, and photoelectron impact ionization is labeled e*.

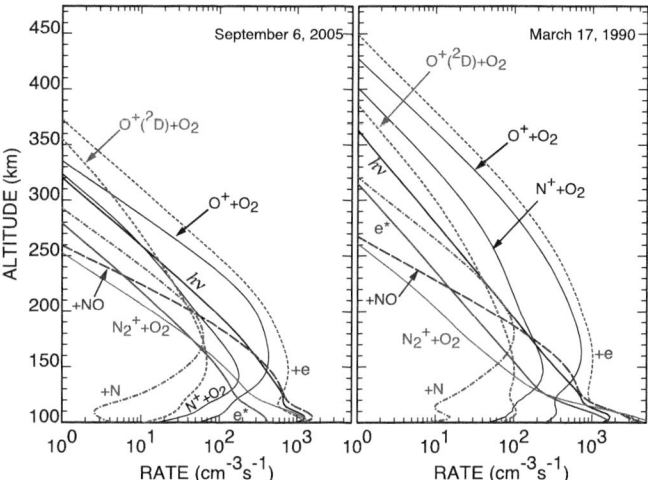

Figure 8. O_2^+ production (solid lines) and loss (broken lines) rates for 6 September 2005 and 17 March 1990. The sources and sinks are labeled with the reacting species. Photoionization is labeled $h\nu$, and photoelectron impact ionization is labeled e*.

at high altitudes and increases in importance with solar activity due to the higher electron densities.

2.5. O_2^+ Sources and Sinks

Figure 8 shows the principal O_2^+ production (solid lines) and loss rates (broken lines) for noon on 6 September 2005 and 17 March 1990 for Millstone Hill. The dominant source above 150 km is the reaction $O^+ + O_2$. In contrast to the case of O^+ and N_2^+, photoelectrons are not a dominant source in the E region. This is because photoionization is more important for O_2^+ in the E region. Solar photons with wavelengths above 913 Å can penetrate with little attenuation to the E region because they are not absorbed by O and N_2. The main O_2^+ sink is electron recombination, but charge exchange with NO is also important below 150 km.

2.6. NO^+ Sources and Sinks

Figure 9 shows the principal NO^+ production (solid lines) and loss rates (broken lines) for noon on 6 September 2005 and 17 March 1990 for Millstone Hill. Photoionization and photoelectron impact ionization of NO are not important sources of NO^+ above 100 km. At solar minimum, the dominant sources are $N_2^+ + O$ below 250 km and $O^+ + N_2$ above 300 km. However, vibrational excitation of N_2 causes a very large solar cycle increase in the $O^+ + N_2$ source to make it the largest source above 250 km at solar maximum. The $O_2^+ + NO$ reaction becomes important at low altitudes for both solar maximum and minimum. Electron recombination is the only NO^+ sink.

2.7. N^+ Sources and Sinks

Figure 10 shows the principal N^+ production (solid lines) and loss rates (broken lines) for noon on 6 September 2005 and 17 March 1990 for Millstone Hill. The dominant sources of N^+ are photodissociative ionization of N_2 below 250 km and $O^+ + N(^2D)$ at higher altitudes. Dissociative ionization of N_2 by photoelectrons is a small source of N^+ except at very low altitudes. The photodissociative ionization threshold is 510 Å and maximizes around 250 Å. Thus, there is

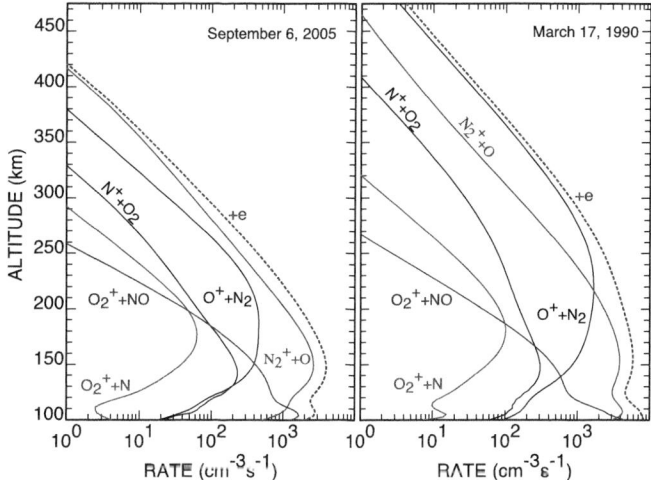

Figure 9. NO^+ production (solid lines) and loss (broken lines) rates for 6 September 2005 and 17 March 1990. The sources and sinks are labeled with the reacting species.

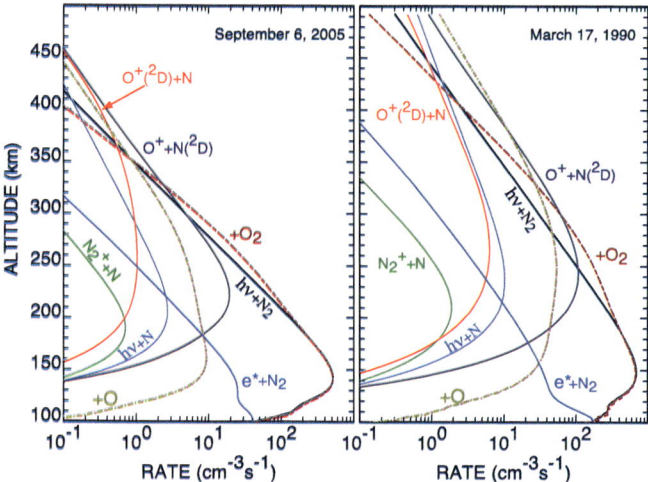

Figure 10. N^+ production (solid lines) and loss (broken lines) rates for 6 September 2005 and 17 March 1990. The sources and sinks are labeled with the reacting species. Photoionization is labeled $h\nu$, and photoelectron impact ionization is labeled e*.

considerable overlap with the wavelengths (<400 Å) that are important for the ionosphere photoelectron flux. This explains why photoelectron impact ionization does not dominate photoionization below 150 km. Photoionization of N is a small source of N^+, and photoelectron impact ionization of N is negligible.

The main N^+ sink is the reaction with O_2, which has several branches that lead to different products (see Table 1). Only the total O_2 loss rate is shown in Figure 10. Figures 8 and 9 show that the $N^+ + O_2$ reaction is a minor source for O_2^+ and NO^+. The reaction of N^+ with O becomes an important sink at higher altitudes especially at solar maximum, but is a negligible source of O^+. N^+ is in approximate local chemical equilibrium below ~400 km, but diffusion becomes increasingly important for N^+ above ~350 km at solar minimum and above ~450 km at solar maximum.

3. CONCLUSIONS

This paper uses a comprehensive ionosphere model to examine the solar cycle variation of the chemistry of O^+, N_2^+, NO^+, O_2^+, and N^+ between 100 and 500 km altitude in the Earth's ionosphere at equinox. All the ion densities increase with solar activity, but the O^+ and N^+ densities increase much more than the molecular ion densities. This is because increases in the molecular ion production rates are counterbalanced by the increased losses due to increased electron and neutral densities. Electron recombination is particularly important because the molecular ion loss rates are proportional to the electron density, and the peak electron density increases by almost an order of magnitude from solar minimum to solar maximum.

In the F_1 region of the ionosphere between 120 and 200 km, solar cycle increases in ion production rate due to increasing solar EUV intensity are largely offset by increased attenuation from the increased O and N_2 column densities. The solar cycle variation of the molecular ion densities is largest in the region due to the large solar cycle increases in the soft X-ray irradiances that create a large increase in photoelectron flux. In general, the sources and sinks of the ions scale with solar activity as would be expected from the changes in solar EUV intensity with some notable exceptions. For example, the loss rates involving electrons are faster at solar maximum above about 200 km because of the large increase in electron density. Also, the reaction rate of O^+ with N_2 increases much more than expected due to the increased reaction rate from vibrational excitation of N_2.

The situation is more complex in summer because the solar cycle variation in electron density above 200 km may not be as great. This means that the solar cycle variation of the molecular ion densities will be greater in summer than at equinox above 200 km. The winter solar cycle relative changes are similar to equinox changes above 200 km. There will be seasonal differences below 200 km, but these are mainly related to the substantial differences in absorption of the solar irradiance due to the different solar zenith angles.

Acknowledgments. This research was supported by NASA grant NNX09J76G and NSF grant AGS-1048350 to George Mason University.

REFERENCES

Abdou, W. A., D. G. Torr, P. G. Richards, M. R. Torr, and E. L. Breig (1984), Results of a comprehensive study of the photochemistry of N_2^+ in the ionosphere, *J. Geophys. Res.*, 89(A10), 9069–9079.

Anicich, V. G. (1993), Evaluated bimolecular ion-molecule gas phase kinetics of positive ions for use in modelling planetary atmospheres, cometary comae, and interstellar clouds, *J. Phys. Chem. Ref. Data*, 22, 1469–1569.

Bates, D. R. (1989), Theoretical considerations regarding some inelastic atomic collision processes of interest in aeronomy: Deactivation and charge transfer, *Planet. Space Sci.*, 17, 363–368.

Chang, T., D. G. Torr, P. G. Richards, and S. C. Solomon (1993), Reevaluation of the O+(^{2}P) reaction rate coefficients derived from Atmosphere Explorer C observations, *J. Geophys. Res.*, 98(A9), 15,589–15,597.

Constantinides, E. R., J. H. Black, A. Dalgarno, and J. H. Hoffman (1979), The photochemistry of N^+ ions, *Geophys. Res. Lett.*, 6(7), 569–572.

Dalgarno, A. (1979), A. E. reaction rate data, *Rep. AFGL-TR 790067*, Air Force Geophys. Lab., Bedford, Mass.

Dotan, I., and A. A. Viggiano (1999), Rate constants for the reactions of O^+ with NO as a function of temperature (300–1400 K), *J. Chem. Phys.*, *110*, 4730–4733.

Dotan, I., P. M. Hierl, R. A. Morris, and A. A. Viggiano (1997), Rate constants for the reactions of N^+ and N_2^+ with O_2 as a function of temperature (300–1800 K), *Int. J. Mass Spectrom. Ion Processes*, *167/168*, 223–230.

Fahey, D. W., I. Dotan, F. C. Fehesenfeld, D. L. Albritton, and L. A. Viehland (1981), Energy dependence of the rate constant of the reaction $N^+ + NO$ at collision energies 0.04 to 2.5 eV, *J. Chem. Phys.*, *74*, 3320–3323.

Ferguson, E. E. (1973), Rate constants of thermal energy binary ion-molecule reactions of aeronomic interest, *At. Data Nucl. Data Tables*, *12*(2), 159–178.

Fox, J. L., and K. Y. Sung (2001), Solar activity variations of the Venus thermosphere/ionosphere, *J. Geophys. Res.*, *106*(A10), 21,305–21,335.

Glosik, J., A. B. Rakshit, N. D. Twiddy, N. G. Adams, and D. Smith (1978), Measurement of the rates of the reaction on ground and metastable excited states of O_2^+, NO^+ and O^+ with atmospheric gases at thermal energy, *J. Phys. B At. Mol. Phys.*, *11*, 3365, doi:10.1088/0022-3700/11/19/013.

Goldan, P. D., A. L. Schmeltekopf, F. C. Fehsenfeld, H. I. Schiff, and E. E. Ferguson (1966), Thermal energy ion-neutral reaction rates. II. Some reactions of ionospheric interest, *J. Chem. Phys.*, *44*, 4095–4103.

Hierl, P. M., I. Dotan, J. V. Seeley, J. M. Van Doren, R. A. Morris, and A. A. Viggiano (1997), Rate constants for the reactions of O^+ with N_2 and O_2 as a function of temperature (300–1800 K), *J. Chem. Phys.*, *106*(9), 3540–3544.

Johnsen, R., and M. A. Biondi (1980), Charge transfer coefficients for the $O^+(^2D) + N_2$ and $O^+(^2D) + O_2$ excited ion reactions at thermal energy, *J. Chem. Phys.*, *73*, 190–193.

Li, X., Y. L. Huang, G. D. Flesch, and C. Y. Ng (1997), A state-selected study of the ion-molecule reactions $O^+(^4S, ^2D, ^2P) + N_2$, *J. Chem. Phys.*, *106*, 1373–1381.

McFarland, M., D. L. Albritton, F. C. Fehsenfeld, E. E. Ferguson, and A. L. Schmeltekopf (1974), Energy dependence and branching ratio of the $N_2^+ + O$ reaction, *J. Geophys. Res.*, *79*(19), 2925–2926.

McLaughlin, B. M., and K. L. Bell (1998), Electron-impact excitation of the fine-structure levels $(1s^2 2s^2 2p^{3 4}S_{3/2}^0, ^2D_{5/2,3/2}^0, ^2P_{3/2,1/2}^0)$ of singly ionized atomic oxygen, *J. Phys. B At. Mol. Phys.*, *31*, 4317–4329.

Mehr, F. J., and M. A. Biondi (1969), Electron temperature dependence of recombination of O_2^+ and N_2^+ ions with electrons, *Phys. Rev.*, *181*, 264–271.

Midey, A. J., and A. A. Viggiano (1999), Rate constants for the reaction of O_2^+ with NO from 300 to 1400 K, *J. Chem. Phys.*, *110*, 10,746–10,748.

O'Keefe, A., G. Mauclaire, D. Parent, and M. T. Bowers (1986), Product energy disposal in the reaction of $N^+(^3P)$ with $O_2(X^3\Sigma)$, *J. Chem. Phys.*, *84*, 215–219.

Picone, J. M., A. E. Hedin, D. P. Drob, and A. C. Aikin (2002), NRLMSISE-00 empirical model of the atmosphere: Statistical comparisons and scientific issues, *J. Geophys. Res.*, *107*(A12), 1468, doi:10.1029/2002JA009430.

Peterson, W. K., E. N. Stavros, P. G. Richards, P. C. Chamberlin, T. N. Woods, S. M. Bailey, and S. C. Solomon (2009), Photoelectrons as a tool to evaluate spectral variations in solar EUV irradiance over solar cycle timescales, *J. Geophys. Res.*, *114*, A10304, doi:10.1029/2009JA014362.

Richards, P. G. (2011), Reexamination of ionospheric photochemistry, *J. Geophys. Res.*, *116*, A08307, doi:10.1029/2011JA016613.

Richards, P. G., and W. K. Peterson (2008), Measured and modeled backscatter of ionospheric photoelectron fluxes, *J. Geophys. Res.*, *113*, A08321, doi:10.1029/2008JA013092.

Richards, P. G., J. A. Fennelly, and D. G. Torr (1994), EUVAC: A solar EUV Flux Model for aeronomic calculations, *J. Geophys. Res.*, *99*(A5), 8981–8992.

Richards, P. G., T. N. Woods, and W. K. Peterson (2006), HEUVAC: A new high resolution solar EUV proxy model, *Adv. Space Res.*, *37*(2), 315–322, doi:10.1016/j.asr.2005.06.031.

Richards, P. G., R. R. Meier, and P. J. Wilkinson (2010), On the consistency of satellite measurements of thermospheric composition and solar EUV irradiance with Australian ionosonde electron density data, *J. Geophys. Res.*, *115*, A10309, doi:10.1029/2010JA015368.

Scott, G. B. L., D. A. Fairley, D. B. Milligan, C. G. Freeman, and M. J. McEwan (1999), Gas phase reactions of some positive ions with atomic and molecular oxygen and nitric oxide at 300 K, *J. Phys. Chem. A*, *103*, 7470–7473.

Solomon, S. C. (2006), Numerical models of the E-region ionosphere, *Advances in Space Research*, *37*, 1031–1037, doi:10.1016/j.asr.2005.09.040.

Solomon, S. C., and L. Qian (2005), Solar extreme-ultraviolet irradiance for general circulation models, *J. Geophys. Res.*, *110*, A10306, doi:10.1029/2005JA011160.

St.-Maurice, J.-P., and D. G. Torr (1978), Nonthermal rate coefficients in the ionosphere: The reactions of O^+ with N_2, O_2, and NO, *J. Geophys. Res.*, *83*(A3), 969–977.

Torr, D. G., and M. R. Torr (1979), Chemistry of the thermosphere and ionosphere, *J. Atmos. Terr. Phys.*, *41*, 797–839.

Vejby-Christensen, L., D. Kella, H. B. Pedersen, and L. H. Anderson (1998), Dissociative recombination of NO^+, *Phys. Rev. A*, *57*, 3627–3634.

Woods, T. N., et al. (2008), XUV photometer system (XPS): Improved solar irradiance algorithm using Chianti spectral models, *Sol. Phys.*, *250*(2), 235–267, doi:10.1007/s11207-008-9196-6.

Zipf, E. C. (1980), The dissociative recombination of vibrationally excited N_2^+ ions, *Geophys. Res. Lett.*, *7*(9), 645–648.

Zipf, E. C., P. J. Espy, and C. F. Boyle (1980), The excitation and collisional deactivation of metastable $N(^2P)$ atoms in auroras, *J. Geophys. Res.*, *85*(A2), 687–694.

P. G. Richards, School of Physics, Astronomy and Computational Sciences, George Mason University, Fairfax, VA 22030, USA. (pgrichds@gmail.com)

Energetics and Composition in the Thermosphere

A. G. Burns, W. Wang, S. C. Solomon, and L. Qian

National Center for Atmospheric Research, Boulder, Colorado, USA

The thermosphere is defined by the high temperatures that occur in its upper regions. These high temperatures result from the absorption of energetic EUV radiation combined with weak in situ cooling. Heat is eventually lost from the upper thermosphere primarily through downward heat conduction. This heating and cooling is modified through a variety of processes, the most important of which, at low and middle latitudes, is heating by compression and cooling by expansion. These thermal processes help to drive the dynamics of the thermosphere, which, in turn, are a major cause of changes in neutral composition. In this paper, we discuss the global changes in heating that arise as a result of high-latitude energy inputs during geomagnetic storms and the ways that we can gain a better understanding of these processes. One of the scientific results of this paper was that low- and middle-altitude heating during a simulated geomagnetic storm was mainly caused by the propagation and dissipation of gravity waves generated near the auroral zone.

1. INTRODUCTION

The thermosphere is characterized by its heat as its name suggests. The hottest temperatures in the Earth's atmosphere occur in this region. Upper thermospheric temperatures typically reach 1200–1400 K in the summer daytime at solar maximum (e.g., see Naval Research Laboratory-Mass Spectrometer Incoherent Scatter model (MSIS)) [*Picone et al.*, 2002] and have been seen to reach 2500 K in Dynamics Explorer 2 (DE 2) [*Hoffman et al.*, 1981] Wind and Temperature Spectrometer (WATS) [*Spencer et al.*, 1981] by the first author of this paper.

The coldest region of the atmosphere, the mesopause, lies below the thermosphere. This creates strong vertical temperature gradients in the lower thermosphere (from ~100 to ~150 km), which, in conjunction with viscosity and the drag of ions on the neutrals, results in a very stable region where no neutral turbulence is seen at above about 105 km. Above about 120 km, viscosity prevents shears on the scale of turbulence to exist at all [e.g., *Rishbeth and Garriott*, 1969]. Turbulent mixing is thus lacking in the thermosphere, so the individual species' densities decrease with altitude with a scale height that is dependent on that species' mass. Any change in the dynamics in the upper atmosphere leads not only to changes in temperature but also to changes in neutral composition.

The basic thermal and compositional structure of the upper atmosphere is shown in Figure 1 (from National Research Laboratory-NRL MSIS) [*Picone et al.*, 2002]. Figure 1 describes atmospheric structure for solar maximum conditions (the differences in solar radiation over a solar cycle only have a major effect in the thermosphere and ionosphere not on the middle and lower atmosphere). The thermal structure in this plot was described above, as was the decrease in density. The three major neutral species in the thermosphere are shown here. Molecular nitrogen (N_2) and molecular oxygen (O_2) have the same mass mixing ratios (mmrs, the ratio of the mass density of the species to the total mass density) near the bottom of the thermosphere as they do at the ground. At higher altitudes, the densities of these species fall off at a rate that primarily depends on their individual scale heights. Large O mmrs are not found in the

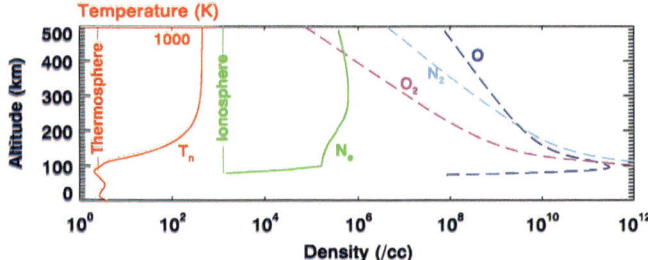

Figure 1. Temperature and density variations with altitude in the thermosphere (from National Research Laboratory- Mass Spectrometer Incoherent Scatter model) [*Picone et al.*, 2002].

lower atmosphere. They occur in the upper atmosphere because O_2 is dissociated by EUV and UV radiation in the lower thermosphere. Equilibrium is reached because the atomic oxygen subsequently diffuses downward to regions where collisions are frequent enough to cause the atomic oxygen to recombine to produce molecular oxygen [e.g., *Rishbeth and Garriott*, 1969; *Burns et al.*, 1989].

Note that electron densities in the upper atmosphere are approximately 1/10,000,000 of those of the neutral density in the E region near 100 km and 1/1000 of those at the F_2 peak (near 300 km). Thus, the upper atmosphere is a weakly ionized gas. Despite these low relative densities, the ionosphere can have a profound effect on the thermosphere (e.g., through Joule heating and ion drag).

Figure 1 represents the mean state of the upper atmosphere. In reality, the region is a dynamic place that is driven primarily by external forcing. Solar EUV and UV radiation primarily affects the upper thermosphere by ionizing atomic oxygen. The resulting secondary electrons collide with the ions and the neutrals heating them and causing further ionization. The primary path for this heat transfer in the upper thermosphere is the electrons heating the ions through collisions and the ions then colliding rapidly with the neutrals, heating them in turn [*Roble et al.*, 1987; *Roble*, 1995]. This radiation varies with solar cycle and solar rotation and can also change as a result of solar flares [e.g., *Qian et al.*, 2006].

Another form of external energy input comes from magnetosphere-ionosphere-thermosphere coupling at high latitudes. Particles flow from the Sun as well as radiation. This solar wind interacts with the magnetosheath that bounds the magnetosphere, resulting in particle precipitation around the auroral oval and creating an electric potential pattern in the high latitudes. Both the speed and the magnetic field imbedded in the solar wind vary greatly, leading to large variations in the magnetosphere and, thus, in the potential and particle precipitation in the upper atmosphere, which, in turn, drive changes in the energetic, dynamics (including waves) and composition of this region.

A third form of forcing results from waves propagating up from below. These waves have a great range of spatial and temporal scales from the gravity waves, whose breaking causes the turbulence in the lowest regions of the thermosphere to the upper atmosphere effects of Sudden Stratospheric warmings, the effects of which have been described in the F_2 region ionosphere [*Liu and Roble*, 2002]. The ways that external forcing changes temperatures will be the main theme of this paper.

The next section details some of the data and models that have been used to develop our understanding of the upper atmosphere. Section 3 describes the equations that determine the compositional and thermal structure of the thermosphere. The National Center for Atmospheric Research Thermosphere Ionosphere Electrodynamics Model (NCAR-TIEGCM) [*Richmond et al.*, 1992], which has been used to develop the results of this paper, is also described in this section. The causes of the thermal structure are considered in section 4, and a summary is given in section 5.

2. DATA AND MODELS THAT HAVE INCREASED OUR UNDERSTANDING OF THE UPPER ATMOSPHERE

The thermosphere was first detected using rocket flights [*Rocket Panel*, 1952]. Since then, the region has been probed in numerous ways. The following is a short précis of those techniques that are most relevant to understanding thermal and compositional variations.

The 3-D structure of the thermosphere was revealed using satellite drag measurements [e.g., *Jacchia and Slowey*, 1964]. This technique is still in use today and is currently being used to improve empirical models of the thermosphere [*Picone et al.*, 2002] and to study the recent solar minimum [*Qian et al.*, 2006; *Emmert et al.*, 2008; *Solomon et al.*, 2011]. Total neutral density has also been obtained from accelerometers. These instruments are also used for satellite guidance and continue to be flown to this day. Recent examples of their use include studying neutral density peaks near the cusp [e.g., *Crowley et al.*, 2010], investigating middle-latitude density peaks [e.g., *Lei et al.*, 2010] and analyzing the effects of flares on the thermosphere [e.g., *Qian et al.*, 2011].

Another type of instrument has been used in space to measure neutral composition: the mass spectrometer. The first of these instruments was flown in low Earth orbit in the 1960s, and they continued to be flown regularly in this region through to the early 1980s [*Carignan et al.*, 1981], but have seldom been attempted since. They provided a wealth of information about composition in the upper thermosphere.

Compositional variations have also been investigated by FUV imaging. Global ultraviolet imaging [*Paxton et al.*,

1999] data have been used in many studies [e.g., *Zhang and Paxton*, 2008; *Crowley et al.*, 2008] to study processes like geomagnetic storm-driven changes in composition.

Temperature measurements from space are much rarer. Temperatures were first deduced from neutral densities using scale heights and assuming diffusive equilibrium [e.g., *Jacchia and Slowey*, 1964]. The first successful direct measurements by satellites were made by the Spherical Fabry-Perot Interferometer (FPI) that flew on board OGO 6 [*Blamont and Luton*, 1972]. These measurements were later followed by those made by the WATS [*Spencer et al.*, 1981], which flew on board the DE 2 satellite and the Wind Imaging Interferometer (WINDII) [*Shepherd et al.*, 1993] on board the UARS satellite, which provided temperatures up to about 200 km.

Ground-based instruments have also provided information about temperature changes in the upper atmosphere. FPIs have been used to deduce neutral temperatures in the upper atmosphere for many years [e.g., *Killeen et al.*, 1995]. A limited number of these instruments are currently deployed in a widespread array of sites, ranging from Resolute in the Arctic [e.g., *Wu et al.*, 2008] through the equatorial regions [e.g., *Meriwether*, 2006] to the Antarctic [e.g., *Hernandez et al.*, 1990]. In recent years, an imaging FPI has been developed in Alaska [*Conde and Smith*, 1998] that has allowed both spatial and temporal information about temperature changes to be measured from the ground. Incoherent scatter radar data can also be used to infer neutral temperature data [e.g., *Hedin*, 1983].

There are two main types of models: semiempirical models and theoretical first principle models. The former are developed by gathering a vast array of data and then applying physical relationships to interpolate for states or regions for which there are no data. The Jacchia models [e.g., *Jacchia*, 1971] exemplified the first successful application of this technique to the thermosphere. They are still used today to provide operational satellite drag information for orbital prediction [e.g., *Bowman*, 2005]. As neutral composition data became available, a new generation of semiempirical models became available that included this information. The prime example of this new generation of models is the MSIS suite of models [*Hedin*, 1983, 1987, 1991; *Picone et al.*, 2002]. They are probably the most commonly used thermospheric models today. NRL-MSIS was used to produce Figure 1 of this paper.

A number of theoretical first principle models have been developed for the thermosphere including the NCAR-TIEGCM [*Richmond et al.*, 1992] and other related NCAR models [e.g., *Roble and Ridley*, 1994], and the coupled thermosphere-ionosphere model [e.g., *Fuller-Rowell et al.*, 1996]. All of these models solve a coupled set of equations including, but not limited to, the momentum equations, thermodynamic equations, and the continuity equation. The latter two equations are described in the next section.

Theoretical first principle models like these have an advantage over empirical models. Empirical model can only describe the thermosphere, whereas theoretical, first principle models can both describe the thermosphere and explain it [*Rishbeth*, 2007]. Thus, theoretical, first principle models can be further interrogated to help to understand the physics and chemistry that drive the variations that are seen in the models [*Killeen and Roble*, 1984]. One diagnostic processor is used in this paper: it describes the terms for the neutral energy equation [*Burns et al.*, 1992, 1995]. The equations that underlie these diagnostics are described in the following section.

3. THE CONTINUITY EQUATION, THE ENERGY EQUATION, AND BOUNDARY CONDITIONS

The equations that drive the composition equations are described in this section. Their formulation is that used in the NCAR-TIEGCM. Other models use similar equations although there are possibly slight differences in the parameterizations, in the form of the equations and the grids used.

The NCAR-TIEGCM [*Roble et al.*, 1988; *Richmond et al.*, 1992] is a first principle upper atmospheric general circulation model that solves the Eulerian continuity, momentum, and energy equations for the coupled thermosphere-ionosphere system. It uses pressure surfaces as the vertical coordinate and extends in altitude from approximately 97 to 600 km, although the upper value varies over the solar cycle. The normal resolution of the model is 5° in latitude and longitude, and 0.5 scale height in altitude; a 2 min time step is employed. Tidal forcing at the lower boundary is specified using the migrating diurnal and semidiurnal tides specified by the global scale wave model (GSWM) [*Hagan and Roble*, 2001], and semiannual and annual density periodicities can be obtained by applying seasonal variation of the eddy diffusivity coefficient at the lower boundary [*Qian et al.*, 2009].

The major species continuity equation (equation (2)) is solved for the mmrs of O_2 and O [*Dickinson et al.*, 1984]; the N_2 mmr is then calculated using equation (2), which assumes that the mmrs of the three major species adds up to 1.

$$\frac{\partial \widetilde{\Psi}}{\partial t} = -\frac{e^Z}{\tau}\frac{\partial}{\partial Z}\left[\frac{\overline{m}}{m_{N_2}}\left(\frac{T_{00}}{T_n}\right)^{0.25}\widetilde{\alpha}^{-1}L\widetilde{\Psi}\right] + S - R + e^Z\frac{\partial}{\partial Z}\left[K(Z)e^{-Z}\frac{\partial}{\partial Z}\widetilde{\Psi}\right] - \left(\vec{V}\cdot\nabla\widetilde{\Psi} + w\frac{\partial \widetilde{\Psi}}{\partial Z}\right) \quad (1)$$

$$\Psi_{N_2} = 1 - \Psi_O - \Psi_{O_2} \qquad (2)$$

The term on the left of equation (1) is the time rate of change of mmr. The first term on the right is molecular diffusion, the second (S) and third (R) terms represent the production of O by the dissociation of O_2 and the loss of O by three-body recombination and vice versa for O_2. The next term is an eddy diffusion term. In the NCAR-TIEGCM, the eddy diffusion coefficient drops off exponentially with height above the bottom boundary at 97 km. The final two terms represent the horizontal and vertical advection terms that dominate the short-term changes (e.g., geomagnetic storm–related ones) in composition in the upper thermosphere.

Roble et al. [1988] described the thermodynamic equation that is used to calculate the temperature:

$$\frac{\partial T}{\partial t} = \frac{ge^Z}{P_0 C_p} \frac{\partial}{\partial Z} \left\{ \frac{K_T}{H} \frac{\partial T}{\partial Z} + K_E H^2 C_P \rho \left[\frac{g}{C_P} + \frac{1}{H} \frac{\partial T}{\partial Z} \right] \right\}$$
$$- \mathbf{V} \cdot \nabla T - w \left(\frac{\partial T}{\partial Z} + \frac{RT}{C_P \bar{m}} \right) + \frac{(Q-L)}{C_P} \qquad (3)$$

The left-hand side is the rate of temperature change. The right-hand side contains a number of physically meaningful processes: thermal conduction (first term), eddy heat conduction (second term), heat transport by the horizontal winds (third term), heating as the result of compression and cooling by expansion (all of the last term on the top line), and a number of heat production and loss terms (the Q-L terms) given by *Roble et al.* [1987] including, among others, heating by solar radiation through heat transfer from electrons to the ions and neutrals, heating as a result of exothermic ion and neutral chemical reactions, Joule heating, and radiational cooling by NO, $O(^3P)$, and CO_2.

Diffusive equilibrium is assumed for the top boundary conditions of the composition of the neutral gas, the neutral winds and temperatures are assumed to have no vertical gradients, and an upward flux is imposed on O^+ density during daytime. The bottom boundary conditions for *T*, *U*, and *V* at ~97 km are imposed using the migrating diurnal and semidiurnal tides specified by the GSWM [*Hagan et al.*, 1999] in normal operations. These tidal variations are imposed on a mean neutral temperature of 181 K. Several other conditions are imposed at the bottom boundary including an O_2 mmr of 0.22, an N_2 mmr of 0.78, no vertical gradient of O, $N(^4S)$ is in diffusive equilibrium, and an NO density of 8×10^6 cm^{-3}.

The physical processes represented by equations (1) and (3) can be further analyzed by the techniques suggested by *Killeen and Roble* [1984]. The decomposition of the continuity equation was described by *Burns et al.* [1989] and applied by *Burns et al.* [1991, 2004]. The decomposition of the thermodynamics equation has not been fully described, but it was used by *Burns et al.* [1992, 1995]. Because the decomposition of the thermodynamics equation has not been studied as fully as that of the composition, the remainder of this paper will be used to discuss the application of this postprocessor to changes in the low- and middle-latitude temperatures during a geomagnetic storm.

4. THE CAUSES OF TEMPERATURE VARIATIONS IN THE LOW AND MIDDLE LATITUDES OF THE UPPER THERMOSPHERE DURING GEOMAGNETIC STORMS

Temperature changes in the low and middle latitudes during geomagnetic storms have been studied much less than composition changes. The main reason for this has been the sparseness of direct measurements of temperature in the thermosphere, although some papers using ground- and space-based observations of storms have been published.

Ground-based studies have relied primarily on neutral temperatures inferred from incoherent scatter radar [e.g., *Salah et al.*, 1976] measurements and direct measurements by FPIs [e.g., *Niciejewski et al.*, 1994]. The former are limited by the small number of stations and will not be considered further here. Two examples of studies that addressed the issue of storm time heating at low and middle latitudes were *Biondi and Sipler* [1985] and *Biondi and Meriwether* [1985], who used FPIs to find that nighttime temperatures increased at these latitudes during storms.

Some direct measurements have also been made from space. The first were made by an FPI [*Blamont and Luton*, 1972], who showed that temperatures could increase by 80 K in low-latitude regions during geomagnetic storms. Later measurements from space confirmed and extended these results. DE 2 WATS [*Spencer et al.*, 1981] data indicated that temperature increases of up to 200 K near midnight and up to about 100 K near noon occurred at low and middle latitudes [*Burns et al.*, 1992, 1995], and WINDII [*Shepherd et al.*, 1993] showed similar results [*Wiens et al.*, 2002].

This paper will not further review these results, nor will it detail previous modeling efforts. Instead, we will undertake a brief description of how and why low- and middle-latitude temperatures changed in the vertical plane during the AGU storm on 15–16 December 2006. Space prevents the inclusion of a plot of geomagnetic conditions during this storm [e.g., see *Wang et al.*, 2010] for these conditions. The salient points that relate to the following temperature plots are that the initial phase began just after 12:00 UT on the 15th, the main phase began a little before 00:00 UT on the 16th, and the main phase had ended by 12:00 UT on the 16th.

Figure 2 shows the temperatures at 09:00 UT for the day prior to the storm (a) and for the day of the main phase (b) that were modeled by the NCAR-TIEGCM. Overall, there is an increase of temperature everywhere, but there are particularly strong increases in temperature around the auroral oval and immediately to the early morningside of the oval. These increases are primarily the result of Joule heating in the oval and direct advection of heat into the early morning hours by the horizontal neutral winds (not shown). The tail is like that seen in the N_2/O ratio, which also occurs in the early morning hours [Hedin and Carignan, 1985], and it is formed by the same process (horizontal advection), but the temperature tail is nowhere near as extensive in either latitude or LT. This difference occurs because the prime recovery process (downward heat conduction) for temperature is much faster than the one for composition (molecular diffusion). Although it is not obvious in this plot, the temperature increases are generally greater at night than in the daytime, a result that is consistent with previous studies [Burns et al., 1992, 1995; Wiens et al., 2002].

Both these studies and earlier studies [e.g., Tauesch et al., 1971] gave rise to an explanation for the majority of the low- and middle-latitude heating that is seen during geomagnetic storms. Heat is input into the high latitudes, primarily as a result of Joule heating around the auroral oval. This heating gives rise to changes in temperature gradient and subsequently circulation. As the circulation changes [e.g., Roble et al., 1979] from a summer to winter one (in the solstices) or an equator to pole one (in the equinoxes) into at least a partially pole to equator one, convergence increases in the low and middle altitudes, which generates stronger downward winds (or equivalently, weaker upward ones) in this region, increasing temperatures through stronger compressional heating (or weaker cooling by expansion). Burns et al. [1995] also demonstrated that the in situ diurnal tide was also weakened by this process, as heat is preferentially transferred from the high latitudes to the middle latitudes in the early morning hours.

This was the understanding with which we undertook the case study used for this paper. However, Figure 3 shows a much more dynamic picture. Instead of being dominated by changes in the global circulation, the changes in temperature in the main phase of the storm are dominated by synoptic-scale changes in circulation that we associate with large-scale gravity waves. These waves are characterized in Figure 3 by regions of convergence (where the arrows appear to get darker as they are more concentrated) and divergence (where the arrows appear to be lighter as they are less concentrated) of neutral winds. High temperatures are associated with the regions of convergence, and low temperatures are associated with the regions of divergence. The location of these regions of convergence and divergence changes with time in a regular way, indicating the passage of waves. Further, there is also heating of the background thermosphere, which may be an indication of changing large-scale circulation, but may also indicate that the waves are nonlinear and are depositing energy throughout the thermosphere as they dissipate.

Figure 4 shows the effects of these waves for two times early in the storm on two of the heating and cooling terms: stability (adiabatic heating and cooling) and downward heat conduction. These two terms are the two major controllers of the temperature changes at low and middle latitudes during geomagnetic storms. The most noticeable feature of Figures 4a and 4b, the two stability term plots, is the dominance of the waves. A bright red band of heating (implying downward winds and, thus, compressional heating) occurs at latitudes of about 40° in both hemispheres in Figure 4a. An hour later, this peak occurs at about 20° (Figure 4b). It has a nearly constant phase with altitude above the $z = 0$ ($z = \ln(p/p0)$ is approximately at 200–220 km in this simulation; $p0$ is 5×10^{-4} μbar)

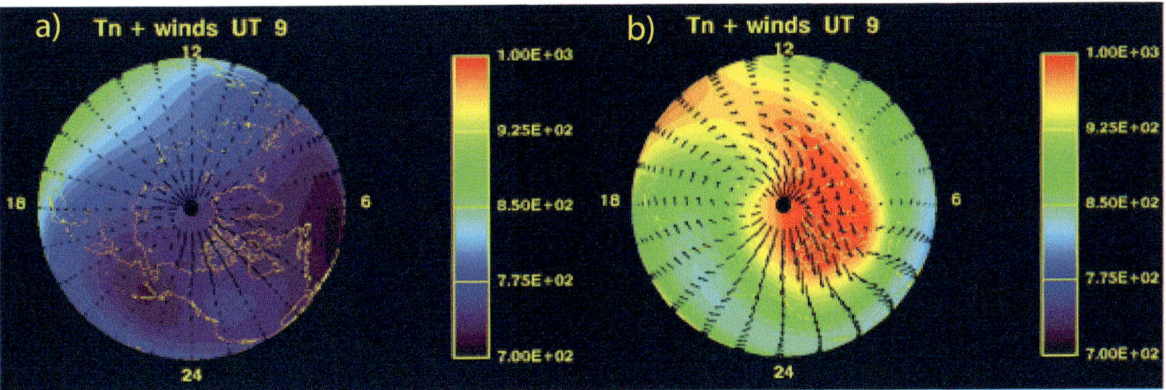

Figure 2. Temperatures and horizontal wind vectors before the storm (a) and during the main phase of the storm (b) at 09:00 UT for the $z = 2$ (~300 km) pressure surface.

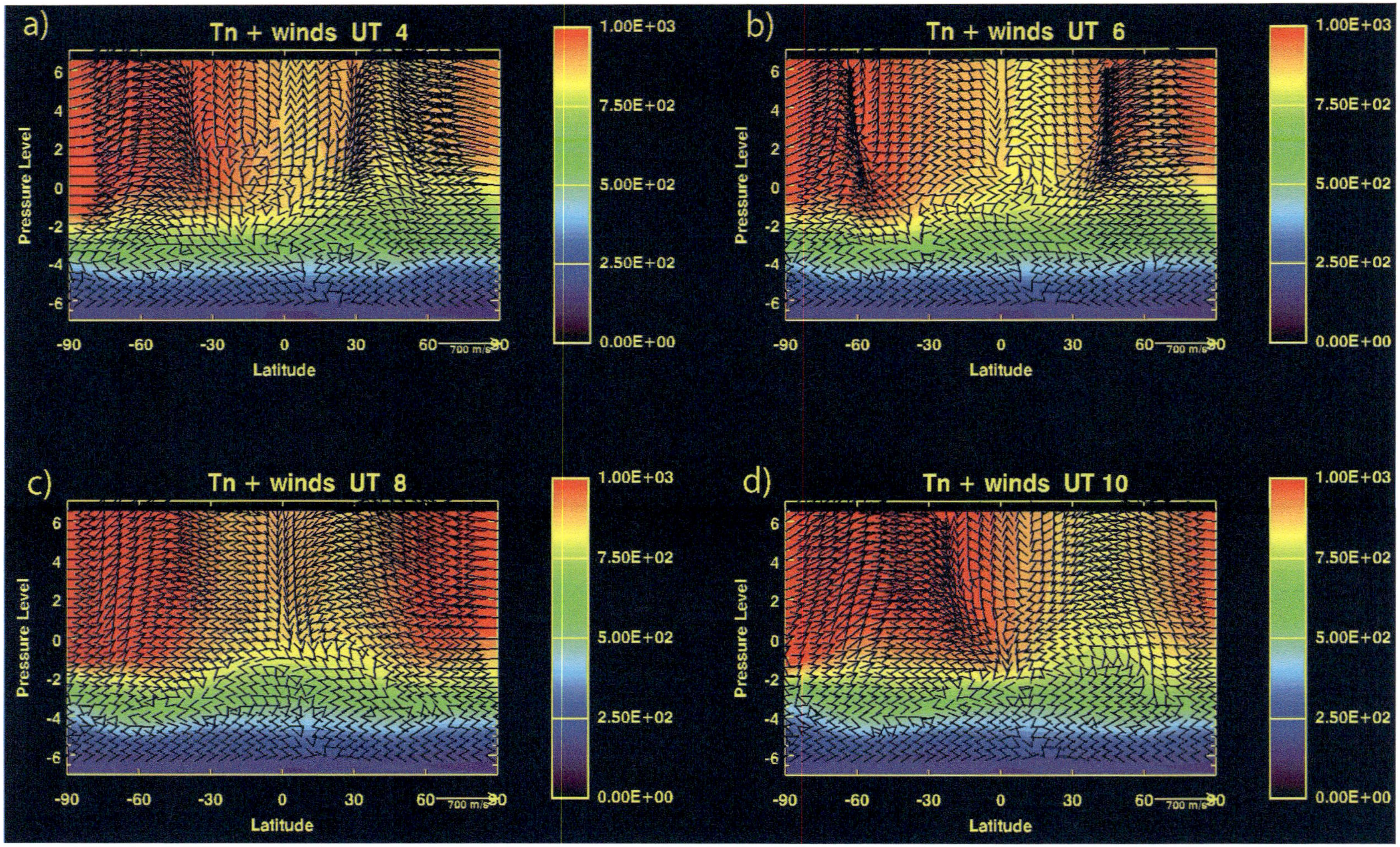

Figure 3. Temperatures and winds at midnight plotted as a function of pressure surface and latitude for 4 UTs, from 0400 to 1000 at midnight local time during the main phase of the storm. Pressure level 0 corresponds to about 280 km at quiet time.

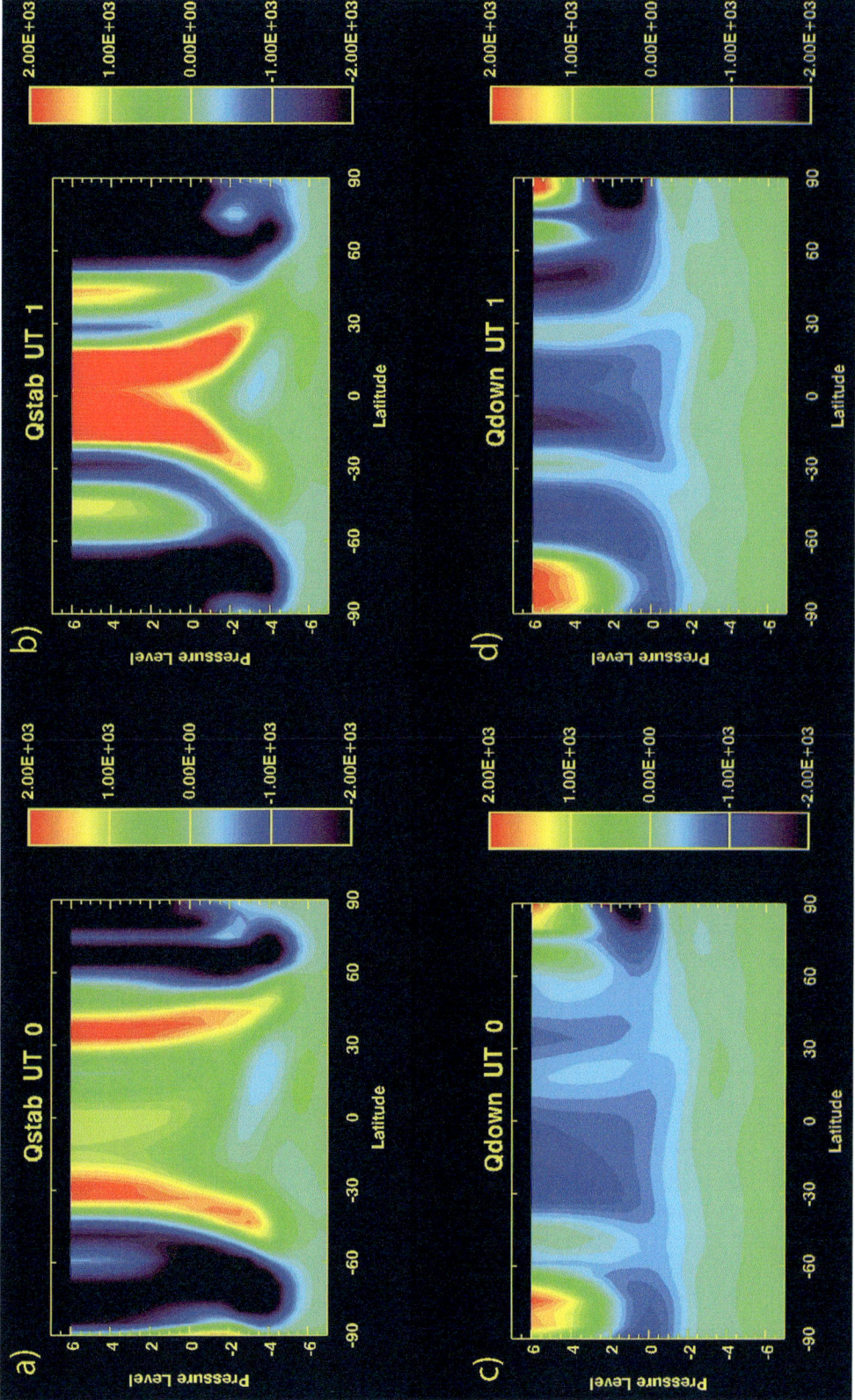

Figure 4. Compressional heating and cooling by expansion (in units of K s^{-1}) for two times during the main phase of the 15–16 December 2006 geomagnetic storm, plotted as latitude versus pressure surface at midnight (a and b). The same for downward heat conduction (c and d).

pressure surface, but apparently moves more slowly at lower altitudes, consistent with *Mayr et al.*'s [1984] calculations for this wave. Downward heat conduction also responds to this wave, albeit with much less amplitude (Figures 4c and 4d). One interesting feature of the "downward" heat conduction is that it actually causes small temperature increases in the upper thermosphere in the trough of the wave (where upward winds cause cooling by expansion) as a result of upward heat conduction. Such behavior is even stronger at high altitudes, where there are not enough collisions in the uppermost parts of the atmosphere to cause much Joule heating, so the shape of the vertical temperature profile drives upward conduction of heat. This behavior does raise the question of whether the boundary condition in which there is no vertical gradient of temperature is appropriate in this case. A possible way to determine this is to move the upper boundary to a higher-pressure surface to determine if the same behavior occurs.

Two more points need to be made. The first is that the characteristics of the propagation of these waves are very different between day and night. They are much weaker in the day than in the night to the point where there is often little evidence of them in the daytime (Hedin, private communication), so if they are a major source of heating at middle and low altitudes during storms, smaller temperature increases should be expected during the daytime. This is in agreement with observations [*Burns et al.*, 1995], but it is also the case if the source of the heating is changed in the general circulation.

The second point is that these are for one coronal mass ejection–driven storm, which occurred near solar minimum. It is possible that the effects of these waves may be disproportionately important in this case as waves seem to be relatively more important at solar minimum (for example, *Ma et al.* [2010] showed that the midnight temperature maximum was more prominent at solar minimum).

5. SUMMARY

We have briefly described the processes that determine the temperature and compositional structure of the thermosphere and outlined the experimental and theoretical efforts that have given rise to this understanding. In addition, we have discussed the processes that give rise to the low- and middle-latitude temperature changes during geomagnetic storms. In the last case, we have found that, for the simulation described, the main source of these temperature changes is the waves that are generated in the auroral regions, rather than the longer-term changes in circulation. We caution that these results apply to the simulation of one storm near solar minimum conditions when it is expected that waves would be more important.

Acknowledgments. This research was supported by the Center for Integrated Space Weather Modeling (CISM), which is funded by the STC program under agreement ATM-0120950, and by NASA grants NNX10AQ49G, NNX08AQ31G, NNX08AQ91G, and NNX10AF21G to the National Center for Atmospheric Research. NCAR is supported by the National Science Foundation.

REFERENCES

Biondi, M. A., and J. W. Meriwether Jr. (1985), Measured response of the equatorial thermospheric temperature to geomagnetic activity and solar flux changes, *Geophys. Res. Lett.*, 12(5), 267–270.

Biondi, M. A., and D. P. Sipler (1985), Studies of equatorial 630.0 nm airglow enhancements produced by a chemical release in the F-region, *Planet. Space Sci.*, 32, 1605–1610.

Blamont, J. E., and J. M. Luton (1972), Geomagnetic effect on the neutral temperature of the *F* region during the magnetic storm of September 1969, *J. Geophys. Res.*, 77(19), 3534–3556.

Bowman, B. (2005), Drag coefficient variability at 175–500 km from the orbit decay analyses of spheres, paper AAS 05-257 presented at the AIAA/AAS Astrodynamics Specialist Conference, Lake Tahoe, Calif.

Burns, A. G., T. L. Killeen, and R. G. Roble (1989), Processes responsible for the compositional structure of the thermosphere, *J. Geophys. Res.*, 94(A4), 3670–3686.

Burns, A. G., T. L. Killeen, and R. G. Roble (1991), A theoretical study of thermospheric composition perturbations during an impulsive geomagnetic storm, *J. Geophys. Res.*, 96(A8), 14,153–14,167.

Burns, A. G., T. L. Killeen, and R. G. Roble (1992), Thermospheric heating away from the auroral oval during geomagnetic storms, *Can. J. Phys.*, 70, 544–552.

Burns, A. G., T. L. Killeen, W. Deng, G. R. Carignan, and R. G. Roble (1995), Geomagnetic storm effects in the low- to middle-latitude upper thermosphere, *J. Geophys. Res.*, 100(A8), 14,673–14,691.

Burns, A. G., T. L. Killeen, W. Wang, and R. G. Roble (2004), The solar-cycle-dependent response of the thermosphere to geomagnetic storms, *J. Atmos. Sol. Terr. Phys.*, 66, 1–14.

Carignan, G. R., B. P. Block, J. C. Maurer, A. E. Hedin, C. A. Reber, and N. W. Spencer (1981) The neutral mass spectrometer on Dynamics Explorer B, *Space Sci. Instrum.*, 5, 429–441.

Conde, M., and R. W. Smith (1998), Spatial structure in the thermospheric horizontal wind above Poker Flat, Alaska, during solar minimum, *J. Geophys. Res.*, 103(A5), 9449–9471, doi:10.1029/97JA03331.

Crowley, G., A. Reynolds, J. P. Thayer, J. Lei, L. J. Paxton, A. B. Christensen, Y. Zhang, R. R. Meier, and D. J. Strickland (2008), Periodic modulations in thermospheric composition by solar wind high speed streams, *Geophys. Res. Lett.*, 35, L21106, doi:10.1029/2008GL035745.

Crowley, G., D. J. Knipp, K. A. Drake, J. Lei, E. Sutton, and H. Lühr (2010), Thermospheric density enhancements in the dayside cusp region during strong B$_Y$ conditions, *Geophys. Res. Lett.*, *37*, L07110, doi:10.1029/2009GL042143.

Dickinson, R. E., E. C. Ridley, and R. G. Roble (1984), Thermospheric general circulation with coupled dynamics and composition, *J. Atmos. Sci.*, *41*, 205–219.

Emmert, J. T., J. M. Picone, and R. R. Meier (2008), Thermospheric global average density trends, 1967–2007, derived from orbits of 5000 near-Earth objects, *Geophys. Res. Lett.*, *35*, L05101, doi:10.1029/2007GL032809.

Fuller-Rowell, T. J., D. Rees, S. Quegan, R. J. Moffett, M. V. Codrescu, and G. H. Millward (1996), A coupled thermosphere-ionosphere model (CTIM), in *STEP Handbook on Ionospheric Models*, edited by R. W. Schunk, pp. 217–238, Utah State Univ., Logan.

Hagan, M. E., and R. G. Roble (2001), Modeling diurnal tidal variability with the National Center for Atmospheric Research thermosphere-ionosphere-mesosphere-electrodynamics general circulation model, *J. Geophys. Res.*, *106*(A11), 24,869–24,882, doi:10.1029/2001JA000057.

Hagan, M. E., M. D. Burrage, J. M. Forbes, J. Hackney, W. J. Randel, and X. Zhang (1999), GSWM-98: Results for migrating solar tides, *J. Geophys. Res.*, *104*(A4), 6813–6827, doi:10.1029/1998JA900125.

Hedin, A. E. (1983), A Revised thermospheric model based on mass spectrometer and incoherent scatter data: MSIS-83, *J. Geophys. Res.*, *88*(A12), 10,170–10,188, doi:10.1029/JA088iA12p10170.

Hedin, A. E. (1987), MSIS-86 thermospheric model, *J. Geophys. Res.*, *92*(A5), 4649–4662, doi:10.1029/JA092iA05p04649.

Hedin, A. E. (1991), Extension of the MSIS thermosphere model into the middle and lower atmosphere, *J. Geophys. Res.*, *96*(A2), 1159–1172, doi:10.1029/90JA02125.

Hedin, A. E., and G. R. Carignan (1985), Morphology of thermospheric composition variations in the quiet polar thermosphere from Dynamics Explorer measurements, *J. Geophys. Res.*, *90*(A6), 5269–5277, doi:10.1029/JA090iA06p05269.

Hernandez, G., R. W. Smith, R. G. Roble, J. Gress, and K. C. Clark (1990), Thermospheric dynamics at the South Pole, *Geophys. Res. Lett.*, *17*(9), 1255–1258, doi:10.1029/GL017i009p01255.

Hoffman, R. A., G. D. Hogan, and R. C. Maehl (1981), Dynamics Explorer spacecraft and ground operations system, *Space Sci. Instrum.*, *5*, 349–367.

Jacchia, L. G. (1971), Revised static models of the thermosphere and exosphere with empirical temperature profiles, *SAO Rep. 332*, Smithson. Astrophys. Obs., Cambridge, Mass.

Jacchia, L. G., and J. Slowey (1964), Temperature variations in the upper atmosphere during geomagnetically quiet intervals, *J. Geophys. Res.*, *69*(19), 4145–4148.

Killeen, T. L., and R. G. Roble (1984), An analysis of the high-latitude thermospheric wind pattern calculated by a thermospheric general circulation model: 1. Momentum forcing, *J. Geophys. Res.*, *89*(A9), 7509–7522.

Killeen, T. L., Y.-I. Won, R. J. Niciejewski, and A. G. Burns (1995), Upper thermosphere winds and temperatures in the geomagnetic polar cap: Solar cycle, geomagnetic activity, and interplanetary magnetic field dependencies, *J. Geophys. Res.*, *100*(A11), 21,327–21,342, doi:10.1029/95JA01208.

Lei, J., J. P. Thayer, and J. M. Forbes (2010), Longitudinal and geomagnetic activity modulation of the equatorial thermosphere anomaly, *J. Geophys. Res.*, *115*, A08311, doi:10.1029/2009JA015177.

Liu, H.-L., and R. G. Roble (2002), A study of a self-generated stratospheric sudden warming and its mesospheric–lower thermospheric impacts using the coupled TIME-GCM/CCM3, *J. Geophys. Res.*, *107*(D23), 4695, doi:10.1029/2001JD001533.

Ma, R., J. Xu, W. Wang, J. Lei, H.-L. Liu, A. Maute, and M. E. Hagan (2010), Variations of the nighttime thermospheric mass density at low and middle latitudes, *J. Geophys. Res.*, *115*, A12301, doi:10.1029/2010JA015784.

Mayr, H. G., I. Harris, F. Varosi, and F. A. Herrero (1984), Global excitation of wave phenomena in a dissipative multi-constituent medium: 1. Transfer function of the Earth's thermosphere, *J. Geophys. Res.*, *89*(A12), 10,929–10,959, doi:10.1029/JA089iA12p10929.

Meriwether, J. W. (2006), Studies of thermospheric dynamics with a Fabry–Perot interferometer network: A review, *J. Atmos. Sol. Terr. Phys.*, *68*, 1576–1589, doi:10.1016/j.jastp.2005.11.014.

Niciejewski, R. J., T. L. Killeen, and Y. Won (1994), Observations of neutral winds in the polar cap during northward IMF, *J. Atmos. Sol. Terr. Phys*, *56*, 285–295, doi:10.1016/0021-9169(94)90036-1.

Paxton, L. J., et al. (1999), Global ultraviolet imager (GUVI): Measuring composition and energy inputs for the NASA Thermosphere Ionosphere Mesosphere Energetics and Dynamics (TIMED) mission, *SPIE Opt. Spectrosc. Techn. Instrum. Atmos. Space Res.*, *3756*, 265–276.

Picone, J. M., A. E. Hedin, D. P. Drob, and A. C. Aikin (2002), NRLMSISE-00 empirical model of the atmosphere: Statistical comparisons and scientific issues, *J. Geophys. Res.*, *107*(A12), 1468, doi:10.1029/2002JA009430.

Qian, L., R. G. Roble, S. C. Solomon, and T. J. Kane (2006), Calculated and observed climate change in the thermosphere, and a prediction for solar cycle 24, *Geophys. Res. Lett.*, *33*, L23705, doi:10.1029/2006GL027185.

Qian, L., S. C. Solomon, and T. J. Kane (2009), Seasonal variation of thermospheric density and composition, *J. Geophys. Res.*, *114*, A01312, doi:10.1029/2008JA013643.

Qian, L., A. G. Burns, P. C. Chamberlin, and S. C. Solomon (2011), Variability of thermosphere and ionosphere responses to solar flares, *J. Geophys. Res.*, *116*, A10309, doi:10.1029/2011JA016777.

Richmond, A. D., E. C. Ridley, and R. G. Roble (1992), A thermosphere/ionosphere general circulation model with coupled electrodynamics, *Geophys. Res. Lett.*, *19*(6), 601–604.

Rishbeth, H. (2007), Thermospheric targets, *Eos Trans. AGU*, *88*(17), 189, doi:10.1029/2007EO170002.

Rishbeth, H., and O. K. Garriott (1969), *Introduction to Ionospheric Physics*, Academic Press, New York.

Roble, R. G. (1995), Energetics of the mesosphere and thermosphere, in *The Upper Mesosphere and Lower Thermosphere: A Review of Experiment and Theory, Geophys. Monogr. Ser.*, vol. 87, edited by R. M. Johnson and T. L. Killeen, pp. 1–21, AGU, Washington, D. C., doi:10.1029/GM087p0001.

Roble, R. G., and E. C. Ridley (1994), A thermosphere-ionosphere-mesosphere-electrodynamics general circulation model (time-GCM): Equinox solar cycle minimum simulations (30–500 km), *Geophys. Res. Lett.*, *21*(6), 417–420.

Roble, R. G., R. E. Dickinson, E. C. Ridley, and Y. Kamide (1979), Thermospheric response to the November 8–9, 1969, magnetic disturbances, *J. Geophys. Res.*, *84*(A8), 4207–4216.

Roble, R. G., E. C. Ridley, and R. E. Dickinson (1987), On the global mean structure of the thermosphere, *J. Geophys. Res.*, *92*(A8), 8745–8758.

Roble, R. G., E. C. Ridley, A. D. Richmond, and R. E. Dickinson (1988), A coupled thermosphere/ionosphere general circulation model, *Geophys. Res. Lett.*, *15*(12), 1325–1328.

Rocket Panel (1952), Pressures, densities and temperatures in the upper atmosphere, *Phys. Rev.*, *88*, 1027–1032.

Salah, J. E., J. V. Evans, D. Alcayde, and P. Bauer (1976), Comparison of exospheric temperatures at Millstone Hill and St. Santin, *Ann. Geophys.*, *32*, 257–266.

Shepherd, G. G., et al. (1993), WINDII, the wind imaging interferometer on the Upper Atmosphere Research Satellite, *J. Geophys. Res.*, *98*(D6), 10,725–10,750.

Solomon, S. C., L. Qian, L. V. Didkovsky, R. A. Viereck, and T. N. Woods (2011), Causes of low thermospheric density during the 2007–2009 solar minimum, *J. Geophys. Res.*, *116*, A00H07, doi:10.1029/2011JA016508. [Printed 117(A2), 2012].

Spencer, N. W., L. E. Wharton, H. B. Niemann, A. E. Hedin, G. R. Carignan, and J. C. Maurer (1981), The Dynamics Explorer wind and temperature spectrometer, *Space Sci. Instrum.*, *5*, 417–428.

Taeusch, D. R., G. R. Carignan, and C. A. Reber (1971), Neutral composition variation above 400 kilometers during a magnetic storm, *J. Geophys. Res.*, *76*(34), 8318–8325.

Wang, W., J. Lei, A. G. Burns, S. C. Solomon, M. Wiltberger, J. Xu, Y. Zhang, L. Paxton, and A. Coster (2010), Ionospheric response to the initial phase of geomagnetic storms: Common features, *J. Geophys. Res.*, *115*, A07321, doi:10.1029/2009JA014461.

Wiens, R. H., V. P. Bhatnagar, and G. Thuillier (2002), Geomagnetic storm heating effects on the low-latitude dayside thermosphere from WINDII observations at equinox, *J. Atmos. Sol. Terr. Phys.*, *64*, 1393–1400.

Wu, Q., D. McEwen, W. Guo, R. J. Niciejewski, R. G. Roble, and Y. I. Won (2008), Long-term thermospheric neutral wind observations over the northern polar cap, *J. Atmos. Sol. Terr. Phys.*, *70*, 2014–2030, doi:10.1016/j.jastp.2008.09.004.

Zhang, Y., and L. J. Paxton (2008). An empirical Kp-dependent global auroral model based on TIMED/GUVI FUV data, *J. Atmos. Sol. Terr. Phys.*, *70*(8), 1231–1242, doi:10.1016/j.jastp.2008.03.008.

A. G. Burns, L. Qian, S. C. Solomon, and W. Wang, National Center for Atmospheric Research, P.O. Box 3000, Boulder, CO 80307-3000, USA. (aburns@ucar.edu)

Numerical Methods in Modeling the Ionosphere

J. D. Huba

Plasma Physics Division, Naval Research Laboratory, Washington, District of Columbia, USA

G. Joyce

Icarus Research, Inc., Bethesda, Maryland, USA

A brief review of numerical methods used in ionospheric modeling is presented. We focus on two numerical schemes to solve the continuity and velocity equations of the ions along the geomagnetic field: the fully implicit method and the semi-implicit method. We describe each method in detail highlighting the advantages and disadvantages of each method. We compare and contrast simulation results for a simplified problem to illustrate their attributes. We also discuss two gridding systems used to model crossfield transport: Lagrangian (moving grid) and Eulerian (fixed grid).

1. INTRODUCTION

The ionosphere is a weakly ionized gas surrounding the Earth in the altitude range ~90–1000 km [*Kelley*, 2009; *Schunk and Nagy*, 2000]. The plasma is primarily produced by solar EUV radiation and consists of both molecular and ion species. It is embedded in a dynamic, multicomponent neutral gas, as well as the Earth's geomagnetic field. Thus, a model of the Earth's ionosphere comprises a number of complex components: photoionization, chemical reactions, plasma transport, ion-neutral coupling, and electrodynamics, and necessarily requires sophisticated numerical algorithms to solve the nonlinear equations that describe the system.

There are several papers that have provided an in-depth description of some of the numerical methods used in modeling the ionosphere [*Schunk and Sojka*, 1996; *Bailey and Balan*, 1996; *Millward et al.*, 1996; *Huba et al.*, 2000].

However, to our knowledge, there are no papers that provide a full description of all the numerical methods used in current ionosphere models or contrast different numerical algorithms used in these models.

A complete description of the numerics used in ionospheric models is certainly beyond the scope of this paper. Instead, we focus on two numerical schemes to solve the continuity and velocity equations of the ions along the geomagnetic field: the fully implicit method and the semi-implicit method. We describe each method in detail highlighting the advantages and disadvantages of each method. We also compare and contrast simulation results for a simplified problem to illustrate their attributes. Finally, we also discuss two gridding systems used in ionosphere models as they apply to cross-field plasma transport: Lagrangian (moving grid) and Eulerian (fixed grid).

2. GENERAL EQUATIONS

The basic plasma equations that are solved to describe ionospheric dynamics are the following:
ion continuity

$$\frac{\partial n_i}{\partial t} + \nabla \cdot (n_i \mathbf{V}_i) = \mathcal{P}_i - \mathcal{L}_i n_i, \qquad (1)$$

ion velocity

$$\frac{\partial \mathbf{V}_i}{\partial t} + \mathbf{V}_i \cdot \nabla \mathbf{V}_i = -\frac{1}{\rho_i}\nabla \mathbf{P}_i + \frac{e}{m_i}\mathbf{E} + \frac{e}{m_i c}\mathbf{V}_i \times \mathbf{B}$$
$$+ \mathbf{g} - \nu_{in}(\mathbf{V}_i - \mathbf{V_n}) - \sum_j \nu_{ij}(\mathbf{V}_i - \mathbf{V}_j), \quad (2)$$

ion temperature

$$\frac{\partial T_i}{\partial t} + \mathbf{V}_i \cdot \nabla T_i + \frac{2}{3}T_i \nabla \cdot \mathbf{V}_i + \frac{2}{3}\frac{1}{n_i k}\nabla \cdot \mathbf{Q}_i = Q_{in} + Q_{ij} + Q_{ie}, \quad (3)$$

electron momentum along **B**

$$0 = -\frac{1}{n_e m_e}b_s \frac{\partial P_e}{\partial s} - \frac{e}{m_e}E_s, \quad (4)$$

electron temperature along **B**

$$\frac{\partial T_e}{\partial t} - \frac{2}{3}\frac{1}{n_e k}b_s^2 \frac{\partial}{\partial s}\kappa_e \frac{\partial T_e}{\partial s} = Q_{en} + Q_{ei} + Q_{phe}. \quad (5)$$

In the above, $\mathcal{P}_i$ is the production term associated with photoionization and chemistry (e.g., charge exchange), $\mathcal{L}_i$ is the loss term associated with chemistry (e.g., charge exchange, recombination), ν_{in} and ν_{ij} are the ion-neutral and ion-ion collision frequencies, $Q_{\alpha\beta}$ are the collisional heating/cooling terms between ion and neutral species, Q_{phe} is the photoelectron heating term, $\mathbf{Q}_i = -k_i\nabla T_i$ is the ion heat flux, K_i and K_e are the ion and electron heat conductivities, s is the coordinate along **B**, $b_s = B_s/B_0$, B_s is the magnetic field (in dipole coordinates), and B_0 is the value of the field at the Earth's surface, $n_e = \sum_i n_i$ is the electron density, $P_{i,e} = n_{i,e}T_{i,e}$ are the ion and electron pressure, g is gravity, V_n is the neutral wind, and **E** is the electric field. Specific definitions of these terms can be found in the work of *Huba et al.* [2000].

The electric field can be specified using an empirical model [*Scherliess and Fejer*, 1999], analytically or self-consistently. The self-consistent electric field is found by solving a potential equation $\nabla\Sigma\nabla\Phi = S(J_\parallel, V_n, g)$, where $\mathbf{E} = -\nabla\Phi$, Σ is the ionospheric conductance, S is the source term associated with the drivers (i.e., high-latitude currents, neutral wind, and gravity). This topic is discussed in this monograph by Richmond and Maute.

One technique to solve equations (1)–(3) is to use a time-splitting method. Schematically, the equations are solved in two steps:

$$\begin{array}{ll} t_0 + \Delta t \to t^* & \text{parallel transport} \\ t^* + \Delta t \to t_1 & \text{perpendicular transport} \end{array} \quad (6)$$

The density, velocity, and temperature are advanced from time t_0 to time t^* for motion parallel to the geomagnetic field. These values are then used to advance the solution to the next time step t_1 for motion perpendicular to the geomagnetic field. The splitting method is used because the equations are solved differently for the parallel and perpendicular motion to the geomagnetic field. This is not necessary but is extremely useful because the magnetic field strongly controls the plasma motion in the very low β ($= 8\pi n T/B^2$) ionosphere: plasma transport along the magnetic is dominated by diffusion, pressure, and gravity, while plasma transport across the magnetic field is dominated by the $E \times B$ drift motion.

3. FIELD-ALIGNED DYNAMICS

We first address two numerical methods to solve the ion equations along the geomagnetic field: the fully implicit and the semi-implicit methods. The continuity and velocity equations that describe the ion motion along the geomagnetic field are

$$\frac{\partial n_i}{\partial t} + b_s^2\frac{\partial}{\partial s}\frac{n_i V_{is}}{b_s} = \mathcal{P}_i - \mathcal{L}_i n_i \quad (7)$$

$$\frac{\partial V_{is}}{\partial t} + V_{is}\frac{\partial V_{is}}{\partial s} = -\frac{1}{n_i m_i}b_s\frac{\partial(P_i + P_e)}{\partial s}$$
$$+ g_s - \nu_{in}(V_{is} - V_{ns}) - \sum_{j(\text{ions})}\nu_{ij}(V_{is} - V_{js}). \quad (8)$$

The subscript s on vector quantities indicates the component of the vector in the s direction.

3.1. Fully Implicit Algorithm

The key assumption to develop an implicit algorithm to solve equations (7) and (8) is to neglect ion inertia in equation (8)

$$0 = -\frac{1}{n_i m_i}b_s\frac{\partial(n_i T_i + n_e T_e)}{\partial s}$$
$$+ g_s - \nu_{in}(V_{is} - V_{ns}) - \sum_{j(\text{ions})}\nu_{ij}(V_{is} - V_{js}), \quad (9)$$

where we have used the definition of pressure $P = nT$. The basic procedure is to solve equation (9) for V_{is} as a function of n_i (and the other variables) and substitute it into equation (7). The time discretization of equation (7) is then written as

$$\left(\frac{1}{\Delta t} + \mathcal{L}_i\right)n_i^{t+\Delta t} + b_s^2\frac{\partial}{\partial s}\frac{n_i^{t+\Delta t}f(n_i^{t+\Delta t},\ldots)}{b_s} = \frac{n_i^t}{\Delta t} + \mathcal{P}_i, \quad (10)$$

where $f(n_i^{t+\Delta t}, \ldots)$ denotes the solution to V_{is}. Except for the right-hand side of equation (10), the ion density n_i is defined at the upper time level $t + \Delta t$; this is the crux of the fully implicit scheme. Defining the spatial discretization as $\partial g/\partial s = (g_{j+1} - g_{j-1})/\Delta s$, one can write equation (10) in tridiagonal form

$$An_{i,j-1}^{t+\Delta t} + Bn_{i,j}^{t+\Delta t} + Cn_{i,j-1}^{t+\Delta t} = D, \quad (11)$$

which can be solved for $n_i^{t+\Delta t}$ using standard numerical algorithms [*Press*, 2003].

The above is a broad overview of the fully implicit differencing technique used in ionospheric modeling. The major difficulty in implementing this algorithm is that the solution to V_{is} involves all ion species because $n_e = \sum_i n_i$ and ion-ion coupling. One strategy to lessen the algebraic complexity of the problem is to only consider transport of the atomic ions (i.e., H^+, He^+, and O^+) and treat the molecular ions (i.e., N_2^+, NO^+, and O_2^+) to be in chemical equilibrium (i.e., $\mathcal{P}_i = \mathcal{L}_i n_i$). In general, this is a reasonable assumption, especially during nominal daytime and nighttime conditions, because of the fast time scales associated with molecular chemistry at altitudes below 200 km. However, in the nighttime solar terminators, the molecular ions can depart from chemical equilibrium because of strong neutral wind shears. One detailed description of an implicit numerical algorithm to solve equations (7) and (9) is given by *Bailey and Balan* [1996].

Another technique to solve equations (7) and (9) implicitly is through iteration. The idea is to solve equation (9) for V_{is} without expanding the electron density into the ion densities. This allows V_{is} to be written in terms of n_i directly. The system of ion equations can then be solved where $n_e = \sum_i n_i^t$ is incorporated into D in equation (11). At each time step, the equations are iterated until n_e no longer changes. This method was used in the 1-D ionosphere model described in the work of *Oran et al.* [1974]. A shortcoming of this method is that there is no guarantee a priori that the solution will converge.

The primary benefit of solving equations (7) and (9) fully implicitly is that a relatively large time step can be used. For example, time steps of 5–15 min are commonly used in ionospheric simulations using fully implicit schemes. On the other hand, there is an issue using this scheme at very high altitudes (e.g., the plasmasphere). To the lowest order, the ion velocity given by equation (9) is proportional to v_{in}^{-1}. At high altitudes, v_{in} becomes very small, and the ion velocity becomes unphysically large.

3.2. Semi-Implicit Algorithm

The Naval Research Laboratory ionosphere models SAMI2 and SAMI3 use a semi-implicit algorithm to solve equations (7) and (8) [*Huba et al.*, 2000]. In this scheme, ion inertia is included in the ion velocity equation. The difference equation for continuity is written as

$$\frac{n_{i,j}^{t+\Delta t} - n_{i,j}^t}{\Delta t} + \frac{(n_i^{t+\Delta t} V_{is}^t)_{j+1/2} - (n_i^{t+\Delta t} V_{is}^t)_{j-1/2}}{\Delta s_j} = \mathcal{P} - n_{i,j}^{t+\Delta t}\mathcal{L}, \quad (12)$$

where $\Delta s_j = (s_{j+1} - s_{j-1})/2$, and $\mathcal{P}$ and $\mathcal{L}$ are evaluated at time t. The density is evaluated at the upper time level $t + \Delta t$ so that the difference scheme is implicit (i.e., backward biased). However, the velocity V_{is} is evaluated at the current time level t so the scheme is only "semi-implicit." This method allows the Courant condition ($\Delta t < \Delta s/V$) to be based upon the advection velocity $V = V_{is}$ and not the sum of the advection velocity and the sound speed $V = V_{is} + C_s$. This technique is useful in modeling subsonic plasmas such as the ionosphere.

The donor cell method is used to compute the fluxes nV at the half steps $j - 1/2$ and $j + 1/2$. The velocity at the half step is calculated as a simple average:

$$V_l = \frac{1}{2}(V_{j-1} + V_j); \quad V_r = \frac{1}{2}(V_j + V_{j+1}). \quad (13)$$

The difference equation is written as

$$\frac{n_{i,j}^{t+\Delta t} - n_{i,j}^t}{\Delta t} + \frac{a_0 n_{i,j-1}^{t+\Delta t} + b_0 n_{i,j}^{t+\Delta t} + c_0 n_{i,j+1}^{t+\Delta t}}{\Delta s_j} = \mathcal{P} - n_{i,j}^{t+\Delta t}\mathcal{L}, \quad (14)$$

where

$$a_0 = \begin{cases} -V_l & V_r > 0 \text{ and } V_l > 0 \\ 0 & V_r < 0 \text{ and } V_l < 0 \\ 0 & V_r > 0 \text{ and } V_l < 0 \\ -V_l & V_r < 0 \text{ and } V_l > 0 \end{cases} \quad (15)$$

$$b_0 = \begin{cases} V_r & V_r > 0 \text{ and } V_l > 0 \\ -V_l & V_r < 0 \text{ and } V_l < 0 \\ V_r - V_l & V_r > 0 \text{ and } V_l < 0 \\ 0 & V_r < 0 \text{ and } V_l > 0 \end{cases} \quad (16)$$

$$c_0 = \begin{cases} 0 & V_r > 0 \text{ and } V_l > 0 \\ V_r & V_r < 0 \text{ and } V_l < 0 \\ 0 & V_r > 0 \text{ and } V_l < 0 \\ V_r & V_r < 0 \text{ and } V_l > 0 \end{cases} \quad (17)$$

The main idea of the donor cell method is that the flux into or out of each cell is determined by the direction of the velocity at the cell boundary. If the velocity at the half step $j - 1/2$ is positive (i.e., $V_l > 0$), then the flux at the boundary is $n_{j-1}V_l$

because plasma from cell $j - 1$ is being transported into cell j. On the other hand, if the velocity at the half step $j - 1/2$ is negative (i.e., $V_l < 0$), then the flux at the boundary is $n_j V_l$ because plasma is carried from cell j into cell $j - 1$.

Finally, equation (14) is written in the form

$$An_{i,j-1}^{t+\Delta t} + Bn_{i,j}^{t+\Delta t} + Cn_{i,j+1}^{t+\Delta t} = D, \quad (18)$$

where

$$A = \frac{a_0}{\Delta s_j}; \quad B = \frac{1}{\Delta t} + \frac{b_0}{\Delta s_j} + \mathcal{L}; \quad C = \frac{c_0}{\Delta s_j}; \quad D = \frac{n_j^t}{\Delta t} + \mathcal{P},$$

and A, B, C, and D are evaluated at time t. The density at the updated time step $t + \Delta t$ is obtained by solving the tridiagonal matrix (18). The boundary condition at each end of the flux tube is $n_i = \mathcal{P}_i/\mathcal{L}_i$. In SAMI2 and SAMI3, the ion velocity and temperature equations are solved in a similar manner.

3.3. Numerical Comparison

We show results of a direct comparison of the fully implicit and semi-implicit numerical algorithms for an ionospheric plasma. For this comparison, we make a number of simplifying assumptions. We consider only one ion species (H$^+$) and neglect ion-ion coupling. We set the neutral wind and $E \times B$ drift to zero and keep the temperature constant in time. The semi-implicit algorithm uses equation (18) coupled with equation (8).

The fully implicit algorithm used here is described as follows. Neglecting ion inertia, the ion velocity can be written as

$$V_{is} = \frac{-b_s}{\nu_{in}} \left(\frac{1}{n_i m_i} \frac{\partial n_i T_i}{\partial s} + \frac{1}{n_e m_i} \frac{\partial n_e T_e}{\partial s} \right) + \frac{g_s}{\nu_{in}} \quad (19)$$

based on equation (9). Substituting equation (19) into equation (7) yields

$$\frac{\partial n_i}{\partial t} + b_s^2 \frac{\partial}{\partial s}\left(-\frac{1}{m_i \nu_{in}}\frac{\partial n_i T}{\partial s} + \frac{n_i g_s}{b_s \nu_{in}}\right) = \mathcal{P}_i - \mathcal{L}_i n_i, \quad (20)$$

where we use $n_e = n_i$ and define $T = T_e + T_i$.

Implicitly differencing equation (20) leads to

$$\frac{n_{i,j}^{t+\Delta t} - n_{i,j}^t}{\Delta t} - \left(\frac{b_s^2}{m_i \nu_{in}}\right)_j \frac{n_{i,j+1}^{t+\Delta t} T_{i,j+1} - 2 n_{i,j}^{t+\Delta t} T_{i,j} + n_{i,j-1}^{t+\Delta t} T_{i,j-1}}{\Delta s^2}$$
$$+ \left(\frac{g_s}{b_s \nu_{in}}\right)_j \frac{n_{i,j+1}^{t+\Delta t} - n_{i,j-1}^{t+\Delta t}}{\Delta s} = \mathcal{P}_i - n_{i,j}^{t+\Delta t} \mathcal{L}_i, \quad (21)$$

where we have assumed uniform spacing in the s direction. Finally, equation (21) is written in the form

$$An_{j-1}^{t+\Delta t} + Bn_j^{t+\Delta t} + Cn_{j+1}^{t+\Delta t} = D, \quad (22)$$

where

$$A = -\left(\frac{b_s^2}{m_i \nu_{in}}\right)_j \frac{T_{i,j-1}}{\Delta s^2} - \left(\frac{b_s g_s}{\nu_{in}}\right)_j \frac{1}{\Delta s};$$

$$B = \left(\frac{b_s^2}{m_i \nu_{in}}\right)_j \frac{2 T_{i,j}}{\Delta s^2} + \frac{1}{\Delta t} + \mathcal{L}_i$$

$$C = -\left(\frac{b_s^2}{m_i \nu_{in}}\right)_j \frac{T_{i,j+1}}{\Delta s^2} + \left(\frac{b_s g_s}{\nu_{in}}\right)_j \frac{1}{\Delta s}; \quad D = \frac{1}{\Delta t} + \mathcal{P}_i.$$

We have modified SAMI2 [*Huba et al.*, 2000] to solve equations (7) and (8) for both the semi-implicit and fully implicit schemes under the simplifying assumptions outlined above. We only consider the solution along a single field line. The geophysical parameters used are $F10.7 = F10.7A = 120$, day-on-year = 91, and $Ap = 21$. The magnetic apex is 19,500 km, 801 grid points are used, and the code is run for 8 days.

In Figure 1, we show the time evolution of the H$^+$ density as a function of altitude for several values of the density ($n_i = $ 25, 50, 100, and 200 cm^{-3}). The black curve is for the semi-implicit scheme and the red for the fully implicit scheme. The time step for the semi-implicit run is 1 s, while for the fully implicit run it is 300 s. In both cases, the ions achieve a diurnal steady state after ~3 days. However, more important, the two solutions are in very good agreement quantitatively, even at altitudes above 10^4 km.

In Figure 2, we show the H$^+$ density as a function of altitude at time 13:25 LT on day 4 (semi-implicit (black) and fully-implicit (red)). It is clear that there is excellent quantitative agreement between the semi-implicit and fully implicit schemes throughout the entire altitude range.

In Figure 3, we show the H$^+$ velocity as a function of altitude at time 13:25 LT on day 4 (semi-implicit (black) and fully-implicit (red)). There is good quantitative agreement for altitudes $\lesssim 1000$ km. However, above 1000 km, the results diverge: the velocity for the semi-implicit case decreases, but the velocity for the fully implicit case increases significantly with altitude. This latter result is consistent with the velocity scaling as ν_{in}^{-1}.

It is interesting that the fully implicit and the semi-implicit schemes are in excellent quantitative agreement for the plasma density in the collisionless, high-altitude ionosphere despite the fact that the velocities are very different.

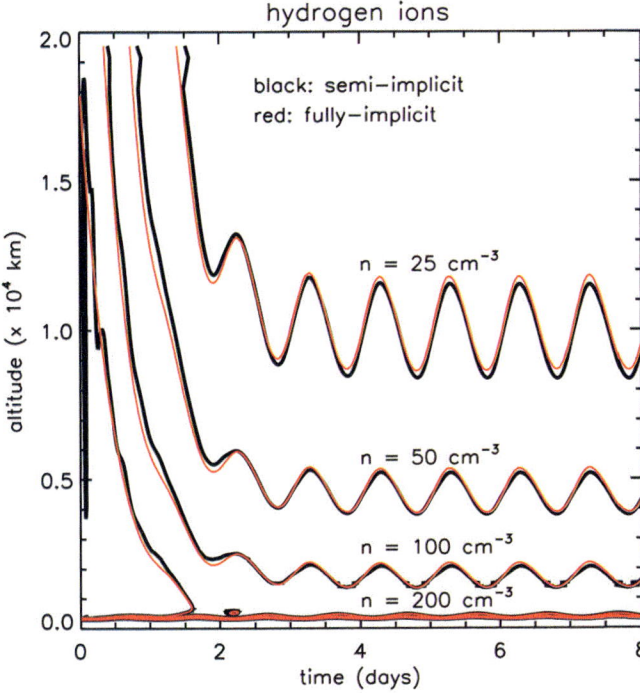

Figure 1. Hydrogen ion density as a function of time and altitude for several values of n_i: semi-implicit (black) and fully implicit (red).

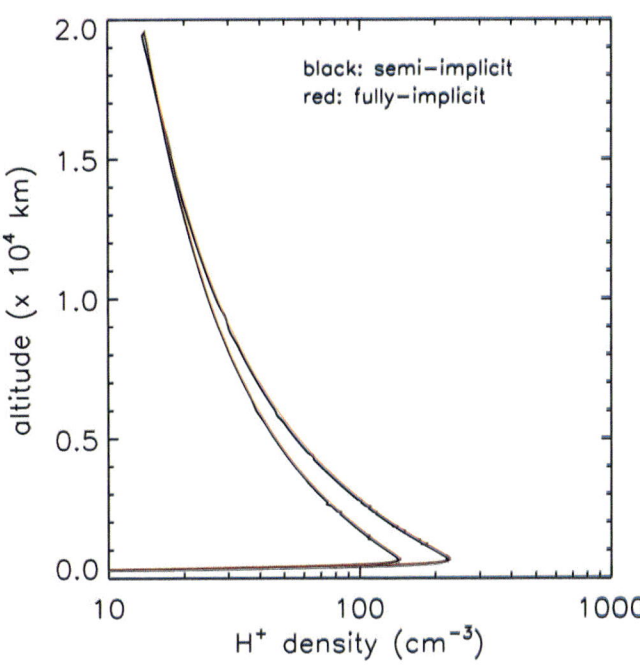

Figure 2. Hydrogen ion density as a function altitude at time 13:25 LT on day 4: semi-implicit (black) and fully implicit (red). Here the higher-density profile is in the Southern Hemisphere, and the lower-density profile is in the Northern Hemisphere.

The reason for this is explained as follows. For the implicit scheme, we solve equation (20). In the high-altitude, collisionless limit, we find that the RHS of equation (20) is very small because there is very little production or loss of H$^+$. Furthermore, since $\nu_{in} \ll 1$, this equation simply blackuces to

$$-\frac{1}{m_i}\frac{\partial n_i T}{\partial s} + \frac{n_i g_s}{b_s} = 0, \qquad (23)$$

which shows that the plasma is in pressure balance with the gravitational force along the magnetic field line as expected. Thus, the fully implicit method can provide an accurate description of the plasma density at high altitude; however, it does not model the plasma velocity accurately because ion inertia is neglected.

4. CROSS-FIELD DYNAMICS

Two techniques used to calculate ion transport across the geomagnetic field caused by the $E \times B$ drift are the Lagrangian method and the Eulerian method. In the Lagrangian method, the motion of "flux tubes" is calculated based on the $E \times B$ drift velocity. The ion density is updated based on conservation of particles. In the Eulerian method, the grid is

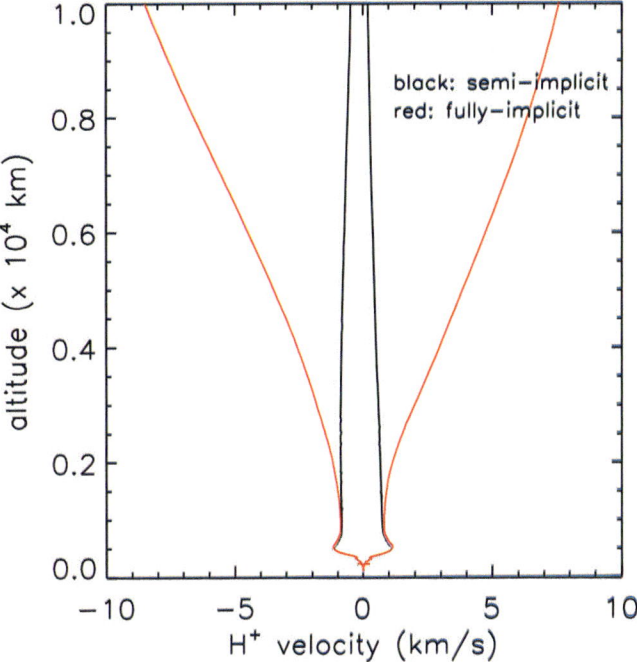

Figure 3. Hydrogen ion velocity as a function altitude at time 13:25 LT on day 4: semi-implicit (black) and fully-implicit (red). Here the negative velocity profile is in the Southern Hemisphere, and the positive velocity profile is in the Northern Hemisphere.

fixed, and the particle flux is calculated at cell boundaries to update the ion density. We briefly describe each numerical technique.

4.1. Lagrangian Grid

We first discuss the Lagrangian scheme. In Figure 4, we consider two flux tubes at times t (position r_1) and $t + \Delta t$ (position r_2) in dipole coordinates (q, p) where $q = (R_E^2/r^2) \cos\theta$, $p = r/(R_E \sin^2\theta)$, and R_E is the radius of the Earth. The flux tube at time t has $p = p_1$; it rises to $p = p_2$ at time $t + \Delta t$ because of an $E \times B$ drift V_E. At time t, p_1 and the s coordinates are known, as well as the physical variables along p_1. The value of p_2 is determined as follows. From the definition of p, we have

$$\Delta p = p_2 - p_1 = (r_2 - r_1)/R_E = \Delta r/R_E \tag{24}$$

along $s = 0$ (i.e., $\theta = \pi/2$). Since $\Delta r = V_E \Delta t$, we find that

$$p_2 = p_1 + V_E \Delta t / R_E. \tag{25}$$

Finally, we note that the magnitude of the geomagnetic field is also known along $p = p_2$.

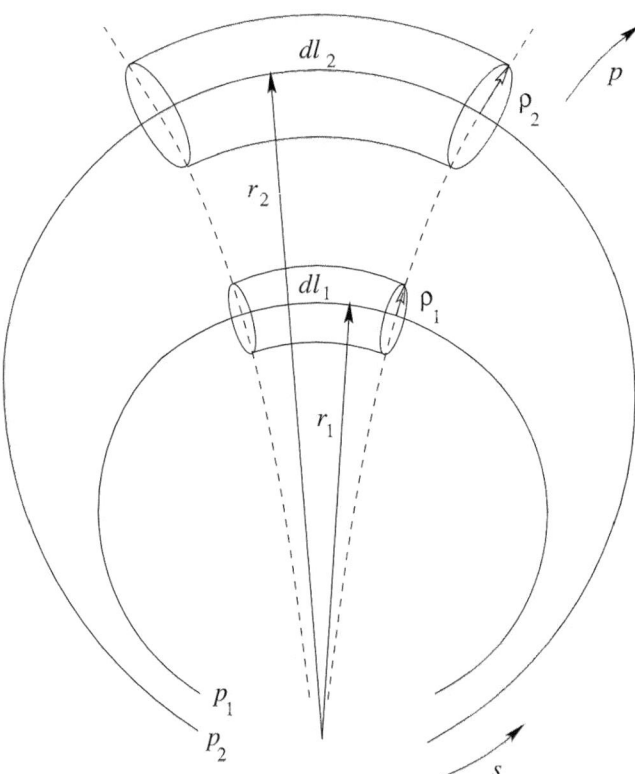

Figure 4. Schematic of "flux tube" motion in the Lagrangian scheme.

We now calculate the updated density along $p = p_2$. The volume of flux tube cell at time t is $\mathcal{V}_1 = \pi \rho_1^2 dl_1$ and at time $t + \Delta t$ is $\mathcal{V}_2 = \pi \rho_2^2 dl_2$, where r is the radius of the flux tube, and dl is the length of the flux tube along the geomagnetic field. The total number of particles in each flux tube cell is conserved so that

$$n_1 \rho_1^2 dl_1 = n_2 \rho_2^2 dl_2. \tag{26}$$

The magnetic flux through each flux tube is also conserved so that

$$\rho_1^2 B_1 = \rho_2^2 B_2. \tag{27}$$

Combining equations (26) and (27), we find that the updated density n_2 at time $t + \Delta t$ is given by

$$n_2 = \frac{dl_1}{dl_2} \frac{B_2}{B_1} n_1. \tag{28}$$

4.2. Eulerian Grid

For the Eulerian (fixed grid) method, we briefly describe the finite volume algorithm. There are other finite-difference algorithms that can be used, but this method has the benefit of being applicable to nonuniform, nonorthogonal meshes. First, we write the transverse component of the continuity equation

$$\frac{\partial n}{\partial t} = -\nabla n \cdot \mathbf{V}_\perp. \tag{29}$$

Next, we integrate over the cell volume

$$\int \frac{\partial n}{\partial t} d^3x = -\int \nabla n \cdot \mathbf{V}_\perp d^3x. \tag{30}$$

Applying Gauss' law, we rewrite equation (30) as

$$\frac{\partial N}{\partial t} = -\oint (n\mathbf{V}_\perp) \cdot \hat{\mathbf{e}}_\mathbf{n} \, d^2x, \tag{31}$$

where $N = \int n d^3x$ is the total number of ions in a cell. Finally, the time advance is carried out

$$N^{t+\Delta t} = N^t - \Delta t \underbrace{\oint (n\mathbf{V}_\perp)^{t+\Delta t/2} \cdot \hat{\mathbf{e}}_\mathbf{n} \, d^2x}_{\text{flux}}, \tag{32}$$

where the flux is calculated at the cell boundary. The density at $n^{t+\Delta t}$ is obtained by dividing $N^{t+\Delta t}$ by the cell volume.

There are a number of algorithms to calculate the flux nV at the cell boundary. One of the simplest is the donor cell method as previously described. This scheme is low order and is numerically diffusive. However, for modeling global ionospheric dynamics, this scheme is adequate because the ionosphere is, in fact, diffusive plasma. On the other hand, high-order transport schemes should be used when modeling ionospheric disturbances that lead to steep density gradients (e.g., equatorial spread F). These schemes will not be discussed here.

5. SUMMARY

A brief discussion of numerical algorithms used in modeling the Earth's ionosphere has been presented. The general methodology is to time split the problem into parallel and perpendicular motion because of the dominant role the geomagnetic field has in organizing the plasma dynamics.

The two techniques discussed for parallel motion are the semi-implicit and fully implicit schemes. The advantage of the fully implicit method is that it allows relatively large time steps (e.g., 5–15 min). The disadvantage is that it can be difficult to implement in a multi-ion plasma. The advantage of the semi-implicit method is that it is relatively easy to implement in a multi-ion plasma. The disadvantage is that it requires relatively small time steps (e.g., 1–30 s). We presented a comparison of the semi-implicit and fully implicit schemes for a simplified, single-ion system. Interestingly, we found that two schemes are in very good agreement for the ion (H^+) density up to 20,000 km despite the fact that the velocities only agree for altitudes less than ~1000 km.

The two methods discussed for cross-field motion are the Lagrangian and Eulerian gridding schemes. The Lagrangian scheme has a dynamic grid and transports flux tubes of plasma. The advantage of this method is that it is relatively easy to implement, and it can use large time steps. The disadvantage is that in complex convection patterns (e.g., during magnetic storms in the polar cap), the flux tubes can become spatially nonuniform, and resolution is lost in some regions. A technique to mitigate this is to interpolate the Lagrangian results onto a fixed grid after each time step. The Eulerian grid uses a fixed mesh. An advantage of this method is that nonuniform meshes can be used with the appropriate transport algorithm (e.g., finite volume method). Disadvantages are that it can be difficult to implement, and small time steps may be required.

Acknowledgments. It is with great sadness that I report that the coauthor of SAMI2/3, Glenn Joyce, passed away in December 2011 during the course of this work (JDH). Glenn's contributions to SAMI2/3 cannot be underestimated; for example, he developed the semi-implicit algorithms used in the models. We thank the referee for several suggestions to improve the manuscript. This research has been supported with NRL 6.1 Base Funds.

REFERENCES

Bailey, G. J., and N. Balan (1996), A low-latitude ionosphere-plasmasphere model, in *STEP: Handbook of Ionospheric Models*, edited by R. W. Schunk, p. 173, Utah State Univ., Logan.

Huba, J. D., G. Joyce, and J. A. Fedder (2000), Sami2 is Another Model of the Ionosphere (SAMI2): A new low-latitude ionosphere model, *J. Geophys. Res.*, *105*(A10), 23,035–23,053.

Kelley, M. (2009), *The Earth's Ionosphere*, 2nd ed., Academic Press, San Diego, Calif.

Millward, G. H., R. J. Moffett, W. Quegan, and T. J. Fuller-Rowell (1996), A coupled thermospheric-ionospheric-plasmasphere model (CTIP), in *STEP: Handbook of Ionospheric Models*, edited by R. W. Schunk, p. 239, Utah State Univ., Logan.

Oran, E. S., T. R. Young, D. V. Anderson, T. P. Coffey, P. C. Kepple, A. W. Ali, and D. F. Strobel (1974), A numerical model of the mid-latitude ionosphere, *NRL Memo. Rep. 2839*, Naval Res. Lab., Washington, D. C.

Press, W. H. (2003), *Numerical Recipes in FORTRAN: The Art of Scientific Computing*, Cambridge Univ. Press, Cambridge, U. K.

Scherliess, L., and B. G. Fejer (1999), Radar and satellite global equatorial F region vertical drift model, *J. Geophys. Res.*, *104*(A4), 6829–6842.

Schunk, R. W., and A. Nagy (2000), *Ionospheres: Physics, Plasma Physics, and Chemistry*, Cambridge Univ. Press, New York.

Schunk, R. W., and J. J. Sojka (1996), USU model of the global ionosphere, in *STEP: Handbook of Ionospheric Models*, edited by R. W. Schunk, p. 153, Utah State Univ., Logan.

J. D. Huba, Code 6790, Naval Research Laboratory, Washington, DC 20375, USA. (huba@ppd.nrl.navy.mil)

G. Joyce, Icarus Research, Inc., Bethesda, MD, USA.

Ionospheric Electrodynamics Modeling

A. D. Richmond and A. Maute

High Altitude Observatory, National Center for Atmospheric Research, Boulder, Colorado, USA

The modeling of quasistatic ionospheric electric fields, currents, and magnetic perturbations associated with thermospheric winds, plasma gravitational and pressure-gradient (G/P) forces, and coupling with the magnetosphere is described, using examples from the National Center for Atmospheric Research thermosphere-ionosphere-electrodynamics general circulation model. A 2-D partial differential equation (PDE) for the electric potential is obtained in magnetic coordinates, with source terms related to the winds, plasma G/P forces, and net field-aligned currents (FAC) from the magnetosphere. Three ways of representing the net FAC in the PDE are discussed: as a specified input FAC distribution, as a FAC distribution derived from a specified distribution of high-latitude electric potential, and as a representation of the FAC from the inner magnetosphere in terms of equivalent magnetospheric conductances. The plasma G/P source is generally weaker than the other sources of electric field and current, but can produce significant effects on the electric field at night and measurable geomagnetic perturbations at satellite heights. The calculation of geomagnetic perturbations below and above the ionosphere is illustrated with simplified calculations that reduce the 3-D current to a horizontal thin current sheet at 110 km altitude, connected to FAC above. This simplified current system can be divided into an "equivalent" current flowing in the sheet, plus a "residual" current composed of the FAC plus divergent sheet current. By definition, the ground-level magnetic perturbations are related only to the equivalent current, while magnetic perturbations above the ionosphere are produced both by the equivalent current and the residual current, with the latter often dominant.

1. INTRODUCTION

Ionospheric electrodynamics is concerned with the electric fields and currents in the ionosphere, the processes that produce them, and their relation to plasma and neutral motions and to perturbations of the geomagnetic field. Plasma motions and the associated electric fields are driven by frictional coupling with neutral winds in the thermosphere, by gravitational and pressure-gradient (G/P) forces on the plasma, and by dynamic processes in the magnetosphere.

Large electron mobility along the magnetic field prevents the development of substantial parallel electric fields, and enforces strong coupling of perpendicular plasma motions and electric fields all along the magnetic field line. Plasma motions perpendicular to the geomagnetic field, produced by forces on the plasma, are transmitted to the plasma all along the field line, inducing transport that affects the ionospheric density. Momentum that is collisionally transferred between the neutrals and the plasma is also transmitted along the geomagnetic field lines through magnetic stresses and associated electric currents, so that ion-drag forces on the neutrals can be exerted in opposite directions at upper and lower heights in the thermosphere, leading to important effects on neutral dynamics, especially in the upper thermosphere, where the mass density is relatively low. The perturbed magnetic field can be readily measured at the ground and in space and is related to the distribution of winds in the thermosphere, to plasma G/P forces, and to the dynamics of the magnetosphere. Geomagnetic observations can, thus, provide information about thermospheric and magnetospheric dynamics and ionospheric plasma density. Knowledge of the geomagnetic perturbations is also important for accurately modeling the Earth's internal magnetic field and its electrical conductivity structure.

The purpose of this article is to discuss several aspects of the modeling of ionospheric electrodynamics: how we can calculate electric fields and currents, plasma velocities, and magnetic perturbations from information about the state of the thermosphere, ionosphere, and magnetosphere. Modeling of ionospheric electrodynamics has traditionally been conducted from an engineering perspective, with electric fields and currents related to each other through Ohm's law and to generation processes caused by movement of the conducting medium through the geomagnetic field, as well as by nonfrictional forces on the plasma. *Vasyliunas* [2012] has recently clarified how, from a physical point of view, the electric fields and currents are passive diagnostics of the plasma forces and motions, and do not, themselves, drive the plasma. Nevertheless, the engineering approach to modeling ionospheric electrodynamics is computationally convenient and gives essentially the correct results for quasistatic conditions with small perturbations to the main geomagnetic field. This is the approach we describe in this article. We use the terminology "ionospheric wind dynamo" or simply "wind dynamo" to describe the process of electric field and current generation by thermospheric winds; "magnetospheric dynamo" to describe the sources of ionospheric electric fields and geomagnetic field–aligned currents arising from processes in the distant magnetosphere (more than one Earth radius above the Earth's surface); and "G/P dynamo" to describe the electrodynamic effects driven by the G/P forces on the ionospheric plasma. We will use one particular model, the National Center for Atmospheric Research thermosphere-ionosphere-electrodynamics general circulation model (NCAR TIE-GCM) [*Qian et al.*, this volume], to focus our description and to illustrate modeling results.

Simulation modeling of ionospheric electrodynamics has been reviewed by *Richmond* [1995a, 2011], while *Le Sager and Huang* [2002a] have discussed the more recent literature on the ionospheric wind dynamo. Let us briefly summarize notable model developments since about 1995. Whereas most earlier models of ionospheric currents used a centered dipole geomagnetic field, *Singh and Cole* [1995] and *Hurtaud et al.* [2007] implemented an eccentric dipole, in order to examine how spatial variations of the geomagnetic field might affect electric fields and currents. The full International Geomagnetic Reference Field (IGRF) was used by *Richmond et al.* [1992], *Le Sager and Huang* [2002a], and *Ren et al.* [2008, 2009]. *Huang et al.* [2010] extended the model of *Le Sager and Huang* [2002a] with the *Tsyganenko* [1989] magnetospheric field and with a specification of the electric potential around the boundary between open and closed geomagnetic field lines. *Takeda* [1996] and *Cnossen et al.* [2011] examined how ionospheric dynamo effects and magnetosphere-ionosphere currents respond to changes in the strength of the geomagnetic field. *Jin et al.* [2008] and *Kawano-Sasaki and Miyahara* [2008] have calculated dynamo effects using winds obtained with ground-to-exobase atmospheric models. *Eccles* [2004], *Alken et al.* [2011], and *Maute et al.* [2012] modeled the electrodynamic effects of the G/P currents. *Huba et al.* [2010a, 2010b] included both wind dynamo and G/P effects in the SAMI3 ionosphere/plasmasphere/electrodynamics model. Various groups have been developing models that couple the ionospheric dynamo with thermospheric and magnetospheric dynamics [e.g., *Richmond et al.*, 1992; *Peymirat et al.*, 1998; *Millward et al.*, 2001; *Raeder et al.*, 2001a; *Ridley et al.*, 2003; *Wiltberger et al.*, 2004; *Wang et al.*, 2004; *Klimenko et al.*, 2006; *Ren et al.*, 2009; *Jin et al.*, 2011; *Pembroke et al.*, 2012]. Calculations of geomagnetic perturbations associated with the modeled ionosphere-magnetosphere currents have sometimes been performed [e.g., *Olsen and Kiefer*, 1995; *Miyahara and Ooishi*, 1997; *Tsunomura*, 1999; *Le Sager and Huang*, 2002b; *Richmond*, 2002; *Doumbia et al.*, 2007; *Ontiveros*, 2009; *Zaka et al.*, 2010; *Alken et al.*, 2011; *Marsal et al.*, 2012].

2. EQUATIONS OF IONOSPHERIC ELECTRODYNAMICS

For global ionospheric electrodynamics, we usually consider only time scales longer than 1 min. We can, therefore, treat the ionospheric electrodynamics as steady state, with divergence-free current density **J** and an electrostatic electric field **E**. (*Vanhamäki et al.* [2005] have illustrated how non-electrostatic effects can sometimes be significant for very rapidly varying fields.) With the use of Ohm's law in the frame of the moving neutrals, **E** and **J** can be written as

$$E = -\nabla \Phi, \quad (1)$$

$$\mathbf{J} = \sigma_P(\mathbf{E}_\perp + \mathbf{u} \times \mathbf{B}) + \sigma_H \mathbf{b} \times (\mathbf{E}_\perp + \mathbf{u} \times \mathbf{B}) + \sigma_\parallel \mathbf{E}_\parallel \mathbf{b} + \mathbf{J}^P, \quad (2)$$

with the Pedersen, Hall, and parallel conductivities σ_P, σ_H, and $\sigma_\parallel$, respectively. The neutral wind vector is denoted by **u**, and **b** is the unit vector along the geomagnetic field **B**. The geomagnetic field **B** can be decomposed into the main field $\mathbf{B}_o$ and a perturbation field $\Delta\mathbf{B}$ produced by ionospheric, magnetospheric, and induced Earth currents. $\Delta\mathbf{B}$ is assumed negligibly small in relation to $\mathbf{B}_o$. $\mathbf{E}_\perp$ and $\mathbf{E}_\parallel$ are the components of **E** perpendicular and parallel to the geomagnetic field, respectively. $\mathbf{J}^P$ is the current driven by plasma G/P forces as discussed in section 6.

Setting the divergence of equation (2) to zero and rearranging terms leads to

$$\nabla \cdot [\sigma_P(\nabla\Phi)_\perp + \sigma_H \mathbf{b}_o \times (\nabla\Phi)_\perp + \sigma_\parallel(\nabla\Phi)_\parallel]$$
$$= \nabla \cdot [\sigma_P \mathbf{u} \times \mathbf{B}_o + \sigma_H \mathbf{b}_o \times (\mathbf{u} \times \mathbf{B}_o) + \mathbf{J}^P]. \quad (3)$$

The conductivity tensor is strongly anisotropic, with $\sigma_\parallel$ several orders of magnitude larger than σ_P and σ_H at all heights above 100 km. *Farley* [1959, 1960] showed that geomagnetic field lines are essentially equipotential above this height for electric fields having scale sizes perpendicular to the geomagnetic field more than several kilometers. *Richmond* [1973] showed that finite $\sigma_\parallel$ has only a relatively small influence on the structure of the electric field in the equatorial electrojet, mainly below 100 km. Therefore, for global-scale electrodynamics, we can assume that geomagnetic field lines are equipotential, and the electric field $\mathbf{E}_\parallel$ is zero. As a consequence, the parallel current density $J_\parallel$ is not determined by Ohm's law, but rather by the condition that the 3-D current has to be divergence-free:

$$J_\parallel = -B_o \int_{s_l}^{s} \frac{\nabla \cdot \mathbf{J}_\perp}{B_o} ds', \quad (4)$$

where s is the distance along a geomagnetic field line, increasing in the direction of $\mathbf{B}_o$, and the integration is carried out from the base of the ionosphere (s_l, where $J_\parallel = 0$) to the point in question (s). $\mathbf{J}_\perp$ is the component of **J** perpendicular to $\mathbf{B}_o$. $J_\parallel$ is defined to be positive in the direction of $\mathbf{B}_o$.

If we assume that the electric potential is the same at magnetically conjugate points in the Southern and Northern Hemispheres, we can reduce equation (3) from three to two dimensions by dividing it by B_o and integrating along the magnetic field all the way from the base of the ionosphere s_l in one hemisphere to the base of the ionosphere in the opposite hemisphere, where $J_\parallel$ also vanishes. This needs to be done in a coordinate system aligned with the geomagnetic field. If a realistic (e.g., IGRF) magnetic field configuration is assumed, Modified Magnetic Apex Coordinates [*Richmond*, 1995b] can be used, in which magnetic longitude is ϕ_m and magnetic latitude is λ_m. The partial differential equation (PDE) for Φ is as follows (modified from Equation (5.23) of *Richmond* [1995b])

$$\frac{1}{R^2 \cos\lambda_m} \left[\frac{\partial}{\partial\phi_m} \left(\frac{\Sigma_{\phi\phi}^T}{\cos\lambda_m} \frac{\partial\Phi}{\partial\phi_m} + \Sigma_{\phi\lambda}^T \frac{\partial\Phi}{\partial|\lambda_m|} \right) + \frac{\partial}{\partial|\lambda_m|} \right.$$
$$\left. \left(\Sigma_{\lambda\phi}^T \frac{\partial\Phi}{\partial\phi_m} - \Sigma_{\lambda\lambda}^T \cos\lambda_m \frac{\partial\Phi}{\partial|\lambda_m|} \right) \right] \quad (5)$$
$$= \frac{1}{R\cos\lambda_m} \left[\frac{\partial(K_{m\phi}^{DT} + K_{m\phi}^{PT})}{\partial\phi_m} + \frac{\partial[(K_{m\lambda}^{DT} + K_{m\lambda}^{PT})\cos\lambda_m]}{\partial|\lambda_m|} \right]$$
$$+ J_{Mr},$$

with $R = R_E + h_R$, where h_R is a reference height at the base of the conducting region. The different Σ quantities are derived from the field line–integrated conductivities, $K_{m\phi}^{DT}$ and $K_{m\lambda}^{DT}$ are derived from the field line–integrated wind dynamo currents, $K_{m\phi}^{PT}$ and $K_{m\lambda}^{PT}$ are derived from the field line–integrated G/P currents (see section 6), and J_{Mr} is the sum of the magnetically conjugate southern and northern upward current densities flowing to the magnetosphere at the top of the ionosphere (for the field line–integrated quantities, see Equations (5.11)–(5.29) of *Richmond* [1995b]). The conductance and current parameters are all scaled by spatially dependent geometric factors relating to the magnetic coordinate system. Equation (5) neglects any component of Φ that is antisymmetric about the magnetic equator, as might arise in the polar caps in association with IMF B_y effects, for example. Techniques can be designed to take into account hemispheric asymmetry of Φ, but this paper does not deal with that problem.

The equatorial boundary condition on Φ requires that the poleward current density summed over the two hemispheres vanishes:

$$-\frac{\Sigma_{\lambda\phi}^T}{R\cos\lambda_m} \frac{\partial\Phi}{\partial\phi_m} - \frac{\Sigma_{\lambda\lambda}^T}{R} \frac{\partial\Phi}{\partial|\lambda_m|} + K_{m\lambda}^{DT} + K_{m\lambda}^{PT} = 0. \quad (6)$$

Once equations (5) and (6) have been solved for Φ, as discussed in the following sections, **E** is calculated from equation (1), and **J** can be calculated from equations (2) and (4).

The right-hand side of equation (5) contains three source terms for the dynamo: the wind dynamo source related to $K_{m\phi}^{DT}$ and $K_{m\lambda}^{DT}$, the G/P source related to $K_{m\phi}^{PT}$ and $K_{m\lambda}^{PT}$, and the magnetospheric source related to J_{Mr}. At high latitudes, the magnetospheric source, discussed in the next section, usually dominates. At middle and low latitudes during magnetically quiet periods, the ionospheric-dynamo source usually dominates. The G/P source can be important at night and in the upper ionosphere, as discussed in section 6.

3. MODELING IONOSPHERE-MAGNETOSPHERE COUPLING

There are different ways in which J_{Mr} can be expressed in equation (5). Most simply, it can be set to zero, so that the solution of equation (5) represents purely the effects of the ionospheric wind and G/P dynamos [e.g., *Takeda and Maeda*, 1980; *Richmond and Roble*, 1987]. It can also be set to zero only within a computation domain at low and middle latitudes, up to a high-latitude boundary at which Φ is specified [e.g., *Iwasaki and Nishida*, 1967; *Senior and Blanc*, 1984; *Peymirat and Richmond*, 1993; *Shen et al.*, 2006; *Huang et al.*, 2010]. In this case, it is implicitly assumed that field-aligned currents (FACs) exist poleward of that boundary, which help determine Φ at the boundary. Poleward of that boundary, Φ is often simply specified as some reference potential Φ^R, rather than being derived from the PDE.

In order to specify a nonzero J_{Mr}, we may equate it to a reference distribution J_{Mr}^R obtained from an empirical model [e.g., *Weimer*, 2000, 2005], a magnetospheric simulation model [e.g., *Wolf et al.*, 1986; *Peymirat et al.*, 1998; *Raeder et al.*, 2001a, 2001b; *Ridley et al.*, 2003; *Wiltberger et al.*, 2004], or observations [e.g., *Marsal et al.*, 2012; B. Anderson et al., AMPERE: Overview and initial results, submitted to *Space Science Reviews*, 2012.] of high-latitude FACS:

$$J_{Mr} = J_{Mr}^R. \quad (7)$$

The areal integral of J_{Mr}^R over the hemisphere must vanish, both to satisfy current continuity and to ensure a realistic numerical solution of equations (2) and (4). *Marsal et al.* [2012] have presented an example of using observed J_{Mr}^R based on AMPERE data, adjusted so that the hemispheric integral vanishes, to solve for the ionospheric electric potential and associated geomagnetic perturbations.

One way to ensure that the hemispheric integral of J_{Mr}^R vanishes is to represent it in the form

$$J_{Mr}^R = \frac{1}{\mu_0 R \cos \lambda_m} \left[\frac{\partial \Delta B_{|\lambda|}^R}{\partial \phi_m} - \frac{\partial (\Delta B_{\phi}^R \cos \lambda_m)}{\partial |\lambda_m|} \right], \quad (8)$$

where μ_0 is the permeability of free space, and $\Delta B_{|\lambda|}^R$ and ΔB_{ϕ}^R are arbitrary functions, with the constraint that $\Delta B_{|\lambda|}^R$ goes to zero at the magnetic equator. With this constraint, the hemispheric integral of equation (8) is automatically zero. $\Delta B_{|\lambda|}^R$ and ΔB_{ϕ}^R can, but need not, represent the hemispherically conjugate sums of horizontal magnetic perturbations above the ionosphere in the magnetic poleward and eastward directions, respectively, mapped to the reference height h_R. Another possibility is that $\Delta B_{|\lambda|}^R/\mu_0$ and $-\Delta B_{\phi}^R/\mu_0$ can represent the magnetic flux tube–integrated eastward and outward/poleward magnetospheric current densities flowing perpendicular to the geomagnetic field, the divergence of which is balanced by the net upward FAC.

J_{Mr} can also be expressed as a quantity proportional to the difference between the ionospheric potential Φ and a reference potential Φ^R [*Peymirat et al.*, 2002]. That is, we can write

$$J_{Mr} = \frac{(1-p)\sigma^R}{pR}(\Phi - \Phi^R), \quad (9)$$

where σ^R is a reference conductivity, which may be constant or spatially varying, and p is a spatially varying parameter which, together with σ^R, controls how strongly Φ is linked to Φ^R. If $p = 1$, J_{Mr} vanishes, as at middle and low latitudes. If $p = 0$, Φ is forced to be identical with Φ^R, in order to avoid infinite J_{Mr}. In that case, we have a perfect voltage generator, and J_{Mr} can no longer be determined from equation (9). In most published TIE-GCM applications, the formulation (9) has been used, with p typically decreasing linearly from 1 to 0 between 60° and 75° magnetic latitude. The version tiegcm1.94.2 discussed in the work of *Qian et al.* [this volume] has a dynamically varying boundary depending on the solar wind conditions, chosen such that the penetration electric fields are reasonable. *Peymirat et al.* [2002] showed a case where p did not vanish over the polar region. The reference potential Φ^R may be obtained from an empirical model like those of *Heelis et al.* [1982] or *Weimer* [2005], or from a data assimilation analysis [e.g., *Crowley et al.*, 1989; *Lu et al.*, 1995], or from a theoretical model [e.g., *Wang et al.*, 2004].

Yet another way to represent J_{Mr}, for ionospheric regions connected to the inner magnetosphere, is to use the concept of "equivalent magnetospheric conductances." One significant aspect of magnetosphere-ionosphere electrodynamic coupling is the tendency for the so-called region 2 FACs in the equatorward portion of the auroral region to hinder the development of strong zonal electric fields, fields that would otherwise cause strong radial convection of plasma in the inner magnetosphere, which would require compression and energization of the plasma. *Vasyliūnas* [1972] showed that, in a steady state, this effect on the magnetosphere-ionosphere electric field can be simulated by representing the magnetospheric hot plasma as a Hall conductor, a concept that has been used in simulation models of the interaction [e.g., *Senior and Blanc*, 1984; *Shen et al.*, 2006]. *Peymirat and Richmond* [1993] showed that the concept can be extended to include effects of magnetospheric ion loss by adding to the magnetospheric Hall conductance a zonal Pedersen conductance. We can use the following representation:

$$J_{Mr} = \frac{1}{R^2 \cos\lambda_m} \left[\frac{\partial}{\partial\phi_m} \left(-\frac{\Sigma_{\phi\phi}^M}{\cos\lambda_m} \frac{\partial\Phi}{\partial\phi_m} - \Sigma_H^M \frac{\partial\Phi}{\partial|\lambda_m|} \right) \right.$$
$$\left. + \frac{\partial}{\partial|\lambda_m|} \left(\Sigma_H^M \frac{\partial\Phi}{\partial\phi_m} \right) \right], \qquad (10)$$

where Σ_H^M and $\Sigma_{\phi\phi}^M$ are equivalent magnetospheric Hall and zonal Pedersen conductances. Using a numerical model of inner-magnetospheric convection under average conditions, *Peymirat and Richmond* [1993] estimated longitudinally averaged equivalent magnetospheric conductances as a function of magnetic latitude. Σ_H^M can be equated with twice their equivalent magnetospheric Hall conductance C_i (which was derived for a single hemisphere), while $\Sigma_{\phi\phi}^M$ can be equated with twice their equivalent magnetospheric zonal Pedersen conductance C_r. Their Figure 4 shows values of C_i and C_r up to 71.6° magnetic latitude. These can be extended to higher latitudes by setting C_i to its value at 71.6° and setting C_r to zero. (Under these conditions, the equivalent magnetospheric conductances make no contribution to the FAC above 71.6°.) More research is needed to find out how their results might be generalized to allow for longitudinally dependent equivalent magnetospheric conductances and to be extended to different levels of magnetospheric activity.

Using equations (7), (9), and (10) in equation (5) yields the following generalized PDE for Φ:

$$p\frac{\partial}{\partial\phi_m} \left[\frac{(\Sigma_{\phi\phi}^T + \Sigma_{\phi\phi}^M)}{\cos\lambda_m} \frac{\partial\Phi}{\partial\phi_m} + (\Sigma_{\phi\lambda}^T + \Sigma_H^M) \frac{\partial\Phi}{\partial|\lambda_m|} \right]$$
$$+ p\frac{\partial}{\partial|\lambda_m|} \left[(\Sigma_{\lambda\phi}^T - \Sigma_H^M) \frac{\partial\Phi}{\partial\phi_m} + \Sigma_{\lambda\lambda}^T \cos\lambda_m \frac{\partial\Phi}{\partial|\lambda_m|} \right]$$
$$+ (1-p)\sigma^R R\cos\lambda_m \Phi =$$
$$pR \left[\frac{\partial(K_{m\phi}^{DT} + K_{m\phi}^{PT})}{\partial\phi_m} + \frac{\partial[(K_{m\lambda}^{DT} + K_{m\lambda}^{PT})\cos\lambda_m]}{\partial|\lambda_m|} \right]$$
$$+ pJ_{Mr}^R R^2 \cos\lambda_m - (1-p)\sigma^R R \cos\lambda_m \Phi^R. \qquad (11)$$

Depending on the manner in which one wishes to represent the effects of magnetosphere-ionosphere coupling for a particular study, some or all of the quantities J_{Mr}^R, σ^R, $\Sigma_{\phi\phi}^M$, and Σ_H^M will usually be set to zero over all or part of the hemisphere.

Figure 1 shows an example of the influence of equivalent magnetospheric conductances on the electric potential. Two calculations are made, one on the left (case A) with no wind ($u = 0$) and no equivalent magnetospheric conductance ($\Sigma^M = 0$), and the other on the right (case B), also with no wind but using the equivalent magnetospheric conductance Σ^M of *Peymirat and Richmond* [1993]. For these computations, the potential is imposed over the polar cap, above 71.6° magnetic latitude, by setting $p = 0$ there. Below that boundary, $p = 1$. J_{Mr}^R is set to zero. Adding Σ^M is seen to contract the midlatitude potential contours toward higher latitudes, so that the electric field is more concentrated in the auroral region and is weaker at midlatitudes. However, even for this simulation of low magnetic activity (cross-polar cap potential of 30 kV), there remains some penetration of the polar electric field to middle and low latitudes. On the left, the average of the northern and southern FAC out of the ionosphere, $J_{Mr}/2$, is characteristic of region 1 FAC, but no region 2 FAC is seen. On the right, region 2 FAC appears in the auroral region, with directions opposite to the region 1 FAC at higher latitude.

4. MODELING IONOSPHERIC CONDUCTIVITIES

Realistic modeling of the electric field and current requires an accurate model of the ionospheric conductivities. The upper atmosphere is mainly ionized by solar EUV and soft X-rays and, at high latitudes, by energetic particle precipitation from the magnetosphere [*Rees*, 1989]. During the day, the ionization sources are, in general, regular and well known. The ionization rates depend on the strength of the ionization source, the atmospheric density and composition, and the absorption and ionization cross sections for each species, as a function of EUV/X-ray wavelength or precipitating particle energy. For the TIE-GCM, *Solomon and Qian* [2005] describe the ionization scheme. *Fang et al.* [2008a] found that the *E*-region electron densities in the TIE-GCM are underestimated compared to the International Reference Ionosphere model by approximately 30%–40%. They increased the soft X-ray fluxes (between 8 and 70 Å) by a factor of 4.4 for moderate solar activity. *Solomon* [2006] pointed out that the agreement between the EUVAC model [*Richards et al.*, 1994] used in the TIE-GCM and the TIMED/Solar EUV Experiment data for the soft X-rays is not as good as for other wavelengths due to the difficulty of accurately measuring in the 8–70 Å range.

The nighttime ionization sources are less well described. Partly, this is due to the rarity of nighttime observations [*Wakai*, 1967; *Shen et al.*, 1976; *Trost*, 1979], the higher variability, and the different dominant sources than during daytime. A major source independent of solar conditions is the flux from starlight (911–1026 Å), which varies with latitude and sidereal time [*Strobel et al.*, 1980]. *Strobel et al.* [1980] showed that the hemispheric difference in starlight

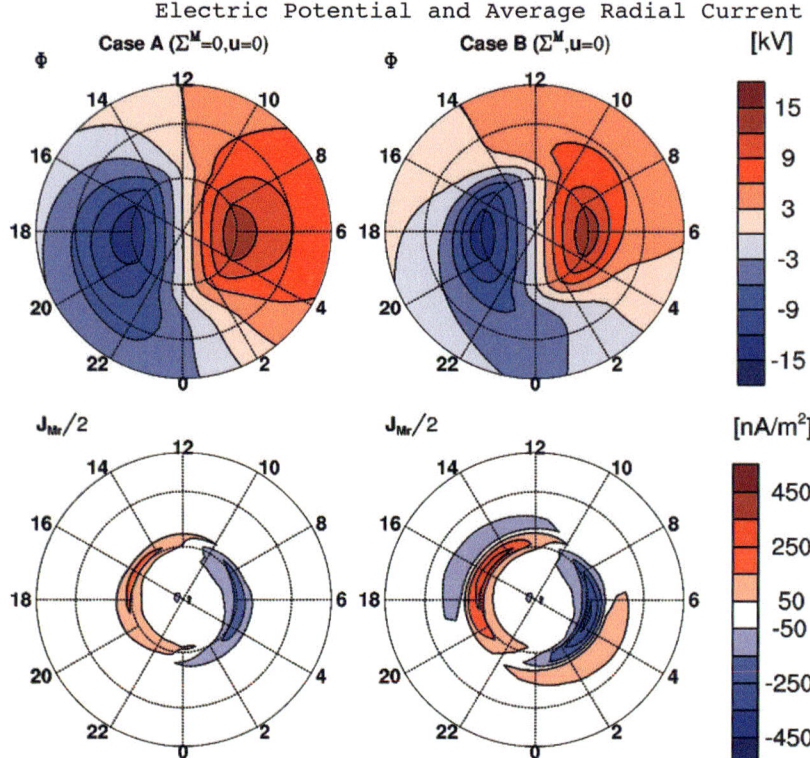

Figure 1. (top) Electric potential and (bottom) hemispherical average of upward current at the top of the ionosphere on 21 December, 4 UT, with 30 kV cross-polar cap potential drop, with no wind dynamo electric field source. (left) Case A does not use equivalent magnetospheric conductances; (right) case B does. The outer circles are 50° magnetic latitude, with magnetic local time shown.

flux is about a factor of 0.38, with the maximum under the southern Milky Way and the Orion regions. The scattering of solar H (Lyα and Lyβ) and He1 lines into the nightside atmosphere by hydrogen and helium in the geocorona is an important source at dawn and dusk [*Strobel et al.*, 1974]. According to the EUVAC model, scattering will increase by a factor of approximately 1.6 from solar minimum to solar maximum. The interplanetary background is much weaker than the stellar and geocoronal sources and is caused by the scattering of solar radiation from interstellar gas at night [*Titheridge*, 2000]. Ablation of meteors is a source of metallic ion, which can affect the nighttime E region and lead to sporadic E layers [*Zhou et al.*, 1999].

The TIE-GCM uses three background ionization sources: starlight (1026 Å), geocorona via the He1 584 Å line, and geocorona via the He2 304 Å line, with the following fluxes: 1.5×10^7, 1.5×10^6, and $1.5 \times 10^6 \frac{\text{photons}}{\text{cm}^2\,\text{s}}$, respectively, independent of location, time of day, season, and solar cycle. At auroral latitudes, particle precipitation is a large and highly variable ionization source, especially at night. Ionospheric simulation models usually use relatively simple parameterizations or empirical models [e.g., *Roble and Ridley*, 1987].

The Pedersen conductivity of the F region is much smaller than that in the E region during the day, but is roughly comparable at night. The F-region conductivity has a much stronger solar cycle variation than the E-region conductivity [e.g., *Takeda and Araki*, 1985], and at the maximum of the solar cycle, the height-integrated F-region Pedersen conductivity is usually much larger at night than the height-integrated E-region conductivity, outside of the auroral zones. The Hall conductivity of the F region is negligible.

5. MODELING THERMOSPHERIC WINDS

Modeling the neutral winds in the E and F regions is important to get the right ionospheric response. The wind can be decomposed into a zonal-mean wind field plus tidal and wave components. The migrating tidal components are important, as *Fesen et al.* [2000] showed using the TIE-GCM to study their effect on the equatorial vertical $\mathbf{E} \times \mathbf{B}$

drift. They found that the semidiurnal migrating tidal component largely determines the magnitude and phase of the daytime drift. *Fang et al.* [2008b] used the TIE-GCM to examine the influence of neutral winds in different altitude regions on the equatorial vertical drift and the magnetic perturbations at the ground and found that the altitude variation of the neutral winds at low latitudes can change the ground magnetic perturbations a few degrees off the equator. *Hagan et al.* [2007] showed that only the inclusion of nonmigrating tides at the lower boundary can reproduce the longitudinal variation of the evening low-latitude ionospheric density observed by IMAGE-FUV [*Immel et al.*, 2006].

The relation between neutral winds and resulting electric fields and drifts is very complex, due to the fact that the electric field is influenced not just by the local wind, but rather by the global wind pattern and conductivity distribution. In general, during the day, the *E*-region wind is more important than the *F*-region wind, but at night, the *F*-region wind can be more important. The relative importance of the *F*-region wind depends on the *F*-region conductivity in relation to the *E*-region conductivity and has a strong solar cycle variation.

6. PLASMA GRAVITY AND PRESSURE-GRADIENT CURRENTS

The G/P current $\mathbf{J}^P$ is much smaller than the neutral wind–driven current, but can have non-negligible effects on nighttime electric fields and on low Earth orbit (LEO) satellite magnetic perturbations. It is given by

$$\mathbf{J}^P = (\rho \mathbf{g} - \nabla[k_B(T_e + T_i)N_e]) \times \mathbf{B}/B^2 = \mathbf{J}(g) + \mathbf{J}(p), \quad (12)$$

with ρ the ion mass density, $\mathbf{g}$ the acceleration of gravity, k_B the Boltzmann constant, T_i and T_e the ion and electron temperatures, and N_e the electron density. We will refer to the component of $\mathbf{J}^P$ related to gravity as $\mathbf{J}(g)$ and that related to the plasma pressure gradient as $\mathbf{J}(p)$.

The G/P terms on the right-hand side of equation (11) are

$$K_{m\phi}^{PT} = |\sin I_m| \int_{s_1}^{s_2} \frac{\mathbf{d}_1 \cdot \mathbf{J}^P}{D} ds' \quad (13)$$

$$K_{m\lambda}^{PT} = \int_{s_1}^{s_2} \frac{\mathbf{d}_2 \cdot \mathbf{J}^P}{D} ds', \quad (14)$$

where

$$\sin I_m = 2\sin \lambda_m (4 - 3\cos^2 \lambda_m)^{-1/2} \quad (15)$$

$$\mathbf{d}_1 = R\cos \lambda_m \nabla \phi_m \quad (16)$$

$$\mathbf{d}_2 = -R\sin I_m \nabla \lambda_m \quad (17)$$

$$D = |\mathbf{d}_1 \times \mathbf{d}_2| \quad (18)$$

and where the integration with respect to distance s along a geomagnetic field line is carried out from the southern s_1 to the northern s_2 foot point of the field line. The foot points are defined by the intersection of the field line with the reference height h_R. For a dipolar field on a spherical Earth, I_m would be the inclination of the geomagnetic field at height h_R, but since it is constant along field lines (except for a sign reversal at the apex), it no longer corresponds to the local inclination at heights above h_R. The base vectors $\mathbf{d}_1$ and $\mathbf{d}_2$ are perpendicular to the geomagnetic field in the magnetic eastward and magnetic downward/equatorward directions, respectively; for a dipolar field on a spherical Earth, they would be unit vectors at h_R, but they decrease in magnitude at higher altitudes along a geomagnetic field line in the same manner as the respective components of $\mathbf{E}_\perp$, owing to the increasing distance between neighboring geomagnetic field lines as we move up along a field line.

Note that $K_{m\phi}^{PT}$ and $K_{m\lambda}^{PT}$ do not depend on the ionospheric conductivity, unlike the wind dynamo quantities $K_{m\phi}^{DT}$ and $K_{m\lambda}^{DT}$. At low latitudes at night, when the latter terms are small, the G/P terms can become relatively important for the electric potential and field.

On large scales, $\mathbf{J}^P$ is mainly in the magnetic east-west direction because plasma pressure gradients are predominantly vertical, as is gravity. However, horizontal plasma pressure gradients and meridional currents can become important on smaller spatial scales, as in equatorial plasma bubbles. Gravity-driven current can also play a role in the development of the Rayleigh-Taylor instability and equatorial spread F, [e.g., *Kelley et al.*, 1981]. Here we focus only on the large-scale currents. *Lühr et al.* [2003] examined magnetic perturbations derived from CHAMP data, in particular, the longitudinal variation of the Appleton anomaly. They also presented a theoretical approximation of the effect of plasma pressure on in situ magnetic perturbations and pointed out its importance when analyzing magnetic field data. *Alken et al.* [2011] calculated the G/P currents and their effect on the electric field and conduction currents. They pointed out that these currents can have a substantial effect in the nighttime. In addition, they showed differences between their modeled magnetic perturbations and the theoretical approximation of *Lühr et al.* [2003] for the diamagnetic effect due to the pressure-gradient current. The effects of G/P current on low-latitude electric fields and drift velocities were examined by *Eccles* [2004], who extended the 2-D flux tube–integrated electrodynamic model of *Haerendel et al.* [1992], and by *Maute et al.* [2012], who used the TIME-GCM, which has the same electrodynamics as the TIE-GCM.

If all terms on the right-hand side of equation (11) are set to zero except the contributions to $K_{m\phi}^{PT}$ and $K_{m\lambda}^{PT}$ from either $\mathbf{J}(g)$ or $\mathbf{J}(p)$, equation (11) can be solved for the potential to obtain the electric field $\mathbf{E}_g$ or $\mathbf{E}_p$ associated solely with $\mathbf{J}(g)$ or $\mathbf{J}(p)$, respectively. Although $\mathbf{J}(g)$ and $\mathbf{J}(p)$ are largest in the F region, $\mathbf{E}_g$ and $\mathbf{E}_p$ drive conduction currents primarily at altitudes where the Pedersen and Hall conductivities are large, which at day is in the E region.

The magnetic eastward component of $\mathbf{J}(g)$, plus the conduction current driven by $\mathbf{E}_g$, which we call $J_\phi(\mathbf{g} + \mathbf{E}_g)$, is shown on the left of Figure 2. The corresponding total pressure-gradient eastward current $J_\phi(\mathbf{p} + \mathbf{E}_p)$ is shown on the right. The simulation is for March (day of year 79) with moderate solar flux ($F10.7 = 120$ W m^{-2} Hz^{-1}) under geomagnetically quiet conditions at 12 UT and 14 magnetic local time.

Because the gravity-driven current $\mathbf{J}(g)$ is proportional to the plasma density, it is largest in the F region, where it is comparable to the ionospheric-dynamo current driven by winds and electric fields at those heights (which are much smaller than the daytime E-region dynamo currents). $\mathbf{J}(g)$ is generally larger at day than at night, so it converges in the evening and diverges after sunrise. The polarization electric field $\mathbf{E}_g$, which develops, is generally westward during the day and early evening and eastward later at night until sunrise. At day, $\mathbf{E}_g$ drives a westward conduction current in the E region that partially balances the eastward F-region current, as seen on the left of Figure 2.

The current driven by plasma pressure gradients, $\mathbf{J}(p)$, is also generally larger at day than at night, but it has opposite directions above and below the peak of the F region, as seen on the right of Figure 2. In the topside ionosphere, where the vertical plasma pressure gradient force tends to balance the gravitational force, $\mathbf{J}(p)$ flows westward and tends to balance $\mathbf{J}(g)$. Below the F-region peak, $\mathbf{J}(p)$ flows eastward and supplements $\mathbf{J}(g)$. The net eastward and westward $\mathbf{J}(p)$ currents largely balance each other, so that $\mathbf{E}_p$ tends to be much smaller than $\mathbf{E}_g$.

Figure 3 shows an example of the relative importance of G/P currents, wind dynamo currents, and quiet-time electric field penetration from the polar cap on vertical plasma drift at the magnetic equator at 12 UT, simulated with the TIME-GCM. For comparison, the empirical model of *Scherliess and Fejer* [1999] is also shown. The wind dynamo is the major driver of the vertical drift. The effects of G/P currents on the vertical drift (the curve labeled *Jpg*) are small during the day, but are non-negligible after sunset and before sunrise. They tend to reduce the strength of the postsunset upward drift and the predawn downward drift. For these magnetically quiet conditions, the influence of penetration electric fields (curve labeled *PE*) is weak. Their contribution to the vertical drift tends to be in the same direction as the wind dynamo contribution and is larger at night than at day.

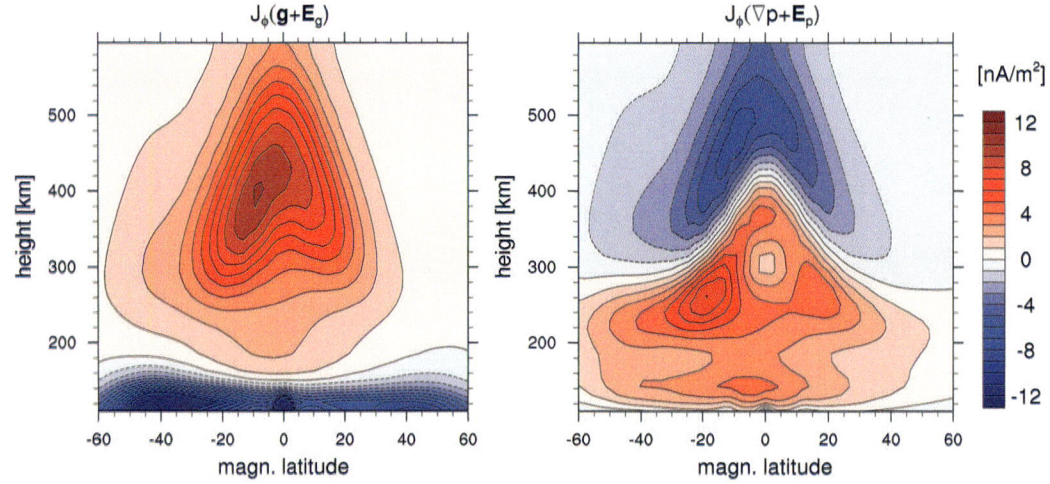

Figure 2. Thermosphere-ionosphere-electrodynamics general circulation model (TIE-GCM) simulations of magnetic eastward current density as a function of magnetic latitude and height, for March equinox, 12 UT, 14 magnetic local time (approximately 30° geographic longitude), $F10.7 = 120 \frac{W}{m^2 \, Hz}$. (left) Due to gravity and associated electric field. (right) Due to plasma pressure gradient and associated electric field.

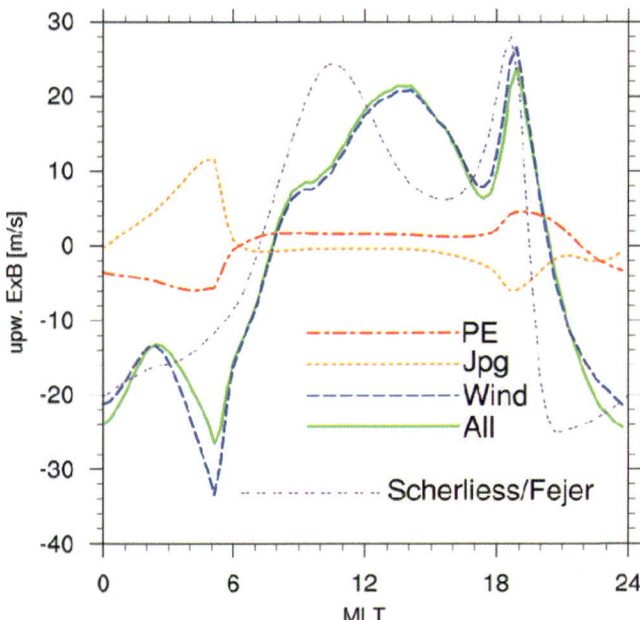

Figure 3. TIME-GCM simulation of upward **E** × **B** drift at 135 km above the geomagnetic equator at 12 UT due to neutral wind (*Wind*; long-dashed blue line), gravity and pressure-gradient current (*Jpg*, short-dashed orange line), penetration electric field (*PE*, dashed-dotted red line), and the sum of all sources (*All*, solid green line). Shown for comparison is the Scherliess/Fejer empirical model [*Scherliess and Fejer*, 1999] (thin purple double-dotted line) (figure reproduced from *Maute et al.* [2012]).

7. SOLVING FOR THE ELECTRIC POTENTIAL

Numerical solution of equations (11) and (6) is carried out by finite differencing to obtain a set of linear equations relating Φ at each grid point to the values at other grid points and to the source terms. These equations can be solved directly or by iterative techniques like the multigrid solver used in the TIEGCM. A number of difficulties can arise in setting up and solving the discretized equations, which we now discuss.

One difficulty is that the quantity $\Sigma_{\lambda\lambda}$ in equations (11) and (6), as defined by *Richmond* [1995b], goes to infinity at $\lambda_m = 0$, while the quantity it multiplies, $(-1/R)\partial\Phi/\partial|\lambda_m|$, represents the poleward electric field and necessarily goes to zero at $\lambda_m = 0$, where the equipotential geomagnetic field lines become horizontal. This difficulty is resolved by defining a new independent variable that is a function of λ_m but goes to zero quadratically rather than linearly with λ_m. For example, $\pm\lambda_m^2/(\lambda_0 + |\lambda_m|)$ could be the new independent variable, where λ_0 is a positive constant. (The TIE-GCM uses a more complex functional relation for its transformed latitude variable.) When equations (11) and (6) are recast using this new latitude variable, it is found that $\Sigma_{\lambda\lambda}$ is multiplied by a function that goes to zero at $\lambda_m = 0$ in such a manner that their product is finite, while the derivative of Φ with respect to the new independent variable does not vanish, and the resultant equations are well-behaved at $\lambda_m = 0$.

A second difficulty is related to the fact that the PDE (11) and the boundary condition (6) are not completely independent when p is set to 1 everywhere. The dependence between equations (11) and (6) is this: the area integral of equation (11) over the hemisphere, between the equator and the pole, is proportional to the line integral of equation (6) around the equator, provided the area integral of J_{Mr} vanishes, as is required by current continuity (no net current flowing into or out of the ionosphere). If these equations are discretized in a manner that conserves current flow between grid elements, the result will be as many linear equations as there are grid points, but the equations will not be linearly independent, so that a unique solution is not possible. A way to resolve this problem is to discard one of the equations obtained for a particular grid point (typically at the pole) and to replace it by $\Phi = \Phi^R$ at that grid point, where Φ^R is a specified potential value. We should note that if the discretization is not carried out in a manner that conserves current flow between grid elements, replacing one of the equations by $\Phi = \Phi^R$ at a particular grid point can result in a numerical solution that has poorly behaved gradients of Φ at that grid point, corresponding to a spurious source or sink of current there.

When the equations are discretized in a current-conserving fashion, the equation for a particular grid point involves neighboring grid points not only directly to the north, south, east, and west but also to the northwest, northeast, southwest, and southeast, arising from gradients in the Hall conductance and from nonorthogonality of the magnetic coordinates. This results in a nine-point stencil for all interior grid points. A possible difficulty may arise concerning the equatorial boundary condition (6), which involves derivatives of Φ both normal and parallel to the boundary. Such mixed-derivative boundary conditions may not be an option in standard elliptic PDE solvers and, therefore, may require a modification of the solver.

A numerical difficulty may arise in solving the discretized versions of equations (11) and (6) when strong gradients of the conductances occur, relative to the grid spacing. The difficulty is exacerbated in regions with a large ratio of Hall to Pedersen conductance like the equatorial electrojet. Depending on the discretization, the influence of the first-order derivatives of the electric potential can dominate over the influence of the second-order derivatives, and terms associated with the Hall conductance ($\Sigma_{\phi\lambda}$, $\Sigma_{\lambda\phi}$) can become dominant with respect to the terms associated with the Pedersen conductance ($\Sigma_{\phi\phi}$, $\Sigma_{\lambda\lambda}$). If numerical problems arise, the condition number of the matrix increases, which leads to relatively higher error in the solution compared to the error in the forcing. A symptom of a high condition number is that iterative solvers might not converge. In principle, this difficulty can be resolved by using a sufficiently fine grid, which increases the influence of the second-order derivatives that stabilize the solution, but in practice, an unreasonably fine grid may be required to ensure numerical stability. In the TIE-GCM, an upwinding scheme is applied to improve the condition number by using one-sided differencing instead of central differencing whenever the diagonal dominance of the matrix is violated. Upwinding schemes make the solution stable, but add numerical diffusion, and they are nonconservative.

8. IONOSPHERIC CURRENT SYSTEM AND GEOMAGNETIC PERTURBATIONS

Having the full 3-D current system in and above the ionosphere permits the calculation of the associated magnetic perturbations by techniques such as spherical harmonic analysis [e.g., *Engels and Olsen*, 1998; *Alken et al.*, 2011]. One difficulty is that the ionospheric model has information only about current within the ionosphere and FAC densities above the ionosphere, which are accurate only at heights up to a few Earth radii where the current geometry is well known, and cross-field currents are negligible. The TIE-GCM treats currents above the ionosphere as purely field aligned all the way to the field line apex, following the IGRF geomagnetic field at all heights. The model closes net FACs (represented by nonzero J_{Mr}) at the apex by assuming that they flow radially to or from infinity. Ring currents, magnetopause currents, and tail currents are ignored.

Magnetic perturbations are produced not only by ionospheric and magnetospheric currents but also by currents induced in the Earth by time-varying external currents [e.g., *Kuvshinov*, 2008], which requires a model of Earth conductivity. Models of various degrees of complexity exist. A very simple model is to treat the Earth as nonconducting down to some depth d, where there is a perfect conductor that prevents any penetration of the magnetic perturbation field to greater depths. The vertical magnetic perturbation vanishes at this depth, where the field is effectively "mirrored." Because of its simplicity, this approximation has been used with the TIE-GCM [e.g., *Doumbia et al.*, 2007], with d typically set to 600 km for quiet-day global magnetic perturbations. *Marsal et al.* [2012] have found that a smaller depth (250 km) often gives better results for rapidly varying smaller-scale magnetic fields.

For purposes of calculating magnetic perturbations above or below the ionosphere, a convenient approximation is to represent the 3-D current density as a 2-D horizontal current sheet of height-integrated horizontal current density **K** at an altitude h_1 (typically 110 km), connected to FAC above the sheet, which has a radial component at the top of the sheet of J_r. Another convenient approximation is to treat the geometry of the FAC as though it flowed along purely dipolar field lines. With these simplifications, we can apply the algorithm of *Richmond* [1974] to calculate ground-level magnetic perturbations in terms of 2-D integrations over the current sheet, instead of 3-D integrations over all space.

Because reducing the 3-D current to a 2-D current sheet plus FAC is only an approximation, there is some freedom in how to allocate low-latitude meridional currents between the current sheet and the FAC. As discussed by *Richmond* [1995b], we can take advantage of this freedom to minimize the latitudinal structure of J_r and of the meridional component of **K** at low latitudes, by specifying a distribution of J_r at low latitudes that varies smoothly in magnetic latitude and by computing the meridional component of **K** in such a manner that current is conserved. Only for geomagnetic field lines that extend above altitudes where we want to simulate observations on LEO satellites do we require J_r to map along field lines in such a way as to give the correct FAC density. On lower field lines, we no longer require this. In practice, this may involve an adjustment of J_r and the sheet current density for field lines that peak below about 300 km, which intersect the current sheet at 110 km around 10° magnetic latitude.

The current can be divided into a nondivergent "equivalent" current flowing only within the sheet, plus a "residual" current composed of a divergent horizontal sheet current connected to the FAC. An example of the equivalent and residual current is shown in Figure 4. By definition, only the

---

Figure 4. (opposite) (left) TIE-GCM simulation of ionospheric currents and (right) geomagnetic perturbations as a function of magnetic local time and magnetic latitude for solar minimum, 8 June, 17 UT (when the subsolar point is at its northernmost magnetic latitude). For currents, arrows show the height-integrated sheet current density, **K**, and colors show the upward component of field-aligned current density at the top of the current sheet, J_{qr}. For magnetic perturbations, arrows show the horizontal components and, colors, the vertical component (positive downward). (top left) The total density, (middle left) the residual current, and (bottom left) the equivalent current. See main text for definitions of residual and equivalent currents. (top right) Magnetic perturbations at 400 km altitude due to equivalent current and (middle right) residual current, and (bottom left) magnetic perturbations at the ground, which by definition are due only to the equivalent current.

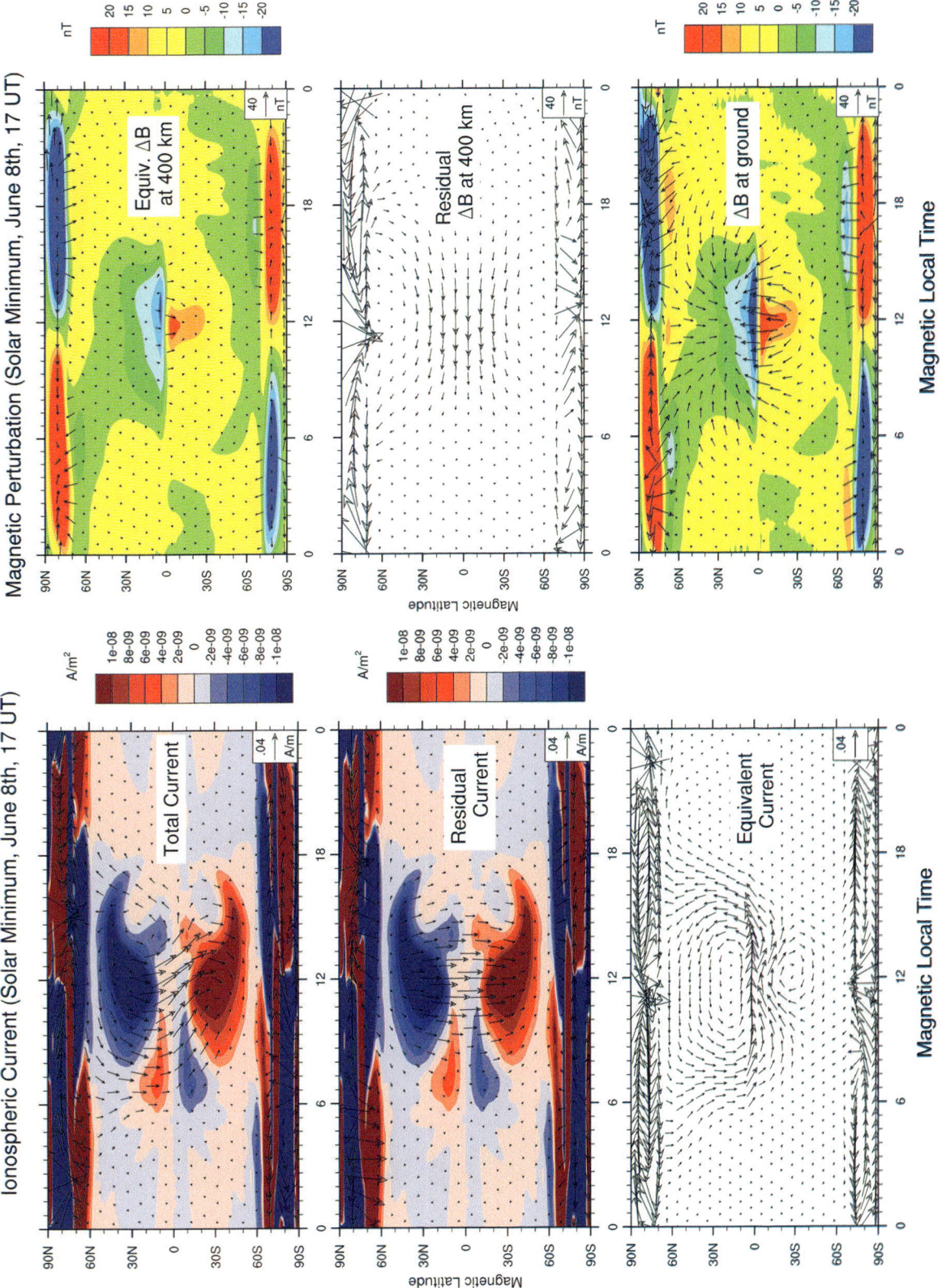

equivalent current component produces any magnetic perturbation on the ground; the residual current produces magnetic perturbations only above the current sheet. To calculate the equivalent current function, *Doumbia et al.* [2007] used spherical harmonics of an order up to $M = 24$ and a degree up to $N = 72$, which is adequate to resolve the equatorial electrojet. When the equivalent current is subtracted from the total sheet current, the remainder sheet current, plus the FAC, gives the residual current. The left side of Figure 4 shows an example from the TIE-GCM. The arrows show the height-integrated components of sheet-current density, while the red and blue colors show the intensity of the vertical component of FAC density at the top of the current sheet. The results are shown in quasidipole (QD) coordinates [*Richmond*, 1995b], and all quantities are scaled to account for the distortion of the true geomagnetic field away from a pure dipole. (In the notation of *Richmond* [1995b], the vector components shown are $K_{q\phi}$ and $K_{q\lambda}$, and the colors show J_{qr}.) The bottom-right plot in Figure 4 shows the ground-level horizontal (arrows) and vertical (colors) magnetic perturbations. Although the magnetic perturbations are calculated assuming dipolar geometry, the computed values are treated as QD components when mapped over the Earth. They include the effects of induced Earth currents, using the simple model of a perfect conductor at 600 km depth. The largest perturbations are at high latitudes, associated with the auroral electrojets, and there is also an enhancement near the magnetic equator associated with the equatorial electrojet.

For calculating the magnetic perturbation above the current sheet, one has to take into account the effects of both the equivalent current and the residual current. Shortly above the current sheet, the magnetic perturbation due to the residual current is $\Delta \mathbf{B}_{res} = -\mu_0 \mathbf{k} \times \mathbf{K}_{res}$, with $\mathbf{K}_{res}$ the residual current and $\mathbf{k}$ the unit upward vector. At higher altitudes, we can assume in the manner of *Doumbia et al.* [2007] that the meridional component of the magnetic perturbation associated with the residual currents is constant in height, while the zonal component is adjusted in height to account for the horizontal component of the FAC flowing between the current sheet and the altitude in question. Since the vertical component of $\Delta \mathbf{B}_{res}$ must vanish at the current sheet (by definition, it is zero below the current sheet, and it must be continuous across the sheet), it is small at LEO altitudes, and we neglect it. The top-right plot in Figure 4 shows the contribution of the equivalent current to the magnetic perturbations at a height of 400 km, while the center-right plot shows the contribution of the residual current. The contribution of the residual current to the horizontal magnetic perturbations tends to be considerably larger than the contribution of the equivalent current. Whereas the horizontal magnetic perturbations associated with the equivalent current are curl-free, those associated with the residual current have a curl directly related to the local vertical component of FAC density. The vertical magnetic perturbations at 400 km are dominated by the equivalent current, especially the electrojets. They tend to be weaker than the vertical perturbations at the Earth's surface.

9. FUTURE DIRECTIONS

Models of ionospheric electrodynamics have been successful in reproducing basic features of observed electric fields and geomagnetic perturbations. Obtaining more accurate results will be important for the development of ionospheric forecast models, for which electrodynamics plays a central role. We can point out some areas where significant improvements to ionospheric electrodynamic models can be made.

The distribution of electrical conductivity is of primary importance for the electric fields and currents. Improved modeling of ion densities, especially in the nighttime ionosphere, will advance our ability to model electric fields. Interactions between the electric fields and the ion densities through plasma transport, heating, and irregularity formation on large and small scales complicate this task.

The primary drivers of ionospheric electrodynamics are thermospheric winds and magnetospheric dynamo processes. As our ability to simulate whole-atmosphere dynamics advances, improvements in the quantitative modeling of winds and their variability can be expected. Magnetospheric modeling is less well developed than is atmospheric modeling, meaning that existing magnetospheric models are limited in their ability to provide realistic electrodynamic inputs to models of ionospheric electrodynamics. There is much room for improvement.

A number of improvements to algorithms for calculating the electric potential will be needed. The potential can be strongly asymmetric between the southern and northern polar caps, and algorithms are needed that can calculate the asymmetric potentials. As models are pushed to higher spatial resolution, improved algorithms that can efficiently take advantage of parallel processing are needed. When ionospheric irregularities are modeled, effects of parallel electric fields need to be taken into account, meaning that full 3-D potential solvers are needed [e.g., *Aveiro et al.*, 2012]. To model rapidly varying electric fields, nonelectrostatic effects will need to be taken into account.

The simplified procedure we presented for calculating geomagnetic perturbations is computationally fast, but cannot be expected to provide highly accurate results, especially at heights within the ionosphere where G/P currents flow. A fully 3-D calculation is needed. Accurate modeling of the effects of induced Earth currents will require time-dependent algorithms that incorporate realistic models of Earth conductivity. Finally, improved ways to account for magnetospheric currents far above the ionosphere, which no longer flow only along the geomagnetic field, and for which the FAC configuration is altered by departures of the field from a simple IGRF configuration, will enable us to simulate geomagnetic perturbations more accurately, especially during magnetically active periods.

Acknowledgments. This work was supported, in part, by NASA grants NNX08AG09G and NNX09AN57G and NSF grants ATM-0836386, AGS-1135446, and AGS-1103149. The National Center for Atmospheric Research is sponsored by the National Science Foundation.

REFERENCES

Alken, P., S. Maus, A. D. Richmond, and A. Maute (2011), The ionospheric gravity and diamagnetic current systems, *J. Geophys. Res.*, *116*, A12316, doi:10.1029/2011JA017126.

Anderson, B., K. Rock, H. Korth, C. L. Waters, D. L. Green, L. P. Dyrud, and R. J. Barnes (2012), AMPERE: Overview and Initial Results, *Space Sci. Rev.*, in press.

Aveiro, H. C., D. L. Hysell, R. G. Caton, K. M. Groves, J. Klenzing, R. F. Pfaff, R. Stoneback, and R. A. Heelis (2012), Three-dimensional numerical simulations of equatorial spread *F*: Results and observations in the Pacific sector, *J. Geophys. Res.*, *117*, A03325, doi:10.1029/2011JA017077.

Cnossen, I., A. D. Richmond, M. Wiltberger, W. Wang, and P. Schmitt (2011), The response of the coupled magnetosphere-ionosphere-thermosphere system to a 25% reduction in the dipole moment of the Earth's magnetic field, *J. Geophys. Res.*, *116*, A12304, doi:10.1029/2011JA017063.

Crowley, G., B. A. Emery, R. G. Roble, H. C. Carlson Jr., and D. J. Knipp (1989), Thermospheric dynamics during September 18–19, 1984: 1. Model simulations, *J. Geophys. Res.*, *94* (A12), 16,925–16,944.

Doumbia, V., A. Maute, and A. D. Richmond (2007), Simulation of equatorial electrojet magnetic effects with the thermosphere-ionosphere-electrodynamics general circulation model, *J. Geophys. Res.*, *112*, A09309, doi:10.1029/2007JA012308.

Eccles, J. V. (2004), The effect of gravity and pressure in the electrodynamics of the low-latitude ionosphere, *J. Geophys. Res.*, *109*, A05304, doi:10.1029/2003JA010023.

Engels, U., and N. Olsen (1998), Computation of magnetic fields within source regions of ionospheric and magnetospheric currents, *J. Atmos. Sol. Terr. Phys.*, *60*, 1585–1592.

Fang, T., A. D. Richmond, J. Y. Liu, A. Maute, C. H. Lin, C. H. Chen, and B. Harper (2008a), Model simulation of the equatorial electrojet in the Peruvian and Philippine sectors, *J. Atmos. Sol. Terr. Phys.*, *70*(17), 2203–2211, doi:10.1016/j.jastp.2008.04.021.

Fang, T. W., A. D. Richmond, J. Y. Liu, and A. Maute (2008b), Wind dynamo effects on ground magnetic perturbations and vertical drifts, *J. Geophys. Res.*, *113*, A11313, doi:10.1029/2008JA013513.

Farley, D. T., Jr. (1959), A theory of electrostatic fields in a horizontally stratified ionosphere subject to a vertical magnetic field, *J. Geophys. Res.*, *64*(9), 1225–1233, doi:10.1029/JZ064i009p01225.

Farley, D. T., Jr. (1960), A theory of electrostatic fields in the ionosphere at nonpolar geomagnetic latitudes, *J. Geophys. Res.*, *65*(3), 869–877, doi:10.1029/JZ065i003p00869.

Fesen, C. G., G. Crowley, R. G. Roble, A. D. Richmond, and B. G. Fejer (2000), Simulation of the pre-reversal enhancement in the low latitude vertical ion drifts, *Geophys. Res. Lett.*, *27*(13), 1851–1854.

Haerendel, G., J. V. Eccles, and S. Çakir (1992), Theory for modeling the equatorial evening ionosphere and the origin of the shear in the horizontal plasma flow, *J. Geophys. Res.*, *97*(A2), 1209–1223, doi:10.1029/91JA02226.

Hagan, M. E., A. Maute, R. G. Roble, A. D. Richmond, T. J. Immel, and S. L. England (2007), Connections between deep tropical clouds and the Earth's ionosphere, *Geophys. Res. Lett.*, *34*, L20109, doi:10.1029/2007GL030142.

Heelis, R. A., J. K. Lowell, and R. W. Spiro (1982), A model of the high-latitude ionospheric convection pattern, *J. Geophys. Res.*, *87*(A8), 6339–6345.

Huang, C.-S., F. J. Rich, O. de La Beaujardiere, and R. A. Heelis (2010), Longitudinal and seasonal variations of the equatorial ionospheric ion density and eastward drift velocity in the dusk sector, *J. Geophys. Res.*, *115*, A02305, doi:10.1029/2009JA014503.

Huba, J. D., G. Joyce, J. Krall, C. L. Siefring, and P. A. Bernhardt (2010a), Self-consistent modeling of equatorial dawn density depletions with SAMI3, *Geophys. Res. Lett.*, *37*, L03104, doi:10.1029/2009GL041492.

Huba, J. D., G. Joyce, J. Krall, C. L. Siefring, and P. A. Bernhardt (2010b), Correction to "Self-consistent modeling of equatorial dawn density depletions with SAMI3", *Geophys. Res. Lett.*, *37*, L20104, doi:10.1029/2010GL045004.

Hurtaud, Y., C. Peymirat, and A. D. Richmond (2007), Modeling seasonal and diurnal effects on ionospheric conductances, region-2 currents, and plasma convection in the inner magnetosphere, *J. Geophys. Res.*, *112*, A09217, doi:10.1029/2007JA012257.

Immel, T. J., E. Sagawa, S. L. England, S. B. Henderson, M. E. Hagan, S. B. Mende, H. U. Frey, C. M. Swenson, and L. J. Paxton (2006), Control of equatorial ionospheric morphology by atmospheric tides, *Geophys. Res. Lett.*, *33*, L15108, doi:10.1029/2006GL026161.

Iwasaki, N., and A. Nishida (1967), Ionospheric current system produced by an external electric field in the polar cap, *Rep. Ionos. Space Res. Jpn.*, *21*, 17–28.

Jin, H., Y. Miyoshi, H. Fujiwara, and H. Shinagawa (2008), Electrodynamics of the formation of ionospheric wave number 4 longitudinal structure, *J. Geophys. Res.*, *113*, A09307, doi:10.1029/2008JA013301.

Jin, H., Y. Miyoshi, H. Fujiwara, H. Shinagawa, K. Terada, N. Terada, M. Ishii, Y. Otsuka, and A. Saito (2011), Vertical connection from the tropospheric activities to the ionospheric longitudinal structure simulated by a new Earth's whole atmosphere-ionosphere coupled model, *J. Geophys. Res.*, *116*, A01316, doi:10.1029/2010JA015925.

Kawano-Sasaki, K., and S. Miyahara (2008), A study on three-dimensional structures of the ionospheric dynamo currents induced by the neutral winds simulated by the Kyushu-GCM, *J. Atmos. Sol. Terr. Phys.*, *70*, 1549–1562.

Kelley, M. C., M. F. Larsen, C. LaHoz, and J. P. McClure (1981), Gravity wave initiation of equatorial spread F: A case study, *J. Geophys. Res.*, *86*(A11), 9087–9100.

Klimenko, M., V. Klimenko, and V. Bryukhanov (2006), Numerical simulation of the electric field and zonal current in the Earth's ionosphere: The dynamo field and equatorial electrojet, *J. Geomagn. Aeron.*, *46*(4), 457–466, doi:10.1134/S0016793206040074.

Kuvshinov, A. (2008), 3-D global induction in the oceans and solid Earth: Recent progress in modeling magnetic and electric fields from sources of magnetospheric, ionospheric and oceanic origin, *Surv. Geophys.*, *29*(2), 139–186, doi:10.1007/s10712-008-9045-z.

Le Sager, P., and T. S. Huang (2002a), Ionospheric currents and field-aligned currents generated by dynamo action in an asymmetric Earth magnetic field, *J. Geophys. Res.*, *107*(A2), 1025, doi:10.1029/2001JA000211.

Le Sager, P., and T. S. Huang (2002b), Longitudinal dependence of the daily geomagnetic variation during quiet time, *J. Geophys. Res.*, *107*(A11), 1397, doi:10.1029/2002JA009287.

Lu, G., A. D. Richmond, B. A. Emery, and R. G. Roble (1995), Magnetosphere-ionosphere-thermosphere coupling: Effect of neutral winds on energy transfer and field-aligned current, *J. Geophys. Res.*, *100*(A10), 19,643–19,659, doi:10.1029/95JA00766.

Lühr, H., M. Rother, S. Maus, W. Mai, and D. Cooke (2003), The diamagnetic effect of the equatorial Appleton anomaly: Its characteristics and impact on geomagnetic field modeling, *Geophys. Res. Lett.*, *30*(17), 1906, doi:10.1029/2003GL017407.

Marsal, S., A. D. Richmond, A. Maute, and B. J. Anderson (2012), Forcing the TIEGCM model with Birkeland currents from the Active Magnetosphere and Planetary Electrodynamics Response Experiment, *J. Geophys. Res.*, *117*, A06308, doi:10.1029/2011JA017416.

Maute, A., A. D. Richmond, and R. G. Roble (2012), Sources of low-latitude ionospheric E × B drifts and their variability, *J. Geophys. Res.*, *117*, A06312, doi:10.1029/2011JA017502.

Millward, G. H., I. C. F. Müller-Wodarg, A. D. Aylward, T. J. Fuller-Rowell, A. D. Richmond, and R. J. Moffett (2001), An investigation into the influence of tidal forcing on F region equatorial vertical ion drift using a global ionosphere-thermosphere model with coupled electrodynamics, *J. Geophys. Res.*, *106*(A11), 24,733–24,744, doi:10.1029/2000JA000342.

Miyahara, S., and M. Ooishi (1997), Variation of *Sq* induced by atmospheric tides simulated by a middle atmosphere general circulation model, *J. Geomagn. Geoelectr.*, *49*, 77–87.

Olsen, N., and M. Kiefer (1995), Geomagnetic daily variations produced by a QBO in thermospheric prevailing winds, *J. Atmos. Terr. Phys.*, *57*, 1583–1589.

Ontiveros, P. (2009), Synthetic magnetogram calculations from magnetosphere-ionosphere coupling models, Ph.D. dissertation, Rice Univ., Houston, Tex.

Pembroke, A., F. Toffoletto, S. Sazykin, M. Wiltberger, J. Lyon, V. Merkin, and P. Schmitt (2012), Initial results from a dynamic coupled magnetosphere-ionosphere-ring current model, *J. Geophys. Res.*, *117*, A02211, doi:10.1029/2011JA016979.

Peymirat, C., and A. D. Richmond (1993), Modeling the ion loss effect on the generation of region 2 field-aligned currents via equivalent magnetospheric conductances, *J. Geophys. Res.*, *98*(A9), 15,467–15,476, doi:10.1029/93JA01034.

Peymirat, C., A. D. Richmond, B. A. Emery, and R. G. Roble (1998), A magnetosphere-thermosphere-ionosphere electrodynamics general circulation model, *J. Geophys. Res.*, *103*(A8), 17,467–17,477, doi:10.1029/98JA01235.

Peymirat, C., A. D. Richmond, and R. G. Roble (2002), Neutral wind influence on the electrodynamic coupling between the ionosphere and the magnetosphere, *J. Geophys. Res.*, *107*(A1), 1006, doi:10.1029/2001JA900106.

Qian, L., A. G. Burns, B. A. Emery, B. Foster, G. Lu, A. Maute, A. D. Richmond, R. G. Roble, S. C. Solomon, and W. Wang (2012) The NCAR TIE-GCM: A community model of the coupled thermosphere/ionosphere system, in *Modeling the Ionosphere/Thermosphere System*, Geophys. Monogr. Ser., doi:10.1029/2012GM001297, this volume.

Raeder, J., Y. Wang, and T. J. Fuller-Rowell (2001a), Geomagnetic storm simulation with a coupled magnetosphere-ionosphere-thermosphere model, in *Space Weather*, Geophys. Monogr. Ser., vol. 125, edited by P. Song, H. J. Singer, and G. L. Siscoe, pp. 377–384, AGU, Washington, D. C., doi:10.1029/GM125p0377.

Raeder, J., Y. L. Wang, T. J. Fuller-Rowell, and H. J. Singer (2001b), Global simulation of magnetospheric space weather effects of the Bastille day storm, *Sol. Phys.*, *204*, 325–338.

Rees, M. (1989), *Physics and Chemistry of the Upper Atmosphere*, 2nd ed., Cambridge Univ. Press, New York.

Ren, Z., W. Wan, Y. Wei, L. Liu, and T. Yu (2008), A theoretical model for mid- and low-latitude ionospheric electric fields in realistic geomagnetic fields, *Chin. Sci. Bull.*, *53*(24), 3883–3890.

Ren, Z., W. Wan, and L. Liu (2009), GCITEM-IGGCAS: A new global coupled ionosphere-thermosphere-electrodynamics model, *J. Atmos. Sol. Terr. Phys.*, *71*, 2064–2076, doi:10.1016/j.jastp.2009.09.015.

Richards, P. G., J. A. Fennelly, and D. G. Torr (1994), EUVAC: A solar EUV flux model for aeronomic calculations, *J. Geophys. Res.*, *99*(A5), 8981–8992.

Richmond, A. D. (1973), Equatorial electrojet – I. Development of a model including winds and instabilities, *J. Atmos. Terr. Phys.*, *35*, 1083–1103.

Richmond, A. D. (1974), The computation of magnetic effects of field-aligned magnetospheric currents, *J. Atmos. Terr. Phys.*, *36*, 245–252.

Richmond, A. D. (1995a), Ionospheric electrodynamics, in *Handbook of Atmospheric Electrodynamics*, vol. 2, edited by H. Volland, pp. 249–290, CRC Press, Boca Raton, Fla.

Richmond, A. D. (1995b), Ionospheric electrodynamics using magnetic apex coordinates, *J. Geomagn. Geoelectr.*, *47*, 191–212.

Richmond, A. D. (2002), Modeling the geomagnetic perturbations produced by ionospheric currents, above and below the ionosphere, *J. Geodyn.*, *33*, 143–156.

Richmond, A. D. (2011), Electrodynamics of ionosphere-thermosphere coupling, in *Aeronomy of the Earth's Atmosphere and Ionosphere*, edited by M. Abdu, D. Pancheva, and A. Bhattacharyya, pp. 191–201, Springer, Dordrecht, The Netherlands.

Richmond, A. D., and R. G. Roble (1987), Electrodynamic effects of thermospheric winds from the NCAR Thermospheric General Circulation Model, *J. Geophys. Res.*, *92*(A11), 12,365–12,376.

Richmond, A. D., E. C. Ridley, and R. G. Roble (1992), A thermosphere/ionosphere general circulation model with coupled electrodynamics, *Geophys. Res. Lett.*, *19*(6), 601–604.

Ridley, A. J., A. D. Richmond, T. I. Gombosi, D. L. De Zeeuw, and C. R. Clauer (2003), Ionospheric control of the magnetospheric configuration: Thermospheric neutral winds, *J. Geophys. Res.*, *108*(A8), 1328, doi:10.1029/2002JA009464.

Roble, R., and E. Ridley (1987), An auroral model for the NCAR thermospheric general circulation model (TGCM), *Ann. Geophys., Ser. A*, *5*(6), 369–382.

Scherliess, L., and B. G. Fejer (1999), Radar and satellite global equatorial F region vertical drift model, *J. Geophys. Res.*, *104*(A4), 6829–6842, doi:10.1029/1999JA900025.

Senior, C., and M. Blanc (1984), On the control of magnetospheric convection by the spatial distribution of ionospheric conductivities, *J. Geophys. Res.*, *89*(A1), 261–284.

Shen, J. S., W. E. Swartz, D. T. Farley, and R. M. Harper (1976), Ionization layers in the nighttime E region valley above Arecibo, *J. Geophys. Res.*, *81*(31), 5517–5526.

Shen, J., M.-Y. Zi, J.-S. Wang, J.-Y. Xu, and S.-L. Liu (2006), Global disturbance currents in the ionosphere during storms, *J. Atmos. Sol. Terr. Phys.*, *68*, 793–802.

Singh, A., and K. Cole (1995), Ionospheric electrodynamic model with an eccentric dipole geomagnetic field, *J. Atmos. Terr. Phys.*, *57*, 795–803.

Solomon, S. (2006), Numerical models of the E-region ionosphere, *Adv. Space Res.*, *37*, 1031–1037.

Solomon, S. C., and L. Qian (2005), Solar extreme-ultraviolet irradiance for general circulation models, *J. Geophys. Res.*, *110*, A10306, doi:10.1029/2005JA011160.

Strobel, D. F., T. R. Young, R. R. Meier, T. P. Coffey, and A. W. Ali (1974), The nighttime ionosphere: E region and lower F region, *J. Geophys. Res.*, *79*(22), 3171–3178.

Strobel, D., C. Opal, and R. Meier (1980), Photoionization rates in the night-time E- and F-region ionosphere, *Planet. Space Sci.*, *28*, 1027–1033.

Takeda, M. (1996), Effects of the strength of the geomagnetic main field strength on the dynamo action in the ionosphere, *J. Geophys. Res.*, *101*(A4), 7875–7880, doi:10.1029/95JA03807.

Takeda, M., and T. Araki (1985), Electric conductivity of the ionosphere and nocturnal currents, *J. Atmos. Terr. Phys.*, *47*(6), 601–609, doi:10.1016/0021-9169(85)90043-1.

Takeda, M., and H. Maeda (1980), Three-dimensional structure of ionospheric currents 1. Currents caused by diurnal tidal winds, *J. Geophys. Res.*, *85*(A12), 6895–6899.

Titheridge, J. (2000), Modelling the peak of the ionospheric E-layer, *J. Atmos. Terr. Phys.*, *62*(2), 93–114, doi:10.1016/S1364-6826(99)00102-9.

Trost, T. F. (1979), Electron concentrations in the E and upper D region at Arecibo, *J. Geophys. Res.*, *84*(A6), 2736–2742.

Tsunomura, S. (1999), Numerical analysis of global ionospheric current system including the effect of equatorial enhancement, *Ann. Geophys.*, *17*, 692–706.

Tsyganenko, N. (1989), A magnetospheric magnetic field model with a warped tail current sheet, *Planet. Space Sci.*, *37*, 5–20.

Vanhamäki, H., A. Viljanen, and O. Amm (2005), Induction effects on ionospheric electric and magnetic fields, *Ann. Geophys.*, *30*, 1735–1746.

Vasyliūnas, V. (1972), The interrelationship of magnetospheric processes, in *Earth's Magnetosphere Processes*, edited by M. McCormac, pp. 29–38, D. Reidel, Hingham, Mass.

Vasyliūnas, V. (2012), The physical basis of ionospheric electrodynamics, *Ann. Geophys.*, *30*, 357–369.

Wakai, N. (1967), Quiet and disturbed structure and variations of the nighttime E region, *J. Geophys. Res.*, *72*(17), 4507–4517.

Wang, W., M. Wiltberger, A. Burns, S. Solomon, T. Killeen, N. Maruyama, and J. Lyon (2004), Initial results from the coupled magnetosphere-ionosphere-thermosphere model: Thermosphere-ionosphere responses, *J. Atmos. Sol. Terr. Phys.*, *66*(15–16), 1425–1441.

Weimer, D. R. (2000), A new technique for the mapping of ionospheric field-aligned currents from satellite magnetometer data, in *Magnetospheric Current Systems*, Geophys. Monogr. Ser., vol. 118, edited by S. Ohtani et al., pp. 381–388, AGU, Washington, D. C., doi:10.1029/GM118p0381.

Weimer, D. R. (2005), Improved ionospheric electrodynamic models and application to calculating Joule heating rates, *J. Geophys. Res.*, *110*, A05306, doi:10.1029/2004JA010884.

Wiltberger, M., W. Wang, A. G. Burns, S. C. Solomon, J. G. Lyon, and C. C. Goodrich (2004), Initial results from the coupled magnetosphere ionosphere thermosphere model: Magnetospheric and ionospheric responses, *J. Atmos. Sol. Terr. Phys.*, *66*(15–16), 1411–1423, doi:10.1016/j.jastp.2004.03.026.

Wolf, R., G. Mantjoukis, and R. Spiro (1986), Theoretical comments on the nature of the plasmapause, *Adv. Space Res.*, *6*(3), 177–186.

Zaka, K. Z., et al. (2010), Simulation of electric field and current during the 11 June 1993 disturbance dynamo event: Comparison with the observations, *J. Geophys. Res.*, *115*, A11307, doi:10.1029/2010JA015417.

Zhou, Q. H., J. D. Mathews, and Q. N. Zhou (1999), Incoherent scatter radar study of the impact of the meteoric influx on nocturnal E-region ionization, *Geophys. Res. Lett.*, *26*(13), 1833–1836.

A. D. Richmond and A. Maute, High Altitude Observatory, National Center for Atmospheric Research, 3080 Center Green Drive, Boulder, CO 80301, USA. (richmond@ucar.edu; maute@ucar.edu)

The NCAR TIE-GCM: A Community Model of the Coupled Thermosphere/Ionosphere System

Liying Qian, Alan G. Burns, Barbara A. Emery, Benjamin Foster, Gang Lu, Astrid Maute, Arthur D. Richmond, Raymond G. Roble, Stanley C. Solomon, and Wenbin Wang

National Center for Atmospheric Research, High Altitude Observatory, Boulder, Colorado, USA

The thermosphere-ionosphere-electrodynamics general circulation model (TIE-GCM) is a community model developed and maintained at the National Center for Atmospheric Research. It also can be run at the NASA Community Coordinated Modeling Center, and is a component of the coupled magnetosphere-ionosphere-thermosphere Model (CMIT). This paper describes the TIE-GCM development history, model elements, model input and output, the equations solved, boundary conditions, and numerical techniques. Some model validation examples are shown, and future improvements and developments are discussed.

1. INTRODUCTION

The thermosphere-ionosphere-electrodynamics general circulation model (TIE-GCM), developed at the National Center for Atmospheric Research (NCAR) High-Altitude Observatory (HAO), is a global 3-D numerical model that simulates the coupled thermosphere/ionosphere system from ~97 to ~600 km altitude. The thermosphere is characterized by a rapid increase of temperature with altitude up to ~300 km above which the atmosphere is nearly isothermal. The major constituents of the thermosphere are O, O_2, and N_2, with O becoming increasingly dominant at higher altitudes until above ~500 km, the exosphere begins, and the lighter species H and He become important. Absorption of solar EUV ionizes the thermospheric constituents and creates the ionosphere. The thermosphere/ionosphere system interacts energetically, dynamically, electrodynamically, and chemically with the magnetosphere above and the mesosphere below.

The TIE-GCM self-consistently solves the fully coupled, nonlinear, hydrodynamic, thermodynamic, and continuity equations of the neutral gas, the ion and electron energy and momentum equations, the ion continuity equation, and neutral wind dynamo. It is an open-source community model and is also available for runs-on-request at the NASA Community Coordinated Modeling Center (CCMC).

2. MODEL DEVELOPMENT HISTORY

The original version of the model, the thermosphere general circulation model (TGCM), was developed by *Dickinson et al.* [1981, 1984] and *Roble et al.* [1982]. Three subsequent major developments were the coupling of the ionosphere to the thermosphere in the thermosphere-ionosphere general circulation model (TIGCM) [*Roble et al.*, 1987, 1988], the implementation of self-consistent electrodynamics in the thermosphere-ionosphere-electrodynamics general circulation model (TIE-GCM) [*Richmond et al.*, 1992; *Richmond*, 1995], and extension downward to include the mesosphere and upper stratosphere in the thermosphere-ionosphere-mesosphere-electrodynamics general circulation model (TIME-GCM) [*Roble and Ridley*, 1994; *Roble*, 1995]. This paper describes the TIE-GCM development branch, but most of this description is applicable to the TIME-GCM as well, particularly aspects related to the ionosphere. The TIE-GCM has been coupled to the Lyon-Fedder-Mobarry (LFM) magnetosphere model [*Lyon et al.*, 2004] to form the coupled magnetosphere-ionosphere-thermosphere model (CMIT) [e.g., *Wang et al.*, 2004; *Wiltberger et al.*, 2004].

Developments to the TIE-GCM since the work of *Richmond et al.* [1992] include the extension of the upper boundary to higher altitude, variable tidal inputs [*Hagan et al.*, 2001], improvements to the electrodynamo calculations [*Richmond and Maute*, this volume], revision of solar irradiance and photoelectron inputs [*Solomon and Qian*, 2005], modification of the upper boundary heat flux [*Lei et al.*, 2007], application of variable eddy diffusivity at the lower boundary [*Qian et al.*, 2009a], revisions to auroral precipitation inputs [*Emery et al.*, 2008] and background (nighttime) ionization rates, and inclusion of the *Weimer* [2005] potential model [*Solomon et al.*, 2012]. Some of these are described in more detail below. Documentation, a user's guide, source code, and output processing software are publically available at http://www.hao.ucar.edu/modeling/tgcm/. The current model release (as of April 2012) is designated version 1.94.2. This model has been in widespread use by universities and other institutions.

3. MODEL DESCRIPTION

3.1. Energy Input and Output

The main external drivers of the thermosphere/ionosphere system are solar UV radiation and geomagnetic forcing, of which the solar radiation is the primary energy input. However, geomagnetic energy input can be comparable to or even larger than the solar radiation input during magnetic storms. Daily averaged solar and geomagnetic energy for solar cycles 21–23 are estimated by *Knipp et al.* [2004]. In addition, the lower atmosphere also transfers momentum and energy to the system near the mesopause. The primary loss of energy is through IR radiative cooling. IR cooling includes CO_2 radiative cooling at 15 µm, which peaks at ~120 km altitude, NO radiative cooling at 5.3 µm, which peaks at ~150 km, and $O(^3P)$ fine-structure cooling at 63 µm, which maximizes above 200 km [*Roble*, 1995]. In addition, energy is transferred internally within the system through vertical heat conduction, eddy diffusion, horizontal diffusion, horizontal and vertical advection, and elastic and inelastic collisions among neutrals, ions, and electrons. The TIE-GCM solves these internal processes physically and self-consistently, as summarized in section 3.2.

The thermosphere absorbs solar irradiance in the soft X-ray (XUV, 1–30 nm), EUV (30–120 nm), and FUV (120–200 nm) ranges. Solar irradiance ionizes, dissociates, and excites the thermospheric constituents. Through these processes, it transfers its energy to the kinetic energy of electrons and ions and to the chemical potential energy of ions and neutrals. The neutral atmosphere is then heated through exothermic ion-neutral and neutral-neutral chemical reactions and through collisions of neutral constituents with electrons and ions. The default input of this forcing for the TIE-GCM is a solar EUV proxy model that is largely based on the EUVAC model [*Richards et al.*, 1994]. The TIE-GCM can also use solar EUV measurements as input. Detailed descriptions of solar input, solar energy deposition, and calculation of photoelectron effects, are provided by *Solomon and Qian* [2005].

Geomagnetic activity results from the interaction between the solar wind and the Earth's magnetic field. Magnetospheric currents flow into the polar regions along the Earth's magnetic field lines and form a magnetosphere-ionosphere current system that drives ionosphere plasma convection and heats the thermosphere through collisions between ions and neutral gases (Joule heating). Energetic charged particles precipitate into the thermosphere/ionosphere along the Earth's magnetic field lines into the auroral region, ionizing thermospheric constituents and heating the thermosphere/ionosphere. In the TIE-GCM, the high-latitude potential imposed by magnetospheric processes is specified by either the *Heelis et al.* [1982] or the *Weimer* [2005] empirical model. The Heelis model is based on the Kp index as input, whereas the Weimer model is parameterized based on solar wind and interplanetary magnetic field data. Energy input associated with auroral particle precipitation is calculated by an analytical auroral model [*Roble and Ridley*, 1987; *Emery et al.*, 2008]. Alternatively, the TIE-GCM can use potential patterns derived from the assimilative mapping of ionospheric electrodynamics (AMIE) procedure [*Richmond and Kamide*, 1988] or from the LFM model, in which mode it becomes a component of the CMIT model. In addition, AMPERE field-aligned currents have also been used to drive the TIE-GCM [*Marsal et al.*, 2012].

The TIE-GCM obtains a self-consistent solution to the low-latitude ionospheric electrodynamo determined by conductances and neutral dynamics, using the method of *Richmond et al.* [1992], in apex coordinates based on the International Geomagnetic Reference Field [*Richmond*, 1995]. It is necessary to merge the electric potentials thus obtained with the externally imposed potentials within each polar cap, which has been accomplished in the past by constraining the electrodynamo solution within the range 60° to 75° magnetic latitude to approach the predefined model using a linear crossover parameter. In the TIE-GCM v. 1.94, the crossover boundaries vary dynamically with the size of the magnetospheric potential pattern, occurring over a 15° range of magnetic latitude starting 5° equatorward of the convection radius boundary [*Solomon et al.*, 2012]. See the work of *Richmond and Maute* [this volume] for more details and recent development on electrodynamics in the TIE-GCM.

The thermosphere/ionosphere is also affected by momentum and energy forcing from various waves in the mesosphere and lower thermosphere (MLT) region. The important waves in the MLT region are gravity waves, tides, and planetary waves. In the TIE-GCM, the amplitudes and phases of tides from the lower atmosphere are provided by the global scale wave model (SGWM) [*Hagan et al.*, 2001]. Effects of turbulent mixing caused by gravity wave breaking are implicitly included by specifying an eddy diffusivity at the lower boundary that varies with altitude and latitude or, alternatively, with a parameterized seasonal variation additionally applied [*Qian et al.*, 2009a].

In the MLT region, NO and CO_2 are excited through collisions with neutral constituents (mainly atomic oxygen) and subsequently radiate or are collisionally quenched. The radiative process becomes dominant as the atmosphere becomes thinner with increasing altitude. NO and CO_2 radiate energy in the IR and, thus, cool the thermosphere/ionosphere system. $O(^3P)$ is also a minor cooling source at high altitude due to the fine-structure radiative transitions. In the TIE-GCM, non-LTE CO_2 cooling at 15 μm is calculated using a cool-to-space approximation, while non-LTE NO cooling at 5.3 μm is calculated using Kockarts's equation, and $O(^3P)$ cooling at 63 μm is calculated using Bates's method [*Roble et al.*, 1987].

3.2. Equations, Boundary Conditions, and Model Output

The TIE-GCM utilizes a spherical coordinate system fixed with respect to the rotating Earth, with latitude and longitude as the horizontal coordinates and pressure surface as the vertical coordinate. The model pressure interfaces are defined as $z = \ln(P_0/P)$, where P_0 is a reference pressure at 5×10^{-4} μbar (5×10^{-5} Pa). The model has a horizontal resolution of 5° × 5°. It has 29 pressure surfaces, ranging from −7 to +7, that cover the altitude range from ~97 to ~600 km (depending on solar activity). The vertical resolution is half a scale height. The default time step is 2 min. The equations solved by the TIE-GCM obtain thermosphere- and ionosphere-dependent variables, i.e., model output, as a function of independent variables. The independent variables are t: universal time, ϕ: latitude, λ: longitude, and z: model pressure interface. The dependent variables are:

Z: geopotential height of model pressure surfaces;
T_N: neutral temperature;
(u, v, w): wind velocity (eastward, northward, vertical (dz/dt) tendency);
(Ψ_{N_2}, Ψ_O, Ψ_{O_2}): major species mass mixing ratio;
($\psi_{N(4S)}$, $\psi_{N(2D)}$, ψ_{NO}): minor species mass mixing ratio;
Φ: electric potential;
T_e: electron temperature;
T_i: ion temperature;
(n_{O^+}, $n_{O_2^+}$, $n_{N_2^+}$, n_{N^+}, n_{NO^+}): ion densities;
n_e: electron density.

Table 1 lists the type of each equation, variables that each equation solves, and physical terms of each equation. Detailed mathematical expressions are available in the model description document at http://www.hao.ucar.edu/modeling/tgcm/. Table 2 lists the boundary conditions for the time-dependent equations in Table 1.

The model can also output many diagnostic variables, such as geometric height, mass density, Joule heating, IR radiative cooling rates, conductivity, ion drift velocities, electric field, the peak altitude and density of the ionospheric F_2 region, and ionospheric currents. Additional diagnostic and derived parameters can be calculated by standard output processing software.

3.3. Numerical Techniques

The TIE-GCM uses an explicit fourth-order centered finite difference for horizontal derivatives, a second-order centered finite difference for time derivatives, and an implicit second-order centered finite difference for vertical derivatives. All time-dependent variables are averaged in time, longitude, and latitude each time step (Shapiro smoothing) to achieve numerical stability. Near the geographic poles, a fast Fourier transform filter is applied in the zonal direction to remove physically unrealistic high-frequency waves and ensure computational stability.

The model is implemented in Fortran 90 as a parallel code using the Message Passing Interface. It can be run on supercomputers, typically using up to 32 processors, but also performs well on typical four to eight processor workstations, and can also be run in single-processor mode on desktop computers or even laptops. It currently achieves performance of better than 600X wall clock on modern workstations.

4. SOME MODEL VALIDATION EXAMPLES

Thermosphere data sets used for validation include neutral density data derived from satellite drag and from CHAMP and GRACE satellites, composition data from TIMED/GUVI, IR cooling data from TIMED/SABER, and neutral wind and temperature data from ground-based Fabry-Perot interferometers. Ionosphere data sets include electron density measurements from Constellation Observing System for Meteorology Ionosphere and Climate (COSMIC), ground-based GPS total electron content data, and ground-based incoherent scatter radar and ionosonde measurements. For the following validation examples, the TIE-GCM was run using the default solar input and Heelis

Table 1. Equations Used in the Thermosphere-Ionosphere-Electrodynamics General Circulation Model (TIE-GCM): The Type of Each Equation, Output Variables Solved by Each Equation, and Physical Terms of Each Equation

Equation	Solve For	Terms of Equation
Horizontal momentum equation	u, v	$\frac{\partial u}{\partial t}$ or $\frac{\partial v}{\partial t}$ = vertical viscosity + Coriolis force + ion drag + horizontal advection + pressure gradient + vertical advection + horizontal diffusion
Vertical momentum equation assuming hydrostatic equilibrium	Z	$\frac{\partial Z}{\partial z} = H$, where H is pressure scale height
Neutral gas continuity equation	w	$\nabla \cdot \vec{V} + e^z \frac{\partial(e^z w)}{\partial z} = 0$
Major species continuity equation	$\Psi_O, \Psi_{O_2}, \Psi_{N_2}$	$\frac{\partial \Psi_O}{\partial t}$ or $\frac{\partial \Psi_{O_2}}{\partial t}$ = vertical molecular diffusion + vertical eddy diffusion + horizontal/vertical advection + chemical production/loss, $\Psi_{N_2} = 1 - \Psi_O - \Psi_{O_2}$
Minor species continuity equation	$\psi_{N(4S)}, \psi_{NO}$	$\frac{\partial \psi}{\partial t}$ = vertical molecular diffusion (including friction with the major species) + vertical eddy diffusion + horizontal advection + vertical advection + chemical production + chemical loss
Minor species continuity assuming photochemical equilibrium	$\psi_{N(2D)}$	production by solar radiation and auroral particle precipitation + production through ion and neutral chemistry = loss through ion and neutral chemistry
Thermodynamic equation (energy equation)	T_n	$\frac{\partial T_n}{\partial t}$ = vertical molecular and eddy heat conduction + horizontal and vertical advection + heating (solar, geomagnetic, and collision with ion and electron) + radiative cooling
Electron energy equation assuming a thermal quasisteady state	T_e	divergence of the electron heat + heating (solar radiance, auroral particle precipitation) = cooling due to collisions with ion and neutral
Ion energy equation assuming quasisteady state	T_i	heating due to electron-ion collisions + Joule heating = cooling due to ion-neutral collisions
Steady state electrodynamo equation	Φ	horizontal divergence of height-integrated current density = field-aligned current from the magnetosphere
Ion continuity equation	n_{O^+}	$\frac{\partial n_{O^+}}{\partial t}$ = transport by neutral wind + transport by ambipolar diffusion + transport by $E \times B$ + chemical production + chemical loss
Ion continuity equation assuming photochemical equilibrium for ions other than O^+	$n_{O_2}, n_{N_2}, n_{N^+}, n_{NO^+}, n_e$	chemical production of each species = chemical loss of each species, n_e is obtained by assuming charge neutrality

model, but incorporating the annual/semiannual eddy diffusivity parameterization described in the work of *Qian et al.* [2009a].

Figure 1a compares TIE-GCM-simulated neutral density to observed density derived from satellite drag data during the declining phase of solar cycle 23. Model simulated density was sampled at the perigees of satellite #12388 for the comparison. Satellite #12388 (Cosmos 1236) has moderately eccentric orbits with perigees varying between 385 and 415 km and apogees varying between 1448 and 1585 km. The satellite perigees precess in latitude and LT, with approximately three latitude cycles and five local time cycles in a year. The corresponding $F10.7$ and magnetic Ap indices are shown in Figure 1b. Solar rotational and solar cycle variations of neutral density, driven by solar radiation, are evident in both the observed and simulated densities. Both the observed and simulated densities also show short-term impulsive variations, consistent with the magnetic Ap index, that is driven by geomagnetic forcing. In addition, annual/semiannual density variations [*Bowman*, 2004; *Emmert and Picone*, 2010] are also discernible, showing as a consistent low-density period near July–August of each year. Comprehensive data-model comparisons of spatial and temporal variations of neutral density can be found in the work of *Qian and Solomon* [2011].

Figure 2 shows the simulated *NO* cooling rate and TIMED/SABER measured *NO* cooling rate [*Mlynczak et al.*, 2008] from 2002 to 2006. The *NO* cooling shown in Figure 2 is global *NO* cooling power, integrated over the globe and in the altitude range from 100 to 200 km. The magnitude and solar cycle variations of the simulated *NO* cooling are consistent with that of the measured *NO* cooling

Table 2. Boundary Conditions of Time-Dependent Equations of the TIE-GCM: Output Variables and Their Lower and Upper Boundary Conditions

Dependent Variables	Lower Boundary Conditions	Upper Boundary Conditions
Z	96 km + perturbation by tides (global scale wave model (GSWM))	N/A
u, v	0 + perturbation by tides (GSWM)	zero vertical flux of horizontal momentum
w	N/A	$w = \nabla \cdot \vec{v}$
T_N	181 K + perturbation by tides (GSWM)	zero vertical heat flux
O_2	fixed mixing ratio of 0.22 (fixed mixing ratio of 0.78 for N_2)	diffusive equilibrium
O	vertical gradient of number density is 0	diffusive equilibrium
$N(4S)$	photochemical equilibrium	diffusive equilibrium
NO	constant density of 8×10^6 cm^{-3}	diffusive equilibrium
O^+	photochemical equilibrium	specified upward or downward flux from the topside ionosphere and plasmasphere
T_e	equal to T_N	specified downward or upward heat flux from the topside ionosphere and plasmasphere

through the declining phase of solar cycle 23, although the model was relatively larger, about 20% on average. In addition, the SABER *NO* cooling rate exhibits a 60 day period corresponding to the solar time sampling of the precession period of the TIMED spacecraft [*Mlynczak et al.*, 2008], with a recurring low *NO* cooling in the second half of each precession cycle that is not in the simulated *NO* cooling rates.

Figure 3 shows data-model comparisons of NmF_2 climatology for 2008. Average NmF_2 between 10:00 and 13:00 LT, calculated using the International Reference Ionosphere (IRI) model [*Bilitza and Reinisch*, 2008] and simulated by the

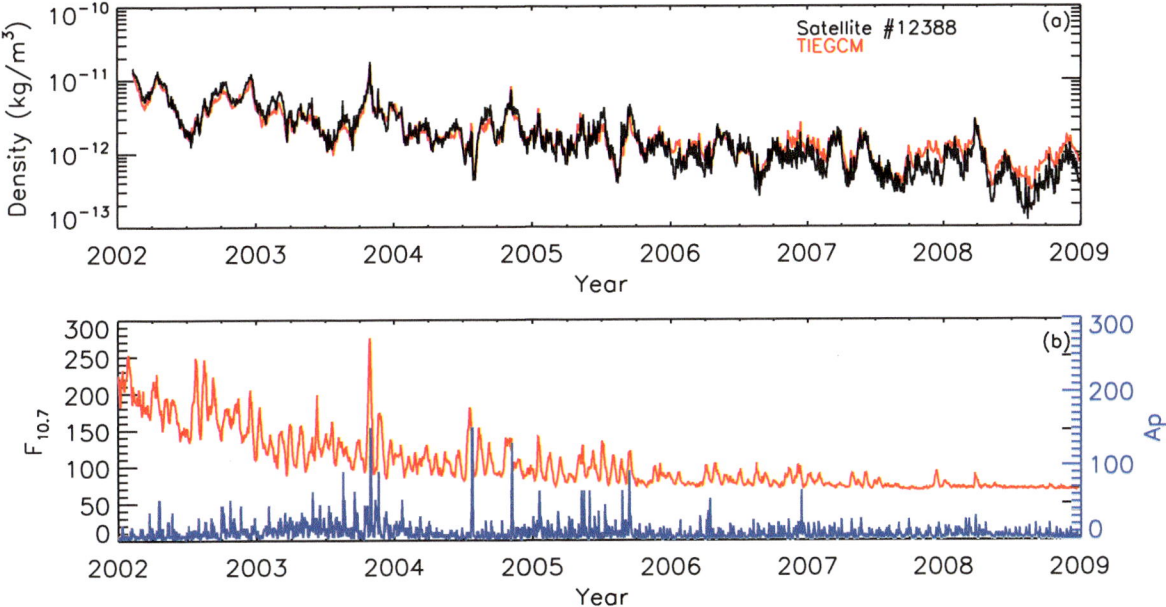

Figure 1. (a) Data-model comparison of thermospheric neutral density at satellite perigees for satellite #12388 from 2002 to 2008. Black, satellite drag-derived density; red: thermosphere-ionosphere-electrodynamics general circulation model (TIE-GCM) simulation. (b) Corresponding $F10.7$ and Ap indices. Satellite #12388 (Cosmos 1236) has moderately eccentric orbits with perigees varying between 385 and 415 km and apogees varying between 1448 and 1585 km. The satellite perigees precess in latitude and LT, with approximately three latitude cycles and five local time cycles in a year.

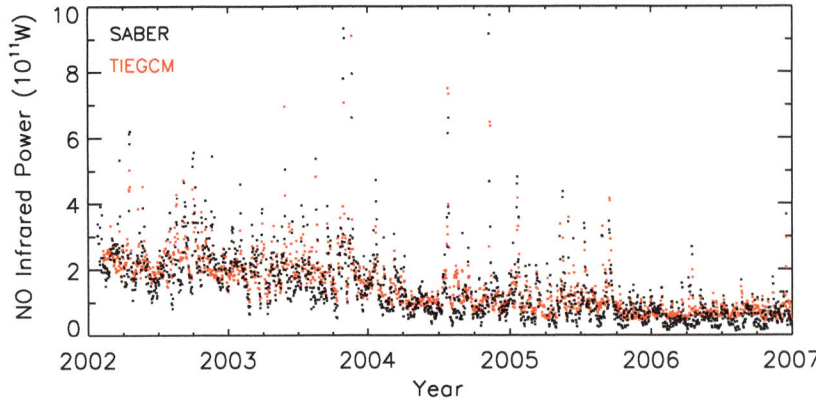

Figure 2. Global *NO* cooling power, integrated over the globe and in the altitude range from 100 to 200 km, from 2002 to 2006. Black, TIMED/SABER; red, TIE-GCM simulation. The estimated uncertainty of SABER NO cooling rates is ~15% (adapted from the work of *Qian et al.* [2010a]).

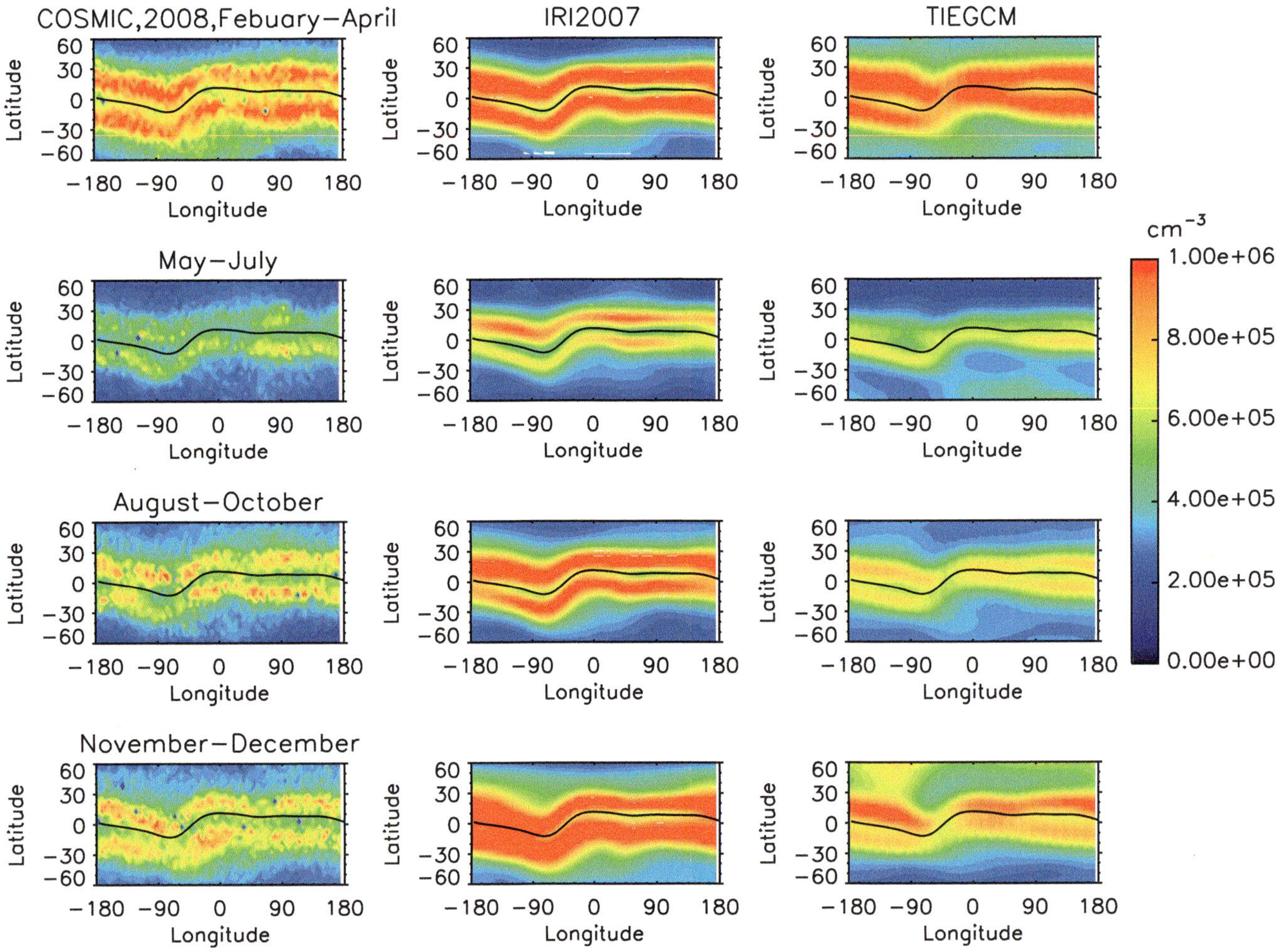

Figure 3. NmF_2 observed by Constellation Observing System for Meteorology Ionosphere and Climate, estimated by the International Reference Ionosphere, and simulated by TIE-GCM, during 2008. NmF_2 is averaged over 10:00–13:00 LT and over the months shown in each panel.

TIE-GCM, are compared to COSMIC-measured NmF_2. Seasonal climatology of TIE-GCM-simulated NmF_2 is consistent with that of the COSMIC-observed seasonal variations. However, the simulated NmF_2 shows a pronounced winter anomaly during the months of November–December that is not in the data. IRI estimated NmF_2 is larger overall than the observed data.

Figure 4 compares TIE-GCM-simulated electron density profiles to Millstone Hill incoherent scatter radar (ISR) data and the profiles estimated by the ISR ionospheric model [*Zhang et al.*, 2005], with profiles observed near equinox (30 March 2007) and at June solstice (21 June 2007) at 12:00 and 15:00 LT. The simulated electron density profiles at different seasons and LTs compare well with the observations.

These are some basic TIE-GCM model validation examples. An evaluation of the TIE-GCM with other models can be found in the work of *Shim et al.* [this volume]. The TIE-GCM, TIME-GCM, and CMIT have been used extensively for ionosphere and thermosphere studies, including studies for geomagnetic storms [e.g., *Burns et al.*, 1992, 1995a, 1995b, 2008; *Crowley et al.*, 2010; *Wang et al.*, 2008, 2010; *Lei et al.*, 2010], solar flares [e.g., *Qian et al.*, 2010b, 2011, 2012], tides [e.g., *Hagan et al.*, 2009; *Pedatella et al.*, 2011], recent solar minimum [e.g., *Solomon et al.*, 2010, 2011], effects of sudden stratospheric warming [e.g., *Liu et al.*, 2010b], equatorial ionosphere [e.g., *Fang et al.*, 2008], IR cooling [e.g., *Lu et al.*, 2010], effect of high-speed solar wind [e.g., *Qian et al.*, 2010a; *Wang et al.*, 2011; *Burns et al.*, 2012; *Solomon et al.*, 2012], data assimilation [e.g., *Lee et al.*, 2012; *Matsuo et al.*, Assimilative thermospheric mass density specification using ensemble Kalman filter, submitted to *Journal of Geophysical Research*, 2012;], and long-term changes [e.g., *Cnossen and Richmond*, 2008; *Qian et al.*, 2009b].

5. FUTURE IMPROVEMENT AND DEVELOPMENT PLANS

The TIE-GCM is an extremely efficient parallel processing model that has been implemented on several computational architectures. The TIE-GCM has fully coupled neutral dynamics and ionospheric electrodynamics, comprehensive photochemistry and thermodynamics, accurate treatment of solar EUV and photoelectron processes, and several

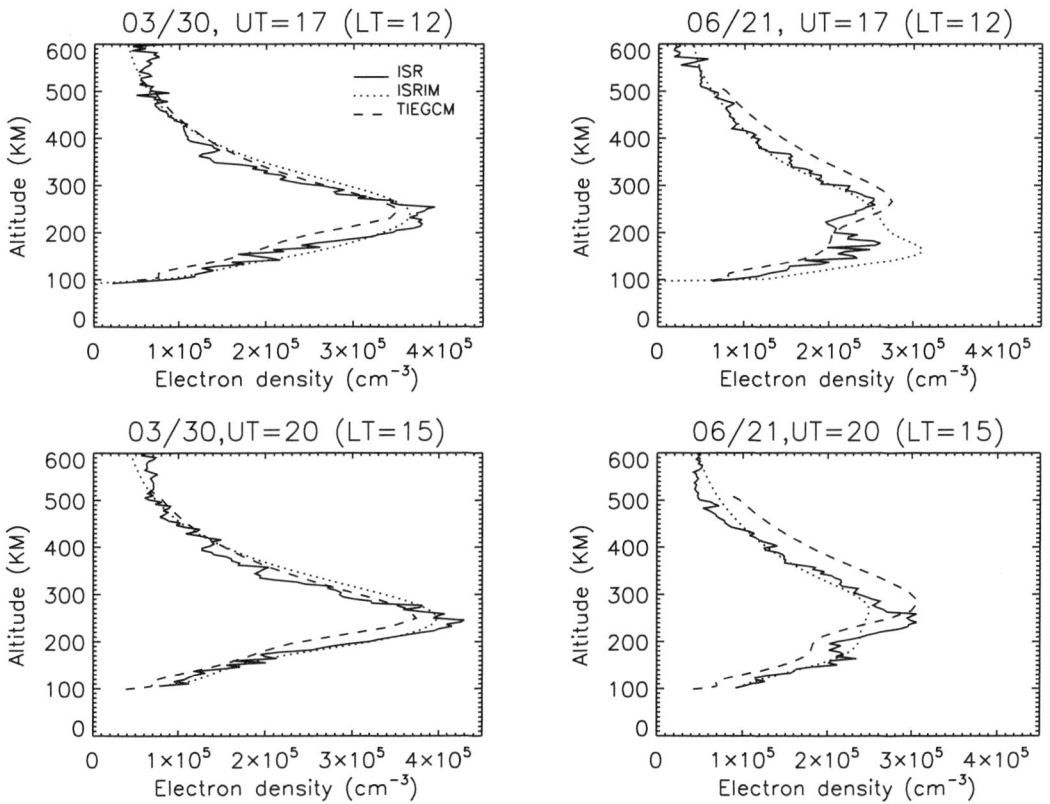

Figure 4. Electron density profiles observed by Millstone Hill incoherent scatter radar (ISR), calculated by the ISR ionospheric model, and simulated by the TIE-GCM, at 12:00 and 15:00 LT, on 30 March 2007 and 21 June 2007.

high-latitude magnetospheric input options. In addition to the Heelis and Weimer models, the TIE-GCM can also use magnetospheric models, or assimilated specifications such as from AMIE, as inputs.

One area of the TIE-GCM that needs improvement is its boundary conditions. The TIE-GCM assumes constant background for Z, T, u, and v (Table 2) at the lower boundary. This simplification causes problems for tides since tidal propagations depend on background. The TIE-GCM currently includes migrating tides, but not nonmigrating tides at its lower boundary. Comparisons between data and TIE-GCM-simulated migrating tides have demonstrated that migrating tides are mostly reasonable in the model [e.g., *Fesen et al.*, 1986; *Hagan et al.*, 2001]. However, the TIE-GCM has troubles to reliably reproduce the peak height for the nonmigrating tides due to the proximity of the lower boundary to the height of the peak nonmigrating tidal amplitudes. It is worth pointing out that there are some differences between the migrating and nonmigrating tides in terms of their generation and source. The nonmigrating tides are almost entirely generated in the troposphere and propagate vertically into the MLT. In contrast, the migrating tides are generated both in the troposphere and in situ in the thermosphere due to solar heating. This is the main reason why the TIE-GCM includes migrating tides, but not nonmigrating tides at its lower boundary. A numerical nudging scheme similar to the method employed in the Whole-Atmosphere Community Climate Model (WACCM) [*Lamarque et al.*, 2011] may be used to address this issue. The lower boundary has also been coupled to the middle atmosphere models such as the Navy Operational Global Atmospheric Prediction System-Advanced Level Physics High Altitude (NOGAPS-ALPHA) forecast/assimilation system using numerical nudging schemes [*Siskind et al.*, 2012]. In addition, the effect of turbulent mixing associated with gravity wave breaking is incorporated using a specified eddy diffusion coefficient. The seasonal varying eddy diffusivity parameterization described in the work of *Qian et al.* [2009a] is available as a standard option; this parameterization was based on a limited satellite drag data set from five satellites over 5 years. It will be improved using more extensive and long-term satellite drag data. Furthermore, the TIE-GCM does not include a plasmasphere model. Consequently, the model specifies topside O^+ flux and electron heat flux at its upper boundary: An ionosphere/plasmasphere model such as SAMI3 [*Huba et al.*, 2000] can be coupled to the TIE-GCM to address this issue.

Another area of improvement is the electron (and ion) energy equations. The TIE-GCM assumes a thermal quasi-steady state of electron temperature, and therefore, the electron energy equation is not time dependent. This assumption is valid below ~300 km where both the electrons and ions respond rapidly to variations in direct thermal forcing and reach steady state. However, 3–16 min is needed to reach steady state between 300 and 600 km, depending on electron density and the type of the thermal forcing. Time-dependent electron and ion energetics have recently been developed for the extended WACCM (WACCM-X) [*Liu et al.*, 2010a] and will be migrated to the TIE-GCM. Also, the electron heating rate calculation will be updated using the parameterization of *Smithtro and Solomon* [2008].

A high-resolution version of the TIE-GCM ($2.5° \times 2.5° \times$ H/4 or twice the current resolution in all three dimensions) has recently been implemented and is currently being tested. This version will better resolve high-latitude inputs and responses, and the vertical structure of tidal perturbations. Other developments currently in progress include a global electric potential calculation based on field-aligned currents [*Marsal et al.*, 2012], further revisions to auroral precipitation inputs, and various performance enhancements. New and future developments will be released to the community and transitioned to the CCMC following validation and verification. The HAO/NCAR team is committed to the continuing development and improvement of the TIE-GCM, to serve the community, and to advance understanding of the thermosphere/ionosphere system.

Acknowledgments. This research was supported by NASA grants NNX08AQ31G and NNX09AJ60G and by the Center for Integrated Space Weather Modeling through NSF agreement ATM-0120950. NCAR is sponsored by the National Science Foundation.

REFERENCES

Bilitza, D., and B. Reinisch (2008), International Reference Ionosphere 2007: Improvements and new parameters, *Adv. Space Res.*, *42*(4), 599–609, doi:10.1016/j.asr.2007.07.048.

Bowman, B. R. (2004), The semiannual thermosphere density variation from 1970 to 2002 between 200–1100 km, paper AAS 2004-174 presented at AAS/AIAA Spaceflight Mechanics Meeting, Maui, Hawaii, 8–12 Feb.

Burns, A. G., T. L. Killeen, and R. G. Roble (1992), Thermospheric heating away from the auroral oval during geomagnetic storms, *Can. J. Phys.*, *70*, 544–552.

Burns, A. G., T. L. Killeen, G. R. Carignan, and R. G. Roble (1995a), Large enhancements in the O/N_2 ratio in the evening sector of the winter hemisphere during geomagnetic storms, *J. Geophys. Res.*, *100*(A8), 14,661–14,671.

Burns, A. G., T. L. Killeen, W. Deng, G. R. Carignan, and R. G. Roble (1995b), Geomagnetic storm effects in the low- to middle-latitude upper thermosphere, *J. Geophys. Res.*, *100*(A8), 14,673–14,691.

Burns, A. G., W. Wang, M. Wiltberger, S. C. Solomon, H. Spence, T. L. Killeen, R. E. Lopez, and J. E. Landivar (2008), An event study to provide validation of TING and CMIT geomagnetic middle-latitude electron densities at the F_2 peak, *J. Geophys. Res.*, *113*, A05310, doi:10.1029/2007JA012931.

Burns, A. G., S. C. Solomon, L. Qian, W. Wang, B. A. Emery, M. Wiltberger, and D. R. Weimer (2012), The effects of corotating interaction region/high speed stream storms on the thermosphere and ionosphere during the last solar minimum, *J. Atmos. Sol. Terr. Phys.*, *83*, 79–87, doi:10.1016/j.jastp.2012.02.006.

Cnossen, I., and A. D. Richmond (2008), Modelling the effect of changes in the Earth's magnetic field from 1957 to 1997 on the ionospheric hmF2 and foF2 parameters, *J. Atmos. Sol. Terr. Phys.*, *70*, 1512–1524.

Crowley, G., D. J. Knipp, K. A. Drake, J. Lei, E. Sutton, and H. Lühr (2010), Thermospheric density enhancements in the dayside cusp region during strong B_Y conditions, *Geophys. Res. Lett.*, *37*, L07110, doi:10.1029/2009GL042143.

Dickinson, R. E., E. C. Ridley, and R. G. Roble (1981), A three-dimensional general circulation model of the thermosphere, *J. Geophys. Res.*, *86*(A3), 1499–1512.

Dickinson, R. E., E. C. Ridley, and R. G. Roble (1984), Thermospheric general circulation with coupled dynamics and composition, *J. Atmos. Sci.*, *41*, 205–219.

Emery, B. A., V. Coumans, D. S. Evans, G. A. Germany, M. S. Greer, E. Holeman, K. Kadinsky-Cade, F. J. Rich, and W. Xu (2008), Seasonal, Kp, solar wind, and solar flux variations in long-term single-pass satellite estimates of electron and ion auroral hemispheric power, *J. Geophys. Res.*, *113*, A06311, doi:10.1029/2007JA012866.

Emmert, J. T., and J. M. Picone (2010), Climatology of globally averaged thermospheric mass density, *J. Geophys. Res.*, *115*, A09326, doi:10.1029/2010JA015298.

Fang, T. W., A. D. Richmond, J. Y. Liu, and A. Maute (2008), Wind dynamo effects on ground magnetic perturbations and vertical drifts, *J. Geophys. Res.*, *113*, A11313, doi:10.1029/2008JA013513.

Fesen, C. G., R. E. Dickinson, and R. G. Roble (1986), Simulation of the thermospheric tides at equinox with the National Center for Atmospheric Research thermospheric general circulation model, *J. Geophys. Res.*, *91*(A4), 4471–4489, doi:10.1029/JA091iA04p04471.

Hagan, M. E., R. G. Roble, and J. Hackney (2001), Migrating thermospheric tides, *J. Geophys. Res.*, *106*(A7), 12,739–12,752, doi:10.1029/2000JA000344.

Hagan, M. E., A. Maute, and R. G. Roble (2009), Tropospheric tidal effects on the middle and upper atmosphere, *J. Geophys. Res.*, *114*, A01302, doi:10.1029/2008JA013637.

Heelis, R. A., J. K. Lowell, and R. W. Spiro (1982), A model of the high-latitude ionospheric convection pattern, *J. Geophys. Res.*, *87*(A8), 6339–6345.

Huba, J. D., G. Joyce, and J. A. Fedder (2000), Sami2 is Another Model of the Ionosphere (SAMI2): A new low-latitude ionosphere model, *J. Geophys. Res.*, *105*(A10), 23,035–23,053, doi:10.1029/2000JA000035.

Knipp, D. J., W. K. Tobiska, and B. A. Emery (2004), Direct and indirect thermosphere heating sources for solar cycle 21-23, *Sol. Phys.*, *224*, 495–505.

Lamarque, J.-F., et al. (2011), CAM-chem: Description and evaluation of interactive atmospheric chemistry in CESM, *Geosci. Model Dev. Discuss.*, *4*, 2199–2278.

Lee, I. T., T. Matsuo, A. D. Richmond, J. Y. Liu, W. Wang, C. H. Lin, J. L. Anderson, and M. Q. Chen (2012), Assimilation of FORMOSAT-3/COSMIC electron density profiles into a coupled thermosphere/ionosphere model using ensemble Kalman filtering, *J. Geophys. Res.*, *117*, A10318, doi:10.1029/2012JA017700.

Lei, J., R. G. Roble, W. Wang, B. A. Emery, and S.-R. Zhang (2007), Electron temperature climatology at Millstone Hill and Arecibo, *J. Geophys. Res.*, *112*, A02302, doi:10.1029/2006JA012041.

Lei, J., J. P. Thayer, A. G. Burns, G. Lu, and Y. Deng (2010), Wind and temperature effects on thermosphere mass density response to the November 2004 geomagnetic storm, *J. Geophys. Res.*, *115*, A05303, doi:10.1029/2009JA014754.

Liu, H.-L., et al. (2010a), Thermosphere extension of the Whole Atmosphere Community Climate Model, *J. Geophys. Res.*, *115*, A12302, doi:10.1029/2010JA015586.

Liu, H.-L., W. Wang, A. D. Richmond, and R. G. Roble (2010b), Ionospheric variability due to planetary waves and tides for solar minimum conditions, *J. Geophys. Res.*, *115*, A00G01, doi:10.1029/2009JA015188.

Lu, G., M. G. Mlynczak, L. A. Hunt, T. N. Woods, and R. G. Roble (2010), On the relationship of Joule heating and nitric oxide radiative cooling in the thermosphere, *J. Geophys. Res.*, *115*, A05306, doi:10.1029/2009JA014662.

Lyon, J. G., J. A. Fedder, and C. M. Mobarry (2004), The Lyon-Fedder-Mobarry (LFM) global MHD magnetospheric simulation code, *J. Atmos. Sol. Terr. Phys.*, *66*(15–16), 1333–1350, doi:10.1016/j.jastp.2004.03.020.

Marsal, S., A. D. Richmond, A. Maute, and B. J. Anderson (2012), Forcing the TIEGCM model with Birkeland currents from the Active Magnetosphere and Planetary Electrodynamics Response Experiment, *J. Geophys. Res.*, *117*, A06308, doi:10.1029/2011JA017416.

Mlynczak, M. G., F. J. Martin-Torres, C. J. Mertens, B. T. Marshall, R. E. Thompson, J. U. Kozyra, E. E. Remsberg, L. L. Gordley, J. M. Russell III, and T. Woods (2008), Solar-terrestrial coupling evidenced by periodic behavior in geomagnetic indexes and the infrared energy budget of the thermosphere, *Geophys. Res. Lett.*, *35*, L05808, doi:10.1029/2007GL032620.

Pedatella, N. M., J. M. Forbes, A. Maute, A. D. Richmond, T.-W. Fang, K. M. Larson, and G. Millward (2011), Longitudinal variations in the *F* region ionosphere and the topside ionosphere-plasmasphere: Observations and model simulations, *J. Geophys. Res.*, *116*, A12309, doi:10.1029/2011JA016600.

Qian, L., and S. C. Solomon (2011), Thermospheric density: An overview of temporal and spatial variations, *Space Sci. Rev.*, *168*, 147–173, doi:10.1007/s11214-011-9810-z.

Qian, L., S. C. Solomon, and T. J. Kane (2009a), Seasonal variation of thermospheric density and composition, *J. Geophys. Res.*, *114*, A01312, doi:10.1029/2008JA013643.

Qian, L., A. G. Burns, S. C. Solomon, and R. G. Roble (2009b), The effect of carbon dioxide cooling on trends in the F2-layer ionosphere, *J. Atmos. Sol. Terr. Phys.*, *71*, 1592–1601, doi:10.1016/j.jastp.2009.03.006.

Qian, L., S. C. Solomon, and M. G. Mlynczak (2010a), Model simulation of thermospheric response to recurrent geomagnetic forcing, *J. Geophys. Res.*, *115*, A10301, doi:10.1029/2010JA015309.

Qian, L., A. G. Burns, P. C. Chamberlin, and S. C. Solomon (2010b), Flare location on the solar disk: Modeling the thermosphere and ionosphere response, *J. Geophys. Res.*, *115*, A09311, doi:10.1029/2009JA015225.

Qian, L., A. G. Burns, P. C. Chamberlin, and S. C. Solomon (2011), Variability of thermosphere and ionosphere responses to solar flares, *J. Geophys. Res.*, *116*, A10309, doi:10.1029/2011JA016777.

Qian, L., A. G. Burns, S. C. Solomon, and P. C. Chamberlin (2012), Solar flare impacts on ionospheric electrodyamics, *Geophys. Res. Lett.*, *39*, L06101, doi:10.1029/2012GL051102.

Richards, P. G., J. A. Fennelly, and D. G. Torr (1994), EUVAC: A solar EUV Flux Model for aeronomic calculations, *J. Geophys. Res.*, *99*(A5), 8981–8992.

Richmond, A. D. (1995), Ionospheric electrodynamics using magnetic apex coordinates, *J. Geomagn. Geoelectr.*, *47*, 191–212.

Richmond, A. D., and Y. Kamide (1988), Mapping electrodynamic features of the high-latitude ionosphere from localized observations: Technique, *J. Geophys. Res.*, *93*(A6), 5741–5759, doi:10.1029/JA093iA06p05741.

Richmond, A. D., and A. Maute (2012), Ionospheric electrodynamics modeling, in *Modeling the Ionosphere-Thermosphere System, Geophys. Monogr. Ser.*, doi:10.1029/2012GM001331, this volume.

Richmond, A. D., E. C. Ridley, and R. G. Roble (1992), A thermosphere/ionosphere general circulation model with coupled electrodynamics, *Geophys. Res. Lett.*, *19*(6), 601–604.

Roble, R. G. (1995), Energetics of the mesosphere and thermosphere, in *The Upper Mesosphere and Lower Thermosphere: A Review of Experiment and Theory, Geophys. Monogr. Ser.*, vol. 87, edited by R. M. Johnson and T. L. Killeen, pp. 1–21, AGU, Washington, D. C., doi:10.1029/GM087p0001.

Roble, R. G., and E. C. Ridley (1987), An auroral model for the NCAR thermosphere general circulation model (TGCM), *Ann. Geophys. Ser. A*, *5*(6), 369–382.

Roble, R. G., and E. C. Ridley (1994), A thermosphere-ionosphere-mesosphere-electrodynamics general circulation model (time-GCM): Equinox solar cycle minimum simulations (30–500 km), *Geophys. Res. Lett.*, *21*(6), 417–420.

Roble, R. G., R. E. Dickinson, and E. C. Ridley (1982), Global circulation and temperature structure of thermosphere with high-latitude plasma convection, *J. Geophys. Res.*, *87*(A3), 1599–1614.

Roble, R. G., E. C. Ridley, and R. E. Dickinson (1987), On the global mean structure of the thermosphere, *J. Geophys. Res.*, *92*(A8), 8745–8758.

Roble, R. G., E. C. Ridley, A. D. Richmond, and R. E. Dickinson (1988), A coupled thermosphere/ionosphere general circulation model, *Geophys. Res. Lett.*, *15*(12), 1325–1328.

Shim, J. S., et al. (2012), Systematic evaluation of ionosphere/thermosphere (IT) models: CEDAR Electrodynamics Thermosphere Ionosphere (ETI) Challenge (2009-2010), in *Modeling the Ionosphere-Thermosphere System, Geophys. Monogr. Ser.*, doi:10.1029/2012GM001293, this volume.

Siskind, D. E., D. P. Drob, J. T. Emmert, M. H. Stevens, P. E. Sheese, E. J. Llewellyn, M. E. Hervig, R. Niciejewski, and A. J. Kochenash (2012), Linkages between the cold summer mesopause and thermospheric zonal mean circulation, *Geophys. Res. Lett.*, *39*, L01804, doi:10.1029/2011GL050196.

Smithtro, C. G., and S. C. Solomon (2008), An improved parameterization of thermal electron heating by photoelectrons, with application to an X17 flare, *J. Geophys. Res.*, *113*, A08307, doi:10.1029/2008JA013077.

Solomon, S. C., and L. Qian (2005), Solar extreme-ultraviolet irradiance for general circulation models, *J. Geophys. Res.*, *110*, A10306, doi:10.1029/2005JA011160.

Solomon, S. C., T. N. Woods, L. V. Didkovsky, J. T. Emmert, and L. Qian (2010), Anomalously low solar extreme-ultraviolet irradiance and thermospheric density during solar minimum, *Geophys. Res. Lett.*, *37*, L16103, doi:10.1029/2010GL044468.

Solomon, S. C., L. Qian, L. V. Didkovsky, R. A. Viereck, and T. N. Woods (2011), Causes of low thermospheric density during the 2007–2009 solar minimum, *J. Geophys. Res.*, *116*, A00H07, doi:10.1029/2011JA016508. [Printed 117(A2), 2012].

Solomon, S. C., A. G. Burns, B. A. Emery, M. G. Mlynczak, L. Qian, W. Wang, D. R. Weimer, and M. Wiltberger (2012), Modeling studies of the impact of high-speed streams and co-rotating interaction regions on the thermosphere-ionosphere, *J. Geophys. Res.*, *117*, A00L11, doi:10.1029/2011JA017417.

Wang, W., M. Wiltberger, A. G. Burns, S. C. Solomon, T. L. Killeen, N. Maruyama, and J. G. Lyon (2004), Initial results from the coupled magnetosphere–ionosphere–thermosphere model: Thermosphere–ionosphere responses, *J. Atmos. Sol. Terr. Phys.*, *66*(15–16), 1425–1441.

Wang, W., J. Lei, A. G. Burns, M. Wiltberger, A. D. Richmond, S. C. Solomon, T. L. Killeen, E. R. Talaat, and D. N. Anderson (2008), Ionospheric electric field variations during a geomagnetic storm simulated by a coupled magnetosphere ionosphere thermosphere (CMIT) model, *Geophys. Res. Lett.*, *35*, L18105, doi:10.1029/2008GL035155.

Wang, W., J. Lei, A. G. Burns, S. C. Solomon, M. Wiltberger, J. Xu, Y. Zhang, L. Paxton, and A. Coster (2010), Ionospheric response to the initial phase of geomagnetic storms: Common features, *J. Geophys. Res.*, *115*, A07321, doi:10.1029/2009JA014461.

Wang, W., J. Lei, A. G. Burns, L. Qian, S. C. Solomon, M.Wiltberger, and J. Xu (2011), Ionospheric day-to-day variability around the whole heliosphere interval in 2008, *Sol. Phys.*, *274*, 457–472, doi:10.1007/s11207-011-9747-0.

Weimer, D. R. (2005), Improved ionospheric electrodynamic models and application to calculating Joule heating rates, *J. Geophys. Res.*, *110*, A05306, doi:10.1029/2004JA010884.

Wiltberger, M., W. Wang, A. Burns, S. Solomon, J. G. Lyon, and C. C. Goodrich (2004), Initial results from the coupled magnetosphere ionosphere thermosphere model: Magnetospheric and ionospheric responses, *J. Atmos. Sol. Terr. Phys.*, *66*, 1411–1423.

Zhang, S.-R., J. M. Holt, A. P. van Eyken, M. McCready, C. Amory-Mazaudier, S. Fukao, and M. Sulzer (2005), Ionospheric local model and climatology from long-term databases of multiple incoherent scatter radars, *Geophys. Res. Lett.*, *32*, L20102, doi:10.1029/2005GL023603.

A. G. Burns, B. A. Emery, B. Foster, G. Lu, A. Maute, L. Qian, A. D. Richmond, R. G. Roble, S. C. Solomon, and W. Wang, National Center for Atmospheric Research, High Altitude Observatory, Boulder, CO, USA. (lqian@ucar.edu)

The Global Ionosphere-Thermosphere Model and the Nonhydrostatic Processes

Yue Deng

Department of Physics, University of Texas at Arlington, Texas, USA

Aaron J. Ridley

Center for Space Environment Modeling, The University of Michigan, Ann Arbor, Michigan, USA

The recently developed global ionosphere-thermosphere model (GITM) uses a 3-D spherical grid that can be stretched in both latitude and altitude. GITM is nontraditional in that it uses an altitude-based grid and does not assume a hydrostatic solution. Using GITM, the primary characteristics of nonhydrostatic effects on the upper atmosphere are investigated. Our results show that after a sudden intense enhancement of high-latitude Joule heating, the vertical pressure gradient force can locally be 25% larger than the gravity force, resulting in a significant disturbance away from hydrostatic equilibrium. This disturbance is transported from the lower-altitude source region to high altitudes through an acoustic wave. The magnitude of the vertical wind perturbation increases with altitude and reaches 150 (250) m s^{-1} at 300 (430) km, which is not typically reproduced by hydrostatic models. The source of nonhydrostatic effects and the sensitivity of the vertical wind and neutral density at low satellite orbits to the energy deposited at low and high thermosphere have been investigated as well. The simulations show that the atmosphere at all altitudes is out of hydrostatic equilibrium after the sudden energy enhancement. But the maximum of the nonhydrostatic effects at high altitudes (300 km) arises from sources below 150 km and propagates vertically through the acoustic wave. The heating above 150 km is responsible for a large increase of the average vertical velocity (40 m s^{-1}) and neutral density (50%) at 300 km and higher altitudes.

1. INTRODUCTION

The ionosphere and thermosphere are two overlapping regions in the upper atmosphere, which are tightly coupled to each other. Both local and global models are very important for a number of upper atmosphere researches and space weather applications. Different types of model have been developed in the community to describe the thermosphere and ionosphere.

The empirical models include the mass spectrometer and incoherent scatter (MSIS) model [*Hedin*, 1983, 1987, 1991] for the thermosphere and the International Reference Ionosphere (IRI) model for the ionosphere. They are spherical harmonic fits to many different satellite and remote observations. The description is based on a large amount of data, and they represent the average conditions for the given inputs. The Global Assimilation of Ionospheric Measurements (GAIM) [*Schunk et al.*, 2004] uses a time-dependent physics-based model of the global ionosphere-plasmasphere and a Kalman filter as a basis for assimilating a diverse set of ionospheric measurements. The primary output of GAIM is a continuous reconstruction of the 3-D electron density distribution.

First-principles-based models simulate the thermosphere and ionosphere by self-consistently determining the density, momentum, and energy from the equations. One of the first global models of this type was the thermosphere general circulation model [*Dickinson et al.*, 1981, 1984]. It solves for mass mixing ratios of the neutral major species O_2, N_2, and O and the minor species, and is a full 3-D code with 5° latitude by 5° longitude by 0.5 scale height altitude cells. The second model of this line is thermosphere-ionosphere general circulation model (TIGCM), which includes a self-consistent ionosphere [*Roble et al.*, 1988]. The O^+ dynamics are considered, while the species O_2^+, N_2^+, NO^+, and N^+ are assumed to be in photochemical equilibrium. *Richmond et al.* [1992] added a self-consistent low-latitude electrodynamics to the TIGCM, resulting in the thermosphere-ionosphere electrodynamic general circulation model (TIEGCM). *Roble and Ridley* [1994] extended the model down into the mesosphere (the thermosphere-ionosphere-mesosphere electrodynamic general circulation model (TIMEGCM)), which added a large amount of chemistry. The minor species transport equation for N(4S) and NO has been extended downward as well. A different first-principles model is the coupled thermosphere ionosphere model (CTIM) [*Fuller-Rowell and Rees*, 1980, 1983; *Rees and Fuller-Rowell*, 1988, 1990]. This model was extended to include the plasmasphere, called the coupled thermosphere ionosphere plasmasphere by modeling the ionosphere along field lines, improving the mass flow between hemispheres and the upper boundary condition on the ion flow.

In this study, we present a relatively newly developed global ionosphere-thermosphere model (GITM) [*Ridley et al.*, 2006]. It differs from other first-principles models in several respects. Some of the important features of GITM include the adjustable resolution, nonuniform grid in the altitude and latitude coordinates, and the dynamics equations solved without the assumption of hydrostatic equilibrium. The primary characteristics of nonhydrostatic process and its influence on the upper atmosphere have been investigated. The source of nonhydrostatic effects and the sensitivity of the vertical wind and neutral density at low satellite orbits to the energy deposited at low and high thermosphere have been studied as well.

2. GLOBAL IONOSPHERE-THERMOSPHERE MODEL

GITM is a 3-D spherical code [*Ridley et al.*, 2006] that models the Earth's thermosphere and ionosphere system using a stretched grid in latitude and altitude. In addition, the number of grid points in each direction can be specified, so the resolution is extremely flexible. GITM explicitly solves for the neutral densities of O, O_2, $N(^2D)$, $N(^2P)$, $N(^2S)$, N_2, NO, H, and He, and ion species $O^+(^4S)$, $O^+(^2D)$, $O^+(^2P)$, O_2^+, N^+, N_2^+, NO^+, H^+, and He^+. The equations for the neutral constituents are based on the Navier-Stokes equations and are listed in the Appendix. The ion momentum equation is solved for assuming steady state, taking into account the pressure, gravity, neutral winds, and external electric fields. GITM has chemistry between the ions and neutrals, ions and electrons, and neutral and neutrals. The primary source of ions on the dayside is due to the solar EUV, and the cross sections for the chemical reactions are specified by *Torr et al.* [1979]. The chemistry within GITM includes all of the chemical equations in the work of *Rees and Fuller-Rowell* [1988]. Alternatively, the chemical equations in the work of *Torr et al.* [1979] can be used. Ion production rates due to auroral electron precipitation are derived from the formulation described by *Frahm et al.* [1997], and the partitioning of the electron precipitation ionization rates are from *Rees* [1989].

One major difference between GITM and other thermosphere codes is the use of an altitude grid instead of a pressure grid. The altitude spacing is done automatically using scale heights and specifying a lower and upper boundary and the number of grid points. Meanwhile, an altitude-dependent gravity acceleration ($\vec{g}$) has been used in GITM [*Deng et al.*, 2008b]. The model is fully parallel using a block-based 2-D domain decomposition with latitudinal and longitudinal ghost cells bordering the blocks [*Oehmke and Stout*, 2001; *Oehmke*, 2004]. This allows parallel computation over the entire domain, exchanging information only between each iteration. GITM uses the Message Passing Interface standard to allow for platform independence.

The simulations can be initiated using MSIS and IRI neutral and ion densities and temperatures for the given date and time. For more theoretical simulations, the code can be initiated with each latitude and longitude having the same height profile from MSIS and IRI. GITM can be restarted from a previous run, so simulations of very long time periods can be conducted without any break in the physics. Having the ability to both restart and start from a base state allows maximum flexibility in the usefulness of the code. The model can use a dipole or the International Geomagnetic Reference Field (IGRF) magnetic field with the Apex coordinate system [*Richmond*, 1995]. This allows experiments ranging from highly idealized magnetic field topological cases to realistic magnetic field cases. The IGRF magnetic field can be exchanged with other systems, so simulations of paleomagnetic conditions could be examined, for example. GITM solves the nonhydrostatic equations and can model problems where the hydrostatic approximation is not accurate. This feature of the model is discussed in section 3.

GITM can be coupled to a large number of models of the high-latitude ionospheric electrodynamics. For example, we can run GITM using results from the assimilative mapping of ionospheric electrodynamics technique [*Richmond and Kamide*, 1988; *Richmond*, 1992] as high-latitude drivers in realistic, highly dynamic time periods. We can also use the *Weimer* [1996], *Foster et al.* [1983], *Heppner and Maynard* [1987], and *Ridley et al.* [2000] electrodynamic potential patterns and the *Hardy et al.* [1987] or *Fuller-Rowell and Evans* [1987] particle precipitation patterns for more idealized conditions. GITM is also part of the University of Michigan's Space Weather Modeling Framework [*Tóth et al.*, 2005], so it can be coupled with a global MHD model [*Powell et al.*, 1999] of the magnetosphere. This allows investigation of the coupling between the thermosphere-ionosphere and the magnetosphere systems [e.g., *Ridley et al.*, 2003].

3. NONHYDROSTATIC PROCESSES

Many theoretical thermosphere/ionosphere models have been developed since the 1970s, including TGCM [*Dickinson et al.*, 1981], CTIM [*Fuller-Rowell and Rees*, 1980], and their later variants. One common assumption used in these models is the hydrostatic equilibrium, under which the pressure gradient force is balanced with the gravity force in the vertical direction:

$$\frac{\partial P}{\partial r} = -\rho g, \quad (1)$$

where P is the pressure, r is the radial distance, ρ is the mass density, and g is the gravitation acceleration. If the hydrostatic approximation is relaxed, the vertical momentum can be expanded to:

$$\frac{\partial u_r}{\partial t} + \mathbf{u} \cdot \nabla u_r + \frac{1}{\rho}\frac{\partial P}{\partial r} = -g + F_f + F_c, \quad (2)$$

where u_r is the vertical component of the neutral wind, $\mathbf{u}$ is the neutral wind vector, F_f contains the forces due to ion-neutral and neutral-neutral friction (when each constituent is solved independently), and F_c contains the centrifugal and Coriolis forces. Compared with equation (1), equation (2) has more terms to bring the nonhydrostatic effects in the system, which can propagate vertically. Including the hydrodynamic processes is beneficial to truly understand both Earth and other planetary atmospheres. Some studies have been done in the low atmosphere (troposphere) using nonhydrostatic models, such as the Weather Research and Forecasting model. Using the GITM, the nonhydrostatic effect on the upper atmosphere has been investigated and quantified as well [*Deng et al.*, 2008a, 2011].

Many papers have reported strong vertical winds (more than 100 m s^{-1}) in observations from DE 2 satellite [*Innis and Conde*, 2002] and Fabry-Perot interferometers (FPI) measurements [*Smith and Hernandez*, 1995; *Aruliah et al.*, 2005]. Such large vertical winds cause strong disturbance of neutral density in the upper atmosphere, which can dramatically alter low-altitude satellite orbits through increasing the atmospheric drag on the satellites. The proposed drivers for the large vertical wind include localized heating [*Price et al.*, 1995], divergence in the horizontal wind [*Smith and Hernandez*, 1995], and acoustic-gravity waves [*Innis and Conde*, 2002]. However, the vertical winds in hydrostatic general circulation models (GCMs), calculated from the divergence of the horizontal wind field, are usually less than 50 m s^{-1} and much smaller than the observed values. Currently, there is no conclusive interpretation about the large vertical winds in the observations, and the nonhydrostatic process associated to the imbalance between gravity and the gradient pressure is one possible mechanism.

Fully nonhydrostatic GCM formulations are also beneficial for whole atmosphere modeling by allowing the acoustic waves [*Akmaev*, 2011]. Upper atmospheric nonhydrostatic models [*Chang and St.-Maurice*, 1991; *Deng et al.*, 2008a, 2011] showed that sudden enhancements of Joule heating during geomagnetically disturbed periods generate intensive acoustic waves associated with strong and highly variable vertical winds. Meanwhile, acoustic waves propagating from the lower atmosphere have also been shown to deposit substantial amounts of energy in the thermosphere [*Hickey et al.*, 2001; *Rind*, 1977]. In the work of *Rind* [1977], infrasound of 0.2 Hz known as microbaroms, generated by interfering ocean waves, propagates into the lower thermosphere and deposits energy of the order of 0.33 W kg^{-1}, the same as that estimated for gravity wave dissipation and capable of producing a heating of at least 30 K d^{-1}. Meanwhile, the gravity waves allowed to propagate vertically in a hydrostatic atmosphere may become internally reflected or ducted under the same conditions in a more realistic nonhydrostatic case [*Akmaev*, 2011], which indicates that the hydrostatic approximation potentially results in overestimation of the gravity wave vertical momentum flux and momentum deposition rates. The nonhydrostatic process may play an important role in the high-frequency gravity wave propagation.

3.1. Results: Buoyancy Acceleration and Vertical Winds

GITM has been run for 30 h of simulation time, reaching a quasisteady state with quiet geomagnetic conditions (interplanetary magnetic field (IMF) $B_z = -1$ nT, hemispheric power = 3 GW, and $F10.7 = 100 \times 10^{-22}$ w m^{-2}Hz^{-1}). The IMF B_z is then changed to -20 nT with the other input parameters remaining the same, and 00 UT in Figure 1 represents the time when B_z drops to -20 nT. The cross polar cap potential (CPCP) correspondingly increases from 45 to 158 kV in the Southern Hemisphere, as shown in Figure 1. In our simulation, the high-latitude electric potential is specified by the Weimer05 empirical model, which is driven by the solar wind conditions and does not include the time lap between the IMF and CPCP. Owing to the correlation between IMF B_z and the energy input from the magnetosphere to the ionosphere [*Deng and Ridley*, 2007], the hemispheric-integrated Joule heating increases abruptly by 19 times and creates a significant disturbance in the thermosphere. After 1 h, IMF B_z returns to the previous value (-1 nT), driving the CPCP and Joule heating to change back to the quiet condition.

Temporal variations of the buoyancy acceleration $\left(-\dfrac{1}{\rho}\dfrac{\partial P}{\partial r}-g\right)$ and the vertical wind distribution in the

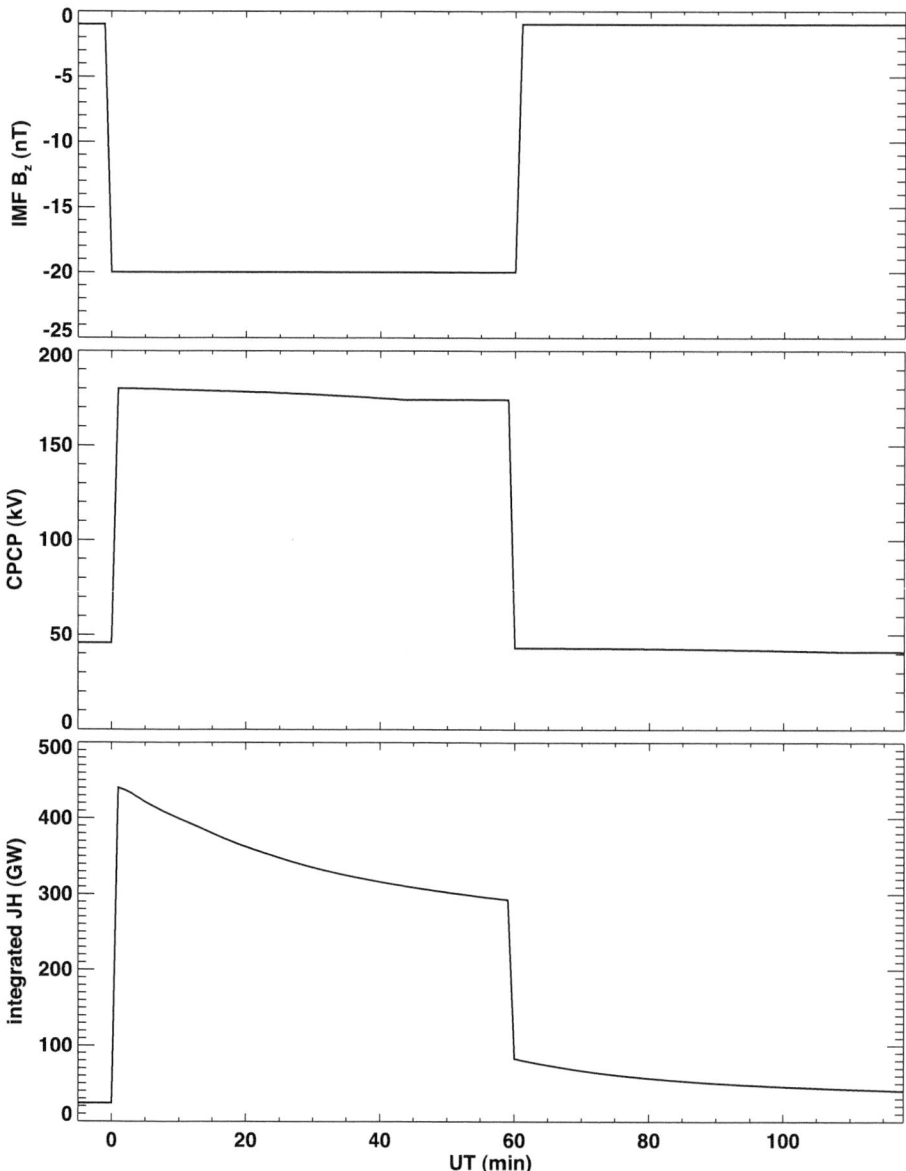

Figure 1. Temporal variation of (top) interplanetary magnetic field B_z, (middle) CPCP and (bottom) hemispheric integrated Joule heating in the Southern Hemisphere.

Southern Hemisphere at 300 km altitude are shown in Figure 2. Both buoyancy acceleration and vertical wind are minimal at time 0, which represents the quiet time. At the third minute, the buoyancy acceleration shows a large upward value at locations with enhanced Joule heating. However, at the fifth and seventh minutes, the buoyancy acceleration changes to be downward and maximizes at positions that correspond with regions of large upward wind seen at the third minute. From the 15th minute to the 55th minute, the buoyancy acceleration returns to the quiet value and changes a little. In general, the buoyancy acceleration occurs immediately after the energy enhancement, but is short-lived since hydrostatic equilibrium quickly reasserts itself despite continued forcing. In response to the variation of buoyancy acceleration, the vertical wind increases dramatically during the first 5 min and decreases after that. Since the vertical wind change is equal to the time integration of the acceleration, the maximum vertical wind (fifth minute) happens just after the maximum buoyancy acceleration (third minute). From the 15th to the 55th minute, the variation of the vertical wind is decelerating slowly with time. The dayside continues to have relatively large vertical winds compared with the quiet time (0 min), even though the buoyancy acceleration is almost zero.

In order to investigate the vertical propagation of the perturbation, the temporal variations of the altitude profiles of the buoyancy acceleration and vertical wind at a specific position (77.5°S, 22.5°E) is presented in Figure 3. It is close to 06:00 LT and is shown in Figure 2a. This particular

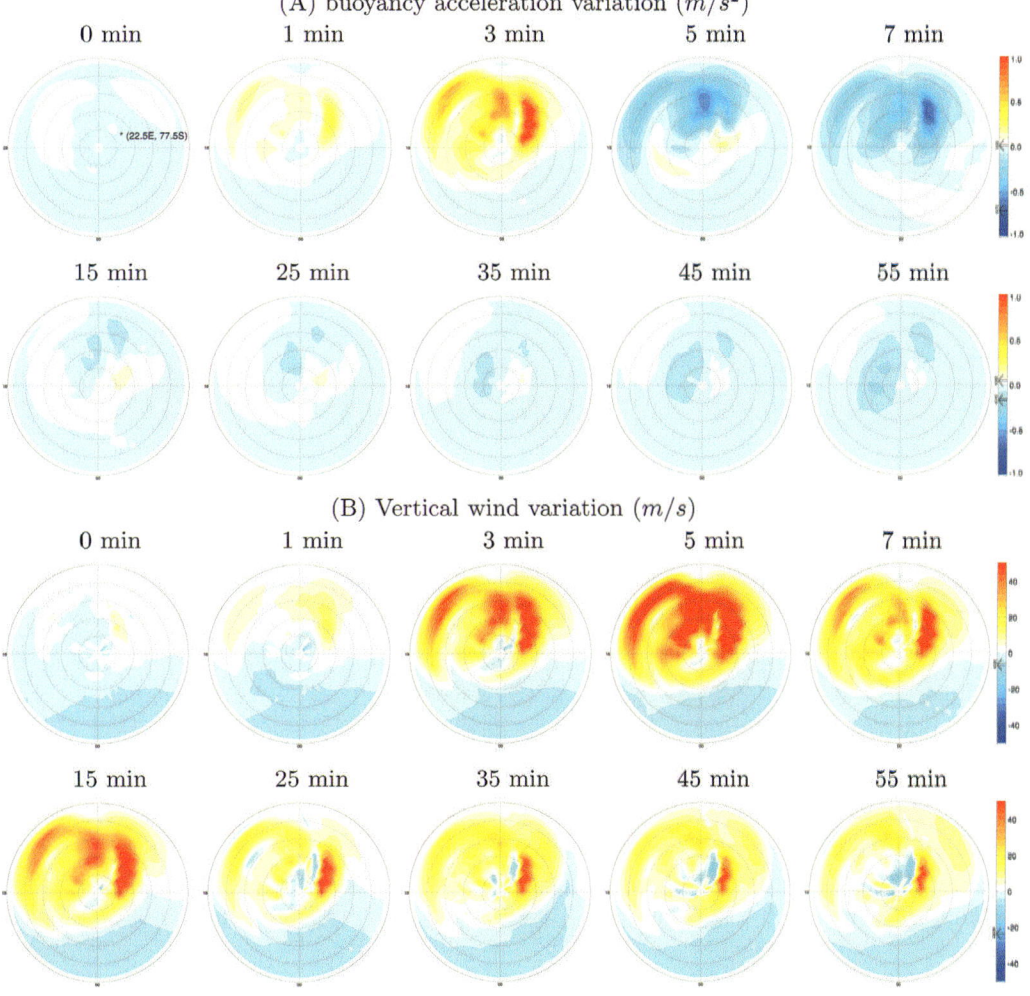

Figure 2. Temporal variation of the distributions of (a) buoyancy acceleration $\left(-\frac{1}{\rho}\frac{\partial P}{\partial r}-g\right)$ (m s^{-2}) and (b) vertical wind (m s^{-1}) in the Southern Hemisphere at 300 km altitude. The outside ring is −40°, and the time is indicated at the top of each distribution. (left) Dusk and (top) noon.

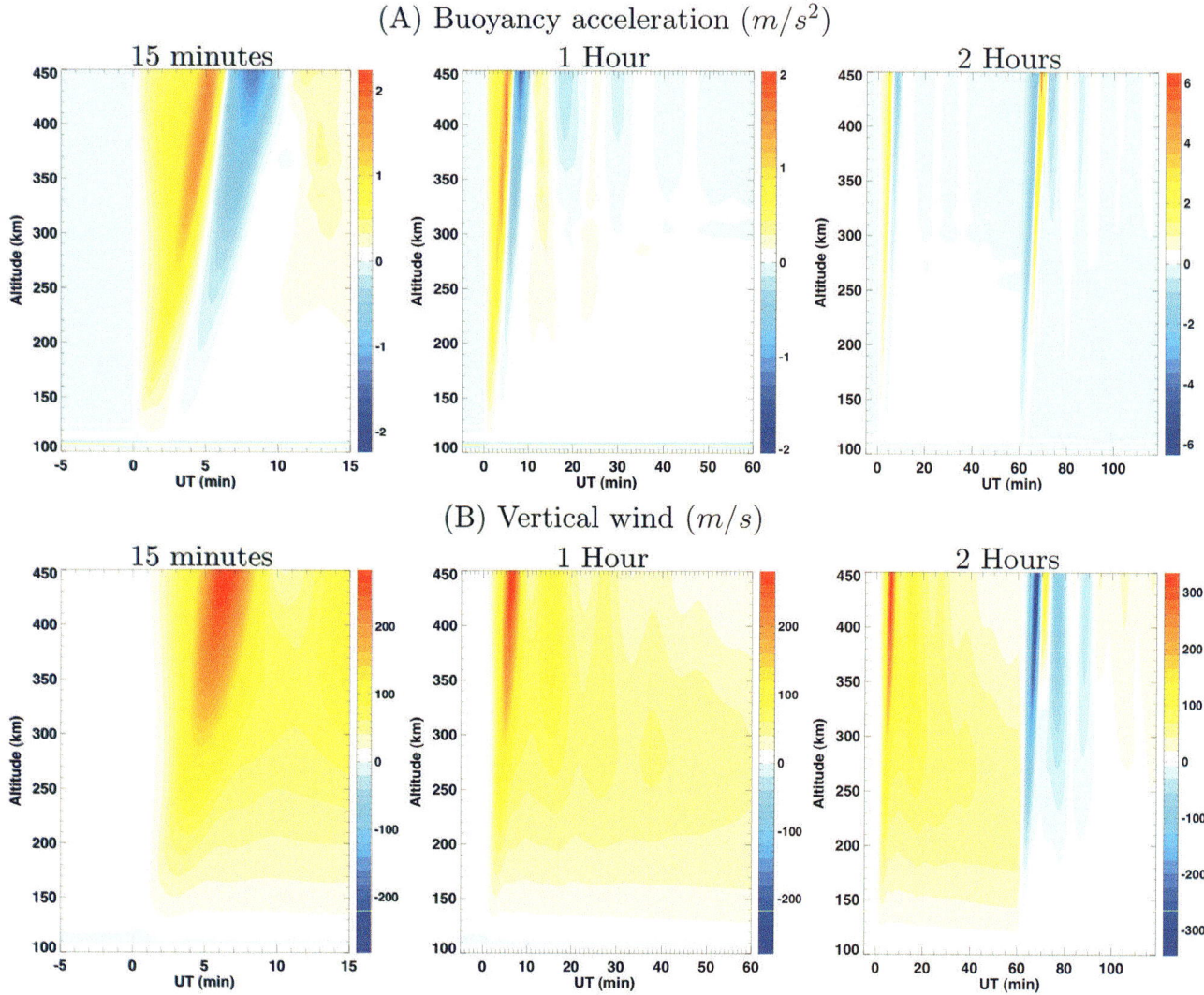

Figure 3. The time versus altitude distribution of (a) buoyancy acceleration (m s^{-2}) and (b) vertical wind (m s^{-1}) at 77.5°S, 22.5°E during 15 min, 1 h, and 2 h time intervals. The location has been shown in Figure 2a.

location has been chosen because it is in the region with the maximum Joule heating [*Deng and Ridley*, 2007] and buoyancy acceleration and the longest period of large upward neutral wind, as shown in Figure 2b. Figure 3a shows that during the first minute, the buoyancy acceleration increases at all altitudes. Additionally, there is a clear positive disturbance propagating from low altitudes to high altitudes during the 1st–6.5th minutes with the propagation speed of 1000 m s^{-1}. It is well known that the peak altitude of the Joule heating is close to 120 km altitude [*Deng and Ridley*, 2007; *Thayer et al.*, 1995]. Owing to the conservation of the perturbation energy and the exponential decrease of the mass density, the disturbance caused by the sudden enhancement of energy input is transported to high altitudes with an ever-magnifying amplitude. While the maximum buoyancy acceleration at 200 km altitude is close to 0.6 m s^{-2}, at 430 km, it can reach 2 m s^{-2}, which is close to 25% of gravity acceleration at this altitude (8.7 m s^{-2}). The lifetime of the disturbance also shows some altitude dependence. At 150 km, the buoyancy force is disturbed from the first to the fifth minute, while at 400 km, it expands to the 1st to the 10th minute. Overall, the buoyancy acceleration returns back to a small value (<0.2 m s^{-2}) after the 11th minute.

The large value of the buoyancy disturbance accelerates the thermosphere vertically, resulting in a significant vertical neutral wind. Figure 3b shows that the magnitude of vertical neutral wind at 430 km altitude reaches 250 m s^{-1} at the sixth minute, which is 1.5 min after the buoyancy acceleration maximum. These large vertical winds are very unlikely to be seen in hydrostatic models, but actually they have been observed in observations, such as by DE 2 [*Innis and Conde*, 2002] and FPIs [*Smith and Hernandez*, 1995; *Aruliah et al.*, 2005]. One possible reason for this large vertical wind is the nonhydrostatic effect, which is caused by the sudden intense enhancement of the energy. The buoyancy acceleration is negative during the 6th to 10th minutes at 430 km; thus, the upward vertical wind actually decreases with time after the sixth minute. From the 10th minute, while the buoyancy acceleration is almost zero, there is still a strong upward vertical wind with the magnitude close to 100 m s^{-1}.

Owing to the large vertical wind, the diffusive dissipation caused by the molecular viscosity and the thermal conduction might conceivably be significant. However, the difference between two runs with and without the vertical viscosity (not shown) in the momentum equation is quite small. This is because molecular viscosity acceleration is equal to $\eta \nabla 2 \vec{U}$, where η is the kinematic viscosity coefficient, and $\vec{U}$ is the velocity. When the velocity is treated as a vertically propagating wave, the viscous damping effect is dependent on $\eta k_z^2 = \eta \frac{4\pi^2}{\lambda_z^2}$, where k_z is the wave number, and λ_z is the wavelength in the vertical direction. Since the vertical wavelength of the acoustic wave is large (200 km as shown in Figure 2), $\eta \frac{4\pi^2}{\lambda_z^2}$ is close to 2 orders of magnitude smaller than the wave frequency (ω = 0.02 rad s^{-1} shown later) around 300 km altitude, and the viscous damping is not significant. Damping by thermal conduction is of the same order as viscous damping [*Pitteway and Hines*, 1963].

The middle panel in Figure 3b shows that the vertical wind decreases slowly with time during the 1 h period when B_z stays fixed at −20 nT. The lifetime of large vertical wind (>50 m s^{-1}) is close to the period of the enhanced forcing (1 h) and much longer than that of the buoyancy acceleration disturbance (<11 min), which is caused by the variation of the pressure gradient. While the nonhydrostatic phenomena in the force term can only be sustained for a short time in the system, the vertical neutral winds lives longer and significantly influences the upper atmosphere. The upward vertical wind lifts the atmosphere and leads to a very large mass density increase at a fixed altitude. For example, at 400 km altitude, the density increases by a factor of 2 (not shown), which dramatically alters the drag on low-altitude satellites.

In the 1960s, some studies were done of acoustic gravity waves in the thermosphere, especially about the wave propagation from nuclear explosions [*Pfeffer and Zarichny*, 1962] and the ion drag effect on propagation [*Hines and Hooke*, 1970]. However, vertically propagating acoustic waves cannot be resolved in hydrostatic models because force balance between the pressure gradient and gravity has been assumed and used to replace the vertical momentum equation. GITM solves the whole vertical momentum equation and technically has the capability to simulate vertically propagating acoustic wave. The simulation results show that the disturbance propagation before the 10th minute is highly likely to be an acoustic wave from the phase speed, propagating direction, and frequency. As calculated earlier, the phase speed is 1000 m s^{-1}, which is very close to the acoustic speed (a_s). Both the energy and phase propagate upward, which is consistent with the dispersion relationship of acoustic waves. The frequency is calculated from the time interval between the negative and positive disturbance peaks, which occurred at times of 5.75 and 8.5 min at 430 km altitude, respectively. The difference between them represents the half period, and the frequency ($\omega = 2\pi/T$) is equal to 0.02 rad s^{-1}. The Brunt-*Väisälä* frequency $\left(N = \sqrt{(\gamma-1)\frac{g^2}{a_s^2}}\right)$ is close to 0.0075 rad s^{-1} at altitudes above 200 km. Hence, $\omega \approx 3N$, which is in the correct frequency range of an acoustic wave. When the variation of the buoyancy acceleration and the vertical wind in a 1 h time period are examined (middle column of Figure 3), there are some oscillations after the 10th minute, which happen simultaneously at different altitudes and do not propagate in altitude. The oscillation frequency is close to the Brunt-*Väisälä* frequency, and they are more likely to be a buoyancy oscillation, instead of an acoustic wave. It is clear that after the sudden enhancement of the energy input, oscillations with different frequencies have been triggered as a ringing effect. To resolve the horizontal-propagating acoustic waves, the horizontal resolution of the simulation should be close to subdegree, which is much higher than 5° by 5° used in this work. Therefore, the horizontal propagation has been neglected in this study.

3.2. Results: Altitudinal Dependence

After a sudden enhancement of the ion convection, Joule heating increases at all altitudes due to the enhancement of the difference between the ion drift and neutral wind velocity. The thermospheric response at higher altitudes is thus affected by energy changes at both E and F region altitudes through the thermal expansion and atmospheric upwelling [*Prölss*, 1981; *Lühr et al.*, 2004]. It is, however, not clear whether the vertical wind and the neutral density variation in the low Earth satellite orbits is primarily caused by the energy deposited at low altitudes or at high altitudes. In this study, GITM has been employed to investigate the source of nonhydrostatic effects [*Deng et al.*, 2008a] and sensitivity of the vertical wind and neutral density at satellite orbits to the energy deposited at both low and high altitudes in the aurora region after a significant heating increase.

In order to separate the impact of energy input at low and high altitudes, we have compared two idealized runs with (case 1) or without (case 2) Joule heating enhancement above the cutting altitude at 150 km. In case 1, Joule heating is calculated with the enhanced ion convection at all altitudes. In case 2, Joule heating below 150 km has been calculated in the same way as case 1, while Joule heating above 150 km is specified with the quiet time values. Case 2 represents the situation when the heating enhances at low altitudes alone. The difference between case 1 and case 2 is similar to the case when the heating enhances at high altitudes alone (not shown). We set the cutting altitude at 150 km because it roughly separates the E region and the F region. Figures 4 and 5 show the temporal variation of the altitudinal distribution of Joule heating, buoyancy acceleration, vertical velocity, and neutral density at 77.5°S, 22.5°E in both cases. The line plots represent the temporal variations at 300 km altitude.

Figure 4a shows that the maximum Joule heating in case 1 is at 120 km altitude and increases from 7.9×10^{-8} to 1.3×10^{-6} W m^{-3} when the ion convection increases. The Joule heating at 300 km is almost 2 orders of magnitude smaller and also increases with the ion convection. Our calculation shows that approximately 25% of the total Joule heating is deposited above 150 km, and 75% is deposited below that level. For example, at 06:10 UT, the total altitude-integrated Joule heating is close to 0.03 W m^{-2}, and the integrated Joule heating from the lower boundary to 150 km is ~0.022 W m^{-2}. The total Joule heating in case 2 is ~25% smaller than that in case 1 due to the method difference. The buoyancy acceleration $\left(-\frac{1}{\rho}\frac{\partial P}{\partial r}-g\right)$, the difference between the pressure gradient force per unit mass and gravity acceleration, is equal to zero under the hydrostatic assumption, which approximately holds before 06:00 UT, as shown in Figure 4b. But it increases at all altitudes after the energy enhancement for both cases, and the value is larger than 2.0 m s^{-2} near the upper boundary. The gravitational acceleration at 400 km altitude is ~8.7 m s^{-2}, and the pressure gradient force per unit mass is thus ~10.7 m s^{-2}, which is 23% larger than the gravity acceleration. Both cases show that after the sudden enhancement of Joule heating, a strong disturbance of buoyancy acceleration propagates vertically with exponentially increasing magnitude resulting in the maximum disturbance at high altitudes. The significant difference between the two cases is the disturbance before the positive maximum. Owing to the heating above 150 km, there is a clear positive disturbance between 06:00 and 06:03 UT at 300 km in case 1, but not in case 2. The difference at 06:03 UT is more than 1.0 m s^{-2}, which results in large differences in the vertical velocity.

The vertical velocity is related to a temporal integration of the vertical acceleration. Figure 5a shows that the maximum vertical wind at 300 km in case 1 is above 150.0 m s^{-1} and almost 3 times larger than that in case 2. Both cases have similar temporal variations at 300 km, and the biggest difference between them is the 1 h average vertical velocity, which is 43.6 m s^{-1} in case 1 and 3.5 m s^{-1} in case 2. This difference indicates that the heating above 150 km (25% of the total energy) and the corresponding nonhydrostatic processes have a stronger influence on the atmosphere at high altitudes than the heating below 150 km (75% of the total energy). The heating at high altitudes sets up a large vertical wind, which is overlapped by a disturbance propagating from the lower altitudes. While the acoustic wave propagating vertically from lower altitudes can cause a large buoyancy acceleration at 300 km, it mainly imposes some temporal variation on the vertical velocity. Figure 5b shows that case 1 also has a much larger density increase than case 2, and the average difference between them at 300 km altitude is ~50% of the background value. Since the thermosphere/ionosphere is a nonlinear system, the variation of neutral density causes additional changes in the system, such as the absorption of solar irradiation, neutral dynamics, and ionospheric density, which in turn feedback to the neutral density. In this paper, the time scale we are concerned with is the first hour after the energy enhancement, and all secondary effects on the neutral density are not discussed. Figures 4b and 5a show a reflection at the upper boundary, which happens in most simulations with general boundary conditions when no addition wave damping layer has been added on top. It is numerical and not physically meaningful. Since the neutral density exponentially decreases with altitude, and the wave is damped quickly when propagating downward, this reflection is smaller than the disturbance propagating from lower altitudes and will not change the main conclusion of our study.

Figures 4 and 5 only represent the results at one specific location (77.5°S, 22.5°E). In order to examine the neutral density variation in the whole high-latitude region, the polar distributions at 07:00 UT for both cases have been shown in Figure 6. The top panel depicts the neutral density and neutral wind vectors at 130 km altitude in the Southern Hemisphere. Clearly, there are a density maximum peak at −55° latitude and 14 LT and a minimum peak at −75° latitude and 11 LT in case 1. The density maximum and minimum peaks are overlapped with the vortex centers very well in both cases. This two-peak structure in the neutral density has been reported in both observations [*Kwak et al.*, 2009] and simulations [*Crowley et al.*, 2006] and explained as the results of the neutral dynamics. When the ion convection is enhanced, the increase of cyclonic (anticyclonic) wind

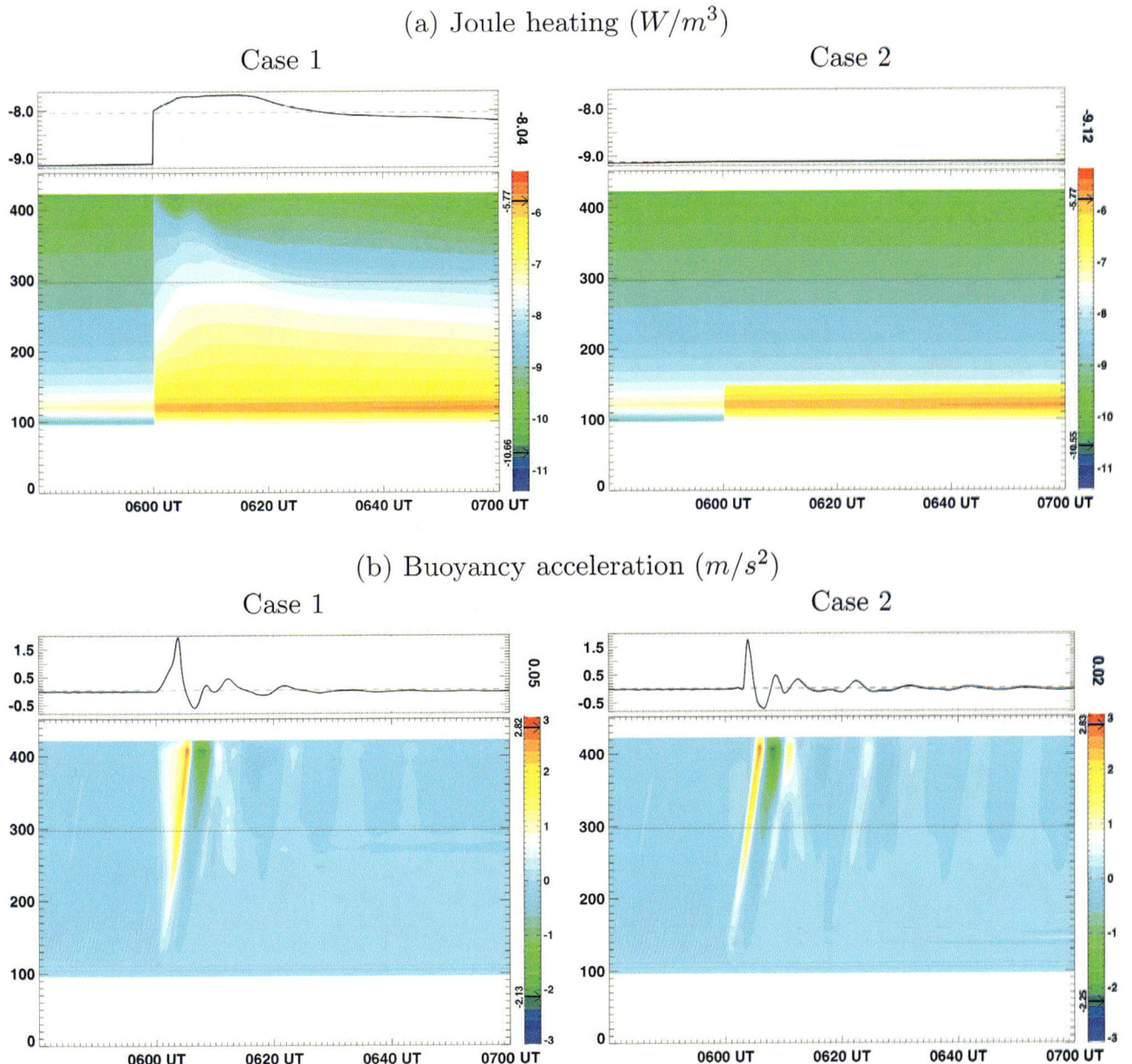

Figure 4. (a) The temporal variation of the altitudinal profiles of Joule heating per unit volume (W m^{-3}) at (77.5°S; 22.5°E) for (left) case 1 and (right) case 2. The numbers and arrows attached to the color bars indicate the minimum and maximum values in the color contour. (top) The line plots show the variations at 300 km altitude, and (right) the numbers represent the average value of the corresponding line plots. (b) is the same as (a), except for the buoyancy acceleration $\left(-\frac{1}{\rho}\frac{\partial P}{\partial r} - g\right)$ (m s^{-2}). The Joule heating (a) is plotted on a logarithmic scale, and the buoyancy acceleration (b) is plotted on a linear scale.

vortex on the dawnside (duskside) tends to have a low (high) pressure and density at its center.

At 300 km altitude, the two-peak structure is persistent, and two cases have similar distributions, except that the magnitude of the neutral density in case 2 is consistently smaller than that in case 1. The percentage difference of the neutral density compared with the background run, in which the IMF B_z is kept as constant ($B_z = -1.0$ nT), has been calculated. As shown in Figure 6c, the percentage difference of neutral density in the whole polar region is

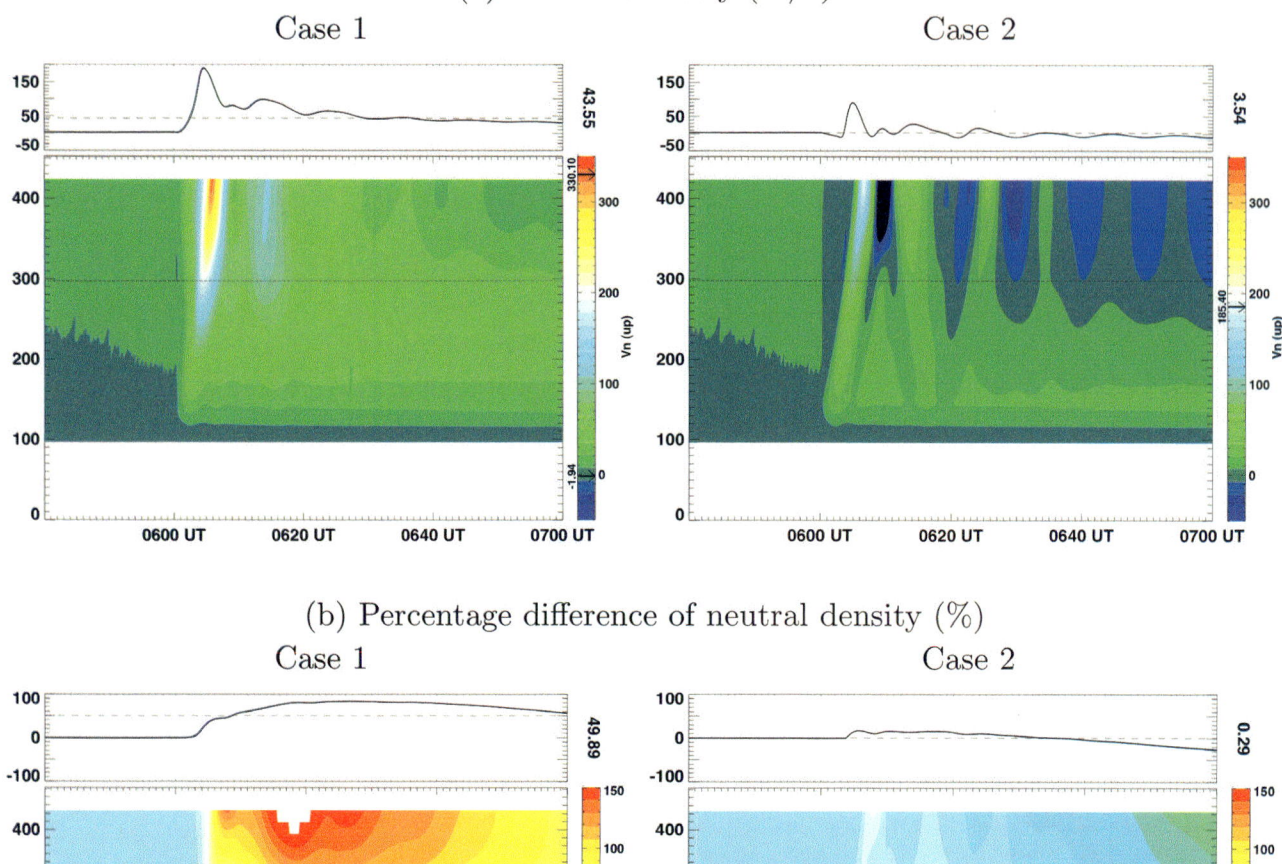

Figure 5. The same as Figure 4, except for the vertical velocity (m s^{-1}) (a), and the percentage difference of neutral density compared with the value in the background case (b), respectively.

positive in case 1. The distribution mimics the structure shown in Figure 6a, with a minimum peak at 11 LT and a maximum peak at 13 LT. Case 2 has a very similar distribution as case 1 except that the magnitude is, on average, 50% smaller. In case 2, in order to separate the effect of Joule heating from other mechanisms, Joule heating above 150 km has been specified as the background values, and all other parameters change with IMF variation. This method creates inconsistency between the Joule heating and convection.

Figure 6. (opposite) (a) Neutral density (kg m^{-3}) distribution at 130 km altitude at 07:00 UT in the Southern Hemisphere. The neutral wind velocity vectors are plotted out on top of the neutral density, and the outside ring is −40°C. (b) Neutral density (kg m^{-3}) distribution at 300 km altitude at 07:00 UT in the Southern Hemisphere. The logarithmic scale has been used for (a) and (b). (c) The same as (b) except for the percentage difference of neutral density at 300 km compared with the background case.

(a) Neutral density and neutral wind distribution (130 km)

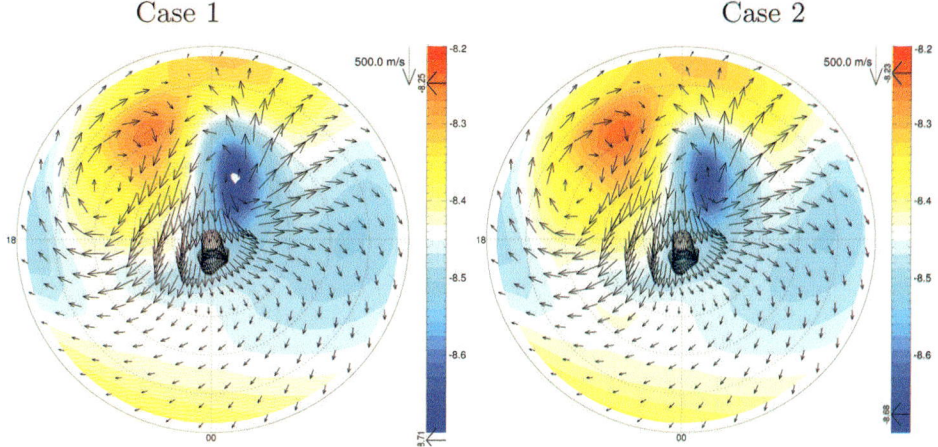

(b) Neutral density distribution (300 km)

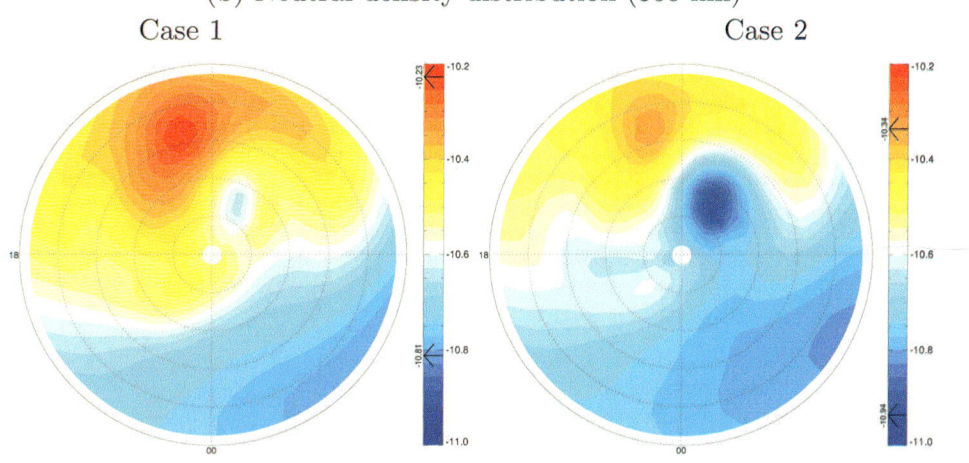

(c) Percentage difference of neutral density (300 km)

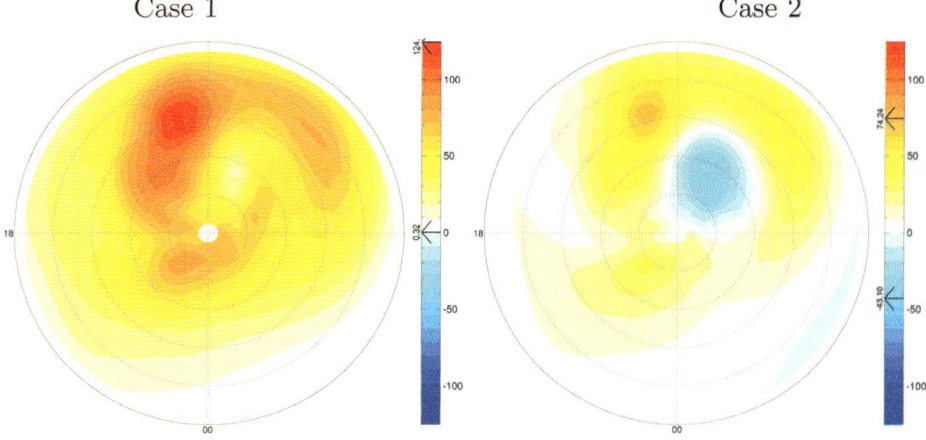

After about 20 min of forcing, Figure 5 shows a gradual downward trend in density in both case 1 and case 2. Figure 6 indicates that part of this trend is likely to be dependent on the particular location chosen and whether or not it is experiencing the cooling effect from the divergent vortex on the dawnside. This modulation of density by the dynamics of the vortices is comparable with the large differences in density between case 1 and case 2 at the specific location (77.5°S, 22.5°E).

4. CONCLUSION AND DISCUSSION

The GITM is a relatively newly developed 3-D, parallel, spherical code that models the Earth's thermosphere and ionosphere system using a stretched altitude grid. The model allows for nonhydrostatic solutions, resolving sound and gravity waves. The altitude-dependent gravitational acceleration has been applied in the model. GITM is an extremely flexible code allowing different models of high-latitude electric fields, auroral particle precipitation, solar EUV inputs, and particle energy deposition to be used. The magnetic field can be represented by an ideal dipole magnetic field or a realistic APEX magnetic field. Many of the source terms can be controlled (switched on and off, or values set) by an easily readable input file. The initial state can be set in three different ways: (1) using an ideal atmosphere, where the user inputs the densities and temperature at the bottom of the atmosphere; (2) using MSIS and IRI; and (3) restarting from a previous run.

Using GITM, which solves the complete vertical momentum equation, the primary characteristics of nonhydrostatic effects on the upper atmosphere are investigated. Our results show that after a sudden intense enhancement of high-latitude Joule heating, the vertical pressure gradient force can locally be 25% larger than the gravity force, resulting in a significant disturbance away from hydrostatic equilibrium. This disturbance is transported from the lower altitude source region to high altitudes through an acoustic wave. The magnitude of the vertical wind perturbation increases with altitude and reaches 150 (250) m s^{-1} at 300 (430) km during the disturbance. The upward neutral wind lifts the atmosphere and raises the neutral density at high altitudes by more than 100%. While the time scale of the buoyancy acceleration perturbation is around 5–10 min in this case, the large vertical wind (above 50 m s^{-1}) at 300 km altitude lasts for a significantly longer time and depends on the lifetime of the forcing. These large vertical winds are observed and are not typically reproduced by hydrostatic models of the thermosphere and ionosphere.

After a sudden enhancement of the ion convection, the Joule heating increases at all altitudes due to the increase of the difference between the ion drift and neutral wind. The thermospheric response at low satellite orbits (400–500 km) is thus subject to the energy variation locally, i.e., at high thermosphere, and the vertical wave propagation from the energy injection at low thermosphere. GITM has been employed to investigate the source of nonhydrostatic processes and the sensitivity of the vertical wind and neutral density at high altitudes to the energy deposited at low and high thermosphere. Through comparing the simulations with and without the Joule heating enhancement above 150 km altitude, the impact of the heating at low and high altitudes to the high-altitude thermosphere has been separated. The numerical simulations show that the atmosphere at all altitudes is out of hydrostatic equilibrium after the sudden energy enhancement. But the maximum of the nonhydrostatic disturbance at high altitudes (300 km) arises from sources below 150 km in the form of an acoustic wave propagating vertically. The heating at high altitudes sets up a large vertical wind, which is overlapped by a disturbance propagating from the low altitudes. While the disturbance propagating vertically from low altitudes can cause a large buoyancy acceleration at 300 km, it mainly brings some temporal variation to the vertical velocity. The heating above 150 km is the primary source for a large increase of the average vertical velocity (40 m s^{-1}) and neutral density (50%) at 300 km and higher altitudes.

APPENDIX A: EQUATIONS IN GITM MODEL

While the GITM code was described in detail by *Ridley et al.* [2006], the main equations are again repeated here for completeness. The variables in the model are [*Deng and Ridley*, 2006]:

T	time
θ	north latitude
ϕ	east longitude
r	radial distance measured from the center of the Earth
N_e	electron density
N_s	number density of species s
N	total number density
$\mathcal{N}_s$	$\ln(N_s)$
ρ	neutral mass density
ρ_i	ion mass density
$\mathbf{v}$	ion velocity
v_r	radial component of the ion velocity
$\mathbf{u}$	neutral velocity
u_θ	northward neutral velocity
u_ϕ	eastward neutral velocity
u_r	radial neutral velocity
$u_{r,s}$	radial neutral velocity of species s

T	temperature
$\mathcal{T}$	p/ρ
Ω	angular velocity of the planet
M_s	molecular mass of species s
G	acceleration of gravity
k	Boltzmann constant
ν_{in}	ion-neutral collision frequency
D_{qs}	diffusion coefficient
η	coefficient of viscosity
$\mathbf{E}$	externally generated electric field (i.e.,magnetospheric)
$\mathbf{B}$	magnetic field
B	magnitude of $|\mathbf{B}|$
$\mathbf{b}$	direction of the magnetic field
P_i	ion pressure
P_e	electron pressure
e	electron charge
K_e	eddy diffusion coefficient

A1. CONTINUITY EQUATION

For each species, the vertical continuity equation in spherical coordinates is:

$$\frac{\partial \mathcal{N}_s^V}{\partial t} = -\frac{\partial u_{r,s}}{\partial r} - \frac{2u_{r,s}}{r} - u_{r,s}\frac{\partial \mathcal{N}_s}{\partial r} \quad (A1)$$

The horizontal continuity equation is:

$$\frac{\partial N_s^H}{\partial t} = -N_s\left(\frac{1}{r}\frac{\partial u_\theta}{\partial \theta} + \frac{1}{r\cos\theta}\frac{\partial u_\phi}{\partial \phi} - \frac{u_\theta\tan\theta}{r}\right) - \frac{u_\theta}{r}\frac{\partial N_s}{\partial \theta} - \frac{u_\phi}{r\cos\theta}\frac{\partial N_s}{\partial \phi}. \quad (A2)$$

The source term $\frac{\partial N_s^S}{\partial t}$ for the neutral density of species s includes the eddy diffusion and chemical sources and losses C_s:

$$\frac{\partial N_s^S}{\partial t} = \frac{\partial}{\partial r}\left[N_s K_e\left(\frac{\partial \mathcal{N}_s}{\partial r} - \frac{\partial \mathcal{N}}{\partial r}\right)\right] + C_s. \quad (A3)$$

The total change rate of density is:

$$\frac{\partial N_s}{\partial t} = N_s\frac{\partial \mathcal{N}_s^V}{\partial t} + \frac{\partial N_s^H}{\partial t} + \frac{\partial N_s^S}{\partial t}. \quad (A4)$$

A2. MOMENTUM EQUATIONS

In rotating spherical coordinates, the vertical momentum equation for each species is:

$$\frac{\partial u_{r,s}}{\partial t} + u_{r,s}\frac{\partial u_{r,s}}{\partial r} + \frac{u_\theta}{r}\frac{\partial u_{r,s}}{\partial \theta} + \frac{u_\phi}{r\cos(\theta)}\frac{\partial u_{r,s}}{\partial \phi} + \frac{k}{M_s}\frac{\partial T}{\partial r} + T\frac{k}{M_s}\frac{\partial \mathcal{N}_s}{\partial r} = -g + \mathcal{F}_s + \frac{u_\theta^2 + u_\phi^2}{r} + \cos^2(\theta)\Omega^2 r + 2\cos(\theta)\Omega u_\phi, \quad (A5)$$

$\mathcal{F}_s$ contains the forces due to the ion-neutral [*Rees*, 1989] and the neutral-neutral friction in the vertical direction [*Colegrove et al.*, 1966]:

$$\mathcal{F}_s = \frac{\rho_i}{\rho_s}\nu_{in}(v_r - u_{r,s}) + \frac{kT}{M_s}\sum_{q\neq s}\frac{N_q}{ND_{qs}}(u_{r,q} - u_{r,s}). \quad (A6)$$

The eastward momentum equation is:

$$\frac{\partial u_\phi}{\partial t} + u_r\frac{\partial u_\phi}{\partial r} + \frac{u_\theta}{r}\frac{\partial u_\phi}{\partial \theta} + \frac{u_\phi}{r\cos\theta}\frac{\partial u_\phi}{\partial \phi} + \frac{1}{r\cos\theta}\frac{\partial \mathcal{T}}{\partial \phi} + \frac{\mathcal{T}}{\rho r\cos\theta}\frac{\partial \rho}{\partial \phi} = \frac{\mathcal{F}_\phi}{\rho} + \frac{u_\phi u_\theta\tan\theta}{r} - \frac{u_r u_\phi}{r} + 2\Omega u_\theta\sin\theta - 2\Omega u_r\cos\theta. \quad (A7)$$

The northward momentum equation is:

$$\frac{\partial u_\theta}{\partial t} + u_r\frac{\partial u_\theta}{\partial r} + \frac{u_\theta}{r}\frac{\partial u_\theta}{\partial \theta} + \frac{u_\phi}{r\cos\theta}\frac{\partial u_\theta}{\partial \phi} + \frac{1}{r}\frac{\partial \mathcal{T}}{\partial \theta} + \frac{\mathcal{T}}{\rho r}\frac{\partial \rho}{\partial \theta} = \frac{\mathcal{F}_\theta}{\rho} - \frac{u_\phi^2\tan\theta}{r} - \frac{u_\theta u_r}{r} - \Omega^2 r\cos\theta\sin\theta - 2\Omega u_\phi\sin\theta. \quad (A8)$$

The force terms due to ion-neutral friction and viscosity are:

$$\begin{aligned}\mathcal{F}_\theta &= \rho_i\nu_{in}(v_\theta - u_\theta) + \frac{\partial}{\partial r}\eta\frac{\partial u_\theta}{\partial r} \\ \mathcal{F}_\phi &= \rho_i\nu_{in}(v_\phi - u_\phi) + \frac{\partial}{\partial r}\eta\frac{\partial u_\phi}{\partial r}.\end{aligned} \quad (A9)$$

A3. ENERGY EQUATION

The vertical thermodynamic equation is:

$$\frac{\partial \mathcal{T}^V}{\partial t} = -u_r\frac{\partial \mathcal{T}}{\partial r} - (\gamma - 1)\mathcal{T}\left(\frac{2u_r}{r} + \frac{\partial u_r}{\partial r}\right). \quad (A10)$$

The horizontal thermodynamic equation is:

$$\frac{\partial T^H}{\partial t} = -\frac{u_\phi}{r\cos\theta}\frac{\partial T}{\partial \phi} - \frac{u_\theta}{r}\frac{\partial T}{\partial \theta}$$
$$-(\gamma-1)T\left(\frac{1}{r}\frac{\partial u_\theta}{\partial \theta} + \frac{1}{r\cos\theta}\frac{\partial u_\phi}{\partial \phi} - \frac{u_\theta\tan\theta}{r}\right). \quad (A11)$$

The thermal energy source term is:

$$\frac{\partial T^S}{\partial t} = \frac{k}{c_v\rho}(Q_{EUV} + Q_{NO} + Q_O$$
$$+ \frac{\partial}{\partial r}\left((\kappa_c + \kappa_{eddy})\frac{\partial T}{\partial r}\right) + N_e \bar{m}_i \nu_{in}(\mathbf{v}-\mathbf{u})^2). \quad (A12)$$

where Q_{EUV} is the solar EUV heating, Q_{NO} is the NO cooling, and Q_O is the O cooling. κ_c is the thermal conductivity, and κ_{eddy} is the eddy diffusion conductivity. $\bar{m}_i$ is the average ion mass.

The total temperature change rate is:

$$\frac{\partial \mathcal{T}}{\partial t} = \frac{\partial \mathcal{T}^V}{\partial t} + \frac{\partial \mathcal{T}^H}{\partial t} + \frac{\partial \mathcal{T}^S}{\partial t}. \quad (A13)$$

A4. EQUATIONS FOR IONS

The ion momentum equation can be solved for the ion velocity:

$$\mathbf{v} = (\mathbf{u}\cdot\mathbf{b})\mathbf{b} + \frac{1}{\nu_{in}}\left(\mathbf{g}\cdot\mathbf{b} - \frac{\nabla(P_i+P_e)\cdot\mathbf{b}}{\rho_i}\right)\mathbf{b}$$
$$+ \frac{\rho_i\nu_{in}\mathbf{A} + eN_e\mathbf{A}\times\mathbf{B}}{\rho_i^2\nu_{in}^2 + e^2N_e^2B^2}, \quad (A14)$$

where

$$\mathbf{A} = \rho_i\mathbf{g}_\perp + eN_e\mathbf{E}_\perp - \nabla(P_i+P_e)_\perp + \rho_i\nu_{in}\mathbf{u}_\perp, \quad (A15)$$

and $\perp$ is defined as the perpendicular direction to the magnetic field.

The ion energy equation is

$$T_i = T_n + \frac{M_n}{3k}(\mathbf{v}-\mathbf{u})^2. \quad (A16)$$

The electron energy equation is the same as in the work of *Schunk and Nagy* [1978].

Acknowledgments. This research was supported by NSF through grant ATM0955629.

REFERENCES

Akmaev, R. A. (2011), Whole atmosphere modeling: Connecting terrestrial and space weather, *Rev. Geophys.*, 49, RG4004, doi:10.1029/2011RG000364.

Aruliah, A. L., E. M. Griffin, A. D. Aylward, E. A. K. Ford, M. J. Kosch, C. J. Davis, V. S. C. Howells, S. E. Pryse, H. R. Middleton, and J. Jussila (2005), First direct evidence of meso-scale variability on ion-neutral dynamics using co-located tristatic FPIs and EISCAT radar in Northern Scandinavia, *Ann. Geophys.*, 23, 147–162.

Chang, C. A., and J.-P. St.-Maurice (1991), Two-dimensional high-latitude thermospheric modeling – A comparison between moderate and extremely disturbed conditions, *Can. J. Phys.*, 69, 1007–1031.

Colegrove, F. D., F. S. Johnson, and W. B. Hanson (1966), Atmospheric composition in the lower thermosphere, *J. Geophys. Res.*, 71(9), 2227–2236.

Crowley, G., T. J. Immel, C. L. Hackert, J. Craven, and R. G. Roble (2006), Effect of IMF B_Y on thermospheric composition at high and middle latitudes: 1. Numerical experiments, *J. Geophys. Res.*, 111, A10311, doi:10.1029/2005JA011371.

Deng, Y., and A. J. Ridley (2006), Dependence of neutral winds on convection E-field, solar EUV, and auroral particle precipitation at high latitudes, *J. Geophys. Res.*, 111, A09306, doi:10.1029/2005JA011368.

Deng, Y., and A. J. Ridley (2007), Possible reasons for underestimating Joule heating in global models: *E* field variability, spatial resolution, and vertical velocity, *J. Geophys. Res.*, 112, A09308, doi:10.1029/2006JA012006.

Deng, Y., A. D. Richmond, A. J. Ridley, and H.-L. Liu (2008a), Assessment of the non-hydrostatic effect on the upper atmosphere using a general circulation model (GCM), *Geophys. Res. Lett.*, 35, L01104, doi:10.1029/2007GL032182.

Deng, Y., A. J. Ridley, and W. Wang (2008b), Effect of the altitudinal variation of the gravitational acceleration on the thermosphere simulation, *J. Geophys. Res.*, 113, A09302, doi:10.1029/2008JA013081.

Deng, Y., T. J. Fuller-Rowell, R. A. Akmaev, and A. J. Ridley (2011), Impact of the altitudinal Joule heating distribution on the thermosphere, *J. Geophys. Res.*, 116, A05313, doi:10.1029/2010JA016019.

Dickinson, R. E., E. C. Ridley, and R. G. Roble (1981), A three-dimensional general circulation model of the thermosphere, *J. Geophys. Res.*, 86(A3), 1499–1512.

Dickinson, R. E., E. C. Ridley, and R. G. Roble (1984), Thermospheric general circulation with coupled dynamics and composition, *J. Atmos. Sci.*, 41, 205–219.

Foster, J. C., J.-P. St.-Maurice, and V. J. Abreu (1983), Joule heating at high latitudes, *J. Geophys. Res.*, 88(A6), 4885–4897.

Frahm, R. A., J. D. Winningham, J. R. Sharber, R. Link, G. Crowley, E. E. Gaines, D. L. Chenette, B. J. Anderson, and T. A. Potemra (1997), The diffuse aurora: A significant source of ionization in the middle atmosphere, *J. Geophys. Res.*, *102* (D23), 28,203–28,214.

Fuller-Rowell, T. J., and D. S. Evans (1987), Height-integrated Pedersen and Hall conductivity patterns inferred from the TIROS-NOAA satellite data, *J. Geophys. Res.*, *92*(A7), 7606–7618.

Fuller-Rowell, T. J., and D. Rees (1980), A three-dimensional, time-dependent, global model of the thermosphere, *J. Atmos. Sci.*, *37*, 2545–2567.

Fuller-Rowell, T. J., and D. Rees (1983), Derivation of a conservative equation for mean molecular weight for a two-constituent gas within a three-dimensional, time-dependent model of the thermosphere, *Planet. Space Sci.*, *31*, 1209–1222.

Hardy, D. A., M. S. Gussenhoven, R. Raistrick, and W. J. McNeil (1987), Statistical and functional representations of the pattern of auroral energy flux, number flux, and conductivity, *J. Geophys. Res.*, *92*(A11), 12,275–12,294.

Hedin, A. E. (1983), A revised thermospheric model based on mass spectrometer and incoherent scatter data: MSIS-83, *J. Geophys. Res.*, *88*(A12), 10,170–10,188.

Hedin, A. E. (1987), MSIS-86 thermospheric model, *J. Geophys. Res.*, *92*(A5), 4649–4662.

Hedin, A. E. (1991), Extension of the MSIS Thermosphere Model into the middle and lower atmosphere, *J. Geophys. Res.*, *96*(A2), 1159–1172.

Heppner, J. P., and N. C. Maynard (1987), Empirical high-latitude electric field models, *J. Geophys. Res.*, *92*(A5), 4467–4489.

Hickey, M. P., G. Schubert, and R. L. Walterscheid (2001), Acoustic wave heating of the thermosphere, *J. Geophys. Res.*, *106*(A10), 21,543–21,548, doi:10.1029/2001JA000036.

Hines, C. O., and W. H. Hooke (1970), Discussion of ionization effects on the propagation of acoustic-gravity waves in the ionosphere, *J. Geophys. Res.*, *75*(13), 2563–2568.

Innis, J. L., and M. Conde (2002), Characterization of acoustic–gravity waves in the upper thermosphere using Dynamics Explorer 2 Wind and Temperature Spectrometer (WATS) and Neutral Atmosphere Composition Spectrometer (NACS) data, *J. Geophys. Res.*, *107*(A12), 1418, doi:10.1029/2002JA009370.

Kwak, Y.-S., A. D. Richmond, Y. Deng, J. M. Forbes, and K.-H. Kim (2009), Dependence of the high-latitude thermospheric densities on the interplanetary magnetic field, *J. Geophys. Res.*, *114*, A05304, doi:10.1029/2008JA013882.

Lühr, H., M. Rother, W. Köhler, P. Ritter, and L. Grunwaldt (2004), Thermospheric up-welling in the cusp region: Evidence from CHAMP observations, *Geophys. Res. Lett.*, *31*, L06805, doi:10.1029/2003GL019314.

Oehmke, R., and Q. Stout (2001), Parallel adaptive blocks on the sphere, paper presented at the 10th SIAM Conf. Parallel Processing for Scientific Computing, SIAM, Portsmouth, Va.

Oehmke, R. C. (2004), High performance dynamic array structures, Ph.D. thesis, Univ. of Mich., Dep. of Electr. Eng. and Comput. Sci., Ann Arbor.

Pfeffer, R. L., and J. Zarichny (1962), Acoustic gravity wave propagation from nuclear explosions in the Earth's atmosphere, *J. Atmos. Sci.*, *19*, 256–263.

Pitteway, M. L. V., and C. O. Hines (1963), The viscous damping of atmospheric gravity waves, *Can. J. Phys.*, *41*, 1935–1948.

Powell, K. G., P. L. Roe, T. J. Linde, T. I. Gombosi, and D. L. D. Zeeuw (1999), A solution-adaptive upwind scheme for ideal magnetohydrodynamics, *J. Comput. Phys.*, *154*, 284–309.

Price, G. D., R. W. Smith, and G. Hernandez (1995), Simultaneous measurements of large vertical winds in the upper and lower thermosphere, *J. Atmos. Terr. Phys.*, *57*, 631–643.

Prölss, G. W. (1981), Latitudinal structure and extension of the polar atmospheric disturbance, *J. Geophys. Res.*, *86*(A4), 2385–2396, doi:10.1029/JA086iA04p02385.

Rees, D., and T. J. Fuller-Rowell (1988), Understanding the transport of atomic oxygen in the thermosphere using a numerical global thermospheric model, *Planet. Space Sci.*, *36*, 935–948.

Rees, D., and T. J. Fuller-Rowell (1990), Numerical simulations of the seasonal/latitudinal variations of atomic oxygen and nitric oxide in the lower thermosphere and mesosphere, *Adv. Space Res.*, *10*(6), 83–102.

Rees, M. H. (1989), *Physics and Chemistry of the Upper Atmosphere*, Cambridge Univ. Press, New York.

Richmond, A. D. (1992), Assimilative mapping of ionospheric electrodynamics, *Adv. Space Res.*, *12*, 59–68.

Richmond, A. D. (1995), Ionospheric electrodynamics using magnetic apex coordinates, *J. Geomagn. Geoelectr.*, *47*, 191–212.

Richmond, A. D., and Y. Kamide (1988), Mapping electrodynamic features of the high-latitude ionosphere from localized observations: Technique, *J. Geophys. Res.*, *93*(A6), 5741–5759.

Richmond, A. D., E. C. Ridley, and R. G. Roble (1992), A thermosphere/ionosphere general circulation model with coupled electrodynamics, *Geophys. Res. Lett.*, *19*(6), 601–604.

Ridley, A. J., G. Crowley, and C. Freitas (2000), An empirical model of the ionospheric electric potential, *Geophys. Res. Lett.*, *27*(22), 3675–3678.

Ridley, A. J., A. D. Richmond, T. I. Gombosi, D. L. De Zeeuw, and C. R. Clauer (2003), Ionospheric control of the magnetospheric configuration: Thermospheric neutral winds, *J. Geophys. Res.*, *108*(A8), 1328, doi:10.1029/2002JA009464.

Ridley, A. J., Y. Deng, and G. Toth (2006), The global ionosphere-thermosphere model, *J. Atmos. Sol. Terr. Phys.*, *68*, 839–864.

Rind, D. (1977), Heating of the lower thermosphere by the dissipation of acoustic waves, *J. Atmos. Terr. Phys.*, *39*, 445–456.

Roble, R. G., and E. C. Ridley (1994), A thermosphere-ionosphere-mesosphere-electrodynamics general circulation model (time-GCM): Equinox solar cycle minimum simulations (30–500 km), *Geophys. Res. Lett.*, *21*(6), 417–420.

Roble, R. G., E. C. Ridley, A. D. Richmond, and R. E. Dickinson (1988), A coupled thermosphere/ionosphere general circulation model, *Geophys. Res. Lett.*, *15*(12), 1325–1328.

Schunk, R. W., and A. F. Nagy (1978), Electron temperatures in the F region of the ionosphere: Theory and observations, *Rev. Geophys.*, *16*(3), 355–399.

Schunk, R. W., et al. (2004), Global Assimilation of Ionospheric Measurements (GAIM), *Radio Sci.*, *39*, RS1S02, doi:10.1029/2002RS002794.

Smith, R. W., and G. Hernandez (1995), Vertical winds in the thermosphere within the polar cap, *J. Atmos. Terr. Phys.*, *57*, 611–620.

Thayer, J. P., J. F. Vickrey, R. A. Heelis, and J. B. Gary (1995), Interpretation and modeling of the high-latitude electromagnetic energy flux, *J. Geophys. Res.*, *100*(A10), 19,715–19,728.

Torr, D. G., et al. (1979), An experimental and theoretical study of the mean diurnal variation of O^+, NO^+, O_2^+, and N_2^+ ions in the mid-latitude F_1 layer of the ionosphere, *J. Geophys. Res.*, *84*(A7), 3360–3372.

Tóth, G., et al. (2005), Space Weather Modeling Framework: A new tool for the space science community, *J. Geophys. Res.*, *110*, A12226, doi:10.1029/2005JA011126.

Weimer, D. R. (1996), A flexible, IMF dependent model of high-latitude electric potentials having "Space Weather" applications, *Geophys. Res. Lett.*, *23*(18), 2549–2552.

Y. Deng, Department of Physics, University of Texas at Arlington, TX, USA. (yuedeng@uta.edu)

A. J. Ridley, Center for Space Environment Modeling, The University of Michigan, Ann Arbor, MI, USA. (ridley@umich.edu)

Traveling Atmospheric Disturbance and Gravity Wave Coupling in the Thermosphere

L. C. Gardner and R. W. Schunk

Center for Atmospheric and Space Sciences, Utah State University, Logan, Utah, USA

The global thermosphere is continually subjected to wave activity that can arise as a result of atmospheric, thermospheric, and ionospheric disturbances. The waves transport mass, momentum, and energy into, out of, and throughout the thermosphere. The waves can be in the form of gravity waves (GWs), tides, or planetary waves. In this study, the propagation and interaction of two GW types were investigated: a shorter wavelength GW and a long wavelength traveling atmospheric disturbance. The shorter wavelength GW was excited in the lower atmosphere at midlatitudes and then propagated upward and northward. Simultaneously, a traveling atmospheric disturbance (TAD) was excited in the northern thermosphere at high latitudes and propagated southward. The two waves then interacted in the midlatitude thermosphere. The shorter wavelength GW and TAD interacted via a superposition of their wave amplitudes, with the shorter wavelength GW amplitude tending to be symmetric about zero when outside the influence of the TAD. However, the shorter wavelength GW amplitude was nonsymmetric when it interacted with the TAD wave perturbations.

1. INTRODUCTION

During the recent extreme solar minimum, it became clear that the thermosphere-ionosphere (T-I) system is affected by planetary, tidal, gravity, and sound waves that propagate upward from the lower atmosphere. Planetary waves are global-scale oscillations that are generated in the troposphere by orographic forcing (mountains, land masses) [*Garcia and Solomon*, 1985; *Laštovička and Pancheva*, 1991; *Chen*, 1992; *Altadill and Apostolov*, 2001; *Xiong et al.*, 2006; *Li et al.*, 2007; *Liu et al.*, 2010]. The waves can be stationary or propagate in the zonal direction with 2, 5, 10, and 16 day periods. Tides are also global-scale oscillations that arise primarily as a result of solar or lunar influences [*Kato*, 1980; *Forbes*, 1984, 1995; *Hagan*, 1993; *Knipp et al.*, 2004; *Zhang et al.*, 2010]. They can migrate with the sun or can be nonmigrating. Their wavelengths are typically several thousand kilometers, with periods of 24 h (diurnal), 12 h (semidiurnal), or any subharmonic of 24 h. Gravity waves (GWs) are generated by disturbances in the troposphere, such as mountain waves, thunderstorms, hurricanes and typhoons, geostrophic adjustments, or ocean wave atmosphere interactions. These waves have wavelengths of 5–1000 km and periods of 5 min to several hours. Sound waves or acoustic waves are waves with the shortest periods. They tend to have periods less than about 5 min and propagate at the speed of sound. As the period of the waves increases, they transition into GWs, which have periods of about 5 min or greater and phase speeds less than the speed of sound.

Planetary waves and tides are easy to incorporate in global T-I models because they are large scale and have relatively long periods, and these features are consistent with the coarse resolution in most global T-I models. However, modeling gravity and sound waves is a challenge because the global resolution needs to be 1 km vertically and 2–10 km horizontally, with less than a minute wave period. As we will show below, we have had success in modeling realistic GWs, waves that match observation and can be simulated within the resolution of the model, with our global T-I model, including GW breaking.

Two waves are of principal concern in the T-I system, and they are traveling atmospheric disturbances (TADs) and upward propagating atmospheric gravity waves (AGWs or GWs). TADs are longer horizontal wavelength waves generated in the high-latitude auroral regions. They have horizontal scales measuring several thousand kilometers, with phase speeds of 500 to 1000 m s^{-1} and periods from 30 min to 3 h [*Hocke and Schlegel*, 1996; *Bruinsma and Forbes*, 2009; *Miyoshi and Fujiwara*, 2008]. The TADs transport density, momentum, and energy from the high-latitude regions to lower latitudes and can even cross the polar cap in the opposite hemisphere if its scale size is large enough [*Bruinsma and Forbes*, 2009; *Gardner and Schunk*, 2010]. GWs occur globally, are generated by a multitude of sources, and deposit their energy locally where they dissipate. Owing to the decrease of density with altitude, wave amplitudes grow with altitude. If the waves grow too fast or get too large, they will break, depositing their energy and causing local turbulence, as seen in the mesosphere/lower thermosphere [*Fritz and Alexander*, 2003].

2. GLOBAL THERMOSPHERE-IONOSPHERE MODEL

The global thermosphere model (GTM) is a time-dependent, high-resolution, first principles–based model of the global thermosphere [*Ma and Schunk*, 1995; *Schunk and Demars*, 2003, 2005]. The model calculates a simultaneous solution of the neutral gas equations of continuity, momentum, energy, and mean mass, which produces global distributions of the mass density, temperature, and all three components of the neutral wind on a variable altitude grid, typically set at 60 to 600 km. The equations are solved in a spherical coordinate system fixed to the Earth using a multidimensional flux-corrected transport (FCT) technique [*Zalesak*, 1979]. The equations take into account the nonlinear inertia term, pressure gradients, the Coriolis force, centripetal acceleration, ion-neutral collisions, advection, vertical and horizontal thermal conduction, vertical and horizontal viscosity, exothermal chemical reactions, auroral heating, solar heating, several local cooling processes, and the displacement between the geographic and geomagnetic poles. The model uses altitude, not a pressure coordinate, in the vertical direction, and *nonhydrostatic equilibrium flows are allowed*. Note that the FCT method is a well-known numerical technique that was designed to handle *subsonic, transonic, and supersonic flow, as well as shock formation*. The spatial resolution in the model is adjustable. Frequently, global simulations are conducted with a horizontal resolution of 0.5° in latitude and 3° in longitude (50 km latitudinal resolution in the polar caps), but simulations have been conducted with 0.1° × 3° (12 km resolution in latitude, pole-to-pole). In the vertical direction (60–600 km), the layers are distributed nonuniformly according to the neutral gas scale height, and we typically use 75 or 90 layers (25 layers is typical in other models). *The model can take into account planetary, tidal, gravity, and sound waves that propagate up from the lower atmosphere*.

The ionosphere affects the momentum and energy balance in the thermosphere via the ion drag and the ion-neutral frictional heating terms in the transport equations. For these simulations, the International Reference Ionosphere (IRI) model [*Bilitza*, 1990] was used to obtain time-dependent, global, ion density distributions. In addition to the IRI background ionosphere, ionization due to auroral particle precipitation was accounted for using the method of *Roble and Ridley* [1987]. Hence, there is no feedback from the thermosphere model to the ionosphere. This approach is acceptable for the present study because our goal is only to elucidate, in a qualitative way, wave interactions in and throughout the thermosphere.

To date, we have conducted global thermosphere-ionosphere simulations that account for various small-scale (~50–100 km) features. We have calculated the thermosphere's response to single and multiple propagating plasma patches, to circular and cigar-shaped plasma patches [*Ma and Schunk*, 1995, 1997a, 2001], to single and multiple sun-aligned polar cap arcs [*Ma and Schunk*, 1997b], to theta aurora [*Demars and Schunk*, 2005], to equatorial plasma bubbles [*Schunk and Demars*, 2003, 2005], to upwelling in the dayside cusp due to ion heating [*Demars and Schunk*, 2007], and to upward propagating waves from the lower atmosphere [*Schunk et al.*, 2008]. The model has also successfully described the supersonic neutral winds that are observed in the polar caps due to strong plasma convection [*Demars and Schunk*, 2008]. Recently, the model has been used to study the generation of TADs during pulsating geomagnetic storms [*Gardner and Schunk*, 2010], the impact that an upward propagating large-scale GW has on the thermosphere [*Gardner and Schunk*, 2011a], and the interaction of upward propagating atmospheric waves and equatorward propagating thermosphere waves [*Gardner and Schunk*, 2011b]. Some of these results are presented in what follows.

3. SIMULATION SETUP

As noted above, the focus of this study was on the propagation and interaction of TADs and GWs. For the TAD simulation, the horizontal resolution of the global T-I model was 2° × 3° latitude/longitude. The altitude range varied from 60 to 620 km, with 100 levels such that there was a

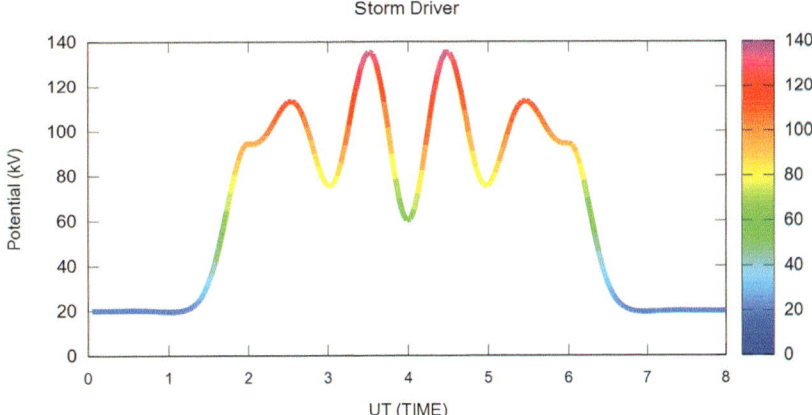

Figure 1. Variation of the polar cap potential (kV) with time that is a reproduction of the pulsating geomagnetic storm of 4 May 1998. From *Gardner and Schunk* [2010].

1 km step size at the lower altitudes that increased to 20 km at the higher altitudes. This resulted in a good vertical resolution in the model throughout the mesosphere/lower thermosphere. A strong pulsating geomagnetic storm was simulated by mimicking an actual geomagnetic storm with an approximately 1 h oscillation superimposed on top of a general rise and decay of the storm.

Figure 1 shows the high-latitude storm driver for the storm described by *Gardner and Schunk* [2010], which has a background cross polar cap potential of 20 kV, a rise to 100 kV at ~01:30 UT and a return to the background value at ~06:30 UT. During the approximately 4 h, the storm is at 100 kV, but superimposed is a 1 h potential oscillation of 40 kV. The potential variation in Figure 1 simulates the pulsing of an actual storm. This storm driver in the high-latitude regions produces TADs with a horizontal wavelength of about 3000 km and a phase speed of nearly 1000 m s^{-1}.

The TADs are generated near the poles, propagate to the poles in the opposite hemispheres, and continue back toward the equator.

Figure 2 is a latitude/longitude plot showing the gravity wave (GW) driver at 03:00 UT near the lower boundary of the model at ~60 km altitude. The GW was generated with a heating function that varied exponentially in a spatial region with a half width of 5° in both latitude and longitude centered at 180° longitude and 30° north latitude. The heating function also had a half width in time of 1 h. The resulting GW has a 2 h period, a horizontal wavelength of 1000 km, and a phase speed of 138 m s^{-1} northward.

As the simulation progressed, the TADs produced at high latitudes in the Northern Hemisphere propagated southward, and the GW produced in the lower atmosphere propagated upward and northward. The waves interacted in the thermosphere as the simulation progressed.

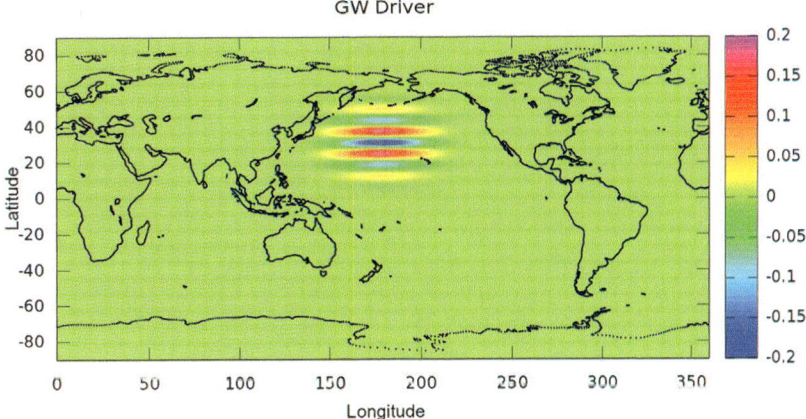

Figure 2. Heating function that was used to generate the gravity wave packet in the lower atmosphere. Adapted from *Gardner and Schunk* [2011a].

104 TAD AND GRAVITY WAVE COUPLING

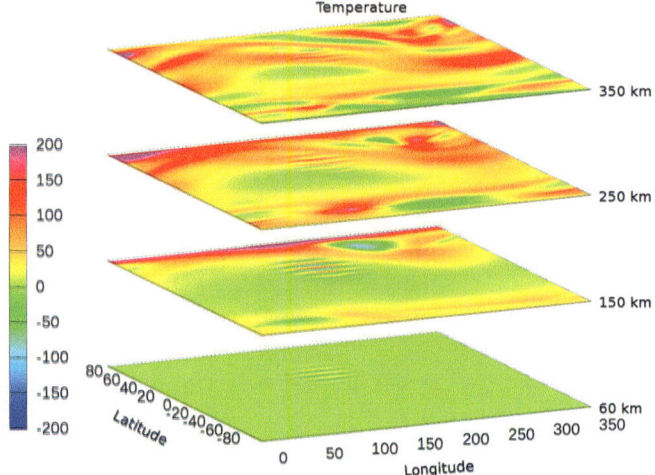

Figure 3. Neutral temperature perturbations versus latitude and longitude at four altitudes (60, 150, 250, and 350 km) and at 04:15 UT.

4. SIMULATION RESULTS

The GW and TAD simulated in this study are both fully modeled in the simulation; there is no parameterization included for unresolved aspects of the waves. This results in a direct representation of the local effects of the GW via the physics included in the model equations. Although the model resolution is insufficient to fully resolve GW breaking, we are more interested in the interaction of the GW with the TAD.

Figure 3 shows neutral temperature perturbations at four altitudes at 04:15 UT. The lower panel in Figure 3 is a 60 km altitude slice that shows the GW driver in temperature, where the wave amplitude is ±20 K. As the wave propagates upward, its amplitude increases due to a decrease in density. With the increasing wave amplitude, there is a possibility that the wave may break if its amplitude gets too large (see for example, the work of *Gardner and Schunk* [2010]), but for this simulation, the wave amplitudes were insufficient to instigate breaking. At 150 km, the GW wave amplitude has increased to ±150 K, and the TAD can be seen at +65° latitude near the GW. At 150 km, the TAD wave speed, which is proportional to the local thermosphere temperature, is slow enough that the two waves have not yet interacted. However, at 250 km, the TAD and GW are just starting to interact. The temperature at this altitude has increased, and therefore, the speed of the TAD has increased. The two waves can be seen to interact at the northward edge of the GW packet by the total increase in the wave amplitude. At the northern edge of the GW packet, the wave amplitude varies from +110 K to −30 K, while at the southern edge the wave amplitude varies from ±30 K. At 350 km, the thermospheric temperature approaches the exospheric temperature, and the TAD propagates at its maximum velocity of nearly

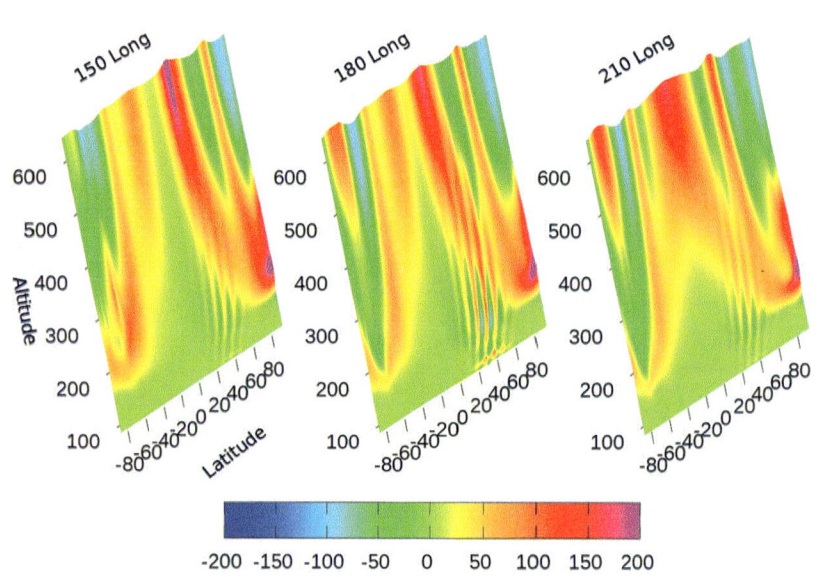

Figure 4. Neutral temperature perturbations versus altitude and latitude at three longitudes (150°, 180°, and 210°) and at 04:15 UT.

1000 m s^{-1}. At this altitude, the GW and TAD completely overlap, and their amplitudes are linearly added.

Further details concerning the TAD and GW interaction are shown in Figure 4, where altitude-latitude slices through the central longitude of the GW are displayed at 04:15 UT. Figure 4 shows the altitude dependence of the two different wave types. The TAD, driven in the thermosphere, has a long horizontal wavelength (3000 km) and a horizontal phase speed that is proportional to the neutral temperature. The horizontal phase speed approaches 1000 m s^{-1} at the upper altitudes where the neutral temperature reaches the exospheric temperature. In contrast, the GW in Figure 3 has a downward phase progression, upward energy transport, and a horizontal wavelength of 1000 km. In Figure 4, the central slice at 180° longitude shows the maximum amplitudes for the GW, with the other two planes, at 150° and 210° longitude, showing decreased amplitudes. As the GW propagates upward, it interacts with the wave front of the TAD in the 40° to 80° latitudinal sector. From 90 km to approximately 200 km, the two waves are independent, and the GW amplitude reaches ±100 K. Above about 200 km, the GW and TAD interact, with their amplitudes summing due to the superposition of the waves. In the 180° longitude slice, the GW amplitude between about 200 and 500 km is shifted in the positive direction due to the positive amplitude in the TAD. The GW wave amplitude varies from −25 to +100 K, showing the superposition of the GW and TAD, where the TAD amplitude is 62 K. A similar superposition of the GW and TAD can be seen in the 150° and 210° longitude slices.

5. SUMMARY

The global thermosphere environment is continually filled with waves from atmospheric, thermospheric, and ionospheric origin, which propagate in all directions. These waves transport mass, momentum, and energy into, out of, and throughout the thermosphere, and they can be in the form of GWs, tides, planetary waves, or TADs. We modeled two of these waves: a GW generated in the lower atmosphere and a TAD generated at high latitudes in the thermosphere. The goal was to study their interaction in the thermosphere. As the GWs propagated upward and northward, and the TAD propagated southward, the two interacted in the midlatitude thermosphere. The GW and TAD interacted via a superposition of their wave amplitudes. The GW amplitude tended to be symmetric about zero when outside the influence of the TAD, but it was nonsymmetric when it interacted with the TAD.

As demonstrated in this study, a GW with a 1000 km horizontal wavelength and a TAD with a 3000 km horizontal wavelength will interact via the superposition principle, and will penetrate through each other. There was little to no wave breaking in the simulation because GW breaking principally occurs in the 90–150 km altitude range when the criteria for wave breaking are exceeded. The criteria for wave breaking were not exceeded for the wave parameters adopted in this study. Future studies will look at the impacts of other wave sources and parameters.

Acknowledgments. This research was supported by NSF Grant 0962544 to Utah State University.

REFERENCES

Altadill, D., and E. M. Apostolov (2001), Vertical propagating signatures of wave-type oscillations (2 and 6.5 days) in the ionosphere obtained from electron-density profiles, *J. Atmos. Sol. Terr. Phys.*, *63*, 823–834.

Bilitza, D. (Ed.) (1990), International Reference Ionosphere 1990, *Rep. NSSDC 90-22*, Natl. Space Sci. Data Cent., Greenbelt, Md.

Bruinsma, S., and J. M. Forbes (2009), Properties of traveling atmospheric disturbances (TADs) inferred from CHAMP accelerometer observations, *Adv. Space Res.*, *43*, 369–376, doi:10.1016/j.asr.2008.10.031.

Chen, P.-R. (1992), Two-day oscillation of the equatorial ionization anomaly, *J. Geophys. Res.*, *97*(A5), 6343–6357.

Demars, H. G., and R. W. Schunk (2005), Effect of the theta aurora on the polar thermosphere, *J. Atmos. Sol. Terr. Phys.*, *67*, 489–499.

Demars, H. G., and R. W. Schunk (2007), Thermospheric response to ion heating in the dayside cusp, *J. Atmos. Sol. Terr. Phys.*, *69*, 649–660.

Demars, H. G., and R. W. Schunk (2008), Modeling and consequences of supersonic winds in the thermosphere, in *Proceedings of the 2008 Ionospheric Effects Symposium*, pp. 381–388, JMG Assoc., Natl. Tech. Inf. Serv., Springfield, Va.

Fritts, D. C., and M. J. Alexander (2003), Gravity wave dynamics and effects in the middle atmosphere, *Rev. Geophys.*, *41*(1), 1003, doi:10.1029/2001RG000106.

Forbes, J. M. (1984), Middle atmosphere tides, *J. Atmos. Terr. Phys.*, *46*, 1049–1067.

Forbes, J. M. (1995), Tidal and planetary waves, in *The Upper Mesosphere and Lower Thermosphere: A Review of Experiment and Theory*, Geophys. Monogr. Ser., vol. 87, edited by R. M. Johnson and T. L. Killeen, pp. 67–87, AGU, Washington, D. C., doi:10.1029/GM087p0067.

Garcia, R. R., and S. Solomon (1985), The effect of breaking gravity waves on the dynamics and chemical composition of the mesosphere and lower thermosphere, *J. Geophys. Res.*, *90*(D2), 3850–3868, doi:10.1029/JD090iD02p03850.

Gardner, L. C., and R. W. Schunk (2010), Generation of traveling atmospheric disturbances during pulsating geomagnetic storms, *J. Geophys. Res.*, *115*, A08314, doi:10.1029/2009JA015129.

Gardner, L. C., and R. W. Schunk (2011a), Large-scale gravity wave characteristics simulated with a high-resolution global thermosphere-ionosphere model, *J. Geophys. Res.*, *116*, A06303, doi:10.1029/2010JA015629.

Gardner, L. C., and R. W. Schunk (2011b), Thermospheric wave interactions in a global thermosphere/ionosphere model, paper presented at the 13th International Ionospheric Effects Symposium, Alexandria, Va.

Hagan, M. E. (1993), Quiet time upper thermospheric winds over Millstone Hill between 1984 and 1990, *J. Geophys. Res.*, *98*(A3), 3731–3739, doi:10.1029/92JA01605.

Hocke, K., and K. Schlegel (1996), A review of atmospheric gravity waves and travelling ionospheric disturbances: 1982–1995, *Ann. Geophys.*, *14*, 917–940.

Kato, S. (1980), *Dynamics of the Upper Atmosphere*, Cent. for Acad. Publ., Tokyo, Japan.

Knipp, D. J., W. K. Tobiska, and B. A. Emery (2004), Direct and indirect thermospheric heating sources for solar cycles 21–23, *Sol. Phys.*, *224*, 495–505.

Laštovička, J., and D. Pancheva (1991), Changes in the characteristics of planetary waves at 80–100 km over central and southern Europe since 1980, *Adv. Space Res.*, *11*, 31–34.

Li, T., C.-Y. She, H.-L. Liu, T. Leblanc, and I. S. McDermid (2007), Sodium lidar–observed strong inertia-gravity wave activities in the mesopause region over Fort Collins, Colorado (41°N, 105°W), *J. Geophys. Res.*, *112*, D22104, doi:10.1029/2007JD008681.

Liu, H.-L., W. Wang, A. D. Richmond, and R. G. Roble (2010), Ionospheric variability due to planetary waves and tides for solar minimum conditions, *J. Geophys. Res.*, *115*, A00G01, doi:10.1029/2009JA015188.

Ma, T.-Z., and R. W. Schunk (1995), Effect of polar cap patches on the polar thermosphere, *J. Geophys. Res.*, *100*(A10), 19,701–19,713.

Ma, T.-Z., and R. W. Schunk (1997a), Effect of polar cap patches on the thermosphere for different solar activity levels, *J. Atmos. Sol. Terr. Phys.*, *59*, 1823–1829.

Ma, T.-Z., and R. W. Schunk (1997b), Effect of sun-aligned arcs on the polar thermosphere, *J. Geophys. Res.*, *102*, 9729–9735.

Ma, T.-Z., and R. W. Schunk (2001), The effects of multiple propagating plasma patches on the polar thermosphere, *J. Atmos. Sol. Terr. Phys.*, *63*, 355–366.

Miyoshi, Y., and H. Fujiwara (2008), Gravity waves in the thermosphere simulated by a general circulation model, *J. Geophys. Res.*, *113*, D01101, doi:10.1029/2007JD008874.

Roble, R. G., and E. C. Ridley (1987), An auroral model for the NCAR thermospheric general circulation model (TGCM), *Ann. Geophys., Ser. A*, *5*, 369–382.

Schunk, R. W., and H. G. Demars (2003), Effect of equatorial plasma bubbles on the thermosphere, *J. Geophys. Res.*, *108* (A6), 1245, doi:10.1029/2002JA009690.

Schunk, R. W., and H. G. Demars (2005), Thermospheric weather due to mesoscale ionospheric structures, in *2005 Ionospheric Effects Symposium*, edited by J. M. Goodman, pp. 799–806, JMG Assoc., Ltd., Springfield, Va.

Schunk, R. W., L. Gardner, L. Scherliess, D. C. Thompson, and J. J. Sojka (2008), Effect of lower atmospheric waves on the ionosphere and thermosphere, paper presented at the 2008 Ionospheric Effects Symposium, JMG Assoc., Natl. Tech. Inf. Serv., Springfield, Va.

Xiong, J., W. Wan, B. Ning, L. Liu, and Y. Gao (2006), Planetary wave-type oscillations in the ionosphere and their relationship to mesospheric/lower thermospheric and geomagnetic disturbances at Wuhan (30.61°N, 114.51°E), *J. Atmos. Sol. Terr. Phys.*, *68*, 498–508.

Zalesak, S. T. (1979), Fully multi-dimensional flux-corrected transport algorithms for fluids, *J. Comput. Phys.*, *31*, 335–362.

Zhang, X., J. M. Forbes, and M. E. Hagan (2010), Longitudinal variation of tides in the MLT region: 2. Relative effects of solar radiative and latent heating, *J. Geophys. Res.*, *115*, A06317, doi:10.1029/2009JA014898.

L. C. Gardner and R. W. Schunk, Center for Atmospheric and Space Sciences, Utah State University, Logan, UT, USA. (robert.schunk@usu.edu)

Air Force Low-Latitude Ionospheric Model in Support of the C/NOFS Mission

Yi-Jiun Su,[1] John M. Retterer,[2] Ronald G. Caton,[1] Russell A. Stoneback,[3] Robert F. Pfaff,[4] Patrick A. Roddy,[1] and Keith M. Groves[2]

In this article, we describe and demonstrate the capabilities of the low-latitude physics-based ionospheric model (PBMOD) developed at the Air Force Research Laboratory to specify radio scintillations using data collected during an April 2009 campaign dedicated to measurements with the Communication/Navigation Outage Forecasting System (C/NOFS). The electric fields/plasma drifts are believed to be the primary driver of the dynamics in the low-latitude ionosphere. With electric field measurements ingested into PBMOD, estimated scintillation strengths (S_4) were comparable with ground measurements at 250 MHz recorded at Ancón, Peru; Christmas Island; and Kwajalein Atoll. These scintillations were associated with upward plasma drifts, although in some places, actual conditions were not precisely determined due to data gaps caused by spurious fields. We also present simulation results obtained from PBMOD driven by four different empirical drift models to specify global ionospheric densities. Discrete longitudinal structures are evident in both averaged density and drift observations. Density outputs from C/NOFS-driven simulations present similar wave 3 or wave 4 structures in geographical longitudes. In contrast, such density structures, likely associated with atmospheric tides, are absent when driving PBMOD with Scherliess-Fejer drifts. Model results have been quantitatively compared with in situ density measurements obtained from C/NOFS, Defense Meteorological Satellite Program, and CHAMP at altitudes ranging from ~350 to 850 km. We found that, on average, the smallest error in modeled densities came from simulations driven by the Ion Velocity Drift Meter drifts. We expect to increase the accuracy of forecasted low-latitude ionospheric densities with more accurate and continuous plasma drift measurements.

1. INTRODUCTION

The ionosphere, an ionized portion of the upper atmosphere, extends from ~60 to several thousand kilometers altitude. The major source of plasmas in the ionosphere is the photoionization of neutral molecules via solar EUV and soft X-ray radiations. The ionized charged particles undergo chemical reactions, recombination, diffusion, and transport [e.g., *Schunk and Nagy*, 2000]. Particularly, plasma diffusion and transport are strongly affected by the Earth's magnetic fields [e.g., *Kelley*, 2009]. Since the beginning of the space age, many ionospheric models have been developed to study different aspects of ionospheric physics. Descriptions of some of those models can be found in the "Handbook of Ionospheric Models" [*Schunk*, 1996] and the

---

[1]Air Force Research Laboratory, Space Vehicles Directorate, Kirtland Air Force Base, New Mexico, USA.

[2]Institute for Scientific Research, Boston College, Chestnut Hill, Massachusetts, USA.

[3]Department of Physics and Center for Space Sciences, University of Texas at Dallas, Texas, USA.

[4]Goddard Space Flight Center, National Aeronautics and Space Administration, Greenbelt, Maryland, USA.

Modeling the Ionosphere-Thermosphere System
Geophysical Monograph Series 201
© 2013. American Geophysical Union. All Rights Reserved.
10.1029/2012GM001268

"Guide to Reference and Standard Ionosphere Models" [*AIAA*, 1999].

Two decades ago, the Air Force Research Laboratory (AFRL) began to plan the Communication/Navigation Outage Forecasting System (C/NOFS) satellite: the first low inclination, low Earth-orbiting satellite mission dedicated to ionospheric irregularities and radio wave scintillations [*de La Beaujadière et al.*, 2004]. A satellite provides only 1-D sampling of the ionospheric parameters and physical quantities. Hence, a global 4-D low-latitude ionospheric model to specify electron density profiles and scintillations in space and time is an important and primary element of the C/NOFS mission supported by the addition of six scientific payloads as well as various ground measurements. In section 2, we describe the Air Force low-latitude ionospheric model and its capabilities. Examples of model runs during an April 2009 campaign are also demonstrated.

After multiple delays, the C/NOFS satellite was launched in April 2008 during the deepest solar minimum on modern record, in spite of the fact that two of the instruments were designed for operation during a solar maximum. Initial results from the C/NOFS mission provided many unexpected scientific discoveries of low-latitude ionosphere during the deep solar minimum. Several papers detailing these results were collected and compiled in a special issue of "Geophysical Research Letters" in 2009. While scientists expected that signatures of low-latitude ionospheric disturbances should be weaker due to the quiet solar wind conditions and weak high-latitude influences, many of these studies have shown that forcing from the thermosphere became more prominent during this deep solar minimum [e.g., *Huang et al.*, 2009, 2011; *Dao et al.*, 2011, *Fejer et al.*, 2010; *Eccles et al.*, 2011; *Su et al.*, 2011]. Using direct and indirect measurements of plasma drifts from the Ion Velocity Drift Meter (IVM) [*Heelis et al.*, 2009] and the Vector Electric Field Instrument (VEFI) [*Pfaff et al.*, 2010] onboard C/NOFS as drivers, we show simulation results demonstrating possible tidal structures obtained from assimilative modeling in section 3.

2. MODEL DESCRIPTION AND OUTPUT

The low-latitude ionospheric physics-based model (PBMOD) has been developed over the last 14 years at AFRL in support of the C/NOFS mission analysis [*Retterer*, 2005; *Retterer et al.*, 2005; *Retterer*, 2010a, 2010b]. PBMOD has been designed to perform ionospheric specification and forecasting of the ambient ionosphere (coarser-scale global structure) and finer-scale localized plasma turbulence (i.e., equatorial spread *F*, bubbles, and etc.). A simplified flowchart describing the overall organization of PBMOD is presented

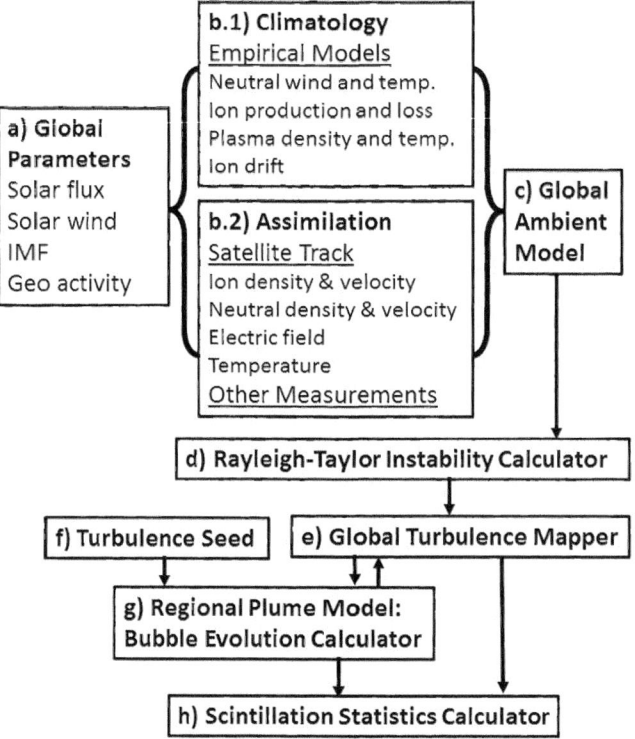

Figure 1. The flowchart of the Air Force physics-based model (PBMOD).

in Figure 1. The model description in this section follows the alphabetic order shown in Figure 1.

(a) Various global parameters recorded on an hourly basis including solar flux, solar wind velocities and densities, interplanetary magnetic fields, and geomagnetic indices are used to set up initial conditions or drivers for various empirical models input into PBMOD.

(b.1) Parameterized by the solar $F_{10.7}$ indices, the *Hinteregger et al.* [1981] reference spectra specify the solar UV fluxes, which control photoproduction of plasma and daytime plasma heating. Production and loss rates of O^+, along with the densities of the molecular ions, are calculated from the local model described by *Jasperse* [1976]. Neutral densities and temperatures are specified by the *Hedin* [1987] Mass Spectrometer Incoherent Scatter (MSIS) model, which is parameterized in terms of the solar $F_{10.7}$ index and the geomagnetic activity index, *Ap*. The horizontal neutral wind components are obtained from the empirical wind model HWM-07 [*Drob et al.*, 2008]. No vertical winds are assumed. The ion and electron temperatures are based on the empirical model published by *Gulyaeva and Titheridge* [2006]. The initial ion densities are given by the FAIM model [*Anderson et al.*, 1989]. Zonal drifts are prescribed by the empirical model of *Fejer et al.* [1991] based on Jicamarca observations, while vertical drifts are specified by *Scherliess and Fejer* [1999] based on observations from Jicamarca and the AE-E satellite. In addition, the disturbance dynamo [*Fejer and Scherliess*, 1997] and penetration electric field effects [*Retterer and Kelley*, 2010] can be optionally included in the empirical drift model.

(b.2) The plasma drift input can now be constrained by in situ measurements of VEFI and IVM on C/NOFS. Recent examples of model output from PBMOD driven by VEFI data have been shown to successfully reproduce large-scale postmidnight density depletions [*Su et al.*, 2009, 2011]. Other satellite or ground measurements when available can be included in a similar fashion. For this effort, we choose to simulate the low-latitude ionosphere using VEFI measurements as a driver during a coordinated April 2009 campaign in the Kwajalein Atoll (KWA). Sample results are shown in Figure 2.

(c) The first-principle ambient model solves continuity equations for both O^+ and H^+ ions as functions of position and time along the magnetic field lines over a range of apex altitudes in the low-latitude ionosphere. Originally based on the LOWLAT model [*Anderson*, 1973; *Preble et al.*, 1994], the tilted dipole field model was replaced by the International Geomagnetic Reference Field model in PBMOD. This ambient model does not consider self-consistent fields of plasma perturbations and, therefore, describes only the background densities without plasma instabilities. The lowest altitude of the ambient model is set at 90 km and extends up to a few thousand kilometers covering both *E* and *F* regions.

(d) The plasma structures obtained from the ambient model are examined for the Rayleigh-Taylor (R-T) instability conditions to be effective. The R-T linear growth rate is calculated based on equations described in Appendix A of *Retterer et al.* [2005] and discussed extensively by *Sultan* [1996] and *Rappaport* [1996]. A 2-D map of integrated growth rates of R-T instability at 06:48UT on day of year (DOY) 118 2009 is shown in Figure 2a, where the color indicates the positive growth rate. In this example, the strongest growth rate is located near Christmas Island (CXI).

(e) Higher-resolution 2- or 3-D plume simulations are then constructed in regions (specified by longitude and time) where the plasmas are unstable (i.e., the R-T growth rate is positive).

(f) In each unstable region, arbitrary small density perturbations are launched triggering bubble formation.

(g) The development and evolution of plasma bubbles is followed using a nonlinear plasma transport code including both continuity and momentum equations in the field-aligned coordinate system and using the current continuity to determine the electric potential [*Retterer*, 2010a]. A snapshot of developing ionospheric bubbles at the magnetic equator from the 2-D plume simulation near the AFRL Scintillation Network Decision Aid (SCINDA) ground station at CXI on DOY 118 is shown in Figure 2b, where the logarithmic density is represented by color according to the color bar on the right.

(e, continued) The global turbulence mapper controls the execution of the plume calculations and then integrates the density along vertical signal paths to obtain a map of total electron content (TEC).

(h) The TEC variation spectrum is then analyzed as a function of east-west wave number, latitude, and UT and is fit with a power law in order to extrapolate to the smaller wavelength range (<1 km). We note that the spatial resolution of the plume model (~10 km) is much larger than the scale size affecting radio scintillations. A linear fit to a power law in wave numbers is a reasonably simple assumption without acquiring high-resolution simulations, which would be unreasonable for forecasting purpose due to current computational power limitations. Finally, a spectral analysis is performed, and the scintillation statistics calculator provides an estimate of the S_4 amplitude-scintillation strength using a phase-screen calculation [*Retterer*, 2010b].

A global map of scintillation is produced as a composite of individual regional maps. S_4 values produced by PBMOD can be compared with direct observations made from SCINDA network ground stations [*Groves et al.*, 1997], a collection of ground-based receivers monitoring the signal strength of satellite-based radio transmissions at more than 60 sites in the equatorial region.

In this paper, we have chosen measurements at ~250 MHz from three SCINDA stations at Ancón, Peru (ANC, Figure 2d),

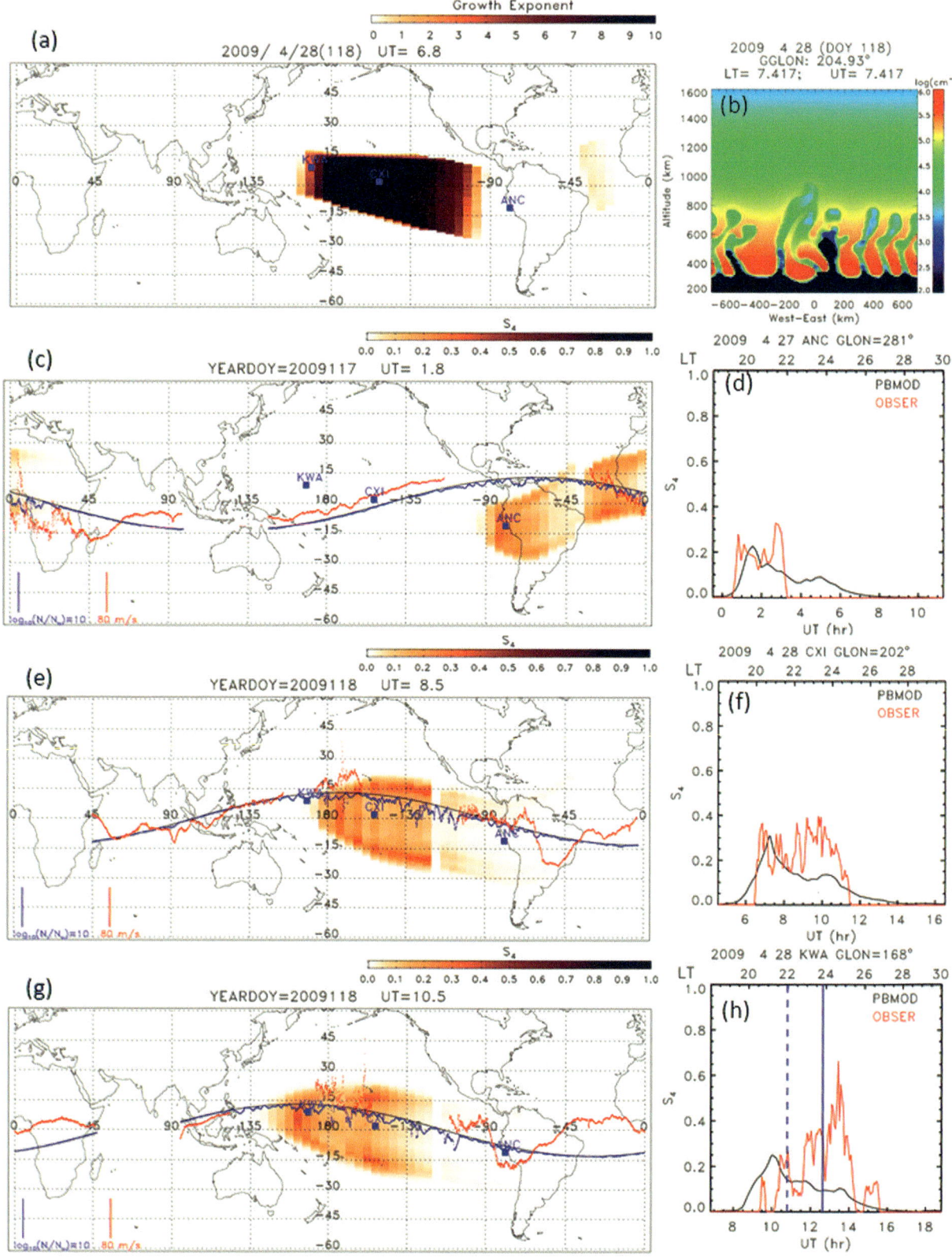

CXI (Figure 2f), and KWA (Figure 2h) to compare with estimated S_4 from PBMOD. These three examples represent good comparisons between model (black) and data (red). In each case, the onset of scintillations occur within 1 h, and the magnitudes of the modeled S_4 values are comparable to the ground VHF measurements with the exception that the model underestimates the scintillation strength at 09:00–11:00 UT at CXI and 12:00–14:00 UT at KWA, and overestimates S_4 between 03:00 and 06:00 UT at ANC. One of the possible reasons for this model-data inconsistency may be due to data gaps in measured drift velocities (discussed at the end of this section). Two-dimensional map-based output of PBMOD estimated S_4 values are shown in Figures 2c, 2e, and 2g. In general, the estimated scintillation regions from PBMOD are associated with observed density perturbations. Incoherent scatter returns on the Advanced Research Project Agency (ARPA) Long-range Tracking and Identification Radar on KWA during the April 2009 campaign also demonstrated density perturbations at the bottom layer of the F region (Figure 3) during the same period as that shown in Figure 2h on DOY 118. Weak scintillation, S_4 of 0.25 (vertical dashed blue line in Figure 2h), was observed at 10:47 UT during an off-perpendicular scan shown in Figure 3b, while stronger scintillation level, S_4 of 0.4 (vertical blue line in Figure 2h), was measured at 12:37 UT displayed in Figure 3d. The formation of irregularities in Figure 3b eventually pierced through the weak F layer in Figure 3d. Unfortunately, spurious fields were seen in VEFI data during parts of this period resulting in significant data gaps in the in situ drift data (seen in each orbit shown in Figures 2c, 2e, and 2g); however, the available data in the scintillation regions indicate upward plasma drifts. For simulation purposes, the gaps were filled in with drift data from the previous orbit in the similar longitude and/or climatology drifts. We should note that gaps have been filled in with upward drifts shown in Figures 2c, 2e, and 2g. The scintillation regimes shown in Figures 2d, 2f, and 2h are associated with upward drifts.

3. A STATISTICAL STUDY OF LOW-LATITUDE IONOSPHERIC DENSITIES

An important driver in the determination of low-latitude ionospheric plasma density and scintillation is the plasma drift. Prior to the C/NOFS mission, PBMOD was driven by an empirical drift model [*Scherliess and Fejer*, 1999] (Figure 4a), referred to as *Scherliess-Fejer* (SF) or CLIMO in this paper. Two instruments, IVM and VEFI, on board the C/NOFS satellite were designed to directly or indirectly measure the plasma drifts. In the statistical study presented here, our intent is to compare the background densities obtained from the model to satellite measurements using four different empirical drift models (shown in Figure 4). Although our study covers ~2 years during solar minimum, we have selected to present results from the northern fall of 2009 as a representation of the entire period due to the limitations on the article length. Because the instrument was designed for a solar maximum mission, IVM collects more accurate data in a plasma environment dominated by O^+ ions. At solar minimum conditions, IVM drift data are reliable at satellite altitudes below 550 km. The satellite requires 67 days of the procession period to cover 24 h LTs and 360° geographic longitude (GLON) to be useful for driving global simulations. The 67 day averaged IVM data is shown in Figure 4b, where the resolution in each LT and GLON bin is 0.25 h and 15°, respectively. Similarly, the VEFI data were averaged over a 3 month period for each season (Figure 4c). IVM and VEFI data sets were combined and averaged to form the fourth drift model shown in Figure 4d. We noted that although both zonal (east-west) and meridional (up-down) drifts are inputs to PBMOD, the low-latitude ionospheric density is much more sensitive to the meridional drifts than zonal drifts.

Simulations of PBMOD were performed every 10 days for seven seasons from June 2008 to February 2010 using yearly averaged $F10.7$ values as input. The density outputs from PBMOD with various plasma drift drivers have been compared to ion density observations near the equator ($-13.8° <$ GLAT $< 13.8°$) from CHAMP at ~350 km, C/NOFS between 400 and 800 km, and Defense Meteorological Satellite Program (DMSP) at ~850 km. A simple description of density measurements by CHAMP, C/NOFS, and DMSP is provided in a recent paper by *Su et al.* [2011]. Density outputs of PBMOD and in situ density measurements for each season were binned according to LT (ΔLT = 1 h), GLON (ΔGLON = 15°), and altitude (ALT, ΔALT = 80 km), i.e., N (nLT = 24, nGLON = 24, nALT = 7). Binned densities between 490 and

Figure 2. (opposite) (a) Integrated Rayleigh-Taylor (R-T) instability growth rate. (b) Plume simulation of bubble formation corresponding to the positive R-T growth rate (a). (c, e, g) Estimated S_4 values from PBMOD, where the black curve represents the Communication/Navigation Outage Forecasting System (C/NOFS) orbit, the blue line represents the density perturbations observed by Planar Langmuir Probe (PLP), and the red line represents the meridional drift obtained from Vector Electric Field Instrument (VEFI). The red curve above and below the orbit indicates upward and downward drift, respectively. (d, f, h) Comparison between PBMOD-estimated S_4 (black) and observed S_4 at ~250 MHz (red) from three Scintillation Network Decision Aid (SCINDA) stations at Ancón, Peru, Christmas Island, and Kwajalein Atoll (KWA), respectively. Three SCINDA stations are also marked as blue squares (a, c, e, g).

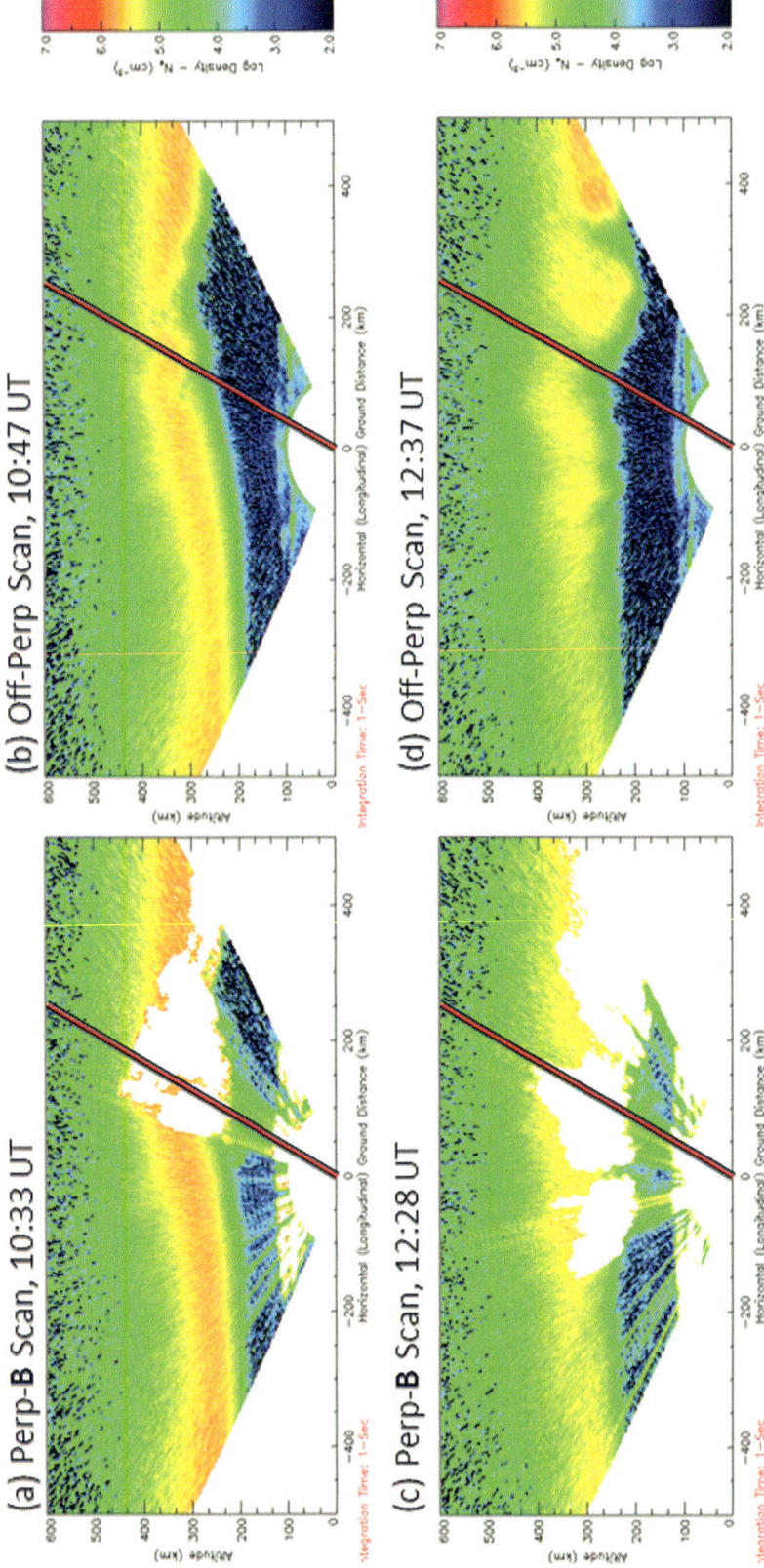

Figure 3. Log electron densities are displayed as a function of altitude and horizontal distance from incoherent scatter returns on Advanced Research Project Agency (ARPA) Long-range Tracking and Identification Radar on the evening of 28 April 2009. Coherent (Bragg) backscatter from irregularities in UHF scans with the radar beam pointing perpendicular to the local magnetic field (Perp-**B**) at 300 km, scans (a) and (c), are unshaded (white). The look angle of the SCINDA VHF data link is marked with a red line. For the Off-Perp scans (b) and (d), the steerable 46 m dish is tilted to the north approximately 6° from perpendicularity with **B**. The absence of Bragg scatter provides a complete picture of the density depletions across a 1000 km longitudinal region centered on KWA.

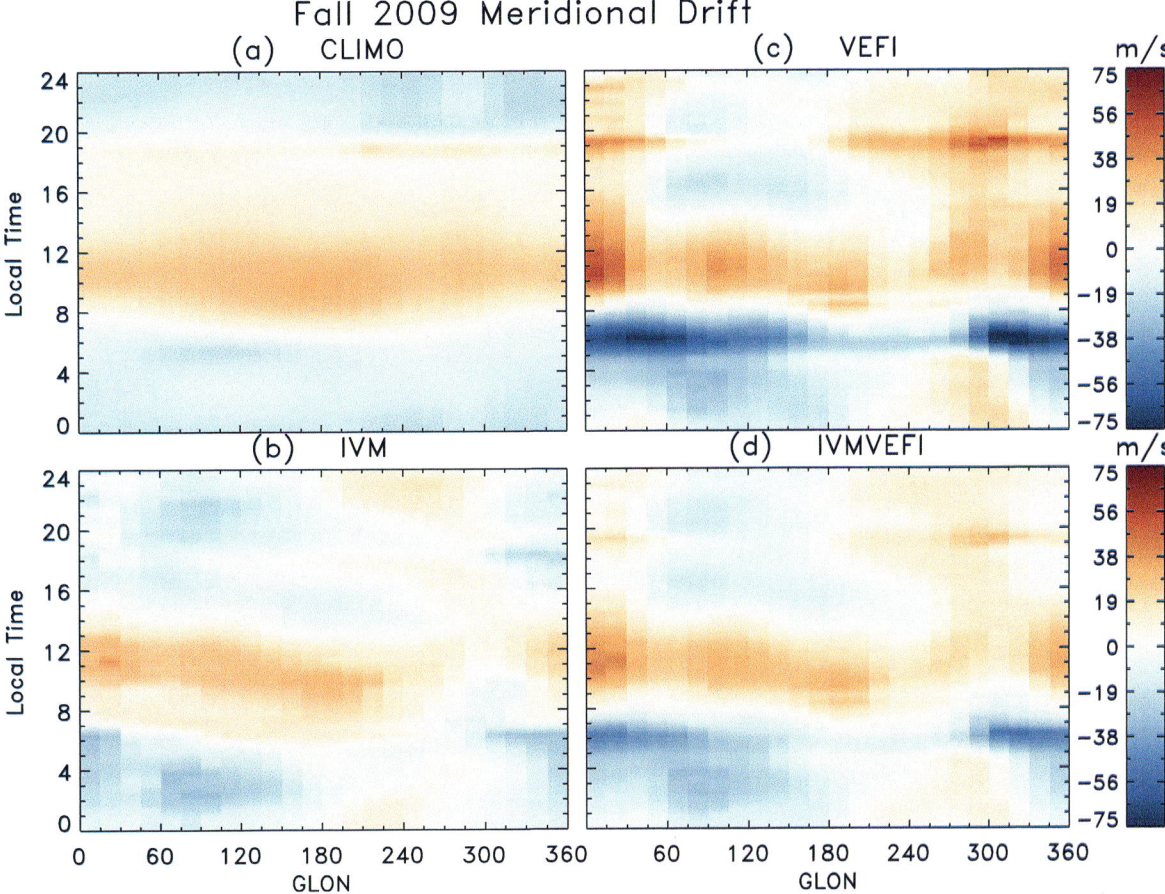

Figure 4. Simulations of PBMOD during the northern fall of 2009 were driven by four drift models: (a) the *Scherliess-Fejer* (SF) empirical model, (b) the 67 day averaged drifts from Ion Velocity Drift Meter (IVM), (c) the averaged drifts from VEFI, and (d) the averaged drift from both IVM and VEFI, where the red and blue represent upward and downward drifts respectively.

570 km altitude during northern fall 2009 are chosen as an example shown in Figures 5a–5e. A distinct observed feature from Figure 5a is the wave 3 or wave 4 longitude structure. This density feature has not been captured using the CLIMO drift model as the driver for PBMOD (Figure 5b); however, model densities (Figures 5c–5e) using average in situ drift measurements as input resemble this observational longitudinal structure. A similar density structure has also been identified during the northern spring and summer months. This structure is weakest during the northern winter season.

For a quantitative study, the logarithmic value of the density ratio between model output and measurement, $\log_{10}\left(\frac{N_{\text{MOD}}}{N_{\text{OBS}}}\right)$, has been calculated for each bin. This value is positive (negative) when the model overestimates (underestimates) the background ionospheric density using in situ measurements as the ground truth. The summations of all positive values divided by the number of positive grid points, $\sum_{i=1}^{n_+}\log\left(\frac{N_{\text{MOD}}}{N_{\text{OBS}}}>1\right)/n_+$, as a function of altitude are shown on the right-hand side of the dashed line in Figure 5f, while $\sum_{i=1}^{n_-}\log\left(\frac{N_{\text{MOD}}}{N_{\text{OBS}}}<1\right)/n_-$ are plotted on the left. We then define the averaged error as $\sum_{i=1}^{n_++n_-}\text{abs}\left[\log\left(\frac{N_{\text{MOD}}}{N_{\text{OBS}}}\right)\right]/(n_++n_-)$, where n_+ and n_- represent the number of positive and negative grid points, respectively. The model output with the smallest error is marked by a pair of solid circles. In Figure 5f, the model driven by the IVM drifts produces the smallest error (red curves with solid circles) for all altitudes, except at ~350 km. Owing to CHAMP and DMSP satellite orbits, there are 68% coverage in LT-GLON from CHAMP and 25% coverage from DMSP when calculating seasonal averaged density errors in Figure 5f. The same technique was also applied to other seasons (results not shown here). We found that 80% of the smallest error occurs when the

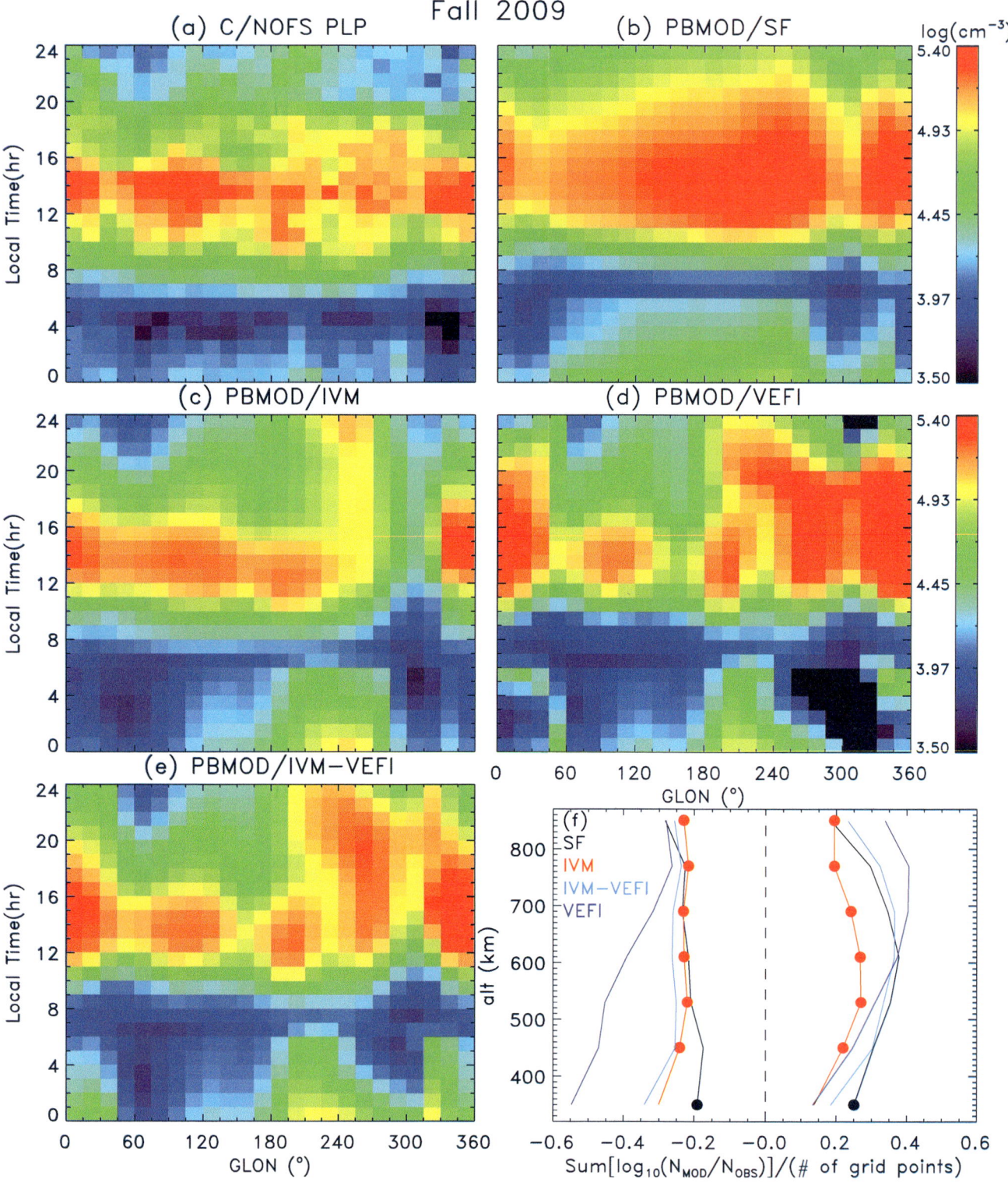

model was driven by IVM measurements, while 10% is based on the CLIMO drifts, and another 10% is from the combination of IVM and VEFI measurements.

4. SUMMARY AND DISCUSSION

This article describes capabilities of the AFRL-PBMOD specifying two key performance parameters as defined by the Air Force Space Command: electron density profiles and scintillations. Using data from an April 2009 campaign dedicated toward C/NOFS measurements, we demonstrate the successful comparisons of scintillation estimation from PBMOD driven by VEFI measurements with ground data at VHF frequencies collected at three SCINDA stations (ANC, CXI, and KWA). In each of these cases, the starting time and the duration of scintillations from the model are comparable to measured S_4 indexes, although the magnitudes might not match precisely. We found these scintillation regions to be associated with upward drifts confirming the R-T instability theory that the nightside ionosphere is unstable due to eastward electric fields (i.e., upward plasma drifts). Although spurious data gaps prevent us from precisely determining the physical conditions, the C/NOFS mission allows us to directly ingest electric field measurements into the model to study ionospheric irregularities for the first time.

In our previous study [*Su et al.*, 2011], we showed that, on average, the densities obtained from PBMOD, when directly ingesting VEFI electric fields, agree better with observed nighttime background densities than when PBMOD was driving with climatological electric fields during June 2008. Although we are not currently able to directly assimilate IVM drifts, as reliable data are obtained only in O^+ rich environments (<550 km altitude during solar minimum), averaged seasonal-dependent plasma drift patterns have been derived for statistical global modeling. Four different drift models are tested in this study: (1) Scherliess-Fejer climatology, (2) IVM, (3) VEFI, and (4) a combination of IVM and VEFI. The density outputs from PBMOD are compared in detail with observations from CHAMP, C/NOFS, and DMSP. The results clearly indicate longitudinal density structures during all seasons, although the signature is the weakest during northern winter. The average drifts derived from C/NOFS measurements shown in Figures 4b–4d also exhibit similar longitudinal structures. On the other hand, this density feature has not been reproduced using the model driven by the climatological drifts because the SF model does not capture this longitudinal velocity structure. Recently, the longitudinal wave 3 or wave 4 signature has been repeatedly reported during solar minimum at the beginning of solar cycle 24 [e.g., *Dao et al.*, 2011; *Su et al.*, 2011; *Huang et al.*, 2011], and it has been associated with atmospheric nonmigrating tides [*Immel et al.*, 2006; *England et al.*, 2006; *Wan et al.*, 2008]. In order to fully understand thermospheric forcing, a knowledge of neutral dynamics is essential. Owing to the increase in solar activity, the Neutral Wind Meter [*Earle et al.*, 2007] on board C/NOFS was turned on in the northern fall of 2011. With in situ neutral wind measurements and increasing coverage of IVM, this new data may shed some light on low-latitude thermosphere-ionosphere dynamics.

In this study, we quantitatively estimate the density errors of PBMOD driven by four types of plasma drifts. From June 2008 to February 2010 (seven seasons) at seven altitude ranges, 80% of the smallest density error occurs when the model was driven by IVM measurements, while 10% is based on the CLIMO drifts, and another 10% is from the combination of IVM and VEFI measurements. In PBMOD, the ionospheric dynamics are much more sensitive to the meridional drift, which reflects the fact that IVM provides more accurate meridional drifts when available, and VEFI measures better zonal drifts. Regardless of environmental or instrumental difficulties in measurements, we expect that a more accurate and continuous plasma drift data set from C/NOFS will be obtained using both VEFI and IVM observations combined through a machine learning technique currently being tested at the University of Texas at Dallas [*Brown et al.*, 2008]. Accurate and continuous measurements of plasma drifts and/or electric fields will increase the accuracy of predicted plasma densities in the low-latitude ionosphere.

Acknowledgments. The C/NOFS mission is supported by the AFRL, the Department of Defense Space Test Program, the National Aeronautics and Space Administration, the Naval Research

Figure 5. (opposite) (a) The C/NOFS-PLP density measurements, (b) the model output of background densities driven by the SF empirical drift model, (c) the model output driven by the averaged IVM drifts, (d) the model output driven by the averaged VEFI measurements, and (e) the model output driven by the averaged drifts from IVM and VEFI measurements at altitude between 490 and 570 km during the northern fall of 2009. (f) Estimated density errors as a function of altitudes from simulations driven from four different drifts: SF (black), IVM (red), VEFI (dark blue), and IVM-VEFI (light blue). The positive values (right of the vertical dashed line) represent that the model overestimates the background density compared to observations, while the negative values (left of the vertical dashed line) indicate an underestimation of the density by the model.

Laboratory, and the Aerospace Corporation. The first author thanks the IVM principle investigator, R. A. Heelis, for supporting the usage of drift data. This research task was supported, in part, by the NASA grant NNH09AK051, the AFOSR grants 11RV04COR and 12RV10COR to AFRL, and the Ionospheric Impacts on RF Systems 6.2 program at AFRL.

REFERENCES

AIAA (1999), *Guide to Reference and Standard Ionosphere Models*, Am. Natl. Stand. Inst., Am. Inst. of Aeronaut. and Astronaut., Inc., Reston, Va.

Anderson, D. N. (1973), A theoretical study of the ionospheric F-region equatorial anomaly, I. theory, *Planet. Space Sci.*, *21*, 409–419.

Anderson, D. N., J. M. Forbes, and M. Codrescu (1989), A fully analytic, low- and middle-latitude ionospheric model, *J. Geophys. Res.*, *94*(A2), 1520–1524.

Brown, M. E., D. J. Lary, A. Vrieling, D. Stathakis, and H. Mussa (2008), Neutral networks as a tool for constructing continuous NDVI time series from AVHRR and MODIS, *Int. J. Remote Sens.*, *29*(24), 7141–7158, doi:10.1080/01431160802238435.

Dao, E., M. C. Kelley, P. Roddy, J. Retterer, J. O. Ballenthin, O. de La Beaujardière, and Y.-J. Su (2011), Longitudinal and seasonal dependence of nighttime equatorial plasma density irregularities during solar minimum detected on the C/NOFS satellite, *Geophys. Res. Lett.*, *38*, L10104, doi:10.1029/2011GL047046.

de La Beaujardière, O., et al. (2004), C/NOFS: A mission to forecast scintillations, *J. Atmos. Sol. Terr. Phys.*, *66*, 1573–1591, doi:10.1016/j.jastp.2004.07.030.

Drob, D. P., et al. (2008), An empirical model of the Earth's horizontal wind fields: HWM07, *J. Geophys. Res.*, *113*, A12304, doi:10.1029/2008JA013668.

Earle, G. D., J. H. Klenzing, P. A. Roddy, W. A. Macaulay, M. D. Perdue, and E. L. Patrick (2007), A new satellite-borne neutral wind instrument for the thermospheric diagnostics, *Rev. Sci. Instrum.*, *78*, 114501, doi:10.1063/1.2813343.

Eccles, V., D. D. Rice, J. J. Sojka, C. E. Valladares, T. Bullett, and J. L. Chau (2011), Lunar atmospheric tidal effects in the plasma drifts observed by the Low-Latitude Ionospheric Sensor Network, *J. Geophys. Res.*, *116*, A07309, doi:10.1029/2010JA016282.

England, S. L., S. Maus, T. J. Immel, and S. B. Mende (2006), Longitudinal variation of the E-region electric fields caused by atmospheric tides, *Geophys. Res. Lett.*, *33*, L21105, doi:10.1029/2006GL027465.

Fejer, B. G., and L. Scherliess (1997), Empirical models of storm time equatorial zonal electric fields, *J. Geophys. Res.*, *102*(A11), 24,047–24,056.

Fejer, B. G., E. R. de Paula, S. A. González, and R. F. Woodman (1991), Average vertical and zonal F region plasma drifts over Jicamarca, *J. Geophys. Res.*, *96*(A8), 13,901–13,906.

Fejer, B. G., M. E. Olson, J. L. Chau, C. Stolle, H. Lühr, L. P. Goncharenko, K. Yumoto, and T. Nagatsuma (2010), Lunar-dependent equatorial ionospheric electrodynamic effects during sudden stratospheric warmings, *J. Geophys. Res.*, *115*, A00G03, doi:10.1029/2010JA015273.

Groves, K. M., et al. (1997), Equatorial scintillation and systems support, *Radio Sci.*, *32*(5), 2047–2064.

Gulyaeva, T. L., and J. E. Titheridge (2006), Advanced specification of electron density and temperature in the IRI ionosphere-plasmasphere model, *Adv. Space Res.*, *38*(11), 2587–2595.

Hedin, A. E. (1987), MSIS-86 thermospheric model, *J. Geophys. Res.*, *92*(A5), 4649–4662.

Heelis, R. A., W. R. Coley, A. G. Burrell, M. R. Hairston, G. D. Earle, M. D. Perdue, R. A. Power, L. L. Harmon, B. J. Holt, and C. R. Lippincott (2009), Behavior of the O+/H+ transition height during the extreme solar minimum of 2008, *Geophys. Res. Lett.*, *36*, L00C03, doi:10.1029/2009GL038652.

Hinteregger, H. E., K. Fukui, and B. R. Gilson (1981), Observational, reference and model data on solar EUV, from measurements on AE-E, *Geophys. Res. Lett.*, *8*(11), 1147–1150.

Huang, C. Y., F. A. Marcos, P. A. Roddy, M. R. Hairston, W. R. Coley, C. Roth, S. Bruinsma, and D. E. Hunton (2009), Broad plasma decreases in the equatorial ionosphere, *Geophys. Res. Lett.*, *36*, L00C04, doi:10.1029/2009GL039423.

Huang, C. Y., S. H. Delay, P. A. Roddy, and E. K. Sutton (2011), Periodic structures in the equatorial ionosphere, *Radio Sci.*, *46*, RS0D14, doi:10.1029/2010RS004569.

Immel, T. J., E. Sagawa, S. L. England, S. B. Henderson, M. E. Hagan, S. B. Mende, H. U. Frey, C. M. Swenson, and L. J. Paxton (2006), Control of equatorial ionospheric morphology by atmospheric tides, *Geophys. Res. Lett.*, *33*, L15108, doi:10.1029/2006GL026161.

Jasperse, J. R. (1976), Boltzmann-Fokker-Planck model for the electron distribution function in the Earth's ionosphere, *Planet. Space Sci.*, *24*, 33–40.

Kelley, M. C. (2009), *The Earth's Ionosphere: Plasma Physics and Electrodynamics*, 2nd ed., Academic Press, London, U. K.

Pfaff, R., et al. (2010), Observations of DC electric fields in the low-latitude ionosphere and their variations with local time, longitude, and plasma density during extreme solar minimum, *J. Geophys. Res.*, *115*, A12324, doi:10.1029/2010JA016023.

Preble, A. J., D. N. Anderson, B. G. Fejer, and P. H. Doherty (1994), Comparison between calculated and observed F region electron density profiles at Jicamarca, Peru, *Radio Sci.*, *29*(4), 857–866.

Rappaport, H. L. (1996), Field line integration and localized modes in the equatorial spread F, *J. Geophys. Res.*, *101*(A11), 24,545–24,551.

Retterer, J. M. (2005), Physics-based forecasts of equatorial radio scintillation for the Communication and Navigation Outage Forecasting System (C/NOFS), *Space Weather*, *3*, S12C03, doi:10.1029/2005SW000146.

Retterer, J. M. (2010a), Forecasting low-latitude radio scintillation with 3-D ionospheric plume models: 1. Plume model, *J. Geophys. Res.*, *115*, A03306, doi:10.1029/2008JA013839.

Retterer, J. M. (2010b), Forecasting low-latitude radio scintillation with 3-D ionospheric plume models: 2. Scintillation calculation, *J. Geophys. Res.*, *115*, A03307, doi:10.1029/2008JA013840.

Retterer, J. M., and M. C. Kelley (2010), Solar wind drivers for low-latitude ionosphere models during geomagnetic storms, *J. Atmos. Sol. Terr. Phys.*, *72*(4), 344–349, doi:10.1016/j.jastp.2009.07.003.

Retterer, J. M., D. T. Decker, W. S. Borer, R. E. Daniell Jr., and B. G. Fejer (2005), Assimilative modeling of the equatorial ionosphere for scintillation forecasting: Modeling with vertical drifts, *J. Geophys. Res.*, *110*, A11307, doi:10.1029/2002JA009613.

Scherliess, L., and B. G. Fejer (1999), Radar and satellite global equatorial *F* region vertical drift model, *J. Geophys. Res.*, *104*(A4), 6829–6842.

Schunk, R. W. (Ed.) (1996), *Solar-Terrestrial Energy Program (STEP): Handbook of Ionospheric Models*, Cent. for Atmos. and Space Sci., Logan, Utah.

Schunk, R. W., and A. F. Nagy (2000), *Ionospheres – Physics, Plasma Physics, and Chemistry*, Cambridge Univ. Press, Cambridge, U. K.

Su, Y.-J., J. M. Retterer, O. de La Beaujardière, W. J. Burke, P. A. Roddy, R. F. Pfaff Jr., G. R. Wilson, and D. E. Hunton (2009), Assimilative modeling of equatorial plasma depletions observed by C/NOFS, *Geophys. Res. Lett.*, *36*, L00C02, doi:10.1029/2009GL038946.

Su, Y.-J., J. M. Retterer, R. F. Pfaff, P. A. Roddy, O. deLaBeaujardière, and J. O. Ballenthin (2011), Assimilative modeling of observed postmidnight equatorial plasma depletions in June 2008, *J. Geophys. Res.*, *116*, A09318, doi:10.1029/2011JA016772.

Sultan, P. J. (1996), Linear theory and modeling of the Rayleigh-Taylor instability leading to the occurrence of equatorial spread *F*, *J. Geophys. Res.*, *101*(A12), 26,875–26,891.

Wan, W., L. Liu, X. Pi, M.-L. Zhang, B. Ning, J. Xiong, and F. Ding (2008), Wavenumber-4 patterns of the total electron content over the low latitude ionosphere, *Geophys. Res. Lett.*, *35*, L12104, doi:10.1029/2008GL033755.

R. G. Caton, P. A. Roddy, and Y.-J. Su, AFRL/RVBXP, BEL Bldg 570, 3550 Aberdeen Ave, S.E., Kirtland AFB, NM 87117, USA. (Yi-Jiun.Su@kirtland.af.mil)

K. M. Groves and J. M. Retterer, Institute for Scientific Research, Boston College, 400 St. Clement's Hall, 140 Commonwealth Ave., Chestnut Hill, MA 02467-3862, USA.

R. F. Pfaff, NASA Goddard Space Flight Center, Code 696, Greenbelt, MD 20771, USA.

R. A. Stoneback, The University of Texas at Dallas, Department of Physics, EC 36, 800 West Campbell Rd, Richardson, TX 75080-3021, USA.

Long-Term Simulations of the Ionosphere Using SAMI3

S. E. McDonald,[1] J. L. Lean,[1] J. D. Huba,[2] G. Joyce,[3] J. T. Emmert,[1] and D. P. Drob[1]

The Naval Research Laboratory (NRL) is conducting an interdisciplinary physics-based space weather model development and validation program called the Integrated Sun-Earth System for the Operational Environment (ISES-OE). The goal of ISES-OE is to improve the quantitative understanding of the space environment, which can disrupt or degrade operational communications and navigation systems and, ultimately, to provide the ability to forecast space weather on multiple time scale, from hours to the 11 year solar cycle. The core ISES-OE model is Another Model of the Ionosphere (SAMI3), NRL's state-of-the-art ionosphere model. As a part of this program, a comprehensive, systematic validation of SAMI3's current capability was conducted to specify the midlatitude to low-latitude ionosphere and its response to heliospheric forcing and thermospheric oscillations. An ensemble of simulation runs is generated using SAMI3 for the solar and geomagnetic conditions during the Whole Heliosphere Interval 2008. The simulations are driven with a solar irradiance model based on Thermosphere, Ionosphere, Mesosphere Energetics and Dynamics (TIMED)/Solar EUV Experiment measurements and with daily Ap and $F10.7$ indices. Thermospheric conditions are specified with NRL's empirical models of the neutral composition and temperature (NRLMSISE-00) and the neutral wind (HWM07). Various input parameters are selected or held constant to quantify their effects on the ionosphere. Additionally, simulations are performed using both empirical electric fields and self-consistently calculated fields. Simulation results are compared with ground- and space-based observations, including ionosondes and GPS-derived global total electron content maps; we illustrate how we are using such comparisons for model validation. Initial results of a multiyear run of SAMI3 over the descending phase of Solar Cycle 23 (2002–2008) are also shown.

1. INTRODUCTION

The geospace environment encompasses vast regions of space from the Sun to the Earth's atmosphere. In order to model and predict the weather of the near-Earth space environment, it is necessary to understand the important coupling mechanisms from the surface of the Sun to the Earth's ionosphere, including its coupling with the atmosphere below. Space weather significantly impacts satellite orbits, satellite communication, and navigation. Decades of research have led to a number of sophisticated physics-based models of various aspects of this region, but only very recently, the space science community has begun to integrate the models in order to understand the coupling between different regions. Simulations that integrate the thermosphere, ionosphere, plasmasphere, and magnetosphere began only in 2001, three decades after numerical models of terrestrial weather integrated the lower atmosphere, oceans, and land.

The purpose of the Naval Research Laboratory's (NRL's) Integrated Sun-Earth System for the Operational Environment

[1] Space Science Division, Naval Research Laboratory, Washington, District of Columbia, USA.
[2] Plasma Physics Division, Naval Research Laboratory, Washington, District of Columbia, USA.
[3] Icarus Research, Inc., Bethesda, Maryland, USA.

Modeling the Ionosphere-Thermosphere System
Geophysical Monograph Series 201
© 2013. American Geophysical Union. All Rights Reserved.
10.1029/2012GM001301

(ISES-OE) program is to quantitatively understand and specify the space environment above 100 km. In contrast to other Sun-Earth integration efforts, a key aspect of ISES is determining the extent to which current models capture the climate of the ionosphere/thermosphere. Until now, complementary long-term simulations using physics-based ionosphere models have not been performed. As part of ISES, an ensemble of multiyear simulations encompassing an entire solar cycle is underway. In conjunction, large databases of drag-derived and remotely sensed neutral thermospheric densities and GPS-derived total electron content (TEC) are being analyzed to provide model validation on time scales as long as the 11 year solar cycle. Several recent studies have taken advantage of long-term measurements of TEC to analyze this past solar cycle [*Afraimovich et al.*, 2008; *Hocke*, 2008, 2009; *Lean et al.*, 2011a, 2011b; *Liu et al.*, 2009].

Figure 1 shows the individual models that compose ISES-OE and the various inputs used to drive the models. The core of ISES-OE consists of Another Model of the Ionosphere (SAMI3) [*Huba et al.*, 2000], NRL's state-of-the-art physics-based 3-D ionosphere model from 85 km to ~8 R_E. SAMI3 is being coupled with NRL's Lyons-Fedder-Mobarry magnetosphere model and an inner magnetosphere module to properly specify the ionospheric boundary conditions. To drive the coupled models, ISES-OE utilizes NRL's Solar Spectral Irradiance (NRLSSI) model of solar EUV spectrum variability based on the Thermosphere, Ionosphere, Mesosphere Energetics and Dynamics (TIMED)/SEE measurements. The ambient solar wind is specified by the NRL Wang-Sheeley model that extrapolates solar surface magnetic fields to the corona and heliosphere, and the coronal mass ejection (CME) impacts by a suite of NRL instruments on Large Angle and Spectrometric Coronagraph (LASCO) and Solar Terrestrial Relations Observatory (STEREO), semiempirical NRL CME propagation models, and from L1 monitoring.

In the results presented here, we report just the simulations of the midlatitude to low-latitude ionosphere. Multiday simulations during the Whole Heliosphere Interval (WHI) 2008 are performed using two versions of SAMI3: (1) SAMI3 with

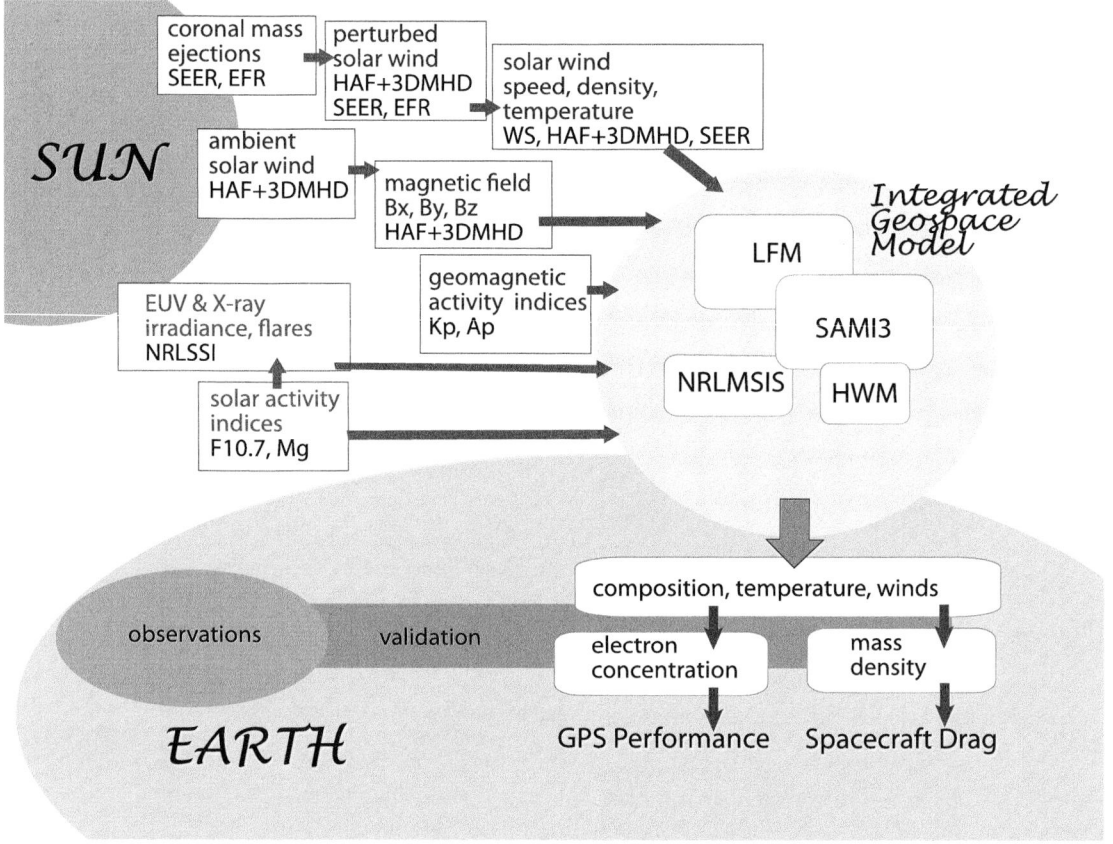

Figure 1. A schematic of models that comprise the Integrated Sun-Earth System for the Operational Environment and the various inputs used to drive the models. Model output, including ionospheric/thermospheric composition, temperature, electric fields, and neutral winds are being validated against observations. Derived products include total electron content (TEC), NmF2, hmF2, and total mass density.

externally specified $E \times B$ drifts and (2) SAMI3 with a potential solver to self-consistently specify electric fields. The results are compared with GPS-derived global TEC maps. We also investigate in more detail the $E \times B$ drifts calculated by the self-consistent SAMI3 and compare these results with an empirical model. Finally, we provide initial results for a multiyear run of the descending phase of Solar Cycle 23 to illustrate the broader range of ISES activity underway.

2. MODEL DESCRIPTION

SAMI3 models the plasma and chemical evolution of seven ion species (H^+, He^+, N^+, O^+, N_2^+, NO^+, and O_2^+). The complete ion temperature equation is solved for three ion species (H^+, He^+, and O^+) as well as the electron temperature equation. The plasma equations in SAMI3 are solved in dipole coordinates and are as follows:

$$\frac{\partial n_i}{\partial t} + \nabla \cdot (n_i \mathbf{V}_i) = p_i - L_i n_i, \quad (1)$$

$$\frac{\partial \mathbf{V}_i}{\partial t} + \mathbf{V}_i \cdot \nabla \mathbf{V}_i = -\frac{1}{\rho_i}\nabla P_i + \frac{e}{m_i}\mathbf{E} + \frac{e}{m_i c}\mathbf{V}_i \times \mathbf{B} \\ + \mathbf{g} - \nu_{in}(\mathbf{V}_i - \mathbf{V}_n) - \Sigma_j \nu_{ij}(\mathbf{V}_i - \mathbf{V}_j), \quad (2)$$

$$\frac{\partial T_i}{\partial t} + \mathbf{V}_i \cdot \nabla T_i + \frac{2}{3}T_i \nabla \cdot \mathbf{V}_i + \frac{2}{3}\frac{1}{n_i k}\nabla \cdot \mathbf{Q}_i = Q_{in} + Q_{ii} + Q_{ie}, \quad (3)$$

$$\frac{\partial T_e}{\partial t} - \frac{2}{3}\frac{1}{n_e k}b_s^2 \frac{\partial}{\partial s}k_e \frac{\partial T_e}{\partial s} = Q_{en} + Q_{ei} + Q_{phe}, \quad (4)$$

where p_i values are the ion production terms, and L_i values are the ion loss terms. The other coefficients and parameters are described by *Huba et al.* [2000]. The magnetic field is modeled as a dipole fit to the International Geomagnetic Reference Field.

For the simulations, we use (1) SAMI3 coupled with the Scherliess-Fejer empirical vertical drift model [*Scherliess and Fejer*, 1999] and (2) SAMI3 with self-consistent electric fields. To distinguish the two models, we refer to them as SAMI3 and SAMI3-SC, respectively. In SAMI3-SC, the potential equation is derived from current conservation ($\nabla \cdot \mathbf{J} = 0$) in dipole coordinates (q, p, ϕ) and is given by

$$\frac{\partial}{\partial p}p\Sigma_{pp}\frac{\partial \Phi}{\partial p} + \frac{\partial}{\partial \phi}\frac{\Sigma_{p\phi}}{p}\frac{\partial \Phi}{\partial \phi} - \frac{\partial}{\partial p}\Sigma_H \frac{\partial \Phi}{\partial \phi} + \frac{\partial}{\partial \phi}\Sigma_H \frac{\partial \Phi}{\partial p} \\ = \frac{\partial F_{pV}}{\partial p} + \frac{\partial F_{\phi V}}{\partial \phi} + \frac{\partial F_{\phi g}}{\partial \phi} - \frac{\partial F_{pg}}{\partial p}, \quad (5)$$

where $\Sigma_{pp} = \int (p\Delta/b_s)\sigma_P ds$, $\Sigma_{p\phi} = \int (1/pb_s\Delta)\sigma_P ds$, $\Sigma_H = \int (1/b_s)\sigma_H dq$, $F_{pV} = \int (B_0/c)r \sin\theta(\sigma_P V_n\phi + \sigma_H V_{np})dq$, $F_{\phi V} = \int (B_0/c)(r_0 \sin^3\theta/\Delta)(-\sigma_P V_{np} + \sigma_H V_n\phi)dq$, $F_{\phi g} = \int (r_E\sin^3\theta/\Delta)(B_0/c)(1/\Omega_i)\sigma_{Hi}g_P ds$, $F_{pg} = \int r \sin\theta(B_0/c)(1/\Omega_i)\sigma_{Pi} ds$, σ_P is the Pedersen conductivity, σ_H is the Hall conductivity, $\Delta = (1 + 3\cos^2\theta)^{1/2}$, $b_s = (r_E^3/r^2)\Delta$, and r_E is the radius of the Earth. The perpendicular electric field $E_\perp = -\nabla\Phi$ is used self-consistently in SAMI3-SC to calculate the $E \times B$ drifts [*Huba et al.*, 2008, 2010]. The SAMI3-SC magnetic field is represented by a tilted dipole. Additional details of the SAMI3-SC model are given by *Huba and Joyce* [this volume].

In both versions of the model, the grid is specified from ±60° magnetic latitude. The thermospheric neutral densities, temperature, and winds are provided by the empirical models NRLMSISE-00 [*Picone et al.*, 2002], HWM07 [*Drob et al.*, 2008], and DWM07 [*Emmert et al.*, 2008]. Though the models have only a one-way coupling with the thermosphere and do not fully simulate feedback effects of ionospheric influence on the ionosphere, such effects, on the time scales of interest, are inherently captured in the empirical models.

The daily solar EUV irradiances from 5 to 105 nm are obtained from the NRLSSI 2C and 3C models, which are based on TIMED/SEE measurements [*Lean et al.*, 2011c]. The major difference between the 2C and 3C models is in how the solar cycle variations are determined; both models use the daily *F*10.7 and *Mg II* indices, but the 3C model includes an additional component, specified by the smoothed *Mg II* index, to replicate the solar cycle variations that SEE measures, which are larger than the 2C model calculates in some wavelength regions. Figure 2a shows the solar irradiance from 5 to 105 nm for the 2C and 3C model during the WHI 2008. Figure 2b shows the *F*10.7 and *Ap* indices, used in SAMI3, and *Bz* for context.

3. SAMI3 COMPARISONS WITH IGS TEC DURING THE WHOLE HELIOSPHERE INTERVAL 2008

For this case study, we chose a 47 day period beginning on 1 March 2008 that encompasses the WHI 1, which extends from 20 March 2008 to 16 April 2008 (solar Carrington Rotation 2068). WHI 2008 is an international coordinated observing and modeling effort to characterize the Sun-Earth system (http://ihy2007.org/WHI.) During the period of interest, the 81 day average solar *F*10.7 flux was 70 solar flux units, though the deepest part of the Solar Cycle 23 minimum did not occur until 2009. The WHI period is characterized by low-level activity resulting from coronal holes on the Sun, which were responsible for geomagnetic activity associated with trailing high-speed streams (HSSs) and corotating interaction regions (CIRs) [*Echer et al.*, 2011; *Gibson et al.*, 2011], and their effects are reflected in the multiple spikes in the *Ap* index (Figure 2b). The storm

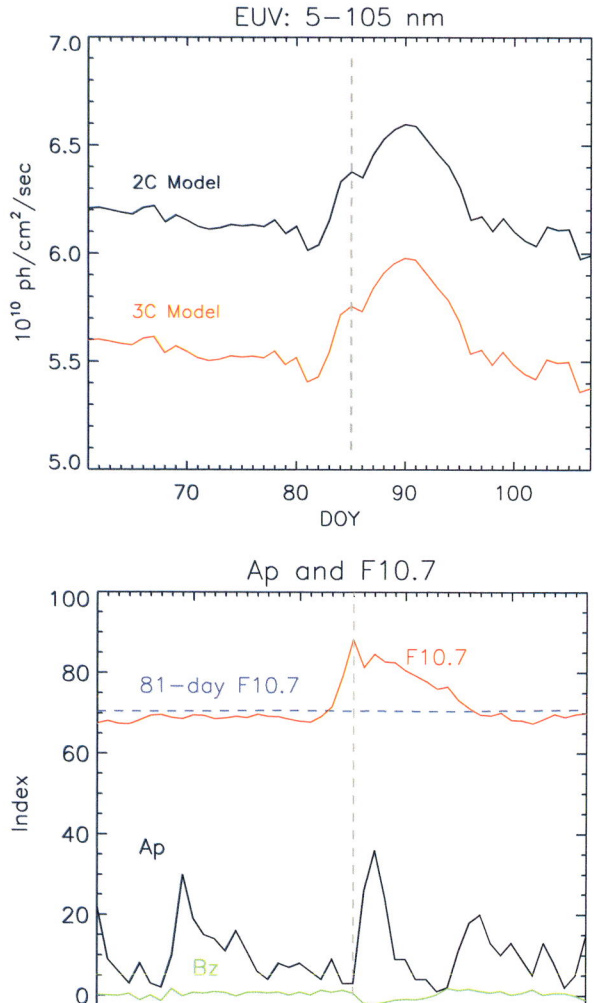

An excellent long-term data product for comparison and validation of the SAMI3 model runs is the International Global Navigation Satellite Systems (GNSS) Service (IGS) global TEC maps, derived from GPS measurements. IGS TEC maps are a weighted mean of maps generated at four different analysis centers including Center for Orbit Determination in Europe, Jet Propulsion Laboratory, European Space Operations Center, and Polytechnical University of Catalonia [*Hernandez-Pajares et al.*, 2009]. The IGS TEC maps are available from 1998 to the present at 2 h resolution on a 2.5° by 5° geographic latitude/longitude grid and can be retrieved from http://cddis.gsfc.nasa.gov.

The IGS cosine-weighted daily mean global TEC is shown in Figure 3 for the WHI 2008. The mean response to the CIRs is on the order of 1 to 3 TECu (1 TECu = 10^{16} electrons m^{-2}), though a geographically localized 2 h response of 30 TECu was measured on day 87 [*Verkhoglyadova et al.*, 2011]. SAMI3 and SAMI-SC simulations are performed over the entire 47 day period. We note that our simulations are performed for the low-latitudes to midlatitudes up to ±60° magnetic latitude and, therefore, do not include high-latitude effects due to the geomagnetic activity during the WHI 2008, with the exception of disturbance winds generated by HWM07/DWM07. In SAMI3, the disturbance winds affect the F region plasma; in SAMI3-SC, these winds also produce a disturbance dynamo. The two SAMI3 runs performed with the NRLSSI 2C and 3C solar irradiances show a similar, but smaller, response to the increases in solar activity. The 3C solar irradiance model brings the

Figure 2. (a) Solar EUV irradiance from the NRL's Solar Spectral Irradiance (NRLSSI) 2C and 3C models. (b) Geomagnetic and solar indices during the WHI 2008 interval. Bz turned negative at the onset of a weak geomagnetic storm due to a corotating interaction region on day 85 (dashed vertical line).

beginning on day 86 was particularly geoeffective as Bz turned southward and took several days to recover. The HSS and CIR events are associated with ~7 and ~9 day periodicities in thermospheric density [*Crowley et al.*, 2008; *Lei et al.*, 2008; *Thayer et al.*, 2008] and electron density in the ionosphere [*Lei et al.*, 2008; *Tulasi Ram et al.*, 2010]. In addition to the HSS events, a CME occurred on day 85 and was associated with an M1.7 class flare [*Nitta*, 2011] that resulted in a jump in the solar EUV irradiance and F10.7. The CME, however, did not affect the solar wind near Earth [*Echer et al.*, 2011].

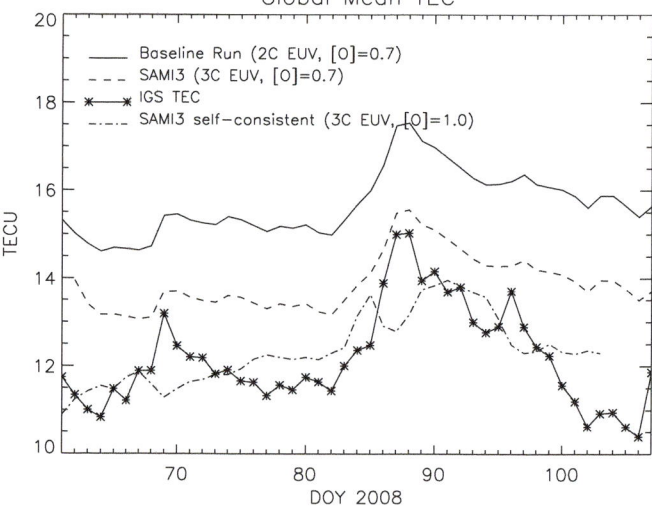

Figure 3. Daily global mean TEC from 1 March through 16 April 2008 is shown for the International Global Navigation Satellite Systems (GNSS) Service (IGS) GPS-derived TEC maps along with that for several A Model of the Ionosphere (SAMI3) simulations.

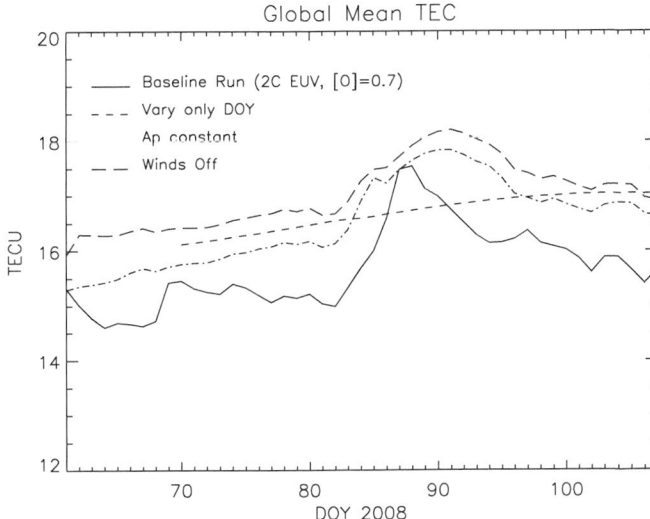

Figure 4. Daily global mean TEC is shown for the SAMI3 simulation using NRLSSI 2C EUV irradiances. The dashed line shows the case in which only the day-of-year was allowed to vary. The dot-dashed line shows the case in which only the Ap index is held constant. The long-dashed line shows the case in which neutral winds are turned off.

SAMI3 TEC into better agreement with the IGS TEC that is achieved with the 2C model, but with a high bias of 2 TECu. It was necessary to reduce the atomic oxygen content from NRLMSISE-00 by a factor of 0.7 in order to further align the SAMI3 TEC absolute level with IGS. This adjustment is consistent with previous findings that NRLMSISE-00 overestimates oxygen content by more than 30% during the recent solar minimum [*Emmert et al.*, 2010]. SAMI3-SC does least well in matching the IGS TEC; this is due to differences in the SAMI3-SC $E \times B$ drifts compared with the Scherliess-Fejer empirical drifts used in SAMI3. These differences are explored in more detail in the next section. First, we provide a more detailed analysis of the SAMI3 results.

To determine the contributing factors to the day-to-day variability in the global TEC, we performed several additional simulations in which various parameters are toggled on and off. In Figure 4, the solid black line corresponds to the "baseline" run shown in Figure 3. With the Ap, $F10.7$, and solar EUV inputs held constant, and only the thermospheric composition allowed to vary from day to day, there is a small increase in the global TEC from day of year (DOY) 70 to 100, which is most likely associated with the semiannual oscillation in thermospheric density that is known to increase in the equinox months. When only the Ap is held constant (dot-dashed line), but all other inputs are allowed to vary from day-to-day, the daily global TEC variation is driven by changes in solar electromagnetic radiation (specified by EUV and $F10.7$) and closely resembles the shape of the variation in the solar EUV irradiance (compare Figure 2), with a peak on day 90. When the Ap and other inputs are allowed to vary, but the effects of the F region neutral winds are turned off, the variation in TEC (long-dashed line) is similar to the Ap constant case. This shows that the F region winds play an important role in determining the day-to-day variation in global TEC associated with weak geomagnetic disturbances driven by CIRs. That the SAMI3 simulations closely match the observed IGS TEC is a good indication that the disturbance wind model (DWM07) within HWM07 is performing well.

We have shown that SAMI3 performs quite well in comparison to measurements of daily averaged global TEC; but not surprisingly upon closer inspection, there are significant differences related to the longitudinal/latitudinal distribution of electron density that manifest on higher temporal and spatial scales. In Figure 5, 2 h global IGS TEC is shown in comparison with the 15 min global SAMI3 TEC. Only several days during the WHI period (days 80–90) are shown in order to emphasize the out-of-phase diurnal variations of SAMI3 and IGS that persist throughout the WHI. Figure 6 illustrates the regional distribution of TEC on days 86–87 responsible for the global differences. TEC at 10, 16, and 22 UT on day 86, and 4 and 10 UT on day 87, are shown for IGS (top row), SAMI3 (middle row), and SAMI3-SC (bottom row). The phase difference in global TEC is very apparent in Figures 5 and 6. On day 86 at 22 UT, the IGS TEC is quite

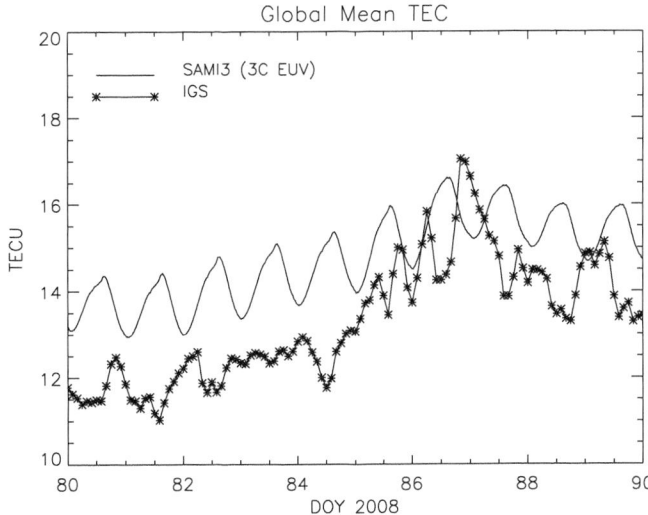

Figure 5. A 2 h global mean IGS TEC is shown for days 80–90 during WHI 2008. The 15 min global mean TEC for SAMI3 using the NRLSSI 3C model is also shown. On most days, there appears to be a 180° phase shift in the diurnal variation of TEC between the data and model.

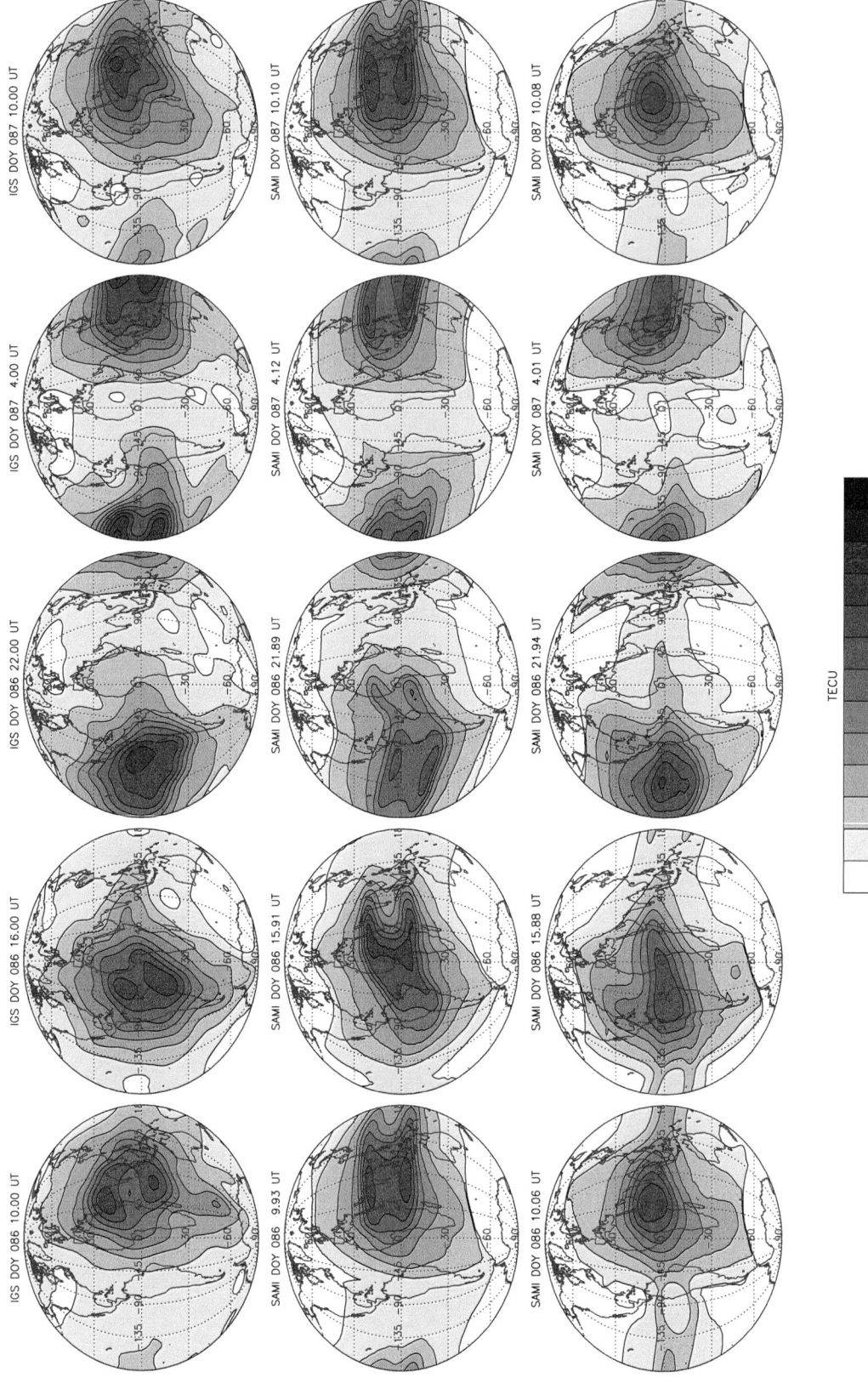

Figure 6. TEC at 10, 18, and 22 UT on day 86 and 10 UT on day 87 during WHI 2008 for (top) IGS, (middle) SAMI3, and (bottom) SAMI-SC. On day 86 at 22 UT, the IGS TEC is quite strong, whereas the SAMI3 TEC is much weaker in the equatorial ionization anomaly (EIA) region. SAMI3-SC shows no significant separation of the EIA crests in comparison with SAMI3.

strong, whereas the SAMI3 TEC is much weaker in the equatorial ionization anomaly (EIA) region. The diurnal variation in TEC is associated with the F region winds; when the winds in SAMI3 are turned off, the daily increase and decrease in global TEC (illustrated in Figure 5) goes away. These comparisons indicate clearly that further investigation and improved representation of winds should improve the SAMI3 model.

Despite the fact that the SAMI3-SC global TEC does not agree with IGS very well, there are times at which the modeled and observed TEC morphology is in good agreement. For example, on day 86 at 22 UT and day 87 at 10 UT, both IGS and SAMI3-SC show a collapsed EIA structure, whereas SAMI3 shows well-separated EIA crests. The IGS TEC maps indicate that there is a good deal of variability in the EIA crests, both day-to-day and longitudinally. At times/longitudes, the anomaly crests are not well separated, the IGS maps are more likely to resemble the SAMI3-SC TEC. That the daily mean TEC more closely resembles the SAMI3 simulations is an indication that the anomaly crests are typically well separated. Previous modeling studies have shown that the separation of the EIA crests is closely related to the strength of the vertical $E \times B$ drifts [*England et al.*, 2008]. It is well known that there are day-to-day and longitudinal variations in the drifts that are associated with lower-atmospheric effects and high-latitude disturbances. Such variation is not captured in climatological models such as HWM07 and the Scherliess-Fejer drift model with the result that SAMI3 shows well-separated anomaly crests at all times. On the other hand, SAMI3-SC shows collapsed crests at all times due to weak daytime vertical $E \times B$ drifts. We would expect that the HWM07 wind climatology would generally reproduce the climatology of the vertical drifts, in which case the SAMI3 and SAMI3-SC TEC would match. That this is not the case suggests that SAMI3-SC is not capturing the E region physics as well as is needed to properly specify the ionosphere.

4. MODELING VERTICAL DRIFTS

One of the most important processes controlling the large-scale distribution of plasma is the generation of a zonal (east-west) electric field that drives a vertical $E \times B$ drift of F region plasma at the equator. To first order, the neutral winds flow from the dayside to the nightside leading to upward plasma drift during the day and downward drift at night. To examine the vertical drifts in SAMI3-SC, we perform a series of runs under solar maximum and minimum conditions for each season and compare the results with the Scherliess-Fejer empirical vertical drift model. Runs are performed using HWM07 and also, for comparison, HWM93 [*Hedin et al.*, 1996; *Huba et al.*, 2009]. Owing to the significant addition of many new data sets and an improved mathematical formulation in HWM07, we expect this wind model to produce a better representation of the vertical drifts than HWM93. However, *Huba et al.* [2009] reported an instance in which HWM93 produced results that were a better match to observations.

Simulations are conducted for day 80 (March), 172 (June), 264 (September), and 355 (December) for the years 2002 (Figure 7) and 2006 (Figure 8). These are all fairly quiet days with $Ap < 21$. In each figure, vertical drifts as a function of longitude and LT are shown for the HWM07 case (left column), the Scherliess-Fejer drift model (middle column), and HWM93 case (right column). We note that HWM does not have an $F10.7$ dependence, so solar cycle differences between the $E \times B$ drifts are due to the differences in thermospheric composition. These simulations indicate that the SAMI3-SC daytime drifts are generally weaker than the Scherliess-Fejer drifts, and this is the case for both versions of HWM. While the SAMI3-SC maximum vertical drifts do not exceed ~15 m s^{-1} during the day, the empirical drifts are generally on the order of ~30 m s^{-1}. The SAMI3-SC drifts are also weaker during the late afternoon and are often downward. At solar maximum, afternoon drifts are downward at all longitudes in March and December, whereas at solar minimum, the afternoon drifts are downward in all seasons. Daytime drifts produced with HWM93 and HWM07 are roughly the same in magnitude and direction, but with differences in longitudinal structure. An exception is September 2006, where the afternoon drifts are upward in the afternoon at some longitudes in the HWM07 case but uniformly downward for HWM93 winds. The significant difference between the SAMI3-SC daytime drifts and that of Scherliess-Fejer offers a primary explanation for why the SAMI3-SC global mean diurnal variation in TEC does not compare well with that of the IGS TEC. As shown in the previous section, weak daytime drifts produce EIA crests that are close to the magnetic equator. F region disturbance winds associated with the geomagnetic activity during the WHI period have little effect on the EIA at such low latitudes; this is the reason the SAMI3-SC global mean TEC resembles that of the SAMI3 simulation in which winds are turned off.

In the evening, the climatological vertical plasma drifts are characterized by a strong postsunset upward drift, typically referred to as the prereversal enhancement (PRE), before the drifts generally turn downward at night. The PRE is the result of complex interactions of the E and F region winds and conductivities. Because of the large upward drifts associated with this PRE of the electric field, it is a significant factor in the development of ionospheric irregularities [*Anderson*

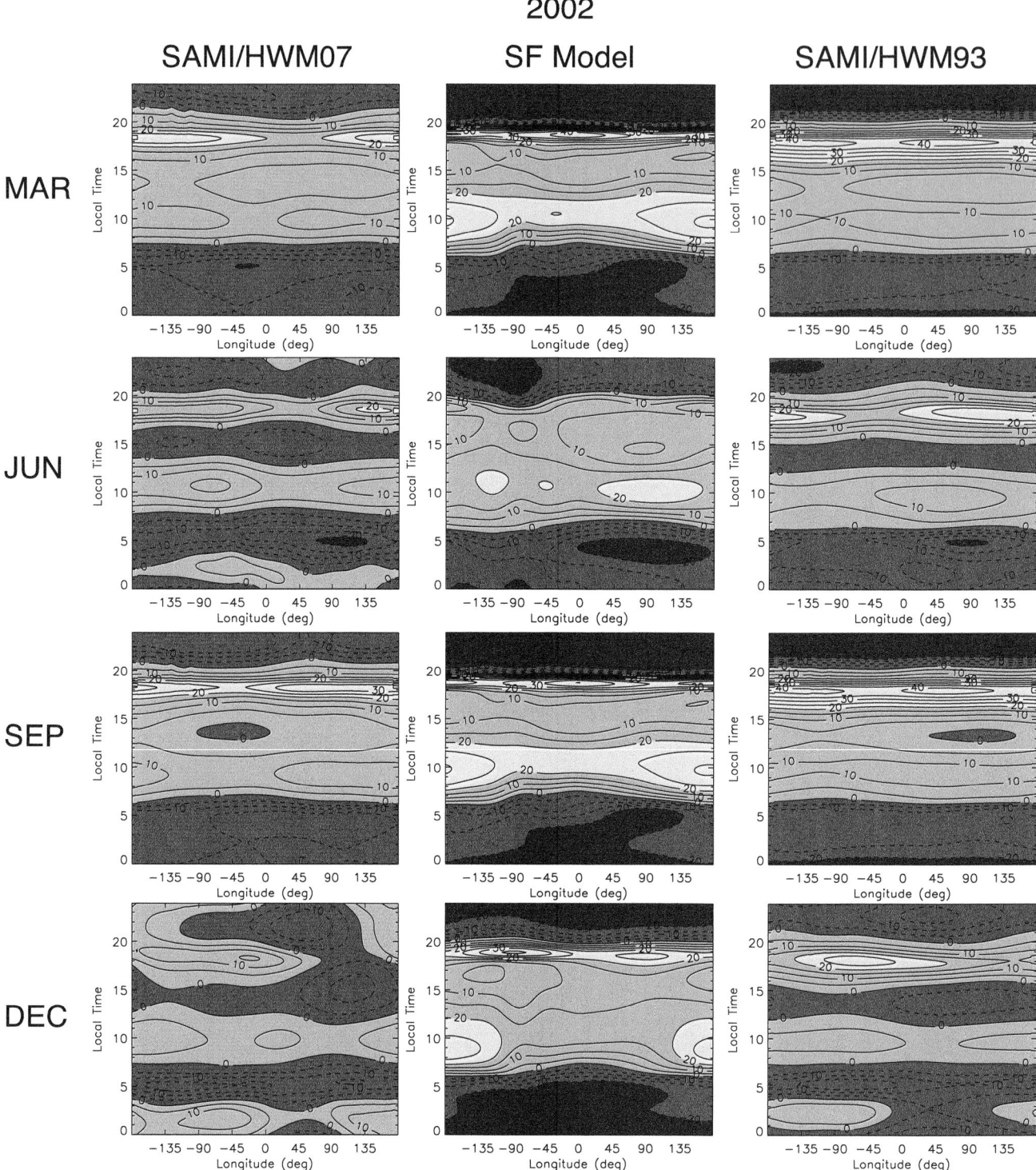

Figure 7. SAMI3-SC vertical plasma drifts are shown for day 80 (March equinox), day 172 (June solstice), day 264 (September equinox), and day 355 (December solstice) for the solar maximum year 2002. SAMI3-SC simulations are performed using (left) HWM07, (right) HWM93 and are compared with the (middle) Scherliess-Fejer vertical drift model.

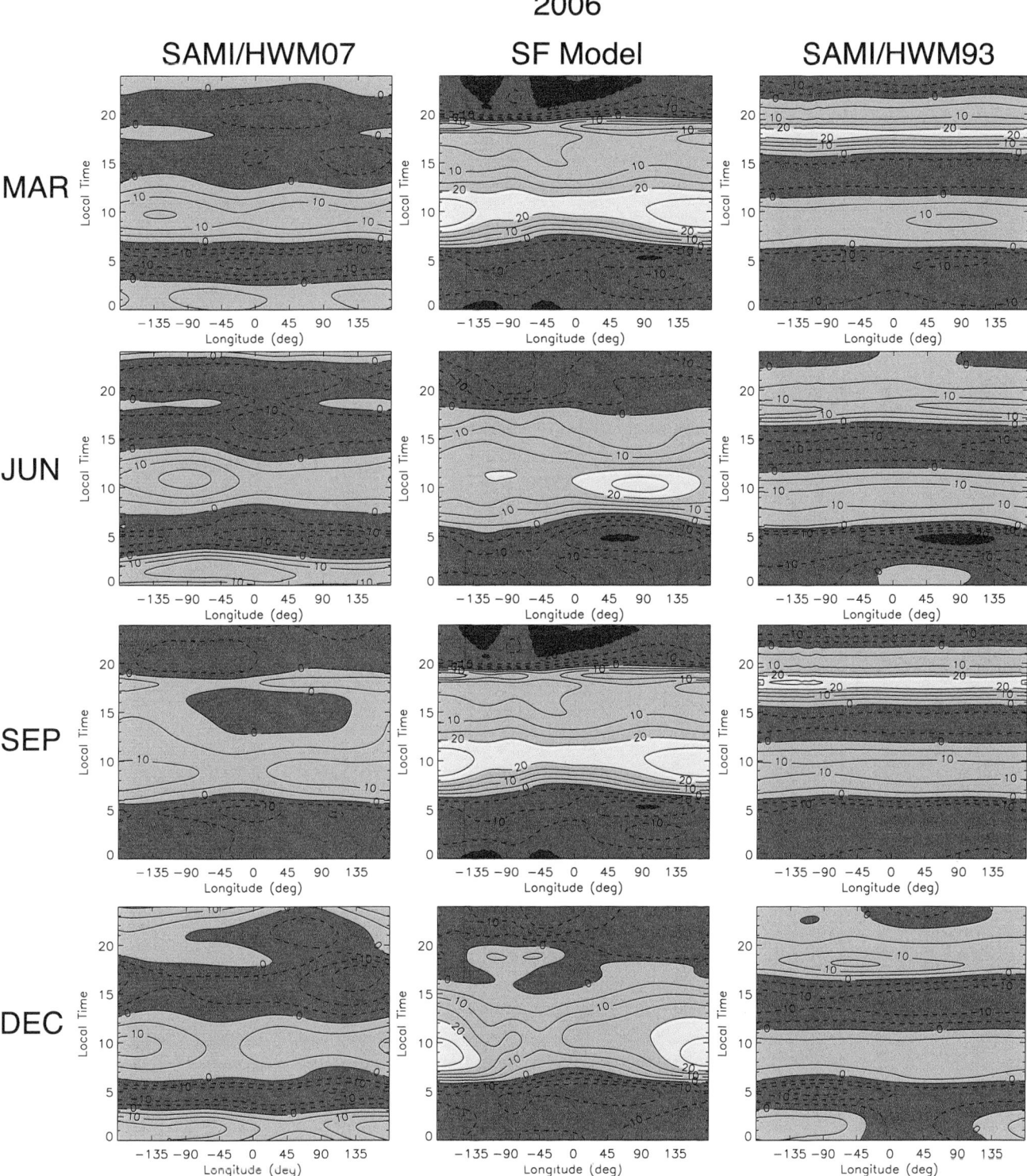

Figure 8. Same as Figure 7, but in this case, the simulations were performed for the year 2006.

et al., 2004; *Basu et al.*, 1996]. There are marked differences between the drifts produced with HWM93 and HWM07. HWM07 generally produces a weaker PRE than HWM93, and at solar minimum, the PRE is not present in most seasons. Additionally, the HWM07 PRE shows greater longitudinal variability and narrower extent in LT than HWM93, especially at the equinoxes. Overall, SAMI3-SC with HWM07 produces a PRE that more closely resembles the climatology, though the peak velocity of the PRE produced with HWM93 is sometimes in better agreement (March equinox, for example).

At nighttime, particularly during the solstices, SAMI3-SC shows postmidnight upward drifts that are not present in the climatology. These upward drifts are more prominent in the HWM07 simulations than in the HWM93 runs. Recent observations of the vertical drifts by Vector Electric Field Instrument (VEFI) and Ion Velocity Meter (IVM) aboard C/NOFS during the extended solar minimum conditions of 2009–2010 indicate the presence of a semidiurnal component in the ion drifts that manifest as downward drift perturbations in the late afternoon and upward drifts at night [*Pfaff et al.*, 2010; *Stoneback et al.*, 2011]. Upward drifts on the order of 20 m s^{-1} in the predawn sector have been observed in some longitudes [*Stoneback et al.*, 2011]. We expect that new observations of vertical drift will improve climatological models, especially at solar minimum. Meanwhile, we continue to work toward improving the *E* region physics of SAMI3-SC.

Our SAMI3-SC simulations with HWM07 and HWM93 underscore the significant impact that neutral winds have on the electric fields and plasma drifts in the ionosphere. But ionospheric conductivity is also important. Preliminary work shows that the O–O$^+$ collision frequency as well as the nighttime sources of ionization have important consequences for the vertical drifts. Furthermore, we find that it is important to sufficiently sample the vertical profile of the *E* region winds, both in the low-latitude and midlatitude regions. In Figure 9, the gray dashed curve shows the *E* × *B* drifts at 12 UT as a function of longitude for DOY 80 in 2006 using a grid with vertical resolution of ~10 km in the midlatitudes. The solid black curve, the "control" run, shows the drifts when a finer grid (~5 km resolution) is used. With the finer grid, the daytime drifts increase, and the postmidnight drifts, though still upward, are reduced. We also experimented with the O–O+ collision frequency by adjusting a multiplicative factor known as the Burnside factor [*Burnside et al.*, 1987; *Salah*, 1993]. The control run uses a Burnside factor of 1.3. Increasing the factor to 1.7 (dot-dashed line) further reduces the postmidnight drift and produces a PRE of the vertical drifts, both in better agreement with the Scherliess-Fejer climatology. Reducing the nighttime photoionization (long-

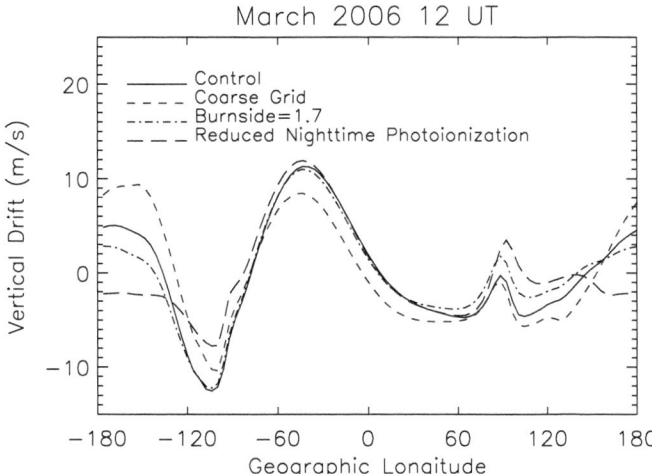

Figure 9. SAMI3-SC vertical drift on day 80 (March 2006) using HWM07. The vertical drift at 12 UT is shown as a function of longitude. Several different runs are performed to assess the sensitivity of the vertical drift to the coarseness of the SAMI3-SC grid (gray dashed line), alteration of the O–O+ collision frequency by increasing the Burnside factor from 1.3 to 1.7 (dot dashed line), and reduction of the nighttime photoionization (long-dashed line).

dashed line) due to stellar sources and scattered sunlight provides even more realistic PRE and nighttime drifts.

These results are in agreement with a study by *Fesen et al.* [2000] in which to produce a PRE in TIEGCM, the nighttime *E* region electron densities had to be reduced below 10^4 cm^{-3} to prevent short circuiting of the *F* region dynamo. We note that *Fesen et al.* [2000] also found that increasing the *E* region semidiurnal tide in the neutral winds increased the magnitude of the daytime drifts. Indeed, increasing the semidiurnal tide in HWM07 has similar effects, but applying such adjustments in our simulations does not appear warranted in the absence of compelling evidence that HWM07 fails to adequately capture this tide.

5. MODELING THE DESCENDING PHASE OF SOLAR CYCLE 23 (2002–2008)

Ongoing ISES work involves systematic study of the ionosphere over longer periods of time in order to investigate the ionospheric responses to solar cycle changes in oscillations in the thermosphere and in solar EUV radiation. Performing such calculations is invaluable for identifying biases in the model. We have performed initial runs using the 2C and 3C EUV models over the descending phase of Solar Cycle 23. Figure 10 shows the daily global TEC as determined by IGS (black line) and SAMI3 (gray line). During the WHI 2008 period, we found SAMI3 to have a high bias in global mean TEC. This bias is apparent throughout the 2002–2008 period

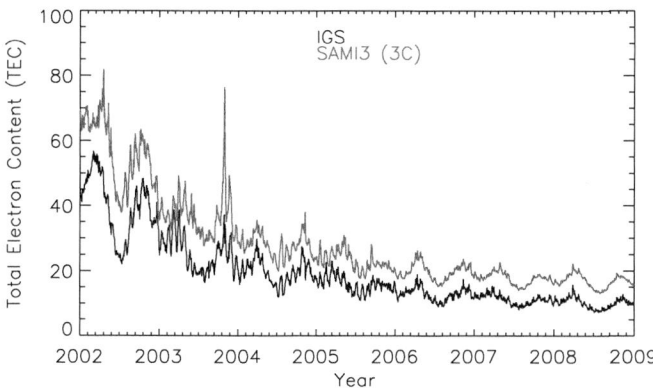

Figure 10. Daily global mean TEC for the descending phase of Solar Cycle 23 (2002 through 2008) are shown for IGS (black) and SAMI3 (gray).

as well and does not have a strong dependence on solar activity. Additional studies using other data sources, such as ionosondes, are being performed to identify possible causes of this bias in the SAMI3 model.

Despite the bias, the calculations show that SAMI3, through NRLMSISE-00, captures the overall annual and semiannual variation in TEC (albeit with reduced annual oscillation amplitude) and the general trend from 2002 through 2008. Solar influences, such as the quasi 27 day solar rotation, are also apparent. A study of the IGS TEC using regression analysis has been performed [*Lean et al.*, 2011a]. The same technique will be applied to the SAMI3 simulations in a future study.

6. SUMMARY OF RESULTS

We have performed multiple ionospheric simulations with SAMI3 during the WHI 2008, using both an empirical vertical drift model and drifts calculated self-consistently from neutral winds. We find that SAMI3, when run with the externally specified Scherliess-Fejer vertical drift model, replicates quite well the daily averaged variation in global mean TEC compared with the IGS GPS-derived TEC maps. During the WHI 2008, variations in TEC associated with CIRs are detectable in the model and are forced primarily through F region disturbance winds in the midlatitudes. Response to the CIRs is not as pronounced as in the IGS TEC, but this is not unexpected for a quiet-time ionospheric model. On shorter time scales, we show that SAMI3 does not reproduce the observed diurnal variation in global mean TEC and identify differences in its regional TEC specification relative to IGS as the probable cause. Further analyses, including validation with additional data sets, are required to better quantify the causes of this difference.

SAMI3-SC, which self-consistently solves for the electric fields, does less well at reproducing the ionospheric variability evident in the IGS TEC. A comparison of the SAMI3-SC vertical $E \times B$ drifts to the Scherliess-Fejer model indicates that the SAMI3 daytime drifts are consistently lower than those of the empirical model; this is true under solar minimum as well as solar maximum conditions. Modeling studies performed by *Fesen et al.* [2000] indicate that the semidiurnal tide in the neutral winds controls the daytime drifts. Clearly, a new comprehensive database of wind measurements is called for in order to properly test the validity of existing neutral wind models. Further analysis of the SAMI3-SC vertical drifts also suggests that the E region physics must also be improved in order to improve the modeled electrodynamics. For example, the $O-O^+$ collision frequency, as well as the nighttime sources of ionization, has important consequences for the postsunset and nighttime vertical drifts.

Finally, our reported SAMI3 simulations during the descending phase of Solar Cycle 23 are initial results of ongoing work to characterize and specify the ionosphere during the entire cycle, including during the two recent solar minimum periods in 1996 and 2008. That we achieve good overall agreement between the modeled and observed IGS TEC suggests that SAMI3 successfully captures long-term changes driven directly by EUV irradiance variations as well as much of the solar cycle modulation of the semiannual and annual oscillations in the global mean TEC. Future work will include simulations of the full solar cycle and comparisons with measurements to better understand ionospheric responses to long-term oscillations in the thermosphere and solar irradiance.

Acknowledgments. This work is supported by the Office of Naval Research.

REFERENCES

Afraimovich, E. L., E. I. Astafyeva, A. V. Oinats, Y. V. Yasukevich, and I. V. Zhivetiev (2008), Global electron content: A new conception to track solar activity, *Ann. Geophys.*, 26(2), 335–344.

Anderson, D. N., B. Reinisch, C. Valladare, J. Chau, and O. Veliz (2004), Forecasting the occurrence of ionospheric scintillation activity in the equatorial ionosphere on a day-to-day basis, *J. Atmos. Sol. Terr. Phys.*, 66(17), 1567–1572, doi:10.1016/j.jastp.2004.07.010.

Basu, S., et al. (1996), Scintillations, plasma drifts, and neutral winds in the equatorial ionosphere after sunset, *J. Geophys. Res.*, 101(A12), 26,795–26,809.

Burnside, R. G., C. A. Tepley, and V. B. Wickwar (1987), The O+-O collision cross-section: Can it be inferred from aeronomical measurements?, *Ann. Geophys. Ser. A*, 5(6), 343–349.

Crowley, G., A. Reynolds, J. P. Thayer, J. Lei, L. J. Paxton, A. B. Christensen, Y. Zhang, R. R. Meier, and D. J. Strickland (2008), Periodic modulations in thermospheric composition by solar wind high speed streams, *Geophys. Res. Lett.*, *35*, L21106, doi:10.1029/2008GL035745.

Drob, D. P., et al. (2008), An empirical model of the Earth's horizontal wind fields: HWM07, *J. Geophys. Res.*, *113*, A12304, doi:10.1029/2008JA013668.

Echer, E., B. T. Tsurutani, W. D. Gonzalez, and J. U. Kozyra (2011), High speed stream properties and related geomagnetic activity during the Whole Heliosphere Interval (WHI): 20 March to 16 April 2008, *Sol. Phys.*, *274*(1–2), 303–320, doi:10.1007/s11207-011-9739-0.

Emmert, J. T., D. P. Drob, G. G. Shepherd, G. Hernandez, M. J. Jarvis, J. W. Meriwether, R. J. Niciejewski, D. P. Sipler, and C. A. Tepley (2008), DWM07 global empirical model of upper thermospheric storm-induced disturbance winds, *J. Geophys. Res.*, *113*, A11319, doi:10.1029/2008JA013541.

Emmert, J. T., J. L. Lean, and J. M. Picone (2010), Record-low thermospheric density during the 2008 solar minimum, *Geophys. Res. Lett.*, *37*, L12102, doi:10.1029/2010GL043671.

England, S. L., T. J. Immel, and J. D. Huba (2008), Modeling the longitudinal variation in the post-sunset far-ultraviolet OI airglow using the SAMI2 model, *J. Geophys. Res.*, *113*, A01309, doi:10.1029/2007JA012536.

Fesen, C. G., G. Crowley, R. G. Roble, A. D. Richmond, and B. G. Fejer (2000), Simulation of the pre-reversal enhancement in the low latitude vertical ion drifts, *Geophys. Res. Lett.*, *27*(13), 1851–1854.

Gibson, S. E., et al. (2011), The Whole Heliosphere Interval in the context of a long and structured solar minimum: An overview from Sun to Earth, *Sol. Phys.*, *274*(1–2), 5–27, doi:10.1007/s11207-011-9921-4.

Hedin, A. E., et al. (1996), Empirical wind model for the upper, middle and lower atmosphere, *J. Atmos. Terr. Phys.*, *58*(13), 1421–1447.

Hernandez-Pajares, M., J. M. Juan, J. Sanz, R. Orus, A. Garcia-Rigo, J. Feltens, A. Komjathy, S. C. Schaer, and A. Krankowski (2009), The IGS VTEC maps: A reliable source of ionospheric information since 1998, *J. Geod.*, *83*(3–4), 263–275, doi:10.1007/S00190-008-0266-1.

Hocke, K. (2008), Oscillations of global mean TEC, *J. Geophys. Res.*, *113*, A04302, doi:10.1029/2007JA012798.

Hocke, K. (2009), Reply to comment by J. T. Emmert et al. on "Oscillations of global mean TEC", *J. Geophys. Res.*, *114*, A01310, doi:10.1029/2008JA013786.

Huba, J. D. and G. Joyce (2012), Numerical methods in modeling the ionosphere, in *Modeling the Ionosphere-Thermosphere System*, *Geophys. Monogr. Ser.*, doi:10.1029/2012GM001306, this volume.

Huba, J. D., G. Joyce, and J. A. Fedder (2000), Sami2 is Another Model of the Ionosphere (SAMI2): A new low-latitude ionosphere model, *J. Geophys. Res.*, *105*(A10), 23,035–23,053.

Huba, J. D., G. Joyce, and J. Krall (2008), Three-dimensional equatorial spread F modeling, *Geophys. Res. Lett.*, *35*, L10102, doi:10.1029/2008GL033509.

Huba, J. D., S. L. Ossakow, G. Joyce, J. Krall, and S. L. England (2009), Three-dimensional equatorial spread F modeling: Zonal neutral wind effects, *Geophys. Res. Lett.*, *36*, L19106, doi:10.1029/2009GL040284.

Huba, J. D., G. Joyce, J. Krall, C. L. Siefring, and P. A. Bernhardt (2010), Self-consistent modeling of equatorial dawn density depletions with SAMI3, *Geophys. Res. Lett.*, *37*, L03104, doi:10.1029/2009GL041492.

Lean, J. L., J. T. Emmert, J. M. Picone, and R. R. Meier (2011a), Global and regional trends in ionospheric total electron content, *J. Geophys. Res.*, *116*, A00H04, doi:10.1029/2010JA016378. [Printed 117(A2), 2012].

Lean, J. L., R. R. Meier, J. M. Picone, and J. T. Emmert (2011b), Ionospheric total electron content: Global and hemispheric climatology, *J. Geophys. Res.*, *116*, A10318, doi:10.1029/2011JA016567.

Lean, J. L., T. N. Woods, F. G. Eparvier, R. R. Meier, D. J. Strickland, J. T. Correira, and J. S. Evans (2011c), Solar extreme ultraviolet irradiance: Present, past, and future, *J. Geophys. Res.*, *116*, A01102, doi:10.1029/2010JA015901.

Lei, J., J. P. Thayer, J. M. Forbes, E. K. Sutton, and R. S. Nerem (2008), Rotating solar coronal holes and periodic modulation of the upper atmosphere, *Geophys. Res. Lett.*, *35*, L10109, doi:10.1029/2008GL033875.

Liu, L., W. Wan, B. Ning, and M.-L. Zhang (2009), Climatology of the mean total electron content derived from GPS global ionospheric maps, *J. Geophys. Res.*, *114*, A06308, doi:10.1029/2009JA014244.

Nitta, N. V. (2011), Observables indicating two major coronal mass ejections during the WHI, *Sol. Phys.*, *274*(1–2), 219–232, doi:10.1007/s11207-011-9806-6.

Pfaff, R., et al. (2010), Observations of DC electric fields in the low-latitude ionosphere and their variations with local time, longitude, and plasma density during extreme solar minimum, *J. Geophys. Res.*, *115*, A12324, doi:10.1029/2010JA016023.

Picone, J. M., A. E. Hedin, D. P. Drob, and A. C. Aikin (2002), NRLMSISE-00 empirical model of the atmosphere: Statistical comparisons and scientific issues, *J. Geophys. Res.*, *107*(A12), 1468, doi:10.1029/2002JA009430.

Salah, J. E. (1993), Interim standard for the ion-neutral atomic oxygen collision frequency, *Geophys. Res. Lett.*, *20*(15), 1543–1546.

Scherliess, L., and B. G. Fejer (1999), Radar and satellite global equatorial F region vertical drift model, *J. Geophys. Res.*, *104*(A4), 6829–6842.

Stoneback, R. A., R. A. Heelis, A. G. Burrell, W. R. Coley, B. G. Fejer, and E. Pacheco (2011), Observations of quiet time vertical ion drift in the equatorial ionosphere during the solar minimum period of 2009, *J. Geophys. Res.*, *116*, A12327, doi:10.1029/2011JA016712.

Thayer, J. P., J. Lei, J. M. Forbes, E. K. Sutton, and R. S. Nerem (2008), Thermospheric density oscillations due to periodic solar wind high-speed streams, *J. Geophys. Res.*, *113*, A06307, doi:10.1029/2008JA013190.

Tulasi Ram, S., C. H. Liu, and S.-Y. Su (2010), Periodic solar wind forcing due to recurrent coronal holes during 1996–2009 and its impact on Earth's geomagnetic and ionospheric properties during the extreme solar minimum, *J. Geophys. Res.*, *115*, A12340, doi:10.1029/2010JA015800.

Verkhoglyadova, O. P., B. T. Tsurutani, A. J. Mannucci, M. G. Mlynczak, L. A. Hunt, A. Komjathy, and T. Runge (2011), Ionospheric VTEC and thermospheric infrared emission dynamics during corotating interaction region and high-speed stream intervals at solar minimum: 25 March to 26 April 2008, *J. Geophys. Res.*, *116*, A09325, doi:10.1029/2011JA016604.

D. P. Drob, J. T. Emmert, J. L. Lean and S. E. McDonald, Space Science Division, Naval Research Laboratory, Washington, DC, USA. (sarah.mcdonald@nrl.navy.mil)

J. D. Huba, Plasma Physics Division, Naval Research Laboratory, Washington, DC, USA.

G. Joyce, Icarus Research, Inc., Bethesda, MD, USA.

Comparative Studies of Theoretical Models in the Equatorial Ionosphere

Tzu-Wei Fang,[1] David Anderson,[1] Tim Fuller-Rowell,[1] Rashid Akmaev,[2] Mihail Codrescu,[2] George Millward,[1] Jan Sojka,[3] Ludger Scherliess,[3] Vince Eccles,[4] John Retterer,[5] Joe Huba,[6] Glenn Joyce,[7] Art Richmond,[8] Astrid Maute,[8] Geoff Crowley,[9] Aaron Ridley,[10] and Geeta Vichare[11]

Two sets of ionospheric models including six non-self-consistent models and five self-consistent models were compared with ionosonde data and the International Reference Ionosphere (IRI). The comparisons were focused on the low latitudes in March equinox under quiet geomagnetic activity and moderate solar activity ($F_{10.7} = 120$). We compared the N_mF_2 and h_mF_2 from all theoretical models, the IRI, and observations at four different LTs at equatorial region in the American sector (75°W). For the non-self-consistent model, we further compared simulation results at Asian sector (120°E). To identify the causes of differences in the non-self-consistent models, runs without neutral wind and/or electric field were conducted at the Asian sector. Results show that the non-self-consistent models are in good agreement with each other and with the IRI especially in the daytime. Large discrepancies are shown in the self-consistent model results, which imply very different electric fields and neutral atmosphere in these models. Also, the daytime N_mF_2 values at the crests of the equatorial anomaly calculated by the self-consistent models are substantially lower than those in the non-self-consistent models. This paper reviews the current status of each theoretical model and examines their capability in simulating the quiet time equatorial ionosphere.

1. INTRODUCTION

The low-latitude ionosphere is characterized by the equatorial ionization anomaly (EIA) produced by the equatorial plasma fountain [*Appleton*, 1946; *Anderson*, 1973; *Rishbeth*, 2000]. The physics of the equatorial ionosphere has been studied intensively using ground-based and satellite measurements in the past few decades. The development of theoretical models has come along with the increase of computer power. These models help us to explain the physics and to strengthen our knowledge of ionospheric morphology. Through solving the first principle equations, most ionosphere-plasmasphere models are capable of reproducing the basic features of low-latitude ionosphere, such

[1]CIRES/University of Colorado, Boulder, Colorado, USA.
[2]NOAA Space Weather Prediction Center, Boulder, Colorado, USA.
[3]Center for Atmospheric and Space Sciences, Utah State University, Logan, Utah, USA.
[4]Space Environment Corporation, Providence, Utah, USA.
[5]Institute for Scientific Research, Boston College, Massachusetts, USA.
[6]Plasma Physics Division, Naval Research Laboratory, Washington, District of Columbia, USA.
[7]Icarus Research, Inc., Bethesda, Maryland, USA.
[8]High Altitude Observatory, National Center for Atmospheric Research, Boulder, Colorado, USA.
[9]Atmospheric and Space Technology Research Associates, San Antonio, Texas, USA.
[10]Center for Space Environment Modeling, University of Michigan, Ann Arbor, Michigan, USA.
[11]Indian Institute of Geomagnetism, Navi Mumbai, India.

as the trough above the magnetic equator and the two crests 10° to 20° away of either side, as well as their diurnal/seasonal variations.

Several different types of ionosphere models are currently used in the community in order to understand the variability in the ionosphere or to reproduce certain phenomena related to space weather events. These include empirical models, which are based on observations, theoretical models that simulate the ionosphere by solving a set of differential equations, and data assimilation models that combine physics-based models and observations to facilitate the capability of forecasting. The theoretical models can be further divided into two categories: the non-self-consistent models and the self-consistent models. The non-self-consistent models require neutral densities and temperatures, neutral winds, and $\mathbf{E} \times \mathbf{B}$ drift velocities as inputs to calculate ion and electron densities and temperatures as a function of altitude, latitude, and LT. The self-consistent models are mostly time-dependent, 3-D, nonlinear models that solve the fully coupled, thermodynamic, and continuity equations of the neutral gas self-consistently with the electron and ion energy, momentum, and continuity equations and output parameters of neutral atmosphere and ionosphere. Even though these theoretical models are developed independently and have different heritages, their structures in the ionosphere share some common elements. For example, they all solve basic equations such as continuity, momentum, and energy equations on a fixed grid, which is defined by specific magnetic coordinates. The design of the grid should have enough spatial resolution to resolve basic ionospheric phenomena. For each model, the lower boundary conditions (e.g., tides from the lower atmosphere) and upper boundary conditions (e.g., plasma fluxes) may need to be specified. Other parameters such as the photoionization, high-latitude electric field, auroral precipitation, reaction rates, etc., are parameterized or are determined by empirical models.

Anderson et al. [1998] compared the daily variations of N_mF_2 and h_mF_2 given by five global models at solar minimum and maximum at Millstone Hill for geomagnetically quiet conditions. The five models included the Coupled Thermosphere-Ionosphere Model (CTIM), the Thermosphere-Ionosphere General Circulation Model (TIGCM), the Time-Dependent Ionospheric Model (TDIM), the Field Line Interhemispheric Plasma model (FLIP), and the Global Theoretical Ionospheric Model (GTIM). The motivation of their study was trying to determine why several physical models consistently underestimated the F region peak electron density, by up to a factor of 2, in the midlatitude, daytime ionosphere at solar maximum. Quite good convergence was found between the mean ionospheric climatology given by different models after adopting the Burnside factor (a scale factor for O-O+ collision frequency) of 1.7 for the collision frequency. The factor is crucial for the diffusion process in the ionosphere and plasmasphere and the impact of neutral winds. Today, however, the evidence suggests the factor is closer to 1.0 [*Nicolls et al.*, 2006]. Tuning the Burnside factor did enhance the agreement between simulation and observation for studies done by *Anderson et al.* [1998], but this solution indicated the lack of physics in these models. After more than a decade from their comparison effort, the current model simulations again do not completely agree with each other and are not able to accurately reproduce observations. Since the development is different between each model, it is important for the community to review the current status of these theoretical models and to improve the relevant physics adopted in simulating the ionosphere and plasmasphere.

The comparisons in the work of *Anderson et al.* [1998] mainly focused on the midlatitudes. Motivated by *Anderson et al.* [1998], a working group, which emphasized the low-latitude ionosphere, was initiated in 2010 to understand the strengths and the limitations of theoretical, time-dependent ionospheric models in representing observed ionospheric structure and variability under moderate solar activity and geomagnetic quiet conditions. To better understand the underlying ionospheric physics and to improve models, a project named "Problems Related to Ionospheric Models and Observation at Equatorial Region (Equatorial-PRIMO)," which coordinates the model-model and model-data comparisons, has been put together to discuss the ongoing work that addresses the physics of equatorial ionosphere at the annual National Science Foundation Coupling Energetics and Dynamics of Atmospheric Regions workshop.

This paper summarizes the first 2 year efforts of Equatorial-PRIMO and its intent to provide an up-to-date and comprehensive comparison of these models and their capability of producing consistent results. The following sections show sensitivity studies that were conducted using a non-self-consistent model to demonstrate the importance of transport processes at different regions of the ionosphere. Details of the models that are participating in Equatorial-PRIMO are described in section 3. The intercomparisons of these models with each other and with the observations at the America sector (75°W) are presented in section 4. Section 5 describes the possible sources for model discrepancies at the Asia sector (120°E). The final section contains the summaries of the project and outlines the future plans of this work.

2. TRANSPORT PROCESSES IN THE IONOSPHERE

The morphology of the F region ionosphere in the equatorial region is strongly controlled by plasma transport processes. These processes have been discussed in several books and publications [e.g., *Rishbeth and Garriott*, 1969; *Rishbeth*, 2000; *Schunk and Nagy*, 2000]. In most of the ionospheric models, the transport term is considered in the continuity equation for the electron and ion density (N)

$$\frac{\partial N}{\partial t} = q - \beta(N) - \nabla(NV),$$

where q is the ion production rate, β is the loss coefficient, and V is the plasma drift velocity. In the above expression, the q and $\beta(N)$ stands for the production and loss terms, respectively. The NV represents the plasma flux due to transport, and its divergence represents the resulting rate of loss per unit volume and unit time. The plasma drift velocity is further controlled by several different transport processes. These include the transport parallel and perpendicular to the magnetic field line. The parallel transport of plasma results from the influences of the neutral wind and plasma diffusion, which is controlled by the gravitational force and the pressure gradient. The perpendicular transport of plasma, especially in the F region, is caused by the electric field through $\mathbf{E} \times \mathbf{B}$ drift in the meridional and zonal directions. Different transport processes can be important at different LTs and for different regions of the ionosphere. How models handle these transport processes is extremely important for successfully reproducing observations. To demonstrate the importance of some of these transport processes in different regions, the Global Ionosphere Plasmasphere model (GIP) model is selected to conduct sensitivity studies. The GIP is a non-self-consistent model that requires neutral atmosphere (density, temperature, and wind velocity) and ionospheric vertical drift as inputs. The fact that these drivers are easily modified makes the non-self-consistent models better choices for this type of study over the self-consistent models. The ionospheric physics and equations in the GIP are based on the Coupled Thermosphere Ionosphere Plasmasphere Electrodynamics model (CTIPe). For the magnetic coordinates, instead of using the tilt dipole structure originally in the CTIPe, the GIP utilizes a Magnetic Apex coordinate system [*Richmond*, 1995] in which a global 3-D grid of magnetic field lines are created by tracing through the full International Geomagnetic Reference Field (IGRF). Important parameters of the GIP are listed in Table 1. More information about the GIP can also be found in the work of *Fang et al.* [2009].

For the sensitivity studies, the zonal and meridional winds from the horizontal wind model (HWM-93) [*Hedin et al.*, 1996], the neutral densities/temperatures from the Naval Research Laboratory (NRL) Mass Spectrometer and Incoherent Scatter Radar Model (MSISE-00) [*Picone et al.*, 2002], and the ionospheric vertical drifts from the empirical model of *Scherliess and Fejer* [1999] are used to drive the GIP. The model is run under moderate solar activity ($F_{10.7}$ = 120) and geomagnetically quiet conditions. The three tests including changes in ionospheric vertical drift at March equinox, changes in neutral wind at June solstice, and changes in the Burnside factor at March equinox are shown

Table 1. Information of the Non-Self-Consistent Models[a]

Model	Output	Altitude Range (km)	Resolution	Magnetic Coordinate	Photoionization
Ionospheric forecast model	N_i (O^+, H^+, NO^+, O_2^+), N_e, T_i, T_e	90–1,600	longitude 5°–15° latitude 2°–5°	Best-fit International Geomagnetic Reference Field (IGRF) dipole for each longitude	EUV flux model for aeronomic calculation (EUVAC)
Ionosphere-plasmasphere model	N_i (O^+, H^+, NO^+, O_2^+, He^+, N_2^+, N^+), N_e, T_i, T_e	90–20,000	longitude 3.75° latitude <1° at low latitude	IGRF dipole	EUVAC
Low-latitude ionosphere sector model	N_i (O^+, H^+, NO^+, O_2^+), N_e, T_i, T_e	90–10,000	single longitude latitude 2°	Best-fit IGRF dipole for each longitude	EUVAC
Physics-based model	N_i (O^+, H^+, NO^+, O_2^+, N_2^+), N_e, T_i, T_e	90–4,000	longitude 7.5° latitude 1°	IGRF apex	Hinteregger fluxes Jasperse Jasperse CSD
Global ionosphere plasmasphere	N_i (O^+, H^+, NO^+, O_2^+, N_2^+, N^+), N_e, T_i, T_e	90–20,000	longitude 4.5° latitude 1°	IGRF apex	fluxes (Tobiska model) cross section [*Torr and Torr*, 1982]
Another Model of the Ionosphere	N_i (H^+, O^+, He^+, N^+, NO^+, N_2^+, O_2^+), N_e, T_i (H^+, O^+, He^+), T_e	90–20,000	Single longitude latitude 1°	IGRF like	EUVAC

[a]N_i and N_e are the ion and electron densities. T_i and T_e are the ion and electron temperatures. CSD stands for Continuous Slowing Down model.

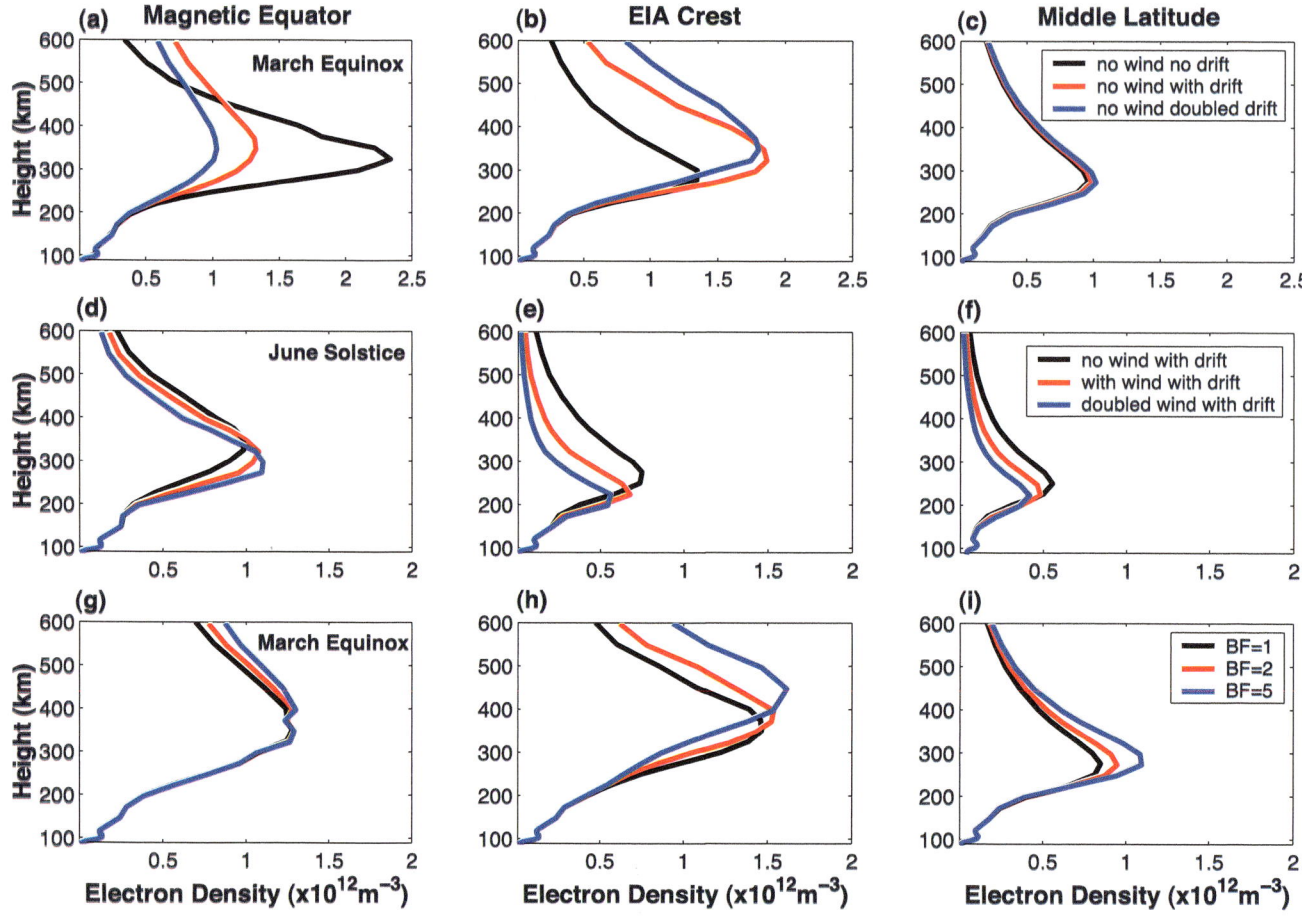

Figure 1. Impacts of vertical drifts at March equinox at the Northern Hemisphere (a–c), neutral winds at June solstice at the Southern Hemisphere (d–f), and Burnside factors at March equinox at the Northern Hemisphere (g–i) on electron density profiles simulated by the global ionosphere plasmasphere model at three different locations. Notice the difference in the x axis scales. See text for more details.

in Figure 1. The height profiles of simulated electron density at the magnetic equator, at 16°, which is close to the peak density of EIA and at middle latitude (42°) along the Jicamarca longitude (75°W) are shown in Figure 1. Simulation results are shown at 14 LT. Overall, the densities are larger in March equinox than in June solstice at all three locations. Figures 1a, 1b, and 1c illustrate the impact of vertical drifts. Black lines stand for the results with no vertical drift and no neutral wind. Red lines are the default run with vertical drifts. Blue lines show model results with twice of vertical drifts as inputs. Without the vertical drifts, the densities are larger at the magnetic equator and decrease with the increasing latitudes. No EIA signature shows up in this simulation. This plasma distribution is mainly due to the slower plasma diffusion at lower altitude and the horizontal structure of magnetic field. With the presence of vertical drifts, a clear EIA signature is produced with larger density near the EIA crests compared to the plasma at the magnetic equator and at higher latitude. The peak density at the crest region is also lifted to a higher altitude. With the doubled vertical drifts, the EIA is strengthened. The enhanced transport process carries the plasma away from the magnetic equator and pushes it further poleward. However, the increase in the vertical drifts hardly has any influence on the density profiles in the midlatitude region.

Figures 1d, 1e, and 1f illustrate the impact of neutral wind. Both zonal and meridional winds are considered in these runs. Notice that results shown for this set of runs are in June solstice because the neutral wind effect can be more significant when the offset of subsolar point and the magnetic equator is larger. Also, the profiles at the EIA crest and midlatitude shown in Figures 1d, 1e, and 1f are located in the Southern Hemisphere. The three simulations cover conditions of no wind (black), default wind (red), and a doubling of the wind (blue). With the increased wind velocity, the peak density increases at the magnetic equator but decreases at the EIA crest and midlatitude. The peak heights reduce with increased wind velocity at all three locations. The reductions in densities and heights at the EIA crest and midlatitude can be explained by the enhanced magnetic southward magnetic meridional wind, which tends to push plasma to the lower altitude where the recombination process is faster.

Figures 1g, 1h, and 1i show the impact of changes in the Burnside factor (BF). In these runs, the factor is set to default (=1, black), 2 (red), and 5 (blue). At both the magnetic equator and midlatitude, increasing the BF only increases the density profiles near and above the F region peak. At all three locations, the densities increase when the BF is larger. Larger BF stands for larger $O-O^+$ collision frequency, which essentially reduces the rate of plasma diffusion parallel to the magnetic field and increase the impact of neutral winds. In the daytime F region, plasma tends to diffuse downward due to gravity and pressure gradient forces. Reducing the diffusion coefficient means larger densities can be maintained at a fixed location in the ionosphere.

These tests demonstrate the importance of the different transport processes (vertical drifts, neutral winds, and plasma diffusion) at crucial locations in the ionosphere. These transport processes are extremely important for ionospheric models, and sensitivity studies help us examine a model in quantitative ways. Applying similar sensitivity studies to different models can provide valuable information for interpreting the similarities and differences among models.

3. PARTICIPATING MODELS IN EQUATORIAL-PRIMO

Two sets of theoretical models are participating in Equatorial-PRIMO. The first set is the non-self-consistent models that include the Space Environment Corporation (SEC) Ionospheric Forecast Model (IFM) [*Schunk et al.*, 1997; *Schunk and Sojka*, 1998], the SEC Ionosphere-Plasmasphere Model (IPM) [*Scherliess et al.*, 2004; *Schunk et al.*, 2004], the SEC Low Latitude Ionosphere Sector Model (LLIONS) [*Anderson*, 1981; *Schunk et al.*, 1997], the Air Force Research Laboratory physics-based model (PBMOD) [*Retterer*, 2005; *Retterer et al.*, 2005], the GIP [*Fang et al.*, 2009], and the NRL SAMI2 is Another Model of the Ionosphere (SAMI2) [*Huba et al.*, 2000]. Important parameters of these non-self-consistent models are listed in Table 1. Under the column "Altitude Range" in Table 1, the first number is the lower-boundary altitude, and the second number is the apex height of the outermost flux tube. No upper-boundary conditions are required for these models. The design of spatial resolutions and the usage of magnetic coordinates in these models reflect their heritages and their development for different purposes. Among these non-self-consistent models, most of them choose the EUV flux model for aeronomic calculations (EUVACs) as their solar flux model except the PBMOD and the GIP.

The other set of ionosphere-thermosphere models is the self-consistent models and includes the NRL SAMI3 is Also a Model of the Ionosphere (SAMI3) [*Huba and Joyce*, 2010], the National Center for Atmosphere Research Thermosphere Ionosphere-Electrodynamics General Circulation Model (TIE-GCM) [*Roble et al.*, 1988; *Richmond et al.*, 1992], the Thermosphere-Ionosphere-Mesosphere-Electrodynamics General Circulation Model (TIME-GCM) [*Roble and Ridley*, 1994] run by the Atmospheric and Space Technology Research Associates, the University of Michigan Global Ionosphere Thermosphere Model (GITM) [*Ridley et al.*, 2006; *Vichare et al.*, 2012], and CTIPe [*Millward et al.*, 1996, 2001]. The SAMI3 does not self-consistently calculate the neutral atmosphere but does solve the global electrodynamics. Therefore, SAMI3 is kept within this set of models. Important parameters of these self-consistent models are listed in Table 2. As shown in Table 2, all these models provide states of the neutral atmosphere and ionosphere but with different altitude coverage. At the lower boundary, different tidal modes are specified in these models. The electron heat and the O^+ fluxes are specified at the upper boundary of the TIE-GCM and the TIME-GCM. The upper boundary conditions for GITM in the ionosphere include: (1) the velocity gradient is continuous, (2) temperature has a zero gradient in both ions and electrons (with a heat magnetospheric heat source for the electrons), and (3) the density is specified so that it decreases with altitude. The SAMI3 and CTIPe calculate plasma along the flux tubes, and no upper boundary conditions are required. The CTIPe and SAMI3 use tilted dipole coordinates, while the TIE-GCM, TIME-GCM, and GITM use Magnetic Apex coordinates. Notice that Tables 1 and 2 only summarize the most common settings for these models. Most models are equipped with different settings in resolutions and boundary conditions. Please see the references of each model for further details.

4. MODEL COMPARISONS IN AMERICAN SECTOR (75°W)

To conduct the simplest comparisons, both sets of models are run in March equinox under moderate solar activity ($F_{10.7} = 120$). The geomagnetic conditions are set to quiet, and the Burnside factor is set to 1. For the non-self-consistent models, the zonal and meridional winds are obtained from the HWM-93, the neutral densities and temperatures are calculated from the NRLMSISE-00, the vertical drifts at the magnetic equator are derived from the empirical climatological model of *Scherliess and Fejer* [1999]. For the

Table 2. Information of the Self-Consistent Models[a]

Model	Output	Lower Boundary Condition	Altitude Range (km)	Ionosphere Resolution	Magnetic Coordinates	Photoionization
Also a model of the ionosphere	H^+, O^+, He^+, N^+, NO^+, N_2^+, O_2^+, N_e, T_i (H^+, O^+, He^+), T_e, Φ	horizontal wind model-93	85–20,000	longitude 3.75° magnetic latitude 1°	tilt dipole	EUVAC
Thermosphere-ionosphere-electrodynamics general circulation	neutral composition, U_n, V_n, T_n, T_i, T_e, N_e, O^+, NO^+, O_2^+, Z, Φ	global-scale wave model 02(GSWM02) migrating diurnal and semidiurnal tides	97 to 450–600	longitude 5° latitude 5°	IGRF apex	EUVAC for <1050 *Woods and Rottman* [2002] for >1050A
Thermosphere-ionosphere-mesosphere-electrodynamics general circulation model	neutral composition, U_n, V_n, W, T_n, T_i, T_e, N_e, O^+, O_2^+, NO^+, N_2^+, N^+, Z, Φ	GSWM migrating diurnal and semidiurnal tides	30 to 450–600	longitude 5° latitude 5°	IGRF apex	EUVAC for <1050 *Woods and Rottman* [2002] for >1050A
Global ionosphere thermosphere model	neutral composition, U_n, V_n, W_n, T_n, V_i, T_i, O^+, O_2^+, NO^+, N_2^+, N^+, T_e, N_e, Φ	GSWM migrating diurnal and semidiurnal tides	100–700	longitude 5° magnetic latitude 1°	IGRF apex	EUVAC Hinteregger's SERF1 model
Coupled thermosphere ionosphere plasmasphere electrodynamics model	neutral compositions, U_n, V_n, T_n, T_i, O^+, H^+, O_2^+, NO^+, N_2^+, N^+, N_e, Φ	migrating semidiurnal tides	thermosphere 80–500 ionosphere 80–10,000	longitude 18° latitude 2°	tilt dipole	EUVAC for <1050 *Woods and Rottman* [2002] for >1050A

[a]N_e is the electron density. T_n, T_i, and T_e are the temperatures for neutral, ion, and electron. U_n and V_n are the zonal and meridional wind velocities, respectively. Z stands for the potential height, while Φ stands for the electric potential.

self-consistent models, their default drivers (inputs and other specifications) are used. Figures 2 and 3 show the comparisons of N_mF_2 and h_mF_2 at 2, 10, 14, and 20 LT at Jicamarca longitude (75°W) from the six non-self-consistent models and four self-consistent models, respectively. Black triangles are averages of N_mF_2 and h_mF_2 observations from two ionosonde stations located at Jicamarca in Peru (magnetic equator) and Tucuman in Argentina (15°S, geomagnetic) between 16 and 26 March in 2004. The mean $F_{10.7}$ during the period is 116, which is very close to the condition for these simulations. For these observations, the standard deviation of scatter is included (thin gray lines). Black lines are results from the International Reference Model (IRI-2007) [*Bilitza and Reinisch*, 2008] in 20 March 2004. The observations and IRI results shown in Figures 2 and 3 are identical.

Figure 2 shows that the differences of N_mF_2 among models are larger in the nighttime than in the daytime. The observed N_mF_2 and h_mF_2 also have a larger spread in the nighttime. The agreement between observation and simulation are better at the magnetic equator than at the EIA crest. Large discrepancies of N_mF_2 in the non-self-consistent models can be found at 14 and 20 LT near the EIA crests. In h_mF_2, the theoretical models produce two crests at 14 LT, while the IRI only shows one peak above the magnetic equator. The differences of model results at 14 and 20 LT near the EIA crests are strongly related to the calculations of transport processes in models. Since the vertical drifts and neutral atmosphere for these non-self-consistent models are from empirical models and are set to the same values, the differences shown in Figure 2 could be due to the transport processes parallel to the magnetic field. At 20 LT, large increases of ionospheric peak density and height compared to those in the daytime are associated with the prereversal enhancement (PRE, the enhanced upward drift near dusk), which is shown in the empirical model and are implemented in all non-self-consistent models. Compared to observations and theoretical models, the IRI shows significantly lower N_mF_2 at the EIA crests and lower h_mF_2 above the magnetic equator at 20 LT. Since IRI is an empirical model based on all available data sources, it is possible that the amount of data at the specific LT and

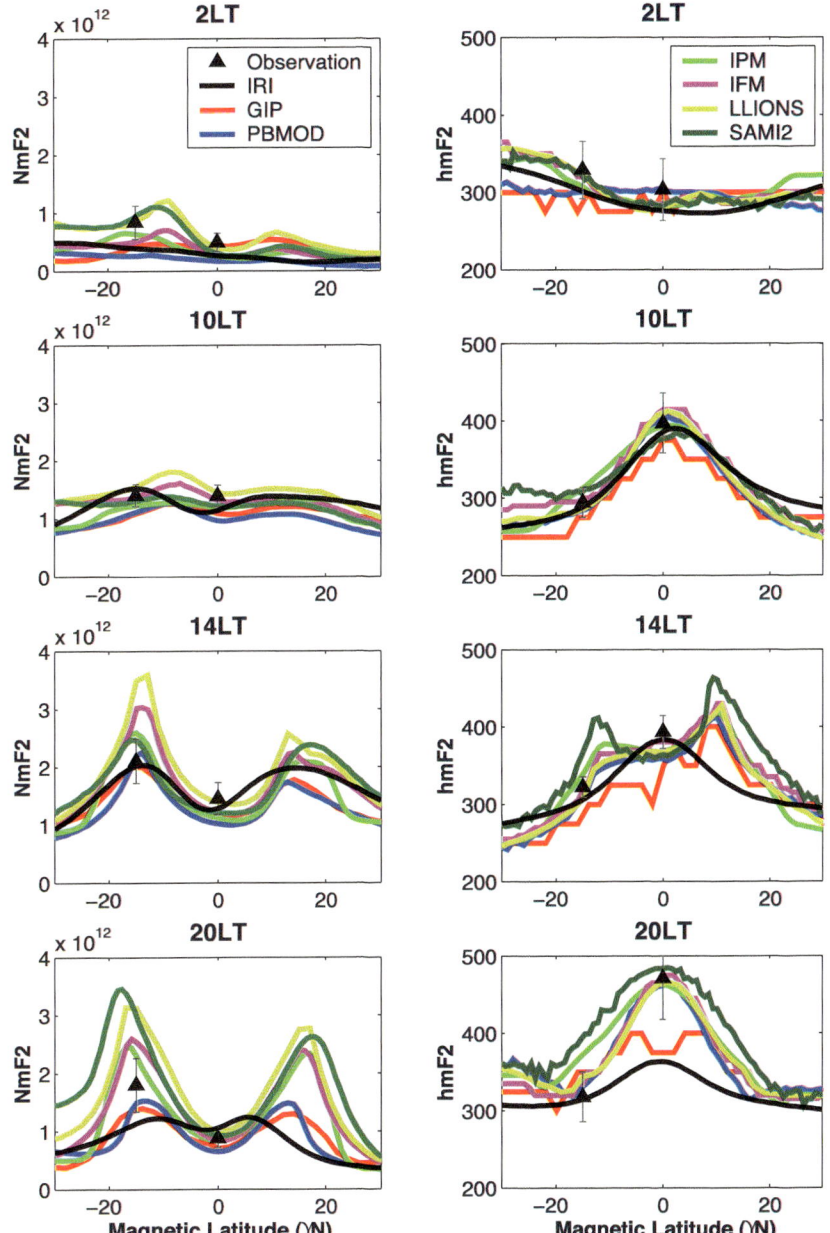

Figure 2. Comparisons of N_mF_2 (m^{-3}, left) and h_mF_2 (km, right) at 2, 10, 14, and 20 LT from six non-self-consistent models (colored lines), International Reference Ionosphere (IRI) (black lines), and observations (triangles). Gray lines are the one standard deviation of the observations.

specific location is not sufficient to capture such rapid change. Similar discrepancy between the IRI and observations during the postsunset period has also been reported in the work of *Lee and Reinisch* [2012].

In Figure 3, the N_mF_2 and h_mF_2 at 10 and 14 LT from the self-consistent models are closer to each other compared to those at other LTs. At 20 LT, the magnitudes and latitudinal distributions of N_mF_2 and h_mF_2 from these models are quite different. The agreement of a particular simulation with IRI and observation can be good or poor for a given model depending on the LT. The daytime EIA crest structures from these self-consistent models are less pronounced compared to the IRI and those non-self-consistent models (Figure 2). The daytime h_mF_2 from the models

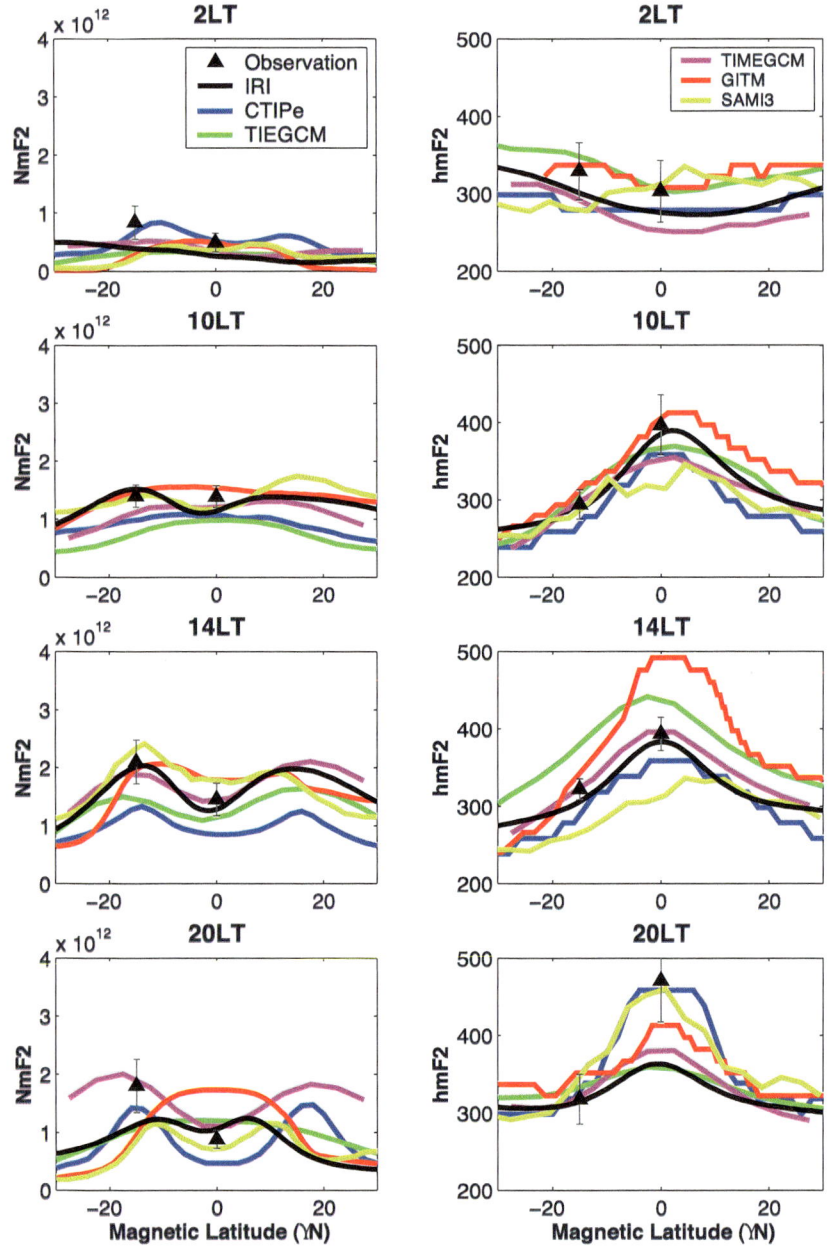

Figure 3. Comparisons of N_mF_2 (m^{-3}, left) and h_mF_2 (km, right) at 2, 10, 14, and 20 LT from five self-consistent models (colored lines), the IRI (black lines), and observations (triangles). Gray lines are the one standard deviation of the observations.

show a peak above the magnetic equator, which agree with IRI. The current version of GITM does not appear to accurately capture the ratio between the N_mF_2 at the EIA and the equator, often producing a larger density at the equator. This is especially true at 2 and 20 LT. The latitudinal distributions of N_mF_2 and h_mF_2 in these models are strongly related to the vertical drifts calculated in each model. However, the magnitudes and phases of the drifts in the daytime and during the PRE (Figure 4) do not seem to explain the differences of N_mF_2 and h_mF_2 at 14 and 20 LT (Figure 3).

Figure 4 shows the comparison of vertical drifts at the magnetic equator from the self-consistent models and from the empirical model of *Scherliess and Fejer* [1999]. In the

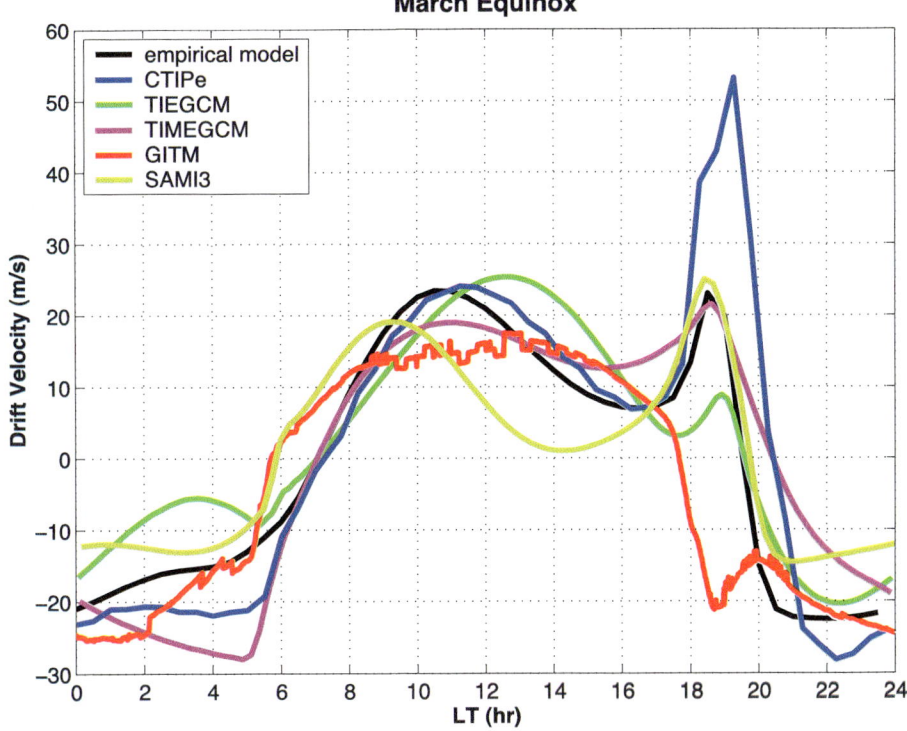

Figure 4. Diurnal variation of vertical drift velocities above the magnetic equator at the American sector simulated by five self-consistent models and the empirical model of *Scherliess and Fejer* [1999].

theoretical models, the peak magnitudes of daytime drifts are about 15–25 m s^{-1}, and the occurrence times of the peak drift range between 11 and 13 LT. The peak daytime drift from the empirical model is about 23 m s^{-1} and occurs at an earlier LT compared to most of the theoretical models. Most models show the PRE at around 19 LT except GITM, which shows a reversal too early and a small increase in drift too late. However, large differences exist in the magnitudes of PRE. The largest PRE is in CTIPe, while the smallest is in TIE-GCM. The magnitudes of PRE in the TIME-GCM and the SAMI3 are comparable with the value from the empirical model. During the night, the differences of drifts are smaller near midnight compared to other times. The drifts shown in Figure 4 are solved using the electrodynamic solver in each model and are strongly controlled by the ionospheric conductivities and neutral winds. The magnitude of PRE is further related to the geometry of the magnetic field and the nighttime E region density in the model [*Fesen et al.*, 2000]. The comparisons of ionospheric parameters in these self-consistent models (Figures 3 and 4) show significant differences among themselves and the observations. The comparisons suggest that the self-consistent neutral atmospheres and temperature structure among these models are likely to be very different.

5. SOURCES OF MODEL DISCREPANCIES

In order to investigate the sources of discrepancies in non-self-consistent models shown in the previous section, the ion-neutral collision frequencies (O-O$^+$), which strongly control the plasma diffusion, and inputs of models, such as wind velocities and vertical drifts, are also compared. The comparison of the ion-neutral collision frequencies shows quantitative agreement (not shown here), which indicates that plasma diffusion in these models should be similar. The comparisons of these inputs (winds and drifts) confirm that the same versions of empirical models are applied in the theoretical models (not shown here).

Furthermore, a set of comparisons that tries to estimate differences in model parameters such as diffusion coefficients and transport processes is also carried out using these non-self-consistent models in the Asian sector (120°E) under the same geophysical conditions. The magnetic equator is located near 10°N geographic latitude at this longitude. Three settings that help us to distinguish model differences are applied in models: (1) no **E** × **B** drift and no neutral wind (case 1), (2) with **E** × **B** drift but no neutral wind (case 2), and (3) with **E** × **B** drift and neutral wind (case 3). The comparisons of N_mF_2 and h_mF_2 at 14 LT from these three

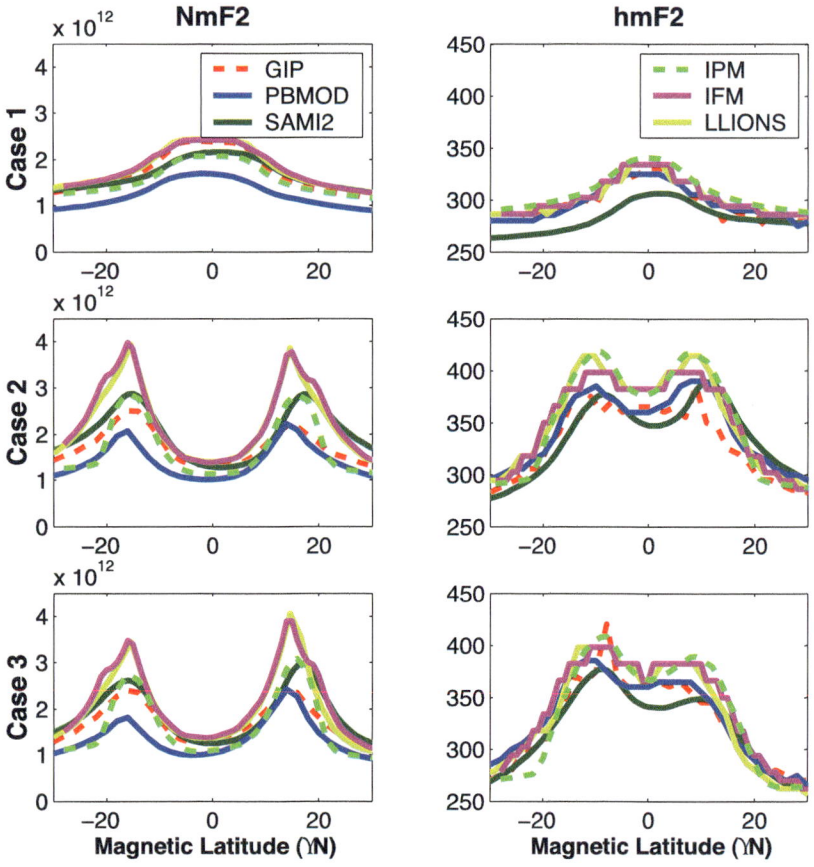

Figure 5. Comparisons of N_mF_2 (m^{-3}, left) and h_mF_2 (km, right) at 14 LT at 120°E under three different cases (top to bottom) from six non-self-consistent models. See section 5 for more details.

cases are demonstrated in Figure 5. The comparison in case 1 (top) provides us the underlying differences in models due to plasma diffusion and the balance between production and loss terms. Since their differences in plasma diffusion are rather small, the comparison suggests that these differences are mainly due to the balance between production and loss terms. Results show that the N_mF_2 in the PBMOD and the h_mF_2 in the SAMI2 are significantly lower at the low latitudes.

Results in case 2 (middle) show that the N_mF_2 among models have larger differences near the EIA crests than above the magnetic equator once the vertical drifts is considered. The N_mF_2 values at the crests in LLIONS and IFM are significantly larger than in the other models, suggesting that the transport processes in these models need to be understood. Also, the latitudinal distribution of EIA is strongly restricted within ±20° magnetic latitude in the IPM, which is particularly due to the height variation of drift that is specified in the model. The simulation results in case 3 (bottom) show slightly smaller N_mF_2 and h_mF_2 near the EIA crests compared to case 2. The overall reduction of N_mF_2 and h_mF_2 at the northern crest is due to the responses of the ionosphere to the imposed meridional wind. The northward wind near the northern crest is capable of pushing the plasma down along the field line to where the recombination is faster and further causes the decrease in plasma density and height. The results of overestimation in N_mF_2 in LLION and IFM and underestimation of PBMOD (N_mF_2) and SAMI2 (h_mF_2) shown in cases 1 and 2 are again repeated in case 3. Through these systematic tests, the differences among the non-self-consistent models can be revealed. Detailed comparisons of other ionospheric parameters and their comparisons with observations will also be carried out in our future workshops.

6. CONCLUSION AND FUTURE PROSPECTS

Comparison results in American sector show that the non-self-consistent model results are generally in good agreement and agree well with IRI results and observations especially in the daytime. The comparisons of model parameters such as diffusion coefficients and transport processes explain part of the differences among the non-self-consistent models. The self-consistent model results are quite different from each other and do not agree with observations. The N_mF_2 values at the EIA crests simulated by the self-consistent models are lower than those from the non-self-consistent model. This result is not able to be explained by the vertical drift since the maximum values of daytime drifts from the self-consistent models do not have significant differences from the empirical model. The phase shifting of daytime drifts in self-consistent models may be caused by different neutral winds at the lower boundary. Differences in the nighttime ionosphere are directly related to the different magnitudes of PREs generated in the models. Since the non-self-consistent models are driven by the same drivers (thermosphere and vertical drift), it is not surprising to see better agreement in ionospheric parameters compared to those from the self-consistent models.

The Equatorial-PRIMO project reviews the current status of these theoretical models and examines their capability in simulating the quiet time equatorial ionosphere, which provides valuable information for the scientific community. This report briefly summarizes comparison results from workshops in the past 2 years. In the coming years, the working group will further focus on identifying other important parameters in the models such as the lower boundary conditions, magnetic coordinates, photoionization, ion/electron temperature model, chemical reaction rates, etc. Besides model simulations, we will also choose appropriate observations that provide information on the equatorial ionosphere and discuss possible model/observation comparisons that will benefit both observers and modelers in understanding the physics of the equatorial ionosphere.

Acknowledgments. The authors wish to thank the National Science Foundation in supporting the Coupling Energetics and Dynamics of Atmospheric Regions Workshop. TWF was supported, in part, by NSF grant ATM-0836386. DNA was supported, in part, by AFOSR grant FA9550-09-0408. JDH was supported by NRL 6.1 Base Funds and NASA. ADR and AM were supported, in part, by NASA grant NNX09AN57G. The National Center for Atmospheric Research is sponsored by the National Science Foundation. Research at UM was supported by NASA grant NNX09AJ59G.

REFERENCES

Anderson, D. N. (1973), A theoretical study of the ionospheric F-region equatorial anomaly, I, Theory, *Planet. Space Sci.*, *21*, 409–419.

Anderson, D. N. (1981), Modeling the ambient, low latitude F-region ionosphere—A review, *J. Atmos. Terr. Phys.*, *43*, 753–762.

Anderson, D. N., et al. (1998), Intercomparison of physical models and observations of the ionosphere, *J. Geophys. Res.*, *103*(A2), 2179–2192.

Appleton, E. V. (1946), Two anomalies in the ionosphere, *Nature*, *157*, 691.

Bilitza, D., and B. W. Reinisch (2008), International Reference Ionosphere 2007: Improvements and new parameters, *Adv. Space Res.*, *42*(4), 599–609, doi:10.1016/j.asr.2007.07.048.

Fang, T.-W., H. Kil, G. Millward, A. D. Richmond, J.-Y. Liu, and S.-J. Oh (2009), Causal link of the wave-4 structures in plasma density and vertical plasma drift in the low-latitude ionosphere, *J. Geophys. Res.*, *114*, A10315, doi:10.1029/2009JA014460.

Fesen, C. G., G. Crowley, R. G. Roble, A. D. Richmond, and B. G. Fejer (2000), Simulation of the pre-reversal enhancement in the low latitude vertical ion drifts, *Geophys. Res. Lett.*, *27*(13), 1851–1854, doi:10.1029/2000GL000061.

Hedin, A. E., et al. (1996), Empirical wind model for the upper, middle and lower atmosphere, *J. Atmos. Terr. Phys.*, *58*, 1421–1447.

Huba, J. D., and G. Joyce (2010), Global modeling of equatorial plasma bubbles, *Geophys. Res. Lett.*, *37*, L17104, doi:10.1029/2010GL044281.

Huba, J. D., G. Joyce, and J. A. Fedder (2000), Sami2 is Another Model of the Ionosphere (SAMI2): A new low-latitude ionosphere model, *J. Geophys. Res.*, *105*(A10), 23,035–23,053, doi:10.1029/2000JA000035.

Lee, C. C., and B. W. Reinisch (2012), Variations in equatorial F_2-layer parameters and comparison with IRI-2007 during a deep solar minimum, *J. Atmos. Sol. Terr. Phys.*, *74*, 217–223.

Millward, G. H., R. J. Moffett, S. Quegan, and T. J. Fuller-Rowell (1996), A coupled thermosphere-ionosphere-plasmasphere model (CTIP), in *Solar-Terrestrial Energy Program: Handbook of Ionospheric Models*, edited by R. W. Schunk, pp. 239–279, Cent. for Atmos. and Space Sci., Utah State Univ., Logan.

Millward, G. H., I. C. F. Müller-Wodarg, A. D. Aylward, T. J. Fuller-Rowell, A. D. Richmond, and R. J. Moffett (2001), An investigation into the influence of tidal forcing on F region equatorial vertical ion drift using a global ionosphere-thermosphere model with coupled electrodynamics, *J. Geophys. Res.*, *106*(A11), 24,733–24,744.

Nicolls, M. J., N. Aponte, S. A. González, M. P. Sulzer, and W. L. Oliver (2006), Daytime *F* region ion energy balance at Arecibo for moderate to high solar flux conditions, *J. Geophys. Res.*, *111*, A10307, doi:10.1029/2006JA011664.

Picone, J. M., A. E. Hedin, D. P. Drob, and A. C. Aikin (2002), NRLMSISE-00 empirical model of the atmosphere: Statistical comparisons and scientific issues, *J. Geophys. Res.*, *107*(A12), 1468, doi:10.1029/2002JA009430.

Retterer, J. M. (2005), Physics-based forecasts of equatorial radio scintillation for the Communication and Navigation Outage Forecasting System (C/NOFS), *Space Weather*, *3*, S12C03, doi:10.1029/2005SW000146.

Retterer, J. M., D. T. Decker, W. S. Borer, R. E. Daniell Jr., and B. G. Fejer (2005), Assimilative modeling of the equatorial ionosphere for scintillation forecasting: Modeling with vertical drifts, *J. Geophys. Res.*, *110*, A11307, doi:10.1029/2002JA009613.

Richmond, A. D. (1995), Ionospheric electrodynamics using magnetic apex coordinates, *J. Geomagn. Geoelectr.*, *47*, 191–212.

Richmond, A. D., E. C. Ridley, and R. G. Roble (1992), A thermosphere/ionosphere general circulation model with coupled electrodynamics, *Geophys. Res. Lett.*, *19*(6), 601–604, doi:10.1029/92GL00401.

Ridley, A. J., Y. Deng, and G. Toth (2006), The global ionosphere–thermosphere model, *J. Atmos. Sol. Terr. Phys.*, *68*, 839–864.

Rishbeth, H. (2000), The equatorial F-layer: Progress and puzzles, *Ann. Geophys.*, *18*, 730–739.

Rishbeth, H., and O. K. Garriott (1969), *Introduction to Ionospheric Physics*, pp. 132–139, Academic Press, New York.

Roble, R. G., and E. C. Ridley (1994), A thermosphere-ionosphere-mesosphere-electrodynamics general circulation model (time-GCM): Equinox solar cycle minimum simulations (30–500 km), *Geophys. Res. Lett.*, *21*(6), 417–420.

Roble, R. G., E. C. Ridley, A. D. Richmond, and R. E. Dickinson (1988), A coupled thermosphere/ionosphere general circulation model, *Geophys. Res. Lett.*, *15*(12), 1325–1328.

Scherliess, L., and B. G. Fejer (1999), Radar and satellite global equatorial F region vertical drift model, *J. Geophys. Res.*, *104*(A4), 6829–6842.

Scherliess, L., R. W. Schunk, J. J. Sojka, and D. C. Thompson (2004), Development of a physics-based reduced state Kalman filter for the ionosphere, *Radio Sci.*, *39*, RS1S04, doi:10.1029/2002RS002797.

Schunk, R. W., and A. F. Nagy (2000), *Ionospheres: Physics, Plasma Physics, and Chemistry*, Cambridge Univ. Press, New York.

Schunk, R. W., and J. J. Sojka (1998), USU time-dependent model of the global ionosphere, in *Guide to Reference and Standard Ionosphere Models*, Rep. ANSI/AIAA G-034-1998, pp. 1–3, Am. Inst. Aeronaut. and Astronaut., Reston, Va.

Schunk, R. W., J. J. Sojka, and J. V. Eccles (1997), Expanded capabilities for the ionospheric forecast model, *Rep. AFRL-VS-HA-TR-98-0001, ADA341230*, Air Force Res. Lab., Space Vehicles Dir., Hanscom AFB, Lincoln, Mass.

Schunk, R. W., et al. (2004), Global Assimilation of Ionospheric Measurements (GAIM), *Radio Sci.*, *39*, RS1S02, doi:10.1029/2002RS002794.

Torr, M. R., and D. G. Torr (1982), The role of metastable species in the thermosphere, *Rev. Geophys. Space Phys.*, *20*(1), 91–144.

Vichare, G., A. J. Ridley, and E. Yigit (2012), Quiet-time low latitude ionospheric electrodynamics in the non-hydrostatic Global Ionosphere–Thermosphere Model, *J. Atmos. Sol. Terr. Phys.*, *80*, 161–172.

Woods, T. N., and G. J. Rottman (2002), Solar ultraviolet variability overtime periods of aeronomic interest, in *Atmospheres in the Solar System: Comparative Aeronomy, Geophys. Monogr. Ser.*, vol. 130, edited by M. Mendillo, A. Nagy, and J. H. White Jr., p. 221, AGU, Washington, D. C.

R. Akmaev and M. Codrescu, NOAA Space Weather Prediction Center, Boulder, CO, USA.

G. Crowley, Atmospheric and Space Technology Research Associates, San Antonio, TX, USA.

V. Eccles, Space Environment Corporation, Providence, UT, USA.

T.-W. Fang, D. Anderson, G. Millward, and T. Fuller-Rowell, CIRES/University of Colorado, Boulder, CO, USA. (tzu-wei.fang@noaa.gov)

J. Huba, Plasma Physics Division, Naval Research Laboratory, Washington, DC, USA.

G. Joyce, Icarus Research, Inc., Bethesda, MD, USA.

A. Maute and A. Richmond, High Altitude Observatory, National Center for Atmospheric Research, Boulder, CO, USA.

J. Retterer, Institute for Scientific Research, Boston College, MA, USA.

A. Ridley, Center for Space Environment Modeling, University of Michigan, Ann Arbor, MI, USA.

L. Scherliess and J. Sojka, Center for Atmospheric and Space Sciences, Utah State University, Logan, UT, USA.

G. Vichare, Indian Institute of Geomagnetism, Navi Mumbai, India.

Systematic Evaluation of Ionosphere/Thermosphere (IT) Models: CEDAR Electrodynamics Thermosphere Ionosphere (ETI) Challenge (2009–2010)

J. S. Shim,[1] M. Kuznetsova,[2] L. Rastätter,[2] D. Bilitza,[3] M. Butala,[4] M. Codrescu,[5] B. A. Emery,[6] B. Foster,[6] T. J. Fuller-Rowell,[5] J. Huba,[7] A. J. Mannucci,[4] X. Pi,[4] A. Ridley,[8] L. Scherliess,[9] R. W. Schunk,[9] J. J. Sojka,[9] P. Stephens,[4] D. C. Thompson,[10] D. Weimer,[11] L. Zhu,[9] D. Anderson,[12] J. L. Chau,[13] and E. Sutton[10]

Systematic quantitative assessment of ionosphere/thermosphere (IT) models is critical in evaluating different modeling approaches, better understanding strengths and weaknesses of the models, and tracking model improvements. The coupling energetics and dynamics of atmospheric regions (CEDAR) community has been leading the CEDAR Electrodynamics Thermosphere Ionosphere (ETI) Challenge since June 2009 in order to address such a need. In the first round of the CEDAR ETI Challenge, we evaluated the performance of about 10 IT models in predicting a number of selected physical parameters, such as vertical ion drifts near the magnetic equator, and electron and neutral densities during the selected nine events, which correspond to three different geomagnetic conditions. To quantify the performance, we obtained four skill scores (RMS error, prediction efficiency, ratio of (maximum (max)-minimum), and ratio of max amplitude). The four skill scores were calculated as a function of geomagnetic activity level and geographic latitude in order to investigate how the model performance depends on them. We found a noticeable dependence of the model performance on geomagnetic activity and latitude. We also observed that the model performance mostly varies with the type of parameters and metrics used.

1. INTRODUCTION

The Earth's upper atmosphere consists of a neutral part, the thermosphere, and an ionized part, the ionosphere. It is a spatially and temporally highly variable region that is controlled by the physical processes driven by external and internal drivers, and the coupling of the ionosphere and thermosphere. A better understanding of the strongly coupled ionosphere-thermosphere (IT) processes is crucial for creating accurate specification and prediction of variations of

[1]GPHI/UMBC, NASA/GSFC, Greenbelt, Maryland, USA.
[2]NASA/GSFC, Greenbelt, Maryland, USA.
[3]School of Physics Astronomy and Computational Science, George Mason University, Fairfax, Virginia, USA.
[4]Jet Propulsion Laboratory, California Institute of Technology, Pasadena, California, USA.
[5]NOAA Space Weather Prediction Center, Boulder, Colorado, USA.
[6]NCAR HAO, Boulder, Colorado, USA.
[7]Plasma Physics Division, Naval Research Laboratory, Washington, District of Columbia, USA.
[8]Space Physics Research Laboratory, University of Michigan, Ann Arbor, Michigan, USA.
[9]CASS, Utah State University, Logan, Utah, USA.
[10]Air Force Research Laboratory, Albuquerque, New Mexico, USA.
[11]Center for Space Science and Engineering Research, Virginia Polytechnic Institute and State University, Blacksburg, Virginia, USA.
[12]CIRES, University of Colorado, Boulder, Colorado, USA.
[13]Radio Observatorio de Jicamarca, Instituto Geofisico del Peru, Lima, Peru.

Modeling the Ionosphere-Thermosphere System
Geophysical Monograph Series 201
© 2013. American Geophysical Union. All Rights Reserved.
10.1029/2012GM001293

the IT system. As a result of extensive theoretical studies of the IT system, various models have been developed in the last several decades [*Schunk et al.*, 2002; *Belehaki et al.*, 2009; *AIAA*, 2010]. The models range from empirical to physics-based data assimilation model. They each have their own limitations in accuracy and reliability of predictions due to, for instance, limited observational data used in empirical modeling and uncertainty in input drivers and simplified equations used in physics-based modeling.

Therefore, there has been an increasing need for systematic quantitative assessment of the models to measure capabilities of different modeling approaches, to better understand their strengths and shortcomings, and to track the model improvements [*Spence et al.*, 2004]. To satisfy this need, the coupling energetics and dynamics of atmospheric regions (CEDAR) community initiated the Electrodynamics Thermosphere Ionosphere (ETI) Challenge with the help of the Community Coordinated Modeling Center (CCMC). In the first round of the CEDAR ETI Challenge (2009–2010), performance of about 10 IT models was assessed using four different scores: RMS error, prediction efficiency, ratio of (maximum (max)-minimum (min)), and ratio of maximum amplitude. In this study, four types of IT models were used: empirical, physics-based ionospheric, coupled IT, and data assimilation models. We compared the model performance in predicting a set of selected physical parameters, such as NmF2, hmF2, vertical drifts, and electron and neutral densities during the selected nine events, which correspond to three different geomagnetic conditions based on maximum Kp value during the events. Measurements used for the comparison are NmF2 and hmF2 from incoherent scatter radars and LEO satellites (COSMIC and CHAMP), electron and neutral densities along the CHAMP trajectory, and vertical ion drifts at Jicamarca near the magnetic equator. The results of the Challenge obtained by using ground-based measurements were reported [*Shim et al.*, 2011], and another report of the Challenge results using space-based measurements is currently in preparation.

In this paper, we present the model performance in predicting electron and neutral densities at the CHAMP orbits and vertical electrodynamic drifts at Jicamarca. The model performance for different geomagnetic conditions and latitude regions is also presented. We provide a model ranking for the same type of models and another ranking that covers all models.

2. OBSERVATIONAL DATA

The in situ electron densities from the planar Langmuir probe (PLP) on board the CHAMP satellite were used as ground truth. CHAMP was orbiting the Earth with an inclination of 87.3° and taking measurements for 10 years since its launch on 15 July 2000 [*Reigber et al.*, 2000]. Owing to the high inclination, the PLP electron density measurements cover almost all latitudes, while all local times are sampled roughly once every 4 months. The PLP takes measurements of electron density at the satellite position every 15 s. The accuracy of the PLP measurements is within 10% [*Liu et al.*, 2007]. The CHAMP PLP data were provided by the Information System and Data Center (http://isdc.gfz-potsdam.de/).

We used the total neutral mass densities inferred from accelerometer measurements on the CHAMP satellite that are available at http://sisko.colorado.edu/sutton/data.html [*Sutton et al.*, 2005]. The neutral densities are 3° latitude average values with a cadence of about 45 s. The absolute uncertainty of the neutral density values is 10%~15% [*Bruinsma et al.*, 2004]. The average error in the neutral density of the selected nine events is about 6%~14%, which corresponds to about $1.5 \times 10^{-13} \sim 3 \times 10^{-13}$ kg m^{-3}, for the altitudes around 340~390 km seen in most of this study. For the study, 1 min average of the observed values of electron and neutral densities were compared with modeled values in the same time frequency.

We also used the vertical drifts derived from the 150 km echoes measured with the Jicamarca Unattended Long-Term Studies of the Ionosphere and Atmosphere (JULIA) coherent scatter radar [*Hysell et al.*, 1997; *Chau and Woodman*, 2004]. The JULIA radar located at Jicamarca, Peru, observes daytime vertical $E \times B$ velocity. The data are recorded every 5 min during the daytime (07:00–18:00 LT). For the selected events (E.2006.348, E.2001.243, E.2007.142, and E.2008.059), the average value of the error does not exceed 1 m s^{-1}. The vertical drifts from JULIA are only available during the day from about 07:00 to 18:00 LT (12:00–23:00 UT). In order to increase data coverage, we also used vertical drifts obtained from Jicamarca-Piura ΔH magnetometer measurements [*Anderson et al.*, 2004] also during the daytime only for four events for which there were no JULIA data available. It is well known that the difference in the magnitudes of the horizontal component (H) between a magnetometer placed directly on the magnetic equator and the one displaced a few degrees away provides the vertical drift in the F region ionosphere. We used modeled and observed vertical drifts with a 5 min temporal resolution.

3. MODELS

We used the model outputs submitted by model developers through the CCMC online submission interface developed for this and other model validation studies. We also used model outputs generated by the CCMC using the IT models hosted at the CCMC [*Webb et al.*, 2009]. Table 1 shows the submissions of outputs using various models, which include

Table 1. Models Submitted for the Coupling Energetics and Dynamics of Atmospheric Regions (CEDAR) Challenge

Model Setting ID	Model/Version	Resolution (lat × lon × alt)
Empirical Models		
1_IRI[a]	IRI-2007, empirical ionospheric model	(50km < alt < 2,000 km)
1_MSIS	Naval Research Laboratory Mass Spectrometer and Incoherent Scatter Extended-00, empirical thermospheric model	
1_JB2008	JB2008, empirical thermospheric density model	
2_JB2008	JB2008 with thermosphere temperature correction derived from the work of *Weimer* [2005] total Poynting fluxes	
Physics-Based Models		
1_SAMI3_HWM93[a] (1_SAMI3_HWM93: only for vertical drift)	Also a Model of the Ionosphere (SAMI3) with the neutral wind model HWM93 [*Hedin et al.*, 1996]	120 × 90 × 160 (90 km < alt < 20,000 km)
1_SAMI3_HWM07	SAMI3 with the neutral wind model HWM07 [*Drob et al.*, 2008]	120 × 90 × 160 (90 km < alt < 20,000 km)
1_USU-IFM[a]	IFM driven by $F10.7$, Kp, and empirical inputs for the thermosphere parameters	60 × 49 × 73 (90 km < alt < 1,600 km)
Coupled IT Models		
1_CTIPE[a]	Coupled thermosphere ionosphere plasmasphere electrodynamics (CTIPe) driven by *Weimer* [2005]	91 × 20 × 15 (~ 90 km < alt < 500 km)
1_GITM	Global ionosphere thermosphere model (GITM)	25 × 50 × 13
3_GITM	GITM with different O^+ reaction rates and collision frequencies, lower boundary conditions, equatorial electrojet, and potential solver from those for 1_GITM	25 × 50 × 13
1_TIE-GCM[a]	TIE-GCM1.93 driven by *Heelis et al.* [1982]	36 × 72 × 29 (~ 90 km < alt < 500 km)
2_TIE-GCM	TIE-GCM1.94 driven by *Weimer* [2005] with dynamic critical colatitudes	36 × 72 × 29 (~ 90 km < alt < 500 km)
Data Assimilation Models		
1_JPL-GAIM	University of Southern California/Jet Propulsion Laboratory (JPL) GAIM with ground-based GPS data ($-55° <$ lat $< 55°$) and COSMIC data	60 × 36 × every 40 km altitude
1_USU-GAIM[a]	USU-GAIM23 with GPS total electron content observations from up to 400 ground stations ($-60° <$ lat $< 60°$)	44 × 24 × 83 (90 km < alt < 1,400 km)

[a]Denotes that the model results are submitted by the Community Coordinated Modeling Center (CCMC) using the models hosted at CCMC. Different model setups are different model setting identification numbers.

empirical, physics-based, coupled IT, and data assimilation models. Multiple output submissions from one model using different input drivers and/or different boundary conditions were distinguished by a unique model setting identifier, for example, 1_JB2008 and 2_JB2008, 1_GITM and 3_GITM, 1_SAMI3_ HWM93 and 1_SAMI3_ HWM07, and 1_TIE-GCM and 2_TIE-GCM. The model setting identifier marked with an alphabet "a" in Table 1 denotes that the model results are submitted by the CCMC.

For the comparison of neutral density at the CHAMP locations, eight submissions of model simulations from five models (JB2008, Naval Research Laboratory Mass Spectrometer and Incoherent Scatter Extended-00 (NRLMSISE-00), coupled thermosphere ionosphere plasmasphere electrodynamics (CTIPe), global ionosphere thermosphere model (GITM), and thermosphere-ionosphere- electrodynamics general circulation model (TIE-GCM)) were analyzed. Nine submissions using eight models (International Reference Ionosphere (IRI), Also a Model of the Ionosphere (SAMI3), Utah State University-Ionosphere Forecast Model (USU-IFM), CTIPe, GITM, TIE-GCM, Jet Propulsion Laboratory-Global Assimilative Ionospheric Model (JPL-GAIM), and USU-GAIM) were included for electron densities. For the comparison of the vertical drift, six submissions of model simulations from four models (IRI, SAMI3, GITM, and TIE-GCM) were analyzed. In the following sections, only brief descriptions of the models used for this study are provided. For more detailed descriptions of the models, please refer to the given references.

3.1. Empirical Models

For the study, one ionospheric and two thermospheric empirical models were used. The IRI [Bilitza, 1990, 2001, 2004; Bilitza and Reinisch, 2007] is the most comprehensive and widely used empirical model for the ionosphere. IRI provides monthly averages of ionospheric density and temperature in the altitude range from 50 to 2000 km for a given location, time, and date. The model has undergone a steady improvement process as new ground and space data became available. We are using IRI-2007 [Bilitza and Reinisch, 2007] for this study. IRI includes a number of different options for specific model parameters and regions, e.g., three different models for the topside electron density and two models for the F peak density. This allows users to test different modeling approaches for their specific application or against their data. For our study, we have relied on the recommended IRI-2007 default settings for all model options.

One of the widely used empirical thermospheric models is NRLMSISE-00, which is based on the earlier models MSIS-86 [Hedin, 1987] and MSISE-90 [Hedin, 1991]. The model covers the altitude range from the ground to the exobase (<1400 km) and provides neutral densities and temperature [Picone et al., 2002]. For the study, the simulation results of NRLMSISE-00 (identified 1_MSIS) were obtained from http://sisko.colorado.edu/sutton/data.html. The Jacchia-Bowman 2008 (JB2008) is another empirical thermospheric model used for the study. The model is developed from the Jacchia-Bowman 2006 model based on Jacchia's diffusion equations [Jacchia, 1965, 1971; Bowman et al., 2007, 2008a, 2008b]. JB2008 uses a correction to the exospheric temperature, due to geomagnetic activity, that is derived from the Dst index and based on the results by Burke [2008]. Another means of calculating the correction to the average exospheric temperature was shown by Weimer et al. [2011]. This alternative method uses the total Poynting fluxes into the polar regions, calculated from empirical models that use the solar wind and IMF measured by the ACE satellite for input values [Weimer, 2005]. Two submissions for the neutral density using JB2008, 1_JB2008 and 2_JB2008, were used (see Table 2). 1_JB2008 is JB2008 run with corrections to global nighttime min exospheric temperature due to auroral heating computed from the Dst index, while 2_JB2008 is JB2008 run with the temperature corrections derived from the total Poynting fluxes [Weimer, 2005; Weimer et al., 2011].

3.2. Physics-Based Ionospheric Models

One of the two physics-based ionospheric models used is SAMI3 based on the 2-D model SAMI2 [Huba et al., 2000, 2008]. SAMI3 calculates the plasma and chemical evolution of ions in the altitude range 85 to 20,000 km. The other model is the USU-IFM [Schunk et al., 1997] based on the USU Time-Dependent Ionosphere Model [Schunk, 1988; Sojka, 1989]. The model covers the altitude range from 90 to 1600 km and all latitudes and longitudes.

Table 2. Setup of the CEDAR Electrodynamics Thermosphere Ionosphere Challenge

Events	Event Name	Date (day of year (DOY)) and Time	F10.7	Kp_max
Strong storms	E.2006.348	14 Dec 2006 (DOY 348) 12:00 UT to 16 Dec (DOY350) 00:00 UT	91	8
	E.2005.243	31 Aug 2005 (DOY 243) 10:00 UT to 01 Sept (DOY244) 12:00 UT	86	7
Moderate storms	E.2001.243	31 Aug 2001 (DOY 243) 00:00 UT to 01 Sept (DOY244) 00:00 UT	192	4
	E.2007.091	01 April 2007 (DOY 091) 00:00 UT to 02 Apr (DOY092) 12:00 UT	72	5
	E.2007.142	22 May 2007 (DOY 142) 12:00 UT to 25 May (DOY145) 00:00 UT	74	6
	E.2008.059	28 Feb 2008 (DOY 059) 12:00 UT to 01 Mar (DOY061) 12:00 UT	69	5
Quiet periods	E.2007.079	20 Mar 2007 (DOY 079) 00:00 UT to 22 Mar (DOY081) 00:00 UT	72	1
	E.2007.190	09 Jul 2007 (DOY 190) 00:00 UT to 10 Jul (DOY191) 00:00 UT	80	0
	E.2007.341	07 Dec 2007 (DOY 341) 00:00 UT to 09 Dec (DOY343) 00:00 UT	80	1

Metrics				Perfect Value
RMS difference	$\text{RMS} = \sqrt{\frac{\sum(x_{obs}-x_{mod})^2}{N}}$		x_{obs} and x_{mod} are observed and modeled values	0
Prediction efficiency (PE)	$\text{PE} = 1 - \frac{\text{RMS}_{mod}}{\text{RMS}_{mean}} = 1 - \sqrt{\frac{\sum(x_{obs}-x_{mod})^2/N}{\sum(x_{obs}-\langle x_{obs}\rangle)^2/N}}$		$\langle x_{obs}\rangle$ is the mean value of the observed measurements	1
Ratio of the maximum (max) change in amplitudes	$\text{ratio (max} - \text{min)} = \frac{(x_{mod})_{max}-(x_{mod})_{min}}{(x_{obs})_{max}-(x_{obs})_{min}}$		$(x_{obs})_{max}$ and $(x_{mod})_{max}$ are the max values of the observed and modeled signals during a certain time window	1
Ratio of the maximum amplitudes	$\text{ratio (max)} = \frac{(x_{mod})_{max}}{(x_{obs})_{max}}$			1

The spatial and temporal resolutions of the IFM are flexible. For this study, 3° × 7.5° latitude-longitude grid was selected. SAMI3 and USU-IFM use empirical inputs for the neutral atmosphere and magnetosphere parameters, e.g., neutral composition, neutral wind, and solar EUV. The neutral winds are obtained from the horizontal wind models (HWMs) [*Hedin et al.*, 1991, 1996; *Drob et al.*, 2008]. Two submissions for the vertical drifts using SAMI3 were included (see Table 1). 1_SAMI3_HWM07 and 1_SAMI3_HWM93 are SAMI3 runs with the neutral wind model HWM07 and HWM93, respectively. The two SAMI3 simulations presented here are for two quiet events (E.2007.079 and E.2007.190) and for two moderate storms (E.2007.091 and E.2007.142). In addition to the two submissions for the vertical drifts made by model developer, we used electron density data from 1_SAMI3_HWM93 submitted by the CCMC. The electron density data in high latitudes (>|50|°) were excluded due to lack of reliability, since SAMI3 does not include high-latitude driving forces, such as the auroral precipitation and the convection electric field pattern.

3.3. Physics-Based Coupled IT Models

For the study, we used three self-consistent coupled IT models, which solve the 3-D momentum, energy, and continuity equations for neutral and ion species: the CTIPe model, the GITM, and the TIE-GCM. The CTIPe [*Fuller-Rowell et al.*, 1996; *Codrescu et al.*, 2000; *Millward et al.*, 2001] consists of four components: global thermosphere, high-latitude ionosphere, midlatitude and low-latitude ionosphere/plasmasphere, and electrodynamical calculation of the global dynamo electric field. The thermospheric code simulates the time-dependent global structure of the wind vector, temperature, and density of the neutral thermosphere. The spatial resolutions of the thermospheric code are 2° in latitude and 18° in longitude, and the vertical direction is divided into 15 levels in logarithm of pressure extending from about 80 to 400~500 km. The GITM [*Ridley et al.*, 2006] uses an altitude grid instead of a pressure grid unlike other global IT models. GITM does not assume hydrostatic equilibrium so that significant vertical flows can develop self-consistently. For the study, two model output submissions (1_GITM and 3_GITM) were made. They differed in the reaction rates of $O^+ + O_2/NO/N_2 \rightarrow O + O_2^+/NO^+/N_2^+$, the equatorial electrojet and potential solver, lower boundary conditions, and the collision frequencies for O^+ to neutral species. The TIE-GCM [*Roble et al.*, 1988; *Richmond et al.*, 1992] uses an imposed electric field model such as the *Weimer* [2005] or *Heelis et al.* [1982] electric potential models at high magnetic latitudes above a critical magnetic colatitude, and a self-consistent dynamo solution equatorward of the second critical colatitude. The critical colatitudes are set to constants or to dynamic values as a function of magnetic activity. It has 29 constant-pressure levels in the vertical, extending from approximately 97 to 500 km in intervals of one-half scale height, and a 5° × 5° latitude-longitude grid, in its base configuration. Two versions of model run, 1_TIE-GCM and 2_TIE-GCM, are submitted by using TIE-GCM1.93 driven by *Heelis et al.* [1982] with constant critical colatitudes and TIE-GCM1.94 driven by *Weimer* [2005] with dynamic critical colatitudes, respectively.

3.4. Physics-Based Data Assimilation Models of the Ionosphere

We used two data assimilation models of the ionosphere. One of them is the GAIM developed by The University of Southern California (USC) and JPL [*Wang et al.*, 2004; *Pi et al.*, 2003]. The JPL/USC-GAIM uses both 4-D variational analysis (4DVAR) and Kalman filter techniques to ingest multiple data sources. Submission of the JPL/USC-GAIM (identified as 1_JPL_GAIM) for the study was generated by assimilating ground-based GPS total electron content (TEC) data (30 s cadence) and COSMIC TEC data (10 s cadence). The ground data from about 200 stations were only assimilated if they came from sites located between ±55° geomagnetic latitude. The spatial resolutions of the model are 10° in longitude, 3° in latitude, and 40 km in altitude. This data set was provided by the JPL team. The other is physics-based data assimilation model of the ionosphere, the GAIM developed at USU [*Schunk et al.*, 2004; *Scherliess et al.*, 2004, 2006]. This model is a Gauss-Markov Kalman filter (GMKF) model, which uses a physics-based model of the ionosphere and a Kalman filter as a basis for assimilating a diverse set of measurements. The physics-based model is the USU-IFM (see section 3.2). The versions of the USU GAIM currently hosted at the CCMC only use GPS observations distributed within ±60° geographic latitude; therefore, the plasma densities at high latitude are provided by the IFM without assimilating any data. We used the USU-GAIM result generated by the CCMC for this study.

4. METRICS

For definite comparisons, the model performance was quantified by using four different metrics (see Table 2). The term "metric" refers to functions that produce one real number (skill score). As defined in Table 2, the RMS error and prediction efficiency (PE) measure how well observed data and modeled values are correlated with each other. Metrics based on ratio are used to quantify the model capability to produce peak values or short-term variations during a certain period of time. Perfect model predictions have RMS differences of 0. Therefore, the closer the RMS error is to 0, the more accurate the model is. PE ranges from negative infinity to 1. The closer PE is to 1, the better is the model performance, while PE of 0 means that the model performance is as accurate as the mean of the observed data, which was considered as a reference model instead of using any empirical model for this study. Negative values of PE indicate that the observed mean is a better predictor of the observations than the model. In order to take account of LT dependence of electron and neutral densities, the daytime PE (06:00–18:00 LT) and the nighttime PE (18:00–06:00 LT) were obtained separately using daytime and nighttime mean values of observations during any given event. The two types of ratio selection were the ratio of the maximum change (difference between max and min values; max − min) and the ratio of the max values of models to those of observations during a certain time interval. Perfect models have a ratio of 1. The ratio of max-min and the ratio of max larger than 1 overestimate max variations and max values.

In the work of *Shim et al.* [2011], it was pointed out that selecting the appropriate length of the time window is crucial to calculate the two ratios, which depend on the time window length. It was found for this study that a suitable length of time window is 1 h, compared to 4 and 7 h for calculating ratios of the vertical drifts at Jicamarca. For the ratio of electron and neutral densities, we selected a 90 min time window length that is close to the period of the CHAMP satellite (about 94 min). Owing to the daytime and nighttime alternation during the 90 min, the ratio of max-min represents the ratio of the difference between daytime max and nighttime min values (diurnal variation).

5. RESULTS

5.1. Electron Density

Tables 3 and 4 show RMS error, PE, ratio of max-min, ratio of max of the model predictions of electron density at the CHAMP orbit, and model ranking. To examine the model performance dependency on latitude, skill scores were calculated for three latitude regions: low ($|lat| < 25°$), middle ($25° < |lat| < 50°$), and high ($|lat| > 50°$) geographic latitudes. Average scores for each geomagnetic activity level was calculated: average of the two strong storms, average of the four moderate storms, and average of the three quiet periods (see Table 2). The model performance for each geomagnetic activity level and latitude regions (total of nine cases) is ranked based on the average values. In Tables 3 and 4, we also show model ranking for the same type of models (subrank) and another ranking that includes all models (rank). First ranked models among all models are shown in bold, and those in the same type of models are shown in italics. Note that 1_SAMI3_HWM93 data at high latitudes were excluded due to lack of reliability, and 1_GITM has data only for one strong (E.2005.243), two moderate storm events (E.2007.091 and E.2007.142), and two events (E.2007.079 and E.2007.190) during quiet conditions. Therefore, comparing 1_GITM with other models based on the averaged performance taken over the events requires caution.

As shown in Table 3, RMS differences for the electron density tend to increase as a function of geomagnetic activity. All models, except 1_GITM and 3_GITM in low latitudes, show the RMS increases due to increasing geomagnetic activity from low (quiet period) to medium (moderate storm) and to high (strong storm) levels. However, a change of geomagnetic activity from medium to high level appears to produce smaller increases in the RMS error than a change from low to medium level activity, especially in high latitudes. PE and ratios (Tables 3–4) also show dependency on geomagnetic activity. Most models tend to have better PE and ratios during storms than quiet times at all latitudes, although the three scores show less systematic dependency on geomagnetic activity than RMS error.

In terms of both RMS error and PE, the empirical model 1_IRI ranks at the top in high latitudes for all three geomagnetic conditions, while in low and middle latitudes, the data assimilation models, 1_JPL-GAIM, and 1_USU-GAIM, and physics based models, TIE-GCMs and 1_USU-IFM, rank top during storms and quiet conditions, respectively. The model rankings for RMS error and PE are similar but not the same (Tables 3 and 4). The ranking for PE is obtained based on the average PE for daytime (06:00–18:00 LT) and nighttime (18:00–06:00 LT), while the RMS error does not have the LT dependence. In addition, PE is normalized by the standard deviation of the observations, whereas the RMS error is not normalized. At high latitudes, the data assimilation models that rank at the top in low and middle latitudes are outranked by 1_IRI, probably due to the limited latitude coverage of data. The USU-GAIM currently hosted at the CCMC only use GPS TEC data between −60 and +60 geographic latitude, and the 1_JPL_GAIM submission for the study was generated by assimilating ground-based GPS TEC data from sites located between ±55° geomagnetic latitude.

As for the ratio of max-min and ratio of max (Table 4), during the strong storms, 1_SAMI3_HWM93 and the data assimilation models show better performance than others in low and higher latitudes, respectively. 1_TIE-GCM shows the best ratios for quiet periods in low and middle latitudes, while 1_IRI ranks at the top for moderate storm at high latitudes. For quiet conditions, most models tend to

Table 3. RMS Errors (10^5 cm^{-3}) and PE in Prediction of Electron Density Along the CHAMP Orbits

Latitude	Model	Strong Storm Rank	Subrank	RMS	Moderate Storm Rank	Subrank	RMS	Quiet Period Rank	Subrank	RMS	Strong Storm Rank	Subrank	PE	Moderate Storm Rank	Subrank	PE	Quiet Period Rank	Subrank	PE
Empirical																			
Low	1_IRI	2	1	2.93	3	1	2.65	7	1	2.03	2	1	0.03	3	1	0.12	8	1	−3.04
Middle	1_IRI	4	1	2.20	3	1	1.53	3	1	0.83	2	1	−0.08	**1**	**1**	**0.16**	5	1	−1.42
High	1_IRI	**1**	**1**	**0.92**	**1**	**1**	**0.79**	**1**	**1**	**0.43**	**1**	**1**	**−0.08**	**1**	**1**	**0.04**	**1**	**1**	**0.17**
Ionospheric																			
Low	1_SAMI3_HWM93	9	2	4.05	7	2	3.70	6	1	1.92	8	2	−0.39	7	2	−0.19	3	2	−0.75
Low	1_USU-IFM	6	1	3.77	4	1	3.05	8	2	2.20	7	1	−0.30	6	1	−0.03	**1**	**1**	**−0.48**
Middle	1_SAMI3_HWM93	8	2	2.85	5	2	2.01	7	1	1.27	7	1	−0.52	7	2	−0.31	6	1	−1.71
Middle	1_USU-IFM	7	1	2.60	4	1	1.94	8	2	1.56	8	2	−0.69	6	1	−0.28	8	2	−2.38
High	1_SAMI3_HWM93	6	1	1.66	7	1	1.88	9	1	1.47	6	1	−1.32	7	1	−1.85	8	1	−1.68
Coupled IT																			
Low	1_CTIPE	7	4	3.82	8	3	3.74	5	3	1.90	6	3	−0.26	8	3	−0.27	6	2	−1.50
Low	1_GITM	10	5	8.37	10	5	7.46	10	5	7.72	10	5	−2.67	10	5	−2.59	10	5	−12.4
Low	3_GITM	5	3	3.60	9	4	4.32	9	4	3.83	9	4	−0.41	9	4	−0.68	9	4	−7.93
Low	1_TIE-GCM	4	2	3.35	6	2	3.35	**1**	**1**	**1.45**	4	2	−0.12	5	2	0.00	4	1	−1.14
Low	2_TIE-GCM	*3*	*1*	*3.19*	*5*	*1*	*3.13*	*4*	*2*	*1.81*	*3*	*1*	*−0.06*	*4*	*1*	*0.07*	*7*	*3*	*−1.85*
Middle	1_CTIPE	5	2	2.33	7	2	2.14	4	3	0.88	5	2	−0.20	8	3	−0.50	3	3	−0.68
Middle	1_GITM	9	4	3.26	6	1	2.13	10	5	1.93	10	5	−0.79	9	4	−0.91	10	5	−6.37
Middle	3_GITM	10	5	3.28	10	5	3.09	9	4	1.81	9	4	−0.78	10	5	−1.10	9	4	−4.17
Middle	1_TIE-GCM	6	3	2.46	9	4	2.82	**1**	**1**	**0.60**	6	3	−0.27	4	1	**−0.13**	**1**	**1**	**−0.16**
Middle	2_TIE-GCM	*3*	*1*	*2.01*	*8*	*3*	*2.53*	*2*	*2*	*0.71*	**1**	**1**	**−0.01**	*5*	*2*	*−0.16*	*2*	*2*	*−0.44*
High	1_CTIPE	*2*	*1*	*1.00*	*4*	*2*	*1.17*	*2*	*2*	*0.53*	*2*	*1*	*−0.27*	*5*	*3*	*−0.85*	*2*	*1*	*0.02*
High	1_GITM	8	4	2.30	8	4	2.25	8	5	1.34	7	3	−1.40	8	4	−3.00	9	5	−2.57
High	3_GITM	9	5	2.62	9	5	2.85	7	4	1.16	8	5	−2.24	9	5	−3.49	7	4	−1.53
High	1_TIE-GCM	7	3	1.71	5	3	1.38	3	2	0.63	4	4	−1.53	3	2	−0.64	3	2	−0.09
High	2_TIE-GCM	5	2	1.24	2	1	0.96	4	3	0.66	2	2	−0.70	2	1	−0.22	4	3	−0.10
Data Assimilation																			
Low	1_JPL-GAIM	**1**	**1**	**2.12**	**1**	**1**	**2.06**	2	1	1.50	**1**	**1**	**0.21**	**1**	**1**	**0.25**	5	1	−1.24
Low	1_USU-GAIM	8	2	3.91	2	2	2.35	3	2	1.57	5	2	−0.26	2	2	0.15	2	2	−0.70
Middle	1_JPL-GAIM	**1**	**1**	**1.80**	2	2	1.34	6	2	1.25	4	1	−0.13	3	2	0.08	7	2	−2.10
Middle	1_USU-GAIM	2	2	1.86	**1**	**1**	**1.09**	5	1	0.89	3	2	−0.13	2	1	0.15	4	1	−0.91
High	1_JPL-GAIM	*3*	*1*	*1.15*	*3*	*1*	*1.16*	*6*	*2*	*1.08*	*3*	*1*	*−0.42*	*4*	*1*	*−0.74*	*6*	*2*	*−1.02*
High	1_USU-GAIM	4	2	1.21	6	2	1.43	5	1	1.01	4	2	−0.54	6	2	−1.08	5	1	−0.86

Table 4. Ratio (Max-Min), and Ratio (Max) in Prediction of Electron Density Along the CHAMP Orbits

Latitude	Model	Strong Storm Rank	Sub-rank	Ratio (Max-Min)	Moderate Storm Rank	Sub-rank	Ratio (Max-Min)	Quiet Period Rank	Sub-rank	Ratio (Max-Min)	Strong Storm Rank	Sub-rank	Ratio (Max)	Moderate Storm Rank	Sub-rank	Ratio (Max)	Quiet Period Rank	Sub-rank	Ratio (Max)
Empirical																			
Low	1_IRI	6	1	0.85	3	1	1.05	5	1	1.23	6	1	0.88	4	1	1.07	5	1	1.26
Middle	1_IRI	4	1	1.29	7	1	1.35	3	1	1.49	4	1	1.26	6	1	1.36	3	1	1.53
High	1_IRI	5	1	0.66	**1**	**1**	**0.93**	**1**	**1**	**1.09**	5	1	0.72	**1**	**1**	**1.07**	2	1	1.23
Ionospheric																			
Low	1_SAMI3_HWM93	**1**	**1**	**0.94**	**1**	**1**	**0.96**	7	1	1.40	2	1	0.96	2	1	0.98	7	1	1.37
	1_USU-IFM	3	2	1.08	5	2	1.13	8	2	1.63	3	2	1.05	6	2	1.12	8	2	1.55
Middle	1_SAMI3_HWM93	*3*	*1*	*1.28*	5	2	1.27	7	1	1.88	3	1	1.23	5	2	1.22	7	1	*1.78*
	1_USU-IFM	6	2	1.40	4	1	1.23	9	2	2.16	5	2	1.32	3	1	*1.19*	9	2	2.07
High	1_SAMI3_HWM93	7	1	1.70	9	1	2.07	9	1	2.51	7	1	1.75	9	1	2.17	9	1	2.63
Coupled IT																			
Low	1_CTIPE	9	4	0.57	8	3	0.65	4	3	1.18	9	4	0.61	8	3	0.72	4	3	1.22
	1_GITM	10	5	1.66	10	5	1.65	10	5	3.16	10	5	1.65	10	5	1.88	10	5	3.08
	3_GITM	8	3	0.74	9	4	0.64	9	4	1.65	8	3	0.76	9	4	0.72	9	4	1.60
	1_TIE-GCM	4	*1*	*0.90*	7	2	0.85	**1**	**1**	**0.99**	4	*1*	*0.91*	7	2	0.88	**1**	**1**	**1.05**
	2_TIE-GCM	5	2	0.87	6	*1*	*0.87*	2	2	1.10	5	2	0.90	3	*1*	*0.93*	3	2	1.17
Middle	1_CTIPE	7	2	0.59	2	*1*	*0.93*	6	3	1.82	7	2	0.59	2	*1*	*0.96*	5	3	1.71
	1_GITM	10	5	2.34	10	5	1.60	10	5	2.66	10	5	2.14	10	5	1.61	10	5	2.55
	3_GITM	9	4	1.64	9	4	1.41	8	4	2.08	9	4	1.54	9	4	1.50	8	4	2.05
	1_TIE-GCM	8	3	1.51	8	3	1.38	**1**	**1**	**0.97**	8	3	1.50	8	3	1.42	**1**	**1**	**1.08**
	2_TIE-GCM	5	*1*	*1.33*	6	2	1.32	2	2	1.13	6	*1*	*1.35*	7	2	1.37	2	2	1.22
High	1_CTIPE	6	3	0.62	3	2	1.16	2	*1*	*1.13*	6	3	0.69	3	2	1.25	**1**	**1**	**1.15**
	1_GITM	10	5	2.44	7	4	1.67	8	5	2.10	9	5	2.53	7	4	1.96	8	5	2.10
	3_GITM	9	4	2.03	8	5	1.74	7	4	1.91	8	4	2.30	8	5	2.12	7	4	1.94
	1_TIE-GCM	2	*1*	*1.06*	4	3	1.19	5	3	0.48	4	2	1.26	4	3	1.42	3	2	0.76
	2_TIE-GCM	4	2	0.79	2	*1*	*0.92*	4	2	0.51	**1**	**1**	**0.91**	2	*1*	*1.08*	4	3	0.73
Data Assimilation																			
Low	1_JPL-GAIM	2	*1*	*0.92*	4	2	0.88	3	*1*	*1.11*	**1**	**1**	**0.97**	5	2	0.93	2	*1*	*1.16*
	1_USU-GAIM	7	2	0.84	2	*1*	*0.96*	6	2	1.27	7	2	0.86	**1**	**1**	**1.00**	6	2	1.31
Middle	1_JPL_GAIM	2	2	1.19	3	2	1.23	5	2	1.72	2	2	1.18	4	2	1.22	6	2	*1.77*
	1_USU-GAIM	**1**	**1**	**1.16**	**1**	**1**	**0.99**	4	*1*	*1.58*	**1**	**1**	**1.14**	**1**	**1**	**1.03**	4	*1*	*1.59*
High	1_JPL_GAIM	2	**1**	**0.97**	5	*1*	*1.23*	3	*1*	*1.44*	3	2	1.21	5	*1*	*1.44*	5	*1*	*1.70*
	1_USU-GAIM	3	2	1.12	6	2	1.55	6	2	1.74	2	1	*1.17*	6	2	1.67	6	2	1.87

overestimate diurnal variations (ratio of max-min > 1) and the daytime maximum (ratio of max > 1) of electron density in all three latitude regions, except for 1_TIE-GCM and 2_TIE-GCM in high latitudes. In low and middle latitudes, 1_GITM, 3_GITM, 1_CTIPE, and 1 USU-IFM show relatively larger differences in the ratios between the three levels of geomagnetic activity than the others.

For predicting electron density, the data assimilation models and the IRI empirical model show better scores than physics-based ionosphere and coupled IT models in low and middle latitudes especially in terms of RMS difference and PE. In high latitudes, the IRI empirical model and physics-based coupled IT models rank higher than the others.

Two physics-based ionospheric models, 1_SAMI3_HWM93 and 1_USU-IFM show similar performance in terms of RMS error and PE. However, 1_USU-IFM produces slightly smaller RMS error during the storms, and 1_SAMI3_HWM93 shows relatively better agreement with observations during the quiet periods for low and middle latitudes and produces better diurnal variations and maximum values of electron density than 1_USU-IFM for most cases.

It appears that 2_TIE-GCM, 1_TIE-GCM, and 1_CTIPE among five coupled IT model submissions perform similarly in producing electron density along the CHAMP tracks, and they show better performance than the other two, 1_GITM and 3_GITM, for most cases. 2_TIE-GCM (driven by Weimer electric potential with dynamic critical crossover latitudes) is slightly better for the storms and worse for the quiet events than 1_TIE-GCM (driven by Heelis electric potential with constant critical crossover latitudes). 1_CTIPE performs worse than 1_TIE-GCM and 2_TIE-GCM in low latitudes, while 1_CTIPE is better for moderate storms in middle latitudes and for the quiet periods in high latitudes.

Differences between performances of the two data assimilation models, 1_JPL-GAIM and 1_USU-GAIM, are hardly observed except for larger RMS errors of 1_USU-GAIM for strong storms in low latitudes and a larger negative PE of 1_JPL-GAIM for quiet events in low and middle latitudes.

5.2. Neutral Density

Tables 5 and 6 show the four scores of the model predictions in neutral density, at the CHAMP orbit, and the ranking of the models. During the strong storms, all models show worst performance in terms of RMS error (Table 5). Highly ranked models, including 1_JB2008, 2_JB2008, 1_MSIS, 1_TIE-GCM, and 2_TIE-GCM, show similar RMS error differences of about $5 \times 10^{-13} \sim 1 \times 10^{-12}$ kg m^{-3} between quiet periods and strong storm events. However, relatively low ranked models such as 1_CTIPE, 1_GITM, and 3_GITM show larger differences up to about 6×10^{-12} kg m^{-3}. The two submissions, 1_JB2008 and 2_JB2008, rank at the top and are followed by another empirical model 1_MSIS. 1_JB2008 has slightly better RMS and PE than 2_JB2008 during strong storm and quiet periods for all three latitudes; however, 2_JB2008 is slightly better than 1_JB2002 for moderate storm. Compared to the three empirical model results, the performances of 1_TIE-GCM and 2_TIE-GCM are rather worse but fairly comparable in terms of RMS. 1_CTIPE produces worse RMS differences and PE than 1_TIE-GCM and 2_TIE-GCM for the storm events, although the RMS and PE of 1_CTIPE are better for the quiet events. 1_GITM and 3_GITM perform worse than the other models especially during the strong storm events. 3_GITM shows the largest RMS errors and negative PE during strong storms. However, for moderate storms, 1_GITM and 3_GITM perform comparably with 1_TIE-GCM and 2_TIE-GCM and better than 1_CTIPE.

As shown in Table 6, among the three empirical models, 1_MSIS shows a better ratio of max-min than JB2008s for strong storms in low latitudes and, for moderate storms, in middle and high latitudes. 2_JB2008 shows the best ratio of max in low latitudes, while 1_JB2008 shows the best ratio of max in middle and high latitudes. During the strong storms, among physics-based coupled IT models, TIE-GCMs that tend to underestimate ratios are better than 1_CTIPE and GITMs, which overestimate ratios. However, during the moderate storm events, the coupled IT models produce similar ratios (less than 1). For quiet periods, 1_CTIPE underestimates the day max, while the GITM and TIEGCM models overestimate daytime max.

The two JB2008s, for quiet periods in the three latitude regions, produce a ratio of max close to 1 and the ratios of max-min smaller than 1. This indicates that JB2008s predict well the daytime maximum neutral density, but overestimate nighttime minimum values. The two GITMs, for moderate storm in all latitudes, show the same features.

5.3. Vertical Drifts at Jicamarca

Tables 7 and 8 show the four scores and the model ranking based on the scores for predicting vertical ion drifts near the magnetic equator at Jicamarca. Note that neither all modeled data nor all observational measurements are available for all events. For instance, there are no observations for the event E.2007.091, and 1_SAMI_HWM07 and 1_SAMI_HWM93 do not have data for the two strong storm cases because they do not have high-latitude model inputs.

All models have the largest RMS error for the strong storms (Table 7). In terms of both RMS error and PE, the empirical model 1_IRI ranks the top followed by the coupled model TIE-GCMs. For all events, all models, except for

Table 5. RMS Errors (10^{-12} kg m^{-3}), and PE in Prediction of Neutral Density Along the CHAMP Orbits

Latitude	Model	Strong Storm				Moderate Storm				Quiet Period				Strong Storm			Moderate Storm			Quiet Period		
		Rank	Subrank	RMS		Rank	Subrank	RMS		Rank	Subrank	RMS		Rank	Subrank	PE	Rank	Subrank	PE	Rank	Subrank	PE
Empirical																						
Low	1_JB2008	**1**	**1**	**1.10**		2	2	0.75		**1**	**1**	**0.25**		**1**	**1**	**0.35**	**1**	**1**	**−0.28**	2	2	−0.29
	2_JB2008	2	2	1.25		**1**	**1**	**0.55**		2	2	0.30		2	2	0.26	2	2	−0.61	**1**	**1**	**−0.05**
	1_MSIS	3	3	1.53		5	3	1.08		3	3	0.56		4	3	0.06	5	3	−1.69	3	3	−1.42
Middle	1_JB2008	**1**	**1**	**1.02**		2	2	0.62		**1**	**1**	**0.25**		**1**	**1**	**0.37**	2	2	0.11	**1**	**1**	**0.22**
	2_JB2008	2	2	1.09		**1**	**1**	**0.46**		2	2	0.29		2	2	0.31	**1**	**1**	**0.31**	2	2	0.03
	1_MSIS	3	3	1.31		5	3	0.96		3	3	0.51		3	3	0.17	5	3	−0.31	3	3	−0.67
High	1_JB2008	**1**	**1**	**1.35**		2	2	0.64		**1**	**1**	**0.24**		**1**	**1**	**0.22**	2	2	0.28	**1**	**1**	**0.25**
	2_JB2008	3	3	1.48		**1**	**1**	**0.60**		2	2	0.30		3	3	0.14	**1**	**1**	**0.31**	2	2	0.01
	1_MSIS	2	2	1.46		5	3	0.95		3	3	0.43		2	2	0.15	5	3	−0.03	3	3	−0.32
Coupled IT																						
Low	1_CTIPE	6	3	2.85		8	5	2.30		4	1	0.81		6	3	−0.63	8	5	−3.26	4	1	−2.26
	1_GITM	7	4	5.37		6	3	1.32		8	5	1.72		7	4	−2.02	6	3	−2.04	8	5	−6.26
	3_GITM	8	5	7.41		7	4	1.38		7	4	1.41		8	5	−3.14	7	4	−2.26	7	4	−4.99
	1_TIE-GCM	5	2	1.72		3	1	0.80		5	2	0.83		5	2	−0.04	3	1	−0.76	5	2	−2.54
	2_TIE-GCM	*4*	*1*	*1.54*		*4*	*2*	*0.83*		*6*	*3*	*1.14*		*3*	*1*	*0.07*	*4*	*2*	*−1.13*	*6*	*3*	*−3.85*
Middle	1_CTIPE	6	3	2.69		8	5	1.88		4	1	0.67		6	3	−0.62	8	5	−1.38	4	1	*−1.13*
	1_GITM	7	4	4.26		6	3	1.24		8	5	1.28		7	4	−1.53	6	3	−0.79	8	5	−3.26
	3_GITM	8	5	6.55		7	4	1.35		7	4	1.09		8	5	−2.91	7	4	−0.92	7	4	−2.65
	1_TIE-GCM	5	2	1.46		3	1	0.75		5	2	0.75		5	2	0.11	3	1	−0.06	5	2	−1.43
	2_TIE-GCM	*4*	*1*	*1.37*		*4*	*2*	*0.79*		*6*	*3*	*1.06*		*4*	*1*	*0.16*	*4*	*2*	*−0.18*	*6*	*3*	*−2.42*
High	1_CTIPE	6	3	2.98		8	5	1.66		4	1	0.56		6	3	−0.72	8	5	−0.84	4	1	−0.76
	1_GITM	7	4	3.83		6	3	1.20		7	4	0.86		7	4	−1.20	6	3	−0.36	8	5	−1.86
	3_GITM	8	5	5.69		7	4	1.29		6	3	0.72		8	5	−2.28	7	4	−0.43	6	3	−1.47
	1_TIE-GCM	5	2	1.72		4	2	0.84		5	2	0.60		5	2	0.00	4	2	0.07	5	2	−0.86
	2_TIE-GCM	*4*	*1*	*1.68*		*3*	*1*	*0.84*		*8*	*5*	*0.89*		*4*	*1*	*0.03*	*3*	*1*	*0.08*	*7*	*4*	*−1.79*

Table 6. Ratio (Max-Min), and Ratio (Max) in Prediction of Neutral Density Along the CHAMP Orbits

Latitude	Model	Strong Storm			Moderate Storm			Quiet Period			Strong Storm			Moderate Storm			Quiet Period		
		Rank	Sub-rank	Ratio (Max-Min)	Rank	Sub-rank	Ratio (Max-Min)	Rank	Sub-rank	Ratio (Max-Min)	Rank	Sub-rank	Ratio (Max)	Rank	Sub-rank	Ratio (Max)	Rank	Sub-rank	Ratio (Max)
	Empirical																		
Low	1_JB2008	3	3	1.12	2	2	1.02	7	3	0.76	2	2	1.07	5	2	1.08	2	2	0.92
	2_JB2008	2	2	1.04	1	1	**1.00**	5	2	0.79	**1**	**1**	**0.95**	4	*1*	*1.05*	**1**	**1**	**0.99**
	1_MSIS	**1**	**1**	**0.99**	3	3	1.05	3	*1*	*0.93*	4	3	0.90	8	3	1.18	3	3	1.15
Middle	1_JB2008	**1**	**1**	**1.02**	2	2	0.99	7	3	0.76	**1**	**1**	**1.03**	5	2	1.07	2	2	0.94
	2_JB2008	2	2	0.93	3	3	0.97	5	2	0.80	3	2	0.92	3	*1*	*1.03*	2	2	**1.01**
	1_MSIS	3	3	0.90	**1**	**1**	**1.01**	**1**	**1**	**1.05**	5	3	0.89	7	3	1.17	3	3	1.18
High	1_JB2008	4	*1*	*0.68*	3	2	0.79	4	2	0.77	**1**	**1**	**0.89**	**1**	**1**	**0.96**	**1**	**1**	**0.99**
	2_JB2008	5	2	0.61	4	3	0.77	3	*1*	*0.81*	5	2	0.78	4	2	0.93	2	2	1.06
	1_MSIS	6	3	0.62	**1**	**1**	**0.82**	6	3	1.27	6	3	0.79	5	3	1.08	4	3	1.24
	Coupled IT																		
Low	1_CTIPE	4	*1*	*1.13*	4	*1*	0.77	2	2	0.63	6	3	1.20	7	5	0.84	5	2	0.75
	1_GITM	7	4	1.50	5	2	0.81	**1**	**1**	**1.02**	7	4	1.67	2	2	0.98	8	5	1.56
	3_GITM	8	5	1.59	6	3	0.82	8	5	1.05	8	5	1.93	3	3	0.97	7	4	1.45
	1_TIE-GCM	6	3	0.65	8	5	0.68	5	4	0.76	5	2	0.86	6	4	0.90	4	*1*	*1.20*
	2_TIE-GCM	5	2	0.74	7	4	0.74	4	3	0.80	3	*1*	*0.91*	**1**	**1**	**1.00**	6	3	1.34
Middle	1_CTIPE	6	3	1.22	8	5	0.70	8	5	0.67	6	3	1.21	8	5	0.78	4	*1*	*0.77*
	1_GITM	7	4	1.67	5	2	0.82	4	3	1.19	7	4	1.64	4	3	0.94	8	5	1.55
	3_GITM	8	5	2.00	4	*1*	*0.89*	6	4	1.20	8	5	1.96	**1**	**1**	**0.99**	7	4	1.48
	1_TIE-GCM	5	2	0.81	7	4	0.73	3	2	0.89	4	2	0.90	6	4	0.90	5	2	1.24
	2_TIE-GCM	4	*1*	*0.90*	6	3	0.78	2	*1*	*0.95*	2	*1*	*0.93*	2	2	0.98	6	3	1.38
High	1_CTIPE	3	3	1.32	8	5	0.64	7	4	0.71	4	3	1.22	8	5	0.74	3	*1*	*0.81*
	1_GITM	7	4	1.49	7	4	0.70	5	3	1.25	7	4	1.47	7	4	0.84	8	5	1.50
	3_GITM	8	5	1.99	2	*1*	*0.81*	8	5	1.30	8	5	1.88	2	*1*	*0.94*	7	4	1.46
	1_TIE-GCM	2	2	0.70	6	3	0.70	**1**	**1**	**1.02**	3	2	0.84	6	3	0.88	5	2	1.29
	2_TIE-GCM	**1**	**1**	**0.82**	5	2	0.72	2	2	1.10	2	*1*	*0.85*	3	2	0.93	6	3	1.46

1_IRI, during the quiet time periods have negative PE, which is calculated using only the available daytime observations. This suggests that the mean of the observations is more accurate than the modeled vertical drifts at Jicamarca during the daytime.

As for the ratio of max-min (Table 8), 1_SAMI3_HWM07, 1_SAMI3_HWM93, and 1_TIEGCM show the best performance in producing max-min for quiet periods, moderate and strong storm cases, respectively. Also, in terms of the ratio of max, 1_TIEGCM and 1_IRI rank the top during quiet periods and storms, respectively. It is evident that maximum values of the modeled vertical drifts and observations show good agreements (close to 1) during the quiet periods, while the models tend to overestimate maximum values during moderate and strong storms.

1_SAMI3_HWM07 has better scores than 1_SAMI3_HWM93 for most cases except for the ratio of max-min during moderate storms. Among the coupled IT model simulations including the two TIE-GCMs and 3_GITM, 2-TIE-GCM is better during the storms in terms of RMS, PE, and ratio of max, while 1_TIE-GCM is better during the quiet periods. For the ratio of max-min, 1_TIE-GCM is better during storms, and 2_TIE-GCM is better for quiet periods.

6. DISCUSSION AND CONCLUSIONS

In order to evaluate the performance of IT models, we calculated four different skill scores: RMS error, PE against the mean of the observations, and ratios of max-min and max in predicting electron and neutral densities and vertical drifts. Measurements used for the comparison were vertical drifts at Jicamarca and CHAMP electron and neutral densities. For the validation study, we selected nine events, which were divided into three geomagnetic levels (strong storms, moderate storms, and quiet periods) according to the maximum Kp value during the events (see Table 2). In addition, we divided the geographic latitudes into three regions: low ($|lat| < 25°$), middle ($25° < |lat| < 50°$), and high ($|lat| > 50°$) latitudes to examine the model performance dependency on the latitude. The model performance of a total of nine cases for each geomagnetic activity level and latitude region was ranked. In this paper, we provided the model ranking for the same type of models (subrank) and another ranking that includes all models (rank).

The results of the study show that the model performance depends on the geomagnetic activity. For the vertical drift prediction, all models show increasing RMS errors with rising geomagnetic activity level. Also, for electron and neutral density prediction, the RMS errors increase as

Table 7. RMS Errors (m s^{-1}) and (PE) in Prediction of Vertical Drift at Jicamarca

Model	Strong Storm				Moderate Storm				Quiet Period			
	Rank	Subrank	RMS	PE	Rank	Subrank	RMS	PE	Rank	Subrank	RMS	PE
Empirical												
1_IRI	1	1	8.22	-0.04	1	1	7.12	-0.20	1	1	3.60	0.18
Ionospheric												
1_SAMI3_HWM93					6	2	20.24	-1.91	5	2	12.24	-1.11
1_SAMI3_HWM07					5	1	17.41	-1.50	4	1	8.84	-1.51
Coupled IT												
3_GITM	4	3	33.30	-3.51	3	2	14.63	-1.60	6	3	13.44	-2.21
1_TIE-GCM	3	2	20.55	-1.75	4	3	17.70	-1.62	2	1	5.91	-0.37
2_TIE-GCM	2	1	13.05	-0.59	2	1	9.12	-0.57	3	2	5.94	-0.38

Table 8. Ratio (Max-Min), and Ratio (Max) in Prediction of Vertical Drift at Jicamarca

Model	Strong Storm			Moderate Storm			Quiet Period			Strong Storm			Moderate Storm			Quiet Period		
	Rank	Sub-rank	Ratio (Max-Min)	Rank	Sub-rank	Ratio (Max-Min)	Rank	Sub-rank	Ratio (Max-Min)	Rank	Sub-rank	Ratio (Max)	Rank	Sub-rank	Ratio (Max)	Rank	Sub-rank	Ratio (Max)
Empirical																		
1_IRI	3	1	0.44	3	1	0.61	3	1	1.22	1	1	1.06	1	1	2.06	3	1	1.07
Ionospheric																		
1_SAMI3_HWM93				**1**	**1**	**0.95**	2	2	0.80				3	2	2.16	6	2	0.62
1_SAMI3_HWM07				*5*	*2*	*0.51*	*1*	*1*	*1.03*				*2*	*1*	*2.10*	*4*	*1*	*0.87*
Coupled IT																		
3_GITM	4	3	2.55	6	3	2.38	6	3	3.21	4	3	3.87	6	3	3.93	5	3	1.34
1_TIE-GCM	**1**	**1**	**0.99**	*2*	*1*	*1.32*	5	2	1.79	3	2	2.27	5	2	3.52	**1**	**1**	**0.99**
2_TIE-GCM	2	2	1.47	*4*	*2*	*1.39*	*4*	*1*	*1.62*	2	1	1.84	*4*	*1*	*2.78*	2	2	0.94

a function of geomagnetic activity for most cases, although the RMS errors of a few models during quiet periods are slightly larger than or similar to those during the storms. The PE and ratios also show dependency on the geomagnetic activity that is less systematic than, and even the opposite of, the RMS's dependency. For example, most models tend to have better PE and ratios during storms than quiet times in low and middle latitudes. The dependency of model performance on the geomagnetic activity for NmF2 and hmF2 prediction and the nonlinearity of the exceptional cases were also illustrated in our earlier paper on IT model evaluation [*Shim et al.*, 2011]. It was pointed out that some possible causes of the nonlinearity are the simultaneous changes in the input drivers such as electric fields, neutral wind, neutral composition, and neutral temperature, which differ from storm to storm, depending on the energy input to the ionosphere from the magnetosphere. Therefore, better knowledge and quantification of the temporal and spatial characteristics of rapid changes in the input drivers in relation to the characteristics of geomagnetic storms are some of the future challenges. With better understanding of these phenomena, IT models could be improved for reproducing weather.

The results of our study also show that model performance depends on the type of metrics and varies with latitude. For example, in the predictions of electron density at the CHAMP track, the ionospheric empirical model, 1_IRI, and the data assimilation models, 1_JPL-GAIM and 1_USU-GAIM, rank at or near the top for RMS error, while the data assimilation models and coupled IT models rank higher in terms of the ratio of max-min especially during the storms. In low and middle latitudes, the data assimilation models are ranked at the top with respect to RMS. However, they are outranked by 1_IRI in high latitudes probably due to the limited latitude coverage of simulation results and observations used for the data assimilation models. The submission of 1_JPL_GAIM for the study was generated by assimilating ground-based GPS TEC data from about 200 stations located between ±55° geomagnetic latitudes, although COSMIC TEC data were also assimilated. 1_USU-GAIM used for the study was obtained using only GPS observations within ±60° geographic latitude. Therefore, the plasma densities at high latitudes are the same as those from the physics-based model USU-IFM, which is used as a background, without any data assimilation.

For thermosphere neutral density predictions, the three empirical model submissions, 1_JB2008, 2_JB2008, and 1_MSIS rank higher for almost all cases even during the storms. However, 1_MSIS shows better ratios of max-min than the JB2008s for strong storm in low latitudes and for moderate storms in middle and high latitudes. 2_JB2008 and 1_JB2008 show the best ratio of max in low and middle/high latitudes, respectively.

For the prediction of vertical drift at Jicamarca, the empirical model 1_IRI, which provides long-term climate trends rather than short-term variations in the weather, shows better performance than the other models in terms of RMS error and PE, while the others produce better ratios of max-min than 1_IRI. This suggests that the physics-based models produce relatively better short-term variations of the ionospheric physical parameters during storms and even quiet periods. Consequently, each one of the models ranks the top at least for one case. Neither data assimilation models nor physics-based coupled IT models are better than the ionospheric models and empirical models for all cases.

From the comparisons among the same types of models, we find that the two physics-based ionospheric models, 1_SAMI3_HWM93 and 1_USU-IFM, show similar performance, in general. Although 1_USU-IFM performs slightly better in predicting electron density during the storms in terms of RMS error, 1_SAMI3_HWM93 produces relatively better diurnal variations (max-min) and maximum values of electron density. Among the five physics-based coupled IT models, it appears that the performance of 2_TIE-GCM, 1_TIE-GCM, and 1_CTIPE are similar to each other and better than 1_GITM and 3_GITM for most cases. For reproducing electron density at the CHAMP orbit, 1_CTIPE performs worse than 1_TIE-GCM and 2_TIE-GCM in low latitudes, while 1_CTIPE does better in middle latitudes for moderate storms and in high latitudes for the quiet periods. In reproducing the neutral density at the CHAMP orbit, 1_CTIPE produces worse (better) RMS and PE than 1_TIE-GCM and 2_TIE-GCM for the storm events (quiet events). 3_GITM shows the largest RMS error and a negative PE during the strong storms. However, for moderate storms, 1_GITM and 3_GITM show comparable performance to 1_TIE-GCM and 2_TIE-GCM, and better performance than 1_CTIPE. In general, the two data assimilation models, 1_JPL-GAIM and 1_USU-GAIM, show similar performance. However, 1_USU-GAIM has larger RMS errors for strong storms in low latitudes, and 1_JPL-GAIM has larger negative PE for quiet events in middle latitudes. Between the two thermospheric empirical models, the JB2008s produce slightly better RMS and PE than 1_MSIS. 1_MSIS shows the best ratio of max-min in low and middle latitudes for all geomagnetic levels.

Furthermore, the models show improvements of their performance when enhanced, and/or more complex input drivers are used. For instance, 1_SAMI3_HWM07 using the upgraded horizontal wind model [Drob et al., 2008] shows better performance in predicting the vertical drifts in terms of RMS error and PE, although it is not as good for the ratio of max-min during moderate storm events as 1_SAMI3_HWM93 driven by the earlier HWM [Hedin et al., 1996]. 2_TIE-GCM (driven by the Weimer [2005] electric potential with dynamic critical crossover latitudes) is better than 1_TIE-GCM (driven by the Heelis et al. [1982] electric potential with constant critical crossover latitudes) in the electron and neutral densities predicted during the storms for all latitudes. However, 2_TIE-GCM does not show systematic improvements for moderate storms and quiet periods. The improvement of 2_TIE-GCM in predicting the ionospheric parameters during strong storms was also noted in the work of Shim et al. [2011]. Comparing 1_GITM and 3_GITM, it is found that 3_GITM has better scores for electron density, whereas 1_GITM has better scores for neutral density. Also, 1_GITM is better for NmF2 prediction, while 3_GITM is better for hmF2 prediction [Shim et al., 2011]. The differences in the input and boundary conditions between 1_GITM and 3_GITM, such as O^+ reaction rates and collision frequencies, lower boundary conditions, equatorial electrojet, and potential solver, affect the performance of GITM. Two JB2008 runs for neutral density, 1_JB2008 (with exospheric temperature corrections derived from the Dst index) and 2_JB2008 (with the temperature corrections derived from Weimer [2005] total Poynting fluxes), show slight differences in their skill scores. 2_JB2008 shows better scores during moderate storm events for most cases, while 1_JB2008 is better for strong storms. Although not all of the model performances with enhanced and/or more complex input drivers are associated with a systematic improvement, the results of the comparison will help further improvement of the models.

This is the first systematic study to quantify the performance of various IT models. There were limitations such as a relatively short duration of events considered, temporal coverage of data, and uncertainty in the measurements. Nevertheless, the results provide a baseline from which the performance of new and improved models can be evaluated. In the future, it would be useful to extend the comparison of coupled IT model simulations with the same or different crucial inputs for the polar ionosphere to study the effects of different drivers on the model performance. Model outputs and observational data used for the validation studies that are or will be done are permanently posted at the CCMC website (http://ccmc.gsfc.nasa.gov) and serve as valuable resources for space science communities in the future.

Acknowledgments. The Jicamarca Radio Observatory is a facility of the Instituto Geofisico del Peru operated with support from the National Science Foundation (NSF) award AGS-0905448 through

Cornell University. The CHAMP neutral density data used in this study are obtained from http://sisko.colorado.edu/sutton/data.html.

REFERENCES

AIAA (2010), *AIAA Guide to Reference and Standard Atmosphere Models (G-003C-2010e)*, Reston, Va.

Anderson, D., A. Anghel, J. Chau, and O. Veliz (2004), Daytime vertical **E** × **B** drift velocities inferred from ground-based magnetometer observations at low latitudes, *Space Weather*, 2, S11001, doi:10.1029/2004SW000095.

Belehaki, A., I. Stanislawska, and J. Lilensten (2009), An overview of ionosphere-thermosphere models available for space weather purposes, *Space Sci. Rev.*, 147(3–4), 271–313, doi:10.1007/s11214-009-9510-0.

Bilitza, D. (1990), International Reference Ionosphere 1990, *NSSDC/WDC-A-R&S 90-22*, 155 pp., Natl. Space Sci. Data Cent., Greenbelt, Md.

Bilitza, D. (2001), International Reference Ionosphere 2000, *Radio Sci.*, 36(2), 261–275.

Bilitza, D. (2004), A correction for the IRI topside electron density model based on Alouette/ISIS topside sounder data, *Adv. Space. Res.*, 33, 838–843.

Bilitza, D., and B. W. Reinisch (2007), International reference ionosphere 2007: Improvements and new parameters, *Adv. Space. Res.*, 42, 599–609.

Bowman, B. R., W. K. Tobiska, F. A. Marcos, and C. Valladares (2007), The JB2006 empirical thermospheric density model, *J. Atmos. Sol. Terr. Phys.*, 70(5), 774–793, doi:10.1016/j.jastp.2007.10.002.

Bowman, B. R., W. K. Tobiska, F. A. Marcos, C. Y. Huang, C. S. Lin, and W. J. Burke (2008a), A new empirical thermospheric density model JB2008 using new solar and geomagnetic indices, paper AIAA 2008-6438 presented at AIAA/AAS Astrodynamics Specialist Conference, Honolulu, Hawaii.

Bowman, B. R., W. K. Tobiska, and M. J. Kendra (2008b), The thermospheric semiannual density response to solar EUV heating, *J. Atmos. Sol. Terr. Phys.*, 70(11–12), 1482–1496, doi:10.1016/j.jastp.2008.04.020.

Bruinsma, S., D. Tamagnan, and R. Biancale (2004), Atmospheric densities derived from CHAMP/STAR accelerometer observations, *Planet. Space Sci.*, 52, 297–312, doi:10.1016/j.pss.2003.11.004.

Burke, W. J. (2008), Storm time energy budgets of the global thermosphere, in *Midlatitude Ionospheric Dynamics and Disturbances*, Geophys. Monogr. Ser., vol. 181, edited by P. M. Kintner Jr. et al., pp. 235–246, AGU, Washington, D. C., doi:10.1029/181GM21.

Chau, J. L., and R. F. Woodman (2004), Daytime vertical and zonal velocities from 150-km echoes: Their relevance to *F*-region dynamics, *Geophys. Res. Lett.*, 31, L17801, doi:10.1029/2004GL020800.

Codrescu, M. V., T. J. Fuller-Rowell, J. C. Foster, J. M. Holt, and S. J. Cariglia (2000), Electric field variability associated with the Millstone Hill electric field model, *J. Geophys. Res.*, 105(A3), 5265–5273, doi:10.1029/1999JA900463.

Drob, D. P., et al. (2008), An empirical model of the Earth's horizontal wind fields: HWM07, *J. Geophys. Res.*, 113, A12304, doi:10.1029/2008JA013668.

Fuller-Rowell, T. J., D. Rees, S. Quegan, R. J. Moffett, M. V. Codrescu, and G. H. Millward (1996), A coupled thermosphere ionosphere model, CTIM, in *STEP Handbook on Ionospheric Models*, edited by R. W. Schunk, p. 217, Utah State Univ., Logan.

Hedin, A. E. (1987), MSIS-86 Thermospheric Model, *J. Geophys. Res.*, 92(A5), 4649–4662.

Hedin, A. E. (1991), Extension of the MSIS Thermosphere Model into the middle and lower atmosphere, *J. Geophys. Res.*, 96(A2), 1159–1172.

Hedin, A. E., et al. (1991), Revised global model of thermosphere winds using satellite and ground-based observations, *J. Geophys. Res.*, 96(A5), 7657–7688.

Hedin, A. E., et al. (1996), Empirical wind model for the upper, middle, and lower atmosphere, *J. Atmos. Terr. Phys.*, 58, 1421–1447.

Heelis, R. A., J. K. Lowell, and R. W. Spiro (1982), A model of the high-latitude ionospheric convection pattern, *J. Geophys. Res.*, 87(A8), 6339–6345.

Huba, J. D., G. Joyce, and J. A. Fedder (2000), Sami2 is Another Model of the Ionosphere (SAMI2): A new low-latitude ionosphere model, *J. Geophys. Res.*, 105(A10), 23,035–23,053.

Huba, J. D., G. Joyce, and J. Krall (2008), Three-dimensional equatorial spread *F* modeling, *Geophys. Res. Lett.*, 35, L10102, doi:10.1029/2008GL033509.

Hysell, D. L., M. F. Larsen, and R. F. Woodman (1997), JULIA radar studies of electric fields in the equatorial electrojet, *Geophys. Res. Lett.*, 24(13), 1687–1690.

Jacchia, L. G. (1965), Static diffusion models of the upper atmosphere with empirical temperature profiles, *Smithson. Contrib. Astrophys.*, 8, 215–257.

Jacchia, L. G. (1971), Semiannual variation in the heterosphere: A reappraisal, *J. Geophys. Res.*, 76(19), 4602–4607.

Liu, H., C. Stolle, and S. Watanabe (2007), Evaluation of the IRI model using CHAMP observations in polar and equatorial regions, *Adv. Space Res.*, 39, 904–909.

Millward, G. H., I. C. F. Müller-Wodarg, A. D. Aylward, T. J. Fuller-Rowell, A. D. Richmond, and R. J. Moffett (2001), An investigation into the influence of tidal forcing on *F* region equatorial vertical ion drift using a global ionosphere-thermosphere model with coupled electrodynamics, *J. Geophys. Res.*, 106(A11), 24,733–24,744, doi:10.1029/2000JA000342.

Pi, X., C. Wang, G. A. Hajj, G. Rosen, B. D. Wilson, and G. J. Bailey (2003), Estimation of E × B drift using a global assimilative ionospheric model: An observation system simulation experiment, *J. Geophys. Res.*, 108(A2), 1075, doi:10.1029/2001JA009235.

Picone, J. M., A. E. Hedin, D. P. Drob, and A. C. Aikin (2002), NRLMSISE-00 empirical model of the atmosphere: Statistical comparisons and scientific issues, *J. Geophys. Res.*, 107(A12), 1468, doi:10.1029/2002JA009430.

Reigber, C., H. Lühr, and P. Schwintzer (2000), CHAMP mission status and perspectives, *Eos Trans. AGU*, *81*(48), Fall Meet. Suppl., Abstract F307.

Richmond, A. D., E. C. Ridley, and R. G. Roble (1992), A thermosphere/ionosphere general circulation model with coupled electrodynamics, *Geophys. Res. Lett.*, *19*(6), 601–604.

Ridley, A. J., Y. Deng, and G. Toth (2006), The global ionosphere-thermosphere model, *J. Atmos. Sol. Terr. Phys.*, *68*, 839–864.

Roble, R. G., E. C. Ridley, A. D. Richmond, and R. E. Dickinson (1988), A coupled thermosphere/ionosphere general circulation model, *Geophys. Res. Lett.*, *15*(12), 1325–1328, doi:10.1029/GL015i012p01325.

Scherliess, L., R. W. Schunk, J. J. Sojka, and D. C. Thompson (2004), Development of a physics-based reduced state Kalman filter for the ionosphere, *Radio Sci.*, *39*, RS1S04, doi:10.1029/2002RS002797.

Scherliess, L., R. W. Schunk, J. J. Sojka, D. C. Thompson, and L. Zhu (2006), Utah State University Global Assimilation of Ionospheric Measurements Gauss-Markov Kalman filter model of the ionosphere: Model description and validation, *J. Geophys. Res.*, *111*, A11315, doi:10.1029/2006JA011712.

Schunk, R. W. (1988), A mathematical model of the middle and high latitude ionosphere, *Pure Appl. Geophys.*, *127*, 255–303.

Schunk, R. W., J. J. Sojka, and J. V. Eccles (1997), Expanded capabilities for the ionospheric forecast model, *Rep. AFRL-VS-HA-TR-98-0001*, Air Force Res. Lab., Hanscom Air Force Base, Bedford, Mass.

Schunk, R. W., L. Scherliess, and J. J. Sojka (2002), Ionospheric specification and forecast modeling, *J. Spacecr. Rockets*, *39*(2), 314–324.

Schunk, R. W., et al. (2004), Global Assimilation of Ionospheric Measurements (GAIM), *Radio Sci.*, *39*, RS1S02, doi:10.1029/2002RS002794.

Shim, J. S., et al. (2011), CEDAR Electrodynamics Thermosphere Ionosphere (ETI) Challenge for systematic assessment of ionosphere/thermosphere models: NmF2, hmF2, and vertical drift using ground-based observations, *Space Weather*, *9*, S12003, doi:10.1029/2011SW000727.

Sojka, J. J. (1989), Global scale, physical models of the F region ionosphere, *Rev. Geophys.*, *27*(3), 371–403.

Spence, H., D. Baker, A. Burns, T. Guild, C.-L. Huang, G. Siscoe, and R. Weigel (2004), Center for integrated space weather modeling metrics plan and initial model validation results, *J. Atmos. Sol. Terr. Phys.*, *66*(15–16), 1499–1507.

Sutton, E. K., J. M. Forbes, and R. S. Nerem (2005), Global thermospheric neutral density and wind response to the severe 2003 geomagnetic storms from CHAMP accelerometer data, *J. Geophys. Res.*, *110*, A09S40, doi:10.1029/2004JA010985.

Wang, C., G. Hajj, X. Pi, I. G. Rosen, and B. Wilson (2004), Development of the Global Assimilative Ionospheric Model, *Radio Sci.*, *39*, RS1S06, doi:10.1029/2002RS002854.

Webb, P. A., M. M. Kuznetsova, M. Hesse, L. Rastaetter, and A. Chulaki (2009), Ionosphere-thermosphere models at the Community Coordinated Modeling Center, *Radio Sci.*, *44*, RS0A34, doi:10.1029/2008RS004108. [Printed 45(1), 2010].

Weimer, D. R. (2005), Improved ionospheric electrodynamic models and application to calculating Joule heating rates, *J. Geophys. Res.*, *110*, A05306, doi:10.1029/2004JA010884.

Weimer, D. R., B. R. Bowman, E. K. Sutton, and W. K. Tobiska (2011), Predicting global average thermospheric temperature changes resulting from auroral heating, *J. Geophys. Res.*, *116*, A01312, doi:10.1029/2010JA015685.

D. Anderson, CIRES, University of Colorado, Boulder, CO, USA.

D. Bilitza, School of Physics Astronomy and Computational Science, George Mason University, Fairfax, VA, USA.

M. Butala, A. J. Mannucci, X. Pi, and P. Stephens, Jet Propulsion Laboratory, California Institute of Technology, Pasadena, CA, USA.

J. L. Chau, Radio Observatorio de Jicamarca, Instituto Geofisico del Peru, Lima, Peru.

M. Codrescu and T. J. Fuller-Rowell, NOAA Space Weather Prediction Center, Boulder, CO, USA.

B. A. Emery and B. Foster, NCAR HAO, Boulder, CO, USA.

J. Huba, Plasma Physics Division, Naval Research Laboratory, Washington, DC, USA.

M. Kuznetsova and L. Rastätter, NASA/GSFC, Greenbelt, MD, USA.

A. Ridley, Space Physics Research Laboratory, University of Michigan, Ann Arbor, MI, USA.

L. Scherliess, R. W. Schunk, J. J. Sojka, and L. Zhu, CASS, Utah State University, Logan, UT, USA.

J. S. Shim, GPHI/UMBC, NASA/GSFC, Greenbelt, MD, USA. (jasoon.shim@nasa.gov)

E. Sutton and D. C. Thompson, Air Force Research Laboratory, Albuquerque, NM, USA.

D. Weimer, Center for Space Science and Engineering Research, Virginia Polytechnic Institute and State University, Blacksburg, VA, USA.

Aspects of Coupling Processes in the Ionosphere and Thermosphere

R. A. Heelis

Hanson Center for Space Sciences, University of Texas at Dallas, Richardson, Texas, USA

An appreciation of the coupling between the charged and neutral constituents in the ionosphere and thermosphere is important to any description of the behavior of each region. In addition, these regions are coupled to the lower atmosphere and to the magnetosphere in a manner that is usually specified by boundary conditions at the lower and upper boundaries, respectively, of the region that is being modeled. Here we discuss some features of observations for which an appreciation of the coupling is required in order to interpret them and some properties of the models that limit our ability to interpret observations. Two specific areas of development are described. One is related to energy balance and the heat supplied from the inner magnetosphere. Another is related to deviations from $E \times B$ drift motion in the lower thermosphere that may contribute to plasma layers that will affect the distribution of ionospheric conductivity.

1. ION-ELECTRON COLLISIONS

It is now well established that energy and momentum exchange between ions and electrons and between the charged and neutral gases in the ionosphere is a critical component to establishing the separate behavior of each of the species [*England et al.*, 2010]. Energy inputs to the magnetosphere, the ionosphere, and the thermosphere originate from the Sun and the Earth's lower atmosphere. However, many internal processes regulate the flow of energy through these various regions, and all computational models attempt to accurately model some or all of these processes in global or regional volumes [*Fuller-Rowell et al.*, 1988; *Wang et al.*, 1999].

Figure 1 describes the principal connections that produce the coupling between different regions of the Earth's space environment. The coupling between the charged and neutral particles in the ionosphere and atmosphere involves not only electromagnetic coupling, which is described by the ion-neutral dynamo interaction [*Richmond*, 1995], but also the chemical interactions between the charged and neutral species [*Crowley et al.*, 2008] and the exchange of momentum and energy that occurs when charged particles move parallel to the magnetic field [*David et al.*, 2011].

Within the atmosphere itself, the dense neutral gas extending through the troposphere and stratosphere is a source of wave motions that propagate upward and interact with larger-scale motions before reaching the lower thermosphere [*Forbes et al.*, 2006]. These processes are described in the yellow circle titled upper-lower atmosphere interactions. The neutral motions subsequently interact with the charged particles producing dynamo currents and electric fields that can influence the charged particles throughout the ionosphere [*Hagan et al.*, 2007]. This ion-neutral dynamo interaction, shown in the blue ellipse, lies inside broader plasma neutral coupling processes, shown in the green ellipse, which include changes in the plasma density and conductivity. Finally, the ionosphere and atmosphere, which together constitute a partially ionized anisotropic conductor, interact with the magnetosphere through magnetic field-aligned currents that regulate the exchanges of electromagnetic energy between the two regions [*Kelley et al.*, 1991]. These are enclosed in the red ellipse.

While the key processes indicated in Figure 1 are not exhaustive, they serve to show how any particular coupling process involves all the others. Thus, while it is possible to

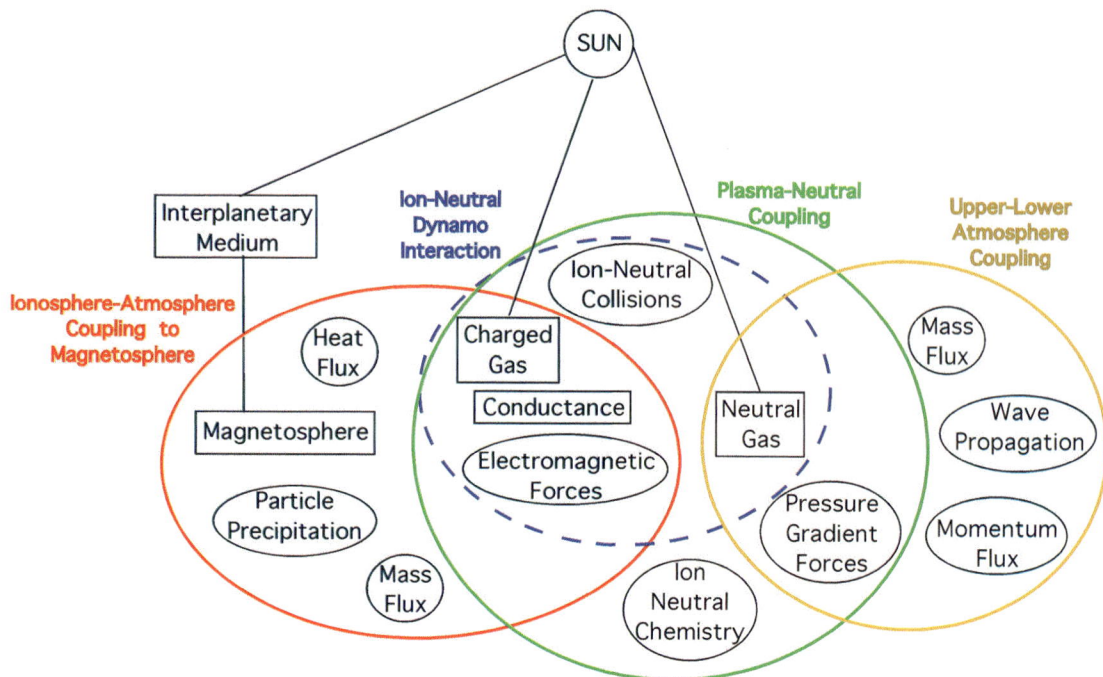

Figure 1. Schematic illustration of coupling processes in the ionosphere, thermosphere, and their relationships to external drivers.

model one component of the coupled system, or one coupling process, it is necessary to consider the limitations imposed by the lack of complete self-consistency and the uncertainties in the specification of the system drivers.

Throughout this monograph, the capabilities of various models will be described, and the features of the systems to which they are most suited will be discussed. In many cases, the adequacy of a given model approach is assessed by comparing model results to observations on occasions when the system drivers can be adequately specified. Computational models are being continually advanced to aid in the interpretation of observations and to expose the most important physical processes that are at work. This advance takes place in an environment where observational capabilities are also advancing. The interchange between these advancing capabilities highlights shortcomings in both measurement and modeling capabilities.

In this paper, while discussing this interchange, we provide an example of present challenges to observations that are required to keep pace with the capabilities of present computational models and one example of observations that challenge the present capabilities of the models.

2. ION ELECTRON COLLISIONS

In the topside *F* region, ionosphere collisions between the ions and electrons are critical in establishing the thermal balance of each species [*Varney et al.*, 2011]. Electrons conduct heat very efficiently in a direction parallel to the magnetic field and have good thermal contact with the ions. At lower altitudes, the ions have good thermal contact with the neutral atmosphere, which has a nearly constant temperature from 250 km up to the exobase. Figure 2 shows the results from the Sami is Another Model of the Ionosphere (SAMI2) model of the ionosphere [*Huba et al.*, 2000] that self-consistently calculates the field-aligned motions and temperatures for the thermal electrons and O^+ and H^+ ions in the presence of a given neutral atmosphere and solar EUV flux. These calculations are performed for equinoxial conditions and low solar activity represented by a 10.7 cm solar flux of 80. The top panel shows the local time variation of the ion temperature in the topside ionosphere near 800 km altitude at solar minimum, for later comparisons with in situ satellite measurements. After sunset, all the charged particle temperatures are the same, and they cool throughout the night. After midnight, the ionosphere and thermosphere have the same temperature, which falls to ~700 K before sunrise. After sunrise, the onset of photoelectron heating produces different temperatures in the ions and electrons. Of importance here is that the O^+ and H^+ temperatures can differ by up to 500 K with H^+ being significantly closer to the electron temperature. As the plasma content of the topside increases during the day, the plasma temperature decreases, and in the topside, the ion and electron temperatures become the same

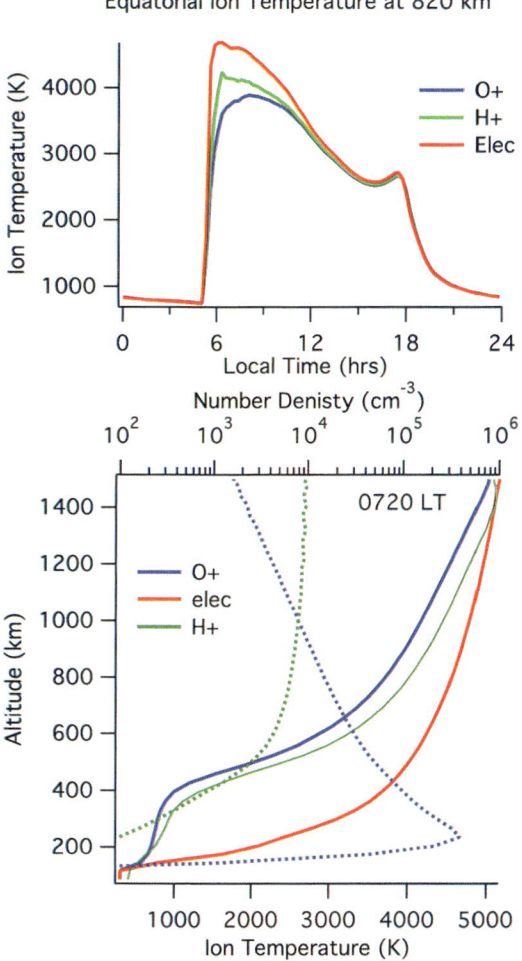

Figure 2. Characteristics of the ion and electron temperatures in the topside ionosphere revealed by the Sami is Another Model of the Ionosphere (SAMI2) model. Note the separation of the constituent ion temperatures and the electron temperature following sunrise.

[*Kakinami et al.*, 2011]. However, it is important to note that in the topside ionosphere after sunrise, the ion temperature is critically dependent on the ion number density and the ion composition. The lower panel of Figure 2 shows an altitude profile of the species temperatures indicated by the solid curves and the lower axis and ion number density, indicated by dotted curves and the upper axis. The O^+ and H^+ temperatures are different over a wide altitude range, which includes the O^+/H^+ transition height. This behavior has critical implications for measurements of the ion temperature that are derived assuming the constituent species have the same temperature. Figure 3 provides an illustration of this issue displayed in measurements of the ion temperature from the DMSP satellite. Shown in this figure are the ion and electron temperatures, the constituent ion concentrations, and the total ion number density for a morningside pass of the DMSP spacecraft across the equatorial region. The electron temperature is consistently higher than the ion temperature, as expected. However, the dominant ion constituent is used to derive the ion temperature. Thus, above the transition height, the temperature represents that of H^+, while below the transition height, the temperature represents that from O^+. Notice that near 18:57 UT and near 19:14 UT, the temperature first increases and then decreases by 500–700 K when the satellite is above the transition height. This is not a spatial gradient in the temperature but a change in the constituent ion that is used to derive it. Such ionospheric behavior presents a challenge to the measurement techniques to consider fitting procedures or observational techniques that allow the signals from O^+ and H^+ to be considered separately in deriving the ion temperature in the postsunrise topside ionosphere.

The presence of spatial gradients in the ion and electron temperature can be rather easily observed with the present observational techniques and, absent the ambiguities due to plasma composition, it remains a challenge to the models to determine the source of such gradients. This challenge is illustrated in Figure 4, which shows the ion and electron temperature, the ion drift velocity, and the total ion concentration observed across the topside high-latitude ionosphere where the major constituent ion is O^+. In the upper panel, a solid thick black line is an estimated baseline for the electron temperature over a region at middle and high latitudes from which significant spatial variations exist. Note that the initial peak in electron temperature, near 13:55 UT is colocated with a minimum in the total ion concentration, while the subsequent increase in the temperature occurs at the poleward extent of the region of sunward flow identified with the auroral zone. The colocation of the electron temperature peak with the total ion/electron concentration minimum suggests that the temperature change could be related to a reduction in the flux tube plasma content while keeping the heat content in the tube the same. However, the latitude profile of the electron temperature does not match well with the latitude profile of the ion density, thus suggesting that an additional heat flux is present at these latitudes. Such a heat flux could be consistent with a location near the edge of the plasmasphere, but self-consistent calculations of the plasma density and the heat balance are required to determine the required heat flux at the top of the ionosphere. The topside temperature profile will also change the topside ion composition. At present, there does not appear to be a satisfactory model that will determine the heat flux in the equatorial plasmapause boundary layer. Thus, this parameter will remain an important variable boundary condition in ionospheric models that should be examined carefully if comparisons between modeled and measured plasma temperatures are made.

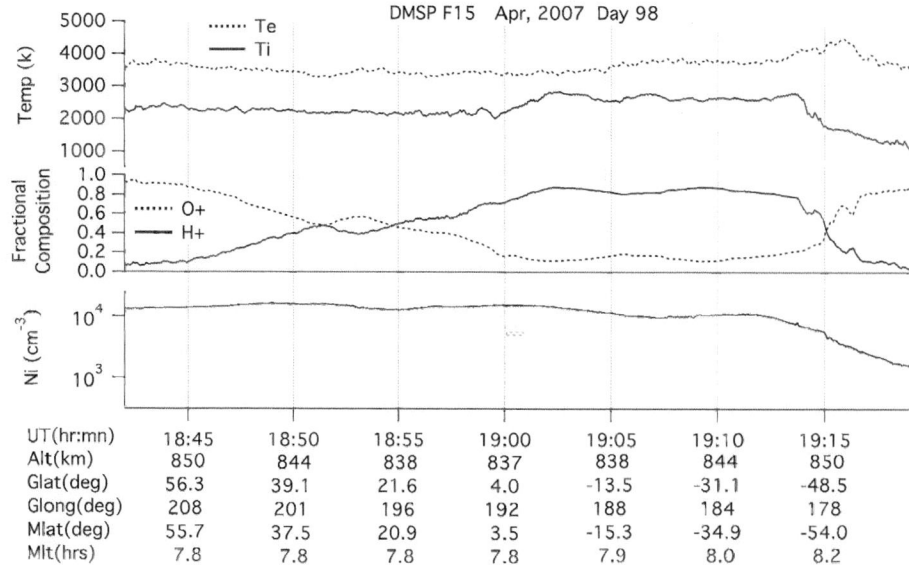

Figure 3. Observations of the ion temperature from DMSP just following sunrise show ion temperatures that differ by 500 K when the dominant ion changes from O^+ to H^+. This apparent change is due entirely to the ionospheric composition.

3. ION-NEUTRAL COLLISIONS IN THE UPPER THERMOSPHERE

Ion-neutral collisions produce a strong coupling between the charged and neutral species in the ionosphere and thermosphere since they are responsible for major changes in the momentum and the energy of both species.

Models of the ionosphere and thermosphere all include the collisional interactions between the ion and neutral gases with different levels of sophistication. In many ionospheric models [e.g., SAMI3], the neutral atmospheric density and motion are specified by empirical models. These model drivers are used to compute the impact of the neutral wind on ion motions parallel to the magnetic field and to determine

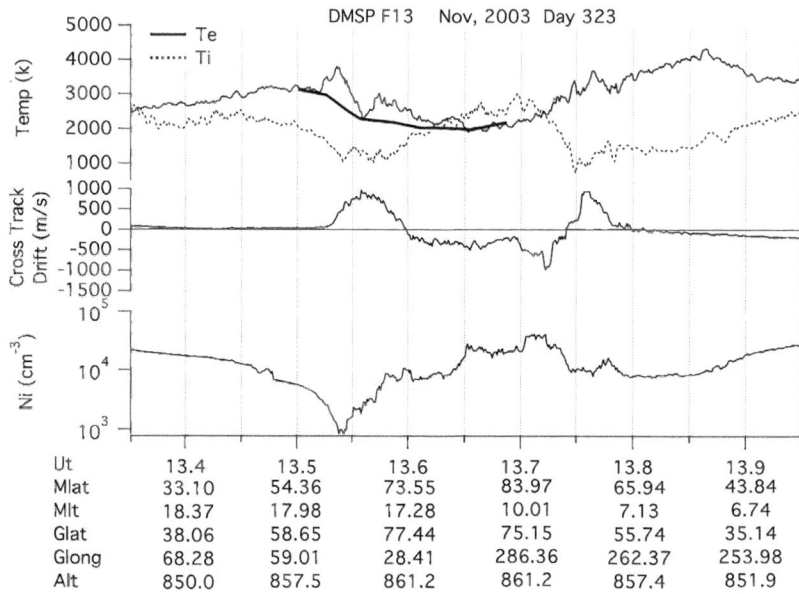

Figure 4. Observations of the electron temperature across the high-latitude region show departures from an expected baseline, estimated by the black line, that are indicative of heat input to the ionosphere. Knowledge of the heat flux at the top of the ionosphere is required to reconcile these measurements.

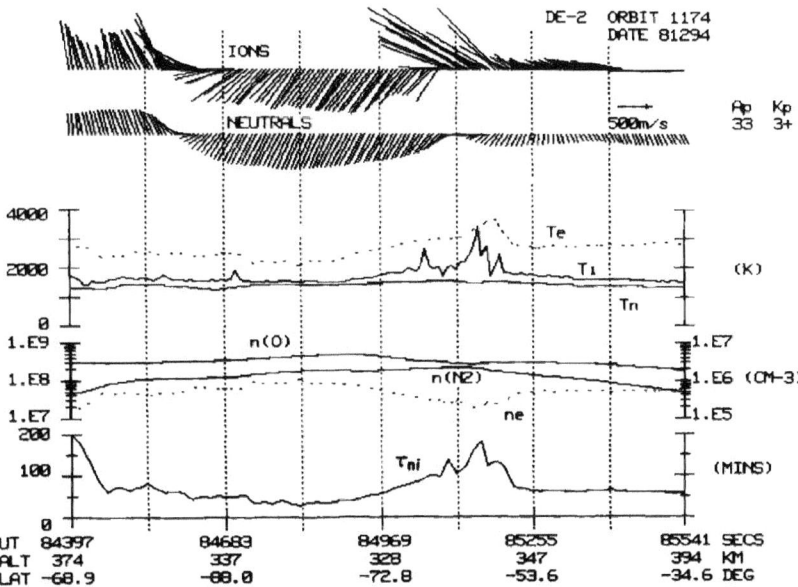

Figure 5. Measurements of the ion and neutral velocities from the Dynamics Explorer satellite show regions near 85,100 s in the top panel where the ion-neutral velocity is very high. In these regions, the ion/electron density, shown in the third panel is decreased, and the time for coupling to the neutrals, shown in the lowest panel, is increased [after *Killeen et al.*, 1984].

the currents perpendicular to the magnetic field that arise in response to the neutral motions perpendicular to the magnetic field. In coupled models of the ionosphere and thermosphere (e.g., Thermosphere Ionosphere Electrodynamics General Circulation Model (TIEGCM)) [*Richmond et al.*, 1992], knowledge of the heat sources and the ion drag are included to compute the ion and neutral motions self-consistently. Both approaches provide valuable insights into the coupled behavior of the charged and neutral species. Figure 5 shows the signatures of ion and neutral flows measured by the DE satellite during a transit across the southern high-latitude region from dusk to dawn [*Killeen et al.*, 1984]. Also shown in the panels below are the plasma and neutral temperature, the neutral and plasma density, and the time scale for responsiveness of the neutral gas to changes in the ion motion imparted through ion-neutral collisions. The similarity in the large-scale convective motions of the plasma and the neutral gas at high latitudes is illustrative of the strong coupling between the ion and neutral gas. Indeed, computational modeling has shown that the large-scale features of ionospheric convection pattern can be stable for many hours and effectively imprint motions with the same scales in the neutral gas [*Fuller Rowell et al.*, 1994] in the upper thermosphere. However, these results also show the effects of relative motions between the plasma and the neutral gas that serve to heat the plasma and to decouple it from the neutral gas. This can be seen quite dramatically between the tick marks at 84,969 and 85,255 s where the difference between the ion and neutral gas velocities are the largest, and the ion and electron temperatures both show local maxima. This signature of frictional heating is well understood and has several other consequences. First, the large relative velocity between the ion and neutral gases changes the charge exchange reaction rate between O^+ and N_2 [*Schunk and Nagy*, 2009]. Second, the recombination rate of the NO^+ is very fast, and thus, the total ion concentration becomes depleted as indicated in the third panel in Figure 5. It is interesting to note that the decrease in the plasma density also decreases the effective drag force on the neutral gas and increases the response time of the neutrals to the ion motion. Thus, in the topside ionosphere, changes in the ion motion can initially drive the neutral gas very effectively. However, the coupling first decreases the velocity difference between the ions and the neutrals. But any frictional heating that results from the velocity difference acts to decrease the plasma density and thus to decrease the coupling that encourages the neutrals to follow the ion motion. The processes, themselves, therefore, intrinsically limit the ion-neutral coupling in the upper thermosphere.

4. ION-NEUTRAL COLLISIONS IN THE LOWER THERMOSPHERE

In the lower thermosphere, the coupling between the ionosphere and thermosphere is quite different because the ion-neutral collision frequency becomes much larger, and the plasma density, in general, is much smaller. In this case, the impact of ion motion on the neutral gas is much smaller and moderated by changes in the neutral gas pressure gradients [*Crowley et al.*, 1996]. For the ions, it is frequently assumed that the ion motion remains related to the electric field by the expression $E = -V \times B$. Then, the formulations that assume the plasma motion perpendicular to the magnetic field is frozen into the magnetic field geometry may be retained. However, there are a number of observations in the ionosphere itself and in the lower thermosphere that suggest there are significant deviations from this assumption that should be investigated.

Figure 6 displays the signatures of a so-called subauroral ion drift (SAID) [*Anderson et al.*, 1993] in the topside ionosphere. Shown in the top panel is the spectral content in the energy flux of precipitating electrons. The middle panel shows the large horizontal plasma drift that appears equatorward of the auroral precipitation. Such a large drift produces frictional heating as described above and a reduction in the plasma density due to the increase in the plasma recombination rate. This latitudinally confined feature is usually contained within a wider plasma depletion commonly referred to as the midlatitude trough. In the lower panel is shown the magnetic field perturbations for which the spatial gradient can be used to recognize regions of upward and downward field-aligned current. They are labeled as regions 1 and 2 in Figure 6. Of particular interest, in the context of modeling, is to note that the very large gradients in the plasma drift and associated electric field occur in the absence of any significant change in the underlying field-aligned current. Thus, current continuity requires that the gradient in the plasma drift be accompanied by equal and opposite gradients in the ionospheric conductivity, such that the gradient in the product (σE) is zero or very small. It is straightforward to assume that the poleward gradient in the conductivity is produced by the auroral precipitation itself, but the equatorward gradient is difficult to establish without invoking transport of ionization in the direction of the electric field itself. This effect is contained in the plasma continuity equation as $dN/dt = -\nabla(NV_\perp) = -N\nabla V_\perp - V_\perp \nabla N$. If it is assumed that $V_\perp = E \times B/B^2$, then $V_\perp$ can be substituted directly into the equation, and the gradient in the density can be accounted for by considering the time derivative following the plasma motion. In this approach, the ionization transport, due to the so-called Pedersen drift of the ions, as originally described by *Banks and Yasuhara* [1978] is not included. In this case, some critical phenomena related to the plasma electrodynamics may remain hidden. For example, in this case, a large field-aligned current would first be required to produce the latitudinally confined rapid subauroral plasma drift. However, the ion motion associated with the closure of this current will rapidly reduce the plasma density due to poleward plasma motion. This reduction in the local plasma

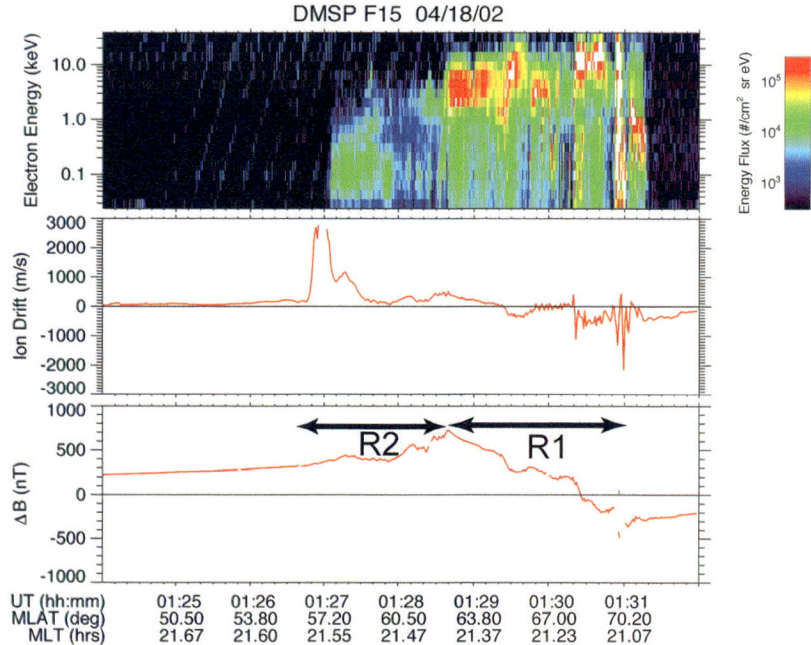

Figure 6. Observations of ion drift and precipitating electrons made from the DMSP satellite show a rapid subauroral flow in the absence of any large field-aligned current. This implies that the conductivity of the medium must behave in a manner that is similar to the drift itself (courtesy, P.C. Anderson).

density would be reflected in a similar behavior for the conductivity. The field-aligned current into and out of the lower thermosphere is given by $J_\| = \Sigma_p \nabla E + E \nabla \Sigma_p$. When E increases with latitude and the conductivity decreases with latitude, these two terms may effectively cancel allowing the SAID to exist in the absence of a large field-aligned current structure as observed.

In the lower thermosphere, the motion of the plasma in a direction that departs from the $E \times B$ direction may also result in changes in plasma density that will not be reproduced in ionospheric models that do not consider such motions. In the lower thermosphere, both the Hall and the Pedersen mobility weight the forces applied to the plasma from the neutral wind. These weightings are taken into account when the flux tube–integrated current perpendicular to the magnetic field is calculated, for inclusion in the solution of the so-called dynamo equation [*Huba et al.*, 2005]. However, if we assume that $V_\perp = E \times B/B^2$, then, the plasma motions perpendicular to the magnetic field produced by gravity, the plasma pressure gradient, and the collisional effect of the wind are not included in the continuity equation. In the presence of altitude variations in the wind, features such as plasma layers can be formed by the plasma motion perpendicular to the magnetic field. Figure 7 shows the effects of altitude variations in the neutral wind on the plasma distribution in the bottomside ionosphere [*Osterman et al.*, 1995]. There are two contributions of importance. The first is the collisional force of the neutral wind applied to the

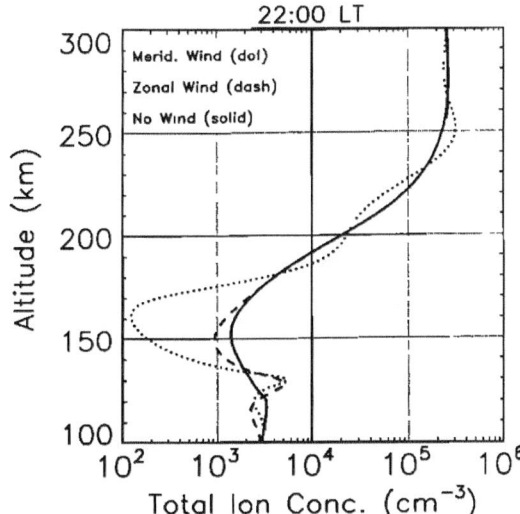

Figure 7. Calculations of the ion number density variation produced by altitude gradients in the neutral wind show that the contribution of vertical drifts due to zonal winds perpendicular to B and vertical drifts due to meridional winds parallel to B are the same at about 130 km altitude [after *Osterman et al.*, 1995].

plasma in the direction parallel to the magnetic field. The effects of this force due to a meridional wind component parallel to the magnetic field are included in all ionospheric models and at middle latitudes, where the magnetic field has a significant inclination; it may give rise to plasma layers in the bottomside F region called intermediate layers [*Roddy et al.*, 2004]. Note, however, that at altitudes near 130 km, the vertical plasma motion induced by zonal winds perpendicular to the magnetic meridian is as effective as the wind parallel to the magnetic meridian, in producing plasma layers. It should be noted that plasma layers formed in this fashion also change the conductivity distribution and could, thus, affect the overall coupling of the winds to higher altitudes through the induced plasma motions.

5. SUMMARY AND CONCLUSIONS

The capabilities of coupled ionosphere-thermosphere models generally allow the major characteristics of both regions to be described. These models depend on a specification of the external drivers to the system, which provides the ability to describe the sensitivity of the results to the drivers [*Lin et al.*, 2009] and to adjust the drivers to reproduce a given observation [*Fuller-Rowell et al.*, 2011]. However, in this work, we have highlighted processes and drivers that appear to be inadequately specified in models. The first involves the energy balance and appropriate partitioning between electrons, O^+ and H^+ ions in the topside ionosphere. This challenge has received attention in recent work [*Varney et al.*, 2011] and would be aided significantly by observations that provide independent measures of the O^+, H^+, and electron temperature. In addition to the internal heat balance in the ionosphere-thermosphere, it is also important to know the heat flux at the top of the thermosphere-ionosphere region. For models for which the upper boundary is a specified altitude or constant atmospheric pressure surface, a specification of the ion electron and neutral temperature is required. For those that include specification of the ionosphere along magnetic flux lines that extend to very large apex heights, then a specification of a heat flux at the equatorial crossing point or at some upper boundary is required. This information requires a detailed consideration of the interaction between hot and cold plasma populations near the edges of the plasmasphere [*Liemohn et al.*, 2000]. Alternatively, an empirical model of the heat flux or the plasma temperatures at an upper boundary would be a valuable tool for the models. Presently, the lack of such an empirical model and the limitations of models that deal with coupling between the thermosphere, ionosphere, and inner magnetosphere make the accurate specification of plasma temperatures in the topside, middle, and high-latitude ionosphere a significant challenge.

An understanding of the coupling between the ion and neutral gases is essential to interpret observations of the dynamics and composition of the charged and neutral species below about 600 km. One of the most commonly studied phenomena that result from this coupling is the ion frictional heating [*Thayer*, 1998]. This mechanism is most effective above the *F* peak, where the ion-neutral collision frequency is large compared to the ion-ion collision frequency or the ion-electron collision frequency, which control the ion-cooling rate. Increases in the ion temperature are also accompanied by a decrease in the ion number density due to the increase in the dissociative change exchange reaction $O^+ + N_2 \Rightarrow NO^+ + N$.

In addition to the energy exchange that produces ion heating, there also exists a momentum exchange between the ion and the neutral gases. The momentum exchange to the neutral gas is dependent on the neutral-ion collision frequency, which in turn is proportional to the total ion number density. Thus, in the *F* region ionosphere, a relative motion between the ion and neutral gas acts to limit the momentum exchange between the two species by increasing the temperature and decreasing the number density of the ions.

In the lower thermosphere, the forces that move the plasma perpendicular to the magnetic field are weighted by the ratio of the ion gyrofrequency to the ion collision frequency. Below about 150 km, this ratio becomes quite small, and the plasma motions are not simply related to the electric field by the expression $E = -V \times B$. In this altitude region below 150 km, the neutral wind may also be dominated by tidal modes with phase velocities that move downward and vertical wavelengths that are comparable to the ionization layer thickness. With sufficient vertical resolution, present models that solve the dynamo equations, like Thermosphere Ionosphere Electrodynamics General Circulation Model (TIEGCM) [*Richmond et al.*, 1992], The Coupled Thermosphere Ionosphere Plasmasphere electrodynamics (CTIPe) model [*Codrescu et al.*, 2008], and SAMI3 [*Huba et al.*, 2005], account for the effects of these variations on the overall drift of the plasma described by an electric field and an equivalent $V_\perp = E \times B/B^2$. However, such winds can produce a local redistribution of the plasma and associated conductivity that may not be considered in the model. There will also exist current systems that circulate locally and which can only be identified by examining the local current density as a function of altitude, latitude, and longitude.

In challenging models to reproduce a given observation, it is apparent that sufficient variability in the input drivers is available to do so. In order to constrain the drivers more effectively, it is necessary to compare measurements of ionospheric and thermospheric parameters simultaneously. For example, most measurements of the ionospheric number density are simultaneously accompanied by measurements of the ion and electron temperature and, in some cases, by the ionospheric composition. Understanding the relationships between these parameters will be essential in a convergent approach to comparing model and measurements. Measurements are always made in a system that is evolving with time, but without a good knowledge of how the system drivers are changing in time. Thus, it is important to use the models to understand the responsiveness of the system to temporal changes in the drivers. Only by knowing the range over which ionospheric parameters change following a change in the input drivers can a measure of the quality of a model-data comparison be made. Knowledge of the responsiveness of the system is also required for the observer, since bounding a parameter by the variability in a given driver is needed to sensibly interpret any given measurement.

As model development moves forward and high-quality measurements become more abundant, we may expect advances in at least two areas. First, data will be used with assimilation techniques to produce a global specification from a sparse set of observations. Second, models will be used to determine the variability in a geophysical parameter attributable to variations in input drivers that are not measured.

Acknowledgments. This work is supported by the University of Texas at Dallas by NASA grant NNX07AT82G. The author thanks P.C. Anderson for providing a figure to describe the configuration of ion drift and current in the SAID region.

REFERENCES

Anderson, P. C., W. B. Hanson, R. A. Heelis, J. D. Craven, D. N. Baker, and L. A. Frank (1993), A proposed production model of rapid subauroral ion drifts and their relationship to substorm evolution, *J. Geophys. Res.*, 98(A4), 6069–6078, doi:10.1029/92JA01975.

Banks, P. M., and F. Yasuhara (1978), Electric fields and conductivity in the nighttime E-region: A new magnetosphere-ionosphere-atmosphere coupling effect, *Geophys. Res. Lett.*, 5(12), 1047–1050, doi:10.1029/GL005i012p01047.

Codrescu, M. V., T. J. Fuller-Rowell, V. Munteanu, C. F. Minter, and G. H. Millward (2008), Validation of the Coupled Thermosphere Ionosphere Plasmasphere Electrodynamics model: CTIPE-Mass Spectrometer Incoherent Scatter temperature comparison, *Space Weather*, 6, S09005, doi:10.1029/2007SW000364.

Crowley, G., J. Schoendorf, R. G. Roble, and F. A. Marcos (1996), Cellular structures in the high-latitude thermosphere, *J. Geophys. Res.*, 101(A1), 211–223, doi:10.1029/95JA02584.

Crowley, G., A. Reynolds, J. P. Thayer, J. Lei, L. J. Paxton, A. B. Christensen, Y. Zhang, R. R. Meier, and D. J. Strickland (2008), Periodic modulations in thermospheric composition by solar wind high speed streams, *Geophys. Res. Lett.*, 35, L21106, doi:10.1029/2008GL035745.

David, M., R. W. Schunk, and J. J. Sojka (2011), The effect of downward electron heat flow and electron cooling processes in the high-latitude ionosphere, *J. Atmos. Sol. Terr. Phys.*, *73*(16), 2399–2409.

England, S. L., T. J. Immel, J. D. Huba, M. E. Hagan, A. Maute, and R. DeMajistre (2010), Modeling of multiple effects of atmospheric tides on the ionosphere: An examination of possible coupling mechanisms responsible for the longitudinal structure of the equatorial ionosphere, *J. Geophys. Res.*, *115*, A05308, doi:10.1029/2009JA014894.

Forbes, J. M., J. Russell, S. Miyahara, X. Zhang, S. Palo, M. Mlynczak, C. J. Mertens, and M. E. Hagan (2006), Troposphere-thermosphere tidal coupling as measured by the SABER instrument on TIMED during July–September 2002, *J. Geophys. Res.*, *111*, A10S06, doi:10.1029/2005JA011492.

Fuller-Rowell, T. J., D. Rees, S. Quegan, R. J. Moffett, and G. J. Bailey (1988), Simulations of the seasonal and universal time variations of the high-latitude thermosphere and ionosphere using a coupled, three-dimensional model, *Pure Appl. Geophys.*, *127*, 189–217.

Fuller-Rowell, T. J., M. V. Codrescu, R. J. Moffett, and S. Quegan (1994), Response of the thermosphere and ionosphere to geomagnetic storms, *J. Geophys. Res.*, *99*(A3), 3893–3914, doi:10.1029/93JA02015.

Fuller-Rowell, T., H. Wang, R. Akmaev, F. Wu, T.-W. Fang, M. Iredell, and A. Richmond (2011), Forecasting the dynamic and electrodynamic response to the January 2009 sudden stratospheric warming, *Geophys. Res. Lett.*, *38*, L13102, doi:10.1029/2011GL047732.

Hagan, M. E., A. Maute, R. G. Roble, A. D. Richmond, T. J. Immel, and S. L. England (2007), Connections between deep tropical clouds and the Earth's ionosphere, *Geophys. Res. Lett.*, *34*, L20109, doi:10.1029/2007GL030142.

Huba, J. D., G. Joyce, and J. A. Fedder (2000), Sami2 is Another Model of the Ionosphere (SAMI2): A new low-latitude ionosphere model, *J. Geophys. Res.*, *105*(A10), 23,035–23,053, doi:10.1029/2000JA000035.

Huba, J. D., G. Joyce, S. Sazykin, R. Wolf, and R. Spiro (2005), Simulation study of penetration electric field effects on the low- to mid-latitude ionosphere, *Geophys. Res. Lett.*, *32*, L23101, doi:10.1029/2005GL024162.

Kakinami, Y., S. Watanabe, J.-Y. Liu, and N. Balan (2011), Correlation between electron density and temperature in the topside ionosphere, *J. Geophys. Res.*, *116*, A12331, doi:10.1029/2011JA016905.

Kelley, M. C., D. J. Knudsen, and J. F. Vickrey (1991), Poynting flux measurements on a satellite: A diagnostic tool for space research, *J. Geophys. Res.*, *96*(A1), 201–207, doi:10.1029/90JA01837.

Killeen, T. L., P. B. Hays, G. R. Carignan, R. A. Heelis, W. B. Hanson, N. W. Spencer, and L. H. Brace (1984), Ion-neutral coupling in the high-latitude *F* region: Evaluation of ion heating terms from Dynamics Explorer 2, *J. Geophys. Res.*, *89*(A9), 7495–7508, doi:10.1029/JA089iA09p07495.

Liemohn, M. W., J. U. Kozyra, P. G. Richards, G. V. Khazanov, M. J. Buonsanto, and V. K. Jordanova (2000), Ring current heating of the thermal electrons at solar maximum, *J. Geophys. Res.*, *105*(A12), 27,767–27,776, doi:10.1029/2000JA000088.

Lin, C. H., A. D. Richmond, G. J. Bailey, J. Y. Liu, G. Lu, and R. A. Heelis (2009), Neutral wind effect in producing a storm time ionospheric additional layer in the equatorial ionization anomaly region, *J. Geophys. Res.*, *114*, A09306, doi:10.1029/2009JA014050.

Osterman, G. B., R. A. Heelis, and G. J. Bailey (1995), Effects of zonal winds and metallic ions on the behavior of intermediate layers, *J. Geophys. Res.*, *100*(A5), 7829–7838, doi:10.1029/94JA03241.

Richmond, A. D. (1995), Modeling equatorial ionospheric electric fields, *J. Atmos. Terr. Phys.*, *57*(10), 1103–1115.

Richmond, A. D., E. C. Ridley, and R. G. Roble (1992), A thermosphere/ionosphere general circulation model with coupled electrodynamics, *Geophys. Res. Lett.*, *19*(6), 601–604, doi:10.1029/92GL00401.

Roddy, P. A., G. D. Earle, C. M. Swenson, C. G. Carlson, and T. W. Bullett (2004), Relative concentrations of molecular and metallic ions in midlatitude intermediate and sporadic-E layers, *Geophys. Res. Lett.*, *31*, L19807, doi:10.1029/2004GL020604.

Schunk, R. W., and A. F. Nagy (2009), *Ionospheres: Physics, Plasma Physics, and Chemistry*, 2nd ed., Cambridge Univ. Press, New York.

Thayer, J. P. (1998), Radar measurements of the electromagnetic energy rates associated with the Dynamic Ionospheric Load/Generator, *Geophys. Res. Lett.*, *25*(4), 469–472, doi:10.1029/97GL03660.

Varney, R. H., D. L. Hysell, and J. D. Huba (2011), Sensitivity studies of equatorial topside electron and ion temperatures, *J. Geophys. Res.*, *116*, A06321, doi:10.1029/2011JA016549.

Wang, W., T. L. Killeen, A. G. Burns, and R. G. Roble (1999), A high-resolution, three-dimensional, time-dependent, nested grid model of the coupled thermosphere–ionosphere, *J. Atmos. Sol. Terr. Phys.*, *61*(5), 385–397.

R. A. Heelis, Hanson Center for Space Sciences, University of Texas at Dallas, Richardson, TX 75083, USA. (heelis@utdallas.edu)

Use of NOGAPS-ALPHA as a Bottom Boundary for the NCAR/TIEGCM

David E. Siskind and Douglas P. Drob

Space Science Division, Naval Research Laboratory, Washington, District of Columbia, USA

We present preliminary results from the National Center for Atmospheric Research thermosphere ionosphere electrodynamics general circulation model (TIEGCM) using a bottom boundary from the Navy Operational Global Atmospheric Prediction System-Advanced Level Physics High Altitude (NOGAPS-ALPHA) model. NOGAPS-ALPHA consists of a forecast module and a data assimilation (DA) system. We use the 6 hourly cycled DA system to initialize a series of forecasts, which are output every hour. The use of hourly output allows the resolution of higher-order tidal modes such as the terdiurnal tide. Results are shown for January 2009. All three migrating tides (diurnal wave 1, semidiurnal wave 2, terdiurnal wave 3) and the eastward traveling nonmigrating diurnal wave 3 tide are present. Results are compared with Sounding of the Atmosphere with Broadband Emission Radiometry (SABER) observations and previously published model results. We evaluate the behavior of these tidal components at the interface between the two models and suggest a method to reduce the discontinuities. Finally, and consistent with previous models, we find a marked decrease in the semidiurnal tide after the dramatic sudden stratospheric warming in late January; however, this does not appear to be offset by a concomitant increase in the migrating terdiurnal component that others have reported.

1. INTRODUCTION

In the past decade, so-called whole atmosphere models have emerged that consider dynamical and chemical coupling between the upper and lower atmosphere. As discussed in a recent review by *Akmaev* [2011], there are several approaches to whole atmosphere modeling. In the first category are true seamless models, which follow the vision originally outlined by *Roble* [2000] and use a single dynamical solver to encompass the domain from the surface to the exobase. Then, there are two other categories of models, which have boundaries somewhere in the atmosphere. For example, there are several vertically extended lower atmosphere climate and weather models with domains from the surface to 100–200 km. Also, there are several thermosphere-ionosphere models with bottom boundaries somewhere in the middle atmosphere. With these latter two cases, achieving a true whole atmosphere modeling capability requires the specification of atmospheric conditions at a boundary. While this introduces some complexity in configuring the model, it can also confer some advantages. For example, a mechanistic model such as the National Center for Atmospheric Research (NCAR) thermosphere ionosphere mesosphere electrodynamics general circulation model (TIME-GCM) [*Roble and Ridley*, 1994] has a bottom boundary in the stratosphere. This allows specification of upward propagating planetary waves or tides [e.g., *Liu et al.*, 2010; *Hagan et al.*, 2009], and thus, self-consistent sensitivity studies can be more readily conducted than with a fully self-consistent seamless model.

Our approach, to use the Navy Operational Global Atmospheric Prediction System-Advanced Level Physics High Altitude (NOGAPS-ALPHA) as a bottom boundary for the thermosphere ionosphere electrodynamics general circulation model (TIEGCM), is similar in some respects to the TIME-GCM, where we have an interface that allows specified meteorological forcing from the lower atmosphere to propagate into the upper atmosphere. However, there are several differences, which are important to document. First, unlike the TIME-GCM formulation of *Roble* [2000], we do not consider downward coupling into the lower atmosphere. Our forcing is one way, only upward. Second, and perhaps most importantly, our interface is near the mesopause ($p = 0.005$ hPa) rather than in the middle stratosphere as in the case of the TIME-GCM. This is important because the dynamics of the mesosphere (50–90 km) is dominated by the breaking of small-scale gravity waves. Typically, these must be parameterized [e.g., *Garcia et al.*, 2007], and this parameterization can often involve ad hoc tuning [e.g., *Eckermann et al.*, 2009]. Depending upon which of several poorly known parameters such as source strength and spectral distribution are used, widely differing responses of the mesosphere and lower thermosphere (MLT) to forcing from the stratosphere are obtained [*Yamashita et al.*, 2010]. By contrast, *Ren et al.* [2011] have recently shown that the requirements for a gravity wave drag parameterization are significantly reduced if mesospheric temperatures are assimilated rather than calculated. This is the approach we adopt here.

The results shown here emphasize the ability of our coupled system to simulate middle and upper atmospheric tides. Tides have long been of great interest because, as discussed by *Forbes* [1995], these large-scale waves "often dominate the meteorology" of the mesosphere and lower thermosphere. We show results for zonal mean winds for several runs of the TIEGCM, we show distribution of the main tidal fields, which have received attention in the literature recently, and finally, we show examples of the response of the thermosphere to the stratospheric warming of late January 2009.

2. MODEL DESCRIPTIONS

2.1. NOGAPS-ALPHA

NOGAPS-ALPHA represents a developmental prototype of the vertical extension of the Navy's operational forecast model. This vertical extension required the adaptation of the vertical coordinate to a middle atmospheric pressure coordinate system [*Eckermann*, 2009] as well as the inclusion of additional stratospheric and mesospheric physical parameterizations, including radiative heating and cooling rates that account for nonlocal thermodynamic equilibrium, ozone and water vapor transport, and photochemistry and gravity wave drag (GWD). We use the non-LTE cooling code of *Fomichev et al.* [1998] and the NCAR GWD parameterization described by *Garcia et al.* [2007]. See the work of *Eckermann et al.* [2009] for details. Analyses are generated by NOGAPS-ALPHA production runs that couple the forecast model to the data assimilation system.

The high-altitude analysis products debuted in the work of *Hoppel et al.* [2008], where temperatures from EOS/MLS and TIMED/SABER were assimilated to 0.01 hPa. The horizontal resolution is defined by spectral triangular truncation to 79 wavenumbers (T79), (about $1.5° \times 1.5°$ latitude-longitude on a Gaussian grid). *Eckermann et al.* [2009] extended this capability to 0.002 hPa (about 88 km), and science results from this product have been reported by *Coy et al.* [2009] and *Siskind et al.* [2010] for sudden stratospheric warming events, by *McCormack et al.* [2009] for the 2 day wave and by *Nielsen et al.* [2010] and *Stevens et al.* [2010] for the summer mesopause. The analysis operates on a 6 hourly update cycle that corresponds to a Nyquist frequency of 12 h. Thus, the use of the analysis fields as a forcing function for a thermosphere-ionosphere model could distort the resultant tidal spectrum. Although *Eckermann et al.* [2009] obtained realistic semidiurnal tidal amplitudes for the middle atmosphere, any tidal frequency greater than semidiurnal (such as terdiurnal) would be aliased onto the diurnal and semidiurnal components. Furthermore, the bottom boundary of the TIEGCM $p = 5.4 \times 10^{-4}$ hPa corresponds identically to the top of the analysis, and this is a problem because the top two levels of the analysis are heavily damped.

Fortunately, the forecast model extends a decade in pressure higher than the analysis and can be output every hour. Like the analysis, the forecast model is typically run at T79, although it can be run at higher resolution [cf. *Siskind et al.*, 2010]. By using sequential initializations of the forecast model every six hours, we can obtain 1 hourly synoptic data (horizontal winds, temperatures, and geopotential height) and resolve diurnal and higher-order tides with minimal aliasing. *Siskind et al.* [2012] presented some preliminary results using the forecast model to drive the TIEGCM, focusing exclusively on the calculation of the zonal mean zonal wind.

This paper presents preliminary results for January 2009. This month was characterized by a historic sudden stratospheric warming that has been the subject of considerable attention in the literature from a wide variety of perspectives [e.g., *Manney et al.*, 2009; *Fuller-Rowell et al.*, 2011; *Coy et al.*, 2011; *Sassi et al.*, 2012]. Figure 1 summarizes the meteorological variability associated with this unusual winter as represented in the NOGAPS-ALPHA analysis [cf. *Eckermann et al.*, 2009]. Note the dramatic change in the zonal wind after 15 January. A coherent pattern of easterlies is seen to descend from the upper mesosphere to the lower stratosphere over a 2 week period. This is important because, while the onset of stratospheric warmings are conventionally

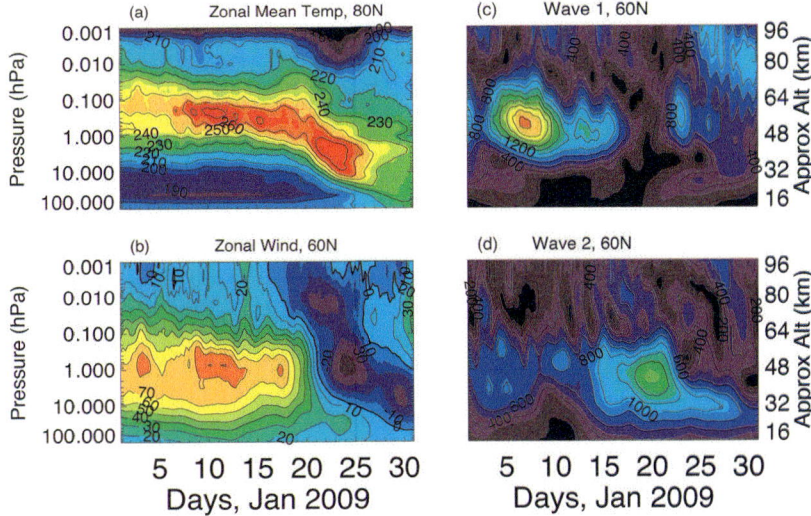

Figure 1. Summary of meteorological variability in January 2009 from the Navy Operational Global Atmospheric Prediction System-Advanced Level Physics High Altitude (NOGAPS-ALPHA) assimilation product of *Eckermann et al.* [2009]. This product is used to initialize a series of six hourly forecasts (see text).

defined by the wind reversal at the 10 mbar level in the midstratosphere, it is known that the mesosphere often responds earlier than the stratosphere [cf. *Coy et al.*, 2011], and furthermore, Figure 1 shows that all four of the meteorological variables in Figure 1 show perturbations at slightly different times. Thus, the peak cooling above 80 km is around 23 January, whereas the zonal wind is beginning to recover to its prewarming state by that date. For planetary waves, there is a high-altitude wave 2 event from the 15th–18th and a high-altitude wave 1 event starting on the 25th. As far as coupling with the thermosphere, it is not yet clear which of these is most relevant and may be different from event to event.

2.2. The TIEGCM

The TIEGCM is a first-principles thermosphere-ionosphere (T-I) general circulation model that extends from a pressure p = 5e-4 hPa (about 91–95 km) to the upper thermosphere. A key feature of the TIEGCM is that it includes a specific calculation of tropical-induced electric fields and resultant ion-neutral coupling in the lower thermosphere [*Richmond et al.*, 1992]. This is in contrast to, for example, the present generation of the WACCM-X model [*Liu et al.*, 2010]. *Fesen et al.* [2000] showed that a realistic diurnal variation of the equatorial ion drift could be achieved, including the presunset reversal of the drift. The version of the TIEGCM that we use is on pressure coordinates with two grid points per scale height and a latitude-longitude resolution of about 5° × 5°. A complete description, including a history of its development, is presented elsewhere in this monograph by *Qian et al.* [this volume].

The bottom boundary interface is at a level $Z = -7$ where $Z = \ln(P_o/P)$, and P_o is 5×10^{-7} hPa. In the standard configuration, this bottom boundary includes migrating diurnal and semidiurnal tides specified by the global-scale wind model (GSWM) [*Hagan et al.*, 2001], together with a fixed, globally averaged (i.e., independent of day of year, latitude, and longitude) background temperature of 181 K and zero background winds. In our applications, we use the zonal (U) and meridional (V) winds, temperatures (T), and geopotential height (z) output from NOGAPS-ALPHA. One issue we discuss below that was not discussed by *Siskind et al.* [2012] is the question of how to minimize possible discontinuities across the bottom boundary. We have found that for tidal studies, such discontinuities are often evident when we simply apply NOGAPS-ALPHA fields at the bottom boundary. We, therefore, considered several approaches to "nudging" the TIEGCM winds and/or temperatures at the first midpoint level, at $Z = -6.75$.

The technique of nudging in atmospheric numerical models was pioneered by the mesoscale modeling community [cf. *Stauffer and Seaman*, 1990, 1994] to improve overall large-scale continuity of mass, momentum, and energy with global synoptic analysis fields in which the mesoscale models are embedded. As defined by *Stauffer and Seaman* [1994], nudging is a continuous data assimilation method that adjusts a model state toward an observed state by adding to one or more of the prognostic equations, artificial tendency terms based on the difference between the two states. The observed state may be independent synoptic-scale analysis field(s) or a number of individual observations. Our approach is closer to analysis nudging, but since we use the one hourly NOGAPS-ALPHA forecast product, rather than the six hourly analysis, as the "observed state," it might be better termed "interpolated analysis nudging."

Provided that the analysis fields are a reasonable representation of the true atmospheric state, the nudging terms effectively add or remove an amount of momentum or energy related to the difference between the computed and observed model state at each time step in order to account for, and even identify, missing model physics. In our case, there are notable differences in the physics of each model. For example, although the TIEGCM includes fourth-order diffusion terms in its momentum and energy equations (similar to Rayleigh friction), it does not include gravity wave drag terms. By contrast, NOGAPS-ALPHA does not properly account for FUV heating and other photochemical effects that can be important in the lower thermosphere. Therefore, when simply using NOGAPS fields as the lower boundary for TIEGCM at $Z = -7.0$, discontinuities can arise at the next midpoint pressure level ($Z = -6.75$) because the dynamical balance of T, U, V, and z are formulated differently in the two models. Through nudging, our approach allows for a tradeoff between the specified dynamical state of NOGAPS (which since it is closely tied to the analysis can be assumed as observationally correct subject to the caveats above) and the dynamical state of TIEGCM (at the lower boundary) without compromising the overall dynamical stability of the TIEGCM model equations.

A discussion of considerations for the determination of optimal nudging schemes for global-scale general circulation models are provided by Zou et al. [1992]. We experimented with several approaches and ultimately found that the best results were by nudging the U and V fields at $Z = -6.75$ with the NOGAPS-ALPHA winds at $p = 4.2 \times 10^{-4}$ hPa using an approximate 1 h relaxation constant. To ensure a smooth transition from an initial TIEGCM model state, driven with a GSWM lower-boundary condition, to our TIEGCM simulations driven with the NOGAPS-ALPHA lower boundary condition, we used a logistic weighting function with an inflection point at $t = 2.5$ days to ramp down the GSWM forcing while ramping up the NOGAPS-ALPHA forcing.

3. MODEL RESULTS

3.1. Monthly Averaged

In the work of Siskind et al. [2012], we showed results for monthly averaged zonal mean winds for two configurations of the TIEGCM. One was for the purely GSWM forced case discussed above, and the second was with the NOGAPS-ALPHA output. The results with NOGAPS-ALPHA forcing were in much better agreement with the Drob et al. [2008] horizontal wind model, and emphasis was placed on the role of the cold summer mesopause in driving an eastward vortex in the lower thermosphere. However, not all the differences between these two models were due to the cold summer mesopause. As we will show below, NOGAPS-ALPHA presents a complex spectrum of tides, both migrating and nonmigrating, to the TIEGCM bottom boundary. Since tidal dissipation is known to enter into the zonal momentum budget, some of the differences shown by Siskind et al. [2012] could be due to differences in tidal forcing rather than simply balancing the latitudinal gradient of temperature. To highlight this, we perform a third calculation, which is similar to the pure GSWM forcing case, but instead of a single globally uniform temperature, we use a time-invariant, but latitudinally varying, zonal mean temperature.

The results from the three TIEGCM simulations are shown in Figure 2. The left-hand panel (pure GSWM forcing) is the same as Figure 3a, the right-hand panel (complete NOGAPS) is the same as Figure 3b. The middle panel (GSWM + zonal mean only) is the new simulation described above. It more clearly isolates the effect of the cold summer mesopause on the zonal mean winds in the thermosphere. It shows an eastward jet in the summer lower thermosphere that peaks at about 30 m s^{-1} and is absent in the baseline case with uniform temperatures. However, the tropical zonal winds are not dramatically different, showing a peak westward flow of about 60 m s^{-1} that is similar to the baseline case. By contrast, the right-hand panel shows significantly weakened tropical winds. This middle simulation is important because it gives a sample of what might be expected if the TIEGCM was driven with a model such as NRLMSIS [Picone et al., 2002], which has reasonably accurate mesospheric temperatures, but only includes the primary migrating tides.

The differences between the three simulations are clarified in Figure 3, which shows difference fields between the three calculations presented in Figure 2. Thus, Figure 2a shows the effect of adding the latitudinally varying temperature. It is most apparent on the summer lower thermospheric jet with some small effects on the tropical winds. Figure 2b shows the added effects of the full NOGAPS-ALPHA variability. Here a large eastward perturbation, centered approximately over the equator, is seen in the zonal wind. Given the latitudinal structure of this perturbation, and based upon arguments presented by Hagan et al. [2009], we suggest that a plausible candidate for this eastward momentum is dissipation of an eastward traveling tide. In the following section, we show how the tides from NOGAPS-ALPHA map onto the TIEGCM.

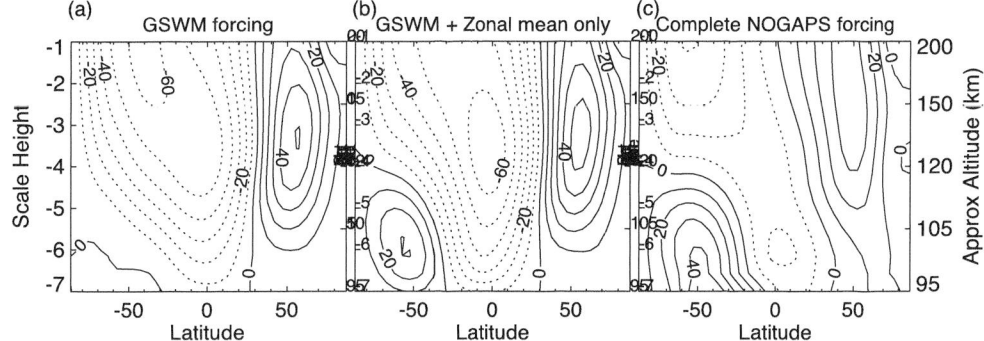

Figure 2. Monthly mean zonal winds from the thermosphere ionosphere electrodynamics general circulation model (TIEGCM) for (a) a bottom boundary with migrating tides from the global-scale wind model (GSWM) model and a globally invariant temperature. (b) The same migrating tides as in (a) but with a time-independent, latitudinally varying zonal mean temperature and winds from NOGAPS-ALPHA and (c) the complete varying wind and temperature fields from the NOGAPS-ALPHA forecast model for January 2009.

3.2. Tides

The study of global-scale oscillations in the middle and upper atmosphere, or tides, has a rich intellectual heritage. The standard theory was developed by *Chapman and Lindzen* [1970]; an excellent modern tutorial on the subject is presented by *Forbes* [1995]. Typically, two classes of tides are distinguished: tides that migrate with the sun's daily heating and nonmigrating tides that are ultimately forced by longitudinally localized releases of heat due to tropospheric convection. Of the migrating tides, three components are presented here: the daily zonal wavenumber one (DW1), the semidiurnal zonal wavenumber two (SW2), and the terdiurnal zonal wavenumber three (TW3). The recent availability of global satellite data sets has spawned an explosion of literature on these tides, particularly the DW1 and SW2 components [e.g., *Forbes et al.*, 2006; *Oberheide et al.*, 2006; *Zhang et al.*, 2006]. We also note that analysis-driven simulations of the terdiurnal tide, a prominent feature of radar winds [*Smith and Ortland*, 2001], are not possible with conventional six hourly cycled analyses. *Wang et al.* [2011] discuss how they circumvent this limitation with incremental updating; here by sampling a series of forecasts each hour, we obtain analogous results. We also show results for the single nonmigrating component, the diurnal eastward wave three (DE3) tide. This has received great attention since the work of *Immel et al.* [2006] and *Hagan et al.* [2007] have linked it to ionospheric variability.

Figure 4 presents monthly T and U fields for the four tidal components discussed above. These amplitudes were obtained from a 2-D fast Fourier transform analysis, where we define diurnal variations as having a period from 0.97 to 1.03 days, semidiurnal is defined as a 0.485 to 0.515 day period, and terdiurnal as having a period of 0.32 to 0.34 days. Our calculation can be placed in context by comparison with other studies of tidal amplitudes in either whole atmosphere

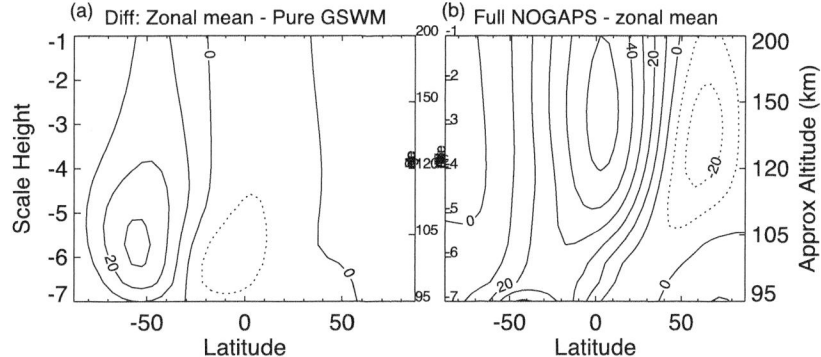

Figure 3. Difference fields for the TIEGCM calculations shown in Figure 1. (a) The middle panel of Figure 1 minus the left-hand panel (zonal mean + GSWM − pure GSWM). (b) The right panel of Figure 1 minus the middle (full NOGAPS − (zonal mean + GSWM).

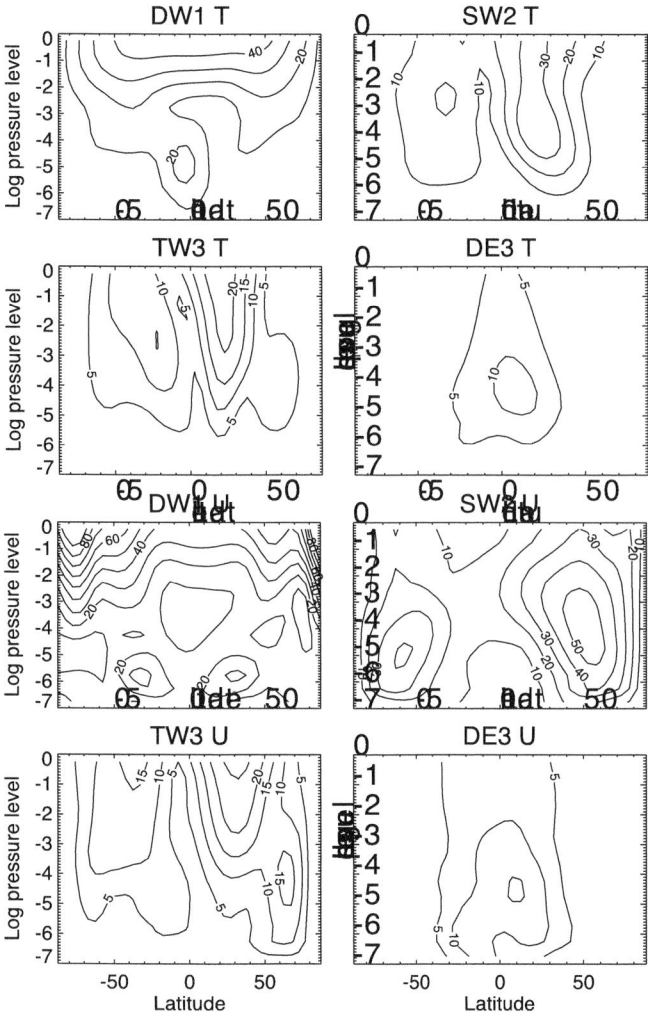

Figure 4. (top 4) Monthly averaged amplitudes of four tidal components for temperature and (bottom 4) zonal wind. The temperature amplitudes are in degrees K, the zonal wind amplitudes are in m s^{-1}. The four components are diurnal westward wave 1, semidiurnal westward wave 2, terdiurnal westward wave 3, and diurnal eastward wave 3. See text for the specific definitions of diurnal, semidiurnal, and terdiurnal.

models [*Akmaev et al.*, 2008] or extended weather models [e.g., *Du and Ward*, 2010]. Thus, Figure 1 of *Akmaev et al.* [2008] shows their calculated DW1 for an entire year at 100 km as a function of latitude. Comparison with our January results for the $Z = -6$ level (roughly equivalent) shows good agreement. Similarly, our SW2 results for $Z = -6$ compare favorably with their Figure 2; our value of 10 K at 40 N is closer to the averaged SABER data they show. Akmaev et al. also show results for DE3 temperatures at 116 km (about $Z = -4.5$ on our plots) and zonal winds at 95 km (roughly our bottom level), which agree well with our results. On the other hand, *Du and Ward* [2010, Figure 3] present TW3 results at 95 and 110 km (about $Z = -5$ for us), which seem to be much greater than what we calculate.

Despite the general agreement with the above-cited references, the above uncertainty with the TW3 results motivates our discussion of possible uncertainties in our calculation due to the nudging we apply to the TIEGCM. We evaluate this by comparing the tidal amplitudes as a function of latitude near the interface between the two models. Figure 5 shows this for T and U for the DW1, SW2, TW3, and DE3 tides. In all eight panels, the dashed curve is the tides from the NOGAPS-ALPHA output at $Z = -7$ level, which is the bottom interface level of the TIEGCM and corresponds directly to NOGAPS-ALPHA output at a pressure of 5.4×10^{-4} hPa. The dotted curves are the TIEGCM tides at $Z = -6.75$ (corresponding to the lowest midpoint level) when only forced by NOGAPS at the bottom interface. As can be clearly seen, there are some notable discontinuities, particularly in the zonal wind DW1 and SW2 components. The solid curves are the TIEGCM tidal components, which result when we additionally nudge the TIEGCM zonal and meridional wind fields at the -6.75 midpoint level with NOGAPS ALPHA output as discussed in section 2.2. Nudging the U and V fields in this matter greatly improves the continuity between the two models. We experimented with nudging the temperature (not shown) and found that it greatly and excessively suppressed the calculated DW1 and DE3 T tides. We, thus, chose to only nudge the winds and not the temperature. We also recognize that we effectively introduced a discontinuity in the northern midlatitude TW3 U tide; however, overall, our calculated tides seem fairly robust to our handling of the bottom boundary. This is demonstrated in Figure 6, which presents altitude profiles of the important tidal components for the two approaches highlighted in Figure 5. For reference, we also overplotted tides calculated from the so-called standard utilization of the TIEGCM, i.e., with GSWM forcing and a global average temperature of 181 K (the same simulation presented in Figure 2a). We should not expect the GSWM-forced tides to agree with the NOGAPS-ALPHA-forced tides since the assumptions underlying GSWM (ozone, Rayleigh friction) are likely very different from that in NOGAPS-ALPHA. The important point is that the two NOGAPS-ALPHA-driven solutions differ by comparably little, especially when compared with the GSWM forced solution. The largest difference is in the peak of the DE3 tide, which varies by about 20%. Ultimately, we suggest that our choice of, and uncertainty associated with, nudging is analogous to some of the tradeoffs that must be made when tuning gravity wave drag (GWD) parameterizations. No one single GWD parameterization can perfectly describe the mesospheric and lower thermosphere circulation and

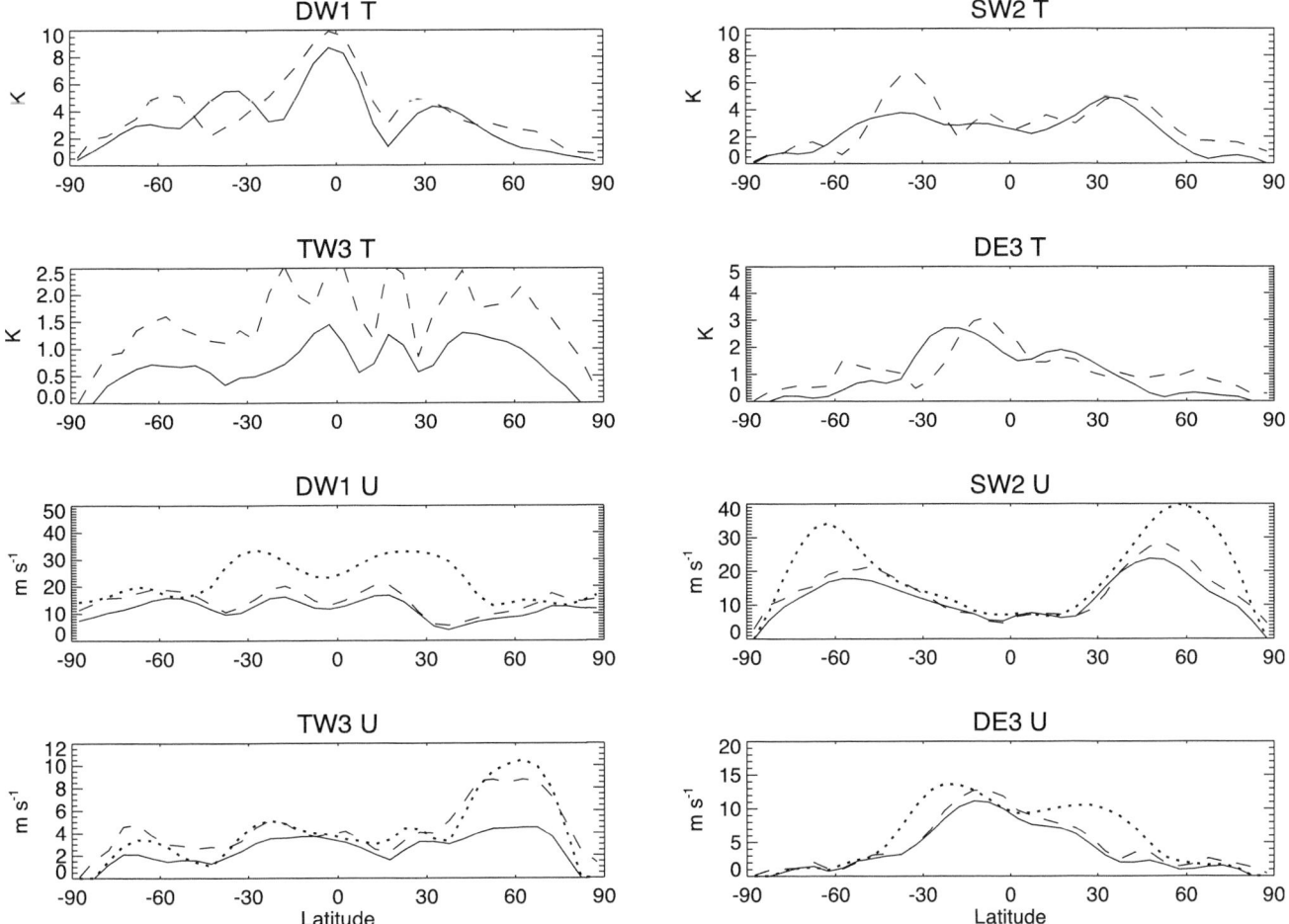

Figure 5. Comparison of monthly averaged latitudinally variation of four tidal components of temperature (*T*) and zonal wind (*U*) tides calculated by NOGAPS-ALPHA and the TIEGCM. The dashed curve in each panel is the particular tidal component calculated from the NOGAPS-ALPHA input to the TIEGCM bottom boundary at the $Z = -7$ log-pressure level. The dotted and solid curves in each panel are the tidal components taken from the TIEGCM *T* and *U* solution at the lowest midpoint level ($Z = -6.75$). The solid curves are from the case where the TIEGCM *U* and *V* fields at $Z = -6.75$ are nudged by NOGAPS; the dotted curves are from the case without nudging. See text for a definition of the TIEGCM pressure levels and their relationship to NOGAPS-ALPHA pressure levels.

temperatures in all circumstances (cf. Figure 1 of *Yamashita et al.* [2010]; also *Eckermann et al.* [2009], section 3 for a discussion of model biases in this regard).

3.3. Changes Due to the January 2009 Sudden Stratospheric Warming

Here we simply show some sample model output to illustrate some of the meteorological changes seen in response to the stratospheric warming of late January. *Fuller-Rowell et al.* [2011] shows that the wind field became more structured in the Northern Hemisphere with higher-order wave modes introduced. This appears to occur with our results as well. Figure 7 shows a comparison of the longitudinal variation of the zonal wind at northern midlatitudes on 10 and 24 January at about 120 km. The predominant wave 2-like pattern on 10 January, reflecting a strong migrating SW2 tide is evident. On 24 January, the wave 2 pattern is gone, and the wind is much more highly structured. However, it is not possible to simply ascribe the change between the two dates as due to a decay in SW2 and growth in TW3 as was seen in Fuller-Rowell's results. Figure 8 shows the 2-D frequency-wavenumber spectrum for the zonal wind for the two 5 day periods, one from the 10th to the 14th (before the warming) and one from the 20th to the 24th (during the warming). The large black blotch at

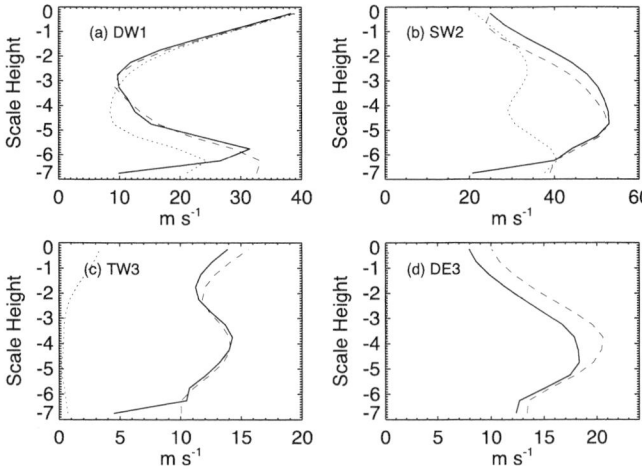

Figure 6. Comparison of monthly averaged profiles of the four zonal wind tides for the two approaches to proscribing the TIEGCM bottom boundary discussed in Figure 5. The solid lines result from nudging the U and V fields; the dashed lines are from the case without nudging. For reference, the migrating tides from a GSWM December solstice solution are presented as the dotted lines. Each panel is an average over a different range of latitudes associated with the peak of the tide. (a) An average from 22.5°N to 32.5°N, (b) an average from 52.5°N to 62.5°N, (c) an average from 52.5°N to 62.5°N, and (d) from 7.5°S to 7.5°N.

wavenumber 2 for two cycles per day in Figure 8a represents the large SW2 component, with an amplitude in excess of 40 m s^{-1}. This is significantly reduced on 24 January, but there is no single dominant feature in the spectrum on this date, although the relative contribution by higher-order waves and

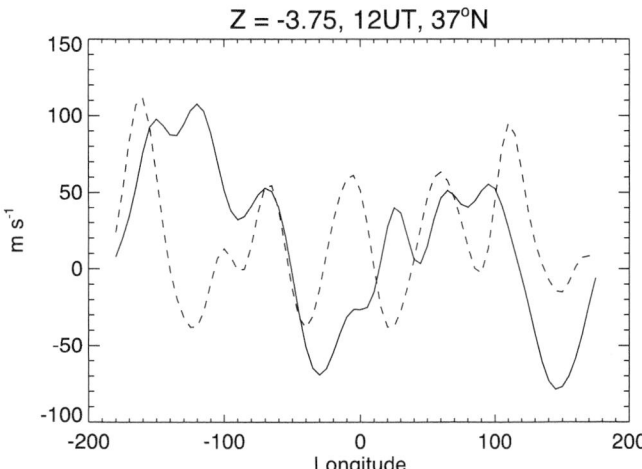

Figure 7. Snapshot of zonal wind variation at $Z = -3.75$ (about 120 km) as a function of longitude at 12 UT for 37°N on 10 January (solid curve) and 24 January (dashed curve).

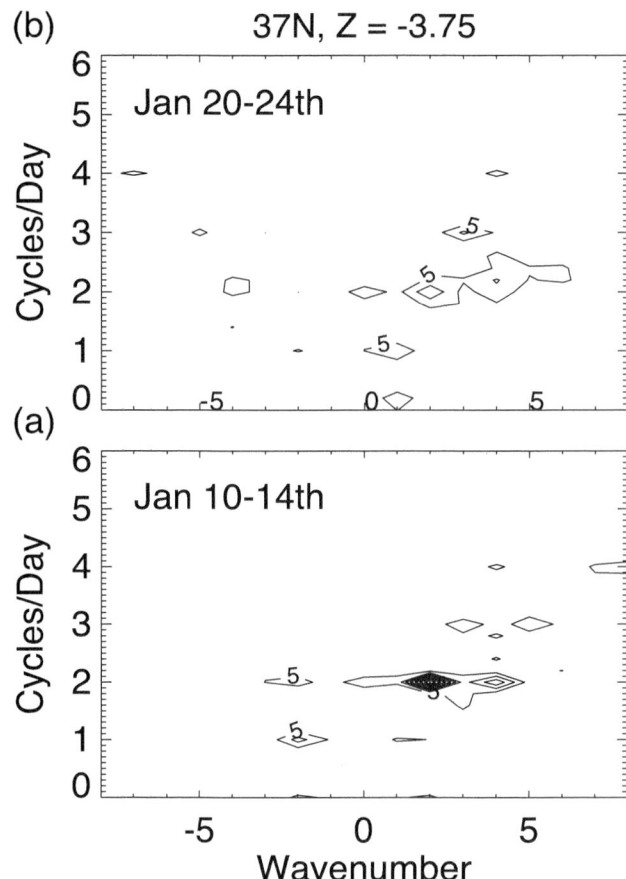

Figure 8. A 2-D spectrum of zonal wind variation for the same location as in Figure 7, for two 5 day intervals (a) for 10–14 January 2009, a 5 day period before the major sudden stratospheric warming, and (b) for 20–24 January 2009, during the warming. The contours are in intervals of 5 m s^{-1}; the peak magnitude of the large blotch at wave 2, 2 cycles d^{-1} in (a) is 40.5 m s^{-1}.

nonmigrating components is clearly evident. *Forbes et al.* [2008] has discussed how a wide range of nonmigrating tides can occur, including up to wave 7. It is likely, although not proven here, that some of the wave-wave interactions discussed by the above authors are occurring here. Thus, although we do not see evidence for a large increase in TW3, clearly, SSW events seem to be associated with numerous complicated tidal interactions, as suggested by previous authors [e.g., *Fuller-Rowell et al.*, 2011; *Wang et al.*, 2011].

4. CONCLUSION

We have provided an introduction into our approach in coupling lower atmospheric model with a thermosphere ionosphere model. Because we use an analysis that extends to higher altitudes than most others, we suggest that using NOGAPS-ALPHA and TIEGCM fills a niche not currently exploited by other systems. We have presented a preliminary analysis as to the relative components of lower atmospheric forcing on the thermospheric zonal mean circulation. For the January 2009 case we show, it appears that the extratropical circulation responds to the latitudinal variation in temperature, most prominently the cold summer mesopause. The tropical circulation appears to correlate with the eastward diurnal tide; however, a more rigorous analysis of the momentum deposition of this tide must be performed to confirm that the strong eastward acceleration we calculate is due to this tide. Second, we show promising results in minimizing discontinuities between NOGAPS-ALPHA and TIEGCM tides at the interface between the two models. Finally, during the 2009 SSW, and consistent with other previous publications, we show a marked decrease in the standard semidiurnal zonal wave 2 pattern in the winter lower thermosphere. However, rather than a simple increase in the migrating terdiurnal tide, our wind pattern appears to be a complicated mix of migrating and nonmigrating components. Future work will extend the coupled TIEGCM-NOGAPS runs for the rest of 2009 and also look at other winters for which we have NOGAPS-ALPHA analyses. Since the truly new component of our coupled system is the nonmigrating tides that we transmit into the thermospheric model, emphasis will be placed on quantifying their role in the composition and momentum budget of the thermosphere and ionosphere.

Acknowledgments. This work was supported by the Office of Naval Research. We acknowledge helpful discussions with Ruth Lieberman, Hanli Liu, Art Richmond, and Stan Solomon.

REFERENCES

Akmaev, R. A. (2011), Whole atmosphere modeling: Connecting terrestrial and space weather, *Rev. Geophys.*, 49, RG4004, doi:10.1029/2011RG000364.

Akmaev, R. A., T. J. Fuller-Rowell, F. Wu, J. M. Forbes, X. Zhang, A. F. Anghel, M. D. Iredell, S. Moorthi, and H.-M. Juang (2008), Tidal variability in the lower thermosphere: Comparison of Whole Atmosphere Model (WAM) simulations with observations from TIMED, *Geophys. Res. Lett.*, 35, L03810, doi:10.1029/2007GL032584.

Chapman, S., and R. S. Lindzen (1970), *Atmospheric Tides*, D. Reidel, Hingham, Mass.

Coy, L., S. D. Eckermann, and K. W. Hoppel (2009), Planetary wave breaking and tropospheric forcing as seen in the stratospheric sudden warming of 2006, *J. Atmos. Sci.*, 66, 495–507, doi:10.1175/2008JAS2784.1.

Coy, L., S. D. Eckermann, K. W. Hoppel, and F. Sassi (2011), Mesospheric precursors to the major stratospheric sudden warming of 2009: Validation and dynamical attribution using a ground-to-edge-of-space data assimilation system, *J. Adv. Model. Earth Syst.*, 3, M10002, doi:10.1029/2011MS000067.

Drob, D. P., et al. (2008), An empirical model of the Earth's horizontal winds fields: HWM07, *J. Geophys. Res.*, 113, A12304, doi:10.1029/2008JA013668.

Du, J., and W. E. Ward (2010), Terdiurnal tide in the extended Canadian Middle Atmospheric Model (CMAM), *J. Geophys. Res.*, 115, D24106, doi:10.1029/2010JD014479.

Eckermann, S. D. (2009), Hybrid σ–p coordinate choices for a global model, *Mon. Weather Rev.*, 137, 224–245.

Eckermann, S. D., K. W., Hoppel, L. Coy, J. P. McCormack, D. E. Siskind, K. Nielsen, A. Kochenash, M. H. Stevens, and C. R. Englert (2009), High-altitude data assimilation system experiments for the Northern Hemisphere summer mesosphere season of 2007, *J. Atmos. Sol. Terr. Phys.*, 71, 531–551.

Fesen, C. G., G. Crowley, R. G. Roble, A. D. Richmond, and B. G. Fejer (2000), Simulation of the pre-reversal enhancement in the low latitude vertical ion drifts, *Geophys. Res. Lett.*, 27 (13), 1851–1854.

Fomichev, V. L., J.-P. Ogibalow, and D. S. Turner (1998), Matrix parameterization of the 15 μm CO_2 band cooling in the middle and upper atmosphere for variable CO_2 concentration, *J. Geophys. Res.*, 103, 11,505–11,528.

Forbes, J. M. (1995), Tidal and planetary waves, in *The Upper Mesosphere and Lower Thermosphere: A Review of Experiment and Theory, Geophys. Monogr. Ser.*, vol. 87, edited by R. M. Johnson and T. L. Killeen, pp. 67–87, AGU, Washington, D. C., doi:10.1029/GM087p0067.

Forbes, J. M., J. Russell, S. Miyahara, X. Zhang, S. Palo, M. Mlynczak, C. J. Mertens, and M. E. Hagan (2006), Troposphere-thermosphere tidal coupling as measured by the SABER instrument on TIMED during July–September 2002, *J. Geophys. Res.*, 111, A10S06, doi:10.1029/2005JA011492.

Forbes, J. M., X. Zhang, S. Palo, J. Russell, C. J. Mertens, and M. Mlynczak (2008), Tidal variability in the ionospheric dynamo region, *J. Geophys. Res.*, 113, A02310, doi:10.1029/2007JA012737.

Fuller-Rowell, T., H. Wang, R. Akmaev, F. Wu, T.-W. Fang, M. Iredell, and A. Richmond (2011), Forecasting the dynamic and electrodynamic response to the January 2009 sudden stratospheric warming, *Geophys. Res. Lett.*, 38, L13102, doi:10.1029/2011GL047732.

Garcia, R. R., D. R. Marsh, D. E. Kinnison, B. A. Boville, and F. Sassi (2007), Simulation of secular trends in the middle atmosphere, 1950–2003, *J. Geophys. Res.*, 112, D09301, doi:10.1029/2006JD007485.

Hagan, M. E., R. G. Roble, and J. Hackney (2001), Migrating thermospheric tides, *J. Geophys. Res.*, 106(A7), 12,739–12,752, doi:10.1029/2000JA000344.

Hagan, M. E., A. Maute, R. G. Roble, A. D. Richmond, T. J. Immel, and S. L. England (2007), Connections between deep tropical clouds and the Earth's ionosphere, *Geophys. Res. Lett.*, 34, L20109, doi:10.1029/2007GL030142.

Hagan, M. E., A. Maute, and R. G. Roble (2009), Tropospheric tidal effects on the middle and upper atmosphere, *J. Geophys. Res.*, *114*, A01302, doi:10.1029/2008JA013637.

Hoppel, K. W., N. L. Baker, L. Coy, S. D. Eckermann, J. P. McCormack, G. Nedoluha, and D. E. Siskind (2008), Assimilation of stratospheric and mesospheric temperatures from MLS and SABER in a global NWP model, *Atmos. Chem. Phys.*, *8*, 6103–6116.

Immel, T. J., E. Sagawa, S. L. England, S. B. Henderson, M. E. Hagan, S. B. Mende, H. U. Frey, C. M. Swenson, and L. J. Paxton (2006), Control of equatorial ionospheric morphology by atmospheric tides, *Geophys. Res. Lett.*, *33*, L15108, doi:10.1029/2006GL026161.

Liu, H.-L., W. Wang, A. D. Richmond, and R. G. Roble (2010), Ionospheric variability due to planetary waves and tides for solar minimum conditions, *J. Geophys. Res.*, *115*, A00G01, doi:10.1029/2009JA015188.

Manney, G. L., M. J. Schwartz, K. Krüger, M. L. Santee, S. Pawson, J. N. Lee, W. H. Daffer, R. A. Fuller, and N. J. Livesey (2009), Aura Microwave Limb Sounder observations of dynamics and transport during the record-breaking 2009 Arctic stratospheric major warming, *Geophys. Res. Lett.*, *36*, L12815, doi:10.1029/2009GL038586.

McCormack, J. P., L. Coy, and K. W. Hoppel (2009), Evolution of the quasi-two day wave during January 2006, *J. Geophys. Res.*, *114*, D20115, doi:10.1029/2009JD012239.

Nielsen, K., D. E. Siskind, S. D. Eckermann, K. W. Hoppel, L. Coy, J. P. McCormack, S. Benze, C. E. Randall, and M. E. Hervig (2010), Seasonal variation of the quasi 5 day planetary wave: Causes and consequences for polar mesospheric cloud variability in 2007, *J. Geophys. Res.*, *115*, D18111, doi:10.1029/2009JD012676.

Oberheide, J., Q. Wu, T. L. Killeen, M. E. Hagan, and R. G. Roble (2006), Diurnal nonmigrating tides from TIMED Doppler Interferometer wind data: Monthly climatologies and seasonal variations, *J. Geophys. Res.*, *111*, A10S03, doi:10.1029/2005JA011491.

Picone, J. M., A. E. Hedin, D. P. Drob, and A. C. Aikin (2002), NRLMSISE-00 empirical model of the atmosphere: Statistical comparisons and scientific issues, *J. Geophys. Res.*, *107*(A12), 1468, doi:10.1029/2002JA009430.

Qian, L., A. G. Burns, B. A. Emery, B. Foster, G. Lu, A. Maute, A. D. Richmond, R. G. Roble, S. C. Solomon, and W. Wang (2012) The NCAR TIE-GCM: A community model of the coupled thermosphere/ionosphere system, in *Modeling the Ionosphere/Thermosphere System, Geophys. Monogr. Ser.*, doi:10.1029/2012GM001297, this volume.

Ren, S., S. Polavarapu, S. R. Beagley, Y. Nezlin, and Y. J. Rochon (2011), The impact of gravity wave drag on mesospheric analyses of the 2006 stratospheric major warming, *J. Geophys. Res.*, *116*, D19116, doi:10.1029/2011JD015943.

Richmond, A. D., E. C. Ridley, and R. G. Roble (1992), A thermosphere/ionosphere general circulation model with coupled electrodynamics, *Geophys. Res. Lett.*, *19*(6), 601–604.

Roble, R. G. (2000), On the feasibility of developing a global atmospheric model extending from the ground to the exosphere, in *Atmospheric Science Across the Stratopause, Geophys. Monogr. Ser.*, vol. 123, edited by D. E. Siskind, S. D. Eckermann, and M. E. Summers, pp. 53–67, AGU, Washington, D. C., doi:10.1029/GM123p0053.

Roble, R. G., and E. C. Ridley (1994), A thermosphere-ionosphere-mesosphere-electrodynamics general circulation model (time-GCM): Equinox solar cycle minimum simulations (30–500 km), *Geophys. Res. Lett.*, *21*(6), 417–420.

Sassi, F., R. R. Garcia, and K. W. Hoppel (2012), Large scale Rossby normal modes during some recent Northern Hemisphere winters, *J. Atmos. Sci.*, *69*, 820–839.

Siskind, D. E., S. D. Eckermann, J. P. McCormack, L. Coy, K. W. Hoppel, and N. L. Baker (2010), Case studies of the mesospheric response to recent minor, major, and extended stratospheric warmings, *J. Geophys. Res.*, *115*, D00N03, doi:10.1029/2010JD014114. [Printed 116(D3), 2011].

Siskind, D. E., D. P. Drob, J. T. Emmert, M. H. Stevens, P. E. Sheese, E. J. Llewellyn, M. E. Hervig, R. Niciejewski, and A. J. Kochenash (2012), Linkages between the cold summer mesopause and thermospheric zonal mean circulation, *Geophys. Res. Lett.*, *39*, L01804, doi:10.1029/2011GL050196.

Smith, A. K., and D. A. Ortland (2001), Modeling and analysis of the structure and generation of the terdiurnal tide, *J. Atmos. Sci.*, *58*, 3116–3134.

Stauffer, D. R., and N. L. Seaman (1990), Use of four-dimensional data assimilation in a limited-area mesoscale model. Part I: Experiments with synoptic-scale data, *Mon. Weather Rev.*, *118*(6), 1250–1277, doi:10.1175/1520-0493(1990)118<1250:UOFDDA>2.0.CO;2.

Stauffer, D. R., and N. L. Seaman (1994), Multiscale four-dimensional data assimilation, *J. Appl. Meteorol.*, *33*(3), 416–434, doi:10.1175/1520-0450(1994)033<0416:MFDDA>2.0.CO;2.

Stevens, M. H., et al. (2010), Tidally induced variations of polar mesospheric cloud altitudes and ice water content using a data assimilation system, *J. Geophys. Res.*, *115*, D18209, doi:10.1029/2009JD01325.

Wang, H., T. J. Fuller-Rowell, R. A. Akmaev, M. Hu, D. T. Kleist, and M. D. Iredell (2011), First simulations with a whole atmosphere data assimilation and forecast system: The January 2009 major sudden stratospheric warming, *J. Geophys. Res.*, *116*, A12321, doi:10.1029/2011JA017081.

Yamashita, C., H.-L. Liu, and X. Chu (2010), Responses of mesosphere and lower thermosphere temperatures to gravity wave forcing during stratospheric sudden warming, *Geophys. Res. Lett.*, *37*, L09803, doi:10.1029/2009GL042351.

Zhang, X., J. M. Forbes, M. E. Hagan, J. M. Russell III, S. E. Palo, C. J. Mertens, and M. G. Mlynczak (2006), Monthly tidal temperatures 20–120 km from TIMED/SABER, *J. Geophys. Res.*, *111*, A10S08, doi:10.1029/2005JA011504.

Zou, X., I. M. Navon, and F. X. Ledimet (1992), An optimal nudging data assimilation scheme using parameter estimation, *Q. J. R. Meteorol. Soc.*, *118*(508), 1163–1186, doi:10.1002/qj.49711850808.

D. P. Drob and D. E. Siskind, Space Science Division, Naval Research Laboratory, 4555 Overlook Avenue SW, Washington, DC 20375, USA. (david.siskind@nrl.navy.mil)

WACCM-X Simulation of Tidal and Planetary Wave Variability in the Upper Atmosphere

H.-L. Liu

High Altitude Observatory, National Center for Atmospheric Research, Boulder, Colorado, USA

A 20 year climate simulation using Community Earth System Model 1/Whole Atmosphere Community Climate Model with thermosphere extension has been made under constant solar and low geomagnetic conditions, and the simulation is analyzed to study the tidal variability in relation to the variability of mean state and planetary waves. On interannual scales, the migrating diurnal and semidiurnal tides (diurnal westward propagating wave 1 (DW1) and semidiurnal westward propagating wave 2 (SW2)) and the nonmigrating diurnal eastward propagating wave 3 component (DE3) in the mesosphere/lower thermosphere (MLT) are modulated by the quasi-biennial oscillations. Correlation analyses are performed between deseasonalized tidal wave amplitudes with winter stratospheric state anomalies for solstitial periods. The correlation between DW1 amplitude at midlatitude and low latitudes and the winter polar stratospheric temperature anomalies is negative in the winter hemisphere and alternates signs over altitudes in the summer hemisphere. SW2 shows a significant positive correlation with the winter polar stratospheric temperature anomalies in the summer stratopause, low latitude and midlatitude in the mesosphere, and lower thermosphere. The correlation alternates signs with altitudes at midlatitudes to high latitudes in the winter hemisphere. The nonmigrating SW1 in the MLT region correlates positively with the planetary wave 1 in the winter stratosphere. DE3 in the equatorial MLT region, where it peaks, does not show any significant correlation with the winter stratosphere anomalies. The short-term variability of these tidal components have time scales of several days, much shorter than the typical time scales of stratospheric planetary wave variability (10–20 days). The magnitude of the day-to-day tidal variability is significant and is persistent throughout the year.

1. INTRODUCTION

The ionosphere and thermosphere (IT) are highly variable on a broad range of time scales, subject to solar and geomagnetic forcings, and perturbations from the lower atmosphere. Of direct relevance to space weather applications is the variability of the F2 region on time scales of several days [e.g., *Rishbeth*, 2006; *Rishbeth and Mendillo*, 2001, and references therein]. The solar and geomagnetic forcings are the most evident drivers of IT variability, due to variability in the solar radiative and particulate outputs and magnetospheric storms and substorms. Another important source of variability is the perturbations from the lower atmosphere. For example, *Forbes et al.* [2000] found that under geomagnetic quiet conditions ($k_p < 1$), the standard deviation of the electron density at the F2 peak (NmF2) was 25%–35% for periods between several hours to 2 days, and 15%–20% for periods between 2 to 30 days. The analysis by *Rishbeth and Mendillo*

Modeling the Ionosphere-Thermosphere System
Geophysical Monograph Series 201
© 2013. American Geophysical Union. All Rights Reserved.
10.1029/2012GM001338

[2001] determined that under medium solar conditions, the geomagnetic activity and perturbations from the lower atmosphere have comparable contributions to the day-to-day F2 region variability, while the solar EUV contribution is relatively small. The perturbations from the lower atmosphere affect the IT most likely through atmosphere waves, including atmospheric thermal tides and lunar tides, planetary waves, and gravity waves. These waves can reach large amplitudes at thermosphere altitudes to compensate for the density decrease and can, thus, cause large perturbations in neutral temperature and winds, and constituents [e.g., *Forbes*, 1995; *Hagan and Forbes*, 2002, 2003; *Fritts and Alexander*, 2003; *Akmaev et al.*, 2009; *Lei et al.*, 2011]. The propagation of these waves is mostly confined below ~200 km because they experience strong dissipation in the middle and upper thermosphere with the rapid increase of molecular viscosity. These waves may affect ionospheric electrodynamics by modulating electric wind dynamo and neutral and plasma transport [*Millward et al.*, 2001; *Forbes*, 2000; *Fesen et al.*, 2000]. For example, a conspicuous wave-4 longitude structure in the equatorial ionospheric anomaly (EIA) is likely caused by the modulation of electrodynamics by the nonmigrating diurnal eastward propagating wavenumber 3 component (DE3) in the ionosphere E region [*Sagawa et al.*, 2005; *Immel et al.*, 2006; *Hagan et al.*, 2009]. These waves show significant variations, due to variation of wave sources, propagation conditions, wave-wave, and wave-mean flow interactions [e.g., *Plumb*, 1983; *Teitelbaum and Vial*, 1991; *Meyer and Forbes*, 1997; *Liu and Hagan*, 1998; *Palo et al.*, 1998; *Liu et al.*, 2004; *Hagan and Roble*, 2001; *Liu et al.*, 2007; *Chang et al.*, 2011; *Yue et al.*, 2012; *Pedatella et al.*, 2012a].

One example when IT displays large short-term variability in response to lower-atmosphere changes is during stratospheric sudden warming (SSW) events [e.g., *Liu and Roble*, 2002; *Goncharenko and Zhang*, 2008; *Chau et al.*, 2012]. During SSWs, quasistationary planetary waves (QSPWs) drive circulation changes in the stratosphere and lower mesosphere, which in turn change the filtering of gravity waves and lead to profound changes of wind, temperature, and transport in the mesosphere and thermosphere [*Liu and Roble*, 2002; *Coy et al.*, 2005; *Ren et al.*, 2008; *Chandran et al.*, 2011]. Although the QSPWs and their forcing on the mean circulation are mostly confined to midlatitude to high latitudes and below the wind reversal (usually in mesosphere and lower thermosphere (MLT) or lower during SSW), their impacts can be global through circulation changes and changes of latitudinal temperature gradient [*Karlsson et al.*, 2009; *Körnich and Becker*, 2010; *Yuan et al.*, 2012], change of planetary waves and gravity waves [*Becker et al.*, 2004; *Becker and Fritts*, 2006], modulation of migrating and non-migrating thermal tidal waves through interaction of tides and QSPWs [*Liu and Roble*, 2002; *Liu et al.*, 2010a; *Chang et al.*, 2009; *Fuller-Rowell et al.*, 2010; *Pedatella and Forbes*, 2010], through change of ozone heating [*Pancheva et al.*, 2003; *Wu et al.*, 2011; *Sridhan et al.*, 2012; *Goncharenko et al.*, 2012], and modulation of lunar semidiurnal tide [*Stening et al.*, 1997]. These tidal variability can lead to ionospheric variability, as evidenced from the recent ionospheric observations during SSW, which showed that anomalies of ion temperature, total electron contents (TEC), plasma drifts, and electrojets during SSW display tidal signatures [*Goncharenko and Zhang*, 2008; *Chau et al.*, 2009; *Goncharenko et al.*, 2010a, 2010b; *Pedatella and Forbes*, 2010; *Fejer et al.*, 2010, 2011; *Park et al.*, 2012; *Yamazaki et al.*, 2012; *Liu and Richmond*, 2013]. Numerical study using the National Center for Atmospheric Research (NCAR) thermosphere-ionosphere-mesosphere-electrodynamics (TIME)-general circulation model (GCM) by *Liu et al.* [2010a] demonstrated that the introduction of an idealized QSPW (wavenumber 1) results in changes of migrating diurnal and semidiurnal tides and excitation of nonmigrating tides, which lead to changes in ionospheric electric field/plasma drift by modulating the E region electric wind dynamo. Using the NOAA Whole Atmosphere Model (WAM) with meteorological data assimilation to drive the Coupled Thermosphere Ionosphere Plasmasphere Electrodynamics Model (CTIPe), *Fuller-Rowell et al.* [2011b] reproduced realistic vertical plasma drift compared with observed ionospheric variability during the 2009 SSW. From the simulation, they found that the terdiurnal tide changes significantly during the 2009 SSW. The change of lunar tide during SSW was recently studied by *Pedatella et al.* [2012c] by using Whole Atmosphere Community Climate Model (WACCM) with the M2 lunar tide implemented [*Pedatella et al.*, 2012b]. From ensemble WACCM simulations, *Pedatella et al.* [2012c] established that the M2 lunar tide is significantly enhanced during SSW. They also found that the migrating solar semidiurnal tide generally increases below ~120 km during SSW. To determine the impacts of the lunar tide during SSW on the ionosphere, *Pedatella et al.* [2012c] performed WACCM-X simulations of SSW (with M2 lunar tide) and used the model output to drive the global ionosphere and plasmasphere (GIP) model. From the GIP simulations, they found that the changes of equatorial vertical plasma drift are similar to the observations when the lunar tide is included in the simulation. On smaller scales, gravity waves can increase or decrease dramatically during SSW, depending on the observational location especially with respect to the polar jet and the phase of the planetary waves [*Wang and Alexander*, 2009; *Yamashita et al.*, 2010a; *Thurairajah et al.*, 2010]. These changes can be caused by the change of propagation conditions due to wind change–associated large-scale flow during SSW, which can, in turn, result in mesosphere and thermosphere changes [e.g., *Liu and Roble*, 2002; *Yamashita et al.*, 2010b]. Large disturbances to the polar vortex can lead to significant gravity wave excitation through spontaneous adjustment [*Wang and Alexander*, 2009; *Yamashita et al.*, 2010a; *Limpasuvan et al.*, 2011]. Their impact on the IT, however, is not known.

Another classical example of ionosphere variability is the equatorial spread F (ESF)/plasma bubble irregularity, which is a phenomenon of great scientific interest and space weather importance. ESF displays large spatial and temporal variability, which makes its prediction challenging. It is critical to understand the processes that control the initiation, development, and variability of the ESF, in order to better forecast such events. Atmospheric gravity waves are a viable candidate for seeding the ESF [*Singh et al.*, 1997]. The sources of gravity waves include wind interaction with orography, convection, frontogenesis, adjustment of unbalanced flow, auroral heating, and secondary generation from gravity wave dissipation (see the work of *Fritts and Alexander* [2003] for an extensive review), and they are highly variable over time and space. The propagation and dissipation are strongly affected by the large-scale flows (mean state, tidal and planetary waves). The development of ESF is also closely associated with large-scale winds, as demonstrated by the numerical study by *Huba et al.* [2009].

These examples underscore the potential significance of the terrestrial weather in IT variability. To study the connection, it is essential for a model to quantify the sources, propagation, and impacts of atmospheric waves. This is one of the motivations for the development of whole atmosphere models in recent years. For a review of the development and status of whole atmosphere modeling, the readers are referred to a review paper by *Akmaev* [2011]. The NCAR Whole Atmosphere Community Climate Model with thermosphere extension (WACCM-X) is one of such models. In this paper, we will present a brief description of the essential model components and capability of WACCM-X. We will then evaluate the planetary wave and tidal variability in the mesosphere and thermosphere and conduct statistical analysis of these waves from extended WACCM-X simulations (20 model years). We will explore, in the presence of large internal variability afforded by a whole atmosphere climate model, if the variability of various PWs and tides are still robustly correlated as indicated in previous mechanistic studies. In this study, we will focus on variability of diurnal westward propagating wave 1 (DW1) and SW2 for the migrating component for obvious reasons. As for nonmigrating tides, we focus on DE3 and the semidiurnal westward propagating wave 1 (SW1) for the variability analysis. DE3 reaches its peak in the equatorial MLT region and can strongly affect ionospheric variability as explained earlier. SW1, resulting from nonlinear interaction between SW2 and QSPW 1 [e.g., *Angelats i Coll and Forbes*, 2002], peaks at midlatitudes to high latitudes. It may reflect the variability of both SW2 and QSPW1, as found in previous studies of SSW [*Chang et al.*, 2009; *Liu et al.*, 2010a; *Wu et al.*, 2011], and can affect ionospheric variability at lower latitudes [*Liu and Richmond*, 2013]. It should be noted that other nonmigrating tides can be strong in the MLT and E region. For example, the DE2 is comparable or even stronger than the DE3 for December–February. It is, however, out of the scope of this study to perform variability analysis of all tidal components. We also note that, although SSW provides an exemplary case for studying atmosphere coupling through waves, the PWs and tides and their variability are ubiquitous throughout a year. A statistical analysis is intended to help understand the wave variability in a more general context.

2. THE THERMOSPHERE EXTENSION OF THE WHOLE ATMOSPHERE COMMUNITY CLIMATE MODEL (WACCM-X)

The WACCM-X is based on the WACCM, which in turn is a superset of the Community Atmospheric Model (CAM). The detailed model features can be found in the work of *Liu et al.* [2010b]. The most recent version of WACCM-X is based upon WACCM4 and CAM4, and it has been integrated in the Community Earth System Model (CESM) and recently released for community use (J.W. Hurrell, et al., The community earth system model: A framework for collaborative research, submitted to *Bulletin of the American Meteorological Society*, 2012). The following is a brief description of the main model features, including information on CESM1 and CAM4 (R.B. Neale, et al., The mean climate of the Community Atmosphere Model (CAM4) in forced SST and fully coupled experiments, submitted to *Journal of Climate*, 2012).

The NCAR CESM version 1 (CESM1) is a coupled model consisting of atmosphere, ocean, land surface, sea and land ice, and carbon cycle components for simulating past, present, and future climates. These components are linked through a coupler that exchanges fluxes and state information among them. A detailed description of CESM1 is given by *Hurrel et al.* [2012]. CAM4 is the default atmosphere component for CESM1. Compared with CAM3, the most notable changes are the inclusion of deep convection effects in the momentum equation [*Richter and Rasch*, 2008] and a dilute approximation in the plume calculation [*Neale et al.*, 2008]. These changes result in improved representation of deep convection [*Gent et al.*, 2011; *Neale et al.*, 2012].

WACCM4 and WACCM4-X can be used in place of CAM4 as the atmospheric component of CESM (version 1.0.4 and later). WACCM4 is the most recent version of WACCM, which is a climate-chemistry general circulation model developed at NCAR. It is based upon the infrastructure of CAM4 with vertical domain extending to 5.9×10^{-6} hPa (~140 km geometric height). The vertical resolution is identical to that of CAM4 up to 100 hPa but is substantially finer above: in the lower stratosphere, the vertical resolution ranges from 1.5 km near the tropopause to about 2 km near the stratopause; in the mesosphere, the vertical resolution is half of the local scale height and remains so up to the model lid. Details of the earlier version, WACCM3, can be found in the work of *Garcia et al.* [2007]. WACCM4 incorporates several improvements and enhancements over the previous version: It can be run coupled to the CESM1 ocean model component (POP2) and sea ice component (CICE); the model's chemistry module has been updated according to the latest JPL-2006 recommendations; a quasibiennial oscillation (QBO) may be imposed (as an option) by relaxing the winds to observations in the Tropics [*Matthes et al.*, 2010]; heating from stratospheric volcanic aerosols is now computed explicitly; the effects of solar proton events are now included; the effect of unresolved orography is parameterized as a surface stress (turbulent mountain stress) leading to an improvement in the frequency of sudden stratospheric warmings, probably by better representing the wind/land surface interaction and, thus, planetary wave generation [*de la Torre et al.*, 2012]; and gravity waves due to convective and frontal sources are parameterized based upon the occurrence of convection and the diagnosis of regions of frontogenesis in the model [*Richter et al.*, 2010]. A newly developed inertial gravity wave parameterization scheme has enabled the internal generation of QBO within WACCM [*Xue et al.*, 2012] and, thus, more consistent resolution of atmosphere variability in the middle and upper atmosphere.

WACCM-X is based on WACCM and, thus, incorporates all the WACCM features mentioned above. It has extended the top boundary to the upper thermosphere (2.5×10^{-9} hPa or ~500 km). As in the regular configuration of WACCM, the chemistry module is interactive with the dynamics through transport and exothermic heating [*Kinnison et al.*, 2007]. In particular, photochemistry associated with ion species (O^+, NO^+, $O2^+$, $N2^+$, and N^+) is part of the chemistry package. Therefore, the neutral and ion species are self-consistently resolved in the model, a unique feature among whole atmosphere models. It should be noted that the ionosphere electrodynamics and plasma transport is still under development. Photolysis rates at wavelength between Lyman-α and 350 nm are calculated according to the work of *Woods and Rottman* [2002] and following the work of *Froehlich* [2000] for wavelengths longer than 350 nm. For wavelengths shorter than Lyman-α (EUV and X-ray), both photolysis and photoionization are considered, and the rates are calculated following the work of *Solomon and Qian* [2005]. The radiative heating calculation in WACCM-X is the same as in WACCM3 [*Marsh et al.*, 2007] with the possibility to parameterize the wavelength-dependent EUV flux either in terms of the solar 10.7 cm radio flux ($F10.7$) or using the method by *Lean and Woods* [2010]. In WACCM-X, the ion drag and Joule heating are calculated according to the work of *Dickinson et al.* [1981] and *Roble et al.* [1982], respectively. The major species diffusion is included using the formulation proposed by *Dickinson et al.* [1981]. To best quantify tidal effects, a lunar tide (M2) module has recently been developed in WACCM and WACCM-X [*Pedatella et al.*, 2012b]. With the extension of the vertical domain to the upper thermosphere, the number of vertical levels is increased to 81 levels (half scale height vertical resolution) or 125 levels (quarter scale height vertical resolution). The former is used in the current study. The standard horizontal resolution used in the WACCM-X simulations is $1.9° \times 2.5°$. The quality of the thermospheric simulation obtained from WACCM-X has been compared with observations, an empirical model (NRLMSISE-00) [*Picone et al.*, 2002] and TIME-GCM. WACCM-X results display complex dynamical behavior from the ground to the model top. In the lower atmosphere and up to about 60 km, the zonal structure of the wind is dominated by planetary waves. Above 60 km, including the wind dynamo region, the dynamic structures are more complex highlighting the large variability of various waves, which is the subject of this paper.

The simulation presented here were conducted using the CESM1 F2000 WACCM-X compset, which is a standard model setup under perpetual solar maximum ($F10.7 = 200$ solar flux unit) and geomagnetically quiet conditions, with sea surface temperature from year 2000. A cyclic QBO is imposed in these simulations. With these conditions, the atmosphere variability in the model is otherwise internally driven. The following analysis is based on 20 years of the model simulation.

3. MODEL RESULTS

3.1. General Features of Model Variability on Various Time Scales

The climatology of mean temperature and winds and some of the major tidal components from WACCM-X was presented in the work of *Liu et al.* [2010b]. The general features of their variability will be examined here. As mentioned in the previous section, the WACCM-X simulations to be analyzed here are made with constant solar EUV and very low Kp, so they do not contribute to the atmospheric variability. We note that in the following analysis, the climatology of a field is formed by first averaging its daily values over the 20 year simulations, and then, a 31 day averaging is applied for each day of the composite year. Figure 1 shows the mean temperature at the North Pole (NP) and 10 hPa and wave number 1 of geopotential height perturbation at 60°N and 1 hPa (around where QSPWs usually peak), over the first January, 1 year and 20 years. For the January, which is from the first year of the simulation, the temperature varies by less than 10 K and is generally within 5 K around the 20 year climatology (dotted line). The NP temperature increases by about 55 K in February and recovers to climatological value in March. Over the 20 perpetual model years, the daily NP temperature varies between 185 and 265 K, in reasonable agreement with, though slightly cooler than, reanalysis results (190–270 K, e.g., http://acdb-ext.gsfc.nasa.gov/Data_-services/met/ann_data.html). The summer NP temperature shows much less variability, and maximum value is ~240 K (about 243 K from reanalysis). Although there is no warming in January of the first year, the wave 1 is very strong and varies between 1000 and 2000 m. It reached 2700 m in February before collapsing during the warming. The wave 1 at the latitude and altitude shown here is predominantly quasistationary planetary wave (QSPW), thus peaking in winter months and becoming small in summer months. The daily value of the amplitude can reach near 3000 m, and the maximums of both the daily and monthly mean values display interannual variability.

From Figure 2, it is seen that the mean stratospheric temperature at the South Pole (SP) from the model is less variable, with the winter values varying around 180 K and reaching as low as 170 K. The spring/summer values are around 255 K with daily values reaching 280 K. While the smaller variability and the summer values are in agreement with the reanalysis results, the winter values from the model are about 10 K cooler than the reanalysis. This cold bias of the southern winter stratosphere is accompanied by an excessively strong polar jet and is a known problem in many GCMs. It is likely caused by the weak or absence of gravity wave forcing in the stratosphere, which is generally neither resolved nor properly parameterized in GCMs [*Austin et al.*, 2003; *McLandress et al.*, 2011] and is responsible for the absence of QBO in most GCMs, as mentioned above. This bias causes less frequent minor SSW in the Southern Hemisphere in the model (Amal Chandran, personal communication). The wave 1 in the model is strong, with both the 20 year climatological and maximum daily values ~500 m larger than their Northern Hemisphere counterparts. The reanalysis, however, indicates that the wave 1 amplitudes in the two hemispheres are comparable. The large wave amplitudes in SH in the model is probably also tied to the weak or absence of gravity wave drag.

The amplitudes of the migrating diurnal (DW1) and semidiurnal (SW2) tidal meridional winds at 10^{-4} hPa (~105 km) over the same time periods as Figure 1 are shown in Figure 3. The latitudes chosen for the plots are where the wave amplitudes usually peak. The DW1 and SW2 meridional winds can change significantly from day to day, with magnitudes comparable to the climatological values of the wave amplitude, even when the stratospheric state is not strongly disturbed (Figure 1a) or around equinox (Figure 1b). These changes seem ubiquitous and persist throughout the year. The phases of the migrating tides also change from day to day and can be as large as 5 and 2 h for DW1 and SW2, respectively (not shown). It is also evident that the day-to-day time scale of the tidal variability is shorter than that of the QSPW 1 (Figures 1d–1e). Given this large variability of tidal amplitudes and phases and the discrepancy between their time scales with those of the PWs, statistical analysis of ensemble simulations is needed to extract relationship between changes of tides, PWs, and lower/middle atmosphere state. On longer time scales, the tidal amplitudes are clearly modulated by QBO. The DW1 is generally stronger/weaker during the westerly/easterly phase of the QBO, while SW2 modulation shows the opposite relationship. These are in agreement with previous observation studies [e.g., *Burrage et al.*, 1995; *Xu et al.*, 2009; *Wu et al.*, 2011]. As noted in the work of *Liu et al.* [2010b], the amplitudes of the DW1 and SW2 are smaller than observations. It is also noted that the seasonal peaks of DW1 in WACCM-X shift to earlier times with altitudes in the MLT region. As a result, the seasonal peaks of DW1 at the height level shown here are around day 50 and 210, about a month earlier than the peaking time around 90 km.

As mentioned in the Introduction, DE3 and SW1 are two prominent nonmigrating components in the lower thermosphere and E region at low latitudes, and midlatitudes to high latitudes, respectively. Their amplitudes at 10^{-4} hPa over the same time periods as discussed above are shown in Figure 4. Like the migrating components, DE3 and SW1 display large variability from day to day. For example, the daily values of DE3 amplitude at the specified level can change between ~20% and 200% of the climatological value within a few days, and the DE3 phase can also change strongly and rapidly. For SW1, the largest amplitude of the meridional wind in the MLT is usually seen in the summer hemisphere at midlatitudes to high latitudes. This agrees with the SW1 from TIDI analysis (including the wave amplitude) by *Wu et al.* [2011]. This is also consistent with previous findings from numerical studies that SW1 from

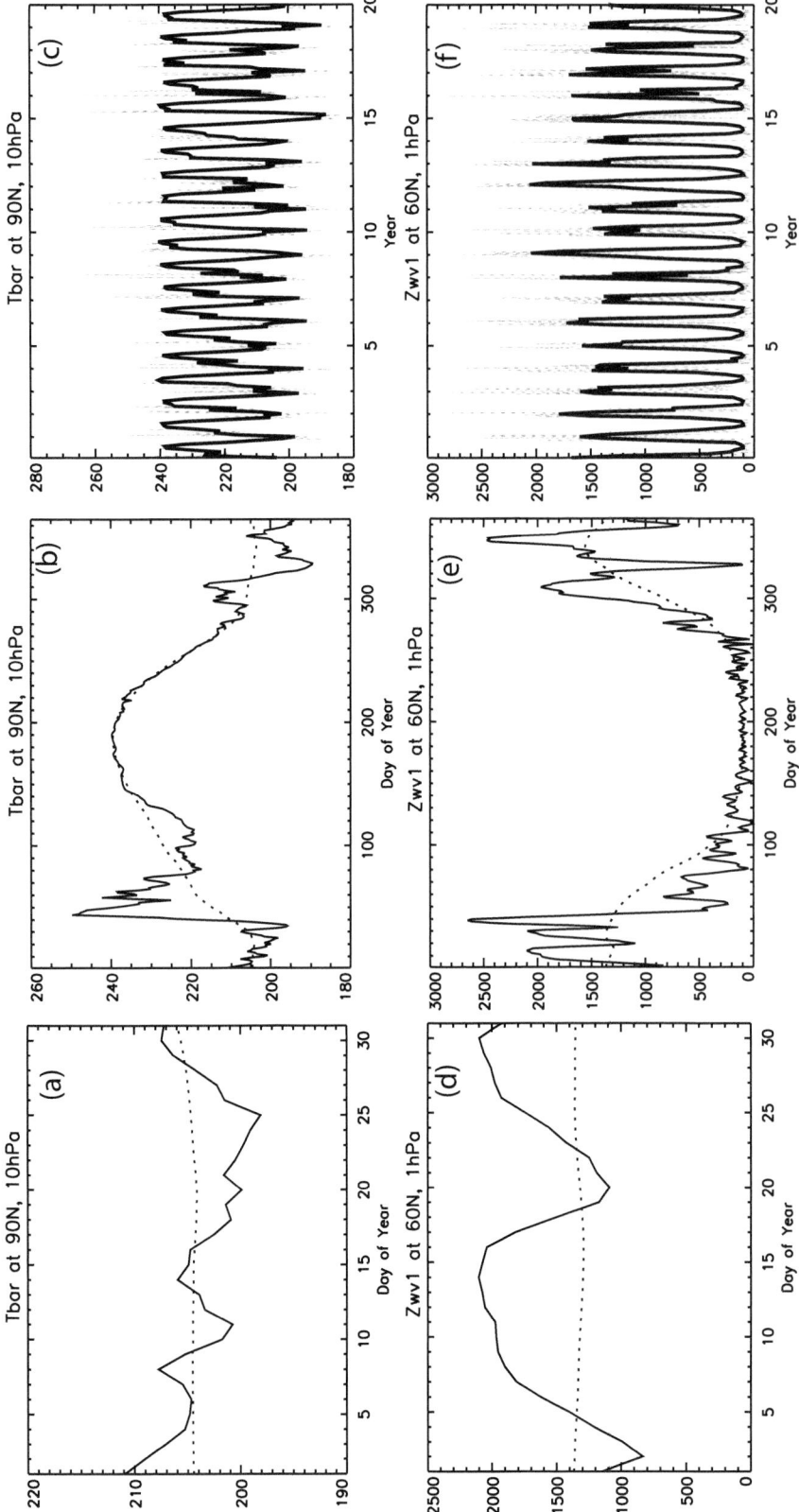

Figure 1. Temperature at 90°N and 10 hPa of (a) January of the first year, (b) first year, and (c) 20 years of Whole Atmosphere Community Climate Model with thermosphere extension (WACCM-X) simulations. The geopotential height amplitude of wave 1 at 60°N and 1 hPa for the same time periods (d–f). (a–b) and (d–e) The solid lines are daily values, and the dotted lines are climatological values from the 20 years of simulation (see text for details). (c) and (f) The solid lines are monthly mean values, and the dashed lines are daily values.

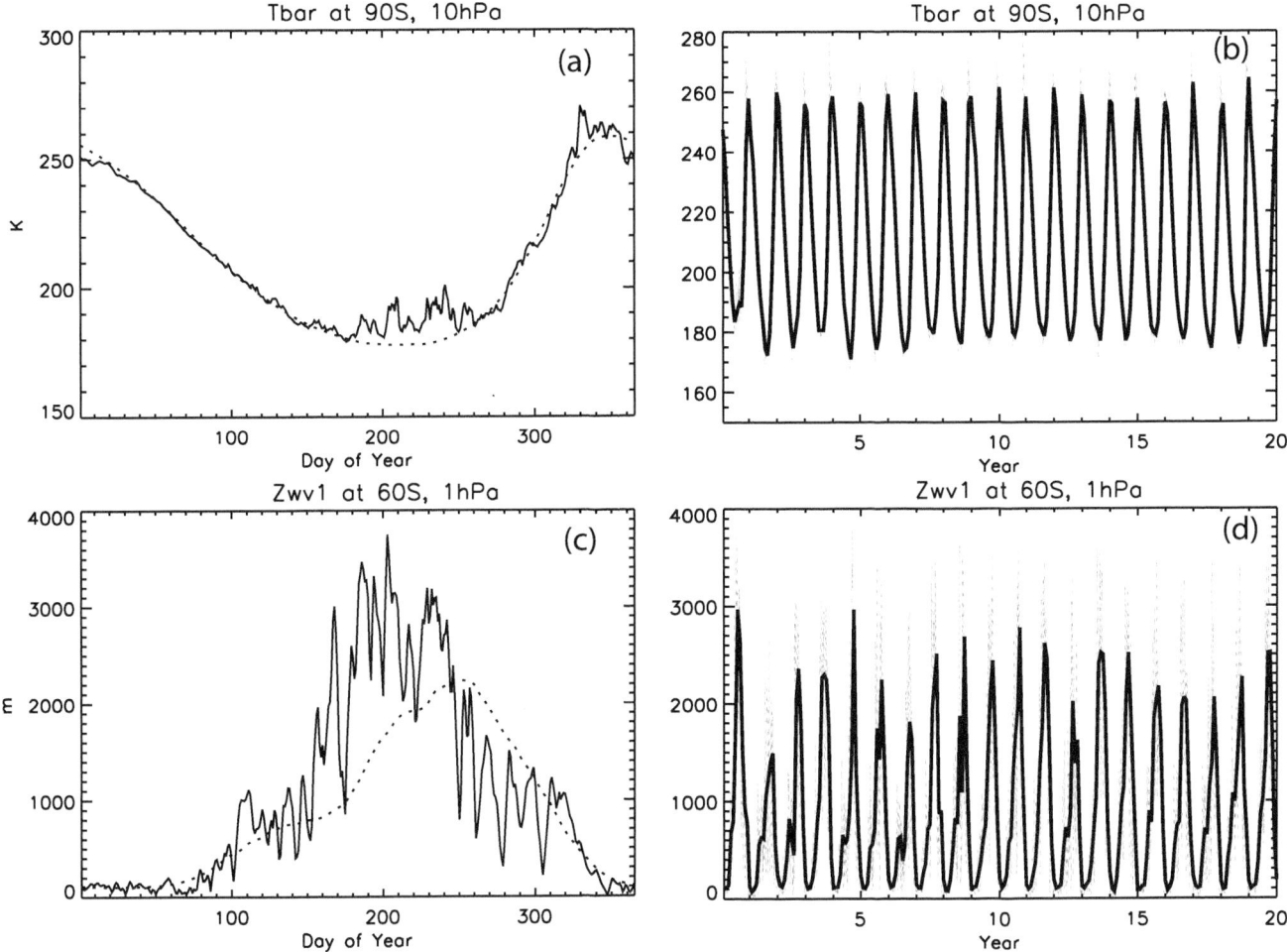

Figure 2. Temperature at 90°S and 10 hPa of (a) first year and (b) 20 years of WACCM-X simulations. The geopotential height amplitude of wave 1 at 60°N and 1 hPa for the same time periods (c–d). (a) and (c) The solid lines are daily values, and the dotted lines are climatological values from the 20 years of simulation (see text for details). (b) and (d) The solid lines are monthly mean values, and the dashed lines are daily values.

SW2 and QSPW1 interaction is usually strongest in the summer hemisphere MLT region [e.g., *Angelats i Coll and Forbes*, 2002; *Chang et al.*, 2009; *Liu et al.*, 2010a]. According to numerical simulations using TIME-GCM [*Liu and Richmond*, 2013], large SW1 at midlatitudes to high latitudes can cause *E* region wind dynamo changes that map down to lower latitudes, and induce significant ionospheric vertical drift changes around dawn under solar minimum conditions. The SW1 maximums during summer seasons, thus, implicate that the upward drift at dawn at low latitudes induced by SW1 would have a semiannual variation with peaks at solstices. This may provide a plausible explanation to the solstitial maximum of upward plasma drift and plasma density depletion at dawn under solar minimum conditions [*de la Beaujardière et al.*, 2009; *Gentile et al.*, 2011]. On

longer time scales, the interannual variation of DE3 is modulated by QBO, and DE3 peak amplitude is generally larger/smaller in the westerly/easterly phase of QBO. The interannual change of SW1, on the other hand, does not show a clear QBO modulation from the figure. We note in passing that the fact that DW1 and DE3, which have opposite zonal phase velocities, show similar QBO-modulating patterns, while DW1 and SW2, whose zonal phase speed are the same, show opposite QBO-modulating patterns suggests that the modulation of wave propagation by zonal wind is probably not the main mechanism here. This is consistent with the previous findings by *Hagan et al.* [1999], who showed that altering mean zonal wind in the Global-Scale Wave Model did not produce the observed QBO modulation of DW1.

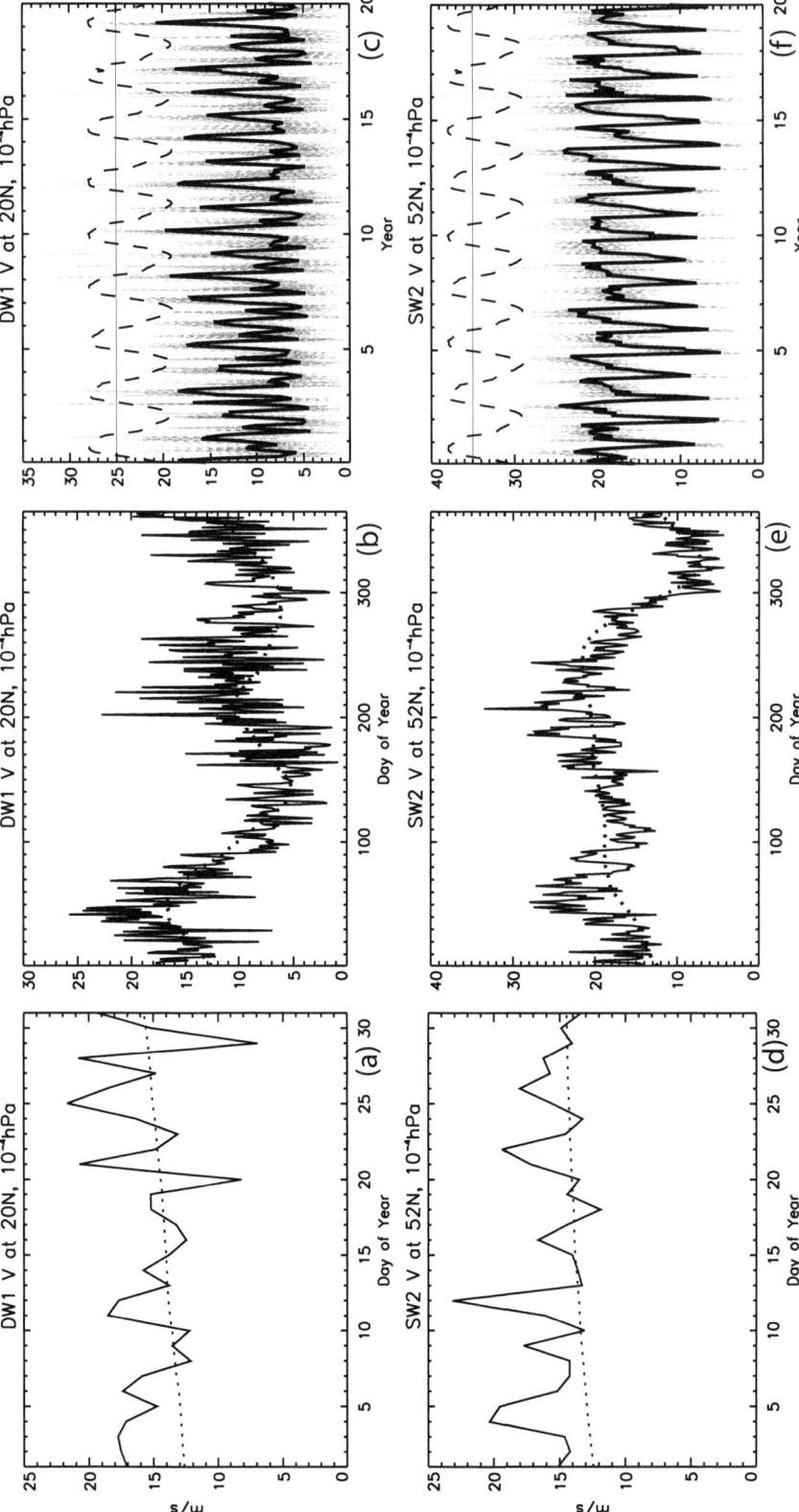

Figure 3. Similar to Figure 1, but for diurnal westward propagating wave 1 (DW1) at 20°N and semidiurnal westward propagating wave 2 (SW2) at 52°N, both at 10^{-4} hPa. (c) and (f) The dark dashed lines are the equatorial zonal mean zonal wind at 30 hPa, shifted by 25 and 35 m s^{-1}, respectively (0 wind denoted by the thin horizontal lines), and scales reduced by fivefold. The quasibiennial oscillation (QBO) phase can be read from the equatorial wind.

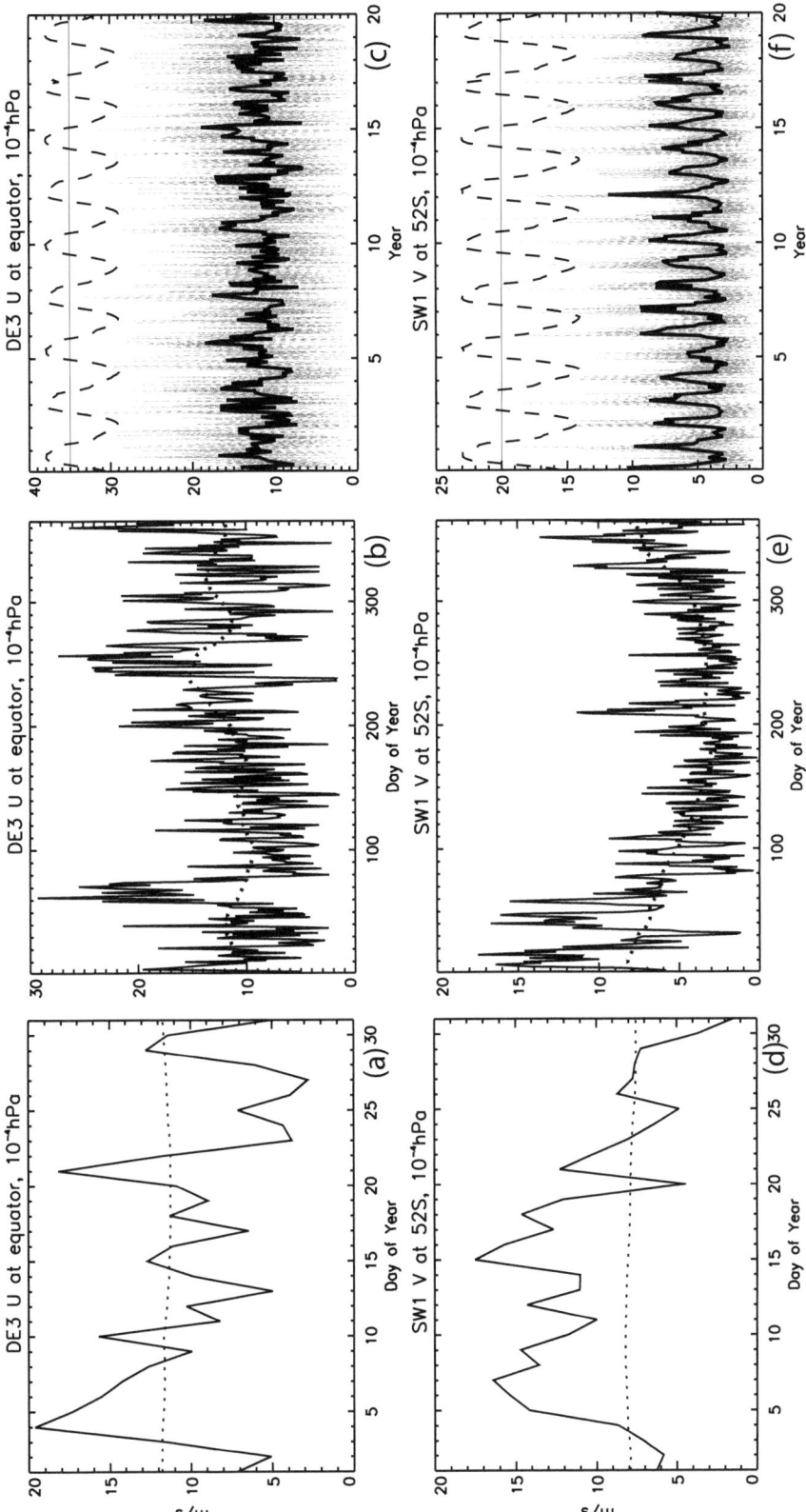

Figure 4. Similar to Figure 3, but for diurnal eastward propagating wave 3 (DE3) at the equator and SW1 at 52°S, both at 10^{-4} hPa. (c) and (f) The dark dashed lines are the equatorial zonal mean zonal wind at 30 hPa, shifted by 35 hPa and 20 m s^{-1}, respectively (0 wind denoted by the thin horizontal lines), and scales reduced by fivefold. The QBO phase can be read from the equatorial wind.

3.2. Correlation of Lower and Upper Atmosphere Around Solstice

As mentioned earlier, the large variability of the atmospheric states and the different time scales associated (both mean state and waves) may obscure the coupling between lower and upper atmosphere. To examine the possible coupling, particularly those suggested by mechanistic studies, but in a more variable environment, correlation studies are conducted on the 20 year WACCM-X simulation results. Given the potentially important role of QSPWs in the coupling process, the correlation calculations here focus around solstitial periods (December–February and June–August) when the QSPWs are the strongest as shown in Figures 1 and 2. *Tan et al.* [2012] demonstrated that the temperature changes of the whole atmosphere show similar spatial patterns in response to temperature anomalies of the polar winter stratosphere regardless of the presence of SSW. We, thus, conduct our correlation analysis over the DJF and JJA periods, without differentiating between SSW and non-SSW periods. Our analysis focuses on the anomalies around the climatological values, so time series are formed by subtracting the climatological values from the respective total fields. As a result, any seasonal effects will be excluded from the correlation analysis.

Figure 5 shows the correlation coefficient of zonal mean temperature (on pressure levels, same in the rest of the paper) with the winter polar temperature at 10 hPa. Figure 5a is essentially the same as the temperature correlation calculation by *Tan et al.* [2012], except that WACCM-X is used here. The whole atmosphere teleconnection pattern below the lower thermosphere from Figure 5a is the same as that obtained by *Tan et al.* [2012], and no further discussion is needed here. In WACCM-X, the positive temperature correlation above the NH winter mesopause at midlatitudes to high latitudes extends to $\sim 10^{-6}$ hPa (220 km). At midlatitudes to low latitudes, the zonal mean temperature above $\sim 10^{-6}$ hPa correlates positively with T (10 hPa, NP) during DJF, with the largest correlation coefficient of 0.3. This is probably caused by a downwelling that converges with an upwelling near 10^{-5} hPa. This positive correlation in the upper thermosphere does not agree with the upper thermosphere cooling at 400 km altitude from the analysis of CHAMP observation of the 2009 SSW event [*Liu et al.*, 2011], but is consistent with the WAM simulation [*Fuller-Rowell et al.*, 2011a]. The disagreement with observations could be due to changes in geomagnetic activity, as pointed out by *Fuller-Rowell et al.* [2011a]. The correlation pattern during SH winter is generally similar, despite the absence of major SSW, low counts of minor SSW, and the cold bias of the SH winter stratosphere in the model. A notable difference is the positive temperature correlation in the winter lower thermosphere is not as prominent as that during NH winter. Because the lower thermosphere temperature change is mainly controlled by the change of gravity wave forcing, and the jet reversal is far less frequent during SH winter, the change in gravity wave filtering (and thus the lower thermosphere temperature variation) is less prominent.

The QSPWs are closely associated with circulation and temperature changes of the middle atmosphere. The temporal and spatial patterns of the relationship and the hemispheric dependence are examined using the WACCM-X output.

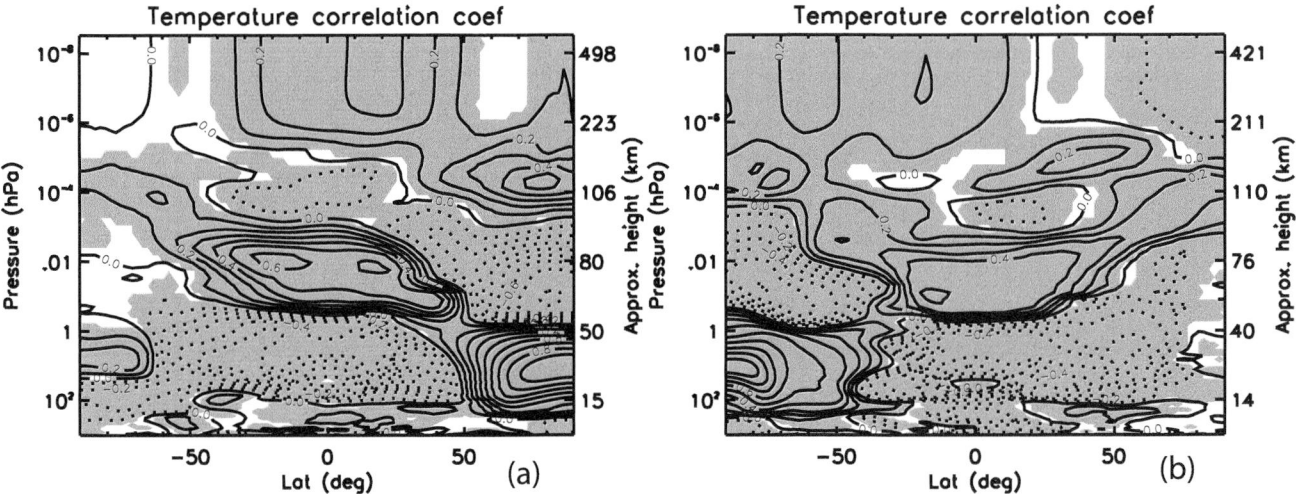

Figure 5. Correlation coefficients between zonal mean temperature with winter polar temperature at 10 hPa, both deseasonalized, for (a) DJF and (b) JJA periods. Solid contour lines are for positive correlation (contour interval: 0.1). Shaded: statistically significant at 95% level.

Figure 6 shows the lagged correlation between wave 1 at 1 hPa (45 km) and 10 hPa (30 km) with T (10 hPa, NP) for DJF and JJA periods. For DJF period and wave 1 at 1 hPa, the largest positive correlation coefficient (~0.3) is found around ~70 N° several days before day 0, while the largest negative correlation coefficient (−0.2) is around ~60°N 10 days after. The maximum values for each latitude shift to smaller lag time/larger lead time at lower latitudes (poleward of 30°N). Thus, the strong wave 1 anomalies at 1 hPa lead the maximum warm anomalies at NP and 10 hPa by several days, followed by minimum wave 1 anomalies about 10 days later. Furthermore, the wave 1 anomalies tend to start at midlatitudes and progress to high latitudes. At 10 hPa, the strong wave 1 anomalies at high latitudes occur at the same time as the warm anomalies therein, while at lower latitudes, the wave 1 anomalies show a lead time of several days. The minimum wave 1 anomalies also occur about 10 days later. This suggests that the occurrence of warm anomalies from the interaction between wave 1 and mean circulation progresses downward and poleward. The minimum wave 1 anomalies following the warm anomalies are associated with the filtering of QSPWs caused by jet reversals during SSW. This is contrasted with the lag correlation for the JJA, when major SSW is absent, and minor SSW is much less frequent (especially with the cold bias). The wave 1 and mean flow interaction during SH winter still causes warm anomalies in the stratosphere along with jet deceleration. But because jet reversal is rare, the QSPWs are not filtered following the warm anomalies. During JJA, the lag correlations in SH still suggest a poleward progression but not a clear downward progression.

The anomalies of migrating diurnal (DW1) and semidiurnal (SW2) tides are correlated with the polar winter temperature anomalies at 10 hPa (Figure 7) (meridional wind used for both components). For both DJF and JJA, the DW1 correlation with temperature is negative at low latitudes to

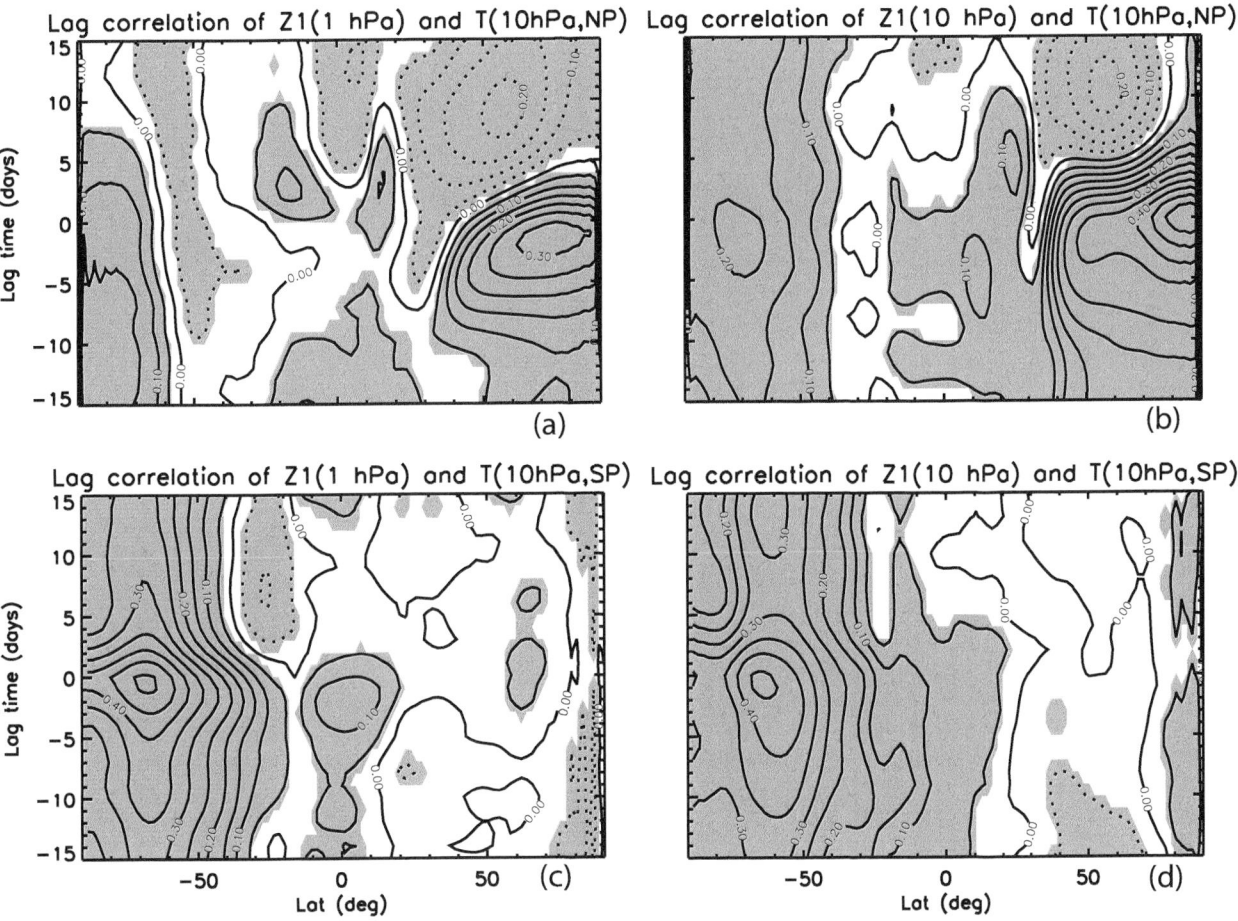

Figure 6. Lag-correlation coefficients between wave 1 at 1 hPa (a and c) and 10 hPa (b and d) with winter polar temperature at 10 hPa for DJF (a and b) and JJA (c and d) periods. Solid contour lines are for positive correlation (contour interval: 0.05). Shaded: statistically significant at 95% level.

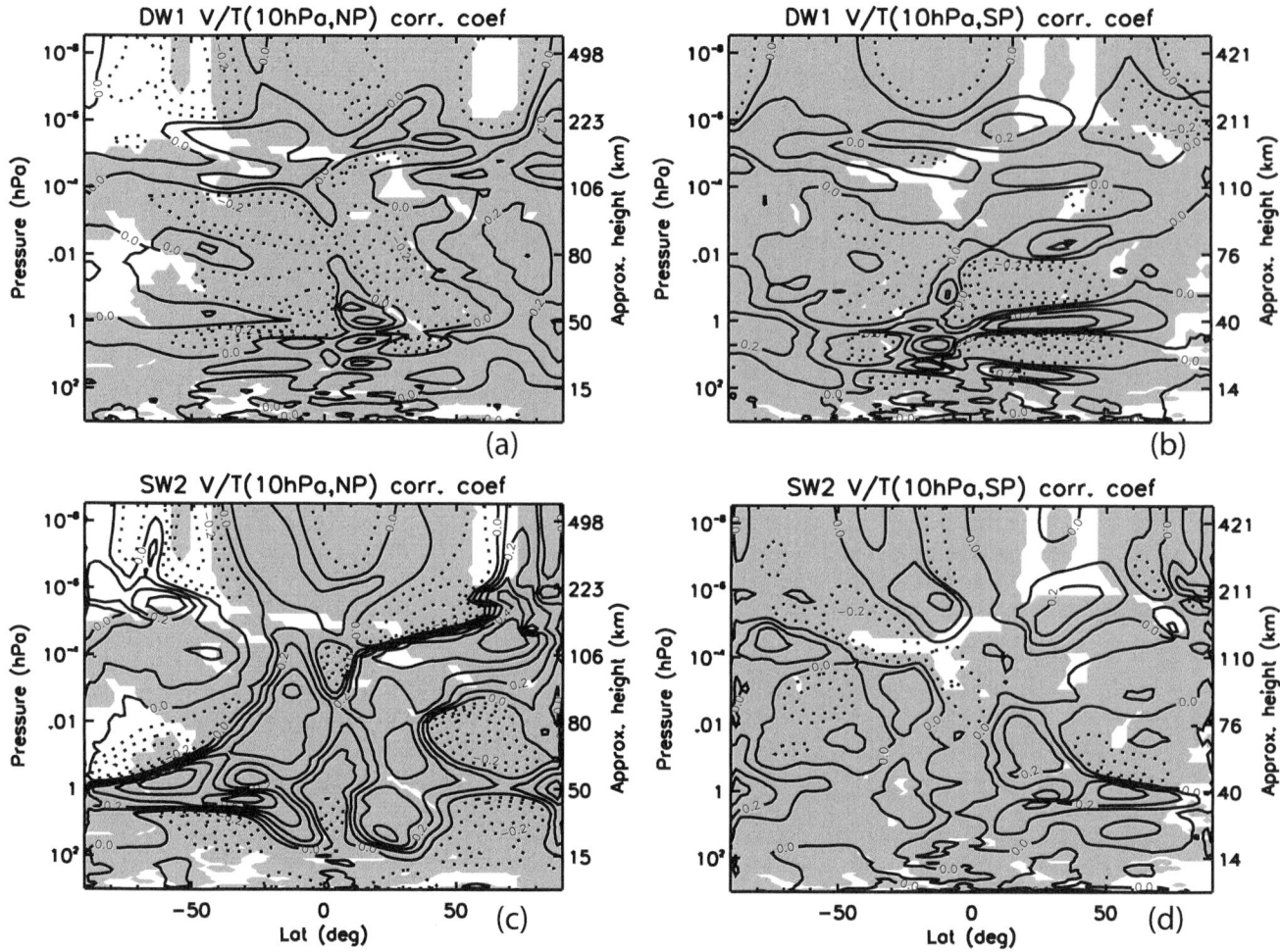

Figure 7. Correlation coefficients between DW1 (a and b) and SW2 (c and d) meridional wind amplitudes with northern (a and c) and southern (b and d) winter polar temperatures at 10 hPa. Solid contour lines are for positive correlation (contour interval: 0.1). Shaded: statistically significant at 95% level.

midlatitudes in the winter MLT (coefficients −0.2 to −0.3) and shows an alternating pattern in the summer hemisphere. These changes are in qualitative agreement with mechanistic model studies for the MLT region [e.g., *Liu et al.*, 2010a] and may result from PW nonlinear interaction with DW1 and modulation of DW1 propagation.

For DJF, the SW2 anomalies in the southern/summer stratopause region have a significant positive correlation (correlation coefficients up to 0.5) with the T (10 hPa, NP) anomalies, and the positive correlation extends to higher altitudes to the lower thermosphere (coefficients also up to 0.5). This is most prominent in the north/winter hemisphere, consistent with the propagation of SW2 from the summer to winter hemisphere. The positive correlation is probably caused by the increase of ozone and ozone heating in the summer stratosphere/stratopause region [*Wu et al.*, 2011;

Sridharan et al., 2012; *Goncharenko et al.*, 2012] during SSW events. At midlatitudes to high latitudes in the winter hemisphere, the correlation alternates signs with altitudes. The negative correlation of SW2, with T (10 hPa, NP) in the stratosphere and mesosphere at mid to high northern/winter latitudes (with correlation coefficients up to −0.4), is due to the decrease of ozone and ozone heating during the warming events (not shown). Immediately above the positive correlation in the northern MLT region, there is a significant negative correlation (−0.3), which is also reported by *Pedatella et al.* [2012c] as a decrease of SW2 during SSW. This negative correlation could be caused by several factors: the decrease of ozone heating in the stratosphere and mesosphere at mid to high (but sunlit) northern/winter latitudes mentioned above, and nonlinear interaction between SW2 and QSPWs [e.g., *Liu et al.*, 2010a], all during the warming events. The

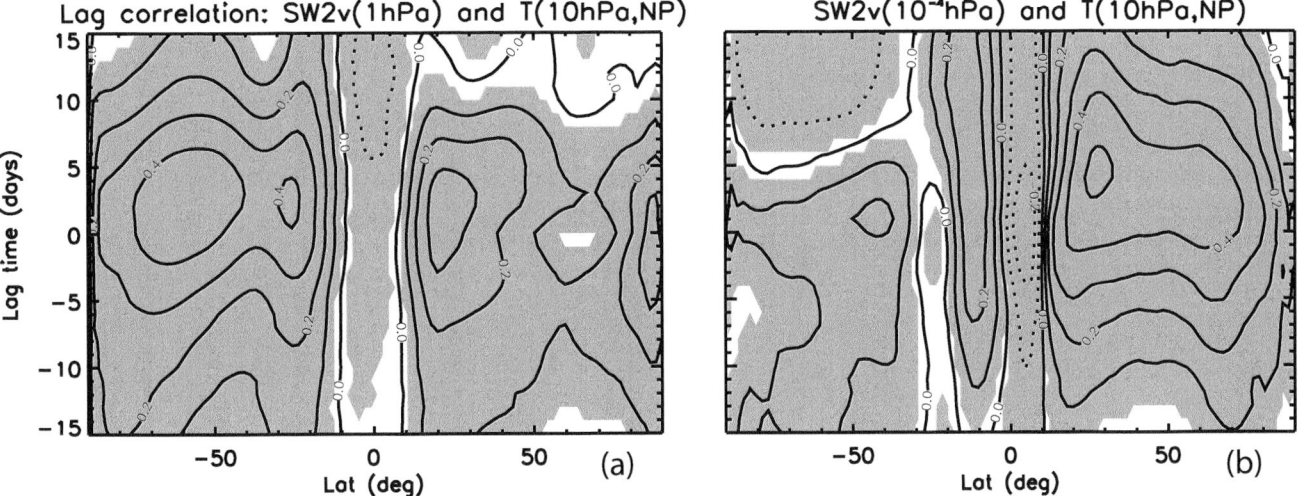

Figure 8. Lag-correlation coefficients between SW2 meridional wind amplitude at (a) 1 hPa and (b) 10^{-4} hPa with temperature at the North Pole and 10 hPa for DJF period. Solid contour lines are for positive correlation (contour interval: 0.1). Shaded: statistically significant at 95% level.

general spatial pattern of the correlation coefficient during JJA is similar to that during DJF, though the values are smaller.

To further study the response of SW2 to temperature anomalies, we examine the lagged correlation between SW2 and T (10 hPa, NP) at both 1 and 10^{-4} hPa (Figure 8). In the SH stratopause region where the SW2 correlates strongly and positively with NH warm anomalies according to Figure 7, the largest response of SW2 (maximum correlation coefficients between 0.4 and 0.5) lags the maximum NP temperature anomalies by ~2 days. In the MLT region, the largest correlation (up to 0.5) is seen in the NH about 2–5 days after the maximum NP temperature anomalies, depending on the latitudes. Similar delays are also seen for the JJA period (not shown).

Correlations of DE3 and SW1 with winter stratospheric state are also examined (Figure 9). The largest correlation between DE3 (temperature) and winter polar stratospheric temperature are found at midlatitudes to high latitudes in the winter hemisphere, with positive correlation in the stratosphere (and also in the lower thermosphere for the northern winter) and negative correlation in the mesosphere. These changes are probably caused by the change of DE3 wave guide at these latitudes during warm anomalies. At equatorial latitudes MLT region where DE3 peaks, however, no clear correlation is seen in these numerical experiments.

In Figure 9, the correlation coefficient between SW1 and wave 1 at 1 hPa is shown, which is generally larger than the correlation coefficient between SW1 and the polar stratospheric winter temperature. This is consistent with the understanding of the SW1 results from the nonlinear interaction between SW2 and wave 1. For both DJF and JJA periods, the largest correlation extends from the winter stratosphere/stratopause to the summer mesopause/lower thermosphere, with maximum correlation of 0.5 for DJF and 0.3 for JJA. This indicates the cross-equatorial propagation of the SW1 from its generation in the winter hemisphere and is consistent with the study of SW1 by *Angelats i Coll and Forbes* [2002], *Chang et al.* [2009], and *Liu et al.* [2010a]. From the lagged correlation analysis (Figure 10), it is seen that the largest MLT correlation lags the strong wave 1 anomalies at 1 hPa by 1–2 days. This fast response reflects the fast propagation of SW1, which has a large vertical wavelength and a relative short period.

4. DISCUSSION AND SUMMARY

Tides and planetary waves from 20 years of WACCM-X simulations display large variability in the middle and upper atmosphere on time scales from interannual to day-to-day. On interannual scales, the amplitudes of migrating diurnal (DW1) and semidiurnal (SW2) tides and the nonmigrating diurnal eastward propagating wave 3 (DE3) tide in the MLT are modulated by the QBO. Both DW1 and DE3 are generally stronger during the westerly phase of QBO, and SW2 is stronger during the easterly phase of QBO. The seasonal variation of these waves in WACCM-X in the mesosphere and thermosphere have been studied and compared with observations before [*Liu et al.*, 2010b]. Another nonmigrating tide, the semidiurnal westward propagating wave 1 (SW1),

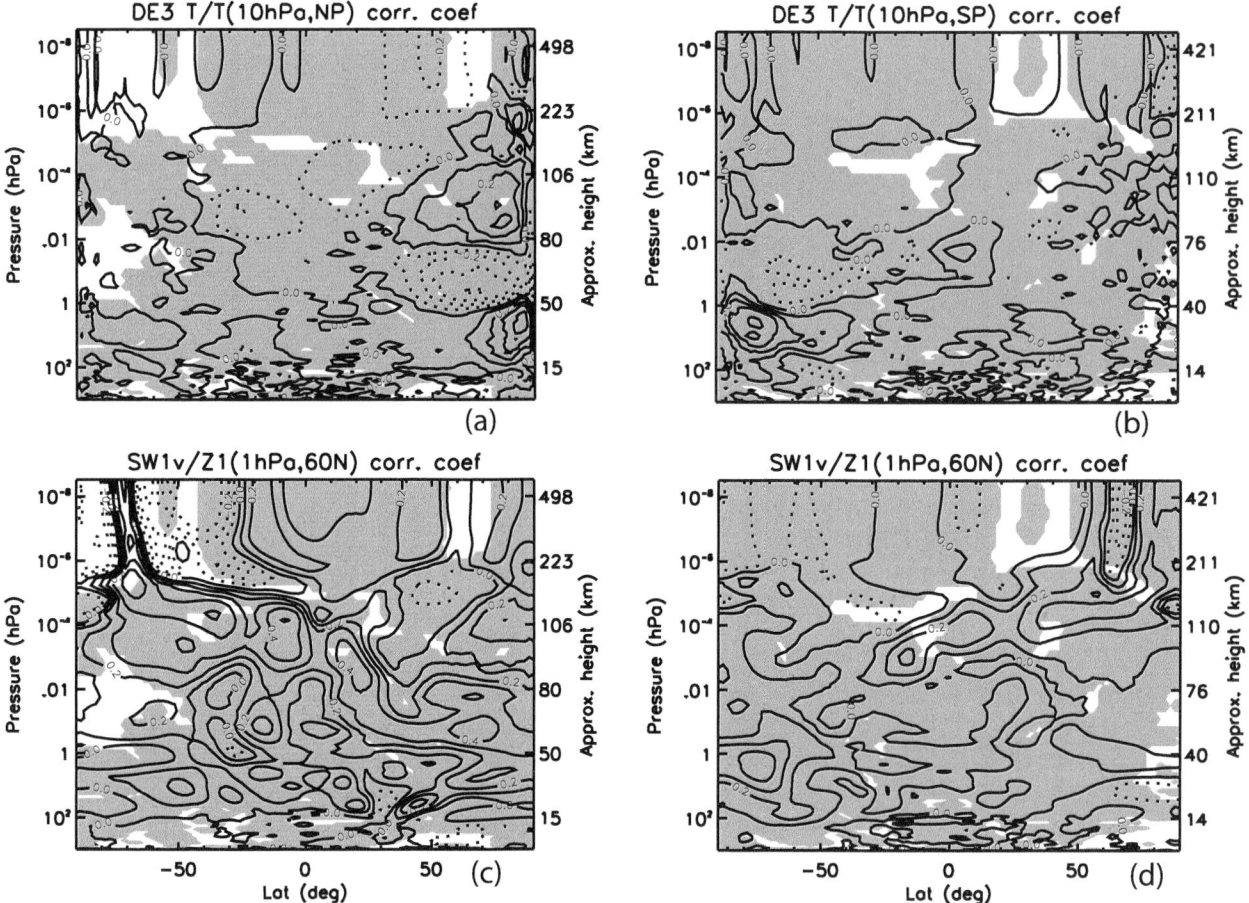

Figure 9. (a–b) Similar to Figures 7a–7b), except for DE3 temperature. Correlation coefficients between SW1 meridional wind amplitudes with wave 1 amplitude at 1 hPa at (c) 60°N and (d) 60°S for DJF and JJA periods, respectively. Solid contour lines are for positive correlation (contour interval: 0.1). Shaded: statistically significant at 95% level.

maximizes at midlatitudes to high latitudes in the summer hemisphere in MLT. It agrees with observations and is excited by nonlinear interaction between QSPW1 and SW2 according to previous studies.

On short time scales, all these waves (DW1, SW2, DE3, and SW1) vary strongly from day to day in the mesosphere and thermosphere. The possible connection between upper atmosphere short-term variability and lower atmosphere control is explored in this study using correlation analysis between deseasonalized time series under solstitial conditions (DJF and JJA), when the QSPWs and their impacts on the middle atmosphere (especially the winter hemisphere) are the strongest. For both solstice periods, the MLT DW1 amplitude at midlatitudes and low latitudes (where DW1 is large) in the winter hemisphere correlates negatively with the winter polar stratospheric temperature anomalies. In the summer hemisphere, the correlation coefficient alternates signs over altitudes. This spatial modulation pattern also extends into the middle and upper thermosphere with statistical significance, though the correlation coefficients are small due to the increasing damping of propagating DW1 components with altitudes. These changes are probably caused by a combination of energy transfer to nonmigrating tides and modulation of DW1 by PWs and/or mean flow changes, and these are in qualitative agreement with previous mechanistic studies. SW2 in the summer stratopause region correlates positively with the winter polar stratospheric temperature anomalies, probably due to the change of ozone and generation of SW2 from ozone heating. The positive correlation extends to the MLT region. The correlation alternates signs with altitudes at midlatitudes to high latitudes in the winter hemisphere, with negative correlation in the mesosphere (50–70 km) and again in the middle thermosphere/F region. This is probably caused by ozone decreases at midlatitudes to high latitudes both in the mesosphere and thermosphere, and nonlinear interaction with QSPWs. The

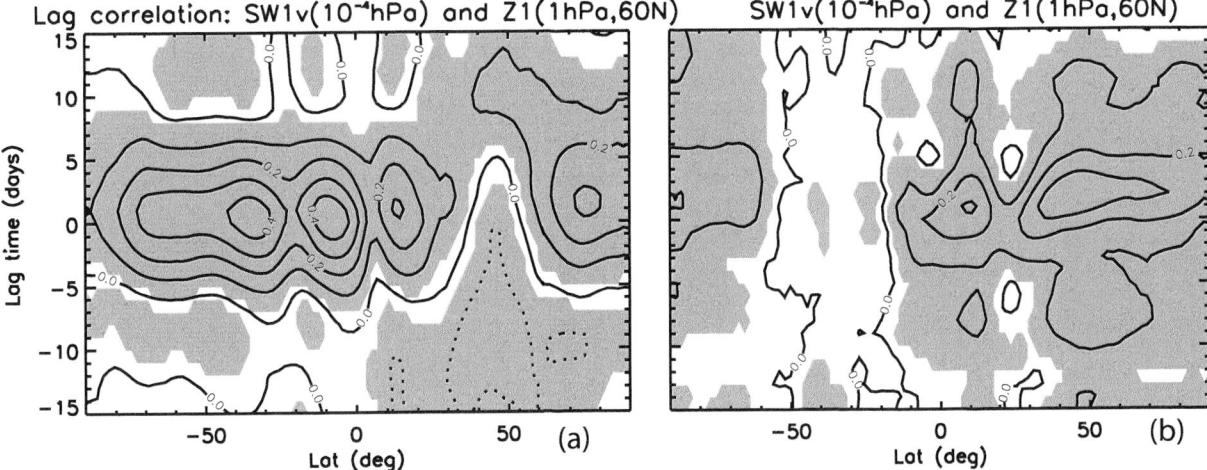

Figure 10. Lag-correlation coefficients between SW1 meridional wind amplitude at 10^{-4} hPa with wave 1 amplitude at 1 hPa at (a) 60°N and (b) 60°S for DJF and JJA periods, respectively. Solid contour lines are for positive correlation (contour interval: 0.1). Shaded: statistically significant at 95% level.

largest SW2 responses in the MLT lag the winter polar stratospheric temperature anomalies by several days. The SW1 in the summer MLT shows the most significant correlation with wave 1 in the winter stratosphere/stratopause region, with a lag of 1–2 days. The SW2 and SW1 correlations discussed above are generally stronger during DJF than JJA, though the spatial patterns are similar. The DE3 in the equatorial MLT region does not show any significant correlation with the polar winter stratospheric temperature anomalies. Although not discussed here, other nonmigrating tidal components (such as DE2 and SE2) can also be large in the thermosphere as resolved by WACCM-X [*Liu et al.*, 2010b] and thus can also contribute to the thermosphere/ionosphere short-term variability.

The teleconnection pattern of the mean state, identified in WACCM by *Tan et al.* [2012], is found to extend into the middle and upper thermosphere, especially during the northern winter season. It should be noted that the current WACCM-X does not include ionospheric electrodynamics or plasma transport. As such, any variability or teleconnection through ion-neutral processes is not accounted for.

Although the correlation analysis suggests a link between tidal variability in the MLT and QSPWs and winter stratosphere state, it is also noted that the time scales of QSPW short-term variation is longer than the tidal variability, with the former being 10–20 days while the latter being a few days. Furthermore, the short-term tidal variability is strong and persistent throughout the year, even when the QSPWs are weak (e.g., at equinoxes). The propagating planetary waves, such as quasi 2 day wave and 5–7 day wave, are known to modulate tides at the respective PW frequencies [e.g., *Ito et al.*, 1986; *Chen*, 1992; *Pedatella et al.*, 2012a]. These waves, however, have clear seasonal dependence. Therefore, the day-to-day tidal variability may not be explained solely in terms of tidal and PW interactions. The cause of the day-to-day tidal variability in the MLT region, as well as its impacts on the ionosphere day-to-day variability, will be explored in a separate study.

Acknowledgments. This work is, in part, supported by the National Science Foundation CEDAR Grant AGS-1138784 and NASA LWS Strategic Capability grant NNX09AJ83G. Resources supporting this work were partially provided by the NASA High-End Computing (HEC) Program through the NASA Advanced Supercomputing (NAS) Division at Ames Research Center. The National Center for Atmospheric Research is sponsored by the National Science Foundation.

REFERENCES

Akmaev, R. A. (2011), Whole atmosphere modeling: Connecting terrestrial and space weather, *Rev. Geophys.*, *49*, RG4004, doi:10.1029/2011RG000364.

Akmaev, R. A., F. Wu, T. J. Fuller-Rowell, and H. Wang (2009), Midnight temperature maximum (MTM) in Whole Atmosphere Model (WAM) simulations, *Geophys. Res. Lett.*, *36*, L07108, doi:10.1029/2009GL037759.

Angelats i Coll, M., and J. M. Forbes (2002), Nonlinear interactions in the upper atmosphere: The $s = 1$ and $s = 3$ nonmigrating semidiurnal tides, *J. Geophys. Res.*, *107*(A8), 1157, doi:10.1029/2001JA900179.

Austin, J., et al. (2003), Uncertainties and assessments of chemistry-climate models of the stratosphere, *Atmos. Chem. Phys.*, *3*, 1–27.

Becker, E., and D. C. Fritts (2006), Enhanced gravity-wave activity and interhemispheric coupling during the MaCWAVE/MIDAS northern summer program 2002, *Ann. Geophys.*, *24*, 1175–1188.

Becker, E., A. Müllemann, F.-J. Lübken, H. Körnich, P. Hoffmann, and M. Rapp (2004), High Rossby-wave activity in austral winter 2002: Modulation of the general circulation of the MLT during the MaCWAVE/MIDAS northern summer program, *Geophys. Res. Lett.*, *31*, L24S03, doi:10.1029/2004GL019615.

Burrage, M. D., M. E. Hagan, W. R. Skinner, D. L. Wu, and P. B. Hays (1995), Long-term variability in the solar diurnal tide observed by HRDI and simulated by the GSWM, *Geophys. Res. Lett.*, *22*(19), 2641–2644.

Chandran, A., R. L. Collins, R. R. Garcia, and D. R. Marsh (2011), A case study of an elevated stratopause generated in the Whole Atmosphere Community Climate Model, *Geophys. Res. Lett.*, *38*, L08804, doi:10.1029/2010GL046566.

Chang, L. C., S. E. Palo, and H.-L. Liu (2009), Short-term variation of the $s = 1$ nonmigrating semidiurnal tide during the 2002 stratospheric sudden warming, *J. Geophys. Res.*, *114*, D03109, doi:10.1029/2008JD010886.

Chang, L. C., S. E. Palo, and H.-L. Liu (2011), Short-term variability in the migrating diurnal tide caused by interactions with the quasi 2 day wave, *J. Geophys. Res.*, *116*, D12112, doi:10.1029/2010JD014996.

Chau, J. L., B. G. Fejer, and L. P. Goncharenko (2009), Quiet variability of equatorial E × B drifts during a sudden stratospheric warming event, *Geophys. Res. Lett.*, *36*, L05101, doi:10.1029/2008GL036785.

Chau, J. L., L. P. Goncharenko, B. G. Fejer, and H.-L. Liu (2012), Equatorial and low latitude ionospheric effects during sudden stratospheric warming events, *Space Sci. Rev.*, *168*, 385–417, doi:10.1007/s11214-011-9797-5.

Chen, P.-R. (1992), Two-day oscillation of the equatorial ionization anomaly, *J. Geophys. Res.*, *97*(A5), 6343–6357.

Coy, L., D. E. Siskind, S. D. Eckermann, J. P. McCormack, D. R. Allen, and T. F. Hogan (2005), Modeling the August 2002 minor warming event, *Geophys. Res. Lett.*, *32*, L07808, doi:10.1029/2005GL022400.

de La Beaujardière, O., et al. (2009), C/NOFS observations of deep plasma depletions at dawn, *Geophys. Res. Lett.*, *36*, L00C06, doi:10.1029/2009GL038884.

de la Torre, L., R. R. Garcia, D. Barriopedro, and A. Chandran (2012), Climatology and characteristics of stratospheric sudden warmings in the Whole Atmosphere Community Climate Model, *J. Geophys. Res.*, *117*, D04110, doi:10.1029/2011JD016840.

Dickinson, R. E., E. C. Ridley, and R. G. Roble (1981), A three-dimensional general circulation model of the thermosphere, *J. Geophys. Res.*, *86*(A3), 1499–1512.

Fejer, B. G., M. E. Olson, J. L. Chau, C. Stolle, H. Lühr, L. P. Goncharenko, K. Yumoto, and T. Nagatsuma (2010), Lunar-dependent equatorial ionospheric electrodynamic effects during sudden stratospheric warmings, *J. Geophys. Res.*, *115*, A00G03, doi:10.1029/2010JA015273.

Fejer, B. G., B. D. Tracy, M. E. Olson, and J. L. Chau (2011), Enhanced lunar semidiurnal equatorial vertical plasma drifts during sudden stratospheric warmings, *Geophys. Res. Lett.*, *38*, L21104, doi:10.1029/2011GL049788.

Fesen, C. G., G. Crowley, R. G. Roble, A. D. Richmond, and B. G. Fejer (2000), Simulation of the pre-reversal enhancement in the low latitude vertical ion drifts, *Geophys. Res. Lett.*, *27*(13), 1851–1854.

Forbes, J. M. (1995), Tidal and planetary waves, in *The Upper Mesosphere and Lower Thermosphere: A Review of Experiment and Theory*, Geophys. Monogr. Ser., vol. 87, edited by R. M. Johnson and T. L. Killeen, pp. 67–87, AGU, Washington, D. C., doi:10.1029/GM087p0067.

Forbes, J. M. (2000), Wave coupling between the lower and upper atmosphere: Case study of an ultra-fast Kelvin wave, *J. Atmos. Sol. Terr. Phys.*, *62*, 1603–1621.

Forbes, J. M., S. Palo, and X. Zhang (2000), Variability of the ionosphere, *J. Atmos. Sol. Terr. Phys.*, *62*, 685–693.

Fritts, D. C., and M. J. Alexander (2003), Gravity wave dynamics and effects in the middle atmosphere, *Rev. Geophys.*, *41*(1), 1003, doi:10.1029/2001RG000106.

Froehlich, C. (2000), Observations of irradiance variations, *Space Sci. Rev.*, *94*, 15–24.

Fuller-Rowell, T., F. Wu, R. Akmaev, T.-W. Fang, and E. Araujo-Pradere (2010), A whole atmosphere model simulation of the impact of a sudden stratospheric warming on thermosphere dynamics and electrodynamics, *J. Geophys. Res.*, *115*, A00G08, doi:10.1029/2010JA015524.

Fuller-Rowell, T., R. Akmaev, F. Wu, M. Fedrizzi, R. A. Viereck, and H. Wang (2011a), Did the January 2009 sudden stratospheric warming cool or warm the thermosphere?, *Geophys. Res. Lett.*, *38*, L18104, doi:10.1029/2011GL048985.

Fuller-Rowell, T., H. Wang, R. Akmaev, F. Wu, T.-W. Fang, M. Iredell, and A. Richmond (2011b), Forecasting the dynamic and electrodynamic response to the January 2009 sudden stratospheric warming, *Geophys. Res. Lett.*, *38*, L13102, doi:10.1029/2011GL047732.

Garcia, R. R., D. R. Marsh, D. E. Kinnison, B. A. Boville, and F. Sassi (2007), Simulation of secular trends in the middle atmosphere, 1950–2003, *J. Geophys. Res.*, *112*, D09301, doi:10.1029/2006JD007485.

Gent, P. R., et al. (2011), The community climate system model version 4, *J. Clim.*, *24*, 4973–4991, doi:10.1175/2011JCLI4083.1.

Gentile, L. C., W. J. Burke, P. A. Roddy, J. M. Retterer, and R. T. Tsunoda (2011), Climatology of plasma density depletions observed by DMSP in the dawn sector, *J. Geophys. Res.*, *116*, A03321, doi:10.1029/2010JA016176.

Goncharenko, L., and S.-R. Zhang (2008), Ionospheric signatures of sudden stratospheric warming: Ion temperature at middle latitude, *Geophys. Res. Lett.*, *35*, L21103, doi:10.1029/2008GL035684.

Goncharenko, L. P., J. L. Chau, H.-L. Liu, and A. J. Coster (2010a), Unexpected connections between the stratosphere and ionosphere, *Geophys. Res. Lett.*, *37*, L10101, doi:10.1029/2010GL043125.

Goncharenko, L. P., A. J. Coster, J. L. Chau, and C. E. Valladares (2010b), Impact of sudden stratospheric warmings on equatorial ionization anomaly, *J. Geophys. Res.*, *115*, A00G07, doi:10.1029/2010JA015400.

Goncharenko, L. P., A. J. Coster, R. A. Plumb, and D. I. V. Domeisen (2012), The potential role of stratospheric ozone in the stratosphere-ionosphere coupling during stratospheric warmings, *Geophys. Res. Lett.*, *39*, L08101, doi:10.1029/2012GL051261.

Hagan, M. E., and J. M. Forbes (2002), Migrating and nonmigrating diurnal tides in the middle and upper atmosphere excited by tropospheric latent heat release, *J. Geophys. Res.*, *107*(D24), 4754, doi:10.1029/2001JD001236.

Hagan, M. E., and J. M. Forbes (2003), Migrating and nonmigrating semidiurnal tides in the upper atmosphere excited by tropospheric latent heat release, *J. Geophys. Res.*, *108*(A2), 1062, doi:10.1029/2002JA009466.

Hagan, M. E., and R. G. Roble (2001), Modeling diurnal tidal variability with the National Center for Atmospheric Research thermosphere-ionosphere-mesosphere-electrodynamics general circulation model, *J. Geophys. Res.*, *106*(A11), 24,869–24,882.

Hagan, M. E., M. D. Burrage, J. M. Forbes, J. Hackney, W. J. Randel, and X. Zhang (1999), QBO effects on the diurnal tide in the upper atmosphere, *Earth Planets Space*, *51*, 571–578.

Hagan, M. E., A. Maute, and R. G. Roble (2009), Tropospheric tidal effects on the middle and upper atmosphere, *J. Geophys. Res.*, *114*, A01302, doi:10.1029/2008JA013637.

Huba, J. D., S. L. Ossakow, G. Joyce, J. Krall, and S. L. England (2009), Three-dimensional equatorial spread F modeling: Zonal neutral wind effects, *Geophys. Res. Lett.*, *36*, L19106, doi:10.1029/2009GL040284.

Immel, T. J., E. Sagawa, S. L. England, S. B. Henderson, M. E. Hagan, S. B. Mende, H. U. Frey, C. M. Swenson, and L. J. Paxton (2006), Control of equatorial ionospheric morphology by atmospheric tides, *Geophys. Res. Lett.*, *33*, L15108, doi:10.1029/2006GL026161.

Ito, R., S. Kato, and T. Tsuda (1986), Consideration of an ionospheric wind dynamo driven by a planetary wave with a two-day period, *J. Atmos. Terr. Phys.*, *48*, 1–13.

Karlsson, B., C. McLandress, and T. G. Shepherd (2009), Inter-hemispheric mesospheric coupling in a comprehensive middle atmosphere model, *J. Atmos. Sol. Terr. Phys.*, *71*, 518–530.

Kinnison, D. E., et al. (2007), Sensitivity of chemical tracers to meteorological parameters in the MOZART-3 chemical transport model, *J. Geophys. Res.*, *112*, D20302, doi:10.1029/2006JD007879.

Körnich, H., and E. Becker (2010), A simple model for the inter-hemispheric coupling of the middle atmosphere circulation, *Adv. Space Res.*, *45*, 661–668.

Lean, J. L., and T. N. Woods (2010), Solar spectral irradiance: Measurements and models, in *Heliophysics: Evolving Solar Activity and the Climates of Space and Earth*, edited by C. J. Schrijver and G. L. Siscoe, pp. 269–298, Cambridge Univ. Press, Cambridge, U. K.

Lei, J., J. M. Forbes, H.-L. Liu, X. Dou, X. Xue, T. Li, and X. Luan (2011), Latitudinal variations of middle thermosphere: Observations and modeling, *J. Geophys. Res.*, *116*, A12306, doi:10.1029/2011JA017067.

Limpasuvan, V., M. J. Alexander, Y. J. Orsolini, D. L. Wu, M. Xue, J. H. Richter, and C. Yamashita (2011), Mesoscale simulations of gravity waves during the 2008–2009 major stratospheric sudden warming, *J. Geophys. Res.*, *116*, D17104, doi:10.1029/2010JD015190.

Liu, H., and M. E. Hagan (1998), Local heating/cooling of the mesosphere due to gravity wave and tidal coupling, *Geophys. Res. Lett.*, *25*(15), 2941–2944.

Liu, H., E. Doornbos, M. Yamamoto, and S. Tulasi Ram (2011), Strong thermospheric cooling during the 2009 major stratosphere warming, *Geophys. Res. Lett.*, *38*, L12102, doi:10.1029/2011GL047898.

Liu, H.-L., and A. D. Richmond (2013), Attribution of ionospheric vertical plasma drift perturbations to large-scale waves and the dependence on solar activity, *J. Geophys. Res.*, doi:10.1002/jgra.50265, in press.

Liu, H.-L., and R. G. Roble (2002), A study of a self-generated stratospheric sudden warming and its mesospheric–lower thermospheric impacts using the coupled TIME-GCM/CCM3, *J. Geophys. Res.*, *107*(D23), 4695, doi:10.1029/2001JD001533.

Liu, H.-L., E. R. Talaat, R. G. Roble, R. S. Lieberman, D. M. Riggin, and J.-H. Yee (2004), The 6.5-day wave and its seasonal variability in the middle and upper atmosphere, *J. Geophys. Res.*, *109*, D21112, doi:10.1029/2004JD004795.

Liu, H.-L., T. Li, C.-Y. She, J. Oberheide, Q. Wu, M. E. Hagan, J. Xu, R. G. Roble, M. G. Mlynczak, and J. M. Russell III (2007), Comparative study of short-term diurnal tidal variability, *J. Geophys. Res.*, *112*, D18108, doi:10.1029/2007JD008542.

Liu, H.-L., W. Wang, A. D. Richmond, and R. G. Roble (2010a), Ionospheric variability due to planetary waves and tides for solar minimum conditions, *J. Geophys. Res.*, *115*, A00G01, doi:10.1029/2009JA015188.

Liu, H.-L., et al. (2010b), Thermosphere extension of the Whole Atmosphere Community Climate Model, *J. Geophys. Res.*, *115*, A12302, doi:10.1029/2010JA015586.

Marsh, D. R., R. R. Garcia, D. E. Kinnison, B. A. Boville, F. Sassi, S. C. Solomon, and K. Matthes (2007), Modeling the whole atmosphere response to solar cycle changes in radiative and geomagnetic forcing, *J. Geophys. Res.*, *112*, D23306, doi:10.1029/2006JD008306.

Matthes, K., D. R. Marsh, R. R. Garcia, D. E. Kinnison, F. Sassi, and S. Walters (2010), Role of the QBO in modulating the influence of the 11 year solar cycle on the atmosphere using constant forcings, *J. Geophys. Res.*, *115*, D18110, doi:10.1029/2009JD013020.

McLandress, C. M., T. G. Shepherd, S. Polavarapu, and S. R. Beagley (2011), Is missing orographic gravity wave drag near 60°s the cause of the stratospheric zonal wind biases in chemistry-climate models?, *J. Atmos. Sci.*, *69*, 802–818.

Meyer, C. K., and J. M. Forbes (1997), A 6.5-day westward propagating planetary wave: Origin and characteristics, *J. Geophys. Res.*, *102*(D22), 26,173–26,178.

Millward, G. H., I. C. F. Müller-Wodarg, A. D. Aylward, T. J. Fuller-Rowell, A. D. Richmond, and R. J. Moffett (2001), An investigation into the influence of tidal forcing on F region equatorial vertical ion drift using a global ionosphere-thermosphere model with coupled electrodynamics, J. Geophys. Res., 106(A11), 24,733–24,744.

Neale, R. B., J. H. Richter, and M. Jochum (2008), The impact of convection on ENSO: From a delayed oscillator to a series of events, J. Clim., 21, 5904–5924.

Palo, S. E., R. G. Roble, and M. E. Hagan (1998), TIME-GCM results for the quasi-two-day wave, Geophys. Res. Lett., 25(20), 3783–3786, doi:10.1029/1998GL900032.

Pancheva, D. V., N. J. Mitchell, H. Middleton, and H. G. Muller (2003), Variability of the semidiurnal tide due to fluctuations in solar activity and total ozone, J. Atmos. Sol. Terr. Phys., 65, 1–19.

Park, J., H. Lühr, M. Kunze, B. G. Fejer, and K. W. Min (2012), Effect of sudden stratospheric warming on lunar tidal modulation of the equatorial electrojet, J. Geophys. Res., 117, A03306, doi:10.1029/2011JA017351.

Pedatella, N. M., and J. M. Forbes (2010), Evidence for stratosphere sudden warming-ionosphere coupling due to vertically propagating tides, Geophys. Res. Lett., 37, L11104, doi:10.1029/2010GL043560.

Pedatella, N. M., H.-L. Liu, and M. E. Hagan (2012a), Day-to-day migrating and nonmigrating tidal variability due to the six-day planetary wave, J. Geophys. Res., 117, A06301, doi:10.1029/2012JA017581.

Pedatella, N. M., H.-L. Liu, and A. D. Richmond (2012b), Atmospheric semidiurnal lunar tide climatology simulated by the Whole Atmosphere Community Climate Model, J. Geophys. Res., 117, A06327, doi:10.1029/2012JA017792.

Pedatella, N. M., H.-L. Liu, A. D. Richmond, A. Maute, and T.-W. Fang (2012c), Simulations of solar and lunar tidal variability in the mesosphere and lower thermosphere during sudden stratosphere warmings and their influence on the low-latitude ionosphere, J. Geophys. Res., 117, A08326, doi:10.1029/2012JA017858.

Picone, J. M., A. E. Hedin, D. P. Drob, and A. C. Aikin (2002), NRLMSISE-00 empirical model of the atmosphere: Statistical comparisons and scientific issues, J. Geophys. Res., 107(A12), 1468, doi:10.1029/2002JA009430.

Plumb, R. A. (1983), Baroclinic instability of the summer mesosphere: A mechanism for the quasi-two-day wave?, J. Atmos. Sci., 40, 262–270.

Ren, S., S. M. Polavarapu, and T. G. Shepherd (2008), Vertical propagation of information in a middle atmosphere data assimilation system by gravity-wave drag feedbacks, Geophys. Res. Lett., 35, L06804, doi:10.1029/2007GL032699.

Richter, J. H., and P. J. Rasch (2008), Effects of convective momentum transport on the atmospheric circulation in the Community Atmosphere Model, version 3, J. Clim., 21, 1487–1499.

Richter, J. H., F. Sassi, and R. R. Garcia (2010), Toward a physically based gravity wave source parameterization in a general circulation model, J. Atmos. Sci., 67, 136–156, doi:10.1175/2009JAS3112.1.

Rishbeth, H. (2006), F-region links with the lower atmosphere?, J. Atmos. Sol. Terr. Phys., 68, 469–478.

Rishbeth, H., and M. Mendillo (2001), Patterns of ionospheric variability, J. Atmos. Sol. Terr. Phys., 63, 1661–1680.

Roble, R. G., R. E. Dickinson, and E. C. Ridley (1982), Global circulation and temperature structure of thermosphere with high-latitude plasma convection, J. Geophys. Res., 87(A3), 1599–1614.

Sagawa, E., T. J. Immel, H. U. Frey, and S. B. Mende (2005), Longitudinal structure of the equatorial anomaly in the nighttime ionosphere observed by IMAGE/FUV, J. Geophys. Res., 110, A11302, doi:10.1029/2004JA010848.

Singh, S., F. S. Johnson, and R. A. Power (1997), Gravity wave seeding of equatorial plasma bubbles, J. Geophys. Res., 102(A4), 7399–7410.

Solomon, S. C., and L. Qian (2005), Solar extreme-ultraviolet irradiance for general circulation models, J. Geophys. Res., 110, A10306, doi:10.1029/2005JA011160.

Sridharan, S., S. Sathishkumar, and S. Gurubaran (2012), Variabilities of mesospheric tides during sudden stratospheric warming events of 2006 and 2009 and their relationship with ozone and water vapour, J. Atmos. Sol. Terr. Phys., 78–79, 108–115, doi:10.1016/j.jastp.2011.03.013.

Stening, R. J., J. M. Forbes, M. E. Hagan, and A. D. Richmond (1997), Experiments with a lunar atmospheric tidal model, J. Geophys. Res., 102(D12), 13,465–13,471.

Tan, B., X. Chu, H.-L. Liu, C. Yamashita, and J. M. Russell III (2012), Zonal-mean global teleconnection from 15 to 110 km derived from SABER and WACCM, J. Geophys. Res., 117, D10106, doi:10.1029/2011JD016750.

Teitelbaum, H., and F. Vial (1991), On tidal variability induced by nonlinear interaction with planetary waves, J. Geophys. Res., 96(A8), 14,169–14,178, doi:10.1029/91JA01019.

Thurairajah, B., R. L. Collins, V. L. Harvey, R. S. Lieberman, and K. Mizutani (2010), Rayleigh lidar observations of reduced gravity wave activity during the formation of an elevated stratopause in 2004 at Chatanika, Alaska (65°N, 147°W), J. Geophys. Res., 115, D13109, doi:10.1029/2009JD013036.

Wang, L., and M. J. Alexander (2009), Gravity wave activity during stratospheric sudden warmings in the 2007–2008 Northern Hemisphere winter, J. Geophys. Res., 114, D18108, doi:10.1029/2009JD011867.

Woods, T. N., and G. J. Rottman (2002), Solar ultraviolet variability over time periods of aeronomic interest, in Atmospheres in the Solar System: Comparative Aeronomy, Geophys. Monogr. Ser., vol. 130, edited by M. Mendillo, A. Nagy, and J. H. White, pp. 221–233, AGU, Washington, D. C., doi:10.1029/130GM14.

Wu, Q., D. A. Ortland, S. C. Solomon, W. R. Skinner, and R. J. Niciejewski (2011), Global distribution, seasonal, and inter-annual variations of mesospheric semidiurnal tide observed by TIMED TIDI, J. Atmos. Sol. Terr. Phys., 73, 2482–2502, doi:10.1016/j.jastp.2011.08.007.

Xu, J., A. K. Smith, H.-L. Liu, W. Yuan, Q. Wu, G. Jiang, M. G. Mlynczak, J. M. Russell III, and S. J. Franke (2009), Seasonal and quasi-biennial variations in the migrating diurnal tide observed by Thermosphere, Ionosphere, Mesosphere, Energetics and Dynamics (TIMED), *J. Geophys. Res.*, *114*, D13107, doi:10.1029/2008JD011298.

Xue, X.-H., H.-L. Liu, and X.-K. Dou (2012), Parameterization of the inertial gravity waves and generation of the quasi-biennial oscillation, *J. Geophys. Res.*, *117*, D06103, doi:10.1029/2011JD016778.

Yamashita, C., H.-L. Liu, and X. Chu (2010a), Gravity wave variations during the 2009 stratospheric sudden warming as revealed by ECMWF-T799 and observations, *Geophys. Res. Lett.*, *37*, L22806, doi:10.1029/2010GL045437.

Yamashita, C., H.-L. Liu, and X. Chu (2010b), Responses of mesosphere and lower thermosphere temperatures to gravity wave forcing during stratospheric sudden warming, *Geophys. Res. Lett.*, *37*, L09803, doi:10.1029/2009GL042351.

Yamazaki, Y., A. D. Richmond, and K. Yumoto (2012), Stratospheric warmings and the geomagnetic lunar tide: 1958–2007, *J. Geophys. Res.*, *117*, A04301, doi:10.1029/2012JA017514.

Yuan, T., B. Thurairajah, C.-Y. She, A. Chandran, R. L. Collins, and D. A. Krueger (2012), Wind and temperature response of midlatitude mesopause region to the 2009 Sudden Stratospheric Warming, *J. Geophys. Res.*, *117*, D09114, doi:10.1029/2011JD017142.

Yue, J., H.-L. Liu, and L. C. Chang (2012), Numerical investigation of the quasi 2 day wave in the mesosphere and lower thermosphere, *J. Geophys. Res.*, *117*, D05111, doi:10.1029/2011JD016574.

H.-L. Liu, High Altitude Observatory, National Center for Atmospheric Research, P.O. Box 3000, Boulder, CO 80307-3000, USA. (liuh@ucar.edu)

Inductive-Dynamic Coupling of the Ionosphere With the Thermosphere and the Magnetosphere

P. Song

Center for Atmospheric Research and Department of Physics, University of Massachusetts Lowell, Massachusetts, USA

V. M. Vasyliūnas

Center for Atmospheric Research and Department of Physics, University of Massachusetts Lowell, Massachusetts, USA

Max-Planck-Institut für Sonnensystemforschung, Katlenburg-Lindau, Germany

The central issue of magnetosphere-ionosphere-thermosphere coupling is the long-range coupling of the system, from the solar wind and magnetosphere to the ionosphere and thermosphere; coupling between the ionosphere and thermosphere is known and appropriately modeled by local coupling via collisions. The long-range coupling is mostly through the electromagnetic force in addition to direct flow. In steady state, the electromagnetic force may be represented by "mapping" of the electric potential along the magnetic field and field-aligned currents with Ohm's law accommodating the local coupling in the ionosphere. To correctly describe the long-range coupling on time scales from longer than few seconds to less than 30 min, the inductive effect (Faraday's law) as well as the dynamic effect (acceleration terms) need to be considered. When the inductive and dynamic effects are included, the physical description of the long-range coupling is drastically changed: the electric potential no longer suffices to describe the electric field, and the electric currents, including Birkeland (magnetic field-aligned) currents, become a secondary derived quantity. All these are replaced by electromagnetic perturbations with associated fluid perturbations, propagating and reflecting between the magnetosphere and ionosphere and from one hemisphere to the other. The perturbations in one part of the ionosphere can propagate into other ionospheric regions, relatively rapidly, which, in turn, couple back to other regions of the magnetosphere. In this article, we discuss the key physical and mathematical differences and compare the physical quantities according to each of the coupling schemes.

1. INTRODUCTION

In a partially ionized gas, the electrons, the ions, and the neutrals each move under the action of different forces. Electromagnetic forces act only on the charged particles but not on the neutrals. The charged and the neutral particles interact through collisions, which transfer energy and momentum from one species to another while converting energy from one type to another. Electrons and ions can be treated together as plasma. Depending on time scales, the plasma and the neutrals may be treated as moving together when the collision times are much shorter than the time scale of interest or as moving separately when the collision times are much longer than the time scale of interest. On intermediate time scales, one species may accelerate or drag the other, while the whole fluid is heated by frictional heating due to collisions among them. Note that collisions are a local process, whereas long-range coupling, which connects the processes and variations in one region to those in another (say, from the magnetosphere to each altitude in the ionosphere) is by electromagnetic forces and by direct flow (if it exists) between the two regions. Magnetosphere-ionosphere/thermosphere (M-IT) coupling is an example of such long-range coupling processes, connecting the highly ionized nearly collisionless magnetospheric plasma to the weakly ionized highly collisional gases of the ionosphere and thermosphere with local coupling (at a given height) between the ionosphere and thermosphere. In this respect, a particular challenge for M-IT coupling theories arises from the rapid increase in collision frequency as one goes from the magnetosphere to the ionosphere/thermosphere, with the result that, for any given perturbation, the dominant process changes from collisionless to strongly collisional over a very short distance, the location and spatial extent of this transition depending on the time scale of the perturbation.

Conventional M-IT coupling models are all based on the classical magnetosphere-ionosphere (M-I) coupling theory, developed by *Vasyliūnas* [1970] and *Wolf* [1970] from earlier models of *Fejer* [1964] and others, which is discussed in section 2.1. Most of the models may be assigned to one of three general categories: an electrostatic ionosphere, a height-integrated ionosphere, and a structured ionosphere with wave propagation. Since the focus of this *Monograph* is the ionosphere/thermosphere (IT) subsystem, we discuss the first category in more detail below.

Most of the models developed in the ionospheric community focus on the region up to no more than a few thousand kilometers above the Earth. Wave propagation or communication time over this relatively small distance is sufficiently short so that one can ignore wave propagation effects, while the full dynamics of the neutral wind can be retained. The most important assumption invoked in these models is that the time-variability of the magnetic field is negligible, hence the inductive term in Faraday's law may be dropped and the electric field assumed curl-free. As a result, the models become electrostatic, which leads to a significant simplification mathematically because the electric potential and the Birkeland current will now suffice to calculate the long-range coupling. The mathematical convenience of using electric fields and currents is elevated to a physical description ("E-J approach") in many discussions. In addition, the time derivative (or dynamic) term in the plasma momentum equation is neglected in most models. This formalism may be adequate to describe the IT system alone (as long as inputs from or coupling to the magnetosphere play no role), although even there, as argued by *Vasyliūnas* [2012], its equations are only conditions for quasisteady state equilibrium and may not provide insight into temporal and causal sequences. For global M-IT coupling, its validity is even more limited: an often overlooked key point is that this scheme, restricted to describing M-IT coupling in quasisteady state equilibrium, cannot be applied to explain transient processes, such as substorms and auroral brightenings, during which the magnetic perturbations and plasma acceleration are not negligibly small.

A key parameter is the time scale that separates the dynamic and the quasisteady state, or equivalently the transient time, required for the coupled M-IT system to change from one quasisteady state to another. Theoretical estimates of this time scale (discussed in section 2.4) depend on the spatial scale of the phenomenon of interest, ranging from a few seconds for purely ionospheric events to 15–30 min for changes on scales of magnetospheric dimensions. The latter is the approximate time scale for many observed important M-IT processes, such as substorms and magnetospheric transients [e.g., *Earle and Kelley*, 1987; *Russell and Ginskey*, 1993, 1995; *Zesta et al.*, 2000; *Bristow et al.*, 2003; *Huang et al.*, 2008, 2010]; the focus of our paper is on time scales from a few minutes to 20–30 min. It is, thus, not surprising that, as pointed out by *Schunk* [2013], conventional ionospheric quasisteady state equilibrium models, e.g., thermosphere ionosphere electrodynamics global circulation model [*Richmond et al.*, 1992], coupled thermosphere ionosphere plasmasphere electrodynamics model [*Fuller-Rowell et al.*, 1996], and the global ionosphere-thermosphere model (GITM) [*Ridley et al.*, 2006], describe well the large-scale slow variations or "climatology" of the ionosphere-thermosphere system but do not work well for the rapidly changing phenomena or "weather." (The SAMI 2 ionospheric model [*Huba et al.*, 2000] does include the time derivatives in the field-aligned component of the momentum equation but does not include the inductive field, so the long-range coupling in the model is still electrostatic.)

On the other hand, many of the M-I coupling models developed in the magnetospheric community, including almost all the global MHD simulation models, retain the inductive term and plasma dynamics in the magnetosphere (see more discussion in the work of *Song et al.* [2009]). They are coupled to the ionosphere through the height-integrated quasisteady state equilibrium Ohm's law, with the ionosphere treated as a height-integrated layer at the lower boundary.

There is an intrinsic inconsistency here: a full inductive and dynamic magnetosphere is being combined with an electrostatic ionosphere. Many of the (magnetospheric) wave models take the same approach, including wave propagation within the magnetosphere and reflection between the magnetosphere and the ionosphere but not within the ionosphere because the "height-integrated" ionosphere has been idealized as having zero extent in altitude. Also, most of those models do not include neutral wind dynamics, except for corotation of the atmosphere as a whole.

As exceptions, there are a few models that include inductive effects as well as plasma dynamics with a structured ionosphere. Taking advantage of powerful Fourier analysis techniques, these models treat time-dependent effects as waves [e.g., *Hughes*, 1974; *Hughes and Southwood*, 1976] and include propagation and reflection effects within the ionosphere. This approach, in general, is able to describe the dynamics of M-I coupling. (Recall that a step function, a mathematical representation of a changing condition, can be decomposed into a wave spectrum, and each frequency can propagate and reflect, but converting the perturbation as a sum of individually propagated and reflected waves from the frequency domain with phase information back to the time domain is by no means a trivial task.)

In contrast to the "E-J" approach mentioned above, the "B-V" approach in which the magnetic field and the plasma bulk flow are treated as the primary quantities has always been a basic principle of MHD [e.g., *Cowling*, 1957; *Dungey*, 1958], and its importance for the magnetosphere and the ionosphere has been urged particularly by *Parker* [1996, 2000, 2007]. A new impetus was provided when *Buneman* [1992] (in a laboratory-plasma paper unknown to the M-IT community) and, independently, *Vasyliūnas* [2001] showed mathematically that in a dynamic process, for a plasma sufficiently dense so that $V_A^2 \ll c^2$, the electric field by itself cannot produce plasma bulk flow, but plasma flow does produce an electric field self-consistently. *Tu et al.* [2008] later confirmed this argument with particle simulations. *Vasyliūnas* [2005a, 2005b, 2011, 2012] investigated the basics of the two approaches extensively and showed that, contrary to widespread assumption, the "B-V" approach is not derived exclusively (or even primarily) from MHD but is, in essence, a consequence of charge quasineutrality. *Song et al.* [2005] and *Vasyliūnas and Song* [2005] derived a general set of equations governing long-range coupling for partially ionized collisional flows with electromagnetic fields, solved simple situations for 1-D steady state, and derived the dispersion relations for wave-type solutions. *Song et al.* [2009] further solved the 1-D time-dependent problem for M-IT coupling and found that, for M-IT coupling, plasma flow and magnetic tension force (rather than electric field and current) are indeed the primary coupling mechanisms. More recently, the new approach and formalism has been applied to study heating processes in the solar chromosphere [*Song and Vasyliūnas*, 2011] and in the terrestrial IT system [*Tu et al.*, 2011].

2. INDUCTIVE-DYNAMIC COUPLING VERSUS ELECTROSTATIC COUPLING

2.1. Differences in Governing Equations

To illustrate the differences between the conventional steady state coupling and the inductive-dynamic coupling, we first examine how each is formulated mathematically. The system consists of plasma, neutral medium, and electromagnetic fields. For simplicity of the comparison, we assume the plasma and neutral medium each consists of multiple species and is represented by its averaged mass and bulk velocity, as well as temperature. The momentum equations for electrons, ions, and neutrals can be combined and rewritten [e.g., *Song et al.*, 2005, 2009; *Vasyliūnas and Song*, 2005; *Vasyliūnas*, 2012] as the generalized Ohm's law, the plasma momentum equation and the neutral momentum equation. Under the approximations listed below, their leading terms are:

$$0 = N_e e(\mathbf{E} + \mathbf{V} \times \mathbf{B}) - \mathbf{J} \times \mathbf{B} - (m_e/e)\nu_e \mathbf{J}, \quad (1)$$

$$\rho \frac{d\mathbf{V}}{dt} = \mathbf{J} \times \mathbf{B} - \rho_i \nu_{in}(\mathbf{V} - \mathbf{u}_n) - \nabla p + \rho \mathbf{g}, \quad (2)$$

and

$$\rho_n \frac{d\mathbf{u}_n}{dt} = \rho_n \nu_{ni}(\mathbf{V} - \mathbf{u}_n) - \nabla p_n + \rho_n \mathbf{g}, \quad (3)$$

where e, ρ_η, p_η, $\mathbf{J}$, $\mathbf{V}$, $\mathbf{u}_n$, $\mathbf{E}$, $\mathbf{B}$, $\mathbf{g}$, N_e, and $\nu_{\eta\xi}$, are the elementary electric charge, the mass density and pressure of species η, the electric current, the bulk velocities of the plasma and neutrals, the electric and the magnetic fields, gravitational acceleration, electron concentration (number density), and the collision frequency between particles of species η and ξ, respectively; $\nu_e \equiv \nu_{ei} + \nu_{en}$. Subscripts i, e, and n denote ions, electrons, and neutrals; densities and pressures without subscripts refer to the plasma (electrons plus ions). Time derivatives in equations (2) and (3) are convective derivatives in the flow of the respective medium. Note that these equations hold locally. The set of equations is completed by the addition of continuity and energy equations and by Faraday's and Ampère's laws

$$\frac{\partial \mathbf{B}}{\partial t} = -\nabla \times \mathbf{E} \quad (4)$$

$$\mathbf{J} = \nabla \times \mathbf{B}/\mu_0, \quad (5)$$

which provide long range forces for global coupling. The principal approximations invoked in deriving equations (1)–(5) are the following:

(a) Neglect of $\partial \mathbf{J}/\partial t$ term and of $\partial \mathbf{E}/\partial t$ (displacement current) term on the left-hand sides of equations (1) and (5), respectively; this applies on time scales that are long compared to inverse plasma frequency and length scales that are large compared to electron inertial length and ensures the validity of charge quasineutrality (for a detailed discussion, see, e.g., the work of *Vasyliūnas* [2005a], and references therein).

(b) Neglect of electron inertia and pressure terms in equation (1) (but **not** necessarily in equation (2)), valid on length scales that are large compared to electron gyroradius.

(c) Neglect of all nonisotropic components of the pressure tensor (including viscous stresses) in equations (2) and (3).

Approximation (a) is the essence of the "B-V" approach; its consequence is that $\mathbf{E}$ and $\mathbf{J}$ must be calculated at a fixed time from equations (1) and (5), respectively, whereas $\mathbf{B}$ and $\mathbf{V}$ evolve in time according to the prognostic equations (4) and (2). As noted above, it applies on sufficiently large time and length scales, which excludes some small-scale ionospheric phenomena (e.g., the Farley-Buneman instability), but includes most aspects of M-IT coupling. Approximations (b) and (c) are made principally for simplicity. Furthermore, the electron-collision (resistive) term in equation (1) is significant only at the lowest altitudes (below approximately 100 km) and can be neglected in the magnetosphere and for many purposes in the ionosphere as well.

Equations (1)–(5) are the general governing equations for M-IT coupling. To be added are continuity and energy equations for each species, to determine densities and pressures, but their discussion is beyond the scope of this paper, in which the key terms of interest are the left-hand-side term in equations (2) and (4), representing the dynamic and the inductive aspects, respectively.

Intuitively, the inductive term in equation (4) should be negligible if the time-varying magnetic field either is sufficiently small (the argument is sometimes made that, in the ionosphere, the perturbed magnetic field is much smaller than the background field) or is varying on a sufficiently long time scale. However, equation (4) (the only equation in the entire set that is exact under all conditions and contains no approximation whatsoever) has only two terms, one on each side; they must, therefore, always be equal, and neither can be small compared to the other. To derive a criterion for neglecting the inductive effect, it is necessary to estimate the electric field from some other equation and compare it to the non-curl-free part of the electric field (which can be calculated by standard techniques, most simply as the induced field in the Coulomb gauge) from equation (4). *Vasyliūnas* [2012] used the ionospheric Ohm's law together with Ampère's law to show that $\nabla \times \mathbf{E}$ can be considered zero if the magnetic field is changing on a time scale that is long compared to

$$\tau = \mu_0 \Sigma_p L, \quad (6)$$

where Σ_p is the Pedersen conductance (height-integrated conductivity), and L is the length scale of the curl. With $\Sigma_p = 10$ mho, $\tau \approx$ s for $L = 200$ km (local ionospheric scale), and $\tau \approx 15$ min for $L = 10\,R_E$ (global magnetospheric scale). The time scale defined by equation (6) has been discussed in several M-I coupling contexts [*Coroniti and Kennel*, 1973; *Holzer and Reid*, 1975; *Vasyliūnas and Pontius*, 2007].

When the inductive effect is neglected, the electric field can be derived from an electrostatic potential. With further neglect of acceleration terms, the governing equations become

$$0 = \mathbf{J} \times \mathbf{B} - \rho v_{in}(\mathbf{V} - \mathbf{u}_n) - \nabla p + \rho \mathbf{g}, \quad (7)$$

$$\mathbf{E} = -\nabla \Phi \quad (8)$$

and

$$\nabla \cdot \mathbf{J} = 0 \quad (9)$$

where Φ is the electrostatic potential; if the resistive term in equation (1) is neglected, magnetic field lines are equipotentials. (Some coupling models introduce additional field-aligned potential drops, mainly in relation to physics of the aurora; these do not change fundamentally the nature of the coupling as long as the potential remains electrostatic.) Equations (8) and (9) follow from equations (4) and (5), in this approximation.

The conventional ionospheric Ohm's law can be derived by combining equations (1) and (7) to eliminate $\mathbf{V}$ and also neglecting the pressure gradient and gravitational forces, to obtain

$$\mathbf{J} = \sigma_p (\mathbf{E}_\perp + \mathbf{u}_n \times \mathbf{B}) + \sigma_H (\mathbf{E} + \mathbf{u}_n \times \mathbf{B}) \times \mathbf{B}/B + \sigma_\parallel \mathbf{E}_\parallel, \quad (10)$$

where $\sigma_p = \dfrac{\rho_i v_{in}}{B^2}\left(1 + \dfrac{v_{in}^2}{\Omega_i^2}\right)^{-1}$ is the Pedersen conductivity, $\sigma_H = -\dfrac{\rho_i \Omega_i}{B^2}\left(1 + \dfrac{v_{in}^2}{\Omega_i^2}\right)^{-1}$ is the Hall conductivity, and $\sigma_\parallel = m_e v_e/N_e e^2$ is the parallel conductivity; see more detailed discussion by *Song et al.* [2001]. Equation (10) superficially looks like (and is often described as) an Ohm's law in the conventional sense of linear relation between current and electric field. Physically, however, **J** given by equation (10) is not an ohmic current but a stress-balance current, a distinction emphasized by *Vasyliūnas and Song* [2005] and *Vasyliūnas* [2011, 2012]. The key assumption on which equation (10) depends is neglect in the plasma momentum equation of all forces except **J** × **B** and collisions with neutrals as well as of all acceleration terms; it does not depend on the electrostatic approximation. When pressure gradient and gravitational forces or dynamic terms are included, additional currents appear, in general, although some of the effects can be accommodated by redefining the conductivities as functions of frequency [*Hughes*, 1974] or even [*Strangeway*, 2012] by introducing a new quantity to replace the electric field. Such modifications preserve the mathematical form of equation (9) but not the physical meaning.

Equations (7)–(10) are the governing equations for electrostatic coupling. If the resistive term in equation (1) is assumed negligible, the Birkeland current cannot be calculated from equation (10) but must be determined from current continuity, equation (9), which relates perpendicular to field-aligned currents:

$$\frac{\partial J_\parallel}{\partial_\parallel} = -\nabla_\perp \cdot \mathbf{J}_\perp = -\nabla_\perp \cdot [\sigma_p(\mathbf{E} + \mathbf{u}_n \times \mathbf{B}) + \sigma_H(\mathbf{E} + \mathbf{u}_n \times \mathbf{B}) \times \mathbf{B}/B]. \quad (11)$$

The electric field between the highly collisional and the collisionless regions is connected by equation (8) along the magnetic field lines, an assumption that is not valid when the inductive effect is significant. Equations (8) and (11) now replace equations (4) and (5) to provide long-range coupling for quasisteady state electrostatic M-I coupling models. Some models introduce additional empirical or semiempirical terms to describe the effects of precipitation particles and Birkeland currents on the conductivities, but these modifications do not change the electrostatic nature of the coupling.

Similarly to the derivation of equation (10), equations (1) and (7) can also be combined to eliminate **J** instead of **V** to obtain

$$0 = e(\mathbf{E} + \mathbf{V} \times \mathbf{B}) - m_i \nu_{in}(\mathbf{V} - \mathbf{u}_n), \quad (12)$$

an equation that is very useful in relating plasma flow and electric field. It has, however, caused considerable confusion: contrary to a common interpretation, the electric field does not produce motion of the plasma but is the result of the motion, linking the electron and ion flows to preserve charge quasineutrality. The description that the electric field produces plasma motion is fundamentally incorrect [*Buneman*, 1992; *Vasyliūnas*, 2001, 2011, 2012; *Tu et al.*, 2008; *Song et al.*, 2009], although it is convenient sometimes and has been used for a long time, as argued by *Kelley* [2009]. When an (external) electric field is imposed onto a plasma, an internal electric field is produced by charge separation, as described in the first chapter of any plasma physics textbook. If the plasma is sufficiently dense, the internal electric field will largely cancel out the imposed one, leaving nearly no electric field inside the plasma.

2.2. Differences in Physical Description of the Coupling Processes

It may be instructive to first consider the coupling process qualitatively in more detail. The time-varying interaction between the solar wind, the magnetosphere, and the ionosphere proceeds primarily by continual creation and relaxation of imbalances between the various force terms in the momentum equation (2). Force imbalance produces plasma motions, which affect all the other quantities, creating a perturbation that can be described to first order as a set of waves, which propagate (and possibly reflect at boundaries, depending on boundary conditions), thereby relaxing the imbalance. The essential difference between electrostatic and dynamic-inductive coupling is that the former neglects all wave effects, whereas the latter explicitly includes them.

When a model, such as SAMI 2 [*Huba et al.*, 2000], includes the dynamic term only but neglects the inductive term, it will contain sonic and gravity waves, produced by force imbalances associated with pressure gradient and gravity terms in equation (2). The sound waves are isotropic, and the gravity waves act primarily in the vertical direction. The two wave modes can be coupled, particularly when the magnetic field is inclined. The wave amplitude increases with altitude unless strong damping is present.

When inductive and dynamic terms are both included in a model, imbalances associated with all the forces, which now include the electromagnetic force, produce MHD waves. With gravity neglected, in a uniform MHD fluid, this gives rise to three wave modes: slow, intermediate or Alfvén, and fast modes. (Note: the term "Alfvén wave," used here for the intermediate mode, is sometimes applied to MHD waves in general.) Different modes not only have different propagation speeds but also different perturbation relations, which can be used to identify the wave mode [*Song et al.*, 1994], as well as to associate particular modes with specific types of changes. The fast mode is relatively more isotropic, i.e., it can easily mitigate the force imbalances both parallel and perpendicular to the magnetic field. The slow and Alfvén modes are more anisotropic and more effective in the direction along the magnetic field. Obliquely propagating Alfvén waves are the only ones (at least in a homogeneous medium) that can carry field-aligned (Birkeland) currents. Between the magnetosphere and the ionosphere, where the plasma beta (ratio of thermal pressure to magnetic pressure) is small, the fast and Alfvén modes are the most effective, and the Alfvén mode is the dominant mechanism in M-I coupling where Birkeland currents play an essential role. For further discussion of the functions of various modes, see the work of *Song and Vasyliūnas* [2010]. For example, the fast mode, since it is the only mode (at least in a homogeneous medium) that propagates in the direction perpendicular to the magnetic field, can directly communicate changes from the subsolar magnetopause to the low-latitude ionosphere. In the high-altitude ionosphere, the fast mode is most efficient in mitigating horizontal force imbalances. Since the fast mode speed is here nearly the same as the Alfvén speed, ~10^3 to 10^4 km s^{-1}, magnetospheric changes are felt nearly instantaneously throughout the whole polar cap as well as into the closed field line regions, for example. The ionosphere can then, in turn, influence other magnetospheric regions where direct intermagnetospheric propagation, especially perpendicular to the magnetic field, might take longer (an example is discussed at the end of section 2.2).

As a simple specific illustration, consider the canonical problem of M-I coupling: plasma flow imposed in the outer magnetosphere (or at the magnetopause). In electrostatic coupling, this is described as an imposed electric field, with the plasma flow calculated from it (E-J description); the electric potential in the magnetosphere (constant along a magnetic field line) is directly mapped into the ionosphere (at every height). As noted in sections 1 and 2.1, this description is not valid in dynamic processes because the inductive effect can no longer be neglected.

In dynamic-inductive coupling, if (according to the B-V description) the electric field is not the primary agent to couple the magnetosphere with the ionosphere/thermosphere, how is coupling achieved? i.e., when the magnetospheric motion changes, how does the ionosphere know about it, and why does the ionosphere/thermosphere move accordingly? (particularly nonintuitive if the magnetospheric motion is entirely parallel to the ionosphere and perpendicular to the magnetic field). Figure 1 (adapted from the discussion by *Song and Vasyliūnas* [2011] of an analogous problem in the solar atmosphere) schematically illustrates the process in a simple 1-D stratified ionosphere/thermosphere with a vertical magnetic field. The plasma and neutrals are initially ($t = 0-$) at rest (or at a common background velocity). A (horizontal) plasma motion (not an electric field!) is imposed on the top boundary at $t = 0+$. With the plasma inside the domain linked by the frozen-in magnetic field, the plasma motion creates a kink on the magnetic field line at the top boundary, which exerts a tension force on the plasma below the boundary and makes it move, creating another kink at the lower altitude, and so on. The net result is a magnetic perturbation front propagating downward along the magnetic field line at the Alfvén speed V_A. Since the neutrals are not subject to the electromagnetic force, a velocity difference between plasma and neutrals is produced as the perturbation front passes. The relative motion causes momentum transfer through ion-neutral collisions, accelerating the neutrals. This is the primary mechanism of momentum coupling from the magnetosphere to the ionosphere. The collisions generate frictional heating of plasma and neutrals and also tend to slow down the plasma motion,

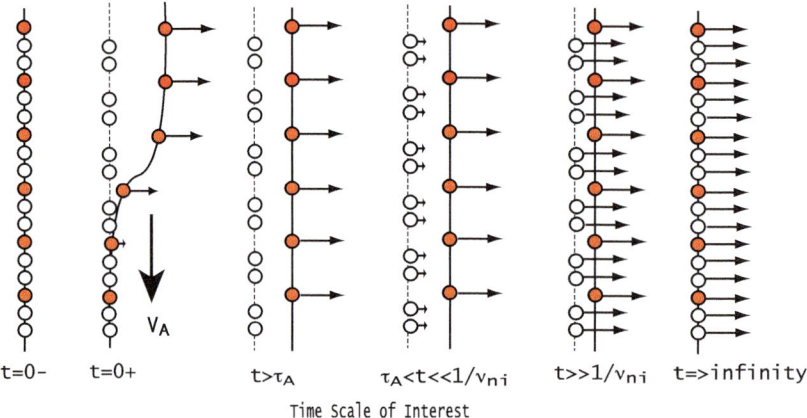

Figure 1. Schematic illustration of the plasma and neutral motion in a partially ionized and magnetized fluid. Closed red dots indicate ions and, open circles, neutrals. The imposed plasma motion at the top boundary produces a magnetic field perturbation that propagates downward, which in turn drives the motion of the plasma further below. Adapted and modified from *Song and Vasyliūnas* [2011].

which, however, continues as long as the motion at the top boundary is sustained. On time scales shorter than $1/\nu_{in}$, where ν_{in} is the ion-neutral collision frequency, the plasma is little affected by the neutrals, and on time scales longer than $t \sim 1/\nu_{ni}$, where ν_{ni} is the neutral-ion collision frequency, the neutrals are catching up with the plasma. In the absence of other forces acting on the neutrals, the system will eventually ($t \to \infty$) reach a state in which the plasma and neutrals have a common velocity equal to the driving velocity. The motion of the plasma creates an electric field, and the distortion of the magnetic field corresponds to a current.

This sequence, depicted in Figure 1, goes from an initial state of rest, through a transient phase of plasma-neutral flow difference, to a final state of equal plasma-neutral flow, all a consequence of an Alfvén wave launched by imposing a small velocity step at the boundary. A small step function can be decomposed into a spectrum of waves (as noted in section 1), and any time profile of perturbations at the upper boundary, including oscillations, can be built up as a series of such small steps. If the perturbations continue without end but always vary more slowly than the neutral-plasma collision time, the system remains in a slowly varying quasisteady state, with a small but nonzero velocity difference between plasma and neutrals and consequent heating by collisions.

The above description is, of course, highly simplified. First, since the perturbation at the top boundary is velocity and not displacement, the field line in the second panel of Figure 1 bends more gradually than is shown in the figure. Second, if the magnetic field is inclined at an angle, there will be a component propagating horizontally, which produces compressible modes. Third, with strictly vertical field and horizontal velocity perturbations, for field-aligned propagation vector, the Alfvén mode and the (compressional) fast mode are indistinguishable. Fourth, horizontal nonuniformity, either in the medium or in the driver, may also generate compressible modes. Velocity shear and field-aligned flow will create additional effects.

In inductive-dynamic coupling, the ionosphere is not merely a passive recipient of driving forces from the magnetosphere but can also impose motions on the magnetosphere, by the process described above but proceeding in the reversed direction in a similar manner to that in the solar chromosphere [*Song and Vasyliūnas*, 2011]. Because the horizontal distances in the ionosphere are much smaller than spatial scales of the magnetosphere, and communication across them is at the fast-mode speed, which in the upper parts of and above the ionosphere is faster than the Alfvén speed in the magnetosphere, it is possible for an imposed magnetospheric change, after propagating to the ionosphere along the local field lines, to propagate quickly across field lines to different regions of the ionosphere, setting the ionosphere there into motion accordingly. This ionospheric motion then propagates back to the magnetosphere at the Alfvén speed and modifies the plasma flow in other magnetospheric regions that have not yet been affected by direct signals from the initially imposed change, as shown in Figure 2.

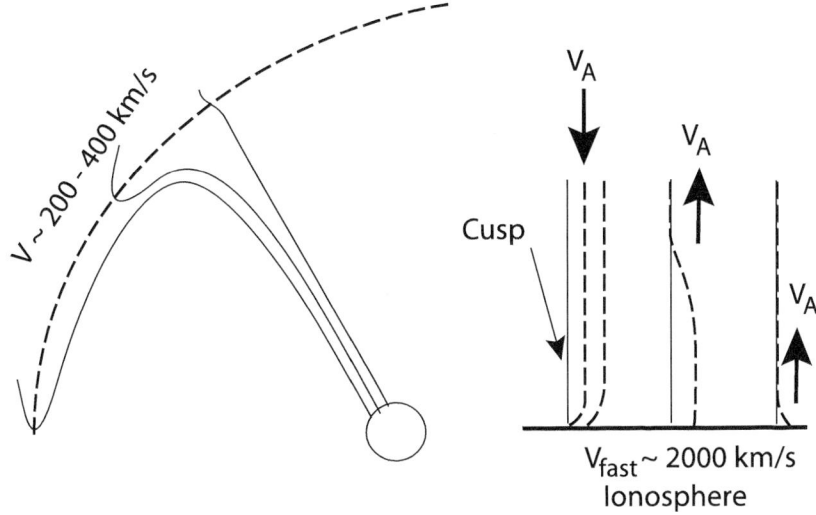

Figure 2. Illustration of the active role of the ionosphere in affecting magnetospheric convection. (left) magnetospheric plasma on open field lines convects along the magnetopause at the Alfvén velocity over a long distance. (right) the corresponding ionospheric feet of the field lines move at fast mode speed over a short distance, imposing motion on adjacent closed field lines. The motion of the ionosphere propagates to the magnetosphere at the Alfvén speed and modifies magnetospheric convection on closed field lines.

2.3. Effect of Reflection

The discussion in section 2.2 provides a basic understanding of how the plasma and neutrals respond in a time-dependent fashion to the solar wind and/or magnetospheric driver, but it describes only the initial phase, with one-way propagation of the perturbation. Afterward, the density gradients of the ionosphere/thermosphere (and to a lesser degree the gradient of the magnetic field) cause partial reflection of the downward propagating perturbation, modifying the basic picture shown in Figure 1. The net result is that the wave is reflected back to the magnetosphere, where it can be re-reflected at the magnetopause or at the conjugate ionosphere. The system, thus, experiences multiple reflections before reaching its steady state. Such a multiple-reflection model has been previously studied by *Holzer and Reid* [1975], who invoked surface reflection from a thin (height-integrated) ionosphere. The actual reflection process is undoubtedly more complicated; since ionospheric parameters vary gradually, reflection should not occur at a simple boundary surface but in a continuous manner over height. As the incident perturbation penetrates deeper down into the ionosphere, the effective wave propagation speed becomes lower, due to the neutral-inertia-loading effect [*Song et al.*, 2005], and the velocity perturbation decreases drastically at and below the altitude where the ion-neutral collision frequency becomes greater than the ion gyrofrequency, and thus, only a small velocity difference is needed if the collision term in equation (2) is to balance the $\mathbf{J} \times \mathbf{B}$ force. That wave reflection is affected by these processes and adds a key ingredient to time-dependent M-IT coupling is an important lesson from our analyses [*Song et al.*, 2009; *Tu et al.*, 2011].

With reflection present, a perturbation of any quantity (velocity, magnetic field, electric field, Poynting vector, etc.) will include contribution from both incident and reflected waves, having different phases. For velocity and magnetic field (the key parameters of inductive-dynamic coupling)

$$\mathbf{V} = \mathbf{V}_{inc} + \mathbf{V}_{ref}, \quad \delta\mathbf{B} = \delta\mathbf{B}_{inc} + \delta\mathbf{B}_{ref}. \quad (13)$$

The velocity and the magnetic perturbations locally are connected by an equation specific to the wave mode, which in regions where collision effects and the Hall term can be neglected takes the simple form of the Walén relation $\delta\mathbf{B} = \pm \mathbf{V}B/V_A = \pm\sqrt{\rho\mu_0}\mathbf{V}$, derived from equations (1), (2), (4), and (5). The negative (positive) sign is for propagation parallel (antiparallel) to the background magnetic field. In general, it is extremely difficult to separate reflected from incident perturbations, either in simulations or in observations, particularly if the reflection occurs in a gradual manner. The reflected perturbation may either enhance or reduce the strength of the incident perturbation, depending on the phase delay. The net perturbation (incident plus reflected) is, thus, expected to be more variable than the incident perturbation alone. Since the ionosphere, as discussed above, acts primarily to reduce the net velocity perturbation, the reflected velocity will tend to subtract from the incident one, and therefore (because of the sign change in the Walén relation), the reflected magnetic field will tend to add to the incident one. This is why reflection may produce overshoots in the magnetic field perturbations, which in turn result in enhanced currents, profound consequences for evaluating the power and heating rate (proportional to the square of the perturbation amplitude). (All these effects, it goes without saying, are absent in the electrostatic models, which contain neither propagation nor reflection.)

2.4. Differences in Physical Quantities

Below, we compare electrostatic coupling described by equations (7)–(11) and inductive-dynamic coupling described by equations (1)–(5), for an initial velocity perturbation V_0 perpendicular to the magnetic field B_0, at a location L_0 away from the ionosphere. Note that collision effects are dominant below 130 km [*Song et al.*, 2001]. A summary of the comparison is given in Table 1.

2.4.1. Communication speed along a magnetic field line.
In the absence of field-aligned bulk flow, communication in long-range coupling is accomplished primarily via electromagnetic coupling described by Maxwell's equations. In quasisteady state models with electrostatic coupling (equation (8)), the coupling is treated as instantaneous as far as the model is concerned (formally, the communication speed is infinite, although it is, of course, admitted that, in reality, the magnetosphere communicates with the ionosphere at the Alfvén speed). For inductive-dynamic coupling, it follows explicitly from equations (2) and (4) that the coupling (communication) speed is the Alfvén speed. For slow variations when $\omega < v_{in}$, the neutral wind affects the waves, by a process referred to as neutral-inertia-loading process [*Song et al.*, 2005]; when collisions are important, the fluid appears heavier than the plasma alone. For perturbations at frequencies less than the ion-neutral collision frequency, the propagation speed decreases continuously with perturbation frequency, and when the frequency becomes less than the neutral-ion collision frequency v_{ni}, approaches $\alpha^{1/2}V_A$ (where $\alpha = \rho_i/(\rho_i + \rho_n)$ is the ionization fraction), i.e., Alfvén speed calculated from total (plasma plus neutrals) mass density. For periods of interest, >1 min^{-1}, the inertia loading effect becomes appreciable below 400 km [*Song et al.*, 2005]. With decreasing altitude, as

Table 1. Comparison of Steady State Coupling and Dynamic Coupling

Quantity	Electrostatic Coupling	Dynamic Coupling
Communication V_{phase}	∞	$V_A \sim \alpha^{1/2} V_A$
Transient time δt	0	$\delta t = \begin{cases} t_A & 2 \text{ min (Alfvén time)} \\ 10 \sim 15 t_A & 30 \text{ min (magnetosphere-ionosphere equilibrium)} \\ 1/\nu_{ni} & 1 \sim 3 \text{h (neutral accel)} \end{cases}$
Reflection	No reflection, $\mathbf{V} = \mathbf{V}_{inc}$, $\delta \mathbf{B}$ not consistent with $\mathbf{V}$	total = $I + R$ for both $\delta \mathbf{B}$ and $\mathbf{V}$, relative sign $\delta \mathbf{B}/\mathbf{V}$ changes on reflection $\mathbf{V} = \mathbf{V}_{inc} - \mathbf{V}_{ref}$, $\delta \mathbf{B} = \pm \delta \mathbf{B}_{inc} \pm \delta \mathbf{B}_{ref}$
Velocity perturbation V	$V = V_0 B_0 D_0 / BD$	$V = (\rho_0/\rho)^{1/4} V_0$ for single Alfvén wave parallel propagation
Magnetic perturbation $\delta \mathbf{B}$	not considered as part of model evolution, calculated from $\mathbf{J}$, not consistent with $\mathbf{V}$	$\delta \mathbf{B} = \pm \mathbf{V} B/V_A = \pm \sqrt{\rho \mu_0}\, \mathbf{V}$ $\|\delta \mathbf{B}\| = (\rho \rho_0 \mu_0^2)^{1/4} V_0 = (\rho/\rho_0)^{1/4} \|\delta \mathbf{B}_0\|$ for single Alfvén wave parallel propagation
Electric field $\mathbf{E}$	$E = E_0 D_0 / D$	$E = (\rho_0/\rho)^{1/4}(B/B_0)^{1/2} E_0$ for single Alfvén wave parallel propagation
Current $\mathbf{J}$	$\mathbf{J}_\perp = \overleftrightarrow{\sigma} \cdot (\mathbf{E} + \mathbf{u}_n \times \mathbf{B}), \nabla \cdot \mathbf{J} = 0$	$\mathbf{J} = \nabla \times \mathbf{B}/\mu_0$, resulting in current conservation
Heating rate q	$q = \sigma_p (\mathbf{E} + \mathbf{u}_n \times \mathbf{B})^2$ $= \left[\left(1 + \frac{\nu_{in}^2}{\Omega_i^2}\right)^{-1} + \frac{\Omega_i \Omega_e}{\nu_e \nu_{in}} \left(\frac{\nu_{en}}{\nu_e} + \frac{\Omega_e^2}{\nu_{en}\nu_e}\right)^{-1} \right]$ $\times \nu_{in} \rho (\mathbf{E}/B + \mathbf{u}_n \times \mathbf{b})^2 \approx \nu_{in} \rho (\mathbf{V} - \mathbf{u}_n)^2$	$q = \mathbf{J} \cdot (\mathbf{E} + \mathbf{V} \times \mathbf{B}) + \nu_{in} \rho (\mathbf{u}_n - \mathbf{V})^2$ $\approx \left(1 + \frac{\nu_e \nu_{in}}{\Omega_e \Omega_i}\right) \nu_{in} \rho \langle (\mathbf{V} - \mathbf{u}_n)^2 \rangle$ $-\frac{1}{2}[\Delta(\rho V^2) + \Delta(\rho_n u_n^2)]/\Delta t$
Poynting vector $\mathbf{S}$	Not considered explicitly, DC part included implicitly in dissipation	$\mathbf{S} = \frac{1}{2} \frac{\delta B^2}{\mu_0} \mathbf{V}_A = \frac{1}{2} \rho V^2 \mathbf{V}_A$ (single wave) Fluctuating component only, DC part may be a problem

the ion-neutral collision frequency increases rapidly and the ionization fraction decreases, the inertia-loading effect can reduce the propagation speed by an order of magnitude.

2.4.2. Transient time. The original formulations of M-I coupling theory [*Fejer*, 1964; *Vasyliūnas*, 1970; *Wolf*, 1970] in terms of quasisteady state electrostatic models were based on the assumption that time scales of interest (in particular, flow times of magnetospheric convection, as stated explicitly by *Vasyliūnas* [1970]) are long compared to wave propagation times within the system. The governing equations (1), (7)–(11) of electrostatic coupling do not contain any time derivatives and, thus, in principle, describe self-consistently only the time-independent solutions for fixed given boundary conditions and system parameters. In practice, however, models are produced (particularly when implemented as numerical simulations) that do evolve with time. This is done by allowing the boundary conditions and/or system parameters to vary with time, e.g., introducing temporal variations at the boundaries of global MHD simulation models, from measurements, or from semiempirical models such as AMIE [*Richmond and Kamide*, 1988]. This approach has been extremely successful in describing many processes in the ionosphere/thermosphere. Nevertheless, the model output still is only a time series of successive but disjointed steady state solutions (the model temporal variations reflecting merely the variability of the input, not the dynamics of coupling) and cannot be considered an adequate approximation to the evolving solutions of the time-dependent equations unless the effective time step of the model is much longer than the time scales of all the transient processes. (It is an illusion to suppose that such models are self-consistently time-dependent and that more dynamic effects can be derived by using higher time resolution data as input or smaller and smaller time steps of the simulations.)

In the electrostatic coupling models, with all transients neglected and perturbations in the magnetosphere mapped instantaneously along field lines to (every altitude of) the ionosphere, the only intrinsic time scales are two time scales associated with collision processes. The ionosphere affects the thermosphere on a time scale $1/v_{ni}$, where v_{ni} is the neutral-ion collision frequency, and the thermosphere affects the ionosphere on a time scale $1/v_{in}$, where v_{in} is the ion-neutral collision frequency. Note that v_{in} and v_{ni} are different quantities; the two collision frequencies are related by momentum conservation during collisions, $\rho_n v_{ni} = \rho_i v_{in}$. As the neutral density is much higher than the plasma density below 800 km [e.g., *Tu et al.*, 2011], it takes a very long time before the neutral wind is affected by the plasma motion, although the plasma motion can be affected by the neutral wind relatively quickly. Neutral dynamics can, therefore, be included in a quasisteady state treatment of the ionosphere. (For plasma, collisions are important when v_{in} is comparable or much greater than the ion gyrofrequency Ω_i.)

In inductive-dynamic coupling, there are three intrinsic time scales: the Alfvén (communication) time, the time for the M-I system (thermosphere not included!) to reach equilibrium or a quasisteady state, and the time for the complete M-IT system to reach equilibrium.

Neglecting collisions, the coupling time from equations (2), (4), and (5) is the Alfvén time $t_A = L/V_A$ during which the magnetic perturbation propagates from the magnetosphere to the ionosphere over a distance L along the magnetic field. The Alfvén time, or magnetosphere-ionosphere communication time, is well known and has been considered by some to be the transient time for M-I coupling. The time for M-I equilibrium, however, is strongly affected by collisions and consequent wave reflection, as discussed in section 2.3.

The condition for M-I equilibrium state is negligible time derivative in equation (2), which implies a current (required to balance the collision term) $J = v_{in}\rho_i V/B$ and associated magnetic perturbation $|\delta \mathbf{B}| = \mu_0 \Delta L J = \mu_0 \Delta L v_{in} \rho_i V/B$ (ΔL is the altitude range over which the current is distributed, and all quantities are evaluated in the ionosphere). Taking the curl of the electric field from the magnetopause to the ionosphere gives, from equations (4) and (1), the time required to produce this magnetic perturbation, in order of magnitude,

$$\delta t \sim \frac{|\delta \mathbf{B}|L}{\Delta E} \sim \left(\frac{\Delta L \Omega_i}{V_A}\right)[(\Omega_i/v_{in})^2 + 1]^{-1/2} t_A \quad (14)$$

(essentially the same time scale as in equation (6)). The dimensionless coefficient of t_A in equation (14) is determined by parameters in the high-collision region and is small when the collision frequency is small. Using typical values of the ionosphere [e.g., *Song et al.*, 2001] gives the transient time for dynamic processes of order of 10~15 t_A, consistent with values quoted by *Song et al.* [2009] and *Tu et al.*, [2011] as well as those obtained by wave analysis [*Lysak and Dum*, 1983]. This is substantially longer than the Alfvén time, indicating that the waves have to reflect several times before the system reaches equilibrium (in agreement with the discussion in section 2.3). For $L \sim 20$ R_E and $V_A \sim 1000$ km s^{-1}, the Alfvén time is 120 s, and the transient time is about 20~30 min; the latter is the observed time scale for many important M-IT processes, such as the substorms and magnetospheric transients [e.g., *Earle and Kelley*, 1987; *Russell and Ginskey*, 1993, 1995; *Zesta et al.*, 2000; *Bristow et al.*, 2003; *Huang et al.*, 2008, 2010]. The time scale (14) is obtained by self-consistently joining long-range coupling with ionosphere-thermosphere local coupling and cannot be derived from local coupling alone such as that modeled by *Strangeway* [2012]. The fundamental reason for this additional time scale is the inertia of the ionosphere, irrelevant when the ionosphere is treated as being in a steady state.

The third time scale, the time for the M-IT system to reach equilibrium, in inductive-dynamic coupling as well as in electrostatic coupling, is the neutral wind acceleration time which is $1/v_{ni}$ (minimum value in the F layer, of an order of a few hours) [e.g., *Song et al.*, 2009]. On time scales longer than a significant fraction of this time, the neutral acceleration cannot be neglected. On still longer time scales, the neutrals and the plasma are tightly coupled, and the three fluids then can be considered a single fluid with collisions as internal dissipation [*Song and Vasyliūnas*, 2011]. The neutral dynamic effects are included in most ionospheric-thermospheric models but neglected in most M-I coupling models.

In the following discussion of magnitudes of the physical quantities, estimates from electrostatic coupling (in equations (15) and (18)) are applicable only to times much longer than the second (M-I equilibrium) transient time scale. For inductive-dynamic processes, estimates in equations (16), (17), and (19) are for a single (incident) wave and thus apply to times below and up to the first (Alfvén-wave communication) transient time scale. For times between the first and the second transient time scales, the quantities become sums of incident and reflected waves and are no longer equal to the values given by the equations. For times much longer than the second transient time scale, the results from inductive-dynamic coupling (after multiple reflections) are essentially the same as those from electrostatic coupling. (All the estimates here are for weakly collisional regions and do not include damping and neutral inertia loading effects, thus do not apply in the lower part of the ionosphere, say, below 130 km.)

2.4.3. Velocity mapping in weakly collisional regions. In electrostatic coupling, the velocity variation along the magnetic field is mapped according to the electric potential mapping (equation (8)), or

$$V = E_0 D_0 / BD = (B_0 D_0 / BD) V_0 \sim (B_0/B)^{1/2} V_0, \quad (15)$$

where D and D_0 are the distance between two field lines in the direction perpendicular to the flow and the (background) magnetic field at two heights. In the ionosphere, since the magnetic field strength does not change significantly over height, the flow velocity does not change as long as it is not affected by collisions. The last expression in equation (15) is an approximation, derived from magnetic flux conservation by ignoring the dependence of the ratio D/D_0 on orientation relative to the magnetic meridian.

In inductive-dynamic coupling, on the other hand, for a single (incident) wave, the velocity may be determined from energy conservation when collisions are weak. The Poynting flux of the perturbations contained in a flux tube is $F = SA \propto S/B = \frac{1}{2} \frac{|\delta \mathbf{B}|^2}{\mu_0} \frac{V_A}{B} = \frac{1}{2} \rho V^2 \frac{V_A}{B}$, where A is the cross section of the flux tube, and the last expression follows from the Walén relation. Energy conservation here implies that F is constant, which gives

$$V = (\rho_0/\rho)^{1/4} V_0. \quad (16)$$

Note the completely different velocity variation along a field line given by the two descriptions, equations (15) and (16). The density dependence of the velocity (slower flow in regions of higher density) during the dynamic stage in inductive-dynamic coupling can be understood as due to plasma inertia.

2.4.4. Magnetic perturbation. In electrostatic coupling, the magnetic perturbation does not play any role in the calculation but is derived from Ampère's law (equation (5)) in order to compare with observations. In inductive-dynamic coupling, the magnetic perturbation for a single wave (in a collisionless or at most weakly collisional region) can be related locally to the velocity perturbation according to the Walén relation discussed in section 2.3. Combined with equation (16), this gives the mapping relation for the magnitude of the magnetic perturbation

$$|\delta \mathbf{B}| = (\rho \rho_0 \mu_0^2)^{1/4} V_0 = (\rho/\rho_0)^{1/4} |\delta \mathbf{B}_0| \quad (17)$$

showing that the magnetic perturbation tends to increase in regions of higher density, also a plasma inertia effect. (Note: the increase at very low frequencies of the effective ion density in the expression by adding part or all of the neutral density due to neutral inertia loading [*Song et al.*, 2009; *Song and Vasyliūnas*, 2011] becomes appreciable only at low altitudes where the Walén relation is no longer a good approximation.)

2.4.5. Electric field mapping along a field line. In electrostatic coupling, the electric field variation along the magnetic field is calculated from the potential,

$$E = E_0 D_0 / D \sim (B/B_0)^{1/2} E_0, \quad (18)$$

where D is the perpendicular distance between two field lines (defined the same way and with the same approximations as in equation (15) for the velocity).

The electric field in inductive-dynamic coupling is calculated from the generalized Ohm's law (1), which gives

$$E = (\rho_0/\rho)^{1/4} (B/B_0)^{1/2} E_0, \quad (19)$$

if electron collisions and Hall effect are neglected, and the velocity is taken from equation (16). Again, the plasma inertia effect results in additional density dependence.

2.4.6. Electric current. In electrostatic coupling, the current is calculated from Ohm's law (equation (10)), with the electric field either given by some other model or obtained by solving equation (11) with the parallel current derived from a global MHD simulation model. The derived temporal variation of the current is intrinsically in conflict with the assumption that the magnetic field is constant. In inductive-dynamic coupling, the current is derived from equation (5), using the calculated magnetic perturbation. The Birkeland (magnetic field-aligned) current is determined automatically by this procedure; no special process or time delay to connect the horizontal currents in the ionosphere with Birkeland currents is needed.

2.4.7. Poynting vector. The importance and measurements of the Poynting vector have been promoted in recent years in the community, which may be considered progress in the understanding of the coupling processes. There are, however, two distinct forms of the Poynting vector. One, the "DC" form is calculated from the (quasi)steady electric and magnetic fields and is obviously the only form applicable in the electrostatic coupling approach. In this form, the Poynting vector is always perpendicular to the (background) magnetic field and, in general, flows into or out of the ionosphere in regions far away from Birkeland currents, even when the energy input to the ionosphere is concentrated in regions where the Birkeland currents are located. The other, the "AC" form, is calculated from perturbation electric and magnetic fields and obviously can be described only in the inductive-dynamic approach. Much of the recent observational evidence for enhanced Poynting-flux energy input into the ionosphere refers to this form.

In inductive-dynamic coupling, the (AC) Poynting vector is $S = \frac{1}{2\mu_0}|\delta\mathbf{B}|^2 V_A = \frac{1}{2}\rho V^2 V_A$ for fluctuating perturbations that obey the Walén relation and propagate all in the same direction along the background magnetic field (which is now, in contrast to the DC form, the direction of the Poynting vector); the factor O comes from averaging over fluctuations. If there are perturbations propagating in both directions (as with incident and reflected waves), the total Poynting vector, being quadratic in the perturbation amplitudes, contains also cross terms of incident and reflected waves. The sign difference between the two in the Walén relation, however, ensures that the cross terms always cancel, independent of any amplitude or phase difference. The calculated (AC) Poynting vector, thus, represents the net flow (up minus down) of electromagnetic energy along the direction of the background magnetic field, as long as the assumptions concerning the Walén relation hold. Energy carried by incident or reflected waves flows down or up, respectively; hence, the measured Poynting vector, which represents the superposition of incident and reflected, is subject to the same uncertainties and ambiguities discussed above for other quantities.

The Poynting vector can be derived from local measurements of the magnetic field and electric field or plasma velocity, and its direction determined from the phase difference between the magnetic and electric or velocity perturbations. The energy flow is downward if the magnetosphere is driving the ionosphere and upward if otherwise. As discussed in section 2.2, ionospheric regions (e.g., low latitudes) that are not directly driven by the external source of magnetospheric convection can still be driven, via fast mode propagation, by ionospheric convection in other regions (e.g., high latitude and polar cap). This secondary ionospheric convection can then locally drive the magnetosphere, a process that sometimes is misleadingly called "penetration electric field." In these regions, the Poynting vector can be upward, similarly to a proposed process for chromospheric heating [*Song and Vasyliūnas*, 2011]. The above description assumes that the driving accelerates the plasma flow. If the driving decelerates the flow instead, e.g., due to a reversal of the IMF direction, the Poynting vector is reversed (often described as a "flywheel effect").

2.4.8. Heating rate. The local heating rate in electrostatic coupling is

$$q = \mathbf{j} \cdot (\mathbf{E} + \mathbf{u}_n \times \mathbf{B}) = \sigma_p(\mathbf{E} + \mathbf{u}_n \times \mathbf{B})^2. \quad (20)$$

This is often referred to as Joule heating, on the basis of the definition of Joule heating as $\mathbf{j} \cdot \mathbf{E}^*$ where $\mathbf{E}^*$ is an electric field. Note, however, that the electric field and, hence, also the Joule heating so defined varies with the chosen frame of reference, whereas the true heating rate does not. There are several frames of reference in our problem: Earth, ionospheric plasma, and thermospheric wind. The conventional treatment chooses the neutral wind frame. *Vasyliūnas and Song* [2005] argued that the Joule heating, as the mechanism that converts electromagnetic energy to thermal energy, should properly be defined in the plasma frame of reference, and they showed that heating given by equation (20) is predominantly collisional and not electromagnetic. That most of the heating occurring in M-IT coupling is frictional and not Joule heating was confirmed by *Tu et al.* [2011]. To obtain the heating rate (equation (20)) observationally, one first uses the measured plasma velocity to calculate the electric field, then the electric potential, the conductivities, the current... etc., each step involving uncertainties and approximations. For weak collisions, the heating rate is

$$q = \left[\left(1 + \frac{v_{in}^2}{\Omega_i^2}\right)^{-1} + \frac{\Omega_i \Omega_e}{v_e v_{in}}\left(\frac{v_{en}}{v_e} + \frac{\Omega_e^2}{v_{en} v_e}\right)^{-1}\right] \\ \times v_{in}\rho(\mathbf{E}/B + \mathbf{u}_n \times \mathbf{b})^2 \approx v_{in}\rho(\mathbf{V} - \mathbf{u}_n)^2. \quad (21)$$

The local heating rate for inductive-dynamic heating is from *Vasyliūnas and Song* [2005],

$$q = \mathbf{J} \cdot (\mathbf{E} + \mathbf{V} \times \mathbf{B}) + v_{in}\rho_i(\mathbf{u}_n - \mathbf{V})^2. \quad (22)$$

For weak collisions and averaged over a long period of time, the heating rate is

$$q \approx \left(1 + \frac{v_e v_{in}}{\Omega_e \Omega_i}\right) v_{in}\rho_i \langle(\mathbf{V} - \mathbf{u}_n)^2\rangle - \frac{1}{2}[\Delta(\rho_i V^2) \\ + \Delta(\rho_n u_n^2)]/\Delta t \sim v_{in}\rho \langle(\mathbf{V} - \mathbf{u}_n)^2\rangle. \quad (23)$$

The heating rates from the two coupling approaches are superficially similar but differ in several respects. First, equation (21) is valid only for quasisteady state and, therefore, does not include heating associated with waves, whereas equation (23) does include wave heating, as indicated by the average signs. In inductive-dynamic coupling, the velocity includes both incident and reflected perturbations; due to reflection, the magnitude of the velocity can be significantly larger than its steady state value during the dynamic period because the average of the velocity perturbations can be close to zero. Second, the physical picture is much clearer for the latter approach: heating is primarily generated by the frictional heating between the plasma and the neutrals. If the velocity difference between the two is measured or otherwise known, the heating rate can be derived relatively easily.

3. CONCLUSIONS AND DISCUSSION

To model global M-IT coupling processes on time scales shorter than about 30 min, the dynamic term in the momentum equation and the inductive term in Faraday's law need to be included. Although the E-J and B-V schemes give the same results in electrostatic and quasisteady state coupling, time-dependent processes can be dealt with only with the B-V scheme because the E-J scheme cannot describe the time evolution self-consistently.

There are major differences between the conventional M-I and M-IT coupling approaches and the inductive-dynamic approach. First, the transition times are estimated differently. Conventionally, the Alfvén time between the magnetosphere and ionosphere or from one hemisphere to the other is considered the transition time scale of dynamic coupling. However, a fully coupled M-IT system cannot complete its transition in one Alfvén time because the ionosphere has its own inertia, and longer time is needed to accelerate the magnetosphere-ionosphere system from one state to another. This transient time is about 10~15 times Alfvén time or typically 20~30 min.

Second, the propagation and reflection of waves, a most important phenomenon during the transition, is not included in the conventional theory. (The term "waves" here refers generally to any type of perturbations.) This process cannot be resolved within the ionosphere/thermosphere if it is treated as a height-integrated layer. This issue is very confusing and has often been overlooked or misunderstood. A question one may ask is: Why is the wave reflection so important? Or, how much does a model miss without including wave reflection? In the presence of incidence and reflection, a physical quantity at a given location and time is the superposition of the two. Depending on the reflection coefficient and the phase difference between the incident and reflected perturbations, the resultant (measured) perturbation can range from as small as zero to as large as twice the incident amplitude both positively and negatively. For a quantity such as energy or heating rate, proportional to the square of the amplitude, the difference can range from zero to a factor 4. Furthermore, because the phase difference between incidence and reflection may vary with time and location, the resulting (measured) signals are actually fluctuating within this range. It is, thus, easy to understand why the observations of a dynamic process (e.g., a substorm) often contain many large-amplitude fluctuations or overshoots, which then may diminish over a transition period after conditions stabilize. The fluctuations can, of course, be averaged out over the transition period, but these intense fluctuations (such as those during substorms or auroral brightenings) are often of great physical significance and constitute the subject of the study. Averaging them out is tantamount to eliminating the phenomenon of primary interest.

Third, the conventional theory does not appropriately describe the ionospheric dynamic responses to the magnetosphere and the coupling within the ionosphere. In fact, because the magnetic field variations in the ionosphere is assumed time constant and the plasma dynamic terms in the perpendicular direction are not considered, MHD modes in the ionosphere itself do not play a role in the theory. In particular, the fast mode, the most effective mechanism to mitigate force imbalance in the horizontal direction, to communicate efficiently between different parts of the ionosphere and to produce ionospheric convection, does not appear in the theoretical description; only some of its effects in the ionosphere are mimicked by the so-called "penetration electric field" postulated for the purpose. The motion of ionospheric plasma in the closed field line region is a result not of imposed electric field in the open field region but because of magnetospheric convection plasma flow there that, due to the requirement of continuity, induces motion in the closed field line regions by action of fast mode waves.

Finally, quantitatively, there are many differences between the electrostatic and dynamic couplings, summarized in Table 1. The coupling speeds, although very different between the two schemes, do not to produce significant differences if the region of interest is only up to a few thousand kilometers above the Earth. The inertia effect of the plasma plays a role in determining the perturbations of the flow velocity, magnetic field, and electric field along the magnetic field. The physical picture for the current is completely different in the two coupling schemes. In quasisteady state coupling, the system is coupled via Birkeland (field-aligned) currents; as the Birkeland currents vary, the Pedersen and Hall currents in the ionosphere are assumed to always vary so as to connect properly to the Birkeland currents. In inductive-dynamic coupling, the currents are carried by the field and flow variations; field-aligned currents are formed in regions where plasma condition and flow vary, as required for self-consistency, and propagate down with magnetic field variations from the magnetosphere to the ionosphere. The Joule heating rate calculated in the conventional theory is actually mostly frictional heating (not Joule heating in the proper physical sense) and neglects the effects of wave heating.

Acknowledgments. The authors thank Dr. J.-N. Tu for comments. This work was supported by the National Science Foundation under grant AGS-0903777 and NASA under grant NNX12AD22G/NSR:3030708.

REFERENCES

Bristow, W. A., G. J. Sofko, H. C. Stenbaek-Nielsen, S. Wei, D. Lummerzheim, and A. Otto (2003), Detailed analysis of substorm observations using SuperDARN, UVI, ground-based magnetometers, and all-sky imagers, *J. Geophys. Res.*, *108*(A3), 1124, doi:10.1029/2002JA009242.

Buneman, O. (1992), Internal dynamics of a plasma propelled across a magnetic field, *IEEE Trans. Plasma Sci.*, *20*, 672–677.

Coroniti, F. V., and C. F. Kennel (1973), Can the ionosphere regulate magnetospheric convection?, *J. Geophys. Res.*, *78*(16), 2837–2851.

Cowling, T. G. (1957), *Magnetohydrodynamics*, Interscience, New York.

Dungey, J. W. (1958), *Cosmic Electrodynamics*, p. 10, Cambridge Univ. Press, London, U. K.

Earle, G. D., and M. C. Kelley (1987), Spectral studies of the sources of ionospheric electric fields, *J. Geophys. Res.*, *92*(A1), 213–224.

Fejer, J. A. (1964), Theory of the geomagnetic daily disturbance variations, *J. Geophys. Res.*, *69*(1), 123–137.

Fuller-Rowell, T. J., D. Rees, S. Quegan, R. J. Moffett, M. V. Codrescu, and G. H. Millward (1996), A coupled thermosphere-ionosphere-plasmasphere model, in *STEP Handbook on Ionospheric Models*, edited by R. W. Schunk, pp. 217–238, Utah State Univ., Logan.

Holzer, T. E., and G. C. Reid (1975), The response of the day side magnetosphere-ionosphere system to time-varying field line reconnection at the magnetopause, 1. Theoretical model, *J. Geophys. Res.*, *80*(16), 2041–2049.

Huang, C.-S., K. Yumoto, S. Abe, and G. Sofko (2008), Low-latitude ionospheric electric and magnetic field disturbances in response to solar wind pressure enhancements, *J. Geophys. Res.*, *113*, A08314, doi:10.1029/2007JA012940.

Huang, C.-S., F. J. Rich, and W. J. Burke (2010), Storm time electric fields in the equatorial ionosphere observed near the dusk meridian, *J. Geophys. Res.*, *115*, A08313, doi:10.1029/2009JA015150.

Huba, J. D., G. Joyce, and J. A. Fedder (2000), Sami2 is Another Model of the Ionosphere (SAMI2): A new low-latitude ionosphere model, *J. Geophys. Res.*, *105*(A10), 23,035–23,053.

Hughes, W. J. (1974), The effect of the atmosphere and ionosphere on long period magnetospheric micropulsations, *Planet. Space Sci.*, *22*, 1157–1172.

Hughes, W. J., and D. J. Southwood (1976), The screening of micropulsation signals by the atmosphere and ionosphere, *J. Geophys. Res.*, *81*(19), 3234–3240.

Kelley, M. C. (2009), *The Earth's Ionosphere: Plasma Physics and Electrodynamics*, 2nd ed., Academic Press, San Diego, Calif.

Lysak, R. L., and C. T. Dum (1983), Dynamics of magnetosphere-ionosphere coupling including turbulent transport, *J. Geophys. Res.*, *88*(A1), 365–380.

Parker, E. N. (1996), The alternative paradigm for magnetospheric physics, *J. Geophys. Res.*, *101*(A5), 10,587–10,625.

Parker, E. N. (2000), Newton, Maxwell, and magnetospheric physics, in *Magnetospheric Current Systems*, *Geophys. Monogr. Ser.*, vol. 118, edited by S. Ohtani et al., pp. 1–10, AGU, Washington, D. C., doi:10.1029/GM118p0001.

Parker, E. N. (2007), *Conversations on Electric and Magnetic Fields in the Cosmos*, Princeton Univ. Press, Princeton, N. J.

Richmond, A. D., and Y. Kamide (1988), Mapping electrodynamic features of the high-latitude ionosphere from localized observations: Technique, *J. Geophys. Res.*, *93*(A6), 5741–5759.

Richmond, A. D., E. C. Ridley, and R. G. Roble (1992), A thermosphere/ionosphere general circulation model with coupled electrodynamics, *Geophys. Res. Lett.*, *19*(6), 601–604.

Ridley, A. J., Y. Deng, and G. Toth (2006), The global ionosphere-thermosphere model, *J. Atmos. Sol. Terr. Phys.*, *68*, 839–864.

Russell, C. T., and M. Ginskey (1993), Sudden impulses at low latitudes: Transient response, *Geophys. Res. Lett.*, *20*(11), 1015–1018.

Russell, C. T., and M. Ginskey (1995), Sudden impulses at subauroral latitudes: Response for northward interplanetary magnetic field, *J. Geophys. Res.*, *100*(A12), 23,695–23,702.

Schunk, R. W. (2013), Ionosphere-thermosphere physics: Current status and problems, in *Modeling the Ionosphere-Thermosphere System*, *Geophys. Monogr. Ser.*, doi:10.1029/2012GM001351, this volume.

Song, P., and V. M. Vasyliūnas (2010), Aspects of global magnetospheric processes, *Chin. J. Space Sci.*, *30*(4), 289–311.

Song, P., and V. M. Vasyliūnas (2011), Heating of the solar atmosphere by strong damping of Alfvén waves, *J. Geophys. Res.*, *116*, A09104, doi:10.1029/2011JA016679.

Song, P., C. T. Russell, and S. P. Gary (1994), Identification of low-frequency fluctuations in the terrestrial magnetosheath, *J. Geophys. Res.*, *99*(A4), 6011–6025, doi:10.1029/93JA03300.

Song, P., T. I. Gombosi, and A. J. Ridley (2001), Three-fluid Ohm's law, *J. Geophys. Res.*, *106*(A5), 8149–8156, doi:10.1029/2000JA000423.

Song, P., V. M. Vasyliūnas, and L. Ma (2005), Solar wind-magnetosphere-ionosphere coupling: Neutral atmosphere effects on signal propagation, *J. Geophys. Res.*, *110*, A09309, doi:10.1029/2005JA011139.

Song, P., V. M. Vasyliūnas, and X.-Z. Zhou (2009), Magnetosphere-ionosphere/thermosphere coupling: Self-consistent solutions for a one-dimensional stratified ionosphere in three-fluid theory, *J. Geophys. Res.*, *114*, A08213, doi:10.1029/2008JA013629.

Strangeway, R. J. (2012), The equivalence of Joule dissipation and frictional heating in the collisional ionosphere, *J. Geophys. Res.*, *117*, A02310, doi:10.1029/2011JA017302.

Tu, J., P. Song, and B. W. Reinisch (2008), On the concept of penetration electric field, *AIP Conf. Proc.*, *974*, 81–85.

Tu, J., P. Song, and V. M. Vasyliūnas (2011), Ionosphere/thermosphere heating determined from dynamic magnetosphere-ionosphere/

thermosphere coupling, *J. Geophys. Res.*, *116*, A09311, doi:10.1029/2011JA016620.

Vasyliūnas, V. M. (1970), Mathematical models of magnetospheric convection and its coupling to the ionosphere, in *Particles and Fields in the Magnetosphere*, edited by B. M. McCormack, pp. 60–71, D. Reidel, Dordrecht, The Netherlands.

Vasyliūnas, V. M. (2001), Electric field and plasma flow: What drives what?, *Geophys. Res. Lett.*, *28*(11), 2177–2180.

Vasyliūnas, V. M. (2005a), Time evolution of electric fields and currents and the generalized Ohm's law, *Ann. Geophys.*, *23*, 1347–1354.

Vasyliūnas, V. M. (2005b), Relation between magnetic fields and electric currents in plasmas, *Ann. Geophys.*, *23*, 2589–2597.

Vasyliūnas, V. M. (2011), Physics of magnetospheric variability, *Space Sci. Rev.*, *158*, 91–118, doi:10.1007/s11214-010-9696-1.

Vasyliūnas, V. M. (2012), The physical basis of ionospheric electrodynamics, *Ann. Geophys.*, *30*, 357–369.

Vasyliūnas, V. M., and D. H. Pontius Jr. (2007), Rotationally driven interchange instability: Reply to André and Ferrière, *J. Geophys. Res.*, *112*, A10204, doi:10.1029/2007JA012457.

Vasyliūnas, V. M., and P. Song (2005), Meaning of ionospheric Joule heating, *J. Geophys. Res.*, *110*, A02301, doi:10.1029/2004JA010615.

Wolf, R. A. (1970), Effects of ionospheric conductivity on convective flow of plasma in the magnetosphere, *J. Geophys. Res.*, *75*(25), 4677–4698.

Zesta, E., H. J. Singer, D. Lummerzheim, C. T. Russell, L. R. Lyons, and M. J. Brittnacher (2000), The Effect of the January 10, 1997, pressure pulse on the magnetosphere-ionosphere current system, in *Magnetospheric Current Systems*, Geophys. Monogr. Ser., vol. 118, edited by S. Ohtani et al., pp. 217–226, AGU, Washington, D. C., doi:10.1029/GM118p0217.

P. Song, Center for Atmospheric Research and Department of Physics, University of Massachusetts Lowell, MA, USA. (Paul_Song@uml.edu)

V. M. Vasyliūnas, Max-Planck-Institut für Sonnensystemforschung, 37191 Katlenburg-Lindau, Germany.

Ionospheric Irregularities: Frontiers

D. L. Hysell and H. C. Aveiro

Earth and Atmospheric Sciences, Cornell University, Ithaca, New York, USA

J. L. Chau

Radio Observatorio de Jicamarca, Instituto Geofísico del Perú, Lima, Peru

Contemporary research into modeling plasma density irregularities and the ionospheric instabilities responsible for them is reviewed. Three classes of waves are singled out for detailed examination. These include gradient drift and Farley-Buneman waves in the equatorial electrojet, collisional interchange instabilities associated with equatorial spread F (ESF), and resistive drift waves, which may be responsible for irregularities in midlatitude sporadic E layers. Three cross-cutting themes drive the analyses. One of these concerns the shortcomings of the often-made equipotential magnetic field line assumption. Another is the role of stochastic forcing in shaping wave behavior, motivated by an overview of stochastic differential equation formalism. The third is the necessity of initial-value analysis for large-scale waves in inhomogeneous flows. The themes are illustrated by analysis and numerical simulations of E and F region ionospheric instabilities.

1. INTRODUCTION

Ionospheric irregularities are features that are not predicted by first-principles, steady state analysis. In practice, they arise from instabilities driven by excess free energy that cannot be rapidly dissipated by binary collisions alone. The free energy can reside in configuration or phase space, and the features range from nearly monochromatic waves to turbulence-like flows. Ionospheric irregularities are most often associated with electrostatic plasma waves that are aspect sensitive with respect to the Earth's magnetic field, although irregularity transients are mediated by electromagnetic fields. Irregularities are found in all of the ionospheric layers and at all latitudes. They both are influenced by and affect the mean state of the ionosphere. They interfere with some diagnostic methods while facilitating others. Their most noticeable role may be their tendency to form diffraction screens, causing radio scintillations and interfering with communication and navigation systems. The ability to forecast scintillations beyond the limits of persistence and climatology remains elusive.

Experimentally, ionospheric irregularities can be detected through radiometric means, including ionospheric sounding, coherent and incoherent scatter, radio scintillation, radio tomography, and synthetic aperture radar. Large-scale plasma density irregularities can also be detected in all-sky optical ionospheric imagery, and lidars can see turbulent upwelling in the mesosphere related to the formation of irregularities in the D region. Finally, plasma and neutral number density, temperature, and electric and magnetic fields measured in situ by instrumented satellites and sounding rockets give precise measurements of fluctuations in state variables in unstable plasmas.

Theoretically, ionospheric irregularities are often approached first through boundary-value analysis of appropriate kinetic or fluid theory models, with the mode(s) with the highest linear growth rates (imaginary part of the eigenfrequencies) presumed to predominate. This kind of analysis can miss essential nonlocal, non-normal, nonlinear, and anomalous effects, however, and produce incomplete results. The most definitive avenue for theoretical investigation of ionospheric irregularities is, therefore, initial-value analysis and numerical simulation. Simulations can be computationally demanding when a broad span of scale sizes is involved or when complete 3-D treatments are required to reproduce natural phenomena, as is often the case. Analysis can still be useful, particularly when the underlying equations can be transformed into well-known ones (e.g., the Navier-Stokes equation, the Taylor-Goldstein equation, Burger's equation, etc.).

Even though ionospheric irregularities have been under scrutiny since the 1930s, fundamental experimental and theoretical results continue to appear in the literature, along with frequent review papers. *Woodman* [2009] and *Kelley et al.* [2011] recently presented comprehensive reviews of plasma irregularities associated with equatorial spread F (ESF), one of the most enduring problems with far-reaching consequences in space plasma physics. The C/NOFS satellite has made in situ density and electric field measurements of ESF irregularities available after a long hiatus of comparable, low-inclination ionospheric satellite missions [*de La Beaujardiere*, 2004], and *Park et al.* [2009] have additionally demonstrated how field-aligned currents associated with ESF can be studied using space-borne magnetometers. Recently, *Pi et al.* [2011] have been able to obtain imagery of low-, middle-, and high-latitude irregularities, including ESF depletions, from synthetic aperture radar (SAR)-derived total electron content measurements. Global models of ESF irregularities can now be constructed, as exemplified by *Huba and Joyce* [2010]. The ionospheric interchange instabilities underlying ESF are closely related to the $\mathbf{E} \times \mathbf{B}$ instabilities responsible for F region coherent scatter at auroral latitudes. *Baker et al.* [2011] recently reviewed how the high-latitude convection pattern can be estimated from HF coherent scatter.

A recurrent issue is the role of gravity waves in initiating ESF. The idea that gravity waves destabilize the postsunset equatorial ionosphere traces to the work of *Kelley et al.* [1981] but has lately received considerable theoretical and modeling attention [e.g., *Vadas and Fritts*, 2004; *Fritts and Vadas*, 2008; *Vadas and Keskinen*, 2010; *Vadas and Liu*, 2009; *Rappaport et al.*, 2009; *Keskinen and Vadas*, 2009; *Tsunoda*, 2010]. Empirical evidence for a connection between gravity wave and ESF occurrence is concurrently being sought [e.g., *Fritts et al.*, 2009; *Takahashi et al.*, 2009; *Abdu et al.*, 2009; *Sreeja et al.*, 2009]. Making the connection is challenging, since it is generally the effects of gravity waves rather than the waves, themselves, that can be directly observed.

Other important classes of irregularities are those found in the equatorial and auroral electrojets and in midlatitude sporadic E layers. Of all the ionospheric instabilities, Farley-Buneman and gradient drift instabilities in the equatorial electrojet are probably the best understood (see the work of *Farley* [2009] for review). This is because they occur in a compact, isolated region of space under generally regular, modest, and measurable forcing conditions. Nevertheless, questions remain, particularly with regard to how Farley-Buneman waves propagate and saturate. A contemporary series of studies has demonstrated that the ion acoustic speed in the electrojet is not isothermal [*St.-Maurice et al.*, 2003; *Hysell et al.*, 2007; *Kagan et al.*, 2008]. Two-dimensional kinetic simulations of Farley-Buneman waves described by *Oppenheim et al.* [2008] and *Dimant and Oppenheim* [2008] recover many of the observed features of the waves, including their tendency to propagate with phase speeds below that predicted by linear theory. Although there has been little computational work on gradient drift waves lately, the fluid simulations described by *Ronchi et al.* [1991] and *Hu and Bhattacharjee* [1999] appear to have captured most of the observed, salient features of large-scale waves propagating in the equatorial electrojet during the day and at night, respectively. *Kagan and St.-Maurice* [2004], *Oppenheim and Dimant* [2004], and *Dimant and Oppenheim* [2004] have expanded the original theory of modified two-stream waves to include electron and ion thermal effects. Recently, *Alken and Maus* [2010] were able to derive empirical estimates of the conductivity of the equatorial electrojet, including wave-driven anomalous corrections, using ground-based radar and space-based magnetometer measurements.

Studying Farley-Buneman waves in the auroral electrojet is complicated by the facts that the forcing is much stronger, more variable, and harder to diagnose. Direct measurements in the waves using instrumented sounding rockets, as analyzed recently by *Hysell et al.* [2008] and *Krane et al.* [2010], offer rare but particularly incisive views into the microphysics of the instabilities at work. (See the works of *Pfaff* [1995], however, for a discussion of the experimental complications involved.) *Makarevich* [2009] and *Uspensky et al.* [2011] reviewed different coherent scatter radar observations of auroral E region irregularities drawn from extensive databases, attempting to tie the gross features to contemporary Farley-Buneman wave theory. A focus of contemporary research is wave heating, the role of parallel electric fields, and the relationship between the ion-acoustic speed and the phase speed of the waves [*St.-Maurice and Kissack*, 2000; *Dimant and Milikh*, 2003; *Bahcivan*, 2007]. *Dimant and Oppenheim* [2011a, 2011b] have shown how anomalous currents associated with Farley-Buneman turbulence at high latitudes effectively increase the Pedersen conductivity and the ionospheric loading of the magnetosphere.

Midlatitude sporadic E ionization layers have been affecting human activity since the earliest days of radio and remain an important means of long-path communications in the HF and VHF bands (e.g., see reviews by *Whitehead* [1972], *Whitehead* [1989], and *Mathews* [1998]). Interest in the electrodynamics of the layers grew in the 1990s after *Riggin et al.* [1986] and *Yamamoto et al.* [1991, 1992] found that they could be populated by intense, small-scale field-aligned irregularities (FAIs) and probed using coherent scatter radar.

Irregularities in sporadic E layers take the form of patchy, elongated, roll-like deformations. Irregularity fronts appear to propagate with periods of 5–10 min, wavelengths of a few tens of kilometers, and directions preferentially toward the southwest or northeast in the Northern Hemisphere, although direction can vary considerably from event to event or even within an event. The fronts are polarized, presenting electric fields often large enough to excite Farley-Buneman instability. FAIs exist throughout the patchy layers even when the condition for Farley-Buneman instability is not met. Irregularities often come in bursts lasting 30–60 min. The irregularities are usually attributed to one of three causes: gravity waves [*Woodman et al.*, 1991; *Didebulidze and Lomidze*, 2010; *Chu et al.*, 2011], neutral instability [*Larsen et al.*, 2007; *Hysell et al.*, 2009], and plasma instability [*Cosgrove and Tsunoda*, 2002, 2004].

A rapidly emerging area of interest concerns medium-scale traveling ionospheric disturbances (MSTIDS) and the associated ionospheric irregularities and dynamos. *Kelley and Miller* [1997] and *Kelley* [2011] argue that the preferred propagation direction for MSTIDs, northeast to southwest in the Northern Hemisphere, is the direction consistent with the minimum Joule heating. *Dymond et al.* [2011] described an unusually comprehensive set of ground- and space-based observations of a midlatitude medium-scale traveling ionospheric disturbance. *Miller et al.* [2009] have shown convincing evidence that TIDs propagating from middle latitudes can instigate ESF, a phenomenon that *Krall et al.* [2011] have been able to reproduce in simulation. *Suzuki et al.* [2009] showed that FAIs are embedded in the phases of MSTIDs as if driven by the associated polarization electric fields under gradient drift instability (see also the works of *Otsuka et al.* [2009] and *Ogawa et al.* [2009]). *Otsuka et al.* [2008] studied statistically the relationship between MSTIDs and disturbed E_s layers. That MSTIDs work together with E region plasma irregularities in coupled instability is an idea that has been investigated extensively for midlatitudes in theory and computation [*Cosgrove and Tsunoda*, 2004;

Tsunoda, 2006; *Cosgrove*, 2007; *Yokoyama et al.*, 2008, 2009]. *Batista et al.* [2008] have argued against E_s layer control of ESF occurrence, however.

Ionospheric irregularities are far too expansive a subject to be covered completely in a single review paper, and no such review will be attempted here. Instead, we report on some emerging ideas about modeling irregularities. These include (1) the importance of nonequipotential magnetic field line treatment, (2) the effects of stochastic forcing on irregularity evolution, and (3) the transient growth of irregularities in sheared flows. These threads will be woven through explicit reviews of Farley-Buneman gradient drift (FBGD) instabilities, instabilities leading to equatorial spread F (ESF), and a potentially important class of drift-wave instabilities that may be operating in equatorial and midlatitude E regions.

2. FUNDAMENTALS

While the plasma irregularities discussed here arise from disparate mechanisms, they can all be understood, in part, in terms of the conservation of electron number density as prescribed by the linearized electron continuity equation,

$$\partial_t n_1/n_o = -\mathbf{v}_{e1}\cdot\nabla_\perp ln n_o - \mathbf{v}_{eo}\cdot\nabla_\perp n_1/n_o - \nabla_\perp\cdot\mathbf{v}_{e1} + \nabla_\parallel(\mu_{e\parallel}\nabla_\parallel\phi_1), \quad (1)$$

where n_o and $\mathbf{v}_{eo}$ are the background number density and electron drift velocity, n_1, $\mathbf{v}_{e1}$, and ϕ_1 are perturbations to the number density, drift velocity, and electrostatic potential, respectively, and $\mu_{e\parallel}$ is the parallel (direct) electron mobility. The perpendicular and parallel subscripts refer to directions with respect to the background magnetic field. For wave growth to occur, at least one of the terms on the right side of equation (1) has to be significant and have the proper phase relationship with the electron density fluctuations. In the E region (transverse), gradient drift waves grow when the electron perturbation $\mathbf{E}\times\mathbf{B}$ drifts are aligned with the background number density gradient (first term on RHS). In the F region, where the ions are also magnetized, the same mechanism produces ionospheric interchange instability. Back in the E region, streaming electrons with finite compressibility are essential components for growing (longitudinal) Farley-Buneman waves (second and third terms on RHS).

The fourth term on the right side of equation (1) represents parallel electron transport and is often incorrectly assumed to be purely stabilizing. In order to evaluate its effect, we can assume a plane wave solution such that the term becomes $+k_\parallel^2|\mu_{e\parallel}|\psi_1$, where k is the wavenumber. The electrostatic potential can be related to the number density through the quasineutrality condition, $\nabla \cdot \mathbf{J} = 0$, where we may write, generically,

$$\mathbf{J} = ne(\mu_\perp \mathbf{E}_\perp + \mu_\parallel \mathbf{E}_\parallel + \mu_\times \hat{b} \times \mathbf{E} - D_\perp \nabla_\perp lnn - D_\parallel \nabla_\parallel lnn), \quad (2)$$

where μ and D refer to the plasma mobility and diffusivity, respectively, the subscripts $\perp$, $\times$, and $\parallel$ refer to the Pedersen, Hall, and direct components, and $\hat{b}$ is a unit vector in the direction of the magnetic field. Note that diamagnetic currents have been neglected here. Furthermore, write $\mathbf{E} = \mathbf{E}_o - \nabla\phi$, segregating the electric field into a background transverse component $\mathbf{E}_o$ that must be calculated self-consistently and a perturbation component $-\nabla\phi_1$ that can vary in all directions. Linearizing and setting the divergence of the current density equal to zero then yields

$$\phi_1 = \frac{-i\mathbf{k}_\perp \cdot \mathbf{E}_o \mu_\perp + i(\mathbf{k}_\perp \times \mathbf{E}_o) \cdot \hat{b}\mu_\times - D_\perp k_\perp^2 - D_\parallel k_\parallel^2}{\mu_\perp k_\perp^2 + \mu_\parallel k_\parallel^2 - ik_\parallel \mu_\parallel / L} \frac{n_1}{n_o}, \quad (3)$$

where L is the parallel conductivity gradient length scale and where all other spatial variations in the transport coefficients have been neglected (see the work of *Hysell et al.* [2002b] for details).

Neglecting the term involving L in the denominator of equation (3) for the moment and considering equation (1), it is clear that the electric field terms affect the propagation of the plasma waves but not their growth, while the terms involving diffusion are purely damping. When L is finite, however, the electric field terms can be such that the fourth term in equation (1) is in phase with the $\partial_t n_1$ term, and instability can result. This instability involves the growth of resistive drift waves and requires finite parallel wavenumbers. Physically, electrons are able to stream along magnetic field lines to maintain charge neutrality, but because of collisions, the electrons overshoot, resulting in growth. This is an overlooked but potentially important modality, since it requires only parallel conductivity gradients and not transverse gradients to function.

Below, we consider a number of irregularity-producing instabilities in more detail. The focus will be on subtle aspects of the instabilities that must be modeled with care if phenomena in nature are to be reproduced accurately.

3. E REGION IRREGULARITIES

The equatorial and auroral electrojets have long been known to be prone to Farley-Buneman [*Farley*, 1963; *Buneman*, 1963] and gradient drift instabilities [*Simon*, 1963; *Hoh*, 1963]. It has since become clear that Farley-Buneman waves also exist within disturbed sporadic E layers at midlatitudes [*Schlegel and Haldoupis*, 1994]. The case for gradient drift waves in sporadic E layers is less clear, although there appear to be no theoretical prohibitions against them, at least for short wavelengths [*Seyler et al.*, 2004]. Farley-Buneman and gradient drift waves are closely intertwined and are often analyzed together and referred to as FBGD waves. Extensive reviews on the subject have been given by *Haldoupis* [1989] and *Sahr and Fejer* [1996]. Below, the current status and some pressing problems are addressed.

3.1. Farley-Buneman Gradient Drift Instability

Farley-Buneman and gradient drift waves are excited in the equatorial and auroral electrojets by rapid electron Hall drifts. Farley-Buneman or modified two-stream waves grow spontaneously when ion inertia causes ions to migrate into the crests and out of the troughs of irregularities in Hall-drifting electron streams faster than diffusion opposes them. These waves are longitudinal. Gradient drift waves grow spontaneously when irregularities in the streams induce transverse $\mathbf{E} \times \mathbf{B}$ drifts in a region of inhomogeneous conductivity such that the irregularities grow through electron interchange. These are transverse waves. Propagation is in the direction of the electron drifts in any case. Whereas the former have dominant wavelengths of a few meters and are cut off at short wavelengths by Landau damping [*Farley*, 1963], the latter predominate at wavelengths of hundreds of meters or kilometers and are cut off at long wavelengths by recombination. Nevertheless, the two instabilities operate together in the equatorial electrojet and can be described by a unified formalism. Appendix A presents a compact fluid theory derivation of the FBGD dispersion relation, which includes the effects of neutral winds, different magnetic aspect angles, and a finite background gradient length scale.

Kudeki et al. [1982] performed pioneering experimental studies of primary gradient drift waves in the electrojet using both spectral analysis and radar interferometry at Jicamarca, observing long-lived, large-scale waves with wavelengths between about 2 and 6 km propagating at about half the electron Hall drift speed. Wavelengths were longer at night than during the day. They attributed the relatively slow phase speeds of the waves to the long-wavelength correction embodied by the k_o term in the dispersion relation and determined that the predominant wavelengths were at least roughly consistent with linear theory with the effects of recombination included. However, *Huba and Lee* [1983] argued that linear theory actually predicts much shorter predominant wavelengths than those observed. They asserted that shear in the electron Hall drift profile must also be considered and has the effect of stabilizing intermediate-scale waves, leaving large-scale waves to predominate.

Soon afterward, *Fu et al.* [1986] reconsidered the role of shear flow. Even though shear flow damps intermediate-scale eigenmodes in boundary value analysis, initial value analysis and numerical simulation predicted strong transient growth at intermediate and small scales. They called the transient response a "quasilocal" mode, something overlooked by boundary value analysis, since the eigenmodes in sheared, non-normal systems are not guaranteed to be orthogonal and complete. Once again, theory seemed to predict dominant waves at intermediate scales, whereas the experimental evidence was for large-scale waves.

Next, *Ronchi et al.* [1989] reinterpreted the findings of *Fu et al.* [1986] in terms of eikonal analysis and the theory of wave trapping. They identified linearly unstable eigenmodes as trapped or convectively unstable waves and transient, quasilocal modes with untrapped or convectively stable modes. Following *St.-Maurice* [1987], *Ronchi et al.* [1989] also considered the anomalous effects associated with small-scale wave turbulence. Including anomalous effects, all gradient drift wave modes, intermediate and large scale, were predicted to be linearly damped, untrapped, transient modes. The explanation for the predominance of large-scale waves now seemed even more distant.

Finally, *Ronchi et al.* [1991] were able to account for the predominance of large-scale gradient drift waves in the electrojet through 2-D simulation including the effects of nonlinear mode coupling. Simulation showed that shear flow stabilizes all wave modes with wavelengths comparable to the shear gradient length scale by convection in Fourier space, which transports the waves to wavenumbers dominated by diffusive dissipation. However, through nonlinear mode coupling, energy from the damped waves is transported back to small wavenumbers. A nonsteady dynamical equilibrium saturated state emerges, characterized by the straining and destruction of large-scale waves by shear and diffusion followed by spontaneous regeneration. Simulations predicted waves with wavelengths of about 2 km and lifetimes of about 30 s.

The nighttime case, meanwhile, was studied by *Hu and Bhattacharjee* [1999]. At night, the plasma density profile in the equatorial electrojet is thought to be structured and jagged [*Prakash et al.*, 1972]. The effect is such as to cause large-scale waves to become trapped (absolutely unstable) or nearly trapped (convectively unstable) in the vicinity of unstable phases of the background density gradient. This is true even when anomalous effects are considered. Trapped, large-scale waves are furthermore able to tunnel from one unstable stratum to another. Intermediate-scale waves, meanwhile, are not affected by the background gradients and remain untrapped (convectively stable). Very long wavelength waves with wavelengths of several kilometers are consequently expected to dominate.

Figure 1 shows radar images acquired at Jicamarca typical of large-scale waves in the daytime and nighttime electrojet. In the images, the pixels convey information about the moments of the underlying Doppler spectrum. Variations in color delineate the phases of the large-scale waves. The waves propagate westward during the day with wavelengths of 1–2 km in the direction of, but considerably more slowly than, the electron drift speed. Tilted wavefronts are a consequence of strong vertical shear in the daytime electron convection speed. Dynamically, the waveforms form, propagate, strain under shear flow, and dissipate, exhibiting a dynamic steady state.

At night, propagation changes direction toward eastward. The vertical span of the unstable region grows, and the wavelength of the waves increases. In this case, the blue- and red-shifted irregularities at either side of the image imply that only a half of a wavelength is contained within the volume illuminated by the radar. Gaps in backscatter in range are indicative of strata that are locally gradient drift stable. The large-scale waves appear to tunnel across these strata, the associated wave packets moving vertically across the entire span of ranges producing echoes even as they propagate horizontally.

Sudan et al. [1973] explained how the large vertical velocities and plasma density gradients associated with large-scale gradient drift waves in the equatorial electrojet can excite vertically propagating, secondary Farley-Buneman waves even when the background zonal electron convection speed is less than the ion acoustic speed. This phenomenon is at work in the example images shown in Figure 1, where the most red- and blue-shifted regions have Doppler shifts near the ion acoustic speed (roughly 350 m s^{-1} in the equatorial electrojet) and narrow spectral widths. Such spectra are commonly referred to as "type 1." Additionally, Farley-Buneman waves can also be directly excited in the upper E region near midday, when the electrojet current is the strongest [*Pfaff et al.*, 1985]. The Doppler spectra from the associated "pure two-stream" irregularities as observed from vertically looking radar are broad and have small Doppler shifts [*Kudeki et al.*, 1987]. These spectra are often referred to as "type 2," as are echoes from gradient drift waves, although the terminology is ambiguous.

Farley-Buneman waves are most unambiguously observed in the equatorial electrojet during strong daytime counter-electrojet conditions, when the background zonal electric field is westward, the electron convection is eastward, and the E region bottomside is gradient drift stable. In this case, only pure two-stream waves emerge and only in narrow

222 IONOSPHERIC IRREGULARITIES: FRONTIERS

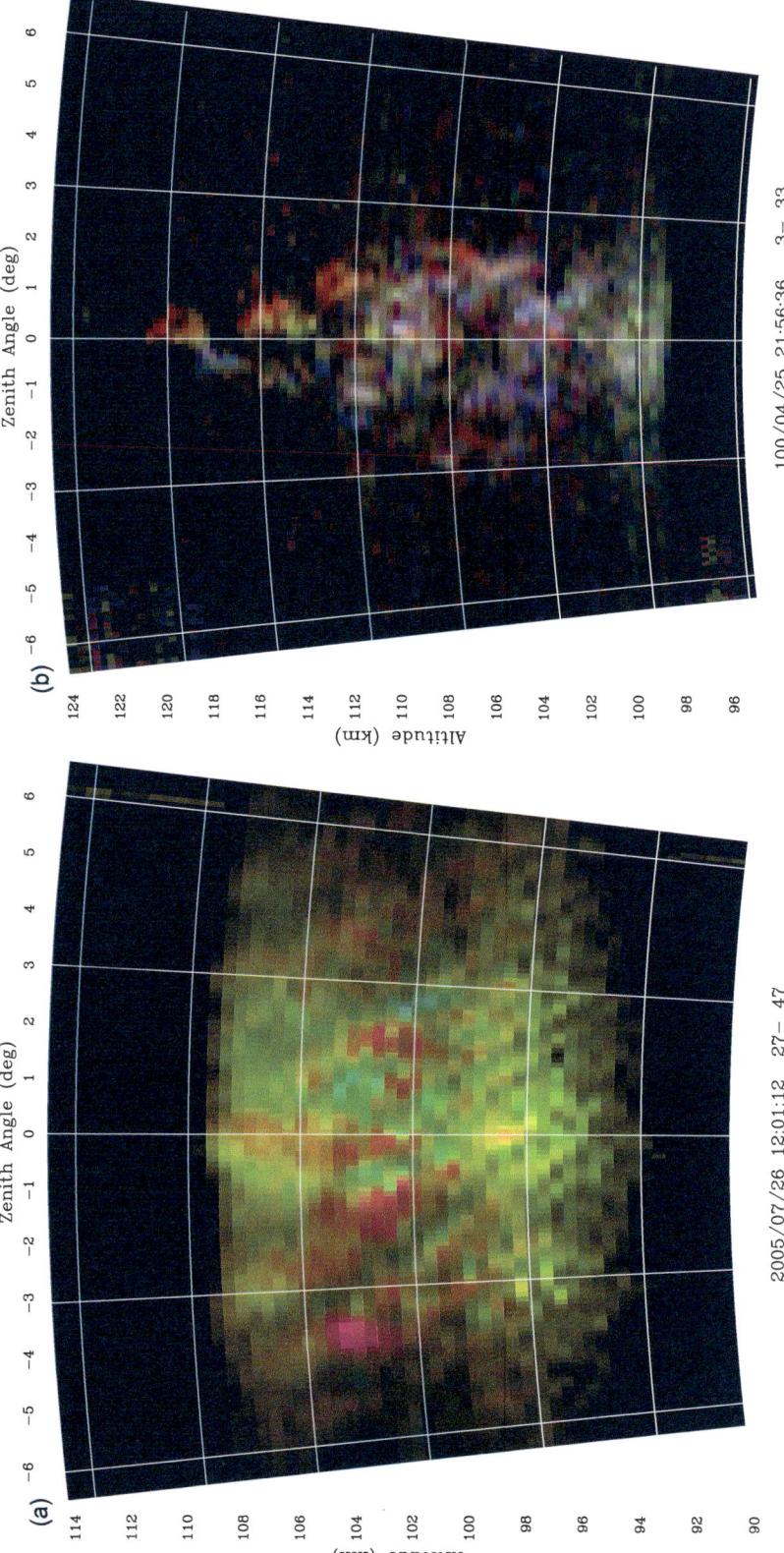

Figure 1. Jicamarca radar images of large-scale gradient drift waves in the (top) daytime and (bottom) nighttime equatorial electrojet. The two images were acquired using different experimental configurations such that the volume imaged was wider in the daytime case than in the nighttime. The brightness, hue, and saturation of the image pixels are related to the scattering intensity, Doppler shift, and spectral width, respectively. Pixel brightness increases with the signal-to-noise ratio on a dB scale. Red (blue) tones are red shifted (blue shifted), while pure (pastel) tones are spectrally narrow (broad).

strata where the convection is supersonic. At Jicamarca, counter electrojet waves have been observed using an antenna with a very broad beam [*Woodman and Chau*, 2002]. Because of the scattering geometry, the range to the echoes was closely related to their zenith angle, which was closely related to the flow angle, the angle between the convection direction and the radar line of sight. (In the presence of large-scale gradient drift waves, a broad span of flow angles is always present within the radar illuminated volume, and so Farley-Buneman wave dependencies on flow angle cannot be distinguished.)

Figure 2 shows spectral moments (Doppler shift and RMS spectral half width) computed for some of the counterelectrojet data obtained on 12 August 2000, described by *Woodman and Chau* [2002]. The curves show how the moments are related to the sine and cosine of the elevation angle plus about 10°. The moments are anisotropic and rotated by about 10° with respect to the flow angle. The asymptotic maximum Doppler shift and RMS half width were 360 and 140 m s^{-1}, respectively, in this case, but vary over time.

The sinusoidal dependencies of the moments on flow angle were recovered by the 2-D particle-in-cell simulations of Farley-Buneman waves described by *Oppenheim et al.* [2008] and *Dimant and Oppenheim* [2008]. The small rotation of the propagation direction of the waves with respect to the flow angle was also recovered and is a consequence of thermal effects [*Oppenheim and Dimant*, 2004]. The same dependencies were also found by comparing VHF imaging radar and sounding rocket observations of Farley-Buneman waves in the auroral zone by *Hysell et al.* [2008], who expanded on the formalism of *Hamza and St-Maurice* [1993a, 1993b] in proposing a theoretical/ phenomenological model for the behavior. Similar behavior has not been observed by fixed-beam HF and VHF radars operating at auroral latitudes [*Makarevich*, 2009], although inferring it without the benefits of radar imaging or the simple geometry available at Jicamarca would be difficult.

The aforementioned observations and simulations point to a more fundamental, long-recognized aspect of Farley-Buneman waves: that they propagate at phase speeds less than the electron convection drift speed but close to the ion acoustic speed [e.g., *Farley*, 2009, and references therein]. While this is inconsistent with the phase speed predicted by linear theory, it is consistent with propagation at the threshold for marginal linear stability (see appendix A). A number of linear, quasilinear, and nonlinear theories have been advanced to explain the phase speed saturation effect. *Sudan* [1983a, 1983b] interpreted phase speed saturation as being due to the effects of random turbulent fluctuations on the electron motion. He derived an anomalous electron diffusivity expression from the kinetic equation for the electrons associated with nonlinear orbit broadening and then attributed the additional transport to an anomalous electron collision frequency, which became part of a modified anisotropy factor ψ^*. In a saturated state, ψ^* grows rapidly as the waves are driven past the threshold for marginal stability such that they never significantly surpass the threshold. *Otani and Oppenheim* [1998] were able to recover phase speed saturation considering only a single triad of nonlinearly interacting modes. Using a creative and independent formalism, *St.-Maurice and Hamza* [2001] analyzed wave speed saturation by considering how 3-D nonplane wave irregularities polarize, turn, and decelerate deterministically.

A frequent point of confusion pertains to the meaning of the "ion acoustic speed" in regard to Farley-Buneman waves. This figure is not simply a static figure but depends on the ratios of specific heat for the electrons and ions, which in turn depend on the altitude of the waves and their wavelength, which is related to the wavelength of the probing radar [*Farley and Providakes*, 1989; *St. Maurice et al.*, 2003; *Hysell et al.*, 2007; *Kagan et al.*, 2008]. Furthermore, wave heating can drastically modify the local electron temperature and the ion acoustic speed with it, particularly in the auroral zone [e.g., *St. Maurice and Laher*, 1985; *St. Maurice*, 1990]. Since parallel electric fields heat the electrons efficiently, heating increases with the increasing RMS magnetic aspect width of the irregularities, which also

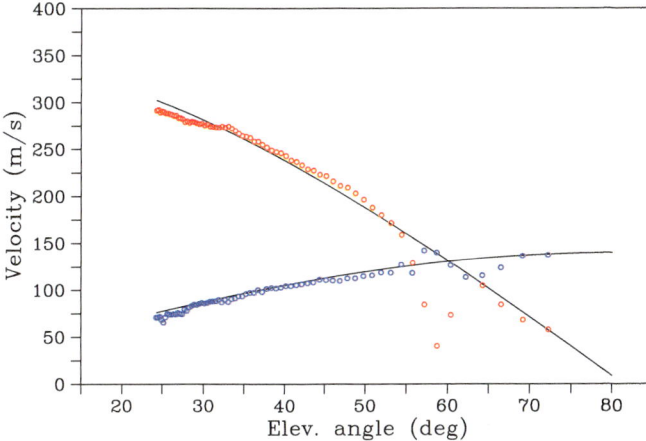

Figure 2. Spectral moments of echoes from pure two-stream waves in the upper electrojet over Jicamarca during counterelectrojet conditions at 10:45 LT Red and blue symbols show the Doppler shift and RMS spectral half width of the echoes, respectively, versus elevation angle. Superimposed lines proportional to the sine and cosine of the elevation angle plus 10° have been added to help guide the eye (see text). The incoherent integration time for the data was 20 s. A null in the antenna radiation pattern near 60° elevation limits data accuracy there.

implies an increasing effective anisotropy factor and a reduced growth rate. The RMS aspect width of Farley-Buneman waves can be estimated by finding the value that implied marginal instability [*St. Maurice*, 1990]. Several studies have explored the systematic relationship between the convection electric field, an effective RMS anisotropy factor, wave heating, the ion acoustic speed, and the Doppler shift of Farley-Buneman waves [*Robinson*, 1986; *Foster and Erickson*, 2000; *Milikh and Dimant*, 2002; *Dimant and Milikh*, 2003; *Bahcivan*, 2007]. The numerical simulations of *Oppenheim et al.* [2008] capture such a systematic relationship, at least in the perpendicular plane.

3.2. Farley-Buneman Turbulence

Here another formalism for investigating the phase speed saturation of Farley-Buneman waves is explored. It is similar in some ways to that of *Sudan* [1983a, 1983b] in attributing saturation to random turbulent fluctuations in the background medium, generated through mode coupling and wave turbulence. The paradigm explored below involves wave scattering off of inhomogeneities in the local index of refraction. The immediate objective is to relate the scattering to the statistical properties of the turbulence. The broader objective of this analysis is to explore a means of assessing the influence that plasma turbulence created by instabilities can have on the mean state of the ionosphere, a phenomenon not limited to Farley-Buneman turbulence but pertinent to all the instabilities discussed in this review (any many that are not discussed).

The formalism is based on stochastic differential equations. Stochastic differential equations were reviewed rather exhaustively by *Van Kampen* [1975], who provides the basis, syntax, and organization for the present analysis. The analysis is 1-D, but the extension to multiple dimensions is straightforward.

Suppose the state vector u for a system evolves under a combination of deterministic and stochastic influences such that

$$\dot{u}(t) = A(t; \varpi)u \quad (4)$$

$$u(0) = a \quad (5)$$

$$A(t; \varpi) = A_o(t) + \alpha A_1(t; \varpi), \quad (6)$$

where $A_o(t)$ is deterministic or "sure," and $A_1(t; \varpi)$ is stochastic, with ϖ denoting one particular member of an ensemble, and with some corresponding probability density function (PDF) $F(\varpi)d\varpi$ being defined. Note that u is a column vector and A, a matrix. Here a is the initial condition for the wavefield, and α is a small expansion parameter. This formulation, which is that of *Van Kampen* [1975], is quite general and could accommodate many linear phenomena of interest in upper atmospheric and space physics.

A conventional way to solve this problem would be through ensemble averages, where a solution is found for each of a large number of trial functions or realizations $A_1(t)$ drawn from a distribution governed by the PDF $F(\varpi)$. The solution vector u could then be found through an average over the ensemble, i.e., $\langle u(t) \rangle$. Depending on the ergodicity of the system and the number of solutions required for adequate statistical confidence, this could be a computationally demanding problem, requiring a complete specification of the random forcing on the system throughout its lifetime for each of many trials. Analytic solutions would seem to be unavailable in general.

The objective of stochastic differential equation analysis is to find a single equation that $\langle u(t) \rangle$ obeys, an equation that depends on a few moments of the function A_1 perhaps, and solve it but once to identify the attendant bias due to the random forcing. Thus, the salient behavior induced by the random forcing can be related systematically to the statistical properties of the forcing without a costly and arduous examination of the detailed microphysics involved, which may be poorly known in any case.

We will approach the problem using Bourret's integral equation [*Bourret*, 1962a, 1962b], which is described in Appendix B. This approach can be applied to systems of ordinary differential equations of any order. While first-order problems can readily be solved, the practical difficulty of the analysis increases essentially with increasing order. The problem of Farley-Buneman turbulence in the linear fluid theory limit is usually posed as a system of three first-order equations: one representing ion continuity, one electron continuity, and one ion momentum. (As electrons can be regarded as massless, the electron momentum equation is algebraic, and the electron velocities can be written explicitly; see appendix A.)

In order to reduce the complexity of the problem, we will describe Farley-Buneman waves using just two first-order linear equations. These are not derived from first principles but are instead contrived to combine to predict a phase velocity for the waves consistent with the full linear theory in the absence of stochastic forcing, viz.

$$v_\phi \equiv V_d/(1 + \psi), \quad (7)$$

where V_d is the electron drift velocity transverse to the magnetic field, and ψ is the anisotropy factor. The system of differential equations can be expressed (using phasor notation) as

$$ikG = \frac{1}{v_\phi}\frac{dF}{dt} \quad (8)$$

$$ikF = n^2 \frac{1}{v_\phi} \frac{dG}{dt} \tag{9}$$

$$n^{-2} = (1 + \alpha\xi(t;\varpi)), \tag{10}$$

where n represents the effective index of refraction of the Farley-Buneman waves and contains a stochastic correction $\xi(t;\varpi)$, a scalar, which is set off by the small expansion parameter α. The correction represents random fluctuations in the background $\mathbf{E} \times \mathbf{B}$ drift speed caused by secondary Farley-Buneman waves and wave turbulence, as seen by the primary wave being analyzed. Clearly, this system reverts to equation (7) when $\alpha = 0$.

The problem stated here is equivalent to the one solved by *Van Kampen* [1975] as an example of the application of Bourret's equation, and we merely reproduce his analysis here. In terms of the original problem, this system of equations can be expressed as:

$$u = \begin{pmatrix} F \\ iG \end{pmatrix} \quad A_o = \begin{pmatrix} 1 \\ -1 \end{pmatrix} \quad B = \begin{pmatrix} 0 \\ -1 \end{pmatrix}$$

$$\dot{u} = (A_o + \alpha \underbrace{\xi(t;\varpi)B}_{A_1})u,$$

where the equations have been simplified using nondimensional time such that $kv_\phi = 1$.

The analysis proceeds in a straightforward way with the recognition that

$$e^{A_o\tau} = I\cos\tau + A_o\sin\tau, \tag{11}$$

where I is the identity matrix. Likewise, the simple forms of the A_o and especially the A_1 matrix for this problem afford a number of additional simplifications. For example, it is easy to show that

$$Be^{A_o\tau}Be^{-A_o\tau} = \begin{pmatrix} 0 & 0 \\ (\sin2\tau)/2 & -(1-\cos2\tau)/2 \end{pmatrix}. \tag{12}$$

At this point, two constants remain to be calculated to populate the stochastic differential equation in equation (B11). These are:

$$c_1 = \int_0^\infty \langle\xi(t)\xi(t-\tau)\rangle\sin2\tau : d\tau \tag{13}$$

$$c_2 = \int_0^\infty \langle\xi(t)\xi(t-\tau)\rangle(1-\cos2\tau)d\tau. \tag{14}$$

Here it may be noted that

$$H(\omega) = \frac{2}{\pi}\int_0^\infty \langle\xi(t)\xi(t-\tau)\rangle\sin\omega\tau d\tau = \frac{1}{\pi}\int_{-\infty}^\infty \frac{S(\omega')}{\omega' - \omega}d\omega', \tag{15}$$

which is to say that the sine transform of the autocorrelation function of a quantity is the Hilbert transform $H(\omega)$ of the corresponding frequency spectrum $S(\omega)$. Consequently, the first of the constants, c_1, is just the Hilbert transform of the spectrum evaluated at normalized frequency 2:

$$c_1 = \frac{\pi}{2}H(2). \tag{16}$$

The cosine transform of the autocorrelation function, meanwhile, is just the spectrum. Therefore, the second of the constants is just the difference of the spectrum evaluated at two frequencies:

$$c_2 = \int_0^\infty \langle\xi(t)\xi(t-\tau)\rangle(1-\cos2\tau)d\tau = \frac{\pi}{2}(S(0)-S(2)). \tag{17}$$

Finally, the stochastic differential equation discussed in the preceding section of this paper can be populated for this problem. The result appears as:

$$\langle\dot{u}\rangle = \left\{\begin{pmatrix} 1 \\ -1 \end{pmatrix} + \frac{\alpha^2}{2}\begin{pmatrix} 0 & 0 \\ c_1 & -c_2 \end{pmatrix}\right\}\langle u\rangle.$$

The two first-order equations can, in the end, be recombined into a more familiar wave equation:

$$\frac{d^2\langle F\rangle}{dt^2} + kv_\phi\frac{\alpha^2 c_2}{2}\frac{d\langle F\rangle}{dt} + k^2v_\phi^2\left(1 - \frac{\alpha^2 c_1}{2}\right)\langle F\rangle = 0, \tag{18}$$

where dimensional time is back in place. (Likewise, the constants c_1 and c_2 as defined above should be evaluated at $\omega = 2kv_\phi = 2kV_d/(1 + \psi)$ using dimensional time.)

At this point, $\langle F\rangle$ may be any of the state variables that describe 1-D Farley-Buneman turbulence, e.g., density, potential, electric field, etc. With $\alpha = 0$, equation (18) describes a simple harmonic oscillator with a frequency $\omega = kv_\phi = kV_d/(1 + \psi)$, consistent with ordinary linear theory for Farley-Buneman waves, as it has been constructed to be. For finite values of α, however, equation (19) describes a damped oscillator with a frequency reduced from its undamped value and with line broadening. The saturated wave frequency and line width will depend on c_1 and c_2, which is to say on the turbulent electric field spectrum. More specifically, the dependence is on the slope and the Hilbert transform of the electric field spectrum. Similar frequency limiting and line broadening are familiar characteristics of waves propagating in turbulent scattering media. This is one intuitive and tractable way of viewing the phase speed saturation of Farley-Buneman waves, which also gives insights into the line broadening that occurs with increasing flow angle.

4. F REGION IRREGULARITIES

Ionospheric irregularities in the F region are mainly attributed to convective instability with similarities to the Rayleigh-Taylor instability in hydrodynamics. Rayleigh-Taylor has an MHD variant and also an electrostatic, two-fluid variant most properly termed the ionospheric interchange instability. An excellent pedagogical review has been given by *Zargham and Seyler* [1989]. In the inertial regime, the underlying equations can be transformed into a system resembling the Navier-Stokes equation. The instability also has an inertialess branch called the collisional interchange instability, which has been analyzed in detail by *Zargham and Seyler* [1987]. Their dispersion relation derivation is similar to the one given for the gradient drift instability in Appendix A except for the fact that the ions are taken to be magnetized along with the electrons. The main drivers for the instability are zonal currents driven by gravity and the background zonal electric field. The collisional interchange instability is sometimes called the "generalized Rayleigh-Taylor" instability and is thought to be responsible for most ESF-related phenomena. In order to evaluate the dispersion relation accurately, given the assumption of equipotential magnetic field lines, an approach incorporating the flux tube–integrated forcing and loading must generally be adopted (G. Haerendel, Theory of equatorial spread F, preprint, Max-Planck Institute für Physik and Astrophysik, Garching, West Germany, 1973)].

In this section, mechanisms including, but not limited to, generalized Rayleigh-Taylor instability working together to produce plasma density irregularities in the equatorial F region leading to equatorial spread F are considered. While some of the conclusions below likely apply to middle and high latitudes, the equatorial ionosphere has unique properties that set ESF apart from other latitudes. The section begins with an analysis of some of those properties and their implications for stability and variability in the region.

It has long been known that strong vertical shear in the zonal plasma drifts exists in the bottomside equatorial F region ionosphere around twilight, where the plasma flow in the bottomside reverses from westward to eastward with increasing altitude even though the thermospheric wind is strictly eastward [*Kudeki et al.*, 1981; *Tsunoda et al.*, 1981]. Several factors contribute to shear flow, including E region dynamo winds, vertical winds and horizontal electric fields on flux tubes with significant Hall conductivity, and currents sourced in the electrojet region near the solar terminator seeking closure (see the works of *Haerendel et al.* [1992] and *Haerendel and Eccles* [1992] and for reviews). Strong vertical current in the bottomside is attendant with the retrograde plasma drifts where the local plasma velocity and neutral wind are antiparallel.

It is also well known that thin layers of plasma irregularities (called bottom-type layers) exist near the region of shear flow, beneath the node [*Woodman and La Hoz*, 1976; *Woodman*, 2009], at the altitude where the valley region meets the bottomside. These layers, which often form at sunset, serve as precursors for more dynamic equatorial spread F events (see Figure 3, which shows a bottom-type layer emerging from a region of sheared plasma flow). *Kudeki and Bhattacharyya* [1999] argued that the irregularities in bottom-type layers are generated by conventional wind-driven gradient drift instabilities, which are readily excited in regions of retrograde plasma drift where zonal conductivity gradients, of the kind that arise across the solar terminator at sunset, also exist.

Hysell et al. [2004] then presented radar imagery supporting the *Kudeki and Bhattacharyya* [1999] hypothesis regarding wind-driven gradient drift instabilities and also showing that bottom-type irregularities usually occur in patches arranged periodically in the zonal direction. They surmised that the periodicity was caused by large-scale waves in the bottomside, the different phases of which being alternately stable and unstable to wind-driven instability. Such waves, if present, could predispose the ionosphere to interchange instabilities and ESF.

The necessity of some kind of ancillary process for initiating ESF is a recurring theme in ESF research. Using the flux tube–integrated formalism, it can readily be shown that, in the absence of unusually strong storm time electric fields at the magnetic equator, the e folding growth time for the collisional plasma interchange instability is seldom less than about 15–20 min in the postsunset bottomside F region and seldom for more than one or two e folding times under moderate solar flux conditions [see, for example, *Sultan*, 1996; *Retterer*, 2005; *Krall et al.*, 2009; *Singh et al.*, 2010]. The time of most rapid wave growth is generally about 1900 SLT, depending on the season, solar cycle, and longitude. How is it possible then for large-scale depletions and radar plumes to start appearing by 2000 solar local time (SLT) and earlier? Nonlinear and/or nonlocal effects may contribute to the rapidity at which ESF develops in nature, e.g., the acceleration of wave growth that occurs as

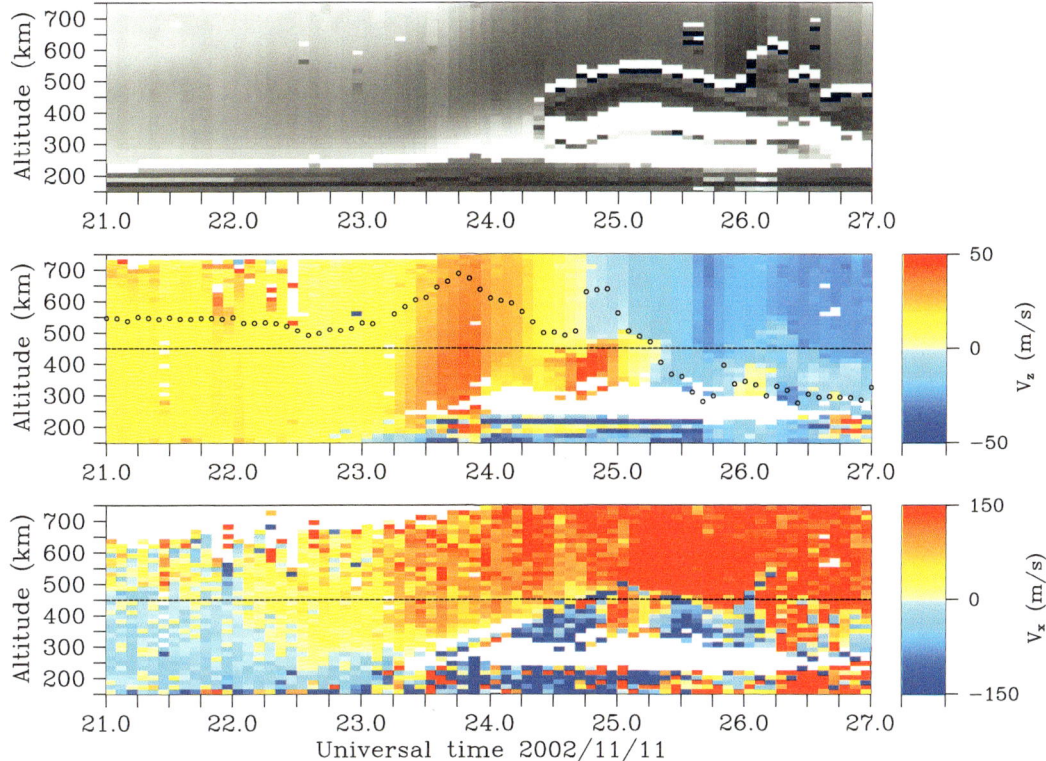

Figure 3. Combined incoherent and coherent scatter radar data from Jicamarca depicting the conditions for onset and evolution of ESF. (top) Plasma number density or, in the case of the strongest echoes, irregularity intensity. (middle) Vertical plasma drifts. (bottom) Zonal plasma drifts.

irregularities ascend to higher altitudes with correspondingly higher growth rates. However, the large-scale waves implied by the bottom-type layers would also seem to be relevant to rapid ESF emergence problem.

The predominance of large-scale waves in the postsunset equatorial ionosphere has been highlighted by *Tsunoda and White* [1981], *Kil and Heelis* [1998], and *Tsunoda* [2005]. As mentioned in the introduction, their source is often attributed to gravity waves [*Röttger*, 1973; *Kelley et al.*, 1981; *Singh et al.*, 1997; *McClure et al.*, 1998]. However, *Hysell and Kudeki* [2004] proposed that retrograde plasma drifts and vertical currents in the ionosphere itself could be the source of the large-scale waves. They followed the formalism developed by *Keskinen et al.* [1988] who assessed the viability of electrostatic Kelvin-Helmholtz instabilities in the auroral F region. That study suggested that ion-neutral collisions damp the instability in the ionosphere. *Hysell and Kudeki* [2004] adapted the work for equatorial application, incorporating the effects of altitude-dependent collisions. They found a collisional branch of the instability that could operate in the collisional regime in regions of strong, retrograde plasma motion. The new found "collisional shear instability" was to Kelvin-Helmholtz what the collisional interchange instability is to Rayleigh-Taylor.

The growth rate of the new instability was found to be potentially several times greater than that of the generalized Rayleigh-Taylor instability under nominal conditions. Boundary value analysis predicted that the growth rate should be a maximum for $kL \sim 1/2$, where k is the horizontal wavenumber, and L is the vertical scale length of the shear. Taking $L \sim 15$ km on the basis of typical observations, therefore, implies a preferred wavelength of about 150–200 km. The associated e folding time was later predicted to be about 5 min. An initial value analysis, meanwhile, suggested that the instability could also exhibit a much shorter dominant wavelength in its early stages, of the order of a few times L in simulation. The transient response of the instability could, therefore, account for the decakilometric large-scale waves in the postsunset ionosphere seen in radar imagery from Jicamarca, while the asymptotic response (the boundary value problem result) could account for the ~150–200 km large-scale waves observed. The transient has been shown to grow even faster than the asymptotic response [*Kudeki et al.*, 2007].

At the most basic level, the generalized Rayleigh-Taylor instability can be understood as plane electrostatic waves propagating in the direction of the background zonal current driven in the equatorial ionosphere by a combination of gravity and an electric field. The current polarizes the plane wave surfaces, and subsequent $\mathbf{E} \times \mathbf{B}$ drifts amplify the wave because the background density is vertically stratified. The growth rate is proportional to the current times two factors of cos (θ), where θ is the angle the wavevector makes with the horizon. One factor comes from the need for the wave to be parallel to the background current and the other from the need to be perpendicular to the background density gradient.

In the collisional shear instability, the background vertical current associated with the imperfectly efficient zonal wind-driven dynamo is the driver. This current can be very large at the base of the F region, where retrograde plasma drifts exist. Now, the growth rate is proportional to the background current times sin (θ) cos (θ), the sine factor entering because the background current is now vertical. The fastest growing waves, therefore, propagate at oblique angles with respect to the horizon. A more careful derivation of the generalized Rayleigh-Taylor and collisional shear instabilities, including flux tube–integration formalism, was given by *Hysell et al.* [2006].

Such calculations, while intuitively informative, are not rigorously correct; the dispersion relation for waves propagating in the direction of the background conductivity gradient is not expected to have plane wave solutions, and a nonlocal analysis like the one given by *Hysell and Kudeki* [2004] is required. That analysis indicated that the growth rate of the instability maximizes near $kL = 0.5$, where k is the wavenumber, and L is the vertical length scale of the shear flow. The underlying equation describing the waves resembled the Taylor-Goldstein equation, which also describes shear instability in hydrodynamics [*Taylor*, 1931; *Goldstein*, 1931]. In the collisional limit, however, the behavior is different, with vertical current replacing shear flow as the main source of free energy. Even so, roll-like deformations at the base of the bottomside characterize the instability, and shear flow is an indispensable feature of a stratified ionosphere supporting vertical current.

Moreover, *Flaherty et al.* [1999] have pointed out that boundary value analysis is also inadequate to describe systems with shear flow, since the associated system of equations is generally non-normal. This means that the eigenmodes are guaranteed to be neither orthogonal nor complete, and large-amplitude transients can exist even in regions of wavevector space where all the eigenmodes are decaying with time. This was the situation confronted by *Ronchi et al.* [1989] dealing with large-scale waves in the equatorial electrojet, and it is also the condition for ESF. Initial value analysis and numerical simulation is required.

The results of a 3-D simulation of equatorial spread F including all the aforementioned instability mechanisms is shown in Figure 4. (For a detailed discussion of the numerical code, see the work of *Aveiro et al.* [2012].) The simulation is cast in magnetic dipole coordinates, (p,q,ϕ), where p is the McIlwain parameter, q is a coordinate that varies along a magnetic field line, and ϕ is the longitude. Controlling numerical diffusion in simulations such as these, which encompass a broad range of spatiotemporal scales is characteristically challenging. For this effort, we employed a flux assignment scheme based on the total variation diminishing (TVD) condition of *Harten* [1983]. This scheme was designed for solving conservation problems with high-order accuracy in such a way as to preserve steepened structures, while minimizing numerical diffusion, dispersion, and nonphysical oscillation. A pedagogical review and extension of the technique has been published by *Trac and Pen* [2003]. That reference describes monotone upwind schemes for conservation laws directly applicable to the ion continuity problem (once recombination has been absorbed into a redefinition of ion density so that the ion continuity equation takes on conservative form, as necessary.) The method we apply combines upward differencing schemes, flux limiting [e.g., *Van-Leer*, 1974], and second-order TVD schemes to minimize both diffusion and dispersion in the time advance. The basic method was extended to three dimensions using the dimensional splitting technique of *Strang* [1968]. In the simulation runs shown here, we use second-order Runge-Kutta time advances with a 10 s step size.

Near the start of the simulation, the background flow is dominated by the evening vortex, indicated by the closed equipotential contours in the equatorial plane [e.g., *Haerendel et al.*, 1992; *Kudeki and Bhattacharyya*, 1999]. Near the uppermost closed contours, strong vertical current flows. This is attendant with the imperfectly efficient F region dynamo and the difference between the zonal neutral wind and the zonal plasma drift at F region altitudes. Strong zonal currents driven mainly by gravity exist at higher altitudes, but only where the background density gradient is modest.

The first waves to emerge appear at the base of the bottomside where the vertical currents are strong. These waves have wavelengths of about 30 km and propagate obliquely with respect to the horizon. They are the transient modes of the collisional shear instability. They are confined to altitudes where vertical current flows, however, and are not responsible for ESF by themselves. They do appear to be responsible for the ubiquitous bottom-type layers seen at Jicamarca and elsewhere, however.

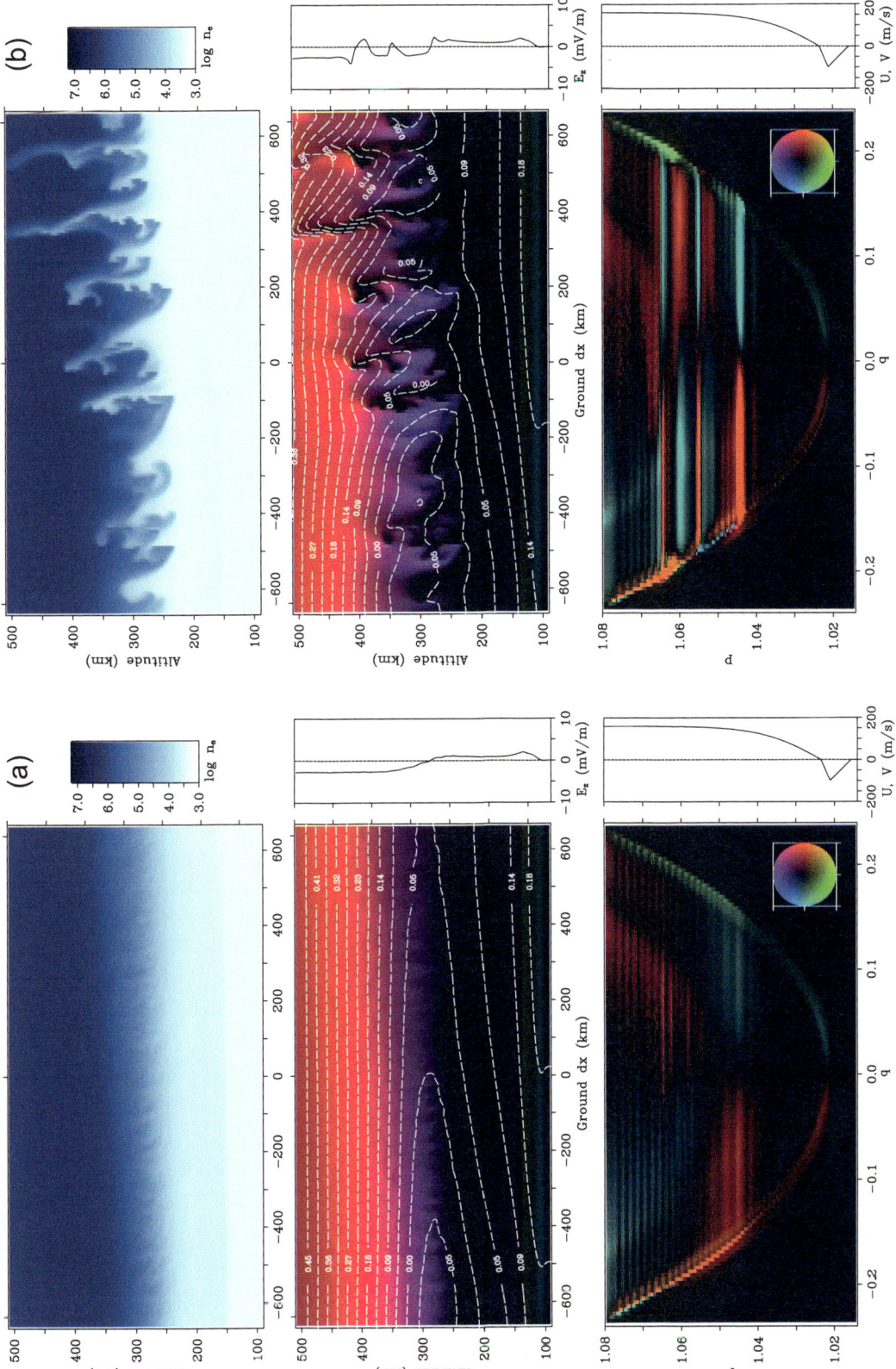

Figure 4. Numerical simulation of plasma irregularities underlying equatorial spread F. (left and right) The groups depict the simulation 20 and 80 min after initialization at 19:15 LT, respectively, under equinox, solar minimum conditions near Jicamarca. (top) Plasma number density in the equatorial plane. (middle) Current density in the equatorial plane with equipotential contours superimposed. (bottom) Current density in the meridional plane. The adjacent line plot shows the assumed zonal neutral wind. The legend for the current densities is given by the color wheel. Maximum scale for the equatorial (meridional) plane is 20 (200) nA m^{-2}. The simulation is initialized with white noise. Note that the contributions to the current density from diamagnetic currents are not included here.

Over time, the original bottom-type irregularities coalesced into waves with wavelengths between 100 and 200 km as the transient response of the instability gave way to the steady state response. Moreover, the simulation was also driven by a background zonal electric field of 0.33 mV m^{-1}, which elevated the bottom-type irregularities to altitudes where substantial zonal current flowed. Afterward, the action of combined collisional shear and generalized Rayleigh-Taylor instability worked to produce large-scale irregularities capable of ascending through the topside. At this point, fully developed ESF was underway. It is noteworthy that the simulations performed recently by *Retterer* [2010a, 2010b] corroborate most of the ESF phenomenology presented here.

Using background electric field measurements from the C/NOFS satellite, *Aveiro et al.* [2012] reproduced ESF depletions very similar to what had been observed by the Altair radar in the summer of 2009. *Aveiro et al.* [2012] was also able to reproduce magnetometer measurements from the CHAMP satellite as it passed through the Kwajalein sector during an ESF event based on the current densities predicted by the simulation under the scenario outlined above. The gross morphology of irregularities in RTI records produced by Jicamarca is also consistent with this scenario [*Aveiro et al.*, 2012].

An important aspect of these simulations is that they do not assume equipotential magnetic field lines. The electrodynamics of elongated, 3-D irregularities differ essentially from those of 2-D ones. Consider that, whereas we expect the ambipolar electric field to be antiparallel to pressure gradients in the direction parallel to B, we expect the electric field to be parallel to pressure gradients in the perpendicular direction. In three dimensions, a complex potential structure surrounds pressure variations that are inseparable. This effect has been studied for meteor trails [*Dimant and Oppenheim*, 2006a, 2006b] but not in the broader context of ionospheric irregularities. Even at the large scales depicted in Figure 4, the plasma convection differs from what would be found using the equipotential approximation.

5. COLLISIONAL DRIFT WAVE INSTABILITIES

In this section, irregularities, which depend on nonequipotential magnetic field lines, are examined. With conventional drift wave, ions move normal to B to preserve charge neutrality, electrons stream along B, and electron inertia leads to overshoot and instability. In the ionosphere, collisional drag replaces inertia in keeping the electron and ion motions out of phase.

Hysell et al. [2002b] considered whether collisional drift waves were responsible for kilometric plasma density irregularities in patchy sporadic E layers. These scales are too small to be related to gravity waves, dynamical instabilities, and E_s layer instabilities and too large to be produced by gradient drift waves [*Seyler et al.*, 2004]. A number of independent sources of information point to the existence of such waves. *Barnes* [1992] inferred the presence of structures with dimensions of about 1 km embedded in larger, drifting plasma clouds from soundings of the layers. Kilometric horizontal structuring was clearly visible in data from a sounding rocket that penetrated a structured E_s layer [*Kelley et al.*, 1995] and in VHF radio scintillations from satellite signals traversing the layers [*Maruyama et al.*, 2000]. *Bernhardt et al.* [2003] observed kilometric structure in optical imagery from sporadic E layers in ionospheric modification experiments at Arecibo.

The dispersion relation for resistive drift waves can be found following the discussion in section 2. The potential equation (3) can be supplemented with the linearized ion continuity equation, written here using similar notation:

$$(-i\omega + i\mathbf{k}_\perp \cdot \mathbf{E}_o \mu_{\perp i} - i(\mathbf{k}_\perp \times \mathbf{E}_o) \cdot \hat{b} \mu_{\times i} + D_\perp k_\perp^2 + D_\| k_\|^2)n_1$$
$$+ n_o \mu_{\perp i} k_\perp^2 \phi_1 = 0. \quad (19)$$

Here the i subscript refers to the ion component of the given transport coefficient. Parallel ion mobility has been neglected, although parallel ion diffusion has not. In the context of irregularities in sporadic E layer "clouds," the unperturbed electric field $\mathbf{E}_o$ includes both the background field and the polarization field created within the cloud.

Combining equation (19) with equation (3) yields the frequency and growth rate of the waves in this local approximation. It can be shown that the highest growth rate is associated with waves with $\mu_\perp k_\perp^2 = \mu_\| k_\|^2$ and with $k_\| L = 1/2$ [*Hysell et al.*, 2002b]. The wave frequency and growth rate in this limit are:

$$\omega_r = \frac{3}{4}\mu_{\perp i}\mathbf{k}_\perp \cdot \mathbf{E}_o + \left(\frac{1}{4}\mu_{\perp i}\frac{\mu_\times}{\mu_\perp} - \mu_{\times i}\right)(\mathbf{k}_\perp \times \mathbf{E}_o) \cdot \hat{b} - \frac{1}{4}\frac{\mu_{\perp i}}{\mu_\perp}$$
$$\times \left(k_\perp^2 D_\perp + k_\|^2 D_\|\right) \quad (20)$$

$$\gamma = \frac{1}{4}\mu_{\perp i}\left[-\mathbf{k}_\perp \cdot \mathbf{E}_o + \frac{\mu_\times}{\mu_\perp}(\mathbf{k}_\perp \times \mathbf{E}_o) \cdot \hat{b}\right] - D_{\perp a}k_\perp^2 - D_{\|a}k_\|^2, \quad (21)$$

where the perpendicular and parallel ambipolar diffusion rates for waves propagating at the assumed magnetic aspect angle are $D_{\perp a} = D_{\perp i} - D_\perp \mu_{\perp i}/4\mu_\perp$ and $D_{\|a} = D_{\|i} - D_\| \mu_{\perp i}/4\mu_\perp$, respectively.

According to equation (21), most rapid growth is anticipated for waves propagating in nearly the $\mathbf{E} \times \mathbf{B}$ direction in ionized layers at altitudes where the Hall mobility is significant. The phase velocity will be nearly in the direction of, but much smaller in magnitude than, the $\mathbf{E} \times \mathbf{B}$ drift speed. In view of the qualifiers leading up to equations (20) and (21), the dominant wavelength will be strongly influenced by L and typically of the order of one to a few kilometers, depending on the vertical depth of the layers.

Figure 5 shows a numerical simulation of collisional drift waves growing in a blob of E region ionization meant to represent a patchy sporadic E layer cloud. The simulation is cast in magnetic dipole coordinates, encompassing a sector of space at midlatitudes in the Northern Hemisphere about 30 km square in the plane perpendicular to B and extending from about 75–200 km in altitude along B. Other simulation parameters are the same as in the ESF runs described earlier.

A background electric field of 0.5 mV m^{-1} directed upward and northward was imposed to drive the simulation. This induced a strong polarization electric field in the center of the cloud with southward, downward, and westward components. The net electric field in the layer (background plus perturbation) was mainly southwestward at the center of the cloud and mainly northward at the periphery. The Hall current was consequently mainly northwestward near the center of the cloud, turning eastward at the periphery. (Electric field rotation such as that seen here is well known to occur in inhomogeneous, anisotropic plasma layers [*Shalimov et al.*, 1998; *Hysell and Drexler*, 2006].) The complicated net current pattern in the plane perpendicular to B in the E layer, which is curiously "S" shaped, closes partially in the F layer via field-aligned currents. Without this closure, the polarization electric field induced in the E layer would be much smaller.

The figure shows the state of the simulation 5 min after initialization at 19:15 LT. Longitudinal, kilometric-scale waves are evident throughout the layer, propagating slowly in the direction opposite the Hall current. The waves follow the S-shaped curve that governs the current. The wavelength, direction of propagation, and initial growth rate of the dominant waves are consistent with the linear dispersion relation derived above. The main nonlinear effect observed in simulation is the ultimate saturation of the wave amplitudes. The relative density perturbations of the waves approach unity, and the associated electric field fluctuations can be larger than the background, driving electric field. Inspection of the current density in the meridional plane shows that the field-aligned currents associated with the wavelike density fluctuations do not flow on surfaces of constant ρ but instead cross magnetic field lines. This is evidence that the waves in question are, in fact, drift waves rather than gradient-drift waves, which would also tend to appear first on the edges of the cloud but not in the center.

Because of their finite parallel wavenumbers, drift waves can extend for long distances along magnetic field lines regardless of their transverse-scale sizes and are, therefore, not strictly E layer phenomena. Although patchy sporadic E layers at midlatitudes provide ideal conditions for collisional drift waves, there appears to be nothing preventing them from operating at high or even low latitudes, wherever the ionospheric conductivity is inhomogeneous and anisotropic and where sufficient background current forcing exists. Preliminary simulations suggest that the waves may be responsible for the irregularities observed at the magnetic equator in the upper E/ valley region at twilight [e.g., *Chau and Hysell*, 2004].

6. SUMMARY AND CONCLUSIONS

This paper reviewed some of the most prevalent ionospheric plasma density irregularities. All involve electrostatic waves, which grow by extracting free energy from their environment, converting it to wave energy, which can rapidly be dissipated. The paper touched on three cross-cutting themes, the first being the shortcomings of irregularity models that assume equipotential magnetic field lines. This assumption neglects the decoupling that occurs at low altitudes where parallel resistivity becomes significant and also sidesteps the essential inseparability of the potential associated with the ambipolar electric field in the directions parallel and perpendicular to the magnetic field. Most obviously, this assumption rules out drift waves, which depend upon finite parallel wavenumbers and which may be important in the ionosphere, particularly in regions where the parallel electron mobility varies rapidly in space.

A second theme was the role of random forcing in shaping ionospheric irregularities. In the case of Farley-Buneman waves and turbulence, the stochastic forcing due to irregularities in the streaming electron flow caused by the waves themselves influences their behavior, limiting their phase speeds and maintaining the condition for marginal linear stability. Using established stochastic differential equation formalisms, the phase speed saturation effect was related to the well-known phenomenon of wave scattering from index of refraction perturbations in a turbulent medium.

The third theme of this paper is the inadequacy of plane-wave analysis, in particular, and boundary-value analysis, in general, for many of the intermediate- and large-scale waves and instabilities important in the ionosphere. The behavior of neither gradient drift waves in the equatorial electrojet nor plasma density depletions in ESF is captured accurately by anything short of initial value analysis via numerical simulation, and conclusions drawn from boundary value analysis should be viewed skeptically. We speculate that threshold conditions based on the free energy available in

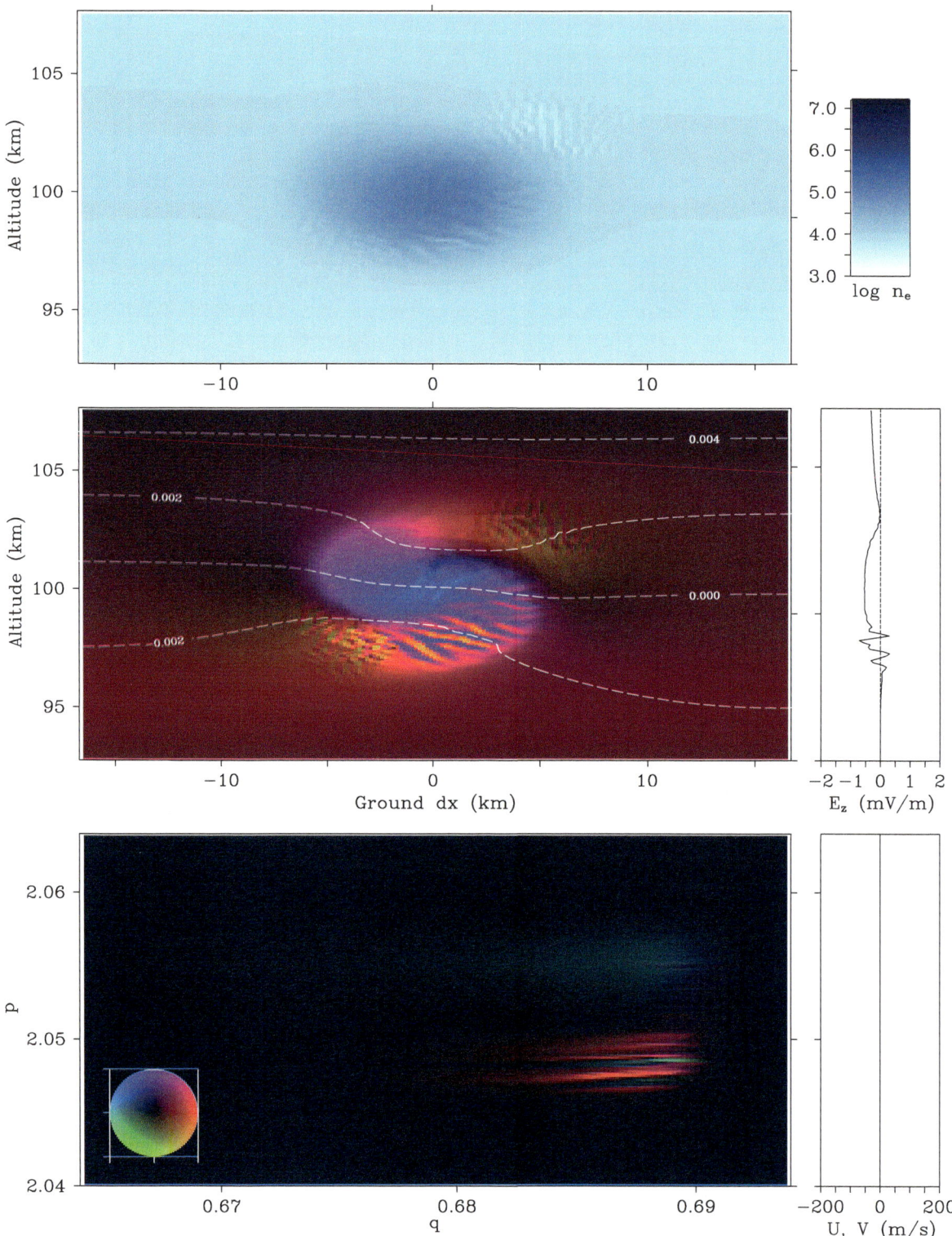

Figure 5. Numerical simulation of drift waves in a patchy sporadic E layer. (top) Plasma number density in a perpendicular-to-B plane passing through the layer. (middle) Current density in the perpendicular-to-B plane with equipotential contours superimposed. The adjacent line plot shows the vertical/meridional electric field component. (bottom) Current density in the meridional plane. The legend for the current densities is given by the color wheel. Maximum scale for the perpendicular-to-B (meridional) plane is 20 (200) nA m^{-2}. The adjacent line plot shows the assumed zonal wind profiles. The simulation is initialized with white noise. Note that the contributions to the current density from diamagnetic currents are not included here.

the ionosphere at a given time may eventually be more expedient forecast tools than the linear growth rate estimates favored in forecast work today.

Ion inertia is critical for Farley-Buneman waves, since inertia is needed for ions to overcome diffusion as they migrate from regions of low to high density. Inertia was neglected for all of the other irregularities discussed. This effectively ruled out the possibility of turbulence in the Kolmogorov or Kraichnan sense, although the latter may be possible in the case of F region irregularities driven by interchange instabilities in the inertial regime.

Absent inertial effects, the dominant nonlinearity in the underlying equations of motion is the one in the convective derivative of the continuity equation. The main effect of this term is usually mode coupling leading to plasma steepening. Steeping is likely important in the production of the small-scale plasma density irregularities detected by coherent scatter radars, although other wave modes may be involved [e.g., *Huba and Ossakow*, 1979, 1981], and kinetic effects generally must also be considered.

Spectral analysis is a common tool applied to the analysis of spacecraft measurements of ionospheric irregularities. Spectral analysis is crucial for analysis of turbulent flows, in which order is generally indiscernible in physical space but evident in wavenumber space, owing to the existence of turbulent cascades, clear separations of scales, etc. Steepened structures, ESF plumes, large-scale gradient drift waves, and related irregularity features, meanwhile, do exhibit order in physical space and so are perhaps best analyzed in that domain.

APPENDIX A: FBGD DISPERSION RELATION

Here we derive the dispersion relation for FBGD waves discussed in the text. The electrons are inertialess and magnetized, and the ions are massive and unmagnetized. Parallel ion motion is neglected throughout. A vertical polarization electric field is presumed to drive a zonal Cowling current and excite the linear instability. The velocities of the ions and electrons can be written explicitly as

$$\mathbf{v}_i = \frac{\Omega_i}{i\omega_i + \nu_{in}} \frac{\mathbf{E}_\perp}{B} - \frac{C_{si}^2}{i\omega_i + \nu_{in}} \nabla_\perp \ln n + v_{io}\hat{x} \quad (A1)$$

$$\mathbf{v}_e = \frac{\mathbf{E} \times \mathbf{B}}{B^2} - \frac{\Omega_e}{\nu_{en}} \frac{\mathbf{E}_\parallel}{B} - \frac{\nu_{en}}{\Omega_e} \frac{\mathbf{E}_\perp}{B} - \frac{\nu_{en}}{\Omega_e^2} C_{se}^2 \nabla_\perp \ln n$$
$$- \frac{C_{se}^2}{\nu_e} \nabla_\parallel \ln n + v_{eo}\hat{x}, \quad (A2)$$

where ν_{jn} is the momentum transfer collision frequency between species j and the neutrals, Ω_j is the cyclotron frequency for species j, and $C_{sj}^2 = \gamma_j K_B T_j/m_j$ with γ_j the ratio of specific heats. Also, v_{io} and v_{eo} are generalized background zonal ion and electron drifts, respectively, driven by a zonal neutral wind or a wind-induced dynamo electric field, as the case may be. The frequency $\omega_i = \omega - k_x v_{io}$ is the intrinsic frequency in the drifting ion frame of reference.

The above are next substituted into the ion and electron continuity equations. Linearization proceeds using the same notational conventions used elsewhere. We define $V_d = E_o/B$ to be the zonal background electron drift velocity in the electrojet. Accordingly, $\omega_e = \omega - k_x(V_d + v_{eo})$. We will assume plane-wave dependencies for the field quantities n and ϕ in time and in the zonal perpendicular (x) and parallel directions but not in the vertical (z) direction. This leads to the following set of differential equations in first-order (perturbation) quantities:

$$i\omega_i n_1 + \frac{C_{si}^2}{i\omega_i + \nu_{in}} k_\perp^2 n_1 + n_o \frac{\Omega_i}{i\omega_i + \nu_{in}} \left(k_x^2 - \frac{\partial^2}{\partial z^2}\right) \frac{1}{B} \phi_1 = 0 \quad (A3)$$

for the ions and

$$i\omega_e n_1 + \frac{\nu_{en}}{\Omega_e^2} C_{se}^2 k_\perp^2 n_1 + \frac{C_{se}^2}{\nu_{en}} k_\parallel^2 n_1 + ik_x \frac{\partial n_o}{\partial z} \frac{1}{B} \phi_1 - n_o \frac{\Omega_e}{\nu_{en}} k_\parallel^2 \phi_1$$
$$- n_o \frac{\nu_e}{\Omega_e} \left(k_x^2 - \frac{\partial^2}{\partial z^2}\right) \frac{1}{B} \phi_1 = 0 \quad (A4)$$

for the electrons. Note that photochemical processes have been neglected for the moment. Since the terms in equations (A3) and (A4) relating to diffusion are only important at small-scale sizes where nonlocal effects are negligible, they have been treated locally, and the notation $k_\perp$ has been introduced to represent the total perpendicular wavenumber at small scales.

Combining the equations above produces

$$\left[i\omega_e + \frac{\nu_e}{\Omega_e^2} C_{se}^2 k_\perp^2 + \frac{C_{se}^2}{\nu_e} k_\parallel^2\right] \left(k_x^2 - \frac{\partial^2}{\partial z^2}\right) \phi_1 - (i\omega_i(i\omega_i + \nu_{in})$$
$$+ C_{si}^2 k_\perp^2) \frac{1}{\Omega_i} \left[ik_x L^{-1} - \frac{\Omega_e}{\nu_{en}} k_\parallel^2 - \frac{\nu_{en}}{\Omega_e} \left(k_x^2 - \frac{\partial^2}{\partial z^2}\right)\right] \phi_1 = 0, \quad (A5)$$

where L is the vertical background density gradient scale length. The often-quoted linear, local dispersion relation for FBGD waves comes from taking the terms in parentheses above to be $k_\perp^2$. After solving the real and imaginary parts of equation (A5), the following expressions for the wave frequency and growth rate (real and imaginary parts of ω) can be found:

$$\omega_r = \frac{k_x(V_d + v_{eo} - v_{io})}{1+\psi} + k_x v_{io} \qquad (A6)$$

$$\gamma = \frac{1}{1+\psi}\left[\frac{\psi}{v_{in}}(\omega_{ir}^2 - k_\perp^2 C_s^2) - \frac{\omega_{ir}}{L}\frac{k_x}{k_\perp^2}\frac{v_{in}}{\Omega_i}\right], \qquad (A7)$$

where $C_s^2 \equiv K_B(\gamma_i T_i + \gamma_e T_e)/m_i$ is the square of the ion acoustic speed, $\psi_o = v_{en}v_{in}/\Omega_e\Omega_i$ and $\psi \equiv \psi_o[1 + (\Omega_e^2/v_{en}^2)(k_\parallel^2/k_\perp^2)]$ is called the anisotropy factor. In the above, $\omega_{ir} = \omega_r - k_x v_{io}$. In deriving these expressions, the growth rate was assumed to be much smaller than the real part of the wave frequency ($\gamma \ll \omega_r$).

Inspection of equation (A7) shows that the instability growth rate is affected by the difference between the electron and ion drift velocities, or by currents. Neutral winds are often cited as influences on stability, whether they depend on how efficiently they generate dynamos and wind-induced electrons drifts, which can significantly limit wind-driven current [*Richmond*, 1973; *Anandarao et al.*, 1977; *Broche et al.*, 1978; *Reddy and Devasia*, 1981; *Hysell et al.*, 2002a].

Since gradient drift wave wavelengths can be comparable to L in the equatorial electrojet, a nonlocal treatment is required to analyze them. Focusing on gradient drift waves, neglecting terms that are quadratic in the wave frequency in equation (A5) and retaining the z derivatives yields the following differential equation for the electrostatic potential:

$$\left[\omega_i k_x L^{-1}\frac{v_{in}}{\Omega_i} + \left(\frac{\psi}{v_{in}}(k_\perp^2 C_s^2 - \omega^2) + i\omega(1+\psi) - ik_x V_d^*\right)\right.$$
$$\left.\times\left(k_x^2 - \frac{\partial^2}{\partial z^2}\right)\right]\phi_1 = 0 \qquad (A8)$$

where $V_d^* = V_d + v_{eo} + \psi v_{io}$. Solving for the nonlocal frequency and growth rate, we can neglect the ω^2 term stemming from ion inertia and unimportant for gradient drift waves and reorganize terms:

$$\frac{\partial^2 \phi_1}{\partial z^2} - k_x^2 \phi_1 \left(\frac{\omega_i L^{-1}(v_{in}/\Omega_i)/k_x}{i\omega(1+\psi) - ikV_d^* + D_\perp k_\perp^2} + 1\right) = 0 \qquad (A9)$$

where the perpendicular diffusion $D_\perp \equiv C_s^2\psi/v_{in}$ has been explicitly defined.

The above expression can be put into a more familiar form with the help of some variable transformations. Expanding the background density profile in a power series about z such that $\frac{1}{n_o}\frac{\partial n}{\partial z} = L^{-1} \rightarrow (1 - z^2/d^2 + \cdots)L_o^{-1}$, where L_o is the density gradient scale length at its shortest, and d is the vertical spatial extent of an ionization layer in the nighttime E region and leads to

$$\frac{\partial^2 \phi_1}{\partial \xi^2} + \left(E - \frac{\xi^2}{4}\right)\phi_1 = 0, \qquad (A10)$$

with $\xi = z/\Delta$, $\Delta^2 = \beta d/2k_x$, $\beta^2 = (i\omega_i(1+\psi) - ikV_d^* + D_\perp k_\perp^2)/(\omega_i L_o^{-1}v_{in}/\Omega_i k_x)$, and $E = \Delta^2 k_x^2(\beta^{-2} - 1)$. Equation (A10) has been written in the form of the quantum oscillator eigenvalue problem and is essentially the same equation derived by *Perkins* [1973] in a nonlocal analysis of the ionospheric interchange instability. The largest growth rate prescribed by equation (A10) comes with the smallest eigenvalue, $E = 1/2$. We can, therefore, solve for the frequency and growth rate of the fastest growing mode in terms of the β^2 parameter, yielding a nonlocal correction to the dispersion relation.

$$\beta^2 = 1 - \frac{1}{2k_x^2 d^2}\left(\sqrt{1 + 4k_x^2 d^2} - 1\right) \qquad (A11)$$

$$\approx 1 - \frac{1}{k_x d}, \qquad k_x^2 d^2 \gg 1. \qquad (A12)$$

The dispersion relation then becomes

$$\omega_i \frac{k_o}{k_x}(1+\psi) = i\omega(1+\psi) - ik_x V_d^* + D_\perp k_\perp^2 \qquad (A13)$$

$$k_o \equiv \frac{v_{in}}{\Omega_i L_o}\frac{\beta^2}{1+\psi}, \qquad (A14)$$

where k_o is similar to the parameter introduced by *Kudeki et al.* [1982] except with the nonlocal correction. This parameter determines the long-wavelength cutoff for the gradient drift instability as is evident from the frequency and growth rate expressions:

$$\omega_r = \frac{k_x(V_d + v_{eo} - v_{io})}{(1+\psi)(1+(k_o/k_x)^2)} + kv_{io} \qquad (A15)$$

$$\gamma = -\omega_{ir}\frac{k_o}{k_x} - \frac{1}{1+\psi}D_\perp k_\perp^2. \qquad (A16)$$

APPENDIX B: BOURRET'S INTEGRAL EQUATION

Bourret [1962a, 1962b] formulated an expansion, which includes provisions for the effects of stochastic forcing in tracking the evolution of a dynamical system. His formalism is not restricted to normal or delta-correlated statistics, although it does assume that the expectation of the stochastic forcing is zero.

The derivation followed below is due to *Van Kampen* [1975], who attributes the main reasoning and conclusions to *Bourret* [1962a, 1962b]. More general means of arriving at the same conclusions are available, as will be pointed out.

If A_o is constant, the solution to sure equation (the equation with $\alpha = 0$) is just

$$u(t) = e^{A_o t} a.$$

If α is sufficiently small, we can expect this solution to have considerable bearing on the behavior of $\langle u(t) \rangle$. This motivates the application of the following unitary transformation:

$$u(t) = e^{A_o t} v(t) \tag{B1}$$

$$A_1(t) = e^{A_o t} V(t) e^{-A_o t}, \tag{B2}$$

which maps the original problem into a new one:

$$\dot{v}(t) = \alpha V(t) v(t) \tag{B3}$$

$$v(0) = a \tag{B4}$$

in which V is now purely stochastic, the sure part of the problem having been transformed away. A solution for the new dependent variable $v(t)$ can be found formally through integration:

$$v(t) = a + \alpha \int_0^t dt' V(t') v(t'), \tag{B5}$$

where a is a constant of integration. This procedure can be repeated iteratively. Repeating it just once and taking the ensemble average then yields:

$$\langle v(t) \rangle = a + \alpha^2 \int_0^t dt' \int_0^{t'} dt'' \langle V(t') V(t'') v(t'') \rangle, \tag{B6}$$

where the time average of V has been taken to be zero in view of the absence of deterministic forcing.

Equation (B6) contains a triple moment involving the dependent variable v being sought and the evolution matrix V, which is related to the original stochastic forcing function through unitary transformation. At this point, it is expedient to assume that the two quantities are statistically uncorrelated. This assumption is not actually essential for finding the central result of this analysis but is consistent with the assumption of small-amplitude fluctuations. The same results can be obtained using the method of cumulant expansion [*Kubo*, 1963; *Van Kampen*, 1974a, 1974b]. Following *Van Kampen* [1975], we can make the assumption here to keep the present, heuristic analysis simple. The resulting equation is

$$\langle v(t) \rangle = a + \alpha^2 \int_0^t dt' \int_0^{t'} dt'' \langle V(t') V(t'') \rangle \langle v(t'') \rangle. \tag{B7}$$

Next, differentiation is applied to evolve a preliminary form of the differential equation for $\langle u(t) \rangle$:

$$\frac{d}{dt} \langle v(t) \rangle = \alpha^2 \int_0^t dt' \langle V(t) V(t') \rangle \langle v(t') \rangle. \tag{B8}$$

Defining the auxiliary variable $\tau = t - t'$ and transforming accordingly then gives

$$\frac{d}{dt} \langle v(t) \rangle = \alpha^2 \int_0^t d\tau \langle V(t) V(t-\tau) \rangle \langle v(t-\tau) \rangle. \tag{B9}$$

At this point, we can assume that the evolution matrix $V(t)$ has a short correlation time, τ_c, such that the temporal translation of the state variable $v(t)$ on the right side of equation (B9) is unimportant and can be neglected for values of $\tau < \tau_c$ for which the evolution matrix is significant. In that case, we have

$$\frac{d}{dt} \langle v(t) \rangle = \alpha^2 \int_0^\infty d\tau \langle V(t) V(t-\tau) \rangle \langle v(t) \rangle. \tag{B10}$$

Finally, we can use the similarity transformation to transform back to the original problem and the original state variable $u(t)$:

$$\frac{d}{dt} \langle u(t) \rangle = \{ A_o + \alpha^2 \int_0^\infty \langle A_1(t) e^{A_o \tau} A_1(t-\tau) \rangle e^{-A_o \tau} d\tau \} \langle u(t) \rangle. \tag{B11}$$

Equation (B11) is the sought-after differential equation obeyed for the ensemble average of the state vector $u(t)$, which evolves according to the original systematic evolution matrix A_o, with a small correction related to the autocorrelation function of the stochastic evolution matrix $A_1(t)$. Notice that very little has been assumed about the statistics underlying the problem. This analysis is not limited to normal statistics, for example.

Acknowledgments. This work was supported by NSF award AGS-1042057 to Cornell University. Additional support was supplied by award FA9550-03-1-0337 from the Air Force Office of Sponsored Research. The Jicamarca Radio Observatory is a facility of the Instituto Geofísico del Perú operated with support from NSF award AGS-0905448 through Cornell. The help of the staff is much appreciated.

REFERENCES

Abdu, M. A., E. A. Kherani, I. S. Batista, E. R. de Paula, D. C. Fritts, and J. H. A. Sobral (2009), Gravity wave initiation of equatorial spread F/plasma bubble irregularities based on observational data from the SpreadFEx campaign, *Ann. Geophys.*, *27*, 2607–2622.

Alken, P., and S. Maus (2010), Relationship between the ionospheric eastward electric field and the equatorial electrojet, *Geophys. Res. Lett.*, *37*, L04104, doi:10.1029/2009GL041989.

Anandarao, B. G., R. Raghavarao, and C. R. Reddi (1977), Electric fields by gravity wave winds in the equatorial ionosphere, *J. Geophys. Res.*, *82*(10), 1510–1512.

Aveiro, H. C., D. L. Hysell, R. G. Caton, K. M. Groves, J. Klenzing, R. F. Pfaff, R. Stoneback, and R. A. Heelis (2012), Three-dimensional numerical simulations of equatorial spread F: Results and observations in the Pacific sector, *J. Geophys. Res.*, *117*, A03325, doi:10.1029/2011JA017077.

Bahcivan, H. (2007), Plasma wave heating during extreme electric fields in the high-latitude E region, *Geophys. Res. Lett.*, *34*, L15106, doi:10.1029/2006GL029236.

Baker, J. B. H., J. M. Ruohoniemi, A. J. Ribeiro, L. B. N. Clausen, R. A. Greenwald, N. A. Frissel, and K. A. Sterne (2011), Super-DARN ionospheric space weather, *IEEE Aerosp. Electron. Syst. Mag.*, *26*, 30–34, doi:10.1109/MAES.2011.6065656.

Barnes, R. I. (1992), An investigation into the horizontal structure of spread-E_s, *J. Atmos. Terr. Phys.*, *54*, 391–399.

Batista, I. S., M. A. Abdu, A. J. Carrasco, B. W. Reinisch, E. R. de Paula, N. J. Schuch, and F. Bertoni (2008), Equatorial spread F and sporadic E-layer connections during the Brazilian Conjugate Point Equatorial Experiment (COPEX), *J. Atmos. Sol. Terr. Phys.*, *70*, 1133–1143.

Bernhardt, P. A., N. A. Gondarenko, P. N. Guzdar, F. T. Djuth, C. A. Tepley, M. P. Sulzer, S. L. Ossakow, and D. L. Newman (2003), Using radio-induced aurora to measure the horizontal structure of ion layers in the lower thermosphere, *J. Geophys. Res.*, *108*(A9), 1336, doi:10.1029/2002JA009712.

Bourret, R. C. (1962a), Propagation of randomly perturbed fields, *Can. J. Phys*, *40*, 782–790.

Bourret, R. C. (1962b), Stochastically perturbed fields, with application to wave propagation in random media, *Nuovo Cimento*, *26*, 1–31.

Broche, P., M. Crochet, and J. Gagnepain (1978), Neutral winds and phase velocity of the instabilities in the equatorial electrojet, *J. Geophys. Res.*, *83*(A3), 1145–1146.

Buneman, O. (1963), Excitation of field aligned sound waves by electron streams, *Phys. Rev. Lett.*, *10*, 285–287.

Chau, J. L., and D. L. Hysell (2004), High altitude large-scale plasma waves in the equatorial electrojet at twilight, *Ann. Geophys.*, *22*, 4071–4076.

Chu, Y. H., C. Y. Wang, S. L. Su, and R. M. Kuong (2011), Coordinated sporadic E layer observations made with Chung-Li 30 MHz radar, ionosonde and FORMOSAT-3/COSMIC satellites, *J. Atmos. Sol. Terr. Phys.*, *73*, 883–894.

Cosgrove, R. B. (2007), Generation of mesoscale F layer structure and electric fields by the combined Perkins and Es layer instabilities, in simulations, *Ann. Geophys.*, *25*, 1579–1601.

Cosgrove, R. B., and R. T. Tsunoda (2002), A direction-dependent instability of sporadic-E layers in the nighttime midlatitude ionosphere, *Geophys. Res. Lett.*, *29*(18), 1864, doi:10.1029/2002GL014669.

Cosgrove, R. B., and R. T. Tsunoda (2004), Instability of the E-F coupled nighttime midlatitude ionosphere, *J. Geophys. Res.*, *109*, A04305, doi:10.1029/2003JA010243.

De La Beaujardiere, O. (2004), C/NOFS: A mission to forecast scintillations, *J. Atmos. Sol. Terr. Phys.*, *66*, 1573–1591.

Didebulidze, G. G., and L. N. Lomidze (2010), Double atmospheric gravity wave frequency oscillations of sporadic E formed in a horizontal shear flow, *Phys. Lett. A*, *354*, 952–959.

Dimant, Y. S., and G. M. Milikh (2003), Model of anomalous electron heating in the E region: 1. Basic theory, *J. Geophys. Res.*, *108*(A9), 1350, doi:10.1029/2002JA009524.

Dimant, Y. S., and M. M. Oppenheim (2004), Ion thermal effects on E-region instabilities: Linear theory, *J. Atmos. Sol. Terr. Phys.*, *66*, 1639–1654.

Dimant, Y. S., and M. M. Oppenheim (2006a), Meteor trail diffusion and fields: 1. Simulations, *J. Geophys. Res.*, *111*, A12312, doi:10.1029/2006JA011797.

Dimant, Y. S., and M. M. Oppenheim (2006b), Meteor trail diffusion and fields: 2. Analytical theory, *J. Geophys. Res.*, *111*, A12313, doi:10.1029/2006JA011798.

Dimant, Y. S., and M. M. Oppenheim (2008), Large-scale 2D and 3D Simulations of plasma turbulence in the lower ionosphere, paper YO3.005 presented at the 50th Annual Meeting of the APS Division of Plasma Physics, Dallas, Tex.

Dimant, Y. S., and M. M. Oppenheim (2011a), Magnetosphere-ionosphere coupling through E region turbulence: 1. Energy budget, *J. Geophys. Res.*, *116*, A09303, doi:10.1029/2011JA016648.

Dimant, Y. S., and M. M. Oppenheim (2011b), Magnetosphere-ionosphere coupling through E region turbulence: 2. Anomalous conductivities and frictional heating, *J. Geophys. Res.*, *116*, A09304, doi:10.1029/2011JA016649.

Dymond, K. F., et al. (2011), A medium-scale traveling ionospheric disturbance observed from the ground and from space, *Radio Sci.*, *46*, RS5010, doi:10.1029/2010RS004535.

Farley, D. T., Jr. (1963), A plasma instability resulting in field-aligned irregularities in the ionosphere, *J. Geophys. Res.*, *68*(22), 6083–6097.

Farley, D. T. (2009), The equatorial E-region and its plasma instabilities: A tutorial, *Ann. Geophys.*, *27*, 1509–1520.

Farley, D. T., and J. Providakes (1989), The variation with T_e and T_i of the velocity of unstable ionospheric two-stream waves, *J. Geophys. Res.*, *94*(A11), 15,415–15,420.

Flaherty, J. P., C. E. Seyler, and L. N. Trefethen (1999), Large-amplitude transient growth in the linear evolution of equatorial spread *F* with a sheared zonal flow, *J. Geophys. Res.*, *104*(A4), 6843–6857.

Foster, J. C., and P. J. Erickson (2000), Simultaneous observations of E-region coherent backscatter and electric field amplitude at F-region heights with the Millstone Hill UHF Radar, *Geophys. Res. Lett.*, *27*(19), 3177–3180.

Fritts, D. C., and S. L. Vadas (2008), Gravity wave penetration into the thermosphere: Sensitivity to solar cycle variations and mean winds, *Ann. Geophys.*, *26*, 3841–3861.

Fritts, D. C., et al. (2009), Overview and summary of the Spread F Experiment (SpreadFEx), *Ann. Geophys.*, *26*, 2141–2155.

Fu, Z. F., L. C. Lee, and J. D. Huba (1986), A quasi-local theory of the E × B instability in the ionosphere, *J. Geophys. Res.*, *91*(A3), 3263–3269.

Goldstein, S. (1931), On the stability of superposed streams of fluids of different densities, *Proc. R. Soc. London, Ser. A*, *132*, 524–548.

Haerendel, G., and J. V. Eccles (1992), The role of the equatorial electrojet in the evening ionosphere, *J. Geophys. Res.*, *97*(A2), 1181–1192.

Haerendel, G., J. V. Eccles, and S. Çakir (1992), Theory for modeling the equatorial evening ionosphere and the origin of the shear in the horizontal plasma flow, *J. Geophys. Res.*, *97*(A2), 1209–1223.

Haldoupis, C. (1989), A review on radio studies of auroral *E*-region ionospheric irregularities, *Ann. Geophys.*, *7*, 239–258.

Hamza, A. M., and J.-P. St-Maurice (1993a), A turbulent theoretical framework for the study of current-driven E region irregularities at high latitudes: Basic derivation and application to gradient-free situations, *J. Geophys. Res.*, *98*(A7), 11,587–11,599.

Hamza, A. M., and J.-P. St-Maurice (1993b), A self-consistent fully turbulent theory of auroral *E* region irregularities, *J. Geophys. Res.*, *98*(A7), 11,601–11,613.

Harten, A. (1983), High resolution schemes for hyperbolic conservation laws, *J. Comput. Phys*, *49*(3), 357–393.

Hoh, F. C. (1963), Instability of Penning-type discharge, *Phys. Fluids*, *6*, 1184–1191.

Hu, S., and A. Bhattacharjee (1999), Gradient drift instabilities and turbulence in the nighttime equatorial electrojet, *J. Geophys. Res.*, *104*(A12), 28,123–28,132.

Huba, J. D., and G. Joyce (2010), Global modeling of equatorial plasma bubbles, *Geophys. Res. Lett.*, *37*, L17104, doi:10.1029/2010GL044281.

Huba, J. D., and L. C. Lee (1983), Short wavelength stabilization of the gradient drift instability due to velocity shear, *Geophys. Res. Lett.*, *10*(4), 357–360.

Huba, J. D., and S. L. Ossakow (1979), On the generation of 3-m irregularities during equatorial spread *F* by low-frequency drift waves, *J. Geophys. Res.*, *84*(A11), 6697–6700.

Huba, J. D., and S. L. Ossakow (1981), On 11-cm irregularities during equatorial spread *F*, *J. Geophys. Res.*, *86*(A2), 829–832.

Hysell, D. L., and J. Drexler (2006), Polarization of elliptic *E* region plasma irregularities and implications for coherent radar backscatter from Farley-Buneman waves, *Radio Sci.*, *41*, RS4015, doi:10.1029/2005RS003424.

Hysell, D. L., and E. Kudeki (2004), Collisional shear instability in the equatorial F region ionosphere, *J. Geophys. Res.*, *109*, A11301, doi:10.1029/2004JA010636.

Hysell, D. L., J. L. Chau, and C. G. Fesen (2002a), Effects of large horizontal winds on the equatorial electrojet, *J. Geophys. Res.*, *107*(A8), 1214, doi:10.1029/2001JA000217.

Hysell, D. L., M. Yamamoto, and S. Fukao (2002b), Simulations of plasma clouds in the midlatitude *E* region ionosphere with implications for type I and type II quasiperiodic echoes, *J. Geophys. Res.*, *107*(A10), 1313, doi:10.1029/2002JA009291.

Hysell, D. L., J. Chun, and J. L. Chau (2004), Bottom-type scattering layers and equatorial spread *F*, *Ann. Geophys.*, *22*, 4061–4069.

Hysell, D. L., M. F. Larsen, C. M. Swenson, and T. F. Wheeler (2006), Shear flow effects at the onset of equatorial spread *F*, *J. Geophys. Res.*, *111*, A11317, doi:10.1029/2006JA011963.

Hysell, D. L., J. Drexler, E. B. Shume, J. L. Chau, D. E. Scipion, M. Vlasov, R. Cuevas, and C. Heinselman (2007), Combined radar observations of equatorial electrojet irregularities at Jicamarca, *Ann. Geophys.*, *25*, 457–473.

Hysell, D. L., G. Michhue, M. F. Larsen, R. Pfaff, M. Nicolls, C. Heinselman, and H. Bahcivan (2008), Imaging radar observations of Farley Buneman waves during the JOULE II experiment, *Ann. Geophys.*, *26*, 1837–1850.

Hysell, D. L., E. Nossa, M. F. Larsen, J. Munro, M. P. Sulzer, and S. A. González (2009), Sporadic *E* layer observations over Arecibo using coherent and incoherent scatter radar: Assessing dynamic stability in the lower thermosphere, *J. Geophys. Res.*, *114*, A12303, doi:10.1029/2009JA014403.

Kagan, L. M., and J.-P. St.-Maurice (2004), Impact of electron thermal effects on Farley-Buneman waves at arbitrary aspect angles, *J. Geophys. Res.*, *109*, A12302, doi:10.1029/2004JA010444.

Kagan, L. M., R. S. Kissack, M. C. Kelley, and R. Cuevas (2008), Unexpected rapid decrease in phase velocity of submeter Farley-Buneman waves with altitude, *Geophys. Res. Lett.*, *35*, L03106, doi:10.1029/2007GL032459.

Kelley, M. C. (2011), On the origin of mesoscale TIDs at midlatitudes, *Ann. Geophys.*, *29*, 361–366.

Kelley, M. C., and C. A. Miller (1997), Electrodynamics of midlatitude spread *F* 3. Electrohydrodynamic waves? A new look at the role of electric fields in thermospheric wave dynamics, *J. Geophys. Res.*, *102*(A6), 11,539–11,547.

Kelley, M. C., M. F. Larsen, C. LaHoz, and J. P. McClure (1981), Gravity wave initiation of equatorial spread *F*: A case study, *J. Geophys. Res.*, *86*(A11), 9087–9100.

Kelley, M. C., D. Riggin, R. F. Pfaff, W. E. Swartz, J. F. Providakes, and C. S. Huang (1995), Large amplitude quasi-periodic fluctuations associated with a midlatitude sporadic E layer, *J. Atmos. Terr. Phys.*, *57*, 1165–1178.

Kelley, M. C., J. J. Makela, O. de La Beaujardière, and J. Retterer (2011), Convective ionospheric storms: A review, *Rev. Geophys.*, *49*, RG2003, doi:10.1029/2010RG000340.

Keskinen, M. J., and S. L. Vadas (2009), Three-dimensional nonlinear evolution of equatorial ionospheric bubbles with gravity wave seeding and tidal wind effects, *Geophys. Res. Lett.*, *36*, L12102, doi:10.1029/2009GL037892.

Keskinen, M. J., H. G. Mitchell, J. A. Fedder, P. Satyanarayana, S. T. Zalesak, and J. D. Huba (1988), Nonlinear evolution of the Kelvin-Helmholtz instability in the high-latitude ionosphere, *J. Geophys. Res.*, *93*(A1), 137–152.

Kil, H., and R. A. Heelis (1998), Global distribution of density irregularities in the equatorial ionosphere, *J. Geophys. Res.*, *103*(A1), 407–417.

Krall, J., J. D. Huba, G. Joyce, and T. Zalesak (2009), Three-dimensional simulation of equatorial spread-F with meridional wind effects, *Ann Geophys.*, *27*, 1821–1830.

Krall, J., J. D. Huba, S. L. Ossakow, G. Joyce, J. J. Makela, E. S. Miller, and M. C. Kelley (2011), Modeling of equatorial plasma bubbles triggered by non-equatorial traveling ionospheric disturbances, *Geophys. Res. Lett.*, *38*, L08103, doi:10.1029/2011GL046890.

Krane, B., J. L. Pecseli, H. Sato, J. Trulsen, and A. W. Wernik (2010), Low-frequency electrostatic waves in the ionospheric E region, *Plasma Sources Sci. Technol.*, *19*(3), 034007, doi:10.1088/0963-0252/19/3/034007.

Kubo, R. (1963), Stochastic Liouville equations, *J. Math. Phys.*, *4*, 174–183.

Kudeki, E., and S. Bhattacharyya (1999), Postsunset vortex in equatorial *F*-region plasma drifts and implications for bottomside spread-*F*, *J. Geophys. Res.*, *104*(A12), 28,163–28,170.

Kudeki, E., B. G. Fejer, D. T. Farley, and H. M. Ierkic (1981), Interferometer studies of equatorial F region irregularities and drifts, *Geophys. Res. Lett.*, *8*(4), 377–380.

Kudeki, E., D. T. Farley, and B. G. Fejer (1982), Long wavelength irregularities in the equatorial electrojet, *Geophys. Res. Lett.*, *9*(6), 684–687.

Kudeki, E., B. G. Fejer, D. T. Farley, and C. Hanuise (1987), The Condor equatorial electrojet campaign: Radar results, *J. Geophys. Res.*, *92*(A12), 13,561–13,577.

Kudeki, E., A. Akgiray, M. A. Milla, J. L. Chau, and D. L. Hysell (2007), Equatorial spread-F initiation: Post-sunset vortex, thermospheric winds, gravity waves, *J. Atmos. Sol. Terr. Phys.*, *69*(17–18), 2416–2427.

Larsen, M. F., D. L. Hysell, Q. H. Zhou, S. M. Smith, J. Friedman, and R. L. Bishop (2007), Imaging coherent scatter radar, incoherent scatter radar, and optical observations of quasiperiodic structures associated with sporadic E layers, *J. Geophys. Res.*, *112*, A06321, doi:10.1029/2006JA012051.

Makarevich, R. A. (2009), Coherent radar measurements of the Doppler velocity in the auroral E region, *Radio Sci. Bull.*, *32*, 33–46.

Maruyama, T., S. Fukao, and M. Yamamoto (2000), A possible mechanism for echo striation generation of radar backscatter from midlatitude sporadic E, *Radio Sci.*, *35*(5), 1155–1164.

Mathews, J. D. (1998), Sporadic E: Current views and recent progress, *J. Atmos. Sol. Terr. Phys.*, *60*, 413–435.

McClure, J. P., S. Singh, D. K. Bamgboye, F. S. Johnson, and H. Kil (1998), Occurrence of equatorial F region irregularities: Evidence for tropospheric seeding, *J. Geophys. Res.*, *103*(A12), 29,119–29,135.

Milikh, G. M., and Y. S. Dimant (2002), Kinetic model of electron heating by turbulent electric field in the E region, *Geophys. Res. Lett.*, *29*(12), 1575, doi:10.1029/2001GL013935.

Miller, E. S., J. J. Makela, and M. C. Kelley (2009), Seeding of equatorial plasma depletions by polarization electric fields from middle latitudes: Experimental evidence, *Geophys. Res. Lett.*, *36*, L18105, doi:10.1029/2009GL039695.

Ogawa, T., N. Nishitani, Y. Otsuka, K. Shiokawa, T. Tsugawa, and K. Hosokawa (2009), Medium-scale traveling ionospheric disturbances observed with the SuperDARN Hokkaido radar, all-sky imager, and GPS network and their relation to concurrent sporadic E irregularities, *J. Geophys. Res.*, *114*, A03316, doi:10.1029/2008JA013893.

Oppenheim, M. M., and Y. S. Dimant (2004), Ion thermal effects on E-region instabilities: 2-D kinetic simulations, *J. Atmos. Sol. Terr. Phys.*, *66*, 1655–1668.

Oppenheim, M. M., Y. Dimant, and L. Dyrud (2008), Large scale simulations of 2D fully kinetic Farley Buneman turbulence, *Ann. Geophys.*, *26*, 543–553.

Otani, N. F., and M. Oppenheim (1998), A saturation mechanism for the Farley-Buneman instability, *Geophys. Res. Lett.*, *25*(11), 1833–1836.

Otsuka, Y., T. Tani, T. Tsugawa, T. Ogawa, and A. Saito (2008), Statistical study of relationship between medium-scale traveling ionospheric disturbance and sporadic E layer activities in summer night over Japan, *J. Atmos. Sol. Terr. Phys.*, *70*, 2196–2202.

Otsuka, Y., K. Shiokawa, T. Ogawa, T. Yokoyama, and M. Yamamoto (2009), Spatial relationship of nighttime medium-scale traveling ionospheric disturbances and *F* region field-aligned irregularities observed with two spaced all-sky airglow imagers and the middle and upper atmosphere radar, *J. Geophys. Res.*, *114*, A05302, doi:10.1029/2008JA013902.

Park, J., H. Luhr, C. Stolle, M. Rother, K. W. Min, and I. Michaelis (2009), The characteristics of field-aligned currents associates with equatorial plasma bubbles as observed by the CHAMP satellite, *Ann. Geophys.*, *27*, 2685–2697.

Perkins, F. (1973), Spread *F* and ionospheric currents, *J. Geophys. Res.*, *78*(1), 218–226.

Pfaff, R. F. (1995), Comparison of wave electric field measurements of two-stream instabilities in the auroral and equatorial electrojets, in *Plasma Instabilities in the Ionospheric E-Region. Proceedings of a Workshop held at the Max-Planck-Institut für*

Aeronomie in Katlenburg-Lindau, Germany, 24–26 Oct., edited by K. Schlegel, p. 19, Cuvillier, Göttingen, Germany.

Pfaff, R. F., M. C. Kelley, B. G. Fejer, N. C. Maynard, L. G. Brace, B. G. Ledley, L. G. Smith, and R. F. Woodman (1985), Comparative in situ studies of the unstable daytime equatorial E region, *J. Atmos. Terr. Phys.*, *47*, 791–811.

Pi, X., A. Freeman, B. Chapman, P. Rosen, and Z. Li (2011), Imaging ionospheric inhomogeneities using spaceborne synthetic aperture radar, *J. Geophys. Res.*, *116*, A04303, doi:10.1029/2010JA016267.

Prakash, S., B. H. Subbaraya, and S. P. Gupta (1972), Rocket measurements of ionization irregularities in the equatorial ionosphere at Thumba and identification of plasma irregularities, *Indian J. Radio Space Phys.*, *1*, 72–80.

Rappaport, Y. G., M. Hayakawa, O. E. Gotynyan, V. N. Ivchenko, A. K. Fedorenka, and Y. A. Selivanov (2009), Stable and unstable plasma perturbations in the ionospheric F region, caused by spatial packet of atmospheric gravity waves, *Phys. Chem. Earth*, *35*, 508–515.

Reddy, C. A., and C. V. Devasia (1981), Height and latitude structure of electric fields and currents due to local east-west winds in the equatorial electrojet, *J. Geophys. Res.*, *86*(A7), 5751–5767.

Retterer, J. M. (2005), Physics-based forecasts of equatorial radio scintillation for the Communication and Navigation Outage Forecasting System (C/NOFS), *Space Weather*, *3*, S12C03, doi:10.1029/2005SW000146.

Retterer, J. M. (2010a), Forecasting low-latitude radio scintillation with 3-D ionospheric plume models: 1. Plume model, *J. Geophys. Res.*, *115*, A03306, doi:10.1029/2008JA013839.

Retterer, J. M. (2010b), Forecasting low-latitude radio scintillation with 3-D ionospheric plume models: 2. Scintillation calculation, *J. Geophys. Res.*, *115*, A03307, doi:10.1029/2008JA013840.

Richmond, A. D. (1973), Equatorial electrojet, I, Development of a model including winds and instabilities, *J. Atmos. Terr. Phys.*, *35*, 1083–1103.

Riggin, D., W. E. Swartz, J. Providakes, and D. T. Farley (1986), Radar studies of long-wavelength waves associated with mid-latitude sporadic E layers, *J. Geophys. Res.*, *91*(A7), 8011–8024.

Robinson, R. T. (1986), Towards a self consistent nonlinear theory of radar auroral backscatter, *J. Atmos. Terr. Phys.*, *48*, 417–422.

Ronchi, C., P. L. Similon, and R. N. Sudan (1989), A nonlocal linear theory of the gradient drift instability in the equatorial electrojet, *J. Geophys. Res.*, *94*(A2), 1317–1326.

Ronchi, C., R. N. Sudan, and D. T. Farley (1991), Numerical simulations of large-scale plasma turbulence in the daytime equatorial electrojet, *J. Geophys. Res.*, *96*(A12), 21,263–21,279.

Röttger, J. (1973), Wavelike structures of large scale equatorial spread F irregularities, *J. Atmos. Terr. Phys.*, *35*, 1195–1206.

Sahr, J. D., and B. G. Fejer (1996), Auroral electrojet plasma irregularity theory and experiment: A critical review of present understanding and future directions, *J. Geophys. Res.*, *101*(A12), 26,893–26,909.

Schlegel, K., and C. Haldoupis (1994), Observation of the modified two-stream plasma instability in the midlatitude E region ionosphere, *J. Geophys. Res.*, *99*(A4), 6219–6226.

Seyler, C. E., J. M. Rosado-Roman, and D. T. Farley (2004), A nonlocal theory of the gradient-drift instability in ionospheric plasmas at mid-latitudes, *J. Atmos. Sol. Terr. Phys.*, *66*(17), 1627–1637.

Shalimov, S., C. Haldoupis, and K. Schlegel (1998), Large polarization electric fields associated with midlatitude sporadic E, *J. Geophys. Res.*, *103*(A6), 11,617–11,625.

Simon, A. (1963), Instability of a partially ionized plasma in crossed electric and magnetic fields, *Phys. Fluids*, *6*, 382–388.

Singh, S., F. S. Johnson, and R. A. Power (1997), Gravity wave seeding of equatorial plasma bubbles, *J. Geophys. Res.*, *102*(A4), 7399–7410.

Singh, S. B., K. Patel, R. P. Patel, A. K. Singh, and R. P. Singh (2010), Modeling of VHF scintillation observed at low latitude, *J. Phys. Conf. Ser.*, *208*(1), 012065, doi:10.1088/1742-6596/208/1/012065.

Sreeja, V., C. Vineeth, T. K. Pant, S. Ravindran, and R. Sridharan (2009), Role of gravity wavelike seed perturbations on the triggering of ESF – A case study from unique dayglow observations, *Ann. Geophys.*, *27*, 313–318.

St.-Maurice, J.-P. (1987), A unified theory of anomalous resistivity and Joule heating effects in the presence of ionospheric E region irregularities, *J. Geophys. Res.*, *92*(A5), 4533–4542.

St.-Maurice, J.-P. (1990), Electron heating by plasma waves in the high-latitude E region ionosphere and related effects: Theory, *Adv. Space Res.*, *10*, 239–249.

St.-Maurice, J.-P., and A. M. Hamza (2001), A new nonlinear approach to the theory of E region irregularities, *J. Geophys. Res.*, *106*(A2), 1751–1759.

St.-Maurice, J.-P., and R. S. Kissack (2000), The role played by thermal feedback in heated Farley-Buneman waves at high latitudes, *Ann. Geophys.*, *18*, 532–546.

St.-Maurice, J.-P., and R. Laher (1985), Are observed broadband plasma wave amplitudes large enough to explain the enhanced electron temperatures of the high-latitude E region?, *J. Geophys. Res.*, *90*(A3), 2843–2850.

St.-Maurice, J.-P., R. K. Choudhary, W. L. Ecklund, and R. T. Tsunoda (2003), Fast type-I waves in the equatorial electrojet: Evidence for nonisothermal ion-acoustic speeds in the lower E region, *J. Geophys. Res.*, *108*(A5), 1170, doi:10.1029/2002JA009648.

Strang, G. (1968), On the construction and comparison of difference schemes, *SIAM J. Numer. Anal.*, *5*, 506–517.

Sudan, R. N. (1983a), Nonlinear theory of type I irregularities in the equatorial electrojet, *Geophys. Res. Lett.*, *10*(10), 983–986.

Sudan, R. N. (1983b), Unified theory of Type I and Type II irregularities in the equatorial electrojet, *J. Geophys. Res.*, *88*(A6), 4853–4860.

Sudan, R. N., J. Akinrimisi, and D. T. Farley (1973), Generation of small-scale irregularities in the equatorial electrojet, *J. Geophys. Res.*, *78*(1), 240–248.

Sultan, P. J. (1996), Linear theory and modeling of the Rayleigh-Taylor instability leading to the occurrence of equatorial spread F, *J. Geophys. Res.*, *101*(A12), 26,875–26,891.

Suzuki, S., K. Hosokawa, Y. Otsuka, K. Shiokawa, T. Ogawa, N. Nishitani, T. F. Shibata, A. V. Koustov, and B. M. Shevtsov (2009), Coordinated observations of nighttime medium-scale traveling ionospheric disturbances in 630-nm airglow and HF radar echoes at midlatitudes, *J. Geophys. Res.*, *114*, A07312, doi:10.1029/2008JA013963.

Takahashi, H., et al. (2009), Simultaneous observation of ionospheric plasma bubbles and mesospheric gravity waves during the SpreadFEx campaign, *Ann. Geophys.*, *27*, 1477–1487.

Taylor, G. I. (1931), Effect of variation in density on the stability of superposed streams of fluids, *Proc. R. Soc. London, Ser. A*, *132*, 499–523.

Trac, H., and U. L. Pen (2003), A primer on Eulerian computational fluid dynamics for astrophysicists, *Astrophysics*, *115*, 303–321.

Tsunoda, R. T. (2005), On the enigma of day-to-day variability in equatorial spread *F*, *Geophys. Res. Lett.*, *32*, L08103, doi:10.1029/2005GL022512.

Tsunoda, R. T. (2006), On the coupling of layer instabilities in the nighttime midlatitude ionosphere, *J. Geophys. Res.*, *111*, A11304, doi:10.1029/2006JA011630.

Tsunoda, R. T. (2010), On seeding equatorial spread *F*: Circular gravity waves, *Geophys. Res. Lett.*, *37*, L10104, doi:10.1029/2010GL043422.

Tsunoda, R. T., and B. R. White (1981), On the generation and growth of equatorial backscatter plumes 1. Wave structure in the bottomside *F* layer, *J. Geophys. Res.*, *86*(A5), 3610–3616.

Tsunoda, R. T., R. C. Livingston, and C. L. Rino (1981), Evidence of a velocity shear in bulk plasma motion associated with the post-sunset rise of the equatorial F-layer, *Geophys. Res. Lett.*, *8*(7), 807–810.

Uspensky, M. V., P. Hanhunen, A. V. Koustov, and K. Kauristie (2011), Volume cross section of auroral radar backscatter and RMS plasma fluctuations inferred from coherent and incoherent scatter data: A response on backscatter volume parameters, *Ann. Geophys.*, *29*, 1081–1092.

Vadas, S. L., and D. C. Fritts (2004), Thermospheric response to gravity waves arising from mesoscale convective complexes, *J. Atmos. Sol. Terr. Phys.*, *66*, 781–804.

Vadas, S. L., and M. J. Keskinen (2010), Correction to "Three-dimensional nonlinear evolution of equatorial ionospheric bubbles with gravity wave seeding and tidal wind effects", *Geophys. Res. Lett.*, *37*, L03101, doi:10.1029/2009GL041216.

Vadas, S. L., and H. Liu (2009), Generation of large-scale gravity waves and neutral winds in the thermosphere from the dissipation of convectively generated gravity waves, *J. Geophys. Res.*, *114*, A10310, doi:10.1029/2009JA014108.

Van Kampen, N. G. (1974a), A cumulant expansion for stochastic linear differential equations. I, *Physica*, *74*, 215–238.

Van Kampen, N. G. (1974b), A cumulant expansion for stochastic linear differential equations. II, *Physica*, *74*, 239–247.

Van Kampen, N. G. (1975), Stochastic differential equations, *Phys. Rep.*, *24*, 171–228.

Van-Leer, B. (1974), Towards the ultimate conservation difference scheme. II. Monotonicity and conservation combined in a second-order scheme, *J. Comput. Phys.*, *14*, 361–370.

Whitehead, J. D. (1972), The structure of sporadic *E* from a radio experiment, *Radio Sci.*, *7*(3), 355–358.

Whitehead, J. D. (1989), Recent work on mid-latitude and equatorial sporadic *E*, *J. Atmos. Terr. Phys.*, *51*, 401–424.

Woodman, R. F. (2009), Spread F – An old equatorial aeronomy problem finally resolved?, *Ann. Geophys.*, *27*, 1915–1934.

Woodman, R. F., and J. L. Chau (2002), First Jicamarca radar observations of two-stream E region irregularities under daytime counter equatorial electrojet conditions, *J. Geophys. Res.*, *107*(A12), 1482, doi:10.1029/2002JA009362.

Woodman, R. F., and C. La Hoz (1976), Radar observations of *F* region equatorial irregularities, *J. Geophys. Res.*, *81*(31), 5447–5466.

Woodman, R. F., M. Yamamoto, and S. Fukao (1991), Gravity wave modulation of gradient drift instabilities in mid-latitude sporadic *E* irregularities, *Geophys. Res. Lett.*, *18*(7), 1197–1200.

Yamamoto, M., S. Fukao, R. F. Woodman, T. Ogawa, T. Tsuda, and S. Kato (1991), Mid-latitude E region field-aligned irregularities observed with the MU radar, *J. Geophys. Res.*, *96*(A9), 15,943–15,949.

Yamamoto, M., S. Fukao, T. Ogawa, T. Tsuda, and S. Kato (1992), A morphological study of mid-latitude *E*-region field-aligned irregularities observed with the MU radar, *J. Atmos. Terr. Phys.*, *54*, 769–777.

Yokoyama, T., Y. Otsuka, T. Ogawa, M. Yamamoto, and D. L. Hysell (2008), First three-dimensional simulation of the Perkins instability in the nighttime midlatitude ionosphere, *Geophys. Res. Lett.*, *35*, L03101, doi:10.1029/2007GL032496.

Yokoyama, T., D. L. Hysell, Y. Otsuka, and M. Yamamoto (2009), Three-dimensional simulation of the coupled Perkins and E_s-layer instabilities in the nighttime midlatitude ionosphere, *J. Geophys. Res.*, *114*, A03308, doi:10.1029/2008JA013789.

Zargham, S., and C. E. Seyler (1987), Collisional interchange instability: 1. Numerical simulations of intermediate-scale irregularities, *J. Geophys. Res.*, *92*(A9), 10,073–10,088.

Zargham, S., and C. E. Seyler (1989), Collisional and inertial dynamics of the ionospheric interchange instability, *J. Geophys. Res.*, *94*(A7), 9009–9027.

H. C. Aveiro and D. L. Hysell, Earth and Atmospheric Sciences, Cornell University, Ithaca, NY 14853, USA. (dlh37@cornell.edu)

J. L. Chau, Radio Observatorio de Jicamarca, Instituto Geofísico del Perú, Lima, Peru.

Three-Dimensional Numerical Simulations of Equatorial Spread F: Results and Diagnostics in the Peruvian Sector

H. C. Aveiro and D. L. Hysell

Earth and Atmospheric Sciences, Cornell University, Ithaca, New York, USA

A 3-D initial boundary value simulation of equatorial spread F (ESF) is described. The simulation advances the plasma number density and electrostatic potential forward in time by enforcing the constraints of quasineutrality and momentum conservation. The simulation evolves under realistic background conditions including bottomside plasma shear flow and vertical current. The results are evaluated using computed numerical proxies for ESF observations, including *in situ* observations from magnetometers on board satellites and remote sensing observations made by coherent/incoherent scatter radar and airglow imagers. The diagnostic codes can be used to validate the numerical simulation, which is evidently able to reproduce the salient characteristics of ESF observed by these instruments.

1. INTRODUCTION

Around sunset, the transfer of dominance from the E region to the F region dynamo creates a shear in the horizontal plasma flow in the equatorial F region bottomside [*Haerendel et al.*, 1992]. Above the shear node, the vertical electric field points downward, and the plasma drifts eastward. Below the node, the vertical electric field points upward, and the plasma drifts westward, even though the neutral flow is everywhere eastward. The imperfectly efficient F region dynamo sets up a vertical current supplied, in part, by field-aligned currents connecting the equatorial F region to the E layer at low latitudes to midlatitudes. Related to this phenomenon, a family of plasma irregularities collectively referred to as equatorial spread F (ESF) disturbs the nighttime equatorial F region (see review by *Woodman* [2009]).

At altitudes below the plasma shear node, the differential motion between plasma and neutrals and associated vertical current provides a source of free energy that can drive collisional shear instability (CSI). During the initial transient phase, the zonal wavelength of the fastest-growing mode (λ) is approximately twice the density gradient scale length $\left(L^{-1} = \frac{1}{n}\frac{\partial n}{\partial z}\right)$, i.e., $\lambda \simeq 2L$. Longer wavelengths dominate during the asymptotic phase, with $\lambda \simeq 4\pi L$ [*Hysell and Kudeki*, 2004]. For typical bottomside gradient-scale lengths, the zonal wavelengths for the transient and asymptotic growth phases are expected to be, respectively, ~30 and 150–200 km. The irregularities produced by CSI develop obliquely at an intermediate angle between zonal and vertical. CSI growth rates are faster than that of generalized Rayleigh-Taylor (gRT) under nominal conditions. Typical growth rates are ~5×10^{-3} s^{-1} or 18 e folds per hour during the transient phase [*Kudeki et al.*, 2007]. CSI, therefore, can account for the scales and rapid growth of the dominant irregularities observed by radar in ESF.

Bottom-type layers are almost ubiquitous over Jicamarca immediately after sunset, and their patchy signature seems to signal the onset of the CSI [*Hysell et al.*, 2004]. Irregularities generated by CSI remain confined to the base of the F region, however, and are, therefore, not space weather effective by themselves. In order to bridge the gap between irregularities at the bottomside and the topside, zonal electric fields are generally needed to propel irregularities toward the F peak. Topside irregularities are driven by the gRT instability near and above the F peak. Typical gRT growth rates at the F peak

Modeling the Ionosphere-Thermosphere System
Geophysical Monograph Series 201
© 2013. American Geophysical Union. All Rights Reserved.
10.1029/2012GM001277

are ~10^{-3} s^{-1} [*Sultan*, 1996; *Hysell et al.*, 2006], which implies no more than three to four e folds per hour. This is a universal instability (i.e., nearly scale independent), and thus, the gRT theory cannot explain by itself the predominant scale sizes of ESF irregularities.

In recent years, efforts to simulate ESF in three dimensions have been made [*Huba et al.*, 2008; *Retterer*, 2010; *Keskinen*, 2010]. These simulations are driven in such a way as to emphasize the role of gRT, while mainly neglecting CSI. More recent work has shown that the two instabilities function together in nature: CSI appears neither to suppress gRT nor merely to seed it; rather, the two mechanisms appear to grow and develop faster when acting together than either when acting alone, exhibiting a greater mixing depth [*Aveiro and Hysell*, 2010].

In this paper, we perform numerical simulations of equatorial spread F, assessing the roles of CSI and gRT and comparing the results with different kinds of space weather diagnostics. The diagnostics in question include magnetometers on board satellites, airglow imagers, and coherent/incoherent scatter radar. The main issue we address here is the degree to which the simulations can reproduce the salient features in remote sensing and in situ measurements of the plasma irregularities in ESF, such as tilting, spacing, layer thickness, altitude range, onset time, and rate of development.

2. NUMERICAL SIMULATIONS

Below, we describe a numerical scheme for the simulation of the plasma density irregularities in equatorial spread F. The 3-D simulation updates the plasma density and electrostatic potential in time assuming quasineutrality and momentum balance. Inertia is neglected in the present incarnation. The characteristics of the neutral atmosphere are imported from empirical models. The background electric field is specified and controls the forcing. Simulations are performed with realistic background conditions including bottomside plasma shear flow and the attendant vertical current.

2.1. Conservation Laws

Neglecting stress, heat flow, and inertia, the fluid continuity equation and momentum equations for electrons and ions are given as:

$$\frac{\partial n_\alpha}{\partial t} + \nabla(n_\alpha \mathbf{v}_\alpha) = P_\alpha - L_\alpha, \qquad (1)$$

$$0 = -\nabla p_\alpha + n_\alpha q_\alpha (\mathbf{E} + \mathbf{v}_\alpha \times \mathbf{B}) + m_\alpha n_\alpha [\mathbf{g} - \nu_{\alpha n}(\mathbf{v}_\alpha - \mathbf{U}) - \sum_{\beta}^{\alpha \neq \beta} \nu_{\alpha \beta}(\mathbf{v}_\alpha - \mathbf{v}_\beta)], \qquad (2)$$

where $p_\alpha = n_\alpha k_B T_\alpha$ is the pressure. The variables n_α, q_α, m_α, T_α, P_α, L_α, and $\mathbf{v}_\alpha$ are number density, electric charge, mass, temperature, production and loss rates, and velocity of the species α, respectively. The loss term includes charge exchange and dissociative recombination for a given species. Nighttime photoionization is neglected, but production due to charge exchange is considered. Here β represents the other ionized bodies that exchange momentum or energy with α. The terms $\nu_{\alpha\beta}$ and $\nu_{\alpha n}$ are the collision frequencies with charged particles and neutrals, respectively. The terms $\mathbf{E}$, $\mathbf{B}$, $\mathbf{U}$, $\mathbf{g}$, and k_B represent the electric field, magnetic field, neutral wind, gravity, and the Boltzmann constant, respectively. The ion composition includes O^+, NO^+, and O_2^+. Equation (1) is not in conservation form, since it includes chemical processes. However, it can be converted to conservative form through variable transformation. Two simplifications are applied to equations (1) and (2): (a) diamagnetic drifts ($\nabla n_\alpha \times \mathbf{B}$) are neglected, since their associated flux divergence is very small compared to the other terms, and (b) Coulomb collisions (the last term in the right-hand side of equation (2)) are treated implicitly and only included in the parallel direction.

2.2. Numerical Scheme

The simulation algorithm performs two computations. First, the self-consistent electric field is found using a 3-D potential solver based on quasineutrality, which can be expressed as

$$\nabla[\hat{\Sigma}\nabla\Phi] = \nabla[\hat{\Sigma}(\mathbf{E}_o + \mathbf{U} \times \mathbf{B}) + \hat{D}\nabla n + \hat{\Xi}\mathbf{g}], \qquad (3)$$

where $\mathbf{E}_o$ is the background electric field, $\hat{\Sigma}$ is the conductivity tensor, $\hat{D}$ includes the diffusivity tensor, $\hat{\Xi}\mathbf{g}$ is gravity-driven current density, and Φ is the electrostatic potential that must arise to preserve a solenoidal current density. Explicit definitions of the tensors are given in the work of *Shume et al.* [2005]. Note that the resulting electric field is the combination of the background and perturbed fields (i.e., $\mathbf{E} = \mathbf{E}_o - \nabla\Phi$). Equation (3) involves an elliptic partial differential equation for the potential in three dimensions. It is solved using the biconjugate gradient stabilized (BiCGStab) method [e.g., *van der Vorst*, 1992] using the algorithms described by *Saad* [1990].

Second, we solve a discretized version of equation (1) for each ion species using a monotone upwind scheme for conservation laws (a pedagogical review can be found in the work of *Trac and Pen* [2003]) directly applicable to the ion continuity problem (once the recombination term has been absorbed into a redefinition of n_α so that equation (1) becomes a conservation equation, as necessary). The characteristic of the neutral atmosphere (densities, temperature, and wind velocity) are updated in time on the basis of inputs from climatological models. The background electric field is specified and partly controls the forcing. The model was constructed using tilted magnetic dipole coordinates (p, q, ϕ), where the tilt is matched to the magnetic declination in the longitude of interest. In our terminology, p represents the McIlwain parameter (L), q is the magnetic colatitude, and ϕ is the longitude [see, e.g., *Hysell et al.*, 2004].

2.3. Empirical Model Drivers

To initialize the model runs, we derive plasma number densities from the parameterized ionospheric model (PIM), a parameterizations of the output of several regional theoretical model outputs generated for different climatological conditions and tuned somewhat to agree with data from Jicamarca and elsewhere [*Daniell et al.*, 1995]. PIM reproduces electron density profiles from Kwajalein and Jicamarca fairly closely if scaled slightly to account for day-to-day variability [*Aveiro and Hysell*, 2010]. The background number density profiles used for initial conditions are intended to reproduce measurements from Jicamarca, including the F region peak height and peak density, the bottomside density scale height, and the density in the valley region.

Plasma mobilities are calculated using neutral composition and temperature estimates from the National Research Laboratory-Mass Spectrometer Incoherent Scatter E00 model [*Picone et al.*, 2002] and ionospheric composition estimates from the International Reference Ionosphere-2007 model [*Bilitza and Reinisch*, 2008]. We take the ionosphere and neutral atmosphere to be in thermodynamic equilibrium after sunset ($T_n = T_e = T_i$). Expressions for the ion-neutral and electron-neutral collision frequencies used to compute conductivities can be taken from the work of *Richmond* [1972].

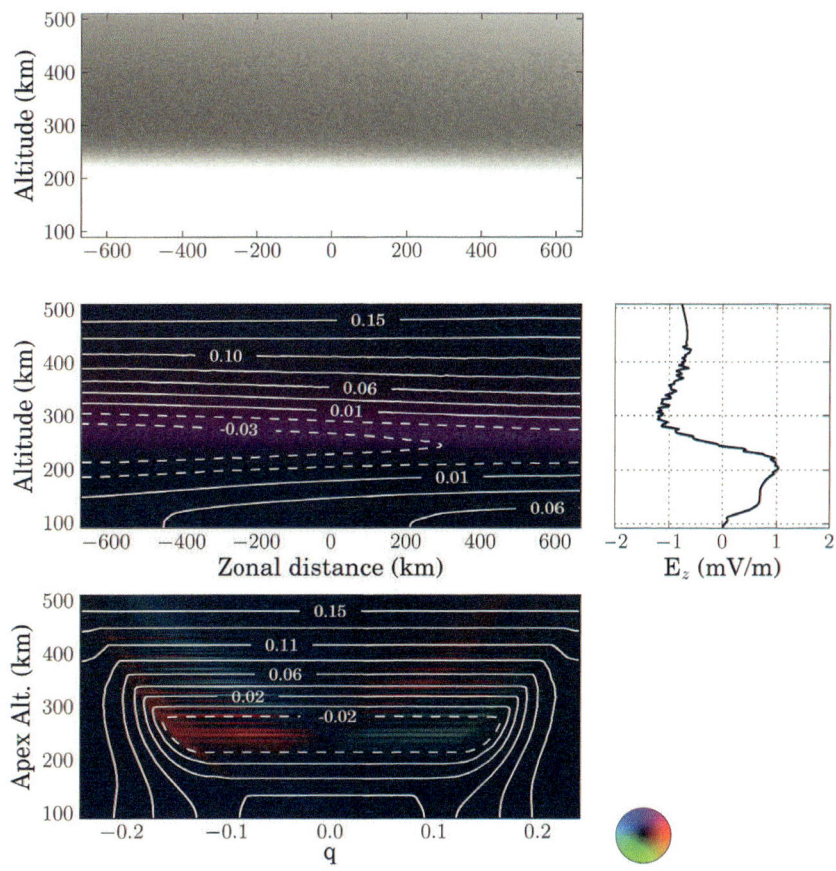

Figure 1. Initial conditions for equatorial spread F model run (i.e., at $t_0 = 00:15$ UT). (top) Plasma number density. (middle) Transverse current density in the equatorial plane with equipotential contours (in kV) superimposed in white (middle right) vertical electric field. (bottom) Current density in the meridional plane. The current densities are vector quantities with magnitudes and directions indicated by the color wheel at the bottom right. The maximum scale is 10 nA m^{-2} in the equatorial plane and 200 nA m^{-2} in the meridional plane. Note that diamagnetic currents are not represented in this diagnostic figure.

Expressions for the Pedersen, Hall, and direct mobilities and diffusivities, themselves, are found in the work of *Kelley* [2009], for example.

For our runs, the zonal neutral winds are obtained from the Horizontal Wind Model-07 [*Drob et al.*, 2008]. As with PIMs, these may be tuned slightly to account for day-to-day variability. The model is sensitive to the wind speeds throughout the thermosphere, since the growth rate of the collisional shear instability depends on the counterstreaming flow of the neutrals and plasma. Meridional and vertical winds were neglected in the simulations presented here.

2.4. Simulation Setup

Our simulation was cast on a rectangular grid $139 \times 133 \times 189$ points wide in (p, q, ϕ) space constructed using a tilted magnetic dipole coordinate system. A cut through the equatorial plane spans altitudes between 90 and 510 km and longitudes between $\pm 6°$. The flux tubes covered by the parallel coordinate all extend to the lower E region. The runs shown below are centered on the dip equator near Jicamarca ($-12.2°$N latitude, $283.1°$E longitude) with background conditions modeled for $F10.7 = 100$. To seed the simulation runs, we added independent Gaussian white noise to the initial number density with a 20% relative amplitude. The local time dependence of the background zonal electric field (E_ϕ) was taken to be a symmetric triangular function that was zero for times before 18 LT and after 21 LT, with a peak of 0.8 mV m^{-1} at 19:30 LT. This zonal electric field time history approximates Jicamarca radar observations made on 8 June 2011 (not shown here). E_ϕ was taken to be constant throughout the F region, but its amplitude is reduced below the valley to suppress the growth of gradient-drift instabilities in the electrojet.

The initial conditions of our simulation at $t_0 = 00{:}15$ UT ($\approx 19{:}15$ LT for the Peruvian sector) are shown in Figure 1. The top panel shows the background plasma density. The middle panel of Figure 1 shows the perpendicular current density with the magnitude and direction indicated by the color disc in the lower panel. There are two major components to the current density that are combined in this depiction. First, a small eastward current driven by gravity exists throughout the region and increases with altitude. In addition, the eastward F region wind drives a vertical current in the region above about 200 km. The vertical current shows a maximum near 300 km and a strong vertical gradient in the bottomside between 280 and 340 km altitude. Below 150 km, the current is influenced by both the local zonal winds and the enhancement of the zonal current in the electrojet. Above the bottomside, the electric field is directed downward, consistent with an eastward $\mathbf{E} \times \mathbf{B}$ drift in the same direction as the dynamo wind. Across the bottomside, the vertical electric field changes direction, producing a sheared zonal $\mathbf{E} \times \mathbf{B}$ drift.

The bottom panel of Figure 1 shows equatorward currents in both hemispheres on field lines that thread the bottomside in the altitude range from 200 to 250 km. These currents feed the upward current driven by the F region zonal wind seen in the middle panel. Above the bottomside, the meridional current is distributed across all the field lines and is poleward in both hemispheres, serving to close the current loop driven by the F region dynamo.

2.5. Coherent/Incoherent Scatter Radar Simulation

Simulated plasma densities in a cut through the magnetic equator are a good proxy for the types of ESF observations made by the ALTAIR radar in the Pacific sector (see, e.g., Figure 1 of *Aveiro et al.* [2012]). Figure 2 shows the results for times ranging from 00:25 to 02:05 hours after the start of the simulation. The top panel of Figure 2 depicts the action of the collisional shear instability. The spatial scales of the irregularities forming at the base of the bottomside range

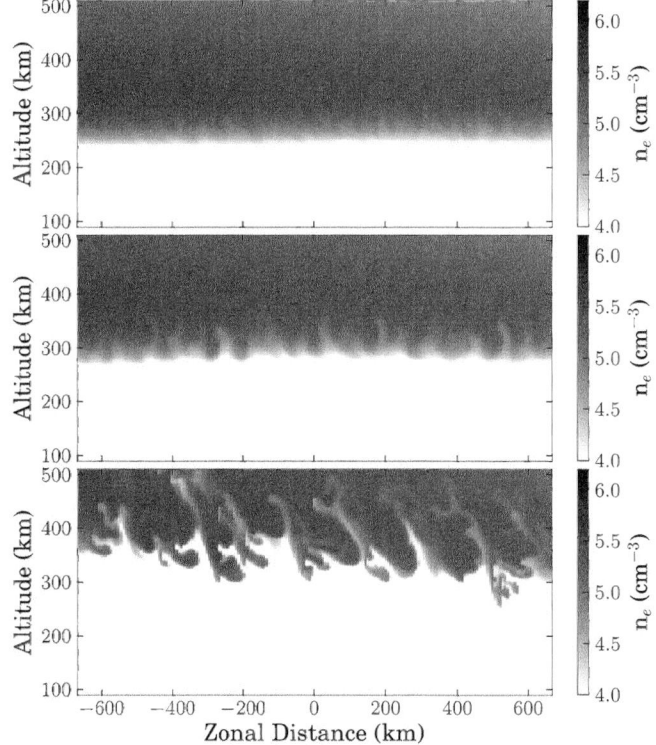

Figure 2. Simulated plasma densities in a cut through the Jicamarca latitudinal plane. The three panels show results for simulation times of 00:25, 00:50, and 02:05 LT, respectively, after a start time $t_0 = 0{:}25$ UT (LT = UT $-$ 5 h at the horizontal center of the simulation).

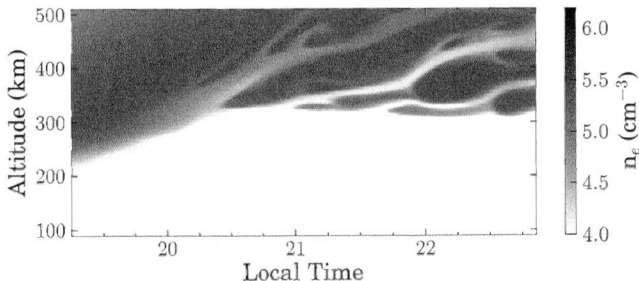

Figure 3. Simulated range-time-intensity maps of electron density observations. In a real observation, incoherent scatter would be contaminated by enhanced coherent scatter from regions of depleted densities.

between 30 and 50 km. The wavefronts are not completely vertical but are tilted (westward). These irregularities were confined to altitudes where the plasma flow is retrograde. The middle panel of Figure 2 shows the vertical development of the irregularities. The plasma depletions had roll-like shapes, but the strong zonal electric field displaced them vertically by ~50 km. The bottom panel of Figure 2 depicts the irregularities after they have crossed the F peak. Bifurcation on the top of the plasma depletions is often evident above the peak. Secondary wind-driven instabilities appear to grow on the western walls of the plasma irregularities. The spatial scales of the irregularities that transcended the F peak range between 100 and 200 km. Had they been observed using a fixed beam coherent scatter radar, the plasma irregularities in Figure 2 would be labeled as bottom-type, bottomside, and topside spread F, respectively.

The Jicamarca 50 MHz radar is not fully steerable like ALTAIR. This limitation has been overcome to some extent with the advent of ionospheric radar imaging [*Hysell*, 1996; *Woodman*, 1997]. However, coherent scatter measured by Jicamarca is still most often presented in a slit camera-like format called range-time-intensity (RTI) maps, which show the temporal behavior of the measured parameter for a span of altitudes (see, e.g., Figure 1 of *Hysell and Burcham* [1998]). Had the simulated ionosphere been sounded by the Jicamarca radar, the resultant RTI map would have been similar to Figure 3 only with enhanced coherent scatter highlighting the regions of depleted density. During the initial hours of the simulation (before 20:30 LT), the bottomside rose up from ~220 to ~300 km altitude in ~1 h, i.e., at an average of ~22 m s^{-1} vertical drift. At the same time, the 50 km thick layer (located between the F peak and valley) created by the collisional shear instability drifted upward with the background ionosphere. Subsequently, irregularities crossed through the F peak and reached the top boundary of the simulation. Some of the topside irregularities formed a "C shape," a feature commonly observed by coherent scatter radars.

2.6. Airglow Imagery Simulation

Airglow emissions corresponding to the simulation runs were computed for the 6300 Å (or oxygen red) line. The volume emission rate (in photons cm^{-3} s^{-1}) is given by [*Link and Cogger*, 1988]

$$V_{6300} = \frac{0.76\beta k_1 n_{O^+} n_{O_2}}{1 + (k_2 n_{N_2} + k_3 n_{O_2} + k_4 n_e + k_5 n_O)/A_{1D}}, \quad (4)$$

where n_α represents the densities for species α, and A_{1D} is the Einstein transition coefficient for the O(^{1}D) state ($A_{1D} = A_{6300} + A_{6364} + A_{6392} = 7.45 \times 10^{-3}$ s^{-1}, $A_{6300} = 5.63 \times 10^{-3}$, $A_{6364} = 1.82 \times 10^{-3}$, $A_{6392} = 8.92 \times 10^{-7}$) [*Link and Cogger*, 1989]. Also, β is the production efficiency of O(^{1}D) from dissociative recombination of O_2^+ and was obtained from the Frank-Condon distribution, Table 13 in the work of *Bates* [1990]. The reactions involving the coefficients k_i are given in Table 1.

A simulated airglow image was obtained through the integration of the volume emission rates along the camera line of sight. For the present computation, the camera was placed virtually in Chilean territory looking north, i.e., observing the airglow emissions and depleted flux tubes due to F region plasma irregularities near Jicamarca. The location of the virtual camera is close to the airglow imaging system at the Cerro Tololo InterAmerican Observatory near La Serena [*Makela and Miller*, 2008]. We do not assume a virtual

Table 1. Chemistry of O(^{1}D) in the Nightglow

Reaction	Rate Coefficient (cm^3 s^{-1})[a]	References
$O^+ + O_2 \rightarrow O_2^+ + O$	$k_1 = 3.23 \times 10^{-12} e^{3.72/\tau_i - 1.87/\tau_i^2}$	
$O(^1D) + N_2 \rightarrow O(^3P) + N_2$	$k_2 = 2.0 \times 10^{-11} e^{111.8/T_n}$	*Link and Cogger* [1988]
$O(^1D) + O_2 \rightarrow O(^3P) + O_2$	$k_3 = 2.9 \times 10^{-11} e^{67.5/T_n}$	
$O(^1D) + e \rightarrow O(^3P) + e$	$k_4 = 1.60 \times 10^{-12} T_e^{0.91}$	
$O(^1D) + O \rightarrow O(^3P) + O$	$k_5 = 2.55 \times 10^{-12}$	*Sobral et al.* [1993]

[a]$\tau_i = T_i/300$.

altitude of emission (see the work of *Tinsley* [1982] for a more complete explanation of the geometry involved in the airglow imagery) and use coordinates representing scanning angles in elevation and azimuth. Since typical camera exposure times are longer than a minute [see, e.g., *Makela and Miller*, 2008], the simulated airglow emissions represent 2 min averages.

Figure 4 shows the postsunset airglow imagery simulation. During the first hours after sunset (Figure 4), stronger emissions appeared on the western half of the image due to recombination. Intermediate-scale irregularities (30–50 km) were barely detectable in the bottomside of the *F* region. At

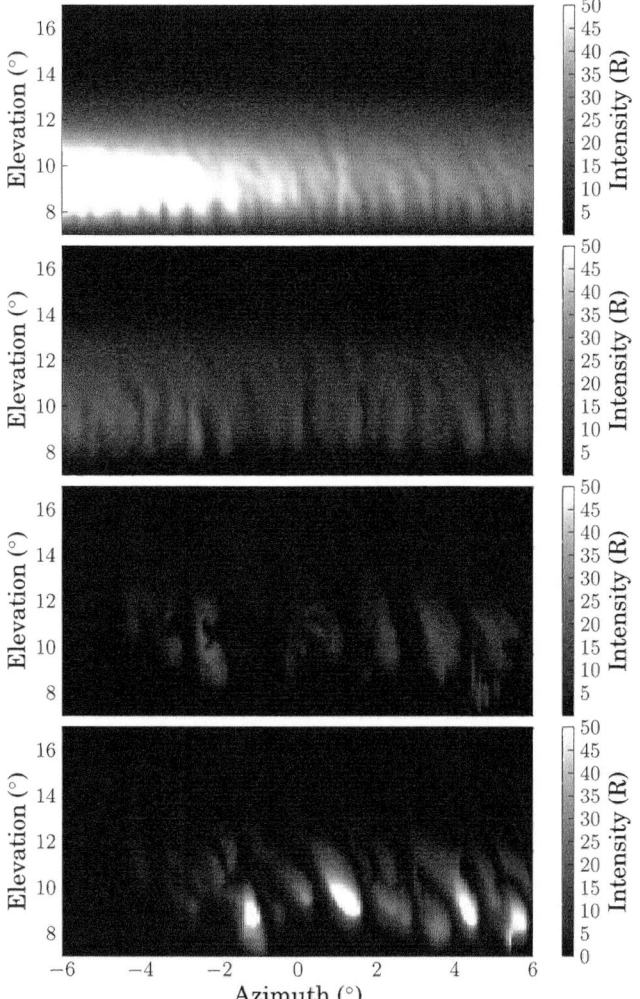

Figure 4. Simulated airglow (in Rayleighs) observed by a camera located south of Jicamarca (looking north). The four panels show results for simulation times of 00:25, 00:50, 02:05, and 03:20 hours, respectively, after a start time t_0 = 09:30 UT (LT = UT + 11 h at the horizontal center of the simulation).

t_0 + 00:50 hours, airglow depletions were present almost everywhere. Large-scale airglow depletions were spaced apart by 100–200 km, but smaller ones were also detectable. At t_0 + 02:05 hours, the emission layers appeared to have drifted upward (i.e., stronger emissions occurred away from the equator). Wide, large-scale depletions were evident with zonal width of tens of kilometers. The last panel (t_0 + 03:20 hours) shows stronger emissions compared to the previous one (t_0 + 02:05 hours). Enhanced brightness close to midnight is similar to what has been shown by *Makela and Miller* [2008]. Bifurcations and secondary instabilities growing on the western walls of the plasma depletions were also evident; however, the waveforms observed in airglow appeared distorted, and fine structure has been lost when compared to the electron density maps in Figure 2. Finally, the apparent zonal drift of the irregularities computed based on the frame-to-frame displacement is ~22 m s^{-1}. Since the background was drifting at ~60 m s^{-1}, this indicates how the irregularities (as measured by airglow) drifted slower than the background ionosphere. This scenario occurs during the first hours after sunset, but as the *E* region density declines, the *F* region dynamo becomes more efficient, and both the plasma irregularities and the background plasma drift zonally at the neutral wind speed.

2.7. In Situ Satellite Probe Simulations

To compute the magnetic induction due to ionospheric currents, we used Ampère's law. Note that, despite being neglected during the computation of the electrostatic potential, diamagnetic currents are included in the diagnostic computation of the magnetic field perturbations. Following the work of *Aveiro et al.* [2011], we neglected the displacement currents and applied the curl operator to Ampère's law, giving $\nabla^2 \mathbf{B} = -\mu_0 \nabla \times \mathbf{J}$, which represents a set of three independent partial differential equations for B_L, B_ϕ, and $B_\parallel$. These were solved by applying Neumann boundary conditions and using the BiCGStab method in a manner similar to the electrostatic potential solution described above. After the computation of the magnetic field, a consistency check with the current system was performed using Ampère's law for magnetostatics to verify agreement.

Figure 5 shows representative diagnostic information for 03:35 UT. Figure 5a shows the plasma number density. By this time, plasma irregularities, spaced by 100–200 km, had reached the top of the simulation.

The right-top color scale is used to indicate the magnitude and direction of the transverse current densities (scale maximum is 20 nA m^{-2}) and transverse magnetic induction (scale maximum is 2.5 nT). Figure 5c shows the transverse current density in the equatorial plane. Diamagnetic currents

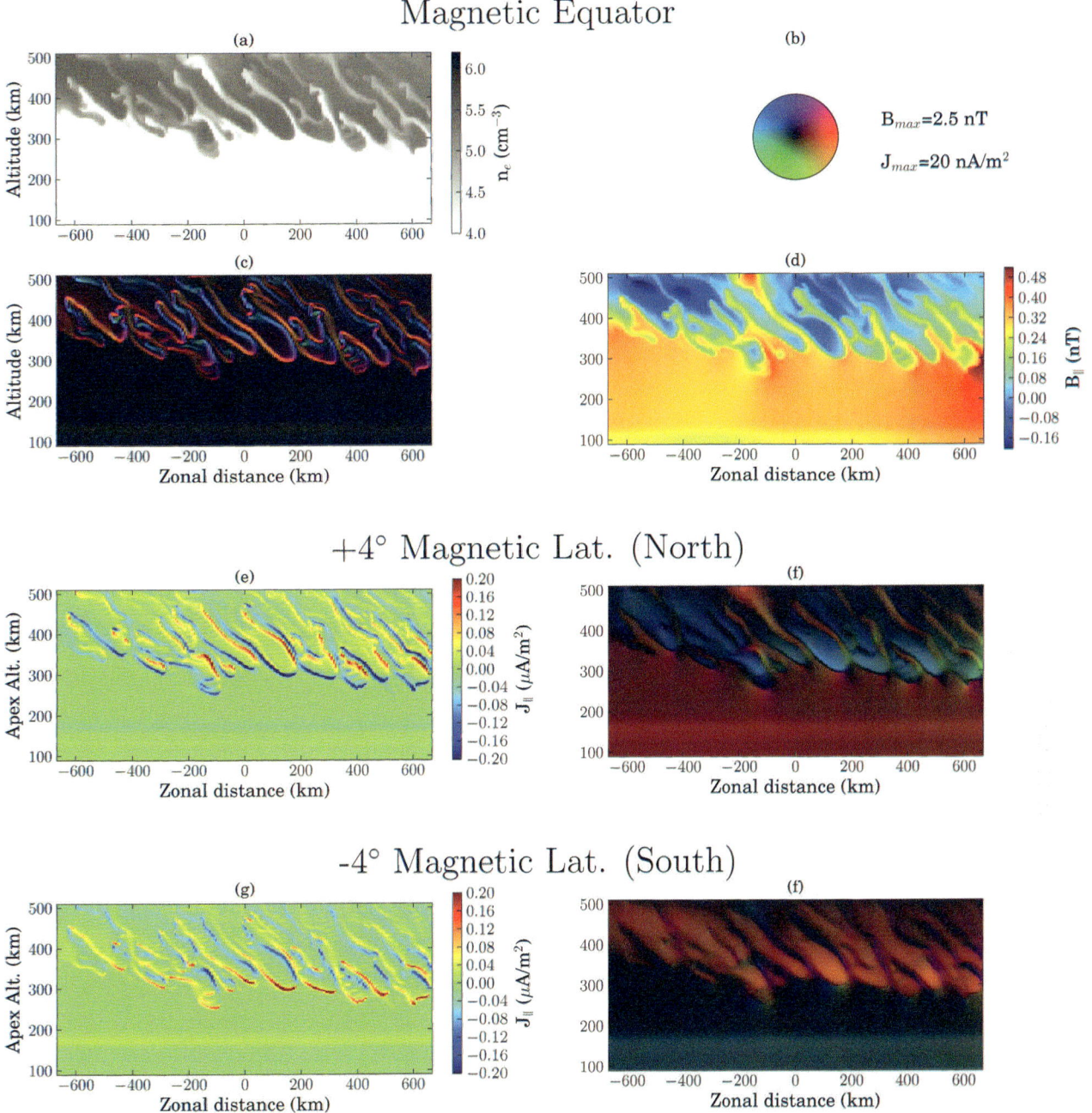

Figure 5. Diagnostic at 03:35 UT: (a) Plasma density in a cut through the equatorial plane on a log scale, (b) color scale of the magnitude and direction of the transverse current densities (scale maximum is 20 nA m^{-2}) and transverse magnetic inductions (scale maximum is 2.5 nT), (c) transverse current densities and (d) parallel magnetic induction in the equatorial plane, (e) parallel current density and (f) transverse magnetic induction at 4° magnetic latitude, and (g) parallel current density and (h) transverse magnetic induction at 4° magnetic latitude.

are observed flowing on the external walls of the plasma depletions in a clockwise direction. If evaluated together with Figure 5d, which indicates parallel magnetic inductions due to ionospheric currents, it can be seen that the currents caused northward deflections of $B_\parallel$ inside the plasma irregularities.

Figures 5e and 5g display the parallel currents at conjugate points ±4° magnetic latitude, respectively, north and south of the magnetic equator. The panels next to the parallel currents indicate the associated transverse magnetic inductions. field-aligned current (FAC) flow poleward (equatorward) on the external west (east) walls of the plasma depletions. Transverse magnetic induction showed a downward (upward) deflection inside the depletions north (south) of the magnetic equator. Transverse magnetic perturbations were negligible in the equatorial plane due to very small FAC and are not shown. The parallel current pattern observed in the simulations agrees with the numerical simulations described by *Aveiro et al.* [2011] and the statistical results of *Stolle et al.* [2006] and *Park et al.* [2009] based on CHAMP satellite data (see, e.g., the Figure 3 of *Aveiro et al.* [2011]).

Had a satellite passed through the simulated ionosphere at 380 km altitude over the Jicamarca sector at about 10:35 UT, the electron density and magnetic field perturbations would have appeared as shown in Figure 6. Magnetic field data were high-pass filtered by a median filter (~700 km) to remove large-scale variations. The simulated results show a plasma depletion at ~±6° magnetic latitude, colocated with a northward $\Delta B_\parallel$ and an upward/westward (downward/eastward) $\Delta B_\perp$ south (north) of the magnetic equator.

3. SUMMARY AND DISCUSSIONS

A 3-D numerical simulation of plasma density irregularities in the postsunset equatorial F region ionosphere leading to ESF was performed to investigate plasma irregularity signatures in different space plasma diagnostics. The simulation of coherent/incoherent scatter observations showed the typical three stages of ESF evolution, from bottom-type to bottomside to topside spread F. Some of the features that the simulated electron density maps share with ALTAIR scans include westward tilted ascending depletions connected to

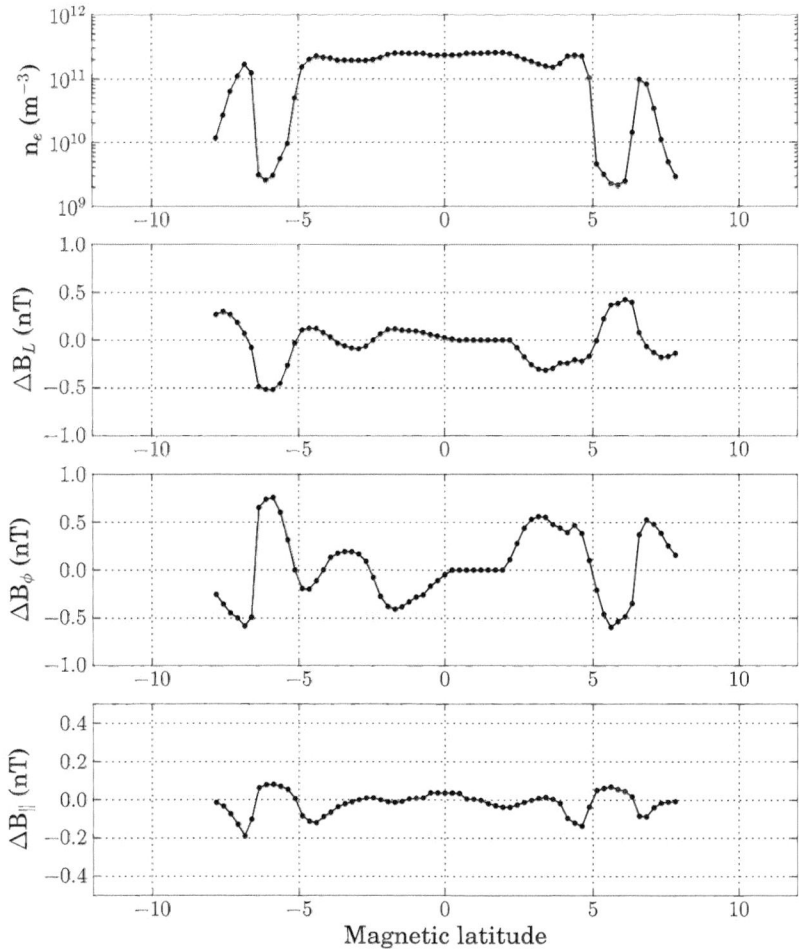

Figure 6. Simulated satellite observations during a meridional pass: (top) plasma density, (middle top) meridional, (middle bottom) zonal, and (bottom) parallel magnetic field perturbations.

the bottomside, periodic spacing of 100–200 km in the zonal direction, bifurcation, secondary instabilities growing on the western walls of the primaries, and rates of development [see, e.g., *Aveiro et al.*, 2012]. Also, the simulated range-time-intensity map recovered the "C shape" structures detected by fixed-beam coherent scatter radars such as the Jicamarca radar. However, the zonal drift of the irregularities seem to be slower than typical observations made by the Jicamarca radar, which may indicate an underestimation of the F region dynamo efficiency in the simulation. Simulations based on field line–integrated electrostatic potential using SAMI3 model showed irregularities drifting zonally approximately at the zonal wind speeds [*Huba and Joyce*, 2010]. Their result suggests an overestimation of the dynamo efficiency due to the absence of secondary instabilities on the western walls of the F region plasma irregularities.

Using simulated plasma and empirical neutral atmosphere composition information, we computed the airglow emission for the 6300 Å (or O^+ red) line. A simulation of airglow observations by a hypothetical camera located a few degrees south of Jicamarca was performed. The results presented here show, for the first time, simulated ground-based airglow observations for the red line emission. Simulations of ground-based airglow measurements for the 5577 Å (or O^+ green) line were described by *Retterer* [2010] and showed bifurcation and tilting of ESF plasma irregularities. Strong emissions were observed on the western half of the image due to recombination during the earliest hours of the night and again close to midnight, similar to what has been reported by *Makela and Miller* [2008]. Intermediate-scale (~30–50 km) irregularities were barely detectable during the first hours after sunset in the simulation; however, those do not seem to be observed in nature by airglow cameras. The comparison between the simulated radar scans (Figure 2) and airglow simulations (Figure 4) showed that details of the plasma irregularities has been lost, and waveforms appeared distorted in airglow observations.

Magnetic perturbations parallel to the main field were mainly due to diamagnetic currents flowing on the external walls of the plasma depletions, and transverse magnetic perturbations were mainly due to FACs. The simulated magnetic field perturbations show good quality with the CHAMP measurements. The ionospheric circulation results agreed with earlier simulations by *Aveiro et al.* [2011] based on CHAMP satellite data. Also, the simulation reproduces the main features of the plasma depletions as derived in a statistical study by *Park et al.* [2009], as, e.g., upward (downward) currents on their external western (eastern) edges.

Overall, the diagnostic codes showed that the numerical simulation scheme described here is able to reproduce many of the most salient characteristics of ESF as observed by a number of different instruments. For the validation of current densities (or magnetic field perturbations), the computation of electrostatic potential must be done in 3-D. Quasi-three-dimensional simulations based on the equipotential field line approach do not provide parallel current information and, thus, cannot be compared to satellite-borne magnetometer data; however, they can still be validated against backscatter radar data and airglow images. Validating simulations against specific observations is difficult since empirical models do not capture day-to-day variability. Thus, simulations can be validated mainly in terms of their ability to reproduce the characteristic behavior of the plasma irregularities, e.g., tilting, spacing, layer thickness, altitude range, onset time, and rate of development.

Acknowledgments. This work was supported by award NNX11A069G from the NASA LWS to Cornell University.

REFERENCES

Aveiro, H. C., and D. L. Hysell (2010), Three-dimensional numerical simulation of equatorial F region plasma irregularities with bottomside shear flow, *J. Geophys. Res.*, *115*, A11321, doi:10.1029/2010JA015602.

Aveiro, H. C., D. L. Hysell, J. Park, and H. Lühr (2011), Equatorial spread F-related currents: Three-dimensional simulations and observations, *Geophys. Res. Lett.*, *38*, L21103, doi:10.1029/2011GL049586.

Aveiro, H. C., D. L. Hysell, R. G. Caton, K. M. Groves, J. Klenzing, R. F. Pfaff, R. Stoneback, and R. A. Heelis (2012), Three-dimensional numerical simulations of equatorial spread F: Results and observations in the Pacific sector, *J. Geophys. Res.*, *117*, A03325, doi:10.1029/2011JA017077.

Bates, D. R. (1990), Oxygen green and red line emission and O_2^+ dissociative recombination, *Planet. Space Sci.*, *38*(7), 889–902.

Bilitza, D., and B. W. Reinisch (2008), International Reference Ionosphere 2007: Improvements and new parameters, *Adv. Space Res.*, *42*, 599–609.

Daniell, R. E., Jr., L. D. Brown, D. N. Anderson, M. W. Fox, P. H. Doherty, D. T. Decker, J. J. Sojka, and R. W. Schunk (1995), Parameterized ionospheric model: A global ionospheric parameterization based on first principles models, *Radio Sci.*, *30*(5), 1499–1510.

Drob, D. P., et al. (2008), An empirical model of the Earth's horizontal wind fields: HWM07, *J. Geophys. Res.*, *113*, A12304, doi:10.1029/2008JA013668.

Haerendel, G., J. V. Eccles, and S. Çakir (1992), Theory for modeling the equatorial evening ionosphere and the origin of the shear in the horizontal plasma flow, *J. Geophys. Res.*, *97*(A2), 1209–1223.

Huba, J. D., and G. Joyce (2010), Global modeling of equatorial plasma bubbles, *Geophys. Res. Lett.*, *37*, L17104, doi:10.1029/2010GL044281.

Huba, J. D., G. Joyce, and J. Krall (2008), Three-dimensional equatorial spread F modeling, *Geophys. Res. Lett.*, *35*, L10102, doi:10.1029/2008GL033509.

Hysell, D. L. (1996), Radar imaging of equatorial F region irregularities with maximum entropy interferometry, *Radio Sci.*, *31*(6), 1567–1578.

Hysell, D. L., and J. D. Burcham (1998), JULIA radar studies of equatorial spread F, *J. Geophys. Res.*, *103*(A12), 29,155–29,167.

Hysell, D. L., and E. Kudeki (2004), Collisional shear instability in the equatorial F region ionosphere, *J. Geophys. Res.*, *109*, A11301, doi:10.1029/2004JA010636.

Hysell, D. L., J. Chun, and J. L. Chau (2004), Bottom-type scattering layers and equatorial spread F, *Ann. Geophys.*, *22*, 4061–4069.

Hysell, D. L., M. F. Larsen, C. M. Swenson, A. Barjatya, T. F. Wheeler, T. W. Bullett, M. F. Sarango, R. F. Woodman, J. L. Chau, and D. Sponseller (2006), Rocket and radar investigation of background electrodynamics and bottom-type scattering layers at the onset of equatorial spread F, *Ann. Geophys.*, *24*, 1387–1400.

Kelley, M. C. (2009), *The Earth's Ionosphere*, Elsevier, New York.

Keskinen, M. J. (2010), Equatorial ionospheric bubble precursor, *Geophys. Res. Lett.*, *37*, L09106, doi:10.1029/2010GL042963.

Kudeki, E., A. Akgiray, M. A. Milla, J. L. Chau, and D. L. Hysell (2007), Equatorial spread-F initiation: Post-sunset vortex, thermospheric winds, gravity waves, *J. Atmos. Sol. Terr. Phys.*, *69* (17–18), 2416–2427.

Link, R., and L. L. Cogger (1988), A reexamination of the O I 6300-Å nightglow, *J. Geophys. Res.*, *93*(A9), 9883–9892.

Link, R., and L. L. Cogger (1989), Correction to "A reexamination of the O I 6300-Å nightglow" by R. Link and L. L. Cogger, *J. Geophys. Res.*, *94*(A2), 1556.

Makela, J. J., and E. S. Miller (2008), Optical observations of the growth and day-to-day variability of equatorial plasma bubbles, *J. Geophys. Res.*, *113*, A03307, doi:10.1029/2007JA012661.

Park, J., H. Lühr, C. Stolle, M. Rother, K. W. Min, and I. Michaelis (2009), The characteristics of field-aligned currents associates with equatorial plasma bubbles as observed by the CHAMP satellite, *Ann. Geophys.*, *27*, 2685–2697.

Picone, J. M., A. E. Hedin, D. P. Drob, and A. C. Aikin (2002), NRLMSISE-00 empirical model of the atmosphere: Statistical comparisons and scientific issues, *J. Geophys. Res.*, *107*(A12), 1468, doi:10.1029/2002JA009430.

Retterer, J. M. (2010), Forecasting low-latitude radio scintillation with 3-D ionospheric plume models: 1. Plume model, *J. Geophys. Res.*, *115*, A03306, doi:10.1029/2008JA013839.

Richmond, A. D. (1972), Numerical model of the equatorial electrojet, *Tech. Rep. AFCRL-72-0668, ERP 421*, Air Force Cambridge Res. Lab., Hanscom AFB, Bedford, Mass.

Saad, Y. (1990), SPARSKIT: A basic tool kit for sparse matrix computations, *Tech. Rep. RIACS-90-20*, Res. Inst. for Adv. Comput. Sci., NASA Ames Res. Cent., Moffett Field, Calif.

Shume, E. B., D. L. Hysell, and J. L. Chau (2005), Zonal wind velocity profiles in the equatorial electrojet derived from phase velocities of type II radar echoes, *J. Geophys. Res.*, *110*, A12308, doi:10.1029/2005JA011210.

Sobral, J. H. A., H. Takahashi, M. A. Abdu, P. Muralikrishna, Y. Sahai, C. J. Zamlutti, E. R. de Paula, and P. P. Batista (1993), Determination of the quenching rate of the $O(^1D)$ by $O(^3P)$ from rocket-borne optical (630 nm) and electron density data, *J. Geophys. Res.*, *98*(A5), 7791–7798.

Stolle, C., H. Lühr, M. Rother, and G. Balasis (2006), Magnetic signatures of equatorial spread F as observed by the CHAMP satellite, *J. Geophys. Res.*, *111*, A02304, doi:10.1029/2005JA011184.

Sultan, P. J. (1996), Linear theory and modeling of the Rayleigh-Taylor instability leading to the occurrence of equatorial spread F, *J. Geophys. Res.*, *101*(A12), 26,875–26,891.

Tinsley, B. A. (1982), Field aligned airglow observations of transequatorial bubbles in the tropical F-region, *J. Atmos. Terr. Phys.*, *44*, 547–557.

Trac, H., and U. L. Pen (2003), A primer on Eulerian computational fluid dynamics for astrophysicists, *Astrophysics*, *115*, 303–321.

Van der Vorst, H. (1992), Bi-CGSTAB: A fast and smoothly converging variant of Bi-CG for the solution of nonsymmetric linear systems, *SIAM J. Sci. Stat. Comput.*, *13*, 631–644.

Woodman, R. F. (1997), Coherent radar imaging: Signal processing and statistical properties, *Radio Sci.*, *32*(6), 2373–2391.

Woodman, R. F. (2009), Spread F – An old equatorial aeronomy problem finally resolved?, *Ann. Geophys.*, *27*, 1915–1934.

H. C. Aveiro and D. L. Hysell, Earth and Atmospheric Sciences, Cornell University, Ithaca, NY 14850, USA. (hca24@cornell.edu; dlh37@cornell.edu)

Density and Temperature Structure of Equatorial Spread F Plumes

J. Krall and J. D. Huba

Plasma Physics Division, Naval Research Laboratory, Washington, District of Columbia, USA

Simulations of simple equatorial spread F (ESF) plumes using the National Research Laboratory SAMI3/ESF 3-D simulation code are analyzed in terms of the momentum and thermal equations governing the ions and electrons. The resulting 3-D plume structure includes regions of enhanced density and temperature. These have been shown to correspond to observed density and temperature features, such as plasma "blobs." We consider the evolution of a plume into a very slowly evolving "fossil plume" state and find that these temperature and density features may persist over time scales that are long compared to the growth time of the ESF plume.

1. INTRODUCTION

Equatorial spread F (ESF), named for the "diffuse echoes from the F region of the ionosphere received continuously at night in equatorial regions over a wide range of wave frequency" [*Booker and Wells*, 1938], has been a subject of active research for the past 35 years [*Ossakow*, 1981; *Hysell*, 2000; *Kelley*, 2009; G. Haerendel, Theory of equatorial spread F, preprint, Max Planck Institute for Extraterrestrial Physics, Munich, Germany, 1974.]. Most recently, sophisticated 3-D computer simulation models of ESF plumes, or "bubbles," have been developed [*Huba et al.*, 2008; *Retterer*, 2010]. Strong electron density gradients associated with ESF affect the propagation of electromagnetic signals, sometimes degrading global communication and navigation systems [*de La Beaujardiere et al.*, 2004; *Steenburgh et al.*, 2008].

In recent work using the SAMI3/ESF code [*Huba et al.*, 2008], we have demonstrated and discovered several basic properties of the Rayleigh-Taylor-like ESF instability. Here we will focus on a subset of these studies that describes the 3-D density and temperature structure within an ESF plume. Central to this is the phenomenon of a "super fountain" [*Huba et al.*, 2009b] within the ESF plume. This is governed by the same physics as the usual equatorial ionospheric fountain that causes the equatorial ionization anomaly (EIA) [*Hanson and Moffett*, 1966; *Anderson*, 1973], except that the driving electric field is much stronger. *Huba et al.* [2009a] showed that the dynamics of the super fountain explain the large variations in temperature found within ESF plumes [*Oyama et al.*, 1988]. The resulting density crests within the plume, analogous to the EIA, can lead to the occurrence of "plasma blobs" [*Oya et al.*, 1986; *Watanabe and Oya*, 1986; *Le et al.*, 2003; *Yokoyama et al.*, 2007; *Krall et al.*, 2010a]. Because ESF plumes often include regions with ion densities equal to or greater than surrounding ionospheric densities, we will refer to them as plumes rather than bubbles in this study.

In the present work, we use results from the SAMI3/ESF 3-D simulation model to illustrate 3-D temperature and density structure of ESF plumes. We will then consider the persistence of this structure as the plume evolves. In a recent study, *Krall et al.* [2010b] showed that the plume stops rising when the field line–integrated mass density within the upper edge of the plume is approximately equal to that of the nearby background ionosphere. This leads to a "fossil" plume state with a low-density bubble near the apex and higher densities at lower altitudes. These lower-altitude, high-density regions can be detected as blobs or as ESF airglow enhancements [*Krall et al.*, 2010b; *Martinis et al.*, 2009].

Below, we will discuss the simulation code and governing equations, followed by examples showing the density and temperature structure within an ESF plume. We will then analyze the results to determine the persistence of temperature and density signatures of an ESF plume. As always, it is hoped that these simulations will aid in the interpretation and analysis of ESF measurements.

2. SAMI3 SIMULATIONS

The NRL 3-D ionosphere code SAMI3/ESF [*Huba et al.*, 2008] is used in this analysis. SAMI3/ESF is a version of the SAMI3 global ionosphere code [*Huba et al.*, 2000; *Huba and Joyce*, 2010] that has been modified for the study of ESF. SAMI3 models the plasma and chemical evolution of seven ion species (H^+, He^+, N^+, O^+, N_2^+, NO^+, and O_2^+). The complete ion temperature equation is solved for three ion species (H^+, He^+, and O^+) as well as the electron temperature equation. The plasma equations solved are as follows:

$$\frac{\partial n_i}{\partial t} + \nabla \cdot (n_i \mathbf{V}_i) = \mathcal{P}_i - \mathcal{L}_i n_i \quad (1)$$

$$\frac{\partial \mathbf{V}_i}{\partial t} + \mathbf{V}_i \cdot \nabla \mathbf{V}_i = -\frac{1}{\rho_i}\nabla P_i + \frac{e}{m_i}\mathbf{E} + \frac{e}{m_i c}\mathbf{V}_i \times \mathbf{B} \\ + \mathbf{g} - \nu_{in}(\mathbf{V}_i - \mathbf{V}_n) - \sum_j \nu_{ij}(\mathbf{V}_i - \mathbf{V}_j) \quad (2)$$

$$0 = -\frac{1}{n_e m_e}\nabla P_e + \frac{e}{m_e}\mathbf{E} + \frac{e}{m_e c}\mathbf{V}_e \times \mathbf{B} \quad (3)$$

$$\frac{\partial T_i}{\partial t} + \mathbf{V}_i \cdot \nabla T_i + \frac{2}{3}T_i \nabla \cdot \mathbf{V}_i + \frac{2}{3}\frac{1}{n_i k}\nabla \cdot \mathbf{Q}_i \\ = Q_{in} + Q_{ii} + Q_{ie} \quad (4)$$

$$\frac{\partial T_e}{\partial t} - \frac{2}{3}\frac{1}{n_e k}b_s^2 \frac{\partial}{\partial s}\kappa_e \frac{\partial T_e}{\partial s} = Q_{en} + Q_{ei} + Q_{phe}. \quad (5)$$

The various coefficients and parameters are specified in the work of *Huba et al.* [2000]. Ion inertia is included in the ion momentum equation for motion along the geomagnetic field, and $\mathbf{E} \times \mathbf{B}$ drifts are computed to obtain motion transverse to the field. Neutral composition and temperature are specified using the empirical NRLMSISE00 (Naval Research Laboratory, Mass Spectrometer Incoherent Scatter radar, Extended in altitude, 2000) model [*Picone et al.*, 2002]. The version of SAMI3 used here and in our other recent ESF studies [*Huba et al.*, 2008, 2009a, 2009b; *Krall et al.*, 2009a, 2009b] is similar to that used in the work of *Huba and Joyce* [2010] in that the perpendicular electric field $\mathbf{E}_\perp = -\nabla \Phi$ is computed self-consistently to determine the $\mathbf{E} \times \mathbf{B}$ drifts. The SAMI3/ESF potential equation is derived from the current conservation ($\nabla \cdot \mathbf{J} = 0$) in dipole coordinates (q, p, ϕ) and is described in the work of *Huba et al.* [2008]. In this study, the potential equation is identical to that of *Krall et al.* [2009b], except that the Hall-conductance terms are neglected relative to Pedersen-conductance terms (e.g., $\Sigma_H = \sigma_H = 0$). The Hall conductance, when included, leads to small zonal asymmetries in the growing plume.

The 3-D simulation model is initialized using results from the 2-D SAMI2 code. SAMI2 is run for 48 h using the desired geophysical conditions. The initial state for the 3-D model corresponds to a time when the ionosphere in the 2-D model has been lifted up by the prereversal enhancement of the zonal electric field. The $\mathbf{E} \times \mathbf{B}$ drift in SAMI2 is prescribed by the Fejer/Scherliess model [*Scherliess and Fejer*, 1999]. The plasma parameters (density, temperature, and velocity) at time 19:20 UT (of the second day) are used to initialize the 3-D model at each magnetic longitudinal plane.

SAMI3 uses a nonorthogonal, irregular grid that conforms to a model geomagnetic field. For this study, the magnetic field is modeled as a dipole field aligned with the Earth's spin axis. The simulation region includes magnetic apex heights from 90 to 2400 km and has an overall longitudinal width of 4° (e.g., ≃460 km). The plasma is, therefore, modeled from hemisphere to hemisphere up to ±31° magnetic latitude. The simulation is centered at geographic longitude 0° so UT and LT are the same. The grid is $(n_z, n_f, n_l) = (101, 202, 96)$ where n_z is the number of grid points along the magnetic field, n_f is the number of "field lines" at each longitude, and n_l is the number of longitude grid points. At the height of the *F* layer, this grid has a resolution of ~6 km × 5 km in altitude and longitude in the magnetic equatorial plane. The simulation region is periodic in longitude (the variation in LT across the 4° width of the simulation is ignored). In essence, we are simulating a narrow "wedge" of the ionosphere in the postsunset period. Unless noted, the ESF instability in each simulation was initiated by imposing a Gaussian-like perturbation in the ion density at $t = 0$: a peak ion density perturbation of 15% centered at 0° longitude with a half width of 0.25°, and at an altitude $z = 300$ km with a half width of 50 km. This seed perturbation extends along the entire flux tube. For these simulations, the neutral winds are set to zero.

3. IONIZATION CRESTS WITHIN AN ESF PLUME

In our first example, we will consider regions of relatively high density that can occur within an ESF plume. It was shown in the work of *Krall et al.* [2010a] that these regions are sometimes detected in situ as plasma blobs. Of particular interest for this example is the field-aligned component of the ion momentum equation, which can be expressed as

$$\frac{\partial V_{is}}{\partial t} + (\mathbf{V}_i \cdot \nabla)V_{is} = -\frac{1}{n_i m_i}b_s\frac{\partial P_i}{\partial s} - \frac{1}{n_i m_i}b_s\frac{\partial P_e}{\partial s} \\ + g_s - \nu_{in}(V_{is} - V_{ns}) - \sum_j \nu_{ij}(V_{is} - V_{js}), \quad (6)$$

where *s* is the coordinate along the field line and where the electron momentum equation, equation (3), has been used to express the polarization electric field in terms of the electron pressure.

Figure 1 shows an example of a simple ESF plume viewed in the plane of the magnetic equator (right panel) and in a constant longitude plane (left panel). In this case, we use the following geophysical conditions: $F10.7 = 75$, $F10.7A = 75$, $Ap = 4$, and day of year 80 (e.g., 21 March 2010). Color-scale plots show values of $\log_{10}(n_e)$, with the location of the left (right) plot being indicated by a white line in the right-hand (left-hand) panel. The edge of the ESF plume, which has an apex height of 1610 km, can be seen in both panels as a discontinuity in the density. The left-hand panel shows the flow of O^+ ions, where velocity vectors indicate the direction of flow *away* from the dot in each case. A dark curve near the center of each plot indicates the H^+/O^+ transition height, where the densities of these two ionospheric constituents are equal. That the bulk of the ESF plume lies below this transition height at all times is indicative of the transport of low-altitude-composition plasma to high altitudes [*Huba et al.*, 2009b]. Here $\mathbf{E} \times \mathbf{B}$ drifts move ions upward across field lines, and gravity moves ions downward along field lines [*Hanson and Bamgboye*, 1984]. As a result, the ESF plume is strongly depleted near its apex, and a pair of ionization crests form within the ESF plume at altitude 440 km and latitude ±21°. These are marked in the left-hand panel by white circles.

For comparison to the ionization crests within the ESF plume, we considered the properties of the background EIA at a position indicated by the red vertical line in the right-hand panel of Figure 1. This is outside the region affected by the localized electric potential of the nearby ESF plume. The positions of the two background EIA crests are indicated by red circles in the left-hand panel. We see that the ESF crests occur at an altitude above that of the background EIA crests on a field line with an apex altitude above that of the background EIA. In Figure 1 (left panel), field lines are indicated by the black dots at the base of each vector; these dots are arrayed along selected field lines.

Past studies of the fountain effect that leads to the EIA [*Hanson and Moffett*, 1966; *Anderson*, 1973] show that the position of the EIA crests along the geomagnetic field is governed by a balance between the downward gravitational force and the upward diffusion force that is exerted by neutrals on the downward moving ions (see equation (6)). The peak in the density occurs where the gravitational and diffusive forces are approximately equal in magnitude. The forces acting on O^+ ions within the ESF plume are shown in Figure 2, where each of the five terms on the right-hand side of equation (6) is plotted in the upper panel. Here the plot is along the field line corresponding to the ionization crests within the ESF plume shown in Figure 1. The lower panel of Figure 2 shows the net force and the electron density. That the ionization crests within the ESF plume are held in place primarily by the downward gravity (F_g) and upward diffusion (F_{in}) forces is confirmed in Figure 2 (top), which shows that the ion and electron pressures are approximately zero where the electron density peaks and that the ion-ion diffusion force F_{ij} is everywhere small. Figure 2 (bottom) shows that the ionization crests are in a near-equilibrium state, with the net force being close to zero in the crest regions. Further results (not shown) show that F_{in} is dominated by collisions between O^+ and O; when the contribution from this interaction is removed, the crests appear at a

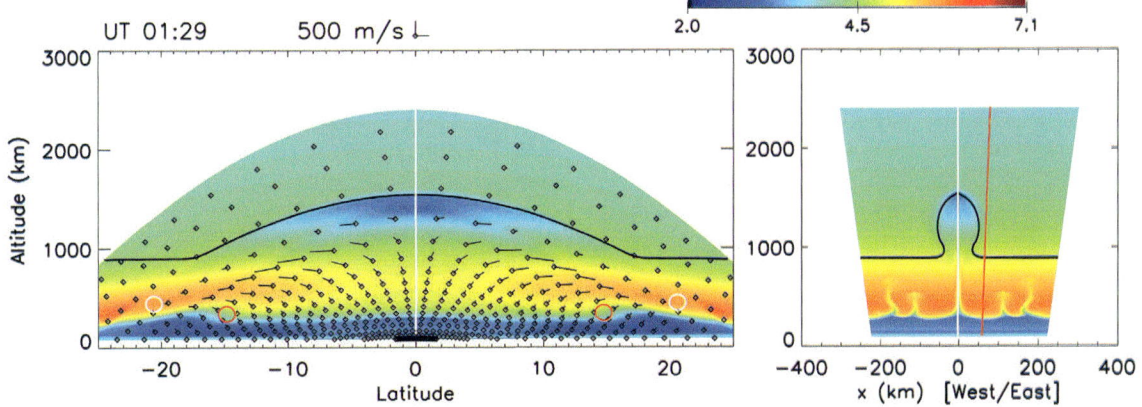

Figure 1. Color-scale plot of $\log_{10}(n_e \text{ cm}^{-3})$ (left) at a fixed longitude and (right) at the magnetic equator. The location of the left-hand plot in the right-hand panel is indicated by a white line and vice versa. Ionization crests are indicated within the ESF plume (white circles) and in the background (red circles). A dark curve near the center of each plot indicates the H^+/O^+ transition height. (left) O^+ ion velocity vectors indicate the direction of flow *away* from the dot [From *Krall et al.*, 2010a, used with permission].

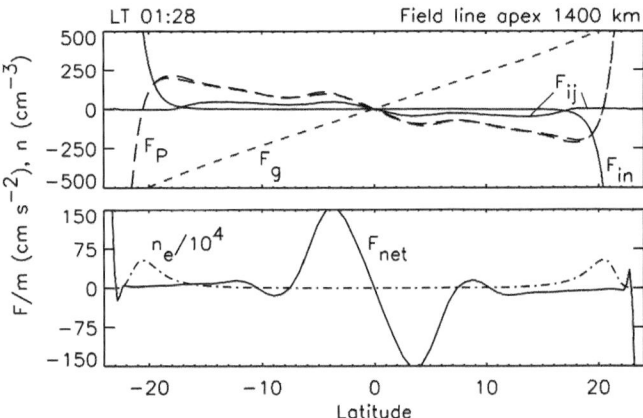

Figure 2. (top) Field-aligned forces, gravity (F_g, dashed line), ion and electron pressures (F_p, long-dashed lines), and diffusion (F_{in}, F_{ij}, solid lines), acting on O$^+$ ions are indicated for a field line corresponding to the ionization crests within the plume. (bottom) Net field-aligned force (solid line) acting on O$^+$ ions is plotted. Electron density (dot-dash) is also shown [From *Krall et al.*, 2010a, used with permission].

much lower altitude. We also considered the effect of ion heating within the plume [*Huba et al.*, 2009a]. This was found to be of little consequence. Indeed, the most significant factor in determining the height of the crests appears to be the downward, field-aligned ion velocity: because the crest height represents a balance between F_g and F_{in}, and because F_{in} increases with velocity, larger velocities lead to higher crest altitudes. As the plume evolves to a point where it is no longer rising, these field-aligned velocities tend to diminish. Further below, we will consider this long-term evolution.

4. THERMODYNAMICS OF AN ESF PLUME

We now consider a simulation first presented by *Huba et al.* [2009a]. This is similar to the first example, except that the geophysical conditions now correspond to more active solar conditions, $F10.7 = 150$, $F10.7A=150$, $Ap = 4$, and day of year 92 (e.g., 7 April 2002), and the initial perturbation is wider, 0.5° longitude, 60 km altitude, and at a higher altitude, 400 km. In Figure 3, we show color-scale plots of the electron, oxygen ion O$^+$, and hydrogen ion H$^+$ temperatures at time 22:19 UT (approximately 2 h after the start of the simulation). The left panels show the temperatures as a function of altitude and latitude, and the right panels as a function of altitude and x (west/east direction). The velocity vectors are shown for O$^+$ and H$^+$ in the left panels. As in Figure 1, the symbol is at the base of the vector, i.e., the velocity is *away* from this point. Finally, the solid line on each electron temperature plot is the H$^+$/O$^+$ transition altitude

(i.e., where the oxygen ion density equals the hydrogen ion density).

The key points of this figure are the following. First, there is a clear transition in the temperatures along the field line that extends from −25° to +25° and reaches a maximum altitude of ~1600 km. This transition corresponds to the low-density plasma bubble boundary [*Huba et al.*, 2009b]. Second, the plasma has cooled inside the bubble; the ion and electron temperatures decrease from ~1050 K outside the bubble to as low as ~400 K inside the bubble at 1500 km in the equatorial plane. The latitudinal extent of the ion cooling is ~±15°, while that of the electrons is ~±10°. The cooling is adiabatic, caused by the increase in volume of the flux tubes as the plasma rises to high altitudes [*Huba et al.*, 2000]. Third, there is heating of the O$^+$ and H$^+$ ions in the altitude range ~400–1000 km at latitudes ~±15°. This is caused by the compression of the ions as they stream down the converging magnetic field lines. The ion vectors in the left panel show the super fountain effect described in the work of *Huba et al.* [2009b]. The ions attain velocities exceeding 1 km s^{-1}. The electrons are heated via collisions with the ions in this region.

The electron temperature increases along the field lines in the upper portion of the plasma bubble (above apex altitudes ~1400 km) more than the ions. This is caused by electron heat conduction, which is more effective than ion heat conduction ($\kappa_e \gg \kappa_i$). We have carried a simulation in which $\kappa_e = 0$, and this high-altitude electron heating did not occur. Finally, relatively intense heating of H$^+$ occurs at an altitude ~300 km at latitudes ~±20°. This interesting artifact is explained in the work of *Huba et al.* [2009a], but is of little interest here because the H$^+$ density is very low at these altitudes.

The cooling and heating of the ions above ~400 km is adiabatic: it is associated with volumetric changes of the plasma as it rises and falls. However, transport of the temperature also plays a role in the distribution of the temperature along the geomagnetic field lines. To show this, we write equation (4) for the ions along the geomagnetic field

$$\frac{\partial T_i}{\partial t} = -\underbrace{\frac{2}{3} T_i b_s^2 \frac{\partial}{\partial s} \frac{V_{is}}{b_s}}_{A} - \underbrace{b_s V_{is} \frac{\partial T_i}{\partial s}}_{B}, \qquad (7)$$

where we neglect thermal conductivity and collisional terms. Here $b_s = B_s/B_0$, where B_s is the geomagnetic dipole field, and B_0 is its value at the Earth's surface at the equator. In Figure 4, we plot the terms A and B shown in equation (7) (dashed lines) and their sum A + B (red line), as well as the difference in the O$^+$ temperature inside and outside of the plasma bubble (blue line). This plot is at time 22:19 UT and

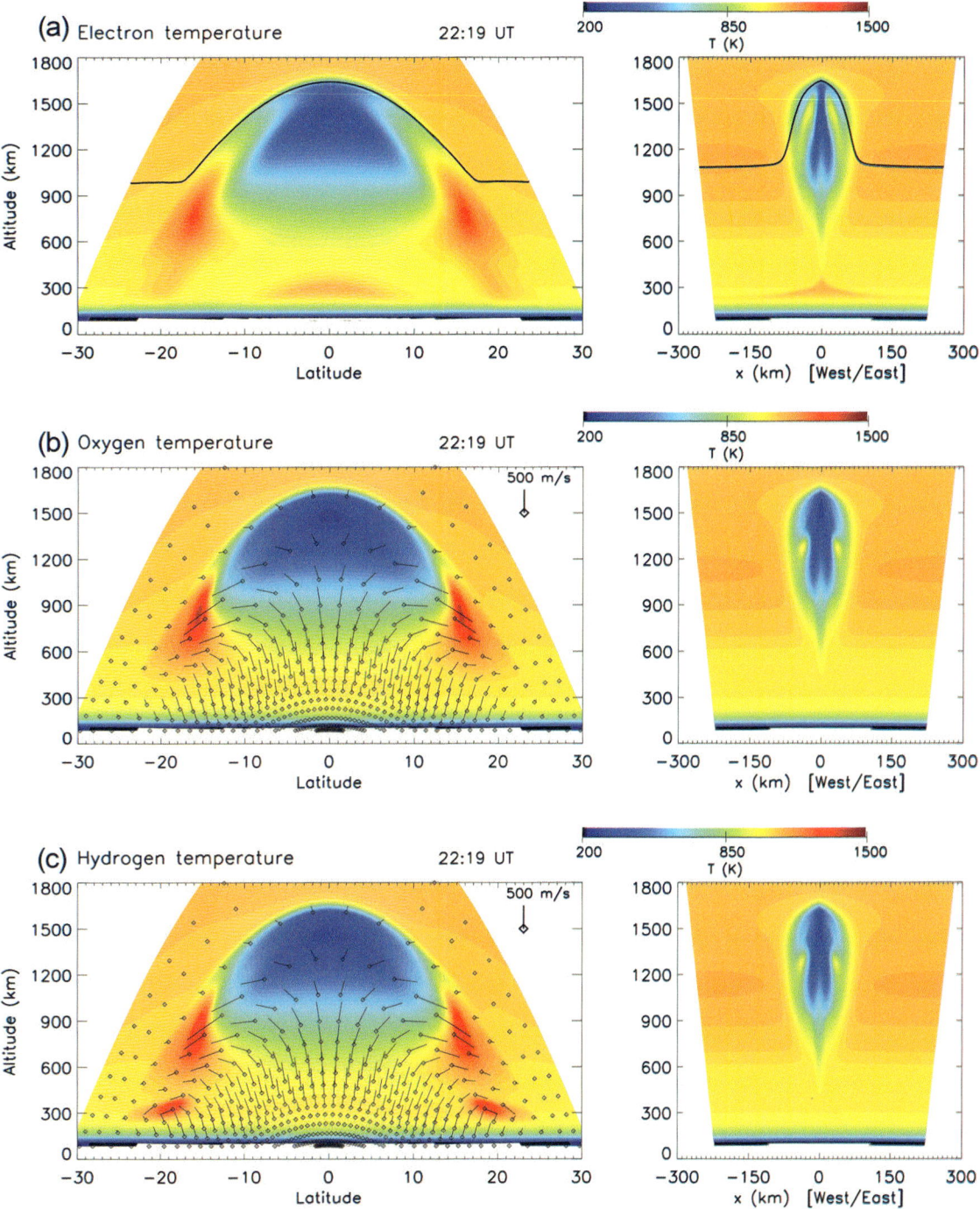

Figure 3. Color-scale plots of the electron, oxygen ion, and hydrogen ion temperatures at time 22:19 UT. (left) The velocity vectors are shown for the oxygen ions O^+ and hydrogen ions H^+. Here the symbol is at the base of the vector, i.e., the velocity is *away* from this point. The solid line in the electron figure is the H^+/O^+ transition altitude [From *Huba et al.*, 2009a, used with permission].

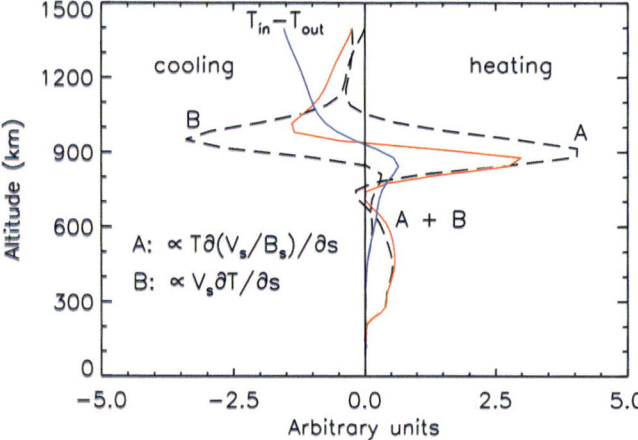

Figure 4. Cooling and heating term comparison [From *Huba et al.*, 2009a, used with permission].

is a function of altitude (km) along a specific field line, one that passes through the O$^+$ temperature enhancement shown in the middle panel of Figure 3. The *x* axis is arbitrary: the purpose is to show where heating (*x* > 0) and cooling (*x* < 0) are expected to occur. Adiabatic heating associated with term A occurs in the altitude range 200–1000 km; cooling occurs above 1000 km. However, the plasma in the bubble is heated in the altitude range 400–900 km and cools above 900 km (blue curve). Cooling above 900 km is associated with term B, the advection of cool ions to lower altitudes along the field. Thus, the distribution of the ion temperature along the magnetic field is not only simply controlled by the adiabatic term A but also by the transport term B. At lower altitudes (below ~400 km), there is no O$^+$ heating; here collisional coupling to the cold neutrals keeps the ions from being heated.

5. TIME EVOLUTION OF AN ESF PLUME

Insight into the time evolution of an ESF plume is provided by *Krall et al.* [2010b], where the growth of a plume to a nearly stationary fossil state is analyzed. In that study, it was found the ESF plume stops rising when the field line–averaged mass density at the leading edge of the plume, $\langle p \rangle$, is approximately equal to that of the nearby background. *Krall et al.* [2010b] helpfully provide conceptual continuity by analyzing the same plume shown in Figure 1, but with the simulation run for an additional 90 min of simulated time. The result, a slowly evolving fossil plume [see *Krall et al.*, 2010b, Figure 2], features very low densities at the apex and higher densities at lower altitudes within the plume. In fact, it resembles Figure 1 (left panel), except with much reduced ion velocities and with the density crests lower by about 100 km.

Figure 5 provides further information on the temporal evolution of the ESF plume of Figure 1. Shown are color-scale plots of the $E \times B$ velocity and electron density as a function of LT (h) and altitude (km) at longitude 0°. Although the vertical $E \times B$ velocity associated with the instability starts at ~21:30 LT, it does not become large (i.e., 500 m s^{-1}) until ~22:15 LT; at this time, it has an impact on the plasma density leading to the development of the upwelling ESF plume as seen in the bottom panel. The large upward velocity of the top of the plume slows significantly at time 23:30 LT (altitude ~1200 km). At time ~03:00 LT, the upward velocity becomes very small (i.e., ~10 m s^{-1}), and the height of the plume approaches an asymptotic value of about 1600 km. Also, at this time, the electron density in the altitude range

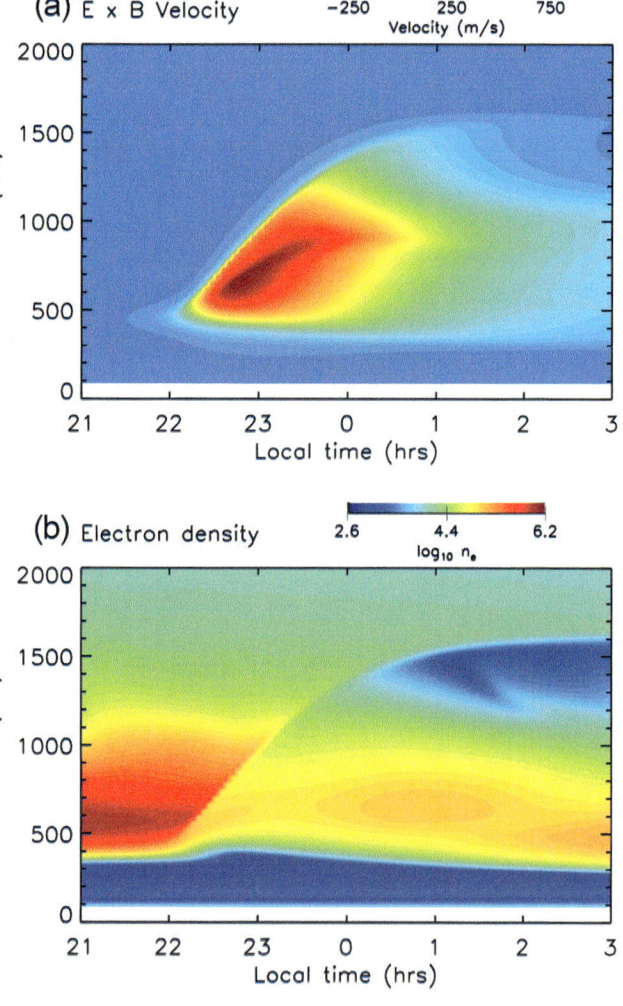

Figure 5. Color-scale plots of the $E \times B$ velocity and electron density (cm^{-3}) as a function of LT and altitude at longitude 0° [From *Krall et al.*, 2010b, used with permission].

1200–1600 km becomes very small because of the "drainage" of the plasma to lower altitudes.

Similar to radar backscatter observations, the plume persists after it has stopped rising ($t > 01{:}00$ LT) for a time long compared to growth time of the plume. In comparing these results to observations [e.g., *Tsunoda*, 1981], we acknowledge that the idealizations of a single-bubble perturbation and zero wind lead to a maximum bubble height of 1600 km, which is unusual for such low values of the $F10.7$ and $F10.7a$ indices. Additional simulations indicate that the maximum bubble height is reduced for nonzero winds [*Krall et al.*, 2009a; *Huba et al.*, 2009c] or in cases where multiple bubbles are seeded [*Krall et al.*, 2010c].

6. DISCUSSION

In this paper, we have reviewed the 3-D temperature and density structure of simulated ESF plumes. As described elsewhere [*Huba et al.*, 2009a; *Krall et al.*, 2010a], these simulations have led to explanations for plasma blobs and ESF temperature measurements. However, measurements of ESF plumes have yet to fully verify the 3-D plume structure. One of the challenges of confirming these findings lies in the fact that plumes evolve with time, while in situ measurements of any given plume occur (at best) once per satellite orbit (about 90 min). One approach would be to analyze a very large number of plume measurements, all during similar geophysical conditions, to build up elements of this 3-D, time-evolving structure. One difficulty would be to distinguish measurements near the outer edge of the plume from those at lower relative altitudes. Another would be to distinguish the early time explosive phase of growth from later stages. Multipoint observations (e.g., a satellite constellation) would clearly be helpful in the effort to verify these results.

Based on the long-term (hours) evolution of our model plumes, we expect that the 3-D density structure, including the low-density apex and the high-density crests, will persist as the plume evolves to obtain a fossil state. We conclude that plasma blobs can be detected in situ long after the plume has stopped rising. On the other hand, Figure 5 shows that velocities within the plume drop significantly within about 3 h of plume initiation. This suggests that the 3-D temperature structure might be observable through combined in situ ion temperature and ion velocity measurements, so long as the data analyzed corresponds to still-active plumes. To properly address this, simulations of ESF must be carried out for comparison to specific sets of observations, using identical geophysical and other conditions.

Acknowledgments. This work was supported by the Naval Research Laboratory 6.1 Base Funds and the NASA LWS program.

REFERENCES

Anderson, D. N. (1973), A theoretical study of the ionospheric F region equatorial anomaly—I. Theory, *Planet. Space Sci., 21,* 409–419.

Booker, H. G., and H. W. Wells (1938), Scattering of radio waves in the F-region of ionosphere, *J. Geophys. Res., 43,* 249–256.

de La Beaujardiere, O., L. Jeong, and the C/NOFS Science Definition Team (2004), C/NOFS: A mission to forecast scintillations, *J. Atmos. Sol. Terr. Phys., 66*(17), 1573–1591.

Hanson, W. B., and D. K. Bamgboye (1984), The measured motions inside equatorial plasma bubbles, *J. Geophys. Res., 89*(A10), 8997–9008.

Hanson, W. B., and R. J. Moffett (1966), Ionization transport effects in the equatorial F region, *J. Geophys. Res., 71*(23), 5559–5572.

Huba, J. D., and G. Joyce (2010), Global modeling of equatorial plasma bubbles, *Geophys. Res. Lett., 37,* L17104, doi:10.1029/2010GL044281.

Huba, J. D., G. Joyce, and J. A. Fedder (2000), Sami2 is Another Model of the Ionosphere (SAMI2): A new low-latitude ionosphere model, *J. Geophys. Res., 105*(A10), 23,035–23,053.

Huba, J. D., G. Joyce, and J. Krall (2008), Three-dimensional equatorial spread F modeling, *Geophys. Res. Lett., 35,* L10102, doi:10.1029/2008GL033509.

Huba, J. D., G. Joyce, J. Krall, and J. Fedder (2009a), Ion and electron temperature evolution during equatorial spread F, *Geophys. Res. Lett., 36,* L15102, doi:10.1029/2009GL038872.

Huba, J. D., J. Krall, and G. Joyce (2009b), Atomic and molecular ion dynamics during equatorial spread F, *Geophys. Res. Lett., 36,* L10106, doi:10.1029/2009GL037675.

Huba, J. D., S. L. Ossakow, G. Joyce, J. Krall, and S. L. England (2009c), Three-dimensional equatorial spread F modeling: Zonal neutral wind effects, *Geophys. Res. Lett., 36,* L19106, doi:10.1029/2009GL040284.

Hysell, D. L. (2000), An overview and synthesis of plasma irregularities in equatorial spread F, *J. Atmos. Sol. Terr. Phys., 62,* 1037–1056.

Kelley, M. C. (2009), *The Earth's Ionosphere: Plasma Physics and Electrodynamics*, Academic, San Diego, Calif.

Krall, J., J. D. Huba, G. Joyce, and S. T. Zalesak (2009a), Three-dimensional simulation of equatorial spread-F with meridional wind effects, *Ann. Geophys., 27,* 1821–1830.

Krall, J., J. D. Huba, and C. R. Martinis (2009b), Three-dimensional modeling of equatorial spread F airglow enhancements, *Geophys. Res. Lett., 36,* L10103, doi:10.1029/2009GL038441.

Krall, J., J. D. Huba, G. Joyce, and T. Yokoyama (2010a), Density enhancements associated with equatorial spread F, *Ann. Geophys., 28,* 327–337.

Krall, J., J. D. Huba, S. L. Ossakow, and G. Joyce (2010b), Why do equatorial ionospheric bubbles stop rising?, *Geophys. Res. Lett., 37,* L09105, doi:10.1029/2010GL043128.

Krall, J., J. D. Huba, S. L. Ossakow, and G. Joyce (2010c), Equatorial spread F fossil plumes, *Ann. Geophys., 28,* 2059–2069, doi:10.5194/angeo-28-2059-2010.

Le, G., C.-S. Huang, R. F. Pfaff, S.-Y. Su, H.-C. Yeh, R. A. Heelis, F. J. Rich, and M. Hairston (2003), Plasma density enhancements associated with equatorial spread F: ROCSAT-1 and DMSP observations, *J. Geophys. Res.*, *108*(A8), 1318, doi:10.1029/2002JA009592.

Martinis, C., J. Baumgardner, M. Mendillo, S.-Y. Su, and N. Aponte (2009), Brightening of 630.0 nm equatorial spread-F airglow depletions, *J. Geophys. Res.*, *114*, A06318, doi:10.1029/2008JA013931.

Ossakow, S. L. (1981), Spread F theories: A review, *J. Atmos. Terr. Phys.*, *43*, 437–452.

Oya, H., T. Takahashi, and S. Watanabe (1986), Observation of low latitude ionosphere by the impedance probe on board the Hinotori satellite, *J. Geomagn. Geoelectr.*, *38*, 111–123.

Oyama, K.-I., K. Schlegel, and S. Watanabe (1988), Temperature structure of plasma bubbles in the low latitude ionosphere around 600 km altitude, *Planet. Space Sci.*, *36*(6), 553–567.

Picone, J. M., A. E. Hedin, D. P. Drob, and A. C. Aikin (2002), NRLMSISE-00 empirical model of the atmosphere: Statistical comparisons and scientific issues, *J. Geophys. Res.*, *107*(A12), 1468, doi:10.1029/2002JA009430.

Retterer, J. M. (2010), Forecasting low-latitude radio scintillation with 3-D ionospheric plume models: 1. Plume model, *J. Geophys. Res.*, *115*, A03306, doi:10.1029/2008JA013839.

Scherliess, L., and B. G. Fejer (1999), Radar and satellite global equatorial F region vertical drift model, *J. Geophys. Res.*, *104*(A4), 6829–6842.

Steenburgh, R. A., C. G. Smithtro, and K. M. Groves (2008), Ionospheric scintillation effects on single frequency GPS, *Space Weather*, *6*, S04D02, doi:10.1029/2007SW000340.

Tsunoda, R. T. (1981), Time evolution and dynamics of equatorial backscatter plumes 1. Growth phase, *J. Geophys. Res.*, *86*(A1), 139–149.

Watanabe, S., and H. Oya (1986), Occurrence characteristics of low latitude ionosphere irregularities observed by impedance probe on board the Hinotori satellite, *J. Geomagn. Geoelectr.*, *38*, 125–149.

Yokoyama, T., S.-Y. Su, and S. Fukao (2007), Plasma blobs and irregularities concurrently observed by ROCSAT-1 and Equatorial Atmosphere Radar, *J. Geophys. Res.*, *112*, A05311, doi:10.1029/2006JA012044.

J. D. Huba and J. Krall, Plasma Physics Division, Naval Research Laboratory, Washington, DC 20375, USA. (jonathan.krall@nrl.navy.mil)

Low-Latitude Ionosphere and Thermosphere: Decadal Observations From the CHAMP Mission

Claudia Stolle

DTU Space, Technical University of Denmark, Copenhagen, Denmark

Huixin Liu

Department of Earth and Planetary Science, Faculty of Science, Kyushu University, Fukuoka, Japan

During its more than 10 year mission, the CHAMP satellite has revealed exciting new observations of the upper F region ionosphere and the thermosphere. Special benefit has been obtained from simultaneous observations of plasma and neutrals, and the magnetic field, which has led to many new findings and insights into the thermosphere-ionosphere system. This chapter reviews these new findings at low and equatorial latitudes, from the point of view of coupling between the atmosphere and the magnetic field and between different atmospheric regions. Gross features in the electron density and temperature and in the neutral density and wind are tackled, such as the equatorial anomaly, the electron temperature morning overshoot, equatorial plasma irregularities, the zonal wind jet, terminator waves, or findings of the wave-4 structure in plasma and neutrals. For specific examples, it is demonstrated that results of physical models have large impact in understanding the new observations and underlying processes. Examples include the distribution of electron temperature during the morning that is suggested to follow the inclination of the geomagnetic field and the longitudinal wave-4 structures in the plasma and neutrals that are believed to be generated by the same source but through different coupling mechanisms. It is also demonstrated that even the most sophisticated models cannot fully reproduce observations of even large-scale ionospheric and thermospheric features.

1. INTRODUCTION

The Earth's ionosphere and thermosphere cover the region from about 100 to 800 km altitude. Different to the lower and middle atmosphere, the upper atmosphere is characterized by gravitationally separated species being partly ionized and free electrons, which allow electrical currents to flow. Knowledge concerning parameters of neutral and plasma density and temperature, winds, and the magnetic and electric fields are crucial to understanding both the thermosphere and ionosphere and their interactions with each other and with other atmospheric regions. During the last few decades, low-orbiting satellite missions have contributed significantly to an increasing amount of in situ observations of various quantities. Among these, the CHAMP satellite [*Reigber et al.*, 2002] has been outstanding with its 10 year long mission, unique orbit altitude of ≤450 km, and simultaneous high-resolution observations of plasma and neutrals.

The increased spatial and temporal coverage of the data has led to several discoveries and has permitted new perspectives on a number of existing research topics. Coupling between the thermosphere and ionosphere with particular regard to results from the CHAMP mission has been reviewed by *Lühr et al.* [2011], with a focus on gross features of the neutral density and wind at 400 km at both low and polar latitudes, the *F* region dynamo, and mass density depletions accompanying equatorial plasma irregularities. Highlights achieved from high-precision magnetic satellite missions including CHAMP have been summarized by *Olsen and Stolle* [2012]. Their paper provides an overview of findings related to ionospheric and magnetospheric current systems derived from magnetometer observations in orbit pointing out the benefit of joint analyses of magnetic field sources both internal and external in origin.

This chapter will concentrate on selected ionospheric and thermospheric processes at low and equatorial latitudes, with emphasis on new features revealed by CHAMP from the perspective of atmospheric coupling. For specific examples, it is demonstrated that results of physical models have large impact in understanding the observations and underlying processes. The chapter starts with a brief summary of the CHAMP mission and relevant data in Section 2. New findings about the ionosphere and thermosphere are described in sections 3 and 4. Conclusions and an outlook are given in Section 5.

2. CHAMP MISSION AND DATA

The CHAMP satellite was launched on 15 July 2000 into a near-circular orbit with an inclination of 87.3°, thus, nearly longitudinal aligned at low latitudes. Successive satellite passes were separated by 23° in geographic longitude. The progress in LT was about 5.5 min d^{-1}; thus, one full LT coverage was obtained in ~130 days. The initial orbital height was 450 km, then decayed gradually to ~320 km in February 2010, and has been last contacted at 150 km on 19 September 2010 when the satellite reentered the atmosphere. The mission was carried out in the declining solar cycle 23 including its deep minimum. The scientific payload was, among others, a planar Langmuir probe (PLP), which provides measurements of the electron density and temperature with 15 s temporal resolution, corresponding to 120 km spatial resolution. Details on the retrieval and validation of these products are available in the works of *McNamara et al.* [2007] and *Rother et al.* [2010]. Electron temperature data are meaningful between 19 February 2002 and 22 February 2010. Data before this period suffer from unfavorable PLP settings, while after that, a 180° rotation of the satellite about the yaw axis made it impossible to perform reliable measurements.

The spacecraft carried a scalar and a fluxgate magnetometer, which are used to retrieve calibrated full vector magnetic field products [e.g., *Maus*, 2007]. Ionospheric currents are inferred from magnetic field observations, after subtracting contributions from the main magnetic field, the lithospheric magnetization, and the magnetosphere. To do so, predictions of geomagnetic field models including all these contributions such as the CHAOS or POMME series (http://www.space.dtu.dk/MagneticFieldModels, http://geomag.org/models) are especially suited. CHAMP electron density and temperature, and magnetic field data are available for open access at http://isdc.gfz-potsdam.de.

With an accelerometer of unprecedented high resolution, CHAMP records the drag force experienced during its flight through the upper atmosphere. From such records, nongravitational acceleration is extracted, which provides measurements of both neutral mass density and cross-track neutral wind via the satellite drag formula. Given the accelerometer's precision of 3×10^{-9} m s^{-2}, the obtained mass density and zonal wind have a precision of 1×10^{-14} kg m^{-3} and 20 m s^{-1}, respectively. CHAMP data provide a sample rate of 0.1 Hz, corresponding to a spatial resolution of ~80 km. A clear-cut description of the essential principles of this method is given in the works of *Liu et al.* [2005, 2006], while more details on the techniques related to raw data processing can be found in the works of *Bruinsma et al.* [2004] and *Doornbos et al.* [2010].

3. OBSERVATIONS OF THE IONOSPHERE

CHAMP plasma and neutral observations have revealed the climatological distribution of key parameters of the upper atmosphere. Figure 1 shows the magnetic LT and latitude variations, and longitudinal averaged electron density (N_e) and temperature (T_e). The data have been selected for low geomagnetic activity ($Kp \leq 3$) and equinox months. Figures 1a and 1b represent an average between 19 February 2002 and 22 February 2010. The mean satellite altitude was 365 km, and the mean solar flux index was $P = 98$, where $P\,[10^{-22}\,\text{Wm}^{-2}\,\text{Hz}^{-1}] = (F10.7 + F10.7\text{A})/2$ [*Richards et al.*, 1994], and $F10.7\text{A}$ is the 81 day average of the $F10.7$ values centered on the day of interest. Corresponding observations of the thermosphere (Figures 1c, 1d) are discussed in Section 4.

3.1. The Equatorial Ionization Anomaly

Electron density (Figure 1a) is lowest late in the night before sunrise; after this, the highest plasma density builds up around the magnetic equator, afterward forming the equatorial ionization anomaly (EIA) recognized as a dual band of enhanced N_e at about 15° north and south of the trough. The EIA is maintained until around midnight. The formation of the EIA is a result of the diurnal variation of the equatorial

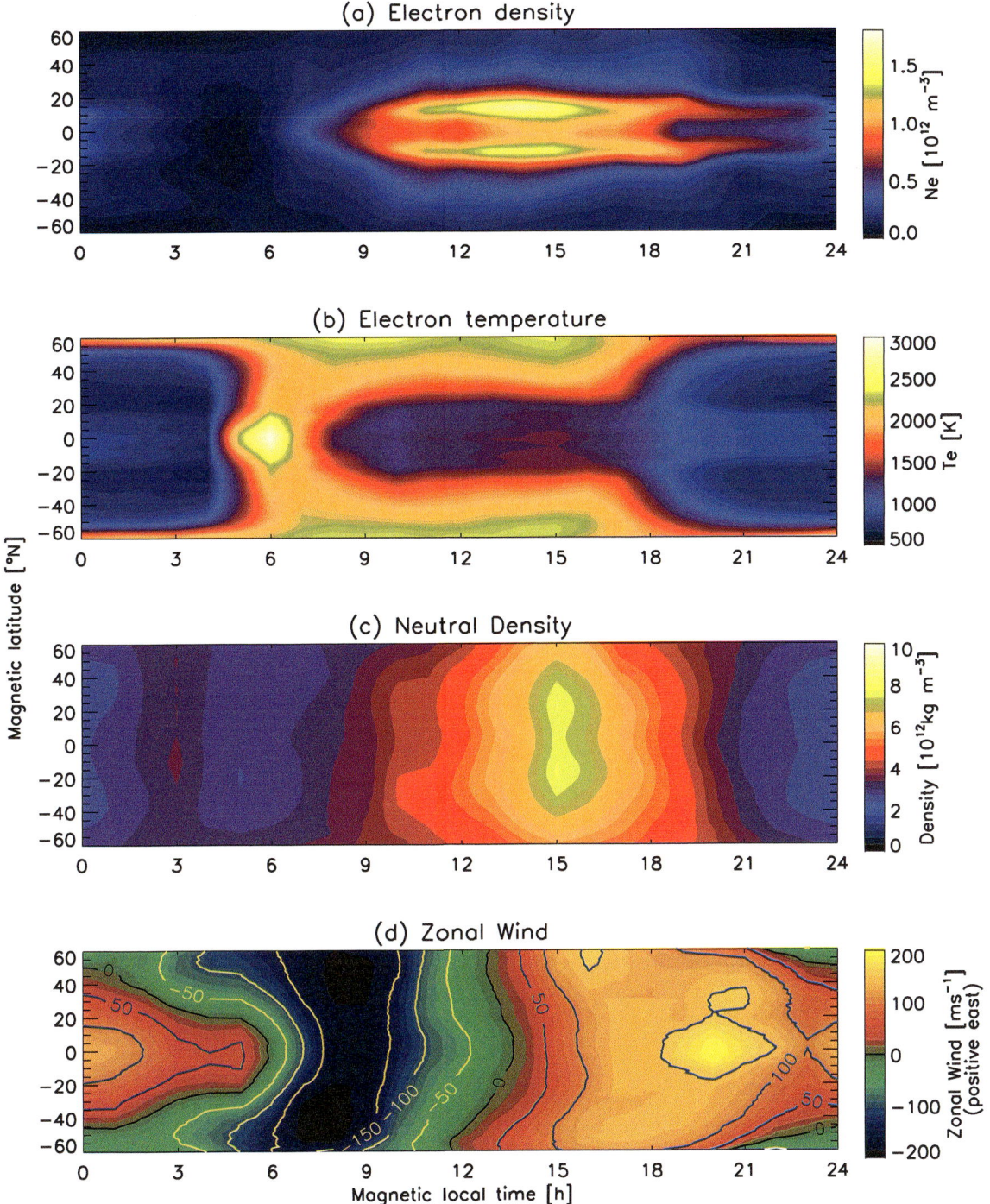

Figure 1. Local time and magnetic latitude variation of CHAMP electron density, temperature, and neutral density and winds for equinox months and longitudinal averaged. Otherwise, see text.

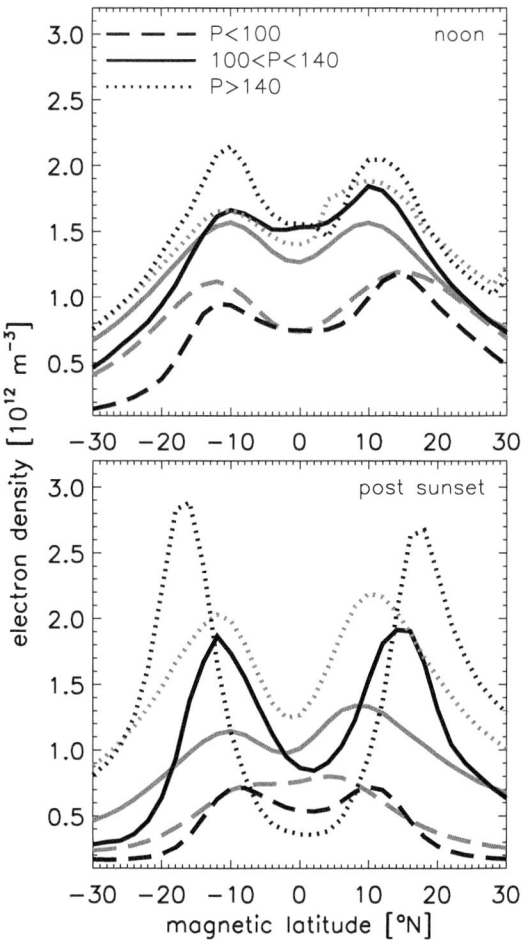

Figure 2. CHAMP N_e observations (black) and corresponding IRI2007 predictions (gray). Mean P values (CHAMP altitude (km)) are 79, 119, 158 (341, 396, 414) for noon, and 74, 121, 184 (339, 400, 392) for postsunset. Otherwise, see text.

zonal electric field, which primarily points eastward during the day [e.g., *Fejer et al.*, 2008]. With the horizontal geomagnetic field at equatorial latitudes, the plasma is lifted up by vertical $E \times B$ drift. Once transported to higher altitudes, the plasma diffuses downward along the geomagnetic field lines into both hemispheres due to gravitational and pressure gradient forces. The EIA has frequently been reviewed in the literature [see, e.g., *Anderson*, 1981]. The zonal electric field often experiences an enhancement after sunset before it turns westward, which is called the prereversal enhancement (PRE). In particular, the PRE has a distinct longitudinal and seasonal variation, and it is strongly correlated with solar flux. On the other hand, the dayside electric field shows only a weak dependence on solar cycle [e.g., *Stolle et al.*, 2008a; *Fejer et al.*, 2008]. Proper modeling of the EIA is a key issue in describing the plasma structure of the low-latitude ionosphere and many related phenomena.

The almost longitudinally aligned CHAMP orbit at low latitudes and the long time series of electron density provide an excellent opportunity to monitor climatological variations of the EIA. Figure 2 shows the latitudinal N_e distribution for $-90°E$ to $-60°E$ longitude (geographic), for data in February, March, April, for three solar flux levels, and for noon (11–15 LT) and postsunset (18–22 LT), respectively, along with the corresponding predictions of the empirical International Reference Ionosphere model, version 2007, (IRI 2007) [*Bilitza and Reinisch*, 2008]. The graphs clearly reveal the EIA structure. The crest-to-trough ratio $\left[\text{CTR} = \dfrac{N_e(\text{north\_crest}) + N_e(\text{south\_crest})}{2N_e(\text{trough})} \right]$ of the EIA derived from CHAMP varies only slightly with solar flux during the day, e.g., 1.44 ($P < 100$), 1.16 ($100 < P < 140$), and 1.44 ($P > 140$). The postsunset EIA CTR increases moderately from 1.34 ($P < 100$) to 2.24 ($100 < P < 140$), but is largely enhanced for higher solar flux reaching 7.78 ($P > 140$). This reflects that the noontime EIA profile is lifted up as a whole with little change in shape, while the EIA shape after sunset changes considerably. *Liu et al.* [2007a] have shown that the equatorial EIA trough during daytime is positively correlated with solar flux, while the postsunset trough N_e shows a positive correlation up to $P \sim 170$ but a negative correlation at higher P values. This reversal can be explained by very deep equatorial troughs that develop for strong upward plasma drifts of 40 m s^{-3} or higher during high solar flux conditions. *Liu et al.* [2007a] also compared low-latitude N_e with predictions of the IRI2001 model and found that the empirical model represents well the EIA structure at all solar flux levels during the day, but largely underestimates the EIA strength for postsunset. Similar conclusions can be drawn from Figure 2. The inappropriate representation of the F_2 layer height at equatorial latitudes in the IRI model has been identified as the major cause for this discrepancy.

EIA simulated by different physical models is discussed within the Problems Related to Ionospheric Models and Observations (PRIMO) project [*Fang et al.*, this volume]. Their simulation conditions correspond approximately to the solid lines in Figure 2. Striking differences between physical models and the IRI2007 predictions (which is taken as the metric reference) occur both during the day and postsunset. While coupled thermosphere/ionosphere models agree well with IRI results during the day, noncoupled models largely overestimate the EIA strengths. At postsunset, noncoupled models agree better with the IRI results, but coupled models predicts EIA structures as flat as during daytime. Results from PRIMO reveal that correctly simulating the EIA is

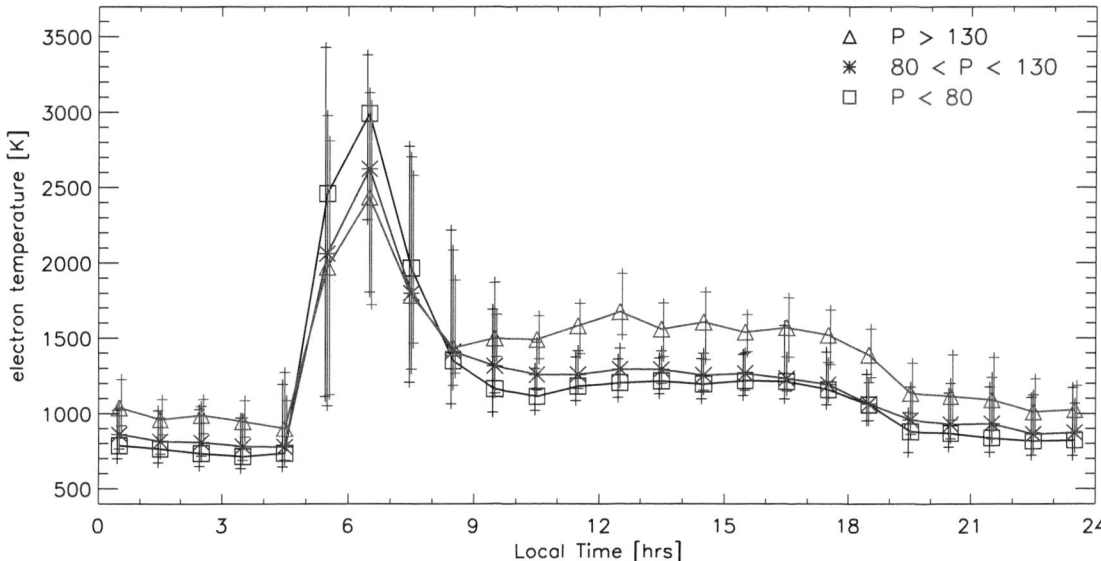

Figure 3. Diurnal variation of CHAMP T_e observations between ±5° dip latitude for three solar flux levels averaged over longitude and season, and leveled to 350 km altitude. Standard deviations are shown as thin vertical lines.

still challenging and is highly sensitive to variations of the modeled electric field and the thermospheric winds, and also to modeled neutral and plasma temperature.

3.2. Equatorial Electron Temperature

Electron temperature (Figure 1b) is low at equatorial latitudes (1000–1500 K) and increased sharply toward higher latitudes (2000–2500 K). The almost steplike increase occurs at ~±20° (~±60°) latitude during the day (night). The electron temperature (T_e) of the equatorial F region has a characteristic diurnal variation. At and just after sunrise, when electron density is still very low, electron temperature increases dramatically. The low-latitude electron temperature decreases again to moderate values coinciding with the gradual increase of electron density at later LTs (compare with Figure 1a). At about 09 LT, T_e has reached its daytime value. This phenomenon is called the electron temperature morning overshoot (MO) and results from excess heat in the low dense electron gas at prompt sunrise [*Da Rosa*, 1966].

Figure 3 shows hourly means of longitudinal averaged T_e at equatorial latitudes (±5° dip latitude) for three solar flux levels. Since different solar flux levels correspond to different heights of the CHAMP satellite, the observations have been leveled to 350 km altitude using IRI2007. The model does not include solar flux variation of T_e; therefore, we do not expect modifications in the observed solar flux trend when applying the leveling. The temperatures during day and night are rather stable in a climatological sense, which is reflected by low standard deviations. At times of the MO, the standard deviation is increased, but meaningful, with a significance of ±30 K. *Stolle et al.* [2011] investigated the distribution and solar flux dependence of T_e observations during local morning and compared the data with results from the field line interhemispheric plasma (FLIP) model [*Richards*, 2004]. The MO with $T_e \geq 3000$ K occurs at the dip equator and then symmetrically decreases to about 1500 K at 20° dip latitude. This shape shows no correlation at all with the flat N_e distribution in this latitudinal and LT sector (see Figures 1a, 1b). FLIP results have shown that heat flux minimizes near the dip equator where the magnetic field is horizontal, heat flux increases by several orders of magnitude away from the equator with increasing magnetic field inclination. Therefore, local heating and cooling are dominant at the dip equator; excess heat cannot be conducted away. As the ion density builds up during the morning, T_e decreases to daytime values due to high electron-ion collisions.

Figure 3 confirms the correlation of T_e with solar flux level during day and night, as has been shown by other data [e.g., *Truhlík et al.*, 2009]. The MO temperature, however, is anticorrelated with solar flux. *Stolle et al.* [2011] found a slope of −14.2 K sfu^{-1} for the MO amplitude at −75°E during December solstice months. The FLIP model predicts a corresponding slope of −13.1 K sfu^{-1} for the same conditions. The anticorrelation of MO temperatures can be explained as follows: The most important local heat source for the thermal electron gas in the F region is through collision with photoelectrons; cooling occurs through Coulomb collisions with ions. The heating rate, Q, is nearly linear in the photoelectron flux, Φ_e, and N_e [*Schunk and Nagy*, 2009, Equations (9.49) and (9.40)]. The cooling rate, L, is quadratic in N_e [*Schunk and Nagy*, 1978, Figure 8a and Equation (48)]. Thus, $Q \propto (N_e, \Phi_e)$ and $L \propto N_e^2$. This means, $\Delta Q < \Delta L$, if $\Delta \Phi_e < \Delta N_e$. Photoelectron flux increases by a factor of 2–3 between low and high solar activity [*Richards and Peterson*, 2008]. AT 06 LT, N_e increases by a factor of 5 from solar flux levels $P < 80$ to $P > 130$ [*Stolle et al.*, 2011, Figure 2]. By this simple approach, Q would increase by a factor of 10–15 and L by a factor of 25, which explains the net cooling with increasing solar flux level. During the day, the electron gas is closely coupled to the high density ion and neutral gas through high collision frequencies, the temperature of both increases with solar flux, as does T_e. At altitudes above 450 km, however, the amplitude of the MO is positively correlated to the solar flux level [*Watanabe and Oyama*, 1996; *Stolle et al.*, 2011]. Non-local heating contributes significantly to thermal electron heating at these altitudes [e.g., *Varney et al.*, 2011] and may play an important role when discussing the height difference, e.g., with subsequent model studies.

3.3. Magnetic Signatures of Equatorial Plasma Irregularities

Over the last several decades, equatorial spread F (ESF) has been used to designate equatorial plasma density irregularities over a large range of scale sizes, which occur regularly in the low-latitude ionosphere shortly after sunset. ESF events are believed to be generated in the bottom side F region when the ionosphere is unstable due to the Rayleigh-Taylor instability. *Kelley* [2009] provides a comprehensive overview of ESF characteristics.

In CHAMP magnetic field and electron density observations, ESFs were first detected by *Lühr et al.* [2002]. Enhancements of the magnetic field magnitude coinciding with plasma depletions have been attributed to diamagnetic currents. *Stolle et al.* [2006] investigated the full vector magnetic data and were able to discriminate between perturbations parallel to the main magnetic field direction (due to diamagnetic currents) and perpendicular to it (due to field-aligned currents). They compiled a comprehensive climatology of the diamagnetic signatures being in good agreement with findings from plasma depletions onboard CHAMP [*Xiong et al.*, 2010] and other satellites, e.g., DMSP [*Huang et al.*, 2001]. In comparison with ROCSAT-1 plasma drift data, *Stolle et al.* [2008b] provided global evidence for the high correlation between ESF occurrence and the PRE vertical plasma drift of about 90%, which has been demonstrated only by local observations so far. At the same time, *Su et al.* [2008] found global evidence of such correlations based on ROCSAT-1 plasma density and drift observations. Larger vertical plasma drift lifts the ionosphere to higher altitudes where it is more susceptible to Rayleigh-Taylor instability due to lower electron density. *Stolle et al.* [2008b] found a longitudinal dependence of the ESF persistence, meaning, its occurrence rate at several hours after sunset. Postmidnight ESFs were detected above the South Atlantic and Africa with occurrence rates up to 30%, but almost never over the Pacific. These results indicate that mechanisms other than the PRE may determine the ESF lifetime. *Xiong et al.* [2012] investigated the behavior of plasma structures with different scale sizes between 15 km and a few hundred kilometers. Structures with scale sizes below 70 km were only detected at times and locations where large PRE values are expected, e.g., during high solar flux levels and above the South Atlantic anomaly between September and December and preferably on high L shells ($L = 1.125$). The occurrence of plasma structures with scale sizes ≥ 150 km does not show such restrictions. The observations indicate that highly structured ESF reach high ionospheric altitudes, and they preferably develop during strong ESF.

Magnetic signatures of field-aligned currents flowing along the ESF walls have been investigated by *Park et al.* [2009]. Until 21 LT, the magnetic deflections show clear hemispheric antisymmetry pointing toward the Earth in the Southern Hemisphere and away from Earth in the Northern Hemisphere. Assuming a prevalent eastward electric field after sunset, they signify a Poynting flux directed poleward and downward along the magnetic field lines. This supports the existence of a high-altitude equatorial electromagnetic source feeding the ESF flux tube, as suggested by *Bhattacharyya and Burke* [2000]. On the other hand, *Aveiro et al.* [2011] compared CHAMP observations with results of a 3-D numerical simulation and were able to reproduce magnetic effects of both diamagnetic and field-aligned currents using a purely electrostatic approach, which may argue against an electromagnetic source. More clarity on this point can be provided by satellite missions that provide observations of both the magnetic and the electric field, such as the C-NOFS satellite or the upcoming Swarm mission.

4. OBSERVATIONS OF THE THERMOSPHERE

In this section, large-scale thermospheric features revealed by CHAMP are described from three perspectives. They are the thermosphere response to solar forcing, to ionospheric forcing, and to lower atmosphere forcing. Figures 1c and 1d present the quiet time latitude-LT distribution of the thermospheric density and wind in geomagnetic coordinates around equinox, for moderate solar flux levels ($P = 150$) and at 400 km altitude. The density increases gradually after sunrise and peaks at about 15 LT. The formation of the peak density at 15 LT instead of local noon is a combined effect of the thermospheric circulation, heat advection, and diffusion, which are further influenced by tides from the lower atmosphere [*Mayr et al.*, 1973]. During about 10–20 LT, the density develops an anomaly-like structure, with peaks at ~±25° magnetic latitude. The neutral wind forms a fast eastward jet at the dip equator in the evening.

4.1. Thermospheric Response to Solar Forcing

Solar forcing includes solar radiation and solar wind. Transient and dramatic changes in solar forcing occur during solar flares and magnetic storms. The thermosphere, with its large mass and high heat capacity, is generally thought to be sluggish in responding to these transient events. However, in the upper thermosphere where the air density is relatively low, early disturbances can be detected by instruments with high sensitivity. This has been demonstrated by several studies using the CHAMP high-resolution measurements [e.g., *Liu and Lühr*, 2005; *Bruinsma et al.*, 2006]. These studies showed that the thermosphere density increases several folds during magnetic storms within a few tens of minutes. Recent efforts to empirically model the thermospheric density response to geomagnetic storms can be found in the work of *Liu et al.* [2010], which showed a close control of the thermospheric density by the merging electric field.

Rapid thermospheric response to solar flares within a few minutes was also observed [*Sutton et al.*, 2006; *Liu et al.*, 2007b]. Taking advantage of simultaneous neutral and plasma observations onboard CHAMP, *Liu et al.* [2007b] revealed that despite a nearly common onset time, the ionospheric perturbation developed much faster and to a larger amplitude than its thermospheric counterpart. For instance, during the 28 October 2003 flare event, the electron density increased by up to 68% within 8 min after the flare, while the thermospheric density reached its maximum enhancement of 50% within 1.5 h after the flare. The underlying process responsible for the flare-induced thermosphere and ionosphere disturbances are apparently different, with the former being mainly driven by a thermosphere expansion due to excessive heating and the latter by changes in equatorial electrodynamics. This difference is manifested in the contrasting latitudinal distribution of the neutral and plasma perturbations, with the neutral perturbation being nearly uniform at latitudes below ±40°, while the plasma is enhanced at equatorial and midlatitudes, but depleted in the EIA crest regions (see *Liu et al.*, 2007b, Figure 2). Plasma-neutral coupling via ion drag restores its dominating role about 2–3 h after the flare bursts, forcing the distribution of both disturbances to become similar. These features reveal a distinctive coupling and decoupling between the flare-induced plasma and neutral perturbations not known before, thus opening a new stage for studies of solar flare effects on the thermosphere-ionosphere system.

On longer time scales, the thermosphere oscillation in response to high-speed solar wind streams is a significant recent finding. Using CHAMP observations in 2005, *Lei et al.* [2008] found a prominent 9 day period in the thermospheric density. These subharmonics of the 27 day solar rotation are caused by recurrent geomagnetic storms driven by fast solar wind streams, which originate from solar coronal holes distributed roughly 120° apart in longitude. Depending on the number of coronal holes and their spatial distribution, the oscillation period varies. These density oscillations demonstrate the close link between the thermosphere and the solar wind. Improved understanding of such oscillations contributes to better predicting the atmospheric conditions.

4.2. Thermospheric Response to Ionospheric Forcing

The gross structures of the neutral density and wind are described well by empirical models like MSIS and HWM. In comparison, CHAMP observations have revealed significant deviations from these gross features, which demonstrate strong magnetic control of the thermosphere via neutral-plasma interactions. Among such deviations, the equatorial mass density anomaly and the equatorial wind jet are two outstanding features.

4.2.1. The equatorial mass density anomaly. The Equatorial Mass density anomaly (EMA) is an interesting and fundamental feature of the Earth's thermosphere in tropical regions [*Liu et al.*, 2005, 2007c]. It is an anomalous latitudinal distribution of the total mass density between about 10 and 20 LT, with its equinox configuration consisting of a density trough near the Earth's dip equator flanked by density crests around ±25° magnetic latitude (see Figure 1c). A similar anomaly in individual constitutes of N_2 and O was noticed by *Philbrick and McIsaac* [1972] in the DE 2 satellite observations. The EMA structure can be viewed as the neutral counterpart of the well-known EIA in the ionosphere (see section 3.1). It resembles fairly well features of the EIA in climatological aspects, e.g., the seasonal and solar cycle variations [*Liu et al.*, 2007c].

Among many factors proposed, three have been demonstrated by numerical simulations to contribute to the EMA formation. These are the meridional ion drag [*Maruyama et al.*, 2003; *Miyoshi et al.*, 2011], the terdiurnal tides [*Miyoshi et al.*, 2011], and the chemical heating fueled by charge exchange between O^+ and O_2 or N_2 [*Fuller-Rowell et al.*, 1997]. The EMA is successfully reproduced by the ground-to-topside model of the Atmosphere and Ionosphere for

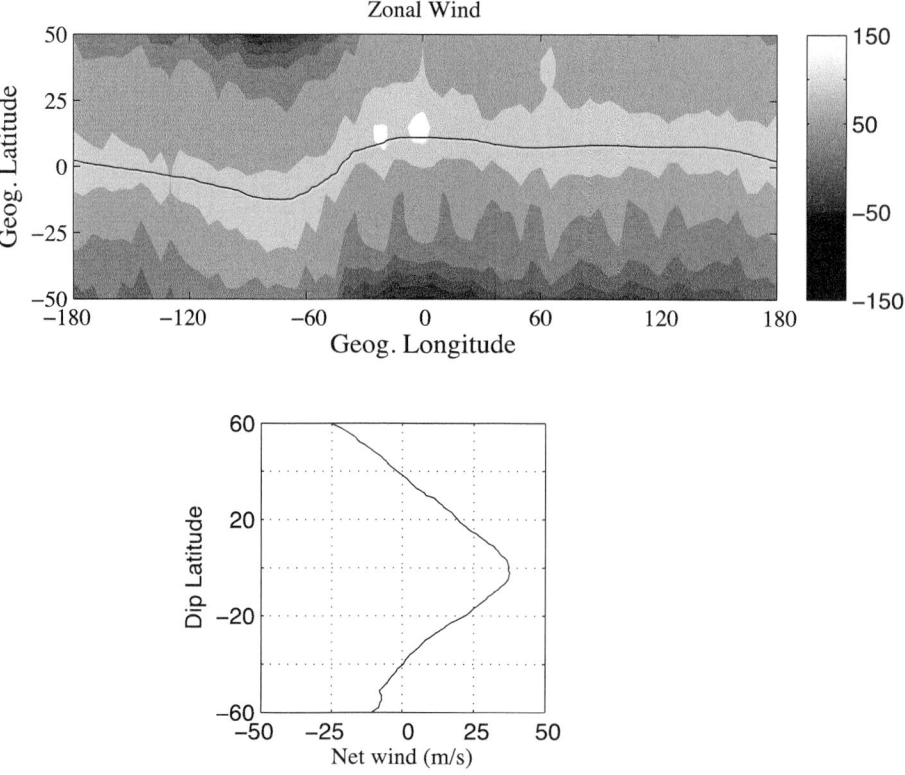

Figure 4. (top) Seasonally averaged zonal wind distribution in geographic coordinates during 18–24 MLT. Note the banded structure along the dip equator (solid line), where the fastest wind flows. (bottom) Seasonally averaged net zonal wind obtained from the neutral wind distribution in Figure 1. Positive eastward (in units of m s^{-1}) [*Liu et al.*, 2009a].

Aeronomy (GAIA) [*Miyoshi et al.*, 2011]. The model simulation confirmed that variations of the ion drag in the meridional direction are the main contributor to the EMA formation, with additional contribution from the terdiurnal tide. The latter suggests effects of lower atmospheric forcing on the EMA. Owing to the dominant role of ion drag, the EMA grows more prominent around equinox and toward solar maximum, in a similar fashion to the EIA. Complicated geomagnetic activity effects on the EMA are discussed in the work of *Lei et al.* [2010]. A quantitative examination of the chemical heating is yet to be carried out.

4.2.2. The equatorial wind jet. The thermospheric zonal wind forms a fast wind jet at the Earth's dip equator instead of the geographic equator as seen in Figure 4. This remarkable feature revealed by the CHAMP satellite with ample samplings [*Liu et al.*, 2009a] corroborates the sparse DE 2 satellite measurements [*Raghavarao et al.*, 1991], making the jet over the dip equator as a credible phenomenon. Eastward wind at the equator maximizes around 20 LT with a speed of 150 m s^{-1}, and westward wind is largest around 08 LT with a speed of -150 m s^{-1} (see Figure 1d). Fast wind jets occur during LTs when the EIA structure exists, being particularly prominent before noon and after sunset. The jet is not well developed in the afternoon due to wind reversal and the small wind speed. The short period of westward wind (06–12 LT) in comparison to that of the eastward wind (at other local times) inevitably leads to a net eastward wind at "magnetic" latitudes below ±35° as derived from CHAMP observations (see Figure 4). This means the upper atmosphere rotates faster than the Earth, which is a phenomenon termed superrotation [*Rishbeth*, 2002]. The superrotation speed is not constant but varies with season and solar cycle, due to corresponding changes in the ion drag [*Liu et al.*, 2006]. The average superrotation is about 22 m s^{-1} for moderate solar flux levels, but increases to 63 m s^{-1} for high solar flux levels.

The mechanism responsible for maximizing the wind at the dip equator is the latitudinal distribution of the zonal ion drag. This has been confirmed with both TIEGCM [*Kondo et al.*, 2011] and GAIA model [*Miyoshi et al.*, 2012], which have successfully reproduced the equatorial wind jet. These works demonstrate that due to the latitudinal distribution of the plasma in the EIA region, the zonal ion drag minimizes at the dip equator where the plasma density is low. On the other hand, the pressure gradient does not vary much between the EIA trough and crest regions. These factors combine to give a stronger wind at the dip equator. It is notable that a fast wind jet also exists at the dip equator in the topside ionosphere, where no EIA exists. This topside wind jet is found to be the direct upward extension of the F region wind jet via strong viscosity [*Kondo et al.*, 2011]. Finally, the F region wind is weak at the equator during non-EIA periods (~05–08 LT). Since the electron density varies little with latitude here, the weakening of the wind may be due to other factors than the ion drag.

4.3. Thermospheric Response to Lower Atmospheric Forcing

4.3.1. The wave-4 structure of the EMA and the neutral wind. Both the EMA and the thermospheric wind exhibit longitudinal structures with four peaks, when observed by sun-synchronous satellites [*Häusler et al.*, 2007; *Liu et al.*, 2009b]. This longitudinal variation is named the "wave-4 structure." It resembles, to a good extent, the wave-4 structure in the EIA [e.g., *Sagawa et al.*, 2005; *Immel et al.*, 2006; *Lühr et al.*, 2007; *Liu and Watanabe*, 2008].

It is important to point out that, despite the seemingly similar longitudinal structure of the EMA and EIA, their excitation processes are different [*Liu et al.*, 2009b]. The excitation of the EIA wave 4 is now accepted to be the modulation of the E region dynamo by eastward traveling nonmigrating diurnal tide with wavenumber 3 (DE3) [*Immel et al.*, 2006; *Hagan et al.*, 2007; *Jin et al.*, 2008], which are excited by deep tropical convection in the troposphere. On the other hand, analysis of simultaneous measurements of the plasma and neutral densities from CHAMP demonstrated that the EMA wave 4 is unlikely to result from the ion drag force associated with the EIA wave 4 because the two are largely out of phase [*Liu et al.*, 2009b]. Direct upward propagation of DE3 to the upper thermosphere appears to be the plausible candidate, since the observed phase shift between the EMA and wind wave 4 agrees well with that predicted by *Oberheide et al.* [2009].

The wave-4 structure in the EMA is well reproduced by the GAIA model [*Miyoshi et al.*, 2012]. Their simulation confirms that ion drag does not act as the major driver for the longitudinal structure. The longitudinal variation of the ion drag is found to be 3 times smaller than, and in the opposite direction to, that of the pressure gradient force. The wave-4 structure of the density is mainly driven by the direct penetration of the DE3. In addition, the eastward propagating semidiurnal tides with wave number 2 (SE2) is found to significantly contribute to the EMA wave-4 formation at middle latitudes.

It is worth noting that the meridional structure of EMA is mainly controlled by the ion drag associated with EIA, while the zonal structure is dominantly controlled by the directed propagation of waves from below. This suggests that the interplay of ionospheric forcing and lower atmospheric forcing on the thermosphere significantly depends on spatial dimension. This makes the global EMA formation a complex but fascinating process.

4.3.2. The terminator wave structures. The solar terminator (ST) is the transition line between day and night. It represents a region of sharp energy change in solar radiation. This consequently leads to strong gradients in the atmospheric density and temperature. In the vicinity of ST, the atmospheric gas is in a nonequilibrium state, giving rise to atmospheric irregularities and inhomogeneities [*Somisikov*, 1995; *Somsikov and Ganguly*, 1995], and atmospheric waves are generated when the ST transverses through the atmosphere as the Earth rotates.

ST-generated waves have been identified in CHAMP density [*Forbes et al.*, 2008] and wind [*Liu et al.*, 2009c] observations. This feature is seen in Figure 5 as banded structures ~30° inclined to the terminator. These ST waves possess wavelengths of ~3000–4500 km and a distinct seasonal rotation of the wavefront. A prominent 90° phase shift exists between the density and wind ST waves in both the meridional and zonal (LT) directions. Clear dawn-dusk asymmetry is seen in Figure 5, with more pronounced wave structures at dusk. This was suggested as a direct effect of different temperature/pressure gradients at the dawn and dusk terminators [*Liu et al.*, 2009c]. The dusk terminator bears a gradient more than 5 times of that near the dawn terminator, hence, being a more efficient wave-generating source. Furthermore, since the mean thermospheric wind is westward near dawn, signatures of the westward traveling terminator waves become weaker.

The GAIA model results demonstrated that the excitation of these ST waves at 400 km is largely due to upward propagating tides [*Miyoshi et al.*, 2009], which are probably generated in the ozone by the moving ST. Thus, thermospheric ST wave structures are not generated locally but are a response to lower atmospheric forcing.

4.3.3. Thermosphere and ionosphere effects during SSW. Stratosphere sudden warming (SSW) is a meteorological event where the stratospheric temperature experiences a rapid and significant rise of more than a few tens of Kelvin in the winter polar region. It is generated by planetary wave-mean wind interaction in the polar region and global-scale meridional circulation [*Matsuno*, 1971]. SSW effects on the atmosphere are global. In the polar region, mesosphere cooling and lower thermosphere warming have been observed

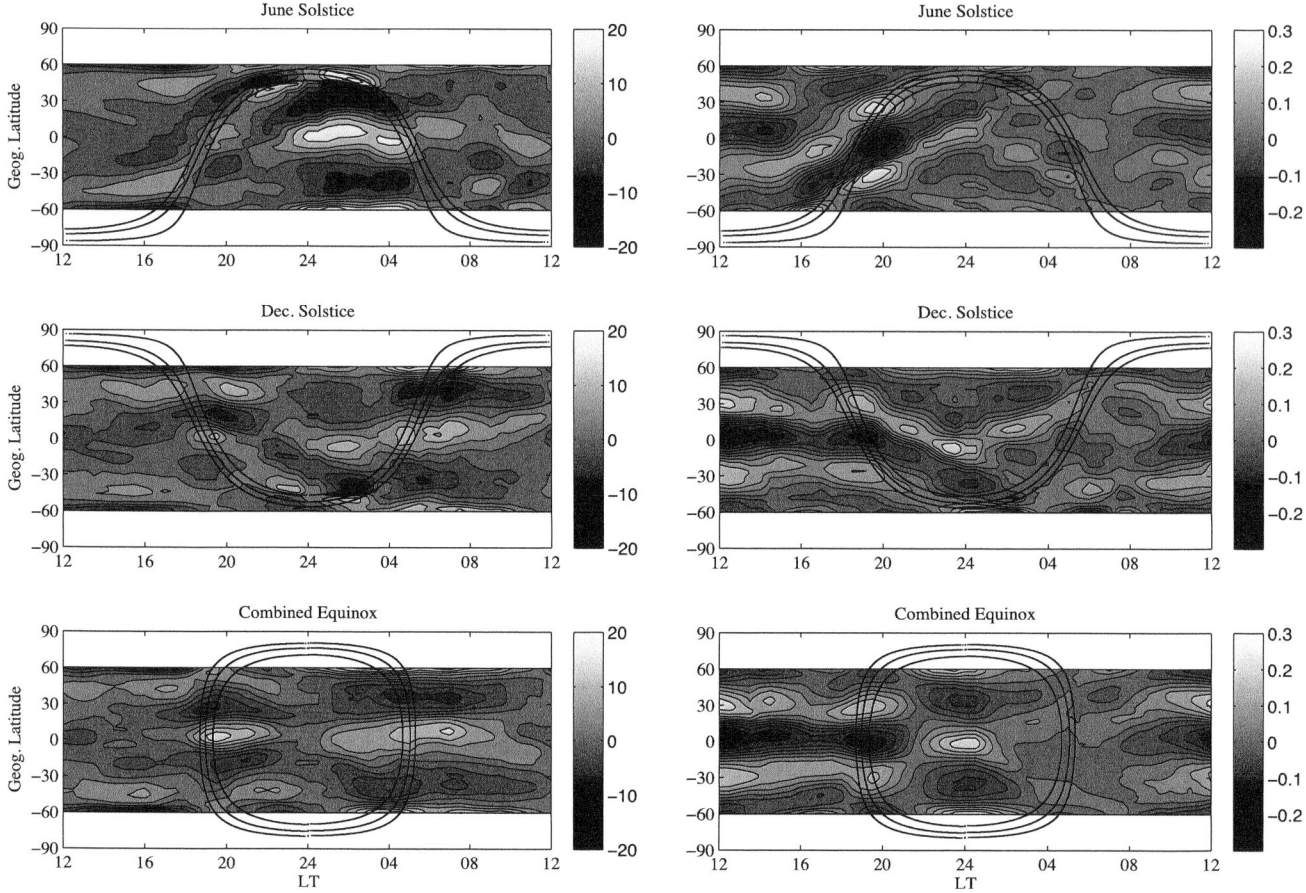

Figure 5. Latitude versus LT distribution of residual zonal wind (left-hand side) and residual density (right-hand side) in different seasons. The zonal wind is in units of m s^{-1}, positive eastward. The residual density is in units of 10^{-12} kg m^{-3}. Solid lines depict the location of the solar terminator at three different altitudes: 100, 200, and 400 km [Liu et al., 2009c].

[Funke et al., 2010; Kuriha et al., 2010]. At middle latitudes, radar observations reveal alternating warming and cooling regions in the E and F region ionosphere [Goncharenko and Zhang, 2008]. At low and equatorial latitudes, significant cooling during SSWs has long been reported in the stratosphere [Fritz and Soules, 1972]. In the ionosphere, at altitudes above ~100 km, a semidiurnal perturbation has been identified during SSW events in various parameters like the vertical plasma drift and the total electron content (TEC) [Chau et al., 2009; Goncharenko et al., 2010]. Analysis of the equatorial electrojet obtained from both CHAMP and ground observations suggests that this modulation results from enhanced lunar semidiurnal tides during SSWs [Fejer et al., 2010; Park et al., 2012; Yamazaki et al., 2012]. However, model results on the role of lunar tides are still at variance with the observations. For example, Fang et al. [2012] reproduced the observed vertical drift pattern without including lunar tidal effects at all.

Liu et al. [2011] reported significant thermosphere cooling observed simultaneously by the CHAMP and GRACE satellites. The GRACE mission consists of two satellites flying initially at ~500 km altitude. Similar to CHAMP, GRACE also provides measurements of thermospheric mass density via accelerometers. Observations of the predawn (04–06 LT) and late afternoon (16–18 LT) sectors during the major SSW in January 2009 were analyzed. After removing geomagnetic activity effects, thermospheric density reduction is found to be largest around 20°S, corresponding to a cooling of ~50 K. As seen in Figure 6, the cooling shows significant longitudinal dependence in terms of its magnitude and timing, with the strongest cooling occurring in the Peruvian sector. On the other hand, given the complex processes related to the SSW, it is possible that the cooling effect is highly LT dependent. This may lead to rather insignificant response when globally averaged. Though the SSW's effect on the upper atmosphere remains an open question, it stimu-

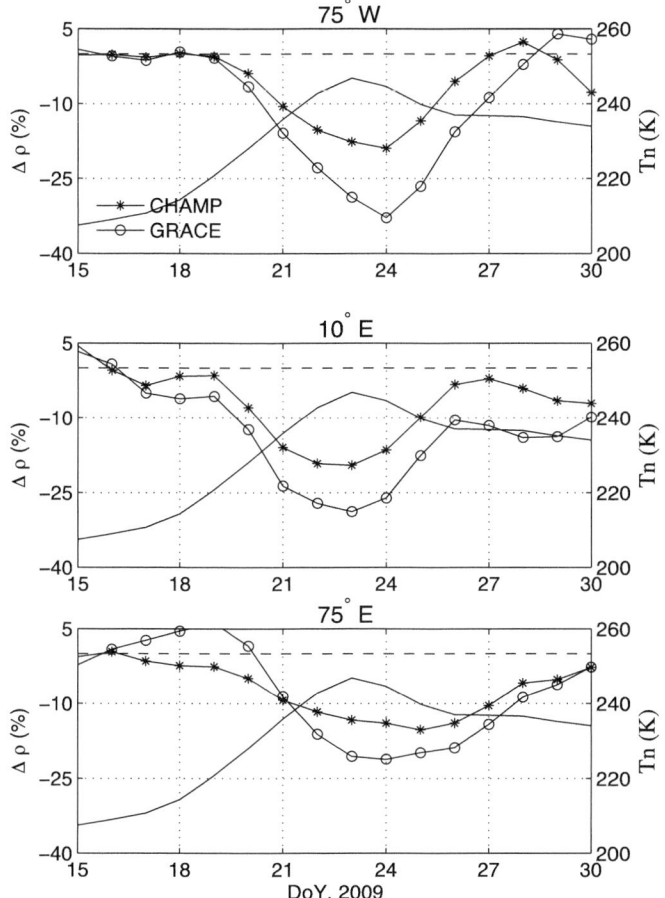

Figure 6. Disturbance (in percentage) of the thermospheric density averaged within 30°S–30°N for the Peruvian sector around 75°W, the European sector around 10°E, and the Indian sector around 75°E. Solid lines are neutral temperature at 10 hPa, indicating the development of the SSW [*Liu et al.*, 2011].

lates new perspectives for the global coupling of the lower and upper atmosphere.

Finally, it is important to be aware that different species contribute to the measured parameter of thermospheric total density, and it effectively results from a combination of energy transfer processes. Although it helps to describe many thermospheric features described in section 4, observations of individual atmospheric constituents and the full wind vector are essential for a better understanding of underlying physical processes.

5. CONCLUSIONS AND OUTLOOK

Ten years of continuous global CHAMP observations of plasma and neutrals, together with observations of the geo-magnetic field, has provided a highly valuable data set at unique altitudes of ≤450 km. This has resulted in detailed insights into a number of topics related to solar-terrestrial interactions and the coupling processes between different atmospheric layers. This paper has highlighted processes at low and equatorial latitudes. It was pointed out that the upper atmosphere should be investigated in the light of an ion/neutral coupling system in the presence of the geomagnetic field.

Physical models have greatly contributed to the understanding of new observations and related generation processes. On the other hand, many models still cannot fully reproduce basic climatological features such as the equatorial ionization or mass density anomaly, which indicates a remaining lack of knowledge of the underlying physics.

Multidisciplinary observations will be continued during the upcoming *Swarm* mission to be launched in 2013 [*Friis-Christensen et al.*, 2008]. *Swarm* is a constellation of three identical polar-orbiting satellites, with two side-by-side flying spacecraft at altitudes ≤460 km and one spacecraft at ~530 km. The mission is targeted at providing electric field and plasma densities and temperatures at 2 Hz sampling rate, and magnetic field and neutral density and zonal wind at 1 Hz sampling rate, corresponding to spatial resolutions of 4 and 8 km, respectively. Such a constellation will allow us to investigate small-scale gradients and variations in the upper atmosphere with less space time ambiguity than before.

Acknowledgments. The CHAMP mission was sponsored by the Space Agency of the German Aerospace Center (DLR) through funds of the Federal Ministry of Economics and Technology, following a decision of the German Federal Parliament (grant code 50EE0944). The data retrieval and operation of the CHAMP satellite was performed by the German Space Operations Center (GSOC).

REFERENCES

Anderson, D. N. (1981), Modeling the ambient, low latitude *F*-region ionosphere – A review, *J. Atmos. Terr. Phys.*, *43*(8), 753–762.

Aveiro, H. C., D. L. Hysell, J. Park, and H. Lühr (2011), Equatorial spread *F*-related currents: Three-dimensional simulations and observations, *Geophys. Res. Lett.*, *38*, L21103, doi:10.1029/2011GL049586.

Bhattacharyya, A., and W. J. Burke (2000), A transmission line analogy for the development of equatorial ionospheric bubbles, *J. Geophys. Res.*, *105*(A11), 24,941–24,950.

Bilitza, D., and B. W. Reinisch (2008), International Reference Ionosphere 2007: Improvements and new parameters, *Adv. Space Res.*, *42*, 599–609.

Bruinsma, S., D. Tamagnan, and R. Biancale (2004), Atmospheric densities derived from CHAMP/STAR accelerometer observations, *Planet. Space Sci.*, *52*, 297–312.

Bruinsma, S., J. M. Forbes, R. S. Nerem, and X. Zhang (2006), Thermosphere density response to the 20–21 November 2003 solar and geomagnetic storm from CHAMP and GRACE accelerometer data, *J. Geophys. Res.*, *111*, A06303, doi:10.1029/2005JA011284.

Chau, J. L., B. G. Fejer, and L. P. Goncharenko (2009), Quiet variability of equatorial E × B drifts during a sudden stratospheric warming event, *Geophys. Res. Lett.*, *36*, L05101, doi:10.1029/2008GL036785.

Da Rosa, A. V. (1966), The theoretical time-dependent thermal behavior of the ionospheric electron gas, *J. Geophys. Res.*, *71*(17), 4107–4120.

Doornbos, E., J. van den Ijssel, H. Lühr, M. Forster, and G. Koppenwallner (2010), Neutral density and crosswind determination from arbitrarily oriented multi-axis accelerometers on satellites, *J. Spacer. Rockets*, *47*(4), 580–589.

Fang, T.-W., T. Fuller-Rowell, R. Akmaev, F. Wu, H. Wang, and D. Anderson (2012), Longitudinal variation of ionospheric vertical drifts during the 2009 sudden stratospheric warming, *J. Geophys. Res.*, *117*, A03324, doi:10.1029/2011JA017348.

Fejer, B. G., J. W. Jensen, and S.-Y. Su (2008), Quiet time equatorial F region vertical plasma drift model derived from ROCSAT-1 observations, *J. Geophys. Res.*, *113*, A05304, doi:10.1029/2007JA012801.

Fejer, B. G., M. E. Olson, J. L. Chau, C. Stolle, H. Lühr, L. P. Goncharenko, K. Yumoto, and T. Nagatsuma (2010), Lunar-dependent equatorial ionospheric electrodynamic effects during sudden stratospheric warmings, *J. Geophys. Res.*, *115*, A00G03, doi:10.1029/2010JA015273.

Forbes, J. M., S. L. Bruinsma, Y. Miyoshi, and H. Fujiwara (2008), A solar terminator wave in thermosphere neutral densities measured by the CHAMP satellite, *Geophys. Res. Lett.*, *35*, L14802, doi:10.1029/2008GL034075.

Friis-Christensen, E., H. Lühr, D. Knudsen, and R. Haagmans (2008), Swarm—An Earth observation mission investigating geospace, *Adv. Space Res.*, *41*(1), 210–216, doi:10.1016/j.asr.2006.10.008.

Fritz, S., and S. D. Soules (1972), Planetary variations of stratospheric temperature, *Mon. Weather Rev.*, *100*, 582–589.

Fuller-Rowell, T. J., M. V. Codrescu, B. G. Fejer, W. Borer, F. Marcos, and D. N. Anderson (1997), Dynamics of the low-latitude thermosphere: Quiet and disturbed conditions, *J. Atmos. Terr. Phys.*, *61*, 1533–1540.

Funke, B., M. López-Puertas, D. Bermejo-Pantaleón, M. García-Comas, G. P. Stiller, T. von Clarmann, M. Kiefer, and A. Linden (2010), Evidence for dynamical coupling from the lower atmosphere to the thermosphere during a major stratospheric warming, *Geophys. Res. Lett.*, *37*, L13803, doi:10.1029/2010GL043619.

Goncharenko, L., and S.-R. Zhang (2008), Ionospheric signatures of sudden stratospheric warming: Ion temperature at middle latitude, *Geophys. Res. Lett.*, *35*, L21103, doi:10.1029/2008GL035684.

Goncharenko, L. P., J. L. Chau, H.-L. Liu, and A. J. Coster (2010), Unexpected connections between the stratosphere and ionosphere, *Geophys. Res. Lett.*, *37*, L10101, doi:10.1029/2010GL043125.

Hagan, M. E., A. Maute, R. G. Roble, A. D. Richmond, T. J. Immel, and S. L. England (2007), Connections between deep tropical clouds and the Earth's ionosphere, *Geophys. Res. Lett.*, *34*, L20109, doi:10.1029/2007GL030142.

Häusler, K., H. Lühr, S. Rentz, and W. Köhler (2007), A statistical analysis of longitudinal dependences of upper thermospheric zonal winds at dip equator latitudes derived from CHAMP, *J. Atmos. Sol. Terr. Phys.*, *69*, 1419–1430.

Huang, C. Y., W. J. Burke, J. S. Machuzak, L. C. Gentile, and P. J. Sultan (2001), DMSP observations of equatorial plasma bubbles in the topside ionosphere near solar maximum, *J. Geophys. Res.*, *106*(A5), 8131–8142.

Immel, T. J., E. Sagawa, S. L. England, S. B. Henderson, M. E. Hagan, S. B. Mende, H. U. Frey, C. M. Swenson, and L. J. Paxton (2006), Control of equatorial ionospheric morphology by atmospheric tides, *Geophys. Res. Lett.*, *33*, L15108, doi:10.1029/2006GL026161.

Jin, H., Y. Miyoshi, H. Fujiwara, and H. Shinagawa (2008), Electrodynamics of the formation of ionospheric wave number 4 longitudinal structure, *J. Geophys. Res.*, *113*, A09307, doi:10.1029/2008JA013301.

Kelley, M. C. (2009), *The Earth's Ionosphere: Plasma Physics and Electrodynamic*, 2nd ed., Academic Press, San Diego, Calif.

Kondo, T., A. D. Richmond, H. Liu, J. Lei, and S. Watanabe (2011), On the formation of a fast thermospheric zonal wind at the magnetic dip equator, *Geophys. Res. Lett.*, *38*, L10101, doi:10.1029/2011GL047255.

Kurihara, J., Y. Ogawa, S. Oyama, S. Nozawa, M. Tsutsumi, C. M. Hall, Y. Tomikawa, and R. Fujii (2010), Links between a stratospheric sudden warming and thermal structures and dynamics in the high-latitude mesosphere, lower thermosphere, and ionosphere, *Geophys. Res. Lett.*, *37*, L13806, doi:10.1029/2010GL043643.

Lei, J., J. P. Thayer, J. M. Forbes, E. K. Sutton, and R. S. Nerem (2008), Rotating solar coronal holes and periodic modulation of the upper atmosphere, *Geophys. Res. Lett.*, *35*, L10109, doi:10.1029/2008GL033875.

Lei, J., J. P. Thayer, and J. M. Forbes (2010), Longitudinal and geomagnetic activity modulation of the equatorial thermosphere anomaly, *J. Geophys. Res.*, *115*, A08311, doi:10.1029/2009JA015177.

Liu, H., and H. Lühr (2005), Strong disturbance of the upper thermospheric density due to magnetic storms: CHAMP observations, *J. Geophys. Res.*, *110*, A09S29, doi:10.1029/2004JA010908.

Liu, H., and S. Watanabe (2008), Seasonal variation of the longitudinal structure of the equatorial ionosphere: Does it reflect tidal influences from below?, *J. Geophys. Res.*, *113*, A08315, doi:10.1029/2008JA013027.

Liu, H., H. Lühr, V. Henize, and W. Köhler (2005), Global distribution of the thermospheric total mass density derived from CHAMP, *J. Geophys. Res.*, *110*, A04301, doi:10.1029/2004JA010741.

Liu, H., H. Lühr, S. Watanabe, W. Köhler, V. Henize, and P. Visser (2006), Zonal winds in the equatorial upper thermosphere: Decomposing the solar flux, geomagnetic activity, and seasonal dependencies, *J. Geophys. Res.*, *111*, A07307, doi:10.1029/2005JA011415.

Liu, H., C. Stolle, M. Förster, and S. Watanabe (2007a), Solar activity dependence of the electron density in the equatorial anomaly regions observed by CHAMP, *J. Geophys. Res.*, *112*, A11311, doi:10.1029/2007JA012616.

Liu, H., H. Lühr, S. Watanabe, W. Köhler, and C. Manoj (2007b), Contrasting behavior of the thermosphere and ionosphere in response to the 28 October 2003 solar flare, *J. Geophys. Res.*, *112*, A07305, doi:10.1029/2007JA012313.

Liu, H., H. Lühr, and S. Watanabe (2007c), Climatology of the equatorial thermospheric mass density anomaly, *J. Geophys. Res.*, *112*, A05305, doi:10.1029/2006JA012199.

Liu, H., S. Watanabe, and T. Kondo (2009a), Fast thermospheric wind jet at the Earth's dip equator, *Geophys. Res. Lett.*, *36*, L08103, doi:10.1029/2009GL037377.

Liu, H., M. Yamamoto, and H. Lühr (2009b), Wave-4 pattern of the equatorial mass density anomaly: A thermospheric signature of tropical deep convection, *Geophys. Res. Lett.*, *36*, L18104, doi:10.1029/2009GL039865.

Liu, H., H. Lühr, and S. Watanabe (2009c), A solar terminator wave in thermospheric wind and density simultaneously observed by CHAMP, *Geophys. Res. Lett.*, *36*, L10109, doi:10.1029/2009GL038165.

Liu, H., E. Doornbos, M. Yamamoto, and S. Tulasi Ram (2011), Strong thermospheric cooling during the 2009 major stratosphere warming, *Geophys. Res. Lett.*, *38*, L12102, doi:10.1029/2011GL047898.

Liu, R., H. Lühr, E. Doornbos, and S.-Y. Ma (2010), Thermospheric mass density variations during geomagnetic storms and a prediction model based on the merging electric field, *Ann. Geophys.*, *28*, 1633–1645.

Lühr, H., S. Maus, M. Rother, and D. Cooke (2002), First in-situ observation of night-time F region currents with the CHAMP satellite, *Geophys. Res. Lett.*, *29*(10), 1489, doi:10.1029/2001GL013845.

Lühr, H., K. Häusler, and C. Stolle (2007), Longitudinal variation of F region electron density and thermospheric zonal wind caused by atmospheric tides, *Geophys. Res. Lett.*, *34*, L16102, doi:10.1029/2007GL030639.

Lühr, H., J. Park, P. Ritter, and H. Liu (2011), In-situ CHAMP observation of ionosphere-thermosphere coupling, *Space Sci. Rev.*, *168*, 237–260.

Maruyama, N., S. Watanabe, and T. J. Fuller-Rowell (2003), Dynamic and energetic coupling in the equatorial ionosphere and thermosphere, *J. Geophys. Res.*, *108*(A11), 1396, doi:10.1029/2002JA009599.

Matsuno, T. (1971), A dynamical model of the stratospheric sudden warming, *J. Atmos. Sci.*, *28*, 1479–1494.

Maus, S. (2007), CHAMP magnetic mission, in *Encyclopedia of Geomagnetism and Paleomagnetism*, edited by D. Gubbins and E. Herrero-Bervera, pp. 59–60, Springer, Heidelberg, Germany.

Mayr, H. G., I. Harris, and H. Volland (1973), Theory of the phase anomaly in the thermosphere, *J. Geophys. Res.*, *78*(31), 7480–7489.

McNamara, L. F., D. L. Cooke, C. E. Valladares, and B. W. Reinisch (2007), Comparison of CHAMP and Digisonde plasma frequencies at Jicamarca, Peru, *Radio Sci.*, *42*, RS2005, doi:10.1029/2006RS003491.

Miyoshi, Y., H. Fujiwara, J. M. Forbes, and S. L. Bruinsma (2009), Solar terminator wave and its relation to the atmospheric tide, *J. Geophys. Res.*, *114*, A07303, doi:10.1029/2009JA014110.

Miyoshi, Y., H. Fujiwara, H. Jin, H. Shinagawa, H. Liu, and K. Terada (2011), Model study on the formation of the equatorial mass density anomaly in the thermosphere, *J. Geophys. Res.*, *116*, A05322, doi:10.1029/2010JA016315.

Miyoshi, Y., H. Fujiwara, H. Jin, H. Shinagawa, and H. Liu (2012), Numerical simulation of the equatorial wind jet in the thermosphere, *J. Geophys. Res.*, *117*, A03309, doi:10.1029/2011JA017373.

Oberheide, J., J. M. Forbes, K. Häusler, Q. Wu, and S. L. Bruinsma (2009), Tropospheric tides from 80 to 400 km: Propagation, interannual variability, and solar cycle effects, *J. Geophys. Res.*, *114*, D00I05, doi:10.1029/2009JD012388. [Printed *116*(D1), 2011].

Olsen, N., and C. Stolle (2012), Satellite geomagnetism, *Annu. Rev. Earth Planet. Sci.*, *40*, 441–465, doi:10.1146/annurev-earth-042711-105540.

Park, J., H. Lühr, C. Stolle, R. M., K. W. Min, and I. Michaelis (2009), The characteristics of field-aligned currents associated with equatorial plasma bubbles as observed by the CHAMP satellite, *Ann. Geophys.*, *27*, 2685–2697.

Park, J., H. Lühr, M. Kunze, B. G. Fejer, and K. W. Min (2012), Effect of sudden stratospheric warming on lunar tidal modulation of the equatorial electrojet, *J. Geophys. Res.*, *117*, A03306, doi:10.1029/2011JA017351.

Philbrick, C. R., and J. P. McIsaac (1972), Measurements of atmospheric composition near 400 km, *Space Res.*, *12*, 743–750.

Raghavarao, R., L. E. Wharton, N. W. Spencer, H. G. Mayr, and L. H. Brace (1991), An equatorial temperature and wind anomaly (ETWA), *Geophys. Res. Lett.*, *18*(7), 1193–1196.

Reigber, C., H. Lühr, and P. Schwintzer (2002), CHAMP mission status, *Adv. Space Res.*, *30*, 129–134.

Richards, P. G. (2004), On the increases in nitric oxide density at midlatitudes during ionospheric storms, *J. Geophys. Res.*, *109*, A06304, doi:10.1029/2003JA010110.

Richards, P. G., and W. K. Peterson (2008), Measured and modeled backscatter of ionospheric photoelectron fluxes, *J. Geophys. Res.*, *113*, A08321, doi:10.1029/2008JA013092.

Richards, P. G., J. A. Fennelly, and D. G. Torr (1994), EUVAC: A solar EUV flux model for aeronomic calculations, *J. Geophys. Res.*, *99*(A5), 8981–8992.

Rishbeth, H. (2002), Whatever happened to superrotation?, *J. Atmos. Sol. Terr. Phys.*, *64*(12–14), 1351–1360, doi:10.1016/S1364-6826(02)00097-4.

Rother, M., K. Schlegel, H. Lühr, and D. Cooke (2010), Validation of CHAMP electron temperature measurements by incoherent scatter radar data, *Radio Sci.*, *45*, RS6020, doi:10.1029/2010RS004445.

Sagawa, E., T. J. Immel, H. U. Frey, and S. B. Mende (2005), Longitudinal structure of the equatorial anomaly in the nighttime ionosphere observed by IMAGE/FUV, *J. Geophys. Res.*, *110*, A11302, doi:10.1029/2004JA010848.

Schunk, R. W., and A. F. Nagy (1978), Electron temperatures in the F region of the ionosphere: Theory and observations, *Rev. Geophys.*, *16*(3), 355–399.

Schunk, R. W., and A. F. Nagy (2009), *Ionospheres*, Cambridge Univ. Press, Cambridge, U. K.

Somsikov, V. M. (1995), On the formation of atmospheric irregularities in the solar terminator region, *J. Atmos. Sol. Terr. Phys.*, *57*, 75–83.

Somsikov, V. M., and B. Ganguly (1995), On the formation of atmospheric inhomogeneities in the solar terminator region, *J. Atmos. Sol. Terr. Phys.*, *57*, 1513–1523.

Stolle, C., H. Lühr, M. Rother, and G. Balasis (2006), Magnetic signatures of equatorial spread F as observed by the CHAMP satellite, *J. Geophys. Res.*, *111*, A02304, doi:10.1029/2005JA011184.

Stolle, C., C. Manoj, H. Lühr, S. Maus, and P. Alken (2008a), Estimating the daytime Equatorial Ionization Anomaly strength from electric field proxies, *J. Geophys. Res.*, *113*, A09310, doi:10.1029/2007JA012781.

Stolle, C., H. Lühr, and B. Fejer (2008b), Relation between the occurrence rate of ESF and the equatorial plasma drift velocity at sunset derived from global observations, *Ann. Geophys.*, *26*, 1–10.

Stolle, C., H. Liu, V. Truhlík, H. Lühr, and P. G. Richards (2011), Solar flux variation of the electron temperature morning overshoot in the equatorial F region, *J. Geophys. Res.*, *116*, A04308, doi:10.1029/2010JA016235.

Su, S.-Y., C. K. Chao, and C. H. Liu (2008), On monthly/seasonal/longitudinal variations of equatorial irregularity occurrences and their relationship with the postsunset vertical drift velocities, *J. Geophys. Res.*, *113*, A05307, doi:10.1029/2007JA012809.

Sutton, E. K., J. M. Forbes, R. S. Nerem, and T. N. Woods (2006), Neutral density response to the solar flares of October and November, 2003, *Geophys. Res. Lett.*, *33*, L22101, doi:10.1029/2006GL027737.

Truhlík, V., D. Bilitza, and L. Třísková (2009), Latitudinal variation of the topside electron temperature at different levels of solar activity, *Adv. Space Res.*, *44*, 693–700.

Varney, R. H., D. L. Hysell, and J. D. Huba (2011), Sensitivity studies of equatorial topside electron and ion temperatures, *J. Geophys. Res.*, *116*, A06321, doi:10.1029/2011JA016549.

Watanabe, S., and K.-I. Oyama (1996), Effects of neutral wind on the electron temperature at a height of 600 km in the low latitude region, *Ann. Geophys.*, *14*, 290–296.

Xiong, C., J. Park, H. Lühr, C. Stolle, and S. Y. Ma (2010), Comparing plasma bubble occurrence rates at CHAMP and GRACE altitudes during high and low solar activity, *Ann. Geophys.*, *28*, 1647–1658, doi:10.5194/angeo-28-1647-2010.

Xiong, C., H. Lühr, S. Y. Ma, C. Stolle, and B. G. Fejer (2012), Features of highly structured equatorial plasma irregularities deduced from CHAMP observations, *Ann. Geophys.*, *30*, 1259–1269, doi:10.5194/angeo-30-1259-2012.

Yamazaki, Y., A. D. Richmond, and K. Yumoto (2012), Stratospheric warmings and the geomagnetic lunar tide: 1958–2007, *J. Geophys. Res.*, *117*, A04301, doi:10.1029/2012JA017514.

H. Liu, Department of Earth and Planetary Science, Faculty of Science, Kyushu University, Fukuoka, Japan.

C. Stolle, DTU Space, Technical University of Denmark, Copenhagen, Denmark. (cst@space.dtu.dk)

Upper Atmosphere Data Assimilation With an Ensemble Kalman Filter

Tomoko Matsuo

Cooperative Institute for Research in Environmental Sciences, University of Colorado, Boulder, Colorado, USA

Space Weather Prediction Center, National Oceanic Atmospheric Administration, Boulder, Colorado, USA

Recent availability of global observations of ionospheric parameters, especially from GPS receivers on low Earth orbiting platforms, has motivated a number of attempts at assimilating ionospheric data. However, assimilation of sparse, irregularly distributed thermosphere observations to global models remains a daunting task. In this paper, we demonstrate the utility of ensemble Kalman filtering (EnKF) techniques to effectively assimilate a realistic set of space- and ground-based observations of the thermosphere and ionosphere into a general circulation model (GCM). An EnKF assimilation procedure has been constructed using the Data Assimilation Research Testbed and the thermosphere-ionosphere electrodynamics GCM, two sets of community software offered by National Center for Atmospheric Research. An important attribute of this procedure is that the thermosphere-ionosphere coupling is self-consistently treated both in a forecast model as well as in assimilation schemes. It effectively facilitates solving the inverse problem of inferring unobserved variables from observed ones, for instance, thermospheric states from better observed ionospheric states. Some results from observing system simulation experiments are shown to demonstrate this point. Furthermore, we discuss some of the issues specific to upper atmospheric EnKF applications and the roles of auxiliary filtering algorithms, such as adaptive covariance inflation and localization of covariance, to cope with these issues.

1. INTRODUCTION

The data assimilation methodology presented in this paper is data driven, yet well informed by first principles, offering a marriage of inductive and deductive techniques. This approach offer an excellent platform from which many system science questions can be addressed, enabling us to bridge the gaps in our understanding of the thermosphere and ionosphere system. While making a diverse set of observations of thermospheric and ionospheric parameters consistent with physical processes included in a first-principles model, a coherent picture of the integrated upper atmosphere emerges. Data assimilation is becoming an integral component of space science basic research.

A number of attempts to assimilate ionospheric data [e.g., *Schunk et al.*, 2004; *Scherliess et al.*, 2006, 2009; *Hajj et al.*, 2004; *Komjathy et al.*, 2010; *Khattatov*, 2010, and references therein] have been motivated by the recent availability of global observations of ionospheric parameters, especially from GPS receivers on low Earth orbiting platforms. However, assimilating sparsely distributed thermosphere observations to global models remains a challenging task [e.g., *Minter et al.*, 2004; *Fuller-Rowell et al.*, 2004, and references therein]. On the other hand, the thermosphere's relatively long-term memory and intricate interactions between neutral and plasma constituents, thermospheric data assimilation has the potential to increase effectiveness of data assimilation in coupled thermosphere-ionosphere dynamical systems, furthering

Modeling the Ionosphere-Thermosphere System
Geophysical Monograph Series 201
© 2013. American Geophysical Union. All Rights Reserved.
10.1029/2012GM001339

our ability to specify and forecast states of the upper atmosphere.

One of this paper's objectives is to demonstrate how the information content of the geospace observing systems can be maximized by taking advantage of the intimate coupling between the thermosphere and ionosphere described in general circulation models (GCMs), with the help of the latest ensemble Kalman filtering (EnKF) techniques [e.g., *Evensen*, 2009]. We use an EnKF data assimilation procedure that has been constructed with the Data Assimilation Research Testbed (DART) [*Anderson et al.*, 2009] and the thermosphere-ionosphere electrodynamics general circulation model (TIEGCM) [*Richmond et al.*, 1992], two sets of community software offered by NCAR. Descriptions and applications of this EnKF data assimilation procedure can be found in the works of *Matsuo and Araujo-Pradere* [2011], *Lee et al.* [2012], and *Matsuo et al.* [2013].

A critical attribute of this EnKF data assimilation procedure is that the feedback between the thermosphere and the ionosphere are self-consistently accounted for in both the analysis and forecast steps of filtering according to coupled dynamics described in the GCM. It facilitates solving the inverse problem of inferring unobserved variables from observed ones, for example, thermospheric states and parameters from better observed ionospheric electron density. Some results from observing system simulation experiments (OSSEs), in which synthetically generated data for a given realistic observing system are assimilated in place of actually observed values, are shown. Note that a control simulation from which synthetic observations are sampled is not part of an ensemble of model forecasts used to obtain the analysis, even though it is simulated by using the same forecast model. With no significant model biases, OSSEs are also called perfect model experiments and allow us to assess the effectiveness of assimilation algorithms and strategies as well as the impact of different types of observations on the assimilation analysis. On the other hand, OSSEs do not address the performance of assimilation algorithms in the presence of realistic model and observation discrepancy. Therefore, the applicability of findings presented here to assimilation experiments with real observations hinges largely on the degree of model biases and the availability of high-quality observations and will be investigated further in the future.

The upper atmosphere is strongly driven by external forcing, and therefore forcing parameter specification in thermosphere-ionosphere GCMs greatly influences the model performance and inevitably the quality of assimilation analysis. Some of the gross inconsistency between observations and model states often results from the misspecification of external drivers that are specified as boundary conditions or parameterized with respect to a handful of adjustable parameters in the models. It is, therefore, a desirable capability of any upper atmosphere data assimilation procedures [e.g., *Pi et al.*, 2003; *Scherliess et al.*, 2009; *Datta-Barua et al.*, 2009], to be able to adjust the model external forcing for given observations. The comparison of filtering experiments with and without the forcing parameter estimations suggests that the quality of assimilation analyses improves significantly when the forcing parameters are successfully estimated.

Finally, we discuss some of the challenges specific to upper atmospheric EnKF applications and the roles of auxiliary assimilation algorithms such as adaptive covariance inflation [e.g., *Anderson and Anderson*, 1999; *Whitaker and Hamill*, 2002; *Anderson*, 2009] and localization of covariance [e.g., *Gaspari and Cohn*, 1999; *Houtekamer and Mitchell*, 2001] to cope with these challenges. Along with the driver estimation mentioned above, another challenge stems from the nature of model error growth, which is almost exclusively driven by external forcing factors in current thermosphere and ionosphere GCMs. In EnKF, the model error growth, gauged as a degree of the spread among ensemble members, is presumed to be representative of the model uncertainty. It is crucial that the probability distribution represented by the ensemble reflects reality. For instance, covariance inflation techniques can rectify the issue of inadequate model error growth by increasing the spread between the ensemble members. Covariance localization techniques can help cope with spurious correlations resulting from sampling and modeling errors.

2. ENSEMBLE KALMAN FILTER

The EnKF can be viewed as solving a problem of sequential Bayesian analysis under the assumption that both forecast and observational errors are normally distributed. Just as in conventional Kalman filters, EnKF algorithms consist of the recursive application of an analysis update step and a forecast step. Some properties of the EnKF used in our application are listed below. For a comprehensive description of the EnKF, the reader should be referred to a textbook by *Evensen* [2009].

2.1. Filter Algorithms

Suppose that the state vector inferred by the filter is denoted by **x** and the observation vector by **y**, and that the relationship of **x** to **y** can be given by

$$\mathbf{y} = \mathbf{H}\mathbf{x} + \boldsymbol{\varepsilon}^r, \qquad (1)$$

where H represents the forward or observation operator and ε^r is the observational error whose covariance is given by a matrix C^r. The standard Kalman filter update formula [e.g., *Jazwinski*, 1970] provides expressions for the assimilation analysis x^a and the analysis error covariance C^a as follows.

$$x^a = x^f + K(y - Hx^f), \quad (2a)$$

$$K = C^f H'(C^r + HC^f H')^{-1}, \quad (2b)$$

where C^f denotes a covariance matrix for the model forecast error. Superscripts **a**, **f**, and **r** refer to variables concerning the assimilation analysis, model forecast, and observational error, while a superscript ' signifies the transpose operation. The analysis error covariance C^a is given by

$$C^a = (I - KH)C^f. \quad (3)$$

Note that the analysis obtained by equation (2) is essentially equivalent to the solution of 3-D variational methods in case of linear H [e.g., *Lorenc*, 1986].

In the EnKF, equations (2) and (3) are not explicitly used to obtain x^a and C^a. Instead, x^a and C^a are represented by the sample statistics of an analysis ensemble $\{x^{a(m)}\}$, where m denotes the mth member of the ensemble, and likewise x^f and C^f are represented by the sample mean and covariance of an ensemble of the model forecasts $\{x^{f(m)}\}$. In other words, $\{x^{a(m)}\}$ is an mth draw from the multivariate normal distribution $\mathcal{N}$ with the mean x^a and the covariance C^a (i.e., $x^{a(m)} \sim \mathcal{N}(x^a, C^a)$) and similarly $x^{f(m)} \sim \mathcal{N}(x^f, C^f)$. A diverse array of ensemble filter algorithms has been developed over the years since first proposed by *Evensen* [1994], resulting in more and more computationally desirable implementations [e.g., *Evensen*, 2009].

Deterministic analysis update schemes, which are implementations of square root ensemble filters (where C^f and C^a are expressed in terms of its matrix square root), handle transformation of the forecast ensemble $\{x^{f(m)}\}$ to the analysis ensemble $\{x^{a(m)}\}$ by strictly preserving the consistency of the sample statistics of $\{x^{a(m)}\}$ with equations (2) and (3) [e.g., *Tippett et al.*, 2003]. In this paper, we use one of deterministic analysis update schemes developed by *Anderson* [2001] among a wide variety of ensemble filter algorithms implemented in DART. Furthermore, DART takes advantage of the sequential nature of the Bayes analysis with respect to observations under the assumption of uncorrelated observational error (i.e., diagonalized observation space) to gain computational efficiency [e.g., *Houtekamer and Mitchell*, 2001; *Anderson*, 2003, and references therein].

In the forecast step, the analysis ensemble at time k $\{x^{a(m)}_k\}$ is integrated forward in time according to the fully nonlinear dynamics of the upper atmosphere to yield the forecast distribution at time $k + 1$, represented by the forecast ensemble $\{x^{f(m)}_{k+1}\}$. The individual forecast member is obtained by

$$x^{f(m)}_{k+1} = \mathcal{F}_k(x^{a(m)}_k). \quad (4)$$

Note that no specific treatment of model error is included in equation (4), and in our application, the model error issue is handled to some extent by artificially inflating the prior variance [e.g., *Anderson and Anderson*, 1999; *Whitaker and Hamill*, 2002].

Like operational assimilation systems, DART includes quality control algorithms to automatically detect and discard observations that are too far away from the forecast ensemble. We can choose to discard observations that are a few standard deviations farther away from x^f than would be expected given the forecast model error C^f and the observational error C^r.

2.2. Application to the Thermosphere and Ionosphere

Description of the EnKF application to the TIEGCM is given in detail in the work of *Matsuo et al.* [2013], so below, some key features are summarized such as constitutions of the state vector **x**, initialization of the forecast ensemble, and implementation of H.

In the version of TIEGCM used here, physical variables are discretized on the model grid of a horizontal resolution of 5° latitude and 5° longitude and a vertical pressure level size of a half-scale height, extending from about 97 km to 500–700 km (depending on solar cycles). Selected subsets of the TIEGCM physical variables, denoted by φ, are included in the state vector **x**. Let us suppose that φ^T represents a vector of temperature values on the model grid and superscript γ_O, γ_{O2}, U, V, and Ne stand, respectively, for atomic oxygen mass mixing ratio, molecular oxygen mass mixing ratio, zonal wind, meridional wind, and electron number density. **x** is, for instance, made up of φ^T, φ^{Ne}, and so on. In the experiments accompanying the forcing estimation, some of the primary model forcing parameters **d** are also included in **x**. Table 1 summarizes the

Table 1. List of the TIEGCM Variables Included in the State Vector **x** for Each Experiment Presented in the Paper[a]

Experiment	**x**
I	$x = [\varphi^{Ne}]$
II IV	$x = [\varphi^{Ne}; \varphi^T; \varphi^U; \varphi^V; \varphi^{\gamma_O}; \varphi^{\gamma_{O2}}]$
III	$x = [\varphi^I; \varphi^{\gamma_O}; \varphi^{\gamma_{O2}}]$
V	$x = [\varphi^T; \varphi^{\gamma_O}; \varphi^{\gamma_{O2}}; d_1]$

[a]Some of the variables are included at both current and previous time steps in **x**. See *Matsuo et al.* [2013] for details.

different compositions of **x** that are used for filtering experiments presented later.

To obtain the initial ensemble, a set of model simulations is launched with different values of primary model forcing **d**, including $F10.7$ index (d_1), cross–polar cap potential drop (d_2), and hemispherical power index (d_3) and integrated for 1–2 weeks with the value of **d** being held unchanged until the ensemble reaches its steady state. Suppose that the initial state vector and the dynamical model are represented by $\mathbf{x}_0$ and $\mathcal{F}$, respectively, and the initial forecast ensemble $\mathbf{x}^{f(m)}_{k=1}$ at time $k = 1$ is given by

$$\mathbf{x}^{f(m)}_{k=1} = \mathcal{F}(\mathbf{x}_0, \mathbf{d}^{(m)}), \qquad (5)$$

where $\mathbf{d}^{(m)} = [d_1^{(m)}, d_2^{(m)}, d_3^{(m)}]^T$ is sampled from the normal distribution $\mathbf{d}^{(m)} \sim \mathcal{N}([\mu_1, \mu_2, \mu_3]^T, \Sigma_d)$. The $F10.7$ index is a measure of the 10.7 cm solar radio flux and is used in the TIEGCM as a proxy for solar EUV flux. The cross–polar cap potential drop and hemispheric power indexes control the magnetospheric driver specified in the form of high-latitude convective electric fields and auroral energetic particle precipitations.

A primary role of the forward (or observation) operator **H** in our application is interpolation of the state vector **x** on the regular model grid to the observation locations. Herein, bilinear interpolation is used in the horizontal direction. In the vertical direction, pressure coordinates are converted to geometric heights using gravity-adjusted geopotential heights and interpolated linearly to the observation height. As for observations of the neutral mass density, which is a function of the temperature, pressure, and mass mixing ratio of major species (i.e., atomic oxygen, molecular oxygen, and molecular nitrogen), **H** encompasses the nonlinear relationship of these physical variables to the mass density. See *Matsuo et al.* [2013] for detail.

3. EFFECTS OF SELF-CONSISTENT TREATMENT OF THE THERMOSPHERE AND IONOSPHERE

Matsuo and Araujo-Pradere [2011] demonstrates that ionospheric data assimilation considerably benefits from self-consistent treatment of thermosphere-ionosphere coupling in the forecast model as well as in assimilation schemes, which can be achieved inherently by using EnKF. In this section, the results from the OSSE designed for a global network of 75 ionosondes (see Table 2 of *Araujo-Pradere et al.* [2002]) are shown. The goal of OSSEs is to examine the benefit of adjusting thermospheric parameters by using electron density observations in the update step on the ionospheric specification. In experiment I, only the electron density is adjusted by observations in the update step. In experiment II, the thermospheric variables, namely, neutral winds, temperature, and compositions, are adjusted along with the electron density, through the sample cross-covariance between electron density observations and thermospheric state variables estimated from the ensemble.

Experiments are conducted under the geomagnetically quiet equinox conditions of 28 March 2002. The assimilation cycle (i.e., how often observations are assimilated) is 30 min, and 64 ensemble members are used. Synthetic observational data are generated by sampling the state $\mathbf{x}^c$ from the control simulation at these ionosonde locations, at every 50 km from 150 to 400 km, and adding the centered Gaussian random fields as the observational error. Here the state $\mathbf{x}^c$ from the control simulation serves as the "truth." The root-mean-squared difference (RMSD) is computed as a standard measure of the differences between the analysis electron density and the "truth," i.e., $\sqrt{\sum_j^J \frac{(\varphi_j^{aNe} - \varphi_j^{cNe})^2}{J}}$, where J denotes the total number of grid points. Note that the size of J is exactly the same for these two experiments, even though the size of the state vector **x** is very different for experiments I and II. Figure 1 shows the RMSD of electron density analyses over the entire model domain from the two filtering experiments: experiment I with the state vector $\mathbf{x} = [\varphi^{Ne}]$ (shown by dashed line) and experiment II with the state vector $\mathbf{x} = [\varphi^{Ne}; \varphi^T; \varphi^U; \varphi^V; \varphi^{\gamma_O}; \varphi^{\gamma_{O2}}]$ (shown by solid line). Note that the overall diurnal cycle of RMSD, seen in both experiments, originates from the longitudinal distribution of ionosonde stations, as data-dense and data-sparse

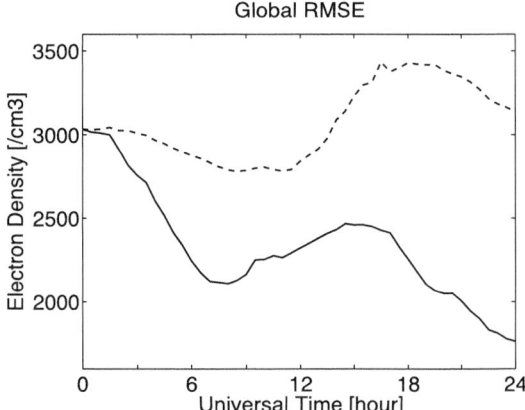

Figure 1. The root-mean-squared difference (RMSD) of the electron density analysis, computed from φ^{Ne} in the analysis state vector $\mathbf{x}^a$ and the "truth" $\mathbf{x}^c$ over the entire model domain, is shown by solid line for the experiment II with the state vector $\mathbf{x} = [\varphi^{Ne}]$ and by dashed line for the experiment I with the state vector $\mathbf{x} = [\varphi^{Ne}; \varphi^T; \varphi^U; \varphi^V; \varphi^{\gamma_O}; \varphi^{\gamma_{O2}}]$ (Modified from Figure 3 of *Matsuo and Araujo-Pradere* [2011]).

parts of the ionosonde observational network rotate under the dayside equatorial ionization anomaly.

Considering that both experiments fully account for the thermosphere-ionosphere coupling described in the TIEGCM in the forecast step, it suggests that self-consistent treatment of the thermosphere and ionosphere in data assimilation schemes play an important role in improving a global ionospheric specification. In spite of sampling error issues associated with the EnKF algorithms, the sample crosscovariance between observations and unobserved physical variables estimated from the ensemble is effective enough to constrain the thermospheric state variables from the observations of electron density, which in turn improves the overall specification of the ionosphere. Whether or not the same statement holds true for experiments with real observations hinges on the quality of the model and the availability of high-quality observations. If the relationship between the thermosphere and the ionosphere is misrepresented in the model, it may degrade the overall assimilation analysis quality by affecting unobserved variables by using observations with erroneously represented correlation.

4. GEOPHYSICAL INFORMATION CONTENT OF DIFFERENT OBSERVATIONS

In this section, the information content of different types of observations is assessed in terms of the global neutral mass density analysis by comparing OSSEs for (1) neutral mass densities obtained from the accelerometer experiment on board the CHAMP satellite [*Reigber et al.*, 2002] and (2) electron density profiles obtained from the radio occultation experiment with the joint US-Taiwan Constellation Observing System for Meteorology, Ionosphere and Climate/Formosa Satellite 3 (COSMIC/FORMOSAT-3) mission [*Rocken et al.*, 2000]. In other words, the effectiveness of sparse but accurate observations of the neutral mass density versus indirect but global observations on the global neutral mass density specification is examined. In these experiments, the forcing parameter value assigned to each ensemble member $\mathbf{d}^{(m)}$ is held unchanged throughout the filtering experiment from initial values, so there is no need to inflate the covariance because the model error growth is predominantly dictated by external forcing.

4.1. CHAMP Mass Density

The CHAMP is a polar orbiting satellite with an orbital period of about 90 min, slowly precessing through different solar LTs over the course of 133 days. Experiment III is designed for the mass density observations obtained along the CHAMP satellite orbit on 28–29 March 2002, by using the state vector given by $\mathbf{x} = [\varphi^T; \varphi^{\gamma_O}; \varphi^{\gamma_{O2}}]$. During this period, the CHAMP satellite provides sampling of the density in 390–450 km and in 5.5 and 18.5 LT sectors. The assimilation window is set to the orbital period of 90 min. See the work of *Matsuo et al.* [2013] for more details.

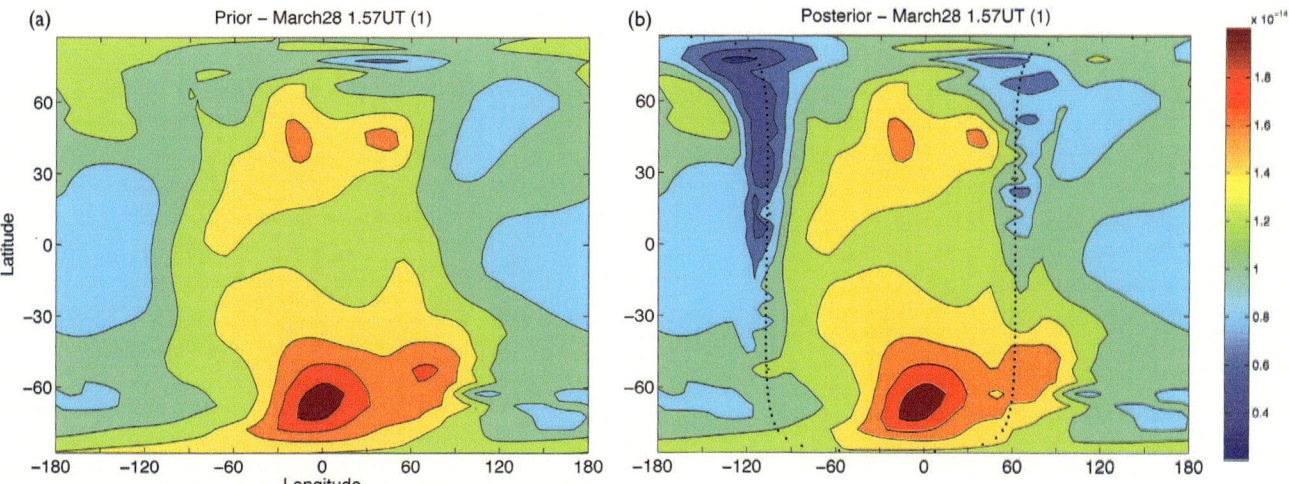

Figure 2. The RMSD of the neutral mass density analysis, computed from neutral density values on the model grid obtained from the analysis state vector $\mathbf{x}^a$ and the "truth" $\mathbf{x}^c$, is displayed in longitude and latitude coordinates. The unit is kg m^{-3}. The contour interval is 0.2×10^{-14}. Black dots indicate the location of the CHAMP satellite track. The RMSD is shown for the analyses (a) before and (b) after the first update step at 1.57 UT on 28 March. (Modified from Figures 2 of *Matsuo et al.* [2013]).

The RMSD is computed from the neutral mass density values on the model grid obtained from the analysis $\mathbf{x}^a$ and the "truth" $\mathbf{x}^c$ over pressure levels 19–24 (corresponding to about 340 to 500 km). Figures 2a and 2b display the longitude and latitude distribution of RMSD: before (a) and after (b) the update step of the first assimilation cycle at 1.57 UT on 28 March. The neutral density specification improvement is limited to the vicinity of satellite orbits. Over the course of the 2 day experiment, the overall RMSD is reduced by a small amount (~8% from the initial level of RMSD). This is because (1) sampling of the neutral mass density along the CHAMP orbit is very limited and (2) the neutral mass density variability is strongly controlled by external forcing, in particular, by solar EUV flux, and forcing parameters are unadjusted in this experiment (the impact of forcing parameter estimation is discussed in section 5).

4.2. COSMIC Electron Density

It is important to examine the possibility of inferring underobserved thermospheric parameters from better observed ionospheric parameters. The initial success of assimilation of the COSMIC/FORMOSAT-3 electron density profiles into the TIEGCM by *Lee et al.* [2012] motivates us to investigate the potential of COSMIC/FORMOSAT-3 electron density profiles to constrain the thermospheric states globally. Experiment IV is, therefore, designed for the COSMIC/FORMOSAT-3 electron density profiles obtained on 8–9 April 2008, using the state vector given by $\mathbf{x} = [\varphi^{Ne}; \varphi^T; \varphi^U; \varphi^V; \varphi^{\gamma_O}; \varphi^{\gamma_{O_2}}]$. The total ensemble number is 90, and the assimilation window is set to 1 h. See *Matsuo et al.* [2013] for more details.

The RMSD is likewise computed from the neutral mass density values on the model grid obtained from the analysis $\mathbf{x}^a$ and the "truth" $\mathbf{x}^c$ over pressure levels 19–24 (corresponding to about 340 to 500 km). Figures 3a and 3b display the RMSD, before (a) and after (b) the update step of the first assimilation cycle at 8 UT on 8 April, in the same fashion as Figure 2. Figures 3c and 3d display the RMSD of the corresponding electron density analyses.

Comparing the RMSD from the pre-assimilation and post-assimilation analysis, shown in Figure 3, the density is well inferred in the vicinity of the COSMIC/FORMOSAT-3 profile locations, mainly from the improved temperature analysis with the help of the crosscovariance between φ^{Ne} (observed) and φ^T (unobserved) in accordance with the thermosphere-ionosphere coupling described in the TIEGCM. The overall RMSD of the neutral mass density is reduced by about 40% from the initial level of RMSD, significantly more than the case with the CHAMP mass density.

In the areas where there are sharp horizontal electron density gradients, especially at the edge of equatorial ionization anomaly, assimilation of the COSMIC/FORMOSAT-3 profiles actually has degrading effects on the neutral density analysis. This is caused by the erroneous correlation between given electron density observations and unobserved variables near the equatorial ionization anomaly boundary, resulting likely from sampling errors. This issue is further discussed in section 6.

These OSSEs suggest that this EnKF assimilation system can facilitate extraction of geophysical information regarding the thermospheric temperature and compositions from the COSMIC/FORMOSAT-3 profiles to the degree that dwarfs the utility of CHAMP neutral mass density observations in terms of global neutral mass density specification. This is, of course, a finding in the context of idealized scenario with no gross discrepancy between the model and observations, and further studies are required to confirm this finding when assimilating real observations in the presence of realistic model biases and observational errors.

5. ESTIMATION OF MODEL PARAMETERS

The upper atmosphere is strongly driven by external forcing. In GCMs of the thermosphere-ionosphere, the description of external forcing is highly parameterized, and the parameter values are prespecified. Because these forcing parameters substantially affect the quality of the model forecast and, therefore, data assimilation analysis, it is desirable to improve the specification of forcing parameters by data assimilation. In this section, we present an example of the forcing parameter estimation by using EnKF.

One of the most straightforward ways to infer unobserved forcing parameters in the EnKF is to augment the EnKF state vector, which is otherwise composed of the model's primary physical variables, with the forcing parameters, and to let these unobserved parameters be updated as part of the state vector according to the sample crosscovariance between state vector elements and observations (i.e., $\mathbf{C}^f \mathbf{H}'$). The next OSSE, experiment V, is performed with the state vector given by $\mathbf{x} = [\varphi^T; \varphi^{\gamma_O}; \varphi^{\gamma_{O_2}}; d_1]$. Note that the forcing parameter d_1 (the $F10.7$ index) is now included as part of the state vector $\mathbf{x}$. Experiment V is designed for the mass density observations along the CHAMP satellite orbit on 28 March 2008 to examine the impact of the estimation of the $F10.7$ index on the neutral mass density specification. See *Matsuo et al.* [2013] for details.

We compare the RMSD of the neutral mass density with and without estimation of the $F10.7$ index d_1, as shown in Figure 4. The initial ensemble is initialized with $\mathbf{d}^{(m)}$ sampled from the Gaussian distribution centered at $[\mu_1, \mu_2, \mu_3] = [175, 45, 16]$. In the control simulation, forcing parameters are set to $[d_1, d_2, d_3] = [200, 45, 16]$, and so the "true" $F10.7$ value is 200. The $F10.7$ d_1 is estimated to be 207. When d_1 is estimated, the $F10.7$ index value assigned to each ensemble

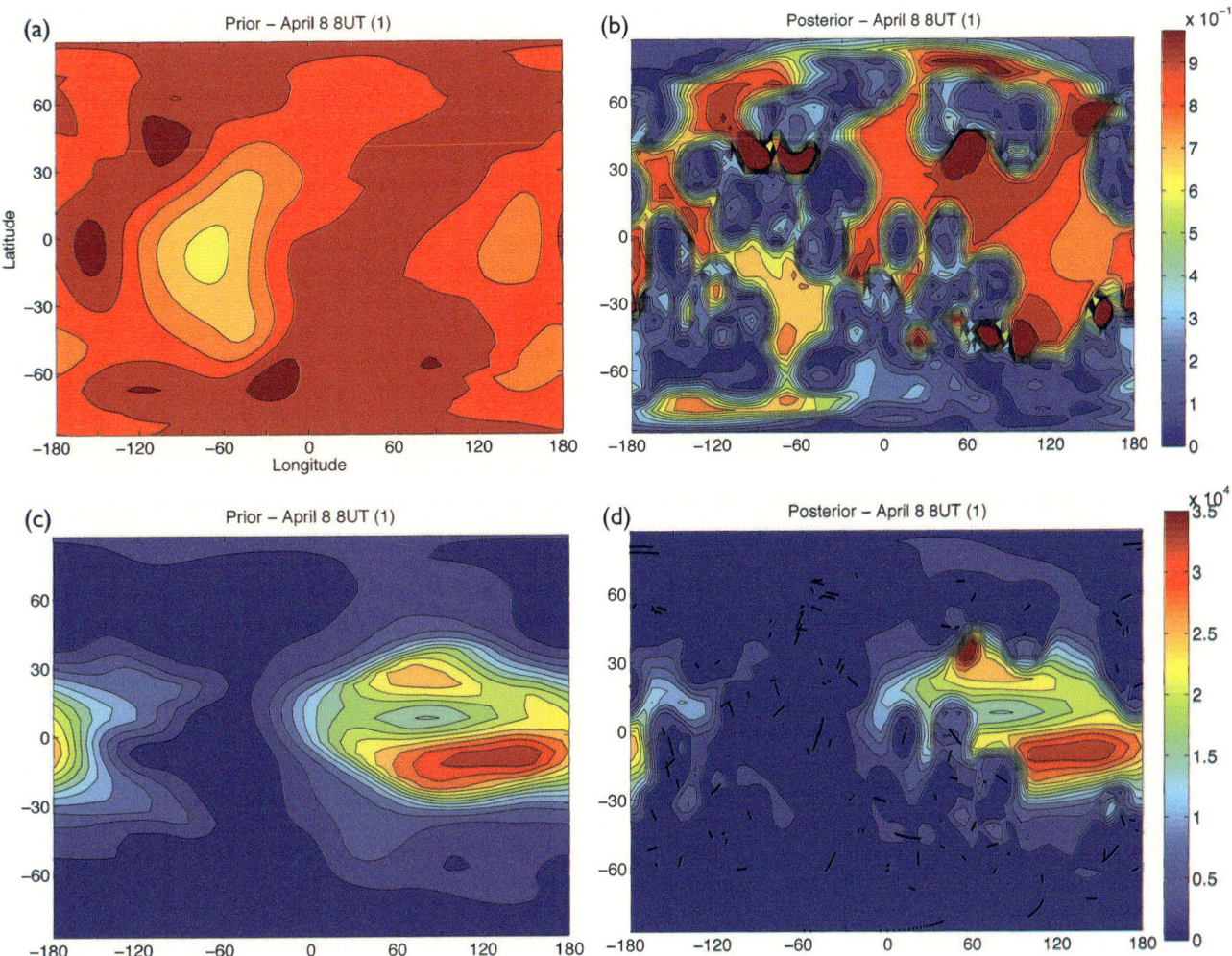

Figure 3. The RMSD of the neutral mass density analysis is displayed in longitude and latitude coordinates in (a) and (b). The unit is kg m^{-3}. The contour interval is 0.75×10^{-14}. The RMSD is shown for the analyses (a) before and (b) after the first update step at 8 UT on 8 April. The color bar on the top applies to (a) and (b). The RMSD of the electron density analysis, computed from corresponding analyses (a) before and (b) after the first update step at 8 UT on 8 April, is displayed. The unit is cm^{-3}. The contour interval is 0.25×10^{4}. The color bar on the bottom applies to (c) and (d) (Modified from Figures 4 of *Matsuo et al.* [2013]).

$d_1^{(m)}$ is adjusted at each assimilation step, but $d_2^{(m)}$ and $d_3^{(m)}$ are held unchanged from the initial setting.

The forcing estimation effectively reduces the model error growth. Because the ensemble spread of a forcing parameter inherently becomes smaller after the update step (i.e., the posterior variance is smaller than the prior variance according to the Bayes' rule). There is no mechanism to recover the ensemble spread in the forecast step, since the dynamical description of forcing parameters is absent in the forecast model. In the experiment presented here, we cope with the reduction of model error growth by artificially inflating the covariance. Specifically, the estimation of d_1 is accompanied by the inflation of the prior covariance by the adaptive covariance inflation technique described in the work of *Anderson* [2007] and of the posterior variance of d_1. Here d_1 is inflated so that the quarter of the initial variance value assigned to $d_1^{(m)}$ is maintained.

Comparing the assimilation analysis obtained after the course of day, an approximately 8% error reduction from the initial level of RMSD is attained from assimilation of CHAMP observations, while an approximately 40% error reduction is attained from assimilation of COSMIC observations. Assimilation of CHAMP observations into the TIEGCM has far greater impact on the neutral mass density specification if the forcing parameter is inferred as well, resulting in 99% of the error reduction. Because of

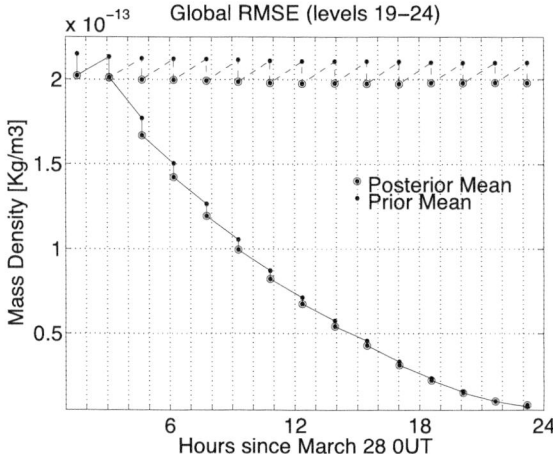

Figure 4. The RMSD of the neutral mass density analysis, with (solid line) and without (dashed line) estimation of d_1, is shown for 8 April 2008. (Modified from Figures 5 of *Matsuo et al.* [2013]).

the strong driver-response relationship between solar EUV flux and the neutral mass density, the estimation of the $F10.7$ index d_1 is particularly effective. Note that the experiment of forcing estimation presented here is highly idealized, and in reality, it is challenging to estimate model parameters for a number of reasons such as the lack of prior uncertainty information for model parameters, the weak or nonlinear relationship between parameters and observations, and the nonuniqueness of solutions.

6. MODEL ERROR ISSUES

Model error issues present serious hurdles for assimilation of actual observations into thermosphere-ionosphere general circulation models. There are two types of model errors that are particularly problematic in EnKF applications: systematic model-observation discrepancies and inadequate model error growth, which is gauged by a degree of the spread among ensemble members in the EnKF.

6.1. Biases

The most contingent types of systematic model-observation discrepancies or model biases involve the erroneous placement of geophysical features with sharp spatial gradients (e.g., the auroral boundary, the equatorial ionization anomaly, day-night terminators). In the upper atmosphere, these features are not fixed geographically, but instead move around considerably with geophysical conditions. It is very difficult to determine the correlation between state variables that are under the influence of very different physical mechanisms. For example, the electron densities inside and outside of the equatorial ionization anomaly region should be correlated very little, if at all, but the sample covariance obtained from the ensemble may suggest a very strong anticorrelation. Spurious correlations resulting from the sampling errors near these geophysical boundaries end up attributing the impact of electron density observations incorrectly and degrading assimilation analysis considerably. Some adaptive methodologies that can identify these geophysical boundaries and localize the sample covariance accordingly will be required to rectify the issues in the future.

6.2. Model Error Growth

In the EnKF, the probability distribution is emulated by the ensemble of model forecasts, and it is important that the ensemble reflects a realistic probability distribution of the upper atmosphere states. In the forecast step, the model error is presumed to grow out of the uncertainty in initial conditions, boundary conditions, and model parameters. Because the upper atmosphere is strongly controlled by external forcing, it is particularly important to incorporate the model error growth associated with external forcing uncertainty. At the moment, the sensitivity of the model to forcing variability and, more importantly, the uncertainty of forcing specification are not well characterized in thermosphere-ionosphere GCMs. Because of the lack of such characterization, the model error growth is inadequately represented by the models. A serious consequence of the insufficient spread among ensemble members is the filter divergence in which the analysis diverges from the "truth" or the observed reality so that the observations no longer make any impact on the analysis. Adaptively inflating covariance [e.g., *Anderson*, 2009] can help cope with issues of model errors to some extent. However, the model error growth issue discussed above will call for close attention from model developers in the future.

7. SUMMARY

This paper summarizes important research and development activities in upper atmosphere data assimilation to advance the capability of first-principles models of the thermosphere and ionosphere beyond the realm of climatology and to increase the geophysical information content of our geospace observing system. Because of a lack of simultaneous global observations of thermospheric parameters, our ability to monitor the global thermospheric conditions is severely limited, which in turn limits our modeling capability. This shortage can be overcome by exploiting our knowledge of the intimate coupling between the thermosphere and ionosphere described in GCMs, with the help of the latest

EnKF techniques. In other words, the observability of thermospheric parameters can be enhanced by using EnKF techniques to solve the inverse problem of inferring thermospheric parameters from better observed ionospheric parameters.

A series of OSSEs are presented to demonstrate the capability of the EnKF data assimilation procedure constructed with two sets of community software offered by NCAR, namely, the TIEGCM and DART. Specific observing systems considered are (1) electron density profiles monitored by a worldwide network of 75 ionosondes (section 3), (2) in situ thermospheric mass density sampled along the CHAMP satellite (sections 4.1 and 5), and (3) electron density profiles obtained from the COSMIC/FORMOSAT-3 mission (section 4.2). The self-consistent treatment of thermosphere-ionosphere coupling in the forecast model as well as in assimilation schemes, both of which can be achieved inherently by using EnKF in conjunction with the TIEGCM, help constrain the thermospheric state variables to be consistent with the ionospheric electron density. Given the ever-enhancing infrastructure for the global navigation satellite system, this is, indeed, a promising prospect for thermospheric data assimilation as well as for ionospheric data assimilation. Because of the thermosphere's relatively long-term memory, thermospheric data assimilation has the potential to increase the effectiveness of ionospheric data assimilation.

This paper also points out the areas that demand further research and development to improve the robustness and effectiveness of EnKF applications to upper atmosphere GCMs. One such area is the estimation of model parameters, especially the ones concerning external forcing, because the upper atmosphere above the mesopause is less sensitive to initial conditions but rather controlled by external forcing. For example, the neutral mass density variability in orbital altitudes is strongly controlled by solar EUV flux. Therefore, the efficacy of forcing parameter estimation considerably affects the quality of data assimilation analysis. In this paper, we show that $F10.7$ index that modulates effects of solar EUV flux in the TIEGCM can be inferred to some extent from the CHAMP neutral mass density by using EnKF, impacting the global neutral density specification considerably. A key to the successful inference of the parameters using EnKF is to cope with the reduction of model error growth associated with the parameter estimation.

The model error issues are major impediments to effective assimilation of upper atmospheric observations into thermosphere-ionosphere GCMs. Systematic model-observation discrepancies or model biases are challenging to all kinds of data assimilation techniques. The EnKF may not be the most robust assimilation procedure when the model bias is significant, and a number of desirable properties of the EnKF data assimilation system are here illustrated in the context of no gross model-observation discrepancies and await further trials in the future. We find the model errors associated with sharp geophysical boundaries to be particularly problematic. Another model error issue specific to EnKF applications is concerned with the model error growth. Most modeling efforts have so far been focused on reproducing the mean atmospheric state for given external driver conditions in a deterministic fashion, and therefore current models do not exhibit adequate model error growth. In the future, the model error growth in connection to variable external forcing needs to be better characterized in GCMs.

Acknowledgments. The author acknowledges the NCAR Institute for Mathematics Applied for Geosciences/Data Assimilation Research Section and High Altitude Observatory/Atmosphere Ionosphere Magnetosphere Section for providing the software support. The work presented in this paper is supported by the Air Force Office of Scientific Research Multidisciplinary University Research Initiative award FA9550-07-1-0565 and the Air Force Research Laboratory award FA9453-12-1-0244. The author thanks T. Mera for his assistance in improving the manuscript.

REFERENCES

Anderson, J. L. (2001), An ensemble adjustment Kalman filter for data assimilation, *Mon. Weather Rev.*, *129*, 2884–2903.

Anderson, J. L. (2003), A local least squares framework for ensemble filtering, *Mon. Weather Rev.*, *131*(4), 634–642.

Anderson, J. L. (2007), An adaptive covariance inflation error correction algorithm for ensemble filters, *Tellus, Ser. A*, *59*, 210–224.

Anderson, J. L. (2009), Spatially and temporally varying adaptive covariance inflation for ensemble filters, *Tellus, Ser. A*, *61*, 72–83, doi:10.1111/j.1600-0870.2008.00361.x.

Anderson, J. L., and S. L. Anderson (1999), A Monte Carlo implementation of the nonlinear filtering problem to produce ensemble assimilations and forecasts, *Mon. Weather Rev.*, *127*(12), 2741–2758.

Anderson, J. L., T. Hoar, K. Raeder, H. Liu, N. Collins, R. Torn, and A. F. Arellano (2009), The Data Assimilation Research Testbed: A community data assimilation facility, *Bull. Am. Meteorol. Soc.*, *90*, 1283–1296, doi:10.1175/2009BAMS2618.1.

Araujo-Pradere, E. A., T. J. Fuller-Rowell, and M. V. Codrescu (2002), STORM: An empirical storm-time ionospheric correction model 1. Model description, *Radio Sci.*, *37*(5), 1070, doi:10.1029/2001RS002467.

Datta-Barua, S., G. S. Bust, G. Crowley, and N. Curtis (2009), Neutral wind estimation from 4-d ionospheric electron density images, *J. Geophys. Res.*, *114*, A06317, doi:10.1029/2008JA014004.

Evensen, G. (1994), Sequential data assimilation with a nonlinear quasi-geostrophic model using Monte Carlo methods to forecast error statistics, *J. Geophys. Res.*, *99*, 10,143–10,162.

Evensen, G. (2009), *Data Assimilation: The Ensemble Kalman Filter*, 2nd ed., 307 pp., Springer, Berlin.

Fuller-Rowell, T. J., C. F. Minter, and M. V. Codrescu (2004), Data assimilation for neutral thermospheric species during geomagnetic storms, *Radio Sci.*, *39*, RS1S03, doi:10.1029/2002RS002835.

Gaspari, G., and S. E. Cohn (1999), Construction of correlation functions in two and three dimensions, *Q. J. R. Meteorol. Soc.*, *125*, 723–757.

Hajj, G. A., B. D. Wilson, C. Wang, X. Pi, and I. G. Rosen (2004), Data assimilation of ground GPS total electron content into a physics-based ionospheric model by use of the Kalman filter, *Radio Sci.*, *39*, RS1S05, doi:10.1029/2002RS002859.

Houtekamer, P. L., and H. L. Mitchell (2001), A sequential ensemble Kalman filter for atmospheric data assimilation, *Mon. Weather Rev.*, *129*, 123–137.

Jazwinski, A. H. (1970), *Stochastic Processes and Filtering Theory*, 376 pp., Academic Press, NewYork.

Khattatov, B. (2010), Assimilation of GPS soundings in ionospheric models, in *Data Assimilation: Making Sense of Observations*, edited by W. Lahoz, B. Khattatov, and R. Ménard, pp. 599–622, Springer, Berlin.

Komjathy, A., B. Wilson, X. Pi, V. Akopian, M. Dumett, B. Iijima, O. Verkhoglyadova, and A. J. Mannucci (2010), JPL/USC GAIM: On the impact of using COSMIC and ground-based GPS measurements to estimate ionospheric parameters, *J. Geophys. Res.*, *115*, A02307, doi:10.1029/2009JA014420.

Lee, I. T., T. Matsuo, A. D. Richmond, J. Y. Liu, W. Wang, C. H. Lin, J. L. Anderson, and M. Q. Chen (2012), Assimilation of FORMOSAT-3/COSMIC electron density profiles into thermosphere/ionosphere coupling model by using ensemble Kalman filter, *J. Geophys. Res.*, *117*, A10318, doi:10.1029/2012JA017700.

Lorenc, A. (1986), Analysis methods for numerical weather prediction, *Q. J. R. Meteorol. Soc.*, *112*, 205–240.

Matsuo, T., and E. A. Araujo-Pradere (2011), Role of thermosphere-ionosphere coupling in a global ionospheric specification, *Radio Sci.*, *46*, RS0D23, doi:10.1029/2010RS004576.

Matsuo, T., I.-T. Lee, and J. L. Anderson (2013), Thermospheric mass density specification using an ensemble Kalman filter, *J. Geophys. Res. Space Physics*, *118*, 1339–1350, doi:10.1002/jgra.50162.

Minter, C. F., T. J. Fuller-Rowell, and M. V. Codrescu (2004), Estimating the state of the thermospheric composition using Kalman filtering, *Space Weather*, *2*, S04002, doi:10.1029/2003SW000006.

Pi, X., C. Wang, G. A. Hajj, G. Rosen, B. D. Wilson, and G. J. Bailey (2003), Estimation of e x b drift using a global assimilative ionospheric model: An observation system simulation experiment, *J. Geophys. Res.*, *108*(A2), 1075, doi:10.1029/2001JA009235.

Reigber, C., H. Lühr, and P. Schwintzer (2002), CHAMP mission status, *Adv. Space Res.*, *30*(2), 129–134, doi:10.1016/S0273-1177(02)00276-4.

Richmond, A. D., E. C. Ridley, and R. G. Roble (1992), A thermosphere/ionosphere general circulation model with coupled electrodynamics, *Geophys. Res. Lett.*, *19*(6), 601–604.

Rocken, C., Y.-H. Kuo, W. Schreiner, D. Hunt, S. Sokolovskiy, and C. McCormick (2000), COSMIC system description, *Terr. Atmos. Oceanic Sci.*, *11*(1), 21–52.

Scherliess, L., R. W. Schunk, J. J. Sojka, D. C. Thompson, and L. Zhu (2006), Utah State University Global Assimilation of Ionospheric Measurements Gauss Markov Kalman filter model of the ionosphere: Model description and validation, *J. Geophys. Res.*, *111*, A11315, doi:10.1029/2006JA011712.

Scherliess, L., D. C. Thompson, and R. W. Schunk (2009), Ionospheric dynamics and drivers obtained from a physics-based data assimilation model, *Radio Sci.*, *44*, RS0A32, doi:10.1029/2008RS004068.

Schunk, R. W., et al. (2004), Global Assimilation of Ionospheric Measurements (GAIM), *Radio Sci.*, *39*, RS1S02, doi:10.1029/2002RS002794.

Tippett, M. K., J. L. Anderson, C. H. Bishop, T. M. Hamill, and J. S. Whitaker (2003), Ensemble square root filters, *Mon. Weather Rev.*, *131*, 1485–1490.

Whitaker, J. S., and T. M. Hamill (2002), Ensemble data assimilation without perturbed observations, *Mon. Weather Rev.*, *130*(7), 1913–1924.

T. Matsuo, 325 Broadway W/NP9, Boulder, CO 80305, USA. (tomoko.matsuo@colorado.edu)

Scientific Investigations Using IDA4D and EMPIRE

G. S. Bust

Atmospheric and Space Technology Research Associates, Boulder, Colorado, USA

John Hopkins University Applied Physics Laboratory, Laurel, Maryland, USA

S. Datta-Barua

Department of Aviation and Technology, San Jose State University, San Jose, California, USA

Over the last several years, two ionospheric data assimilation algorithms have been developed, tested, and applied to various scientific investigations. Ionospheric Data Assimilation Four Dimensional (IDA4D) focuses on estimating the global, 3-D time-evolving distribution of electron density throughout all ionospheric altitudes. Estimating Model Parameters From Ionospheric Reverse Engineering (EMPIRE) takes the output of IDA4D, applies physical constraints, and estimates the underlying internal dynamics that caused the electron density distribution. Recently, improvements to the IDA4D algorithm have been made, including solving for the log of density rather than density. This naturally implies that the estimation becomes a nonlinear iterative algorithm. This allows for easy ingestion of new nonlinear data sources such as UV nighttime radiative recombination observations of either 1356 or 911 Å, which are approximately proportional to line of sight integrals of the square of electron density, time delay (or virtual height) versus frequency observations from ionosondes or digisondes, and bistatic HF and VHF time delay versus frequency observations that are either ground to ground or sky to ground paths. With the new improvements, IDA4D has been recently validated against CHAMP in situ observations of electron density. IDA4D was shown that for densities $\geq 4.0 \times 10^{11} \ e^- \ m^{-3}$, the errors are less than 20%. The new IDA4D algorithm was then combined with EMPIRE to investigate various scientific phenomena that can be better elucidated with simultaneous estimates of plasma density structuring and plasma density transport. In particular, the two algorithms have been used to study the relationship between field-perpendicular and field-parallel ion drifts and the formation and evolution of storm-enhanced density features.

1. INTRODUCTION

Ionospheric imaging and data assimilation methods, and their application to scientific investigations of the ionosphere, have been active areas of research for more than 20 years. A comprehensive review of the state of the field in 2008 can be found in the work of *Bust and Mitchell* [2008]. However, one field that has received less focus is the estimation of external drivers that are responsible for the internal dynamics of the ionosphere-thermosphere system. These include $E \times B$ drifting at high latitudes, equatorial latitudes, and midlatitudes during strong storms and neutral thermospheric winds in the F region of the ionosphere. Many of the outstanding scientific research questions in the areas of the ionosphere and thermosphere require knowledge of the large-scale internal dynamics in order to make significant progress in our understanding. One of the research questions that can be better addressed by simultaneous specification of plasma structuring, electric fields, and neutral winds includes the formation and transport of storm-enhanced density (SED) features and polar cap patches. While exactly how SEDs form, particularly the low-latitude and midlatitude bulge, is not completely understood, it is thought that some combination of transport due to electric fields and neutral winds is responsible. Once formed, the SED is transported poleward, often times all the way to the nightside auroral region. This transport is thought to be due to $E \times B$ transport. In a similar manner, polar cap patches are thought to be transported across the polar cap by $E \times B$ transport. A better understanding of the formation, evolution, and fate of both SEDs and patches could be obtained with simultaneous estimation of plasma structuring, neutral winds, and electric fields.

Another scientific research area is the space-time distribution of Joule heating at high latitudes, including the effects of the Joule heating on changes in the neutral composition with altitude, on launching and propagation of large-scale neutral winds toward the equator, and the transport of compositional changed to other latitude/longitude regions. In order to accurately estimate the Joule heating over the entire high-latitude region, simultaneous local knowledge of the electric fields, neutral winds, and electron density (conductances) is required. Then, in order to understand the propagation effects of high-latitude heating on midlatitudes and low latitudes, knowledge of the neutral winds and plasma structuring is required. A related science research area is the feedback effect of high-latitude conductances and neutral winds upon the electric fields impressed from the magnetosphere and, therefore, on the entire high-latitude current system, which couples the ionosphere to the magnetosphere.

Finally, the formation of small-scale irregularities is controlled by the large or mean state of the ionosphere and thermosphere. Depending on the instability mechanism being studied, this can include the large-scale distribution of electric fields, neutral winds, and gradients in the plasma density. Thus, being able to image these variables simultaneously with observations of small-scale structuring and scintillations could provide insight to the processes that are responsible for the development of instabilities.

However, current data assimilation algorithms [*Mitchell and Spencer*, 2003; *Schunk et al.*, 2004; *Wang et al.*, 2004; *Bust et al.*, 2004; *Bust and Mitchell*, 2008] typically do not self-consistently estimate external drivers, such as neutral winds and electric fields, or internal dynamics, such as electron and ion temperatures, along with the electron density. New data assimilation methods must be developed that couple to a first-principles model and will self-consistently estimate electron density, neutral winds, electric fields, and plasma temperatures throughout the entire ionosphere and plasmasphere, semicontinuously in time. While a number of different methods that estimate model drivers have been developed such as 4DVAR (Four-dimensional variational assimilation) [*Pi et al.*, 2003] and ensemble Kalman filters (EnKFs) [*Scherliess et al.*, 2009; *Matsuo and Araujo-Pradere*, 2011], so far, these methods have only been tested on simulated data, or when applied to real data, the resulting drivers have not been validated.

Over the last several years, two data assimilation algorithms have been developed to allow estimation of both global electron density and dynamical variables such as neutral winds, temperatures, and electric fields. Ionospheric Data Assimilation Four-Dimensional (IDA4D) [*Bust et al.*, 2000, 2001, 2004] is an ionospheric data assimilation algorithm that has been under development for over 12 years. IDA4D provides global 3-D time-evolving maps of ionospheric electron density. Estimating Model Parameters From Ionospheric Reverse Engineering (EMPIRE) [*Datta-Barua et al.*, 2009, 2010, 2011] is a data assimilation algorithm that takes the output of IDA4D and subsequently estimates neutral winds and electric fields. IDA4D and EMPIRE have been used to carry out a number of scientific investigations. A brief review of the development and capabilities of IDA4D and EMPIRE are presented below in sections 2 and 3. Then, scientific results are presented in section 4, while future plans for a new integrated data assimilation algorithm are discussed in section 5.

2. IDA4D

IDA4D has been developed and validated continually for more than 12 years [*Bust et al.*, 2000, 2001, 2004]. IDA4D has been well validated [*Coker et al.*, 2001; *Watermann et al.*, 2002; *Bust and Crowley*, 2007] and used in a number of scientific investigations [*Bust et al.*, 2007; *Bust and Mitchell*, 2008; *Garner et al.*, 2006; *Yin et al.*, 2006; *Mitchell et al.*, 2005]. IDA4D is an objective analysis data assimilation algorithm based closely on the Navy Research Laboratories meteorological 3DVAR algorithm [*Daley and Barker*, 2000]. Temporal evolution of the electron densities are handled by a Guass-Markov Kalman filter. A detailed description of the original algorithm can be found in the work of *Bust et al.* [2004]. Here a brief overview of the original algorithm is provided, then, some new features and capabilities developed over the last few years are discussed.

2.1. Original IDA4D Linear Estimation Method

IDA4D estimation minimizes a cost function J, which includes a χ^2 term and a "regularization" term that represents the misfit between the density and a forecast model density

$$J = 0.5[\vec{y} - \tilde{H}\vec{x}]^T \tilde{R}^{-1}[\vec{y} - \tilde{H}\vec{x}]$$
$$+ 0.5[\vec{x_f} - \vec{x}]^T \tilde{P}_f^{-1}[\vec{x_f} - \vec{x}]. \quad (1)$$

The data and model forecast error covariances are given by $\tilde{R}$ and $\tilde{P}_f$, respectively, and the geometry matrix $\tilde{H}$ relates the discretized grid of electron densities $\vec{x}$ to the vector of observations $\vec{y}$. That is, the relationship between observation y_i and the array density grid values x_k is given as

$$y_i = \sum_k \tilde{H}_{ik} x_k. \quad (2)$$

Taking the gradient of J with respect to $\vec{x}$ and setting the result to zero produces an optimal analyzed estimate of the electron density given as

$$\vec{x_a} = \vec{x_f} + \tilde{P}_f \tilde{H}^T [\tilde{H}\tilde{P}_f \tilde{H}^T + \tilde{R}]^{-1}(\vec{y} - \tilde{H}\vec{x_f}). \quad (3)$$

A formal analyzed error covariance can be derived and is given as

$$\tilde{P}_a = \tilde{P}_f - \tilde{P}_f \tilde{H}^T [\tilde{H}\tilde{P}_f \tilde{H}^T + \tilde{R}]^{-1} \tilde{H}\tilde{P}_f. \quad (4)$$

IDA4D is able to ingest a wide variety of data sources linearly related to electron density. Typical data sources used by IDA4D include (1) ground-based GPS slant total electron content (TEC), (2) ground-based digisonde and incoherent scatter radar (ISR) electron density altitude profiles, (3) Ground-based Doppler orbitography and radiopositioning integrated by satellite (DORIS) TEC, (4) space-based GPS occultation TEC, (5) space-based "topside" GPS TEC, and (6) space-based in situ observations of electron density.

In general, the most available and copious data set is ground GPS TEC. Thus, the performance of most global tomographic imaging and data assimilation algorithms is dependent on the amount and distribution of ground GPS TEC. Figure 1 shows the coverage of GPS TEC lines of sight over the United States of America. The colored contours show the reconstructed electron density for day 99, 2009 at 06:00 UT along the 0° geomagnetic longitude. The white lines are the GPS lines of sight through the ionosphere from the receivers on the ground to the satellites. In order for a line of sight to be visible in the plot, the longitude along the trajectory must be within ±15° of 0° and within ±7.5 min of the reconstruction time. The plot shows that, indeed, there is good coverage over the USA sector, particularly at geomagnetic latitudes >35°, with many crossing rays that are necessary for good tomographic imaging.

While the original implementation of IDA4D produced good results and compared well with individual data sets, there were several areas where improvements could be made. The most important improvements included making the entire algorithm nonlinear, solving for the log of the density and the inclusion of nonlinear data sets such as EUV radiances from 1356 Å. Each of these new capabilities is discussed in greater detail below.

2.2. New IDA4D Nonlinear Estimation Method

Global ionospheric tomographic imaging problems are "mixed-determined" estimation problems that can be ill posed. The linear formulation described by equation (3) permits solution with negative densities, which are not physically realistic. One way around this problem is to add a positivity constraint to the matrix inversion problem. This works well as long as the problem is solved in "model" space, where constraints can be simply added algebraically. However, for the large global imaging problems under consideration, the "model" space can consist of 10^5 unknowns or more, and it is often more computationally efficient to solve the problem in "data" space, where the number of observations are $\sim 10^4$. This is how IDA4D solves the inversion problem (see equation (3)). In addition, the above mathematical formulation makes the assumption that the model and data errors obey Gaussian statistics. However, electron density distributions are closer to a log normal probability distribution. Solving for the log of the density addresses both of these issues. The log of the density ensures the densities will be positive definite, and the mathematical statistical formulation is more suited to solving for the log of the density than the density itself. However, the problem has now become a nonlinear estimation problem and requires an iterative nonlinear estimation scheme. While this approach adds a layer of complexity to the algorithm, there are several advantages to this approach. First, there is no need for additional positivity constraints, which add to complexity of the solution. Second, solving for the log of the density applies equally to both model and data space solutions, and third, once the nonlinear iterative mechanism has been developed and tested, it is straightforward to add nonlinear data sets to IDA4D.

For the nonlinear case, the observations are nonlinearly related to the state variable to be solved for, in this case the log of electron density. Therefore, the nonlinear forward model relating the state variable to the observations must be linearized around the current best estimate of electron density. If we let the forward model for data point i be written as $H^i(\vec{x})$, with $\vec{x} = log(\vec{N_e})$, then, we linearize H^i, for iteration n, as follows:

$$H^i(\vec{x}_n) \approx H^i(\vec{x}_{n-1}) + \sum_k \left(\frac{\partial H^i}{\partial x^k} \bigg|_{x^k_{n-1}} \right)(x^k_n - x^k_{n-1}), \quad (5)$$

where the x^k represents the discretization of the log of the electron density field onto a finite 3-D grid.

The Jacobian matrix elements $\tilde{H}^n_{ik}$ for data point i, grid point k, and iteration n are defined as

$$\tilde{H}^n_{ik} \equiv \left(\frac{\partial H^i}{\partial x^k} \bigg|_{x^k_{n-1}} \right). \quad (6)$$

The estimated analyzed log density solution for the ith iteration is given as

$$\vec{x}^n_a = \vec{x}_f + \tilde{P}_f(\tilde{H}^n)^T [\tilde{H}^n P_f(\tilde{H}^n)^T + \tilde{R}]^{-1} [\vec{y} - \vec{H}(\vec{x}^{n-1}_a) \quad (7)$$
$$+ \tilde{H}^n(\vec{x}^{n-1}_a - \vec{x}_f)].$$

Once the algorithm achieves convergence criteria, the final iteration becomes the analyzed solution, and the analyzed covariance is calculated as before.

2.3. New IDA4D Data Sources

Once the nonlinear iterative algorithm has been formally set up and solved for as in equation (7) above, it becomes possible to include observations that are nonlinearly related to electron density in the IDA4D analysis. Currently, IDA4D accepts three nonlinear data sources: (1) UV night-time radiative recombination observations of either 1356 or 911 Å, which are approximately proportional to line-of-sight integrals of the square of electron density, (2) time-delay (or virtual height) versus frequency observations from ionosondes or digisondes, and (3) bistatic HF and VHF time delay versus frequency observations that are either ground to ground or sky to ground paths.

For each of these observations, the appropriate nonlinear forward model has to be defined and implemented. For the UV data, the forward model is simply an integral along the look direction of the imager of N_e^2, multiplied by a suitable constant. For the upward looking digisonde data, the forward model for the virtual heights is an integral over altitude of the group refractive index obtained from the Appleton-Hartree equation. Finally, for the bistatic HF or VHF observations, a full 3-D ionospheric ray tracer is required that allows for a numerical integration along the (curved) path. In addition to the forward model, for each of these data sets, a Jacobian matrix must be computed for each iteration in a manner similar to that in equations (5) and (6). Once the forward models and Jacobians have been defined properly, a nonlinear iteration solution can be approached using equation (7) above.

2.4. CHAMP Validation of IDA4D

While validation of IDA4D has been performed [*Coker et al.*, 2001; *Watermann et al.*, 2002], these validations have either been comparisons against independent observations of TEC or case studies with ISRs. Recently, a long term multiple-month statistical analysis was carried out comparing in situ observations of electron density from the CHAMP satellite Plasma Languin Probe (PLP) instrument [*Cooke et al.*, 2003] with IDA4D estimates of electron density that are interpolated in space and time onto the CHAMP observation points. This represents a good test of IDA4D accuracy since the CHAMP observations are at a specific altitude in the F region of the ionosphere and a function of latitude and longitude. Over the course of a day, there are >5000 CHAMP observations available for comparison. Initially, 3 months were used for the comparison: April 2007, October 2007, and January 2008. While the initial results were not bad, it was felt that improvements to IDA4D would lead to better comparison results. First, it was realized that the accuracy by which IDA4D could reproduce the input data was limited by the spatial resolution of the grid. IDA4D had been typically using a grid that was 4° × 10° latitude by longitude grid. Requiring a constant density throughout such a large grid cell limits the accuracy of the estimated TEC and, therefore, limits how well the density can be reconstructed. Thus, the resolution was doubled for both latitude and longitude (2° × 5°). Second, the vertical spatial error correlation model was a Gaussian in altitude with the decorrelation lengths input by the user (see the work of *Bust et al.* [2004] for details). However, it is more likely that the model errors are anticorrelated across the peak altitude of the *F* region. This is because what models tend to get wrong is the *height* of the peak in density, more than the *value* of the peak density. Since the CHAMP observations represented point measurements at a specific altitude, it seemed likely that data ingested into the IDA4D at a specific altitude could be misapplied across the *F* region peak. Therefore, a modification was made to the vertical error correlation function as follows

$$\rho(z, z') = sgn(z, z') exp\left[-\frac{(z-z')^2}{LL'}\right], \quad (8)$$

where *sgn* (z, z') is positive if both z, z' are above or below the *F* region peak and negative if one is above and the other below. The decorrelation scale lengths at altitudes z and z' are given by L and L', respectively. These two improvements to IDA4D dramatically improved the comparisons with CHAMP observations. Figure 2 presents statistical comparisons of IDA4D against CHAMP for the entire month of April 2007. In Figure 2, the vertical axis is the percent RMS error between CHAMP and IDA4D. The horizontal axis values represent "threshold" values, for which all CHAMP data greater than or equal to the threshold value was used in the statistical comparison. The horizontal red dashed line is the 30% error level, while the green dashed line represents 25% and the blue dashed line, the 20% error level. It can be seen that for densities $\gtrsim 4 \times 10^{11}$ e$^-$ m^{-3}, the errors are less than 20%. At smaller densities, the IDA4D agreement is much worse. This is due to limitations on the overall accuracy of TEC obtained by IDA4D. Many different validation studies have shown that imaging and data assimilation algorithms only get the residual TEC accuracy to ~2–3 total electron content unit (TECU), 1 TECU = 10^{16} el m^{-2} (see, for example, the work of *Thompson et al.* [2006]). For a 3 TECU error spread over a 500 km path (which represents a slant path through the *F* region), we have an electron density error of ~6 × 10^{10} e$^-$ m^{-3} or larger.

The above developments describe how IDA4D produces global maps of electron density as well as a formal error analysis. However, the global distribution of plasma density is the causal effect and response due to other dynamical drivers and variables such as electric fields and neutral winds. In order to really understand the causal dynamics responsible for the observed plasma distribution, the electron density maps need to be supplemented by spatial and temporally extended maps of $E \times B$ drifts and neutral winds. Since such spatially and temporally extended measurements are not easy to make, and are rarely available, an assimilative algorithm was developed that combines an electron density continuity equation with the time-evolving maps of electron density provided by IDA4D to obtain estimates of the neutral winds and $E \times B$ drifts that are self-consistent with both the continuity equation and the derived densities (and errors) from IDA4D. The resulting algorithm, EMPIRE, is described in the next section. Complete derivations of the algorithm, results, and error analysis can be found in the works of *Datta-Barua et al.* [2009], [2011], and [2010].

3. EMPIRE

EMPIRE is an estimation algorithm that takes the time-evolving global estimate of electron density output from IDA4D as a "derived" data set that can then be used to further estimate the model state parameters in the electron density continuity equation given in the F region of the ionosphere as

$$\frac{dN}{dt} = P - \beta N - \vec{\nabla} \cdot (N \vec{v}_\parallel) - \vec{\nabla} \cdot (N \vec{v}_\perp), \qquad (9)$$

where P is the production, β is the loss term in the F region, $\vec{v}_\parallel$ is the velocity component parallel to the magnetic field direction, and $\vec{v}_\perp$ is the velocity perpendicular to the magnetic field. Here $\frac{dN}{dt}$ and N are from the IDA4D estimates of electron density, and the formal error covariance that is returned from IDA4D is used as the error on the density data and is propagated throughout the EMPIRE estimation algorithm so that output errors are estimated on the other state variables. Since $\frac{dN}{dt}$ is a single scalar variable at each space-time grid point, and there could be up to five unknown parameters in the continuity equation at each grid point (production, β, field-aligned drifts, and the two field-perpendicular components), the number of unknowns must be reduced to have an overdetermined system. This is accomplished by expressing the unknowns using a low-order basis function expansion. For example, if we assume that the field-perpendicular drifts are entirely due to electric fields, which can be derived from an electrostatic potential, then, the potential can be expanded in a low-order power series in magnetic latitude θ and magnetic longitude ϕ (or low-order spherical harmonics) and the coefficients of the series solved for. The implicit assumptions being made in such an expansion are (1) that the higher-resolution spatial-temporal variations observed in the electron density are primarily due to nonlinear mixing of the larger-scale winds, fields, and temperatures and (2) that the observed densities have sensitivity to the larger-scale parameters. In addition, in order for the results to be scientifically useful, the larger-scale (or averaged) values of neutral winds, electric fields, and temperatures that are estimated from such a low-order basis expansion must be of sufficient spatial-temporal resolution in a given scientific investigation. Exactly what the required spatial-temporal resolution is depends on the specific science topic being addressed. For large-scale storm studies, spatial resolution of a few degrees horizontally is sufficient. For studies of the topside distribution of electron temperature (for example), horizontal resolution of thousands of kilometers maybe sufficient. Given that these assumptions hold, we can expand the unknowns in low-order basis functions, gather together terms, and develop an estimation equation as

$$\mathbf{y} = \mathbf{M}\mathbf{x} + \mathbf{a_0} \qquad (10)$$

$$y_i = \frac{\Delta N_i}{\Delta t}, \qquad (11)$$

where y_i is the electron density rate of change observed from ionospheric imaging at the ith grid point, $\mathbf{a_0}$ is the a priori model of any drivers being treated as known quantities through the use of models (empirical or first principle) or other plausibility constraints (such a production and loss are approximately in balance), and $\mathbf{x}$ is the coefficient of functions describing the remaining drivers to be estimated algebraically. The matrix $\mathbf{M}$ maps the state to the observations. The vector of observations $\mathbf{y}$ are extended in space over a large 3-D regional (or possibly global) grid and can be extended over a time period where it is not expected that the other large-scale parameters change dramatically (typically 15–20 min). Typical spatial resolution scales that EMPIRE can resolve are ~5° horizontally in latitude and longitude and ~100 km vertically. The temporal resolution scale is ~15–30 min. When the error covariances on the electron density estimates are taken into account, equation (10) provides an estimation equation for the coefficients of the unknown winds and $E \times B$ drifts, as well as the estimated errors on those drifts.

4. IDA4D AND EMPIRE SCIENTIFIC INVESTIGATIONS

4.1. IDA4D Results

IDA4D has been used for a number of scientific studies over the years. Recently, it has been used to capture the large-scale plasma variations over Antarctica during the minor storm of 5–6 April 2010 and to compare those large-scale variations with small-scale scintillations from a GPS scintillation receiver (see the work of *Kinrade et al.* [2012] for further details). Figure 3 shows the large-scale horizontal plasma structuring over Antarctica at 16:45 UT on this day, with GPS phase scintillations from 16:00–17:00 UT overlaid. In Figure 3, magnetic local noon is at the top of the figure, and magnetic local time (MLT) = 18 on the right-hand side. The white square is the location of the GPS receiver at the South Pole. The white circles represent GPS phase scintillations for different satellites. The largest circles are for phase $\sigma_\phi > 1.0$ radians. The smallest circles represent nonscintillating satellites ($\sigma_\phi < 1.0$). The background model used by IDA4D for this study was International Reference Ionosphere. The IDA4D imaging had GPS TEC data available from ~50 receivers distributed over the South Pole and high latitudes on the mainland, TEC from three DORIS transmitters located on the Antarctic coast, and satellite occultation TEC and topside TEC from seven to nine LEO satellites. Thus, the coverage over Antarctica was actually quite good during this period. (*Kinrade et al.* [2012] shows a coverage plot of data over Antarctica for a 1 h period.) A tongue of ionization (TOI) is entering the polar cap from the afternoon sector, and it appears a large structure has separated from the main tongue just sunward of South Pole (the white square). Whether the large plasma structure is due to a moving patch or soft precipitation in the cusp is not known. What we can see is that GPS phase scintillations occur (the larger circles show significant values of σ_ϕ) all along the poleward edge of the

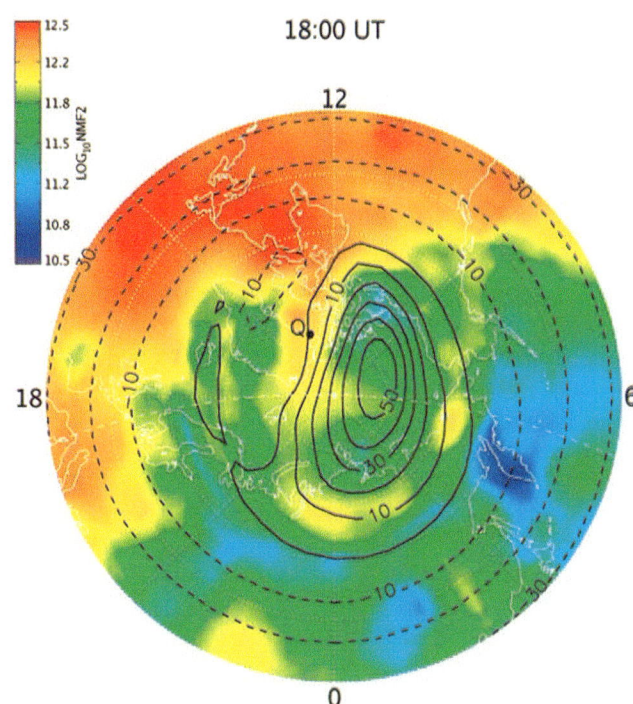

Figure 4. IDA3D electron density distribution as a function of geographic latitude and longitude at 18:00 UT with assimilative mapping of ionospheric electrodynamics convection patterns overlaid. Qanaq is indicated by a black dot. A tongue of ionization extends from the dayside over Qanaq.

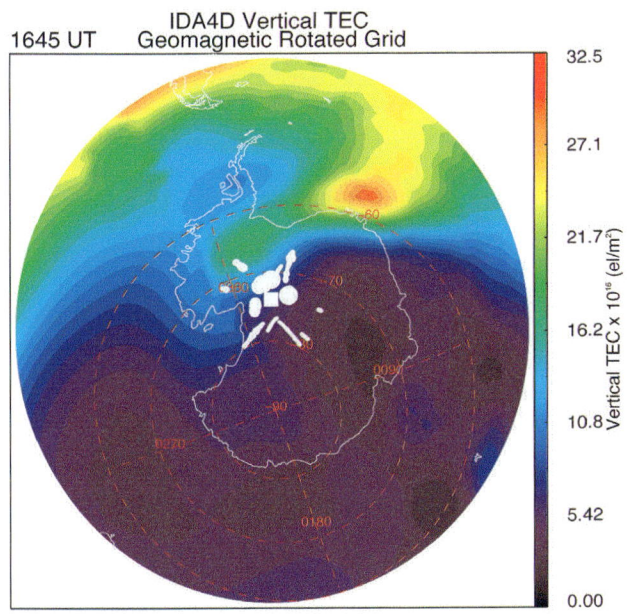

Figure 3. Antarctica 5 April 2010, GPS scintillations from 16:00–17:00 UT, with IDA4D vertical total electron content (TEC) taken at 16:45 UT.

large plasma structure, which indicates the scintillations are associated with the plasma structure in some way.

IDA4D has also been used to investigate the origin, evolution, and fate of polar cap patches [*Bust and Crowley*, 2007]. Figure 4 shows peak electron density in the *F* region obtained from IDA4D across the northern polar cap at 18:00 UT on 12 December 2001. Superimposed on Figure 4 is the assimilative mapping of ionospheric electrodynamics (AMIE) potential pattern with the effects of corotation added. Solid lines are for positive potential, while dashed lines are for negative potential. At lower latitudes, the plasma simply corotates. This AMIE pattern suggests that plasma flows antisunward across the northern face of Greenland near 13 MLT and then on toward Svalbard. What is revealed is the TOI entering the polar cap, breaking up into patches and being transported across the cap as imaged from IDA4D (which makes no use of AMIE or transport), and following very closely the potential patterns from AMIE. This close correspondence between AMIE potential (and derived $E \times B$ drifts) and IDA4D allowed for subsequent "trajectory of patch" analysis, which was at the heart of the investigation.

Finally, IDA4D has been used to investigate several case studies of magnetic storms [*Bust et al.*, 2007; *Bust and*

Crowley, 2008] including the superstorms in October and November 2003, which both showed strong SED features over the USA sector. While both storms have some similar characteristics in the SED, there are differences. Figure 5 shows IDA4D reintegrated vertical TEC for the 30 October 2003 storm (a), and the 20 November 2003 storm (b). The most striking feature of contrast between the two storms is for 30 October, the high-density plasma bulge forms off the west coast of the USA and Mexico, with the orientation aligned with the California/Mexico coastline, while the narrow plume of plasma or TOI stretches back in a southwest to northeast direction up over the pole. For the 20 November case, the bulge forms off the southeastern coast of the USA in the Caribbean, with the TOI stretching from southeast to northwest up over the pole. The second feature of interest is that the 20 November SED forms ~1.5 h earlier in time than the 30 October 30. The third interesting feature is that if the entire time history of the two storms is examined (not shown here), high-density bulge extends to much higher latitudes on 30 October than on 20 November. The 30 October case extends to ~45° geographic latitude, while the 20 November extends to ~35°.

While the above results demonstrate that the global time-evolving images of electron density provided by IDA4D are useful for scientific investigations, it is clear that a more in-depth analysis of the physics could take place if there was also knowledge of the internal dynamics of the system. In particular, it would be useful to have spatial and temporally extended maps of $E \times B$ drifts and neutral winds. In order to improve our understanding of storm dynamics, we turn to results obtained from EMPIRE.

4.2. EMPIRE Results

An EMPIRE analysis was carried out for the 20 November 2003 storm in the local afternoon time period for the USA sector. Figure 7 presents reintegrated vertical TEC from IDA4D during the EMPIRE period of analysis discussed below. The black rectangle defines the EMPIRE horizontal grid. For extreme events such as the superstorms of fall 2003, the hypothesis is that in the region of the SED formation and evolution, the external drivers due to neutral winds and $E \times B$ drifts are going to be more important to the redistribution of plasma than production and loss. The analysis assumed that production approximately equaled loss so that production and loss effects could be ignored, and the field-aligned and field-perpendicular drifts in equation (9) above were estimated. In order to validate EMPIRE and to determine whether the EMPIRE estimated drifts reproduced the actual physics occurring during the storm, the EMPIRE results were compared against vertical drift profiles obtained from Millstone Hill ISR. The Millstone Hill ISR has a fixed

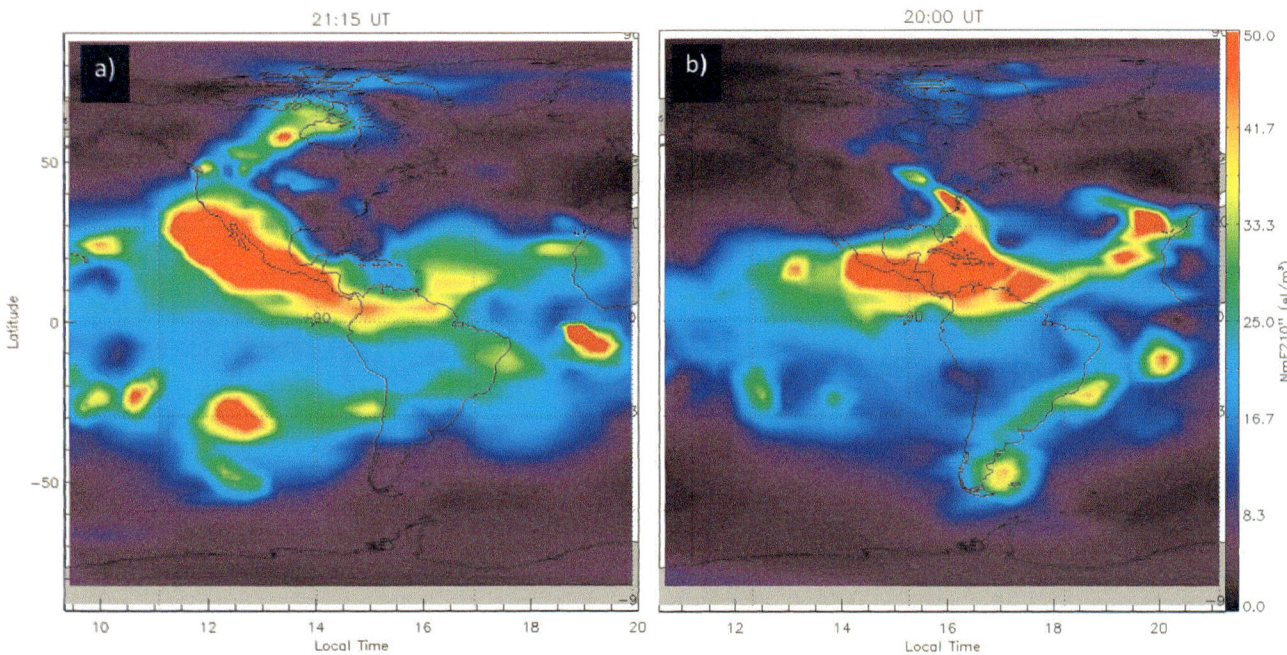

Figure 5. A comparison between the fully formed storm-enhanced density on 30 October and 20 November 2003. Presented are IDA3D horizontal images of peak electron density for (a) 30 October and (b) 20 November. Note that the horizontal axis shows the LT for each longitude sector.

zenith antenna and a steerable antenna (MISA) whose data have been used in previous studies for this particular storm [*Foster and Rideout*, 2005]. In particular, the zenith-pointing antenna observed the SED as it passed by Millstone Hill. Figure 6 presents a comparison between the Millstone Hill vertical drifts and EMPIRE-derived vertical drifts. The results are actually in quite good agreement, particularly in the ~17:50 UT time period when the SED passes by. This validation against the ISR provides confidence that the EMPIRE-derived drifts are reproducing the actual physics to some level of qualitative accuracy.

EMPIRE can be used to investigate the extended spatial-temporal distribution of vertical drifts during the time period of the formation and evolution of the SED. This is useful because in the spatial regions where the SED bulge and plume form, it is thought that the plasma is lifted to higher altitudes by plasma drifts. The Millstone Hill data provide evidence of this at a single latitude and longitude. With EMPIRE, the upward drifts can be estimated over the whole region (see the work of *Datta-Barua et al.* [2011] for details of the analysis). Focusing on the 17:00–19:00 UT time period, Figure 8 presents the vertical drifts as a function of magnetic longitude and latitude at 500 km altitude, which can be compared with the IDA4D results presented in Figure 7. The upward drifts are mapped for (Figure 8a) 17:10 UT, (Figure 8b) 18:10 UT, and (Figure 8c) 18:50 UT. Figure 8a shows that there is an upward vertical flow of ~200–400 m s^{-1} along the eastern edge of the grid region, which corresponds with the leading edge of the SED bulge as it enters the EMPIRE grid region. In particular, there is an upward flow of ≈400 m s^{-1} on the eastern edge of the EMPIRE grid between 48° and 54° magnetic latitude, which is very well correlated with the western edge of the SED plume. Figure 8b presents the upward drifts for the 18:10 UT time period on a scale of −200 (blue) to 200 m s^{-1} (red). While the speeds, ~150 m s^{-1} upward in the southeastern part of the EMPIRE grid and less elsewhere, are not as large as at 17:10 UT, there is still a well-defined relation between the upward drifts and the edge of the SED. Finally, Figure 8c presents upward drifts for the 18:50 UT time period from −200 m s^{-1} (blue) to 200 m s^{-1} (red). Now, the vertical drifts in the northwest region of the EMPIRE grid are strongly upward (~100–150 m s^{-1}) and well correlated with the northwest edge of the SED plume. In addition, the drifts are somewhat positive in the extended northwest section of the grid, falling to ~0 and, then, negative in the eastern and southern parts of the domain. This corresponds fairly well to the middle and trailing edge of the bulge, which has probably already been lifted to above 500 km altitude from the previous time history.

The vertical drifts are a superposition of field-aligned and field-perpendicular drifts. The field-perpendicular drifts are due to $E \times B$ effects and represent the importance of the convective or subauroral polarization stream electric fields to the SED processes. The field-aligned drifts are due to combined effects of neutral winds and diffusion and, therefore,

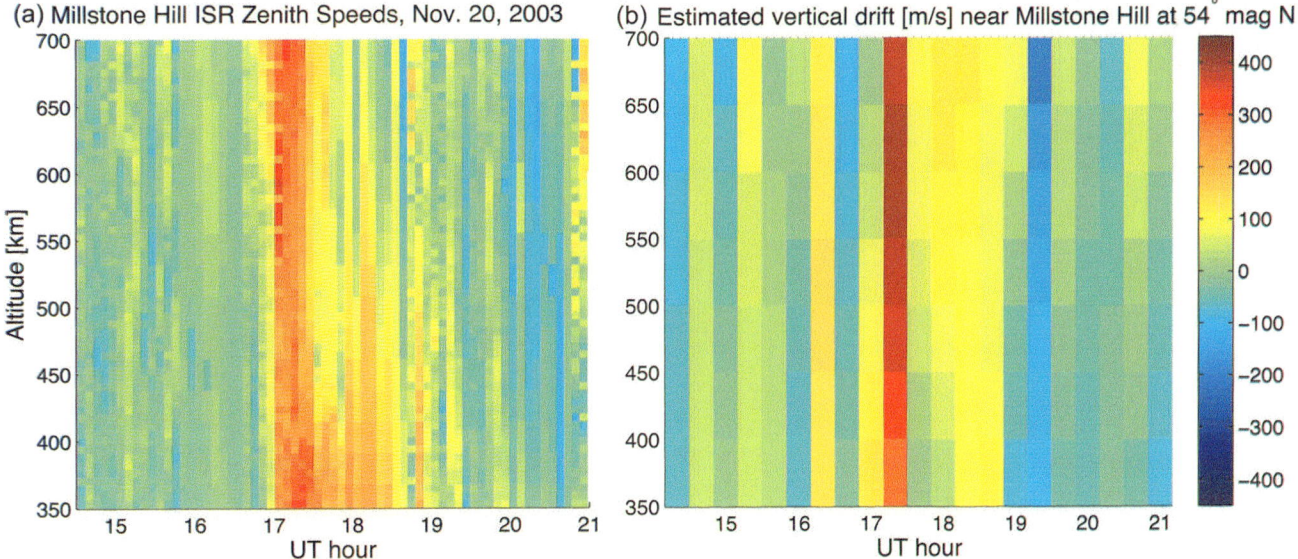

Figure 6. (a) Millstone Hill incoherent scatter radar (ISR) zenith drifts on 20 November 2003 versus time. (b) Estimating Model Parameters From Ionospheric Reverse Engineering (EMPIRE) estimated vertical drifts at 54°N, 1°E magnetic near Millstone Hill ISR location versus time on 20 November 2003.

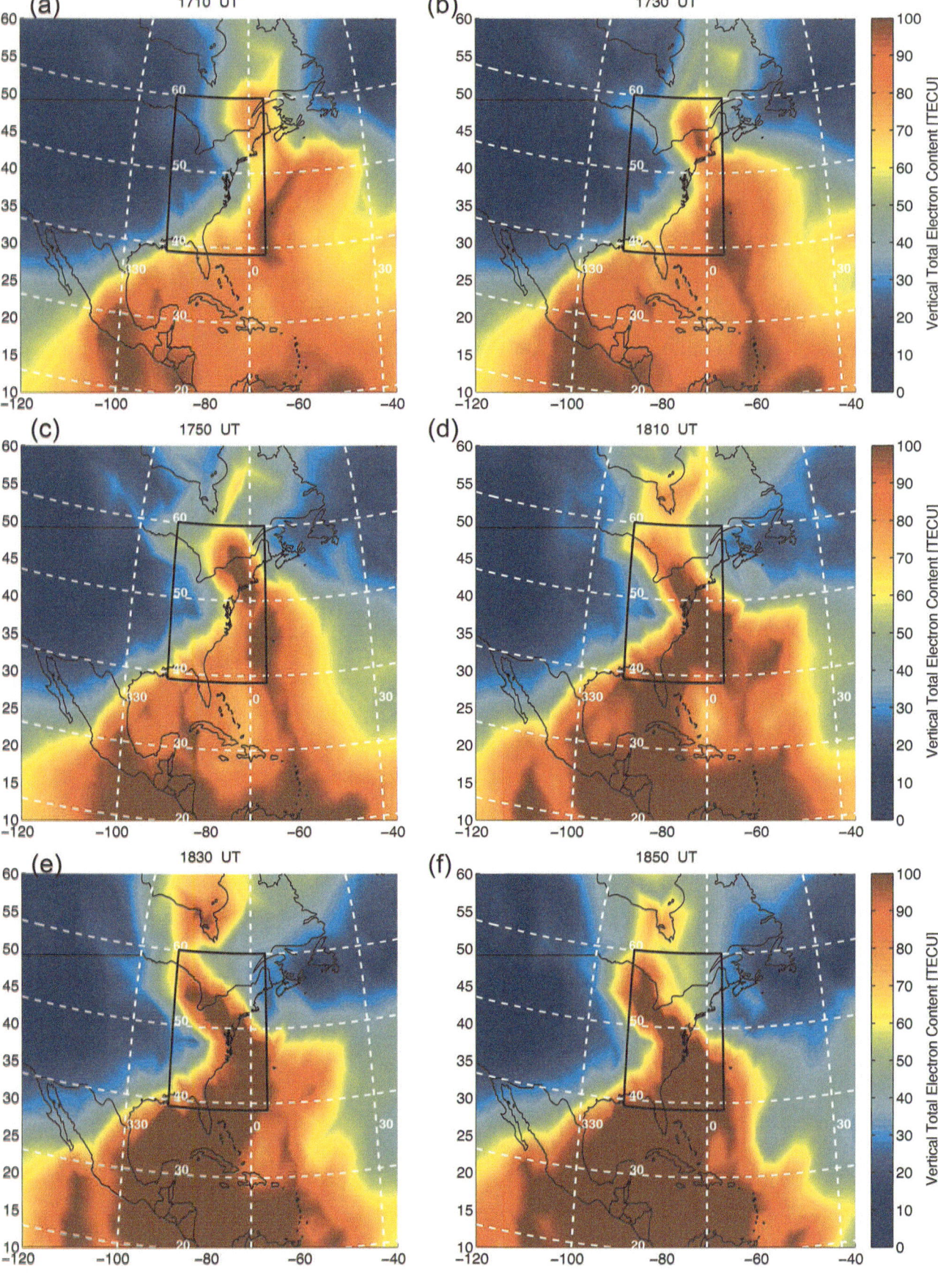

represent more the importance of thermospheric dynamics to the SED processes. With EMPIRE, for the first time, these two effects have been compared and contrasted across an extended spatial region over the several-hour period when the SED passes. Figure 9 presents the same horizontal region and three time periods as Figure 8, only now, the vertical drifts have been decomposed into those due to field-aligned drifts (left-hand column) and those due to $E \times B$ drifts (right-hand column). What is interesting is that for the earlier two times, 17:10 and 18:10 UT, the field-aligned component of the vertical drifts is larger than the $E \times B$ component. It is also located on the eastward edge of the grid, where the SED is first forming and as it drifts westward and northward. Then, at 18:50 UT, the $E \times B$ drift becomes as important as the field-aligned component and has its largest values (if we ignore the values at the top edge of the plot, which have very high errors; see the work of *Datta-Barua et al.* [2010] for details) on the western edge of the grid. This is where it is expected that the SED plume is being transported toward the pole by the large westward and poleward $E \times B$ drifts. Based on these results, it appears that the relative importance of the field-aligned drifts (neutral winds and diffusion) and the $E \times B$ drifts to the upward lifting of the plasma varies in time. At the early times of formation, neutral winds (and possibly diffusion) are at least as important as the $E \times B$ drifts during the development process. At later times, as the SED plume is transported across the pole, $E \times B$ drifts become more important to the subsequent evolution of the SED structure. These tentative results need to be confirmed and additional storms need to be studied.

5. FUTURE RESEARCH: A NEW PLASMASPHERE-IONOSPHERE DATA ASSIMILATION MODEL

In the above sections, two separate data assimilative algorithms have been described: IDA4D and EMPIRE. This work proceeded as separate algorithms and programs in order to be able to understand the effects of data on the results, sensitivity of model estimates to data sources, uniqueness and accuracy of the estimates, and whether there were cases where the system diverged under certain circumstances. Most of that analysis has taken place over the last 5–10 years. This section discusses future research plans for the development of these algorithms can take and how they might be combined together into a complete data assimilation algorithm that simultaneously estimates electron density and internal dynamics of the system, and has the first-principles model capability of predicting forward in time.

This represents the next step forward in enhancing the capability of ionospheric data assimilation to undertake various scientific investigations, as well as provide accurate nowcasts and forecasts for space weather users. There are two extensions that need to be made to IDA4D and EMPIRE in order to have an integrated predictive data assimilation algorithm. First, IDA4D needs to be extended to include the plasmasphere. This will allow for integrated measurements of ground GPS TEC all the way to the GPS satellite altitudes, topside GPS TEC from LEO satellites, and topside in situ measurements of density from satellites such as DMSP. Second, EMPIRE needs to be extended to allow direct measurements of electric fields, neutral winds, temperatures, and composition. This will allow for the optimal use of all data observations available to assimilation.

With these extensions, IDA4D and EMPIRE can be coupled together through a plasmasphere-ionosphere continuity equation model. The continuity equation model is the dynamical model that will be used as the forecast model that takes the current best estimates of the state variables and propagates them forward in time. The other state variables in the continuity equation such as neutral winds, electric fields, and temperatures will come from empirical models. However, these empirical model values will be corrected and updated through the EMPIRE algorithm. The entire linkage between IDA4D, EMPIRE, and the forward model, including the state variable estimates and error covariances, will be handled through an EnKF.

This model allows the EMPIRE estimates of winds, fields, and temperatures to be coupled to the dynamical forecast model so that it uses optimal data-driven estimates of electron densities, winds, electric fields, and temperatures to predict the densities to the next data assimilation time and to make nowcasts and forecasts.

The data assimilation algorithm will work as follows over a complete data assimilation time step: (1) At time t_n, the algorithm produces an analysis of electron density for an EnKF set of predicted ionospheres. (2) The analyzed electron densities are then ingested into EMPIRE along with direct measurements of neutral winds, electric fields, and temperatures. (3) The output of EMPIRE is an ensemble of analyzed electron densities, neutral winds, electric fields, and temperatures. (4) This ensemble set of state variables is ingested into the continuity equation dynamical forecast model, which then predicts the electron densities at the next analysis time, as well as nowcasts and forecasts. (5) Error covariances are estimated, and the assimilation cycle repeats itself. How the new integrated data assimilation algorithm is coupled together

Figure 7. (opposite) IDA4D estimates of vertical TEC over the region of EMPIRE analysis for 20 November 2003 at (a) 17:10, (b) 17:30, (c) 17:50, (d) 18:10, (e) 18:30, and (f) 18:50 UT.

is demonstrated in Figure 10. The output products for the new integrated assimilation model will include global specification of electron density, neutral winds, electric fields, and temperatures.

6. SUMMARY

In order to really understand the physical causes behind observed large- and small-scale ionospheric and thermospheric processes, it is necessary to be able to simultaneously, in a spatially and temporally extended region, obtain estimates of plasma density, composition, dynamics due to winds and electric fields and energetics due to temperatures. New ionospheric data assimilation algorithms that assimilate a variety of different types of data sources and estimate multiple ionospheric and thermosphere state variables show great promise as a diagnostic tool to advance our understanding of the causal physics.

This paper has presented the current status of two data assimilation algorithms, IDA4D and EMPIRE, that together are able to provide spatial and temporally extended estimates of electron density, neutral winds, and electric fields. These algorithms have been used in a variety of scientific studies. In particular, the combined use of EMPIRE and IDA4D in the investigation of the 20 November 2003 superstorm has provided a regional view of the vertical uplifts and drifts as well as the SED plasma structuring as the storm evolves across the United States of America. These new estimates have provided insights into the relative importance and roles of field-aligned and field-perpendicular drifts into the formation and evolution of the SED structure.

IDA4D and EMPIRE were developed separately so that the effects of data and a priori information upon the resulting accuracy and resolution of the estimated state variables could be studied and understood separately for each of these algorithms. The next step involves combining these algorithms together with a "simple-as-possible" first-principles continuity equation forward model of the ionosphere. All of these separate components will be linked together through an EnKF. The new algorithm will be able to ingest direct measurements of winds, fields, temperatures, and composition, which will continue to become more available as more small satellites fly with suites of instruments capable of measuring the ionospheric-thermospheric state along the satellite track. Of course, any such new algorithm must be extensively validated both with simulated and real-world data and must demonstrate that the resulting estimates of the plasma density and internal dynamical state variables are accurate representations of the actual physical state of the ionosphere and thermosphere. However, as new data sources continue to become available, it is anticipated that such integrated data assimilation algorithms will become more and more accurate and will be a useful diagnostic tool for scientific investigations in the ionosphere and thermosphere.

There are a number of scientific topics that are open to investigation by using IDA4D and EMPIRE together in addition to the storm studies described here. An investigation of the extended spatial-temporal distribution of topside ionospheric parameters such as electron and ion temperatures could be carried out at midlatitudes and during quiet times. For such periods, it is expected that the effect on the plasma due to $E \times B$ drifts, or field-aligned neutral winds, will be small, and the continuity equation can be approximated in terms of production, loss, and diffusion. Another scientific study would be to use IDA4D and EMPIRE to estimate the $E \times B$ drifts across the entire high-latitude region as a function of time. As touched upon earlier, if IDA4D and EMPIRE can be used to estimate gradients in the plasma density, $E \times B$ drifts, and neutral wind effects simultaneously, these large-scale values of ionospheric and thermospheric variables can be used as a diagnostic to help sort out what the

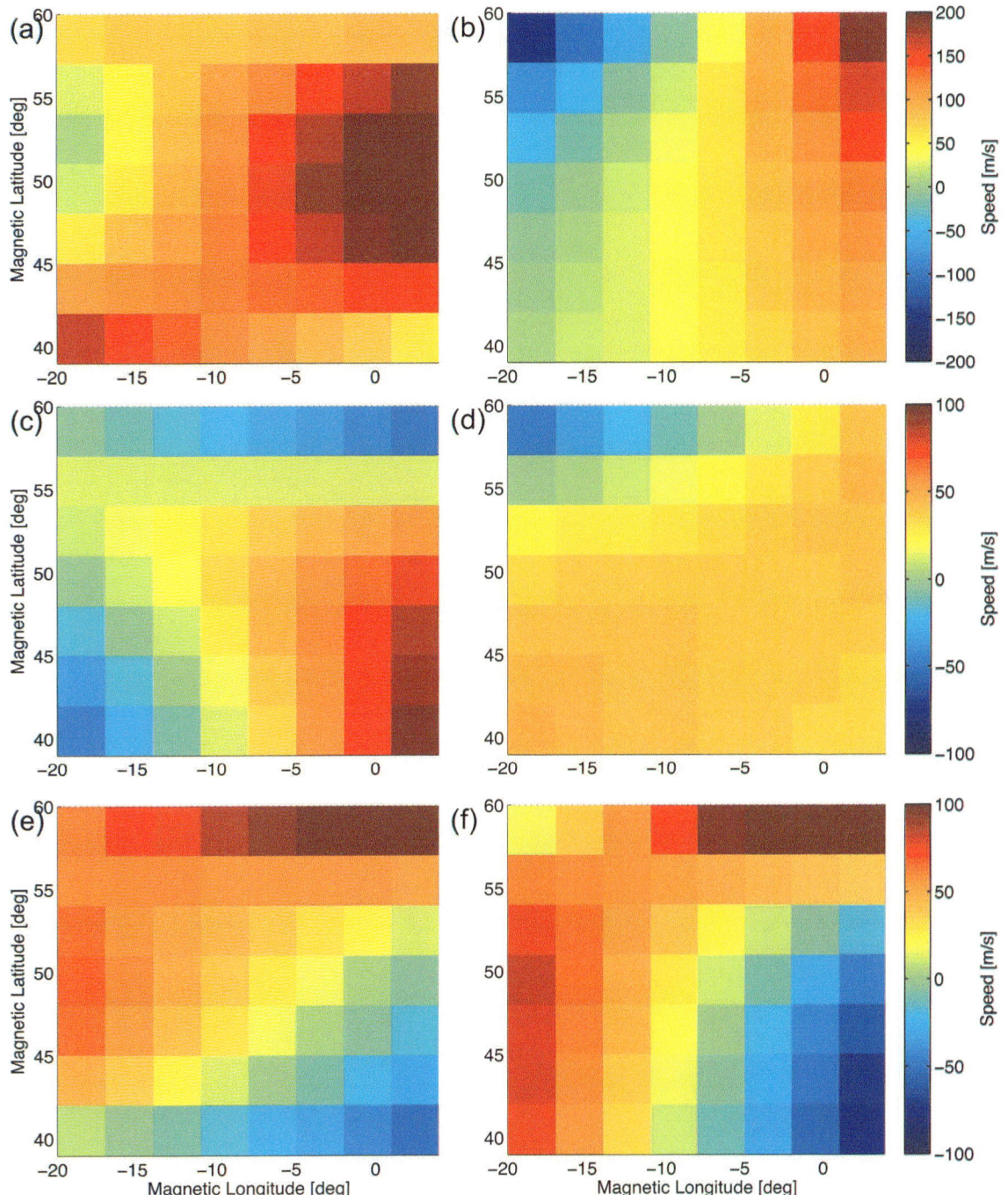

Figure 9. Vertical component of estimated field-aligned drifts at 500 km altitude on 20 November 2003 at (a) 17:10 UT, (c) 18:10 UT, and (e) 18:50 UT. Vertical component of estimated field-perpendicular drifts at (b) 17:10 UT, (d) 18:10 UT, and (f) 18:50 UT

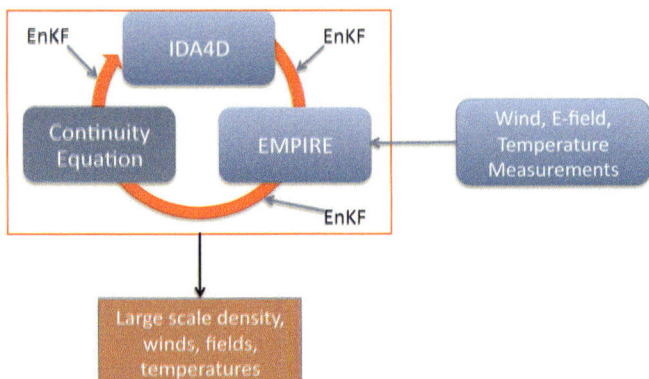

Figure 10. IDA4D-EMPIRE continuity equation ensemble Kalman filter.

instability mechanisms are that are generating observed scintillations (such as from GPS) and also to predict the growth of instabilities at both high and low latitudes.

Finally, there are a number of other data assimilation algorithms that are being developed that can also simultaneously estimate plasma density and some of the other dynamical variables. However, most of those methods have a single, integrated estimation method where various observations are ingested, and some subset of all the state variables and drivers available to the first-principle model are estimated, often through an EnKF. What is unique about the IDA4D and EMPIRE combination is the fact that they are two separate algorithms. In principle, EMPIRE could use densities estimated from any assimilation/imaging method, including 2-D to 3-D scans from ISRs. Such an approach can be considered as complementary to the more complex integrated approach. There are two advantages to such a separated approach.

First, the separate, independent development allows the study of the effects of data and assumptions on each algorithm separately. Particularly for EMPIRE, where we are using estimates of electron density to estimate external drivers, it is important to understand the sensitivity of the drivers to the data and to the approximations made in the basis expansion.

Second, the simultaneous estimation of the global distribution of plasma density and external drivers, such as neutral winds and electric fields, is an inherently difficult estimation problem. The data sources available are mostly sensitive to plasma density. Direct observations of the external drivers are very sparse. Thus, the estimation problem is underdetermined, and the "visibility" of the external drivers to the data is limited. By keeping the algorithms as simple as possible, and only including the minimal necessary physics, we reduce the number of free parameters in this underdetermined "visibility"-limited system. The concept is to only include the minimal amount of physics necessary to (1) adequately investigate the science topic under study and (2) to be able to accurately reproduce the observations. Unnecessary inclusions of model parameters can allow for overfitting of the model parameters to the "noise" in the data. Our approach avoids this problem and allows for investigation of what is the minimal number of physical parameters necessary to represent the overall system.

Acknowledgments. The authors thank Dr. Anthea Coster at Millstone Hill for assistance in use of the Millstone Hill vertical drifts. This work was funded through NSF grants 0939816, 0840650, 0710462, and 0640955. The development of IDA4D was funded, in part, by ONR grant N0001970236.

REFERENCES

Bust, G. S., and G. Crowley (2007), Tracking of polar cap ionospheric patches using data assimilation, *J. Geophys. Res.*, *112*, A05307, doi:10.1029/2005JA011597.

Bust, G. S., and G. Crowley (2008), Mapping the time-varying distribution of high-altitude plasma during storms, in *Midlatitude Ionospheric Dynamics and Disturbances*, Geophys. Monogr. Ser., vol. 181, edited by P. M. Kintner Jr. et al., pp. 91–98, AGU, Washington, D. C., doi:10.1029/181GM10.

Bust, G. S., and C. N. Mitchell (2008), The history, current state, and future directions of ionospheric imaging, *Rev. Geophys.*, *46*, RG1003, doi:10.1029/2006RG000212.

Bust, G. S., D. Coco, and J. Makela (2000), Combined Ionospheric Campaign 1: Ionospheric tomography and GPS total electron count (TEC) depletions, *Geophys. Res. Lett.*, *27*, 2849–2852.

Bust, G. S., D. S. Coco, and T. L. Gaussiran II (2001), Computerized ionospheric tomography analysis of the Combined Ionospheric Campaign, *Radio Sci.*, *36*(6), 1599–1605.

Bust, G. S., T. W. Garner, and T. L. Gaussiran II (2004), Ionospheric data assimilation three-dimensional (IDA3D): A global, multisensor, electron density specification algorithm, *J. Geophys. Res.*, *109*, A11312, doi:10.1029/2003JA010234.

Bust, G. S., G. Crowley, T. W. Garner, T. L. Gaussiran II, R. W. Meggs, C. N. Mitchell, P. S. J. Spenser, P. Yin, and B. Zapfe (2007), Four dimensional GPS imaging of space-weather storms, *Space Weather*, *5*, S02003, doi:10.1029/2006SW000237.

Coker, C., G. Kronschnabl, D. S. Coco, G. S. Bust, and T. L. Gaussiran II (2001), Verification of ionospheric sensors, *Radio Sci.*, *36*, 1523–1529.

Cooke, D., W. Turnbull, C. Roth, A. Morgan, and R. Redus (2003), Ion drift-meter status and calibration, in *First CHAMP Mission Results for Gravity, Magnetic, and Atmospheric Studies*, edited by C. Reigber, H. Lühr, and P. Schwintzer, pp. 212–219, Springer, Berlin.

Daley, R., and E. Barker (2000), The NAVDIS source book 2000: NRL atmospheric variational data assimilation system, *Tech. Rep. NRL/PU/7530-00-418*, Naval Res. Lab., Monterey, Calif.

Datta-Barua, S., G. S. Bust, G. Crowley, and N. Curtis (2009), Neutral wind estimation from 4-D ionospheric electron density images, *J. Geophys. Res.*, *114*, A06317, doi:10.1029/2008JA014004.

Datta-Barua, S., G. Bust, and G. Crowley (2010), Error propagation in ionospheric image-based parameter estimation, paper presented at 23rd International Technical Meeting of The Satellite Division of the Institute of Navigation (ION GNSS 2010), Portland, Oreg.

Datta-Barua, S., G. S. Bust, and G. Crowley (2011), Deducing storm time *F* region ionospheric dynamics from 3-D time-varying imaging, *J. Geophys. Res.*, *116*, A05324, doi:10.1029/2010JA016304.

Foster, J. C., and W. Rideout (2005), Midlatitude TEC enhancements during the October 2003 superstorm, *Geophys. Res. Lett.*, *32*, L12S04, doi:10.1029/2004GL021719.

Garner, T. W., G. S. Bust, T. L. Gaussiran II, and P. R. Straus (2006), Variations in the midlatitude and equatorial ionosphere during the October 2003 magnetic storm, *Radio Sci.*, *41*, RS6S08, doi:10.1029/2005RS003399.

Kinrade, J., C. N. Mitchell, P. Yin, N. Smith, G. S. Bust, M. J. Jarvis, D. J. Maxfield, M. C. Rose, and A. T. Weatherwax (2012), Ionospheric scintillation over Antarctica during the storm of 5–6 April 2010, *J. Geophys. Res.*, *117*, A05304, doi:10.1029/2011JA017073.

Matsuo, T., and E. A. Araujo-Pradere (2011), Role of thermosphere-ionosphere coupling in a global ionospheric specification, *Radio Sci.*, *46*, RS0D23, doi:10.1029/2010RS004576.

Mitchell, C. N., and P. S. J. Spencer (2003), A three-dimensional time-dependent algorithm for ionospheric imaging using GPS, *Ann. Geophys.*, *46*, 687–696.

Mitchell, C. N., L. Alfonsi, G. D. Franceschi, M. Lester, V. Romano, and A. W. Wernik (2005), GPS TEC and scintillation measurements from the polar ionosphere during the October 2003 storm, *Geophys. Res. Lett.*, *32*, L12S03, doi:10.1029/2004GL021644.

Pi, X., C. Wang, G. A. Hajj, G. Rosen, B. D. Wilson, and G. J. Bailey (2003), Estimation of E × B drift using a global assimilative ionospheric model: An observation system simulation experiment, *J. Geophys. Res.*, *108*(A2), 1075, doi:10.1029/2001JA009235.

Scherliess, L., D. Thompson, and R. Schunk (2009), Ionospheric dynamics and drivers obtained from a physics-based data assimilation model, *Radio Sci.*, *44*, RS0A32, doi:10.1029/2008RS004068.

Schunk, R. W., et al. (2004), Global assimilation of ionospheric measurements (GAIM), *Radio Sci.*, *39*, RS1S02, doi:10.1029/2002RS002794.

Thompson, D. C., L. Scherliess, J. J. Sojka, and R. W. Schunk (2006), The Utah State University Gauss-Markov Kalman filter of the ionosphere: The effect of slant TEC and electron density profile data on model fidelity, *J. Atmos. Sol. Terr. Phys*, *68*, 947–958.

Wang, C., G. Hajj, X. Pi, I. G. Rosen, and B. Wilson (2004), Development of the global assimilative ionospheric model, *Radio Sci.*, *39*, RS1S06, doi:10.1029/2002RS002854.

Watermann, J., G. S. Bust, J. P. Thayer, T. Nuebert, and C. Coker (2002), Mapping plasma structures in the high latitude ionosphere using beacon satellite, incoherent scatter radar and ground-based magnetometer observations, *Ann. Geophys.*, *45*, 177–189.

Yin, P., C. N. Mitchell, and G. S. Bust (2006), Observations of the *F* region height redistribution in the storm-time ionosphere over Europe and the USA using GPS imaging, *Geophys. Res. Lett.*, *33*, L18803, doi:10.1029/2006GL027125.

G. S. Bust, Atmospheric and Space Technology Research Associates, Boulder, CO, USA. (Gary.Bust@jhuapl.edu)

S. Datta-Barua, Department of Aviation and Technology, San Jose State University, San Jose, CA, USA.

Customers and Requirements for Ionosphere Products and Services

Rodney Viereck, Joseph Kunches, Mihail Codrescu, and Robert Steenburgh

NOAA Space Weather Prediction Center, Boulder, Colorado, USA

Many space weather impacts on the terrestrial environment result from changes in the ionosphere. Some systems, such as HF radio communication systems, are impacted by enhancements in the D region ionospheric densities. Other systems are impacted by changes in the E and F regions. Other systems, such as single-frequency GPS or Global Navigation Satellite Systems (GNSS), are affected by changes in the total height-integrated electron content. While dual-frequency GPS/GNSS systems provide corrections for many different types of ionospheric variability, they are susceptible to small-scale ionospheric irregularities that cause scintillation of the signal. This report presents an overview of space weather that affects the ionosphere. Ionospheric impacts on various technical systems and the requirements of different customer types for ionospheric services that rely on the National Oceanic and Atmospheric Administration Space Weather Prediction Center (SWPC) for specification and forecasts of the space environment will also be presented. An overview of the current products in operations and in development will be discussed, and a new section within SWPC, the Space Weather Prediction Testbed, will be described. The testbed is where new products and models are tested and validated to assess their performance and their utility to customers and the Forecast Office.

1. INTRODUCTION

Space weather and ionospheric variability impact a number of different types of systems ranging from radio communication to satellite navigation. Users of these systems receive support from the NOAA Space Weather Prediction Center (SWPC) in the form of data, products, and forecasts. To better serve these customers, SWPC has solicited suggestions and advice on what products are needed and at what levels of disturbance various systems are impacted. This paper provides a review of the different types of space weather and how they impact the ionosphere. The impacts on some of the more important technologies and systems are also described. The paper then reviews several models that SWPC currently uses to specify and forecast the environment. And in the last section, the paper describes the Space Weather Prediction Testbed, which was established to facilitate the development, testing, validation, and implementation of new models to improve support to the customers of ionospheric products and services.

2. SPACE WEATHER IMPACTS

Space weather comes in several forms. Nearly all types of space weather affect the ionosphere-thermosphere (I/T) along with the systems that are impacted by ionospheric conditions. In this section, we describe the various space weather categories and their impacts on the ionosphere.

The first sign of impending space weather storms usually comes in the form of a solar flare. During a large flare, the solar X-ray emissions can increase by 4 or 5 orders of magnitude in just a few minutes. These X-rays penetrate the lower ionosphere, enhancing the ionospheric D region density by orders of magnitude. This D region enhancement modifies the reflective properties of the ionosphere for HF radio transmissions, often absorbing or blocking the transmissions altogether. As the lower ionospheric D region electron densities increase, the lower range of HF frequencies (5–10 MHz) is absorbed. As the X-ray flux gets stronger, higher frequencies are absorbed. At high X-ray flux levels, frequencies up to 35–40 MHz are absorbed, and no HF radio transmissions are possible. Other HF communication paths are impacted by changes in the E and F regions. Changes in these layers are driven by ionospheric dynamics more than by solar flares. The densities of the E and F layers contribute significantly to the total electron content (TEC) so this becomes important to GPS users.

While this increase in D region density is significant relative to the background D region densities, it is small compared to the overall ionospheric density and the TEC, which is the height-integrated density of the entire ionosphere, so solar flares have minimal impact on GPS navigation systems.

Sequentially, the next category of space weather that impacts the I/T system comes from the energetic protons that are accelerated near the Sun and arrive at the Earth within a few minutes to hours of the flare-induced X-rays. These protons enter the Earth's upper atmosphere near the polar regions, guided by the solar and terrestrial magnetic fields. As with the X-rays, these energetic protons penetrate to the lower ionosphere and enhance the D region, resulting in the absorption of HF radio waves in the polar regions. This blocks radio HF transmissions. The latitudinal extent of the affected region depends on the state of the magnetosphere; a more disturbed geomagnetic situation allows for protons to have access to lower latitudes.

The third major space weather category is produced when slower-moving coronal mass ejections (CMEs) arrive at the Earth 1 to 4 days after the solar flare erupts. The interaction between the CME and the magnetosphere results in enhanced currents flowing in the ionosphere and the deposition of large amounts of energy in the high-latitude I/T system. The heating enhances mixing, sending molecular species to higher altitudes and driving strong storm time neutral winds. The changes in both neutral composition and winds then drive enhancements and depletions in ionospheric densities that result in sharp gradients. The changes caused by geomagnetic storms affect not only the systems that rely on the ionosphere for normal operation but also the low-orbit satellites, due to increased drag. Often, the heating produces wavelike traveling ionospheric disturbances that result in rapidly changing conditions and the formation of large gradients at midlatitudes. Large geomagnetic storms can cause considerable variations in TEC. These variations have a direct impact on the propagation of GPS/Global Navigation Satellite Systems (GNSS) signals and results in increased errors, especially in the single-frequency systems.

Another ionospheric feature that impacts GPS/GNSS customers is more strongly correlated with normal diurnal behavior rather than major storms. In the E region of the ionosphere, tidal winds drive an eastward dynamo current in the sunlit hemisphere. The resulting eastward electric field, $\mathbf{E}$, is perpendicular to the geomagnetic field, $\mathbf{B}$. The resulting $\mathbf{E} \times \mathbf{B}$ drift drives the plasma upward. The plasma then diffuses along $\mathbf{B}$ field lines under the influence of gravity and pressure gradients. As the Earth rotates into darkness, the eastward electric field increases in response to increasing neutral winds about an hour after sunset. The upward $\mathbf{E} \times \mathbf{B}$ drift intensifies, raising the F layer. Photoionization is greatly diminished after sunset, and the F region is quickly eroded from below as recombination takes place. The combination of lift and bottomside erosion results in steep upward density gradient forces at the base of the F layer, directly opposed by gravity. This configuration is described by Rayleigh-Taylor instability, in which a heavier fluid is supported by a lighter fluid. If perturbed, the fluid turns over and creates low-density structures (bubbles) that rise through the denser fluid. These small-scale structures interfere with the GPS/GNSS signal, causing them to lose lock and provide little or no useful location information. It should be noted that these ionospheric structures also interfere with transionospheric radio communications such as ground-to-space or satellite communications. The structures can create multipath signals that interfere with each other and render the message unintelligible.

The final driver of ionospheric variability does not come from space but from the atmosphere below. It has been shown that planetary waves manifest themselves in the ionospheric airglow [*Immel et al.*, 2006]. Other features of the troposphere such as tides and gravity waves will also propagate upward to the I/T region. Tropospheric forcing of the ionosphere is most important at midlatitudes and low latitudes and may play an important role in the generation of small-scale ionospheric structures that affect the GPS/GNSS systems.

3. SYSTEM IMPACTS AND CUSTOMERS

The two major systems impacted by perturbations in the ionosphere are HF radio propagation and GPS/GNSS satellite navigation systems.

Some examples of industries and activities that rely on accurate and reliable GPS/GNSS services are commercial airlines, precision agriculture, oil exploration, surveying, maritime navigation, and military activities. A recent workshop [*Fisher*, 2011a] and the resulting reports [*Fisher*,

2011b; *Fisher and Kunches*, 2011] highlight the economic impact of GPS/GNSS in today's society. The reports categorize the impacts on GPS/GNSS services in terms of single-frequency errors associated with ionospheric gradients (temporal and spatial) and loss of signal lock associated with small-scale ionospheric structures and scintillation. A growing number of technologies rely on GPS/GNSS for precision timing information. For the most part, the customers require accuracies of only 100 ns or greater, and ionosphere-induced errors in the signal do not impact this user group. A 100 ns timing error is approximately equivalent to 30 m position error although the two are not equivalent. The technologies that require 100 ns timing or better include military munitions and radar, scientific research, and the international time and frequency monitoring facilities. This group of GPS/GNSS users can be impacted by geomagnetic storms and other forms of space weather.

HF radio propagation can be disrupted when the lower part of the ionosphere, the *D* region, is enhanced either by increased X-ray irradiance from solar flares or due to enhanced energetic particle flux associated with solar proton events. The impact of space weather on HF radio communications has been known since the early days of radio in the 1940s. There are many industries and organizations that rely on HF radio. Amateur radio operators are the largest group, and while they use HF radio communications mostly for recreation, they are often asked to assist government agencies, like the Department of Homeland Security and the Federal Emergency Management Administration, during times of crisis. These agencies rely on HF radio communication as a backup to other communication systems during disasters. The commercial airlines and the Federal Aviation Administration rely on HF radio as their primary mode of air-to-ground communication for transoceanic and over-the-poles flights. In most regions of the Earth, airlines use satellite communication as a backup, but this requires a clear line-of-sight between the aircraft and the geosynchronous satellite that is not possible for flights poleward of 82° latitude. For these polar flights, HF radio is the only communication between aircraft and the ground, and regulations do not allow the airlines to fly at these high latitudes when space weather conditions block HF radio signals. Some airlines are exploring alternatives such as a low Earth orbit iridium communication system. The phenomena that disrupt HF radio signals are well understood and easily specified; forecasting *D* region enhancements requires a forecast of X-rays or energetic protons. Accurate predictions of both phenomena continue to elude scientists and forecasters.

A third critical space weather impact within the I/T region is increased neutral density and satellite drag that accompanies increased geomagnetic activity or increased solar EUV irradiance. This space weather impact is important, but it will not be discussed further in this report.

Customer requirements and ionospheric phenomena are listed in Table 1. Some industries, for example, precision agriculture and land surveyors, will be satisfied with products over the continental United States. Most customers of space weather products and services, including aviation, maritime, and oil exploration, have activities that are not limited to the United States, so global products are required. Nearly all industries would like to have accurate forecasts of conditions with multihour or multiday lead times. These requirements, shown in Table 2, are very challenging, and current scientific understanding is not mature enough to meet many of them.

The needs and requirements of customers of ionospheric services are quite varied. In general, most customers would like a forecast of when their systems will be impacted beyond a specific threshold. They would also like an alert that indicates when that threshold has been reached. Some users compensate for ionosphere-induced errors by using local reference stations to improve accuracies. For precision agriculture or precision surveying, the accuracy requirements are often less than 10 cm. For deep sea drilling, the position accuracy may be 100 cm or more. For airlines that must plan in advance, a 24 h flare forecast for HF communication would be very helpful. Unfortunately, there is inadequate understanding of flare physics to provide a 24 h forecast of flare timing and magnitude. Forecasters can only give a probabilistic forecast for flares of a particular magnitude occurring. Current observational techniques and

Table 1. A List of Customer Requirements for the Ionosphere-Thermosphere Environment and the Related Ionospheric Phenomena That Addresses Them

Customer Requirement	Ionospheric Phenomena

Specification and forecasts of GPS/Global Navigation Satellite Systems (GNSS) single-frequency accuracy	• Total electron content (TEC)
	• TEC gradients
	  ◦ Temporal
	  ◦ Spatial

Specification and forecasts of GPS/GNSS dual-frequency availability Ground-to-space radiocommunication	• Scintillation (small-scale structures)

Specification and forecast of HF communication conditions	• *D* region density
	  ◦ Flare enhanced
	  ◦ Proton enhanced
	• *E* and *F* region densities

Table 2. GPS/GNSS Customer Needs and Requirements, Impacted by Space Weather, as Determined Through a Space Weather Prediction Center Survey of Customer Requirements

User Group	United States TEC (USTEC)	US-TEC Forecast	Global TEC	Global TEC Forecast	Scintillation
Agriculture		X	X	X	X
Surveying	X	X	X	X	X
Maritime			X	X	X
Aviation		X	X	X	X
General users	X		X	X	X
GPS system operations			X	X	
Ionospheric correction			X	X	X
Receiver design			X	X	X
Time/frequency	X			X	X

assimilative models may be able to provide alerts as to when the GPS accuracies are being impacted by space weather, but full physics-based models with predictions of all the required drivers (solar EUV irradiance, geomagnetic disturbances, and tropospheric waves), is what is really required for global forecasts of GPS impacts. These models are under development.

4. OPERATIONAL MODELS

The NOAA SWPC has made progress toward meeting the needs of customers whose systems are impacted by the ionosphere. There are currently several products available in operations that address some customer requirements, but there are many that have not been met.

There are several models currently running at SWPC that are considered "operational." In this document, the term "operational" refers to the application of a model or process in the official setting of the Space Weather Forecast Office rather than a model or product in a research or development environment. Operational models typically run in real time and produce outputs for use by forecasters or customers. An operational model has usually undergone rigorous testing and validation such that its performance is reliable and predictable. Operational models have usually been modified with additional internal checking and software documentation such that maintenance, repairs, and upgrades can be made by technical staff rather than researchers.

The global *D* Region Absorption Product is another operational model that provides HF communication users with spatial maps showing the degradation of HF signals at different frequencies between 5 and 35 MHz (see Figure 1). This product requires the current solar X-ray flux, the energetic proton flux at geosynchronous orbit, and the most recent value of the planetary geomagnetic disturbance index, Kp, as input. For GPS/GNSS customers who require estimates of the current conditions over the United States, the U.S. Total Electron Content or USTEC model, shown in Figure 2, was developed [*Mitchell and Spencer*, 2002; *Araujo-Pradere et al.*, 2007]. This product uses the International Reference Ionosphere (IRI) model for the background ionospheric condition and, then, assimilates ground GPS data from around the United States to modify the maps of TEC in real time. This product provides a snapshot of the current ionospheric conditions that impact single-frequency GPS systems.

Other models that are currently in operations or used operationally in the Forecast Office include the Wing Kp prediction model [*Wing et al.*, 2005] and the storm time IRI model [*Fuller-Rowell et al.*, 2001; *Araujo-Pradere et al.*, 2002].

5. THE SPACE WEATHER PREDICTION TESTBED

In response to customer needs and requirements, SWPC is developing new ionospheric products and services and is working to greatly improve forecasts of critical ionospheric conditions for the growing number of customers in these areas. Current activities include the collection of new data sets and the coordination of worldwide ionospheric services through organizations such as the International Space Environment Services and the World Meteorological Organization.

One of the challenges that all operational organizations face is moving scientific discovery into the operational Forecast Office. Most research models require extensive modification and improvement to make them reliable and consistent enough to operate in an operational center such as the SWPC Forecast Office.

To address the issue of transitioning models and products from research to operations, NOAA, the National Weather Service, and SWPC have established the Space Weather Prediction Testbed (SWPT). The concept of a testbed is not new; there are currently nine other testbeds in NOAA supporting operational centers in areas such as aviation, climate,

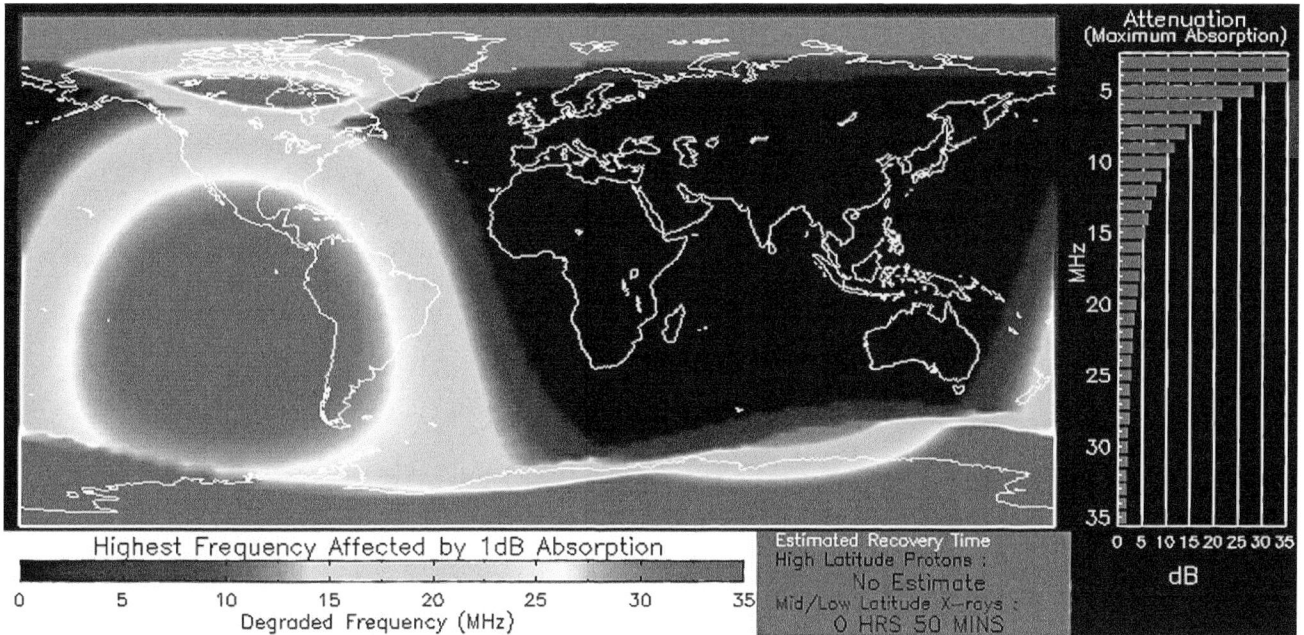

Figure 1. An example of the D Region Absorption Product used for identifying where HF communications will be impacted by solar flares and energetic particle events.

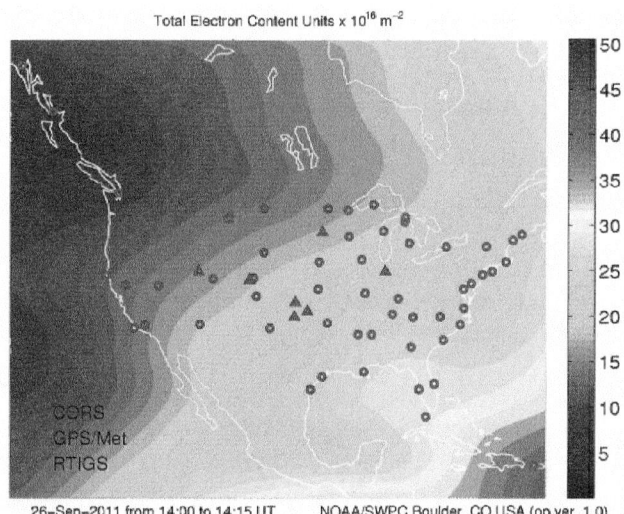

Figure 2. An example plot of the US Total Electron Content (USTEC) product running operationally at Space Weather Prediction Center. This product provides a specification of the current TEC over the United States.

and hurricanes. These testbeds have been established to span the gap between research and operations.

The staff of the SWPT is currently working on four primary projects and several smaller ones.

In the area of solar-heliospheric space weather forecasting, the SWPT has transitioned the Wang-Sheely-Arge (WSA)-Enlil model [*Arge et al.*, 2003; *Odstrcil*, 2003] from research to operations (see Figure 3). This is the first physics-based forecast model to be run operationally at the National Centers for Environmental Prediction computer facility. This model runs every 2 h alongside the weather models that predict traditional meteorological conditions. The WSA-Enlil model provides a 1–4 day forecast of solar wind density and speed at the Earth. This allows the forecasters to more accurately predict the arrival time of CMEs and gives guidance for the magnitude of the resulting geomagnetic storms that are one of the key drivers of I/T variability. One missing component in this model is an accurate estimate of the magnitude and direction of the interplanetary magnetic field within a CME. The geoeffectivness of a particular CME depends strongly on the vertical, or z component, of this field (**Bz**). Without a forecast of **Bz**, the uncertainty of the actual impact of a particular CME is substantially larger than if **Bz** forecasts were accurate and timely.

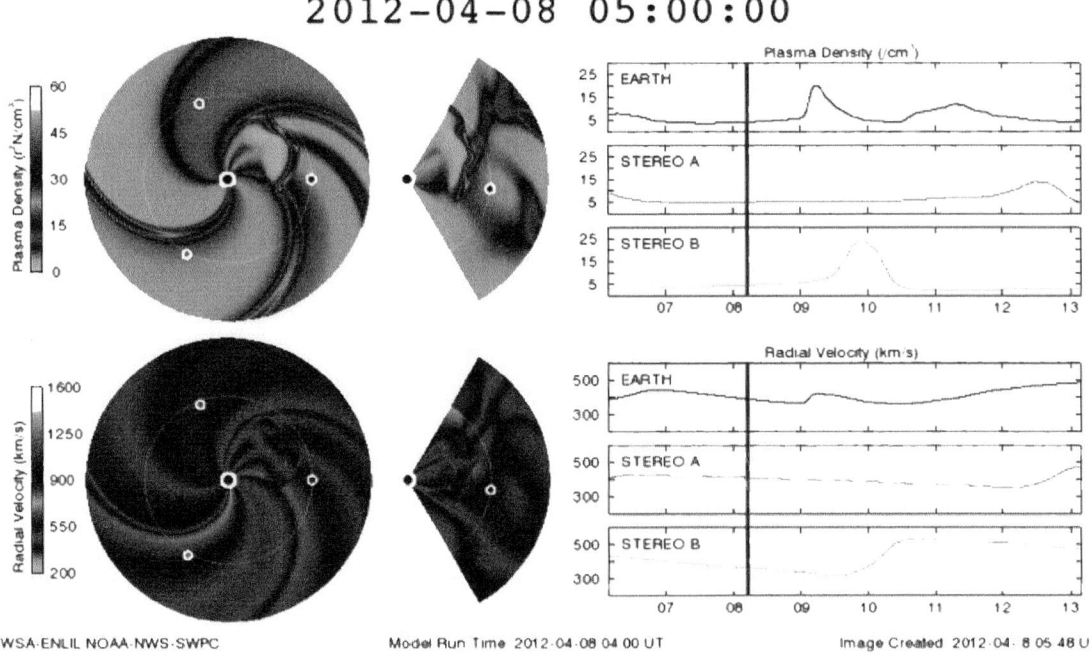

Figure 3. An example plot of the output from the Wang-Sheeley-Arge-Enlil model of the solar heliosphere. The output from this model is a forecast of the (top) solar wind density and (bottom) speed near the Earth.

A second model SWPT is examining for transition to operations, which is the Ovation Prime model developed by *Newell et al.* [2007], [2009], and [2010]. This model, shown in Figure 4, is based on an empirical relationship between solar wind and interplanetary magnetic field conditions at the first Lagrange point (L1) between the Sun and the Earth, and the precipitation auroral electrons and protons in the high latitudes of the Earth. The model provides a short-term (30–45 min) forecast of the particle flux at high latitudes that produce aurora. In addition to the spatial maps of electron and proton fluxes in the high-latitude upper atmosphere, the Ovation Prime model also provides an estimate of the integrated hemispheric power from the particles; this is an input to many ionospheric models. The Ovation model is currently undergoing test and evaluation and is currently available as a test product.

Customers who operate electric power grids have requested higher fidelity, localized specification, and forecasts of the geomagnetic phenomena that impact their systems. In particular, they would like the rate of change in the local magnetic field along the path of individual power lines and the resulting electric fields that cause the unintended currents on these power systems. These changes in the local electromagnetic fields are driven by currents that originate in the magnetosphere. To specify and forecast impacts on the electric power grids, it is necessary to have a magnetospheric model that is driven by solar wind conditions. To address these requirements, NOAA and NASA are currently collaborating through the SWPT and the NASA Community Coordinated Modeling Center (CCMC) to identify and evaluate a magnetospheric model suitable for transition to operations.

There are several physics-based and empirical models of the magnetosphere that are being considered. The CCMC has collected the models and is running them each through a series of space weather events to see how well they perform. The results of this comparison will be used to identify the best candidate for transition from research to operations. The selection criteria are based not only on performance and accuracy during space weather events as well as during quiet and moderate activity levels. It will also be important that the model run in real time and run reliably through periods where the input data are less than optimal, as is often the case when running in real time. The model must also be written and documented clearly in order to make it supportable in an operational setting. It must also run within the constraints of the computer resources that are available for operational support. An additional criteria is that the model developer must be willing to provide support to SWPC during the transition and operation of the model. The model developer must also provide updates and improvements or allow others in the research and operations community to do so. The goal will be to have a magnetospheric model running

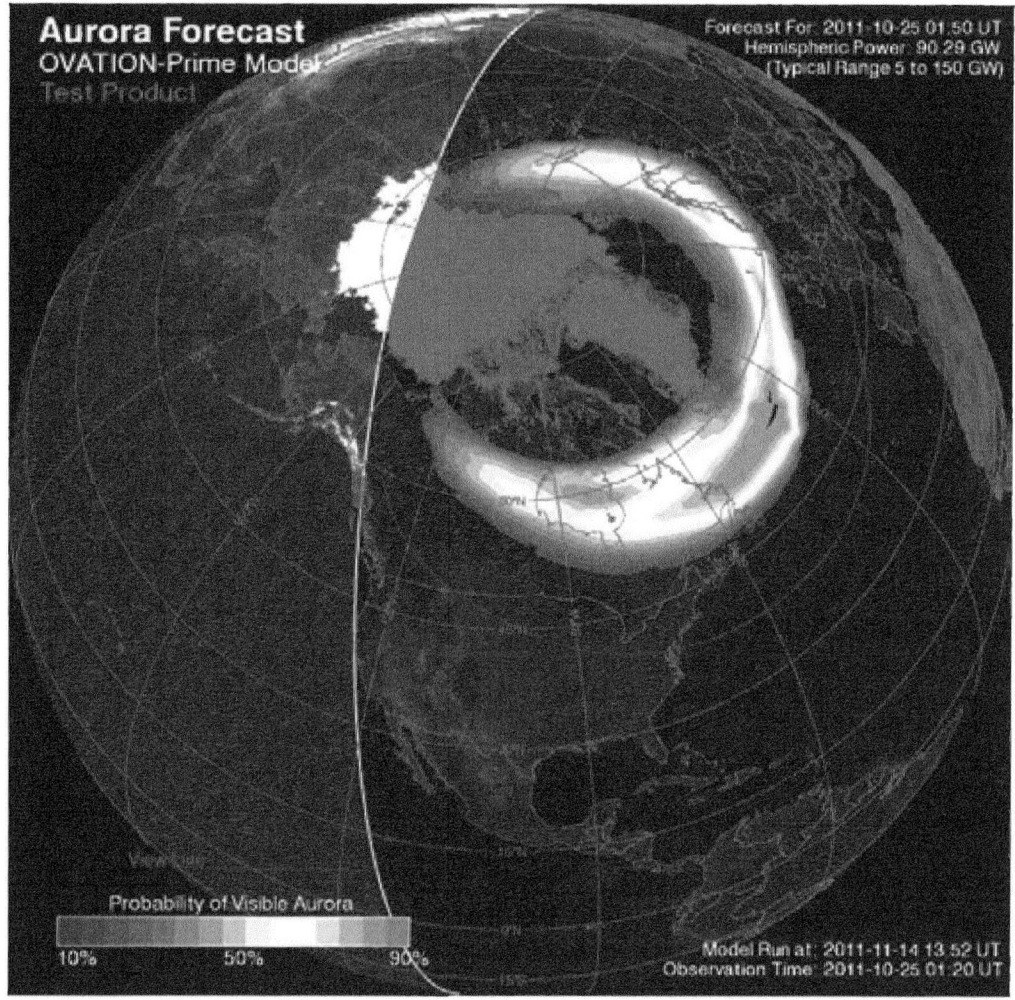

Figure 4. An example image of the Ovation Prime Auroral Forecast model currently under development at the Space Weather Prediction Testbed. This product provides 30–45 min forecasts of the location and intensity of the aurora.

in real time and providing estimates of geomagnetic activity. This model will provide not only products for the electric power industry but also will make better forecasts of the spatial and temporal nature of the energy flux into the ionosphere and thermosphere available. This can then be used to drive models of the I/T system.

The third and most ambitious project at the SWPT is the development of the Whole Atmosphere Model (WAM) coupled to the Ionosphere Plasmasphere Electrodynamics (IPE) model. This effort, under the name Integrated Dynamics in Earth's Atmosphere (IDEA), started with the operational Global Forecast System (GFS) model currently providing tropospheric weather forecasts at the National Weather Service's National Centers for Environmental Prediction. The GFS model was extended from 60 km altitude (the current operational upper limit) up to 600 km, which provides improved specification and forecasts of the neutral density in the thermosphere. The GFS model generates multiday forecasts of the atmospheric conditions, including the tidal and planetary waves that are known to propagate upward to the thermosphere. Details of this model are provided by *Akmaev* [2011]. Current efforts underway by researchers include coupling this model to the IPE model. This will provide a complete ground-to-space model of the atmospheric forcing of the IT system.

It is hoped that in the future, all of the critical drivers of the I/T system will be predicted by operational models at SWPC. The WSA-Enlil model driving the Ovation Prime and geospace models will provide forecasts of geomagnetic storms and their input into the I/T system. The WAM model will provide forecasts of the lower atmosphere and the forcing of the IT system from below. The third and final input is the solar EUV irradiance.

Current activities at the SWPT include the introduction of improved solar EUV observations from the operational NOAA GOES. It is hoped that these new EUV data, along with scientific discoveries from other missions such as the Solar Dynamics Explorer (SDO) and the Extreme Ultraviolet Experiment (EVE) that is on the SDO satellite, will provide much better EUV forcing for the IDEA model and the capability of forecasting solar EUV irradiance several days in advance. This would then provide predictions of the third and final driver of the I/T system.

Other space weather forecasting tools being worked on by the SWPT include the Forecasting Ionospheric Real-Time Scintillation Tool for equatorial ionospheric scintillation forecasts [Redmon et al., 2011] and an expansion of the USTEC model [Araujo-Pradere et al., 2002] of ionospheric TEC from the current coverage of just the Continental United States to include Alaska, and Canada.

One challenge that operational forecasters are currently facing is that customers need scintillation products and services that are not available. In particular, there is a demand for localized dual-frequency GPS availability and forecasts of scintillation impacts at high latitudes. There are a number of applications for precision GPS at high latitudes. The transportation industry is using GPS for polar shipping routes and for low-visibility aircraft and vehicle travel. The oil industry is moving further into the Arctic looking for new oil reserves to tap. This demand is complicated by the fact that GPS reliability becomes more problematic at high latitudes. It is known that there are correlations between geomagnetic activity and GPS reliability. The results of a study done at SWPC can be seen in Figure 5. Figure 5 shows the correlation between GPS cycle slips, a measure of the ionospheric stress on the GPS signal, and the planetary geomagnetic index (Ap). Clearly, the higher the geomagnetic activity level, the higher the number of cycle slips. The research goal, then, is to develop products for high-latitude GPS users that would provide specification and forecasts of the level of ionospheric disturbance and scintillation.

6. CONCLUSION

The NOAA SWPC, the SWPT, continue to work to improve ionospheric products and services, with a goal to provide accurate forecasts of global ionospheric conditions that impact systems and technologies.

Unlike terrestrial weather, the ionosphere is highly driven and responds quickly to three primary external drivers: (1) solar EUV irradiance, (2) high-latitude heating due to geo-

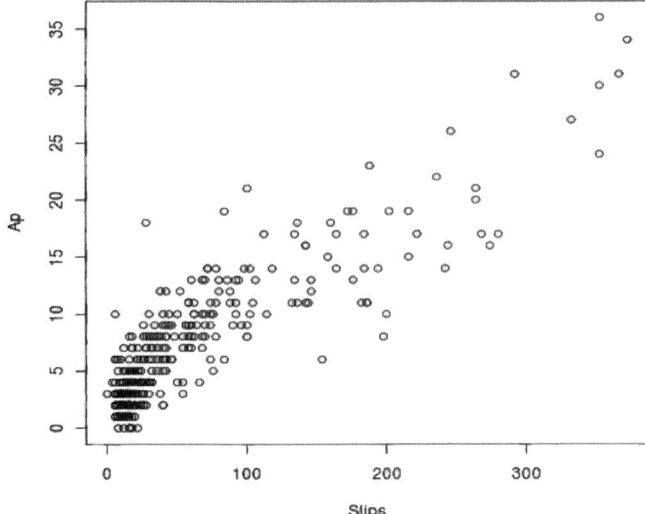

Figure 5. A plot of the relationship between GPS dual-frequency cycle slips and the Ap geomagnetic index.

magnetic storms, and (3) waves, tides, and variability from the lower atmosphere. There is also some level of internal variability and persistence that must be captured when trying to forecast the ionosphere.

Forecasting ionospheric conditions requires reliable forecasts of the three external drivers. Predictions of solar EUV irradiance can be done reasonably through empirical relationships, persistence, and the recurrence of solar active regions on a 27 day solar rotation rate. This, however, does not capture the episodic events associated with solar flares. Forecasting flares is very difficult and is usually done with probabilistic forecasts for achieving specific flare magnitudes over the next 1 to 3 day period. Forecasts of geomagnetic storms have improved in the last few years. The introduction of the WSA-Enlil model into operations has provided 1–4 day forecasts of the arrival of CMEs at the Earth. Predicting the third component of ionospheric forcing requires forecasts of the tropospheric weather and forcing from below. It is now clear that ionospheric models must be coupled with whole atmosphere models to fully capture this third ionospheric driver. It is hoped that coupling an atmospheric model with an ionosphere model will provide new capabilities in specifying and forecasting ionospheric conditions.

The newly formed SWPT has been tasked with developing, validating, and testing new models and model improvements. Taking advantage of scientific advances and new data sets, and making them useful to the operational Space Weather Forecast Office (part of SWPC), is critical to providing customers with new and improved products and services.

Moving research into operations is a primary function of the SWPT. It should also be noted that translating the requests and needs of the customers into requests for new research and scientific development is another function of the SWPT. All of these activities are part of the mission of this organization to span the gap between research and operations.

REFERENCES

Akmaev, R. A. (2011), Whole atmosphere modeling: Connecting terrestrial and space weather, *Rev. Geophys.*, *49*, RG4004, doi:10.1029/2011RG000364.

Araujo-Pradere, E. A., T. J. Fuller-Rowell, and M. V. Codrescu (2002), STORM: An empirical storm-time ionospheric correction model 1. Model description, *Radio Sci.*, *37*(5), 1070, doi:10.1029/2001RS002467.

Araujo-Pradere, E. A., T. J. Fuller-Rowell, P. S. J. Spencer, and C. F. Minter (2007), Differential validation of the US-TEC model, *Radio Sci.*, *42*, RS3016, doi:10.1029/2006RS003459.

Arge, C. N., D. Odstrcil, V. J. Pizzo, and L. R. Mayer (2003), Improved method for specifying solar wind speed near the Sun, *AIP Conf. Proc.*, *679*, 190–193.

Fisher, G. (2011a), Satellite navigation and space weather: Understanding the vulnerability and building resilience. Report of a policy workshop, Am. Meteorol. Soc. Policy Program, Washington, D. C.

Fisher, G. (2011b), Workshop builds strategies to address Global Positioning System vulnerabilities, *Space Weather*, *9*, S01001, doi:10.1029/2010SW000639.

Fisher, G., and J. Kunches (2011), Building resilience of the Global Positioning System to space weather, *Space Weather*, *9*, S12004, doi:10.1029/2011SW000718.

Fuller-Rowell, T., M. Codrescu, and E. Araujo-Pradere (2001), Capturing the storm-time F-region ionospheric response in an empirical model, in *Space Weather*, Geophys. Monogr. Ser., vol. 125, edited by P. Song, H. J. Singer, and G. L. Siscoe, pp. 393–401, AGU, Washington, D. C., doi:10.1029/GM125p0393.

Immel, T. J., E. Sagawa, S. L. England, S. B. Henderson, M. E. Hagan, S. B. Mende, H. U. Frey, C. M. Swenson, and L. J. Paxton (2006), Control of equatorial ionospheric morphology by atmospheric tides, *Geophys. Res. Lett.*, *33*, L15108, doi:10.1029/2006GL026161.

Mitchell, C., and P. Spencer (2002), Development of tomographic techniques for large scale ionospheric imaging, in *Proceedings of the 2002 Ionospheric Effects Symposium*, edited by J. M. Goodman, pp. 601–608, JMG Assoc., Springfield, Va.

Newell, P. T., T. Sotirelis, K. Liou, C.-I. Meng, and F. J. Rich (2007), A nearly universal solar wind-magnetosphere coupling function inferred from 10 magnetospheric state variables, *J. Geophys. Res.*, *112*, A01206, doi:10.1029/2006JA012015.

Newell, P. T., T. Sotirelis, and S. Wing (2009), Diffuse, monoenergetic, and broadband aurora: The global precipitation budget, *J. Geophys. Res.*, *114*, A09207, doi:10.1029/2009JA014326.

Newell, P. T., T. Sotirelis, and S. Wing (2010), Seasonal variations in diffuse, monoenergetic, and broadband aurora, *J. Geophys. Res.*, *115*, A03216, doi:10.1029/2009JA014805.

Odstrcil, D. (2003), Modeling 3-D solar wind structure, *Adv. Space Res.*, *32*, 497–506.

Redmon, R. J., D. Anderson, R. Caton, and T. Bullett (2011), A Forecasting Ionospheric Real-time Scintillation Tool (FIRST), *Space Weather*, *8*, S12003, doi:10.1029/2010SW000582.

Wing, S., J. R. Johnson, J. Jen, C.-I. Meng, D. G. Sibeck, K. Bechtold, J. Freeman, K. Costello, M. Balikhin, and K. Takahashi (2005), Kp forecast models, *J. Geophys. Res.*, *110*, A04203, doi:10.1029/2004JA010500.

M. Codrescu, J. Kunches, R. Steenburgh, and R. Viereck, NOAA Space Weather Prediction Center, Boulder, CO, USA. (Rodney.Viereck@noaa.gov)

Model-Based Inversion of Auroral Processes

Joshua Semeter

Department of Electrical and Computer Engineering and Center for Space Physics, Boston University, Boston, Massachusetts, USA

Matthew Zettergren

Physical Sciences Department, Embry-Riddle Aeronautical University, Daytona Beach, Florida, USA

Self-consistent physical models of an auroral flux tube are capable of calculating optical emissions and ionospheric state variables for a given set of magnetospheric drivers. The collocation of incoherent scatter radar (ISR) and photometric measurements provides a powerful framework for testing the internal consistency of such models. This article presents a framework for using flux tube models in an inverse theoretic sense. Specifically, we consider the problem of estimating the spectrum of primary electrons via inversion of auroral spectral brightnesses in the near infrared (600–900 nm), where prompt and near-prompt emissions of atomic oxygen and molecular nitrogen provide maximum sensitivity to the incident energy spectrum over a limited and contiguous free spectral range. A linear forward model is constructed by applying a set of monoenergetic electron beams to the TRANSCAR model and computing measurable quantities. Inversion is performed using maximum entropy regularization. Internal consistency is tested by comparing observed ISR state variables (N_e, T_e, T_i, V_i), with predictions of the same variables derived by optical inversion. The good agreement suggests that unobserved, but internally consistent, state variables may also be reliably predicted. These include species-specific upflow rates and ion composition changes. Some future directions are suggested.

1. INTRODUCTION

Among the most significant achievements in the field of aeronomy has been the development of self-consistent, first-principles, physical models for the deposition of solar and magnetospheric energy into the ionosphere-thermosphere system [*Strickland et al.*, 1976; *Lummerzheim and Lilensten*, 1994; *Blelly*, 1996]. At their essence, such models take, as input, a source of free energy (e.g., heat flux, electromagnetic flux, precipitation, solar radiation) and compute, as output, changes in the ionosphere-thermosphere (I-T) state (e.g., densities, temperatures, bulk motions, photometric emission rates). A further set of models can be used to map these intrinsic parameters to quantities measured by ground-based instruments. For the case of optical emissions, this involves computing line-of-sight projections through the emitting region, and modeling the response of the optical sensor. For the case of plasma state parameters, these may be resolved as a function of range by an incoherent scatter radar (ISR), a technique that involves the nontrivial task of estimating and fitting the intrinsic fluctuation spectrum of the plasma [*Evans*, 1969].

This article considers the invertibility of this forward process as it pertains to an auroral flux tube. Namely, given a set of heterogeneous ground-based observations, can we reliably quantify the magnetospheric source? Such techniques are important because they can provide a time-dependent description of magnetospheric energy flux in an Earth-fixed reference frame: information that is unobtainable from a rocket or orbital platform.

Our goal is to provide a generic mathematical framework for the combined analysis of optical and ISR observations of the aurora. Our primary focus is on the problem of estimating the flux of auroral primary electrons as a function energy and time. Our approach exploits an electron transport model to produce a set of characteristic response functions over a range of monoenergetic electron beams. Coefficients in this basis correspond to discrete samples of the incident differential number flux spectrum. The major challenge in using ISR and optical measurements collaboratively is time history. Optical spectroscopy provides a measure of energy deposition rate. An ISR, on the other hand, measures the resulting change in plasma state, computed via a continuity relation.

There are a variety of ways to combine this information to quantify magnetospheric drivers. In this chapter, we use measurements of prompt emission features in the near-infrared (NIR) to estimate the incident energy spectrum and then apply these results as a time-dependent boundary condition to our transport model in order to compute the plasma state response which is, in turn, compared with measurements by the ISR. The lack of in situ validation has limited the applicability of such approaches for discovery science. But we argue that the combined use of radar and optical measurements in this way provides a framework for testing the internal consistency of the approach. To the extent that all the physics is included in a self-consistent manner, cross-validation of radar and optical measurements suggests that we may be able to reliably estimate state parameters that are not directly observed by either diagnostic. Such parameters include species-specific ion transport and changes in ionospheric composition.

2. OVERVIEW OF THE FORWARD PROBLEM

The polar ionosphere is subject to fluxes of electromagnetic and kinetic energy flux, which serve to heat, ionize, and excite the neutral gases of the outer atmosphere, producing responses that are readily measured by ground-based radio and optical diagnostics [*Thayer and Semeter*, 2004]. Figure 1 provides a schematic illustration of this forward problem. For purposes of illustration, we restrict our attention here to two spatial dimensions. New diagnostics, such as the AMISR radars [*Semeter et al.*, 2009] and improved models [*Zettergren and Semeter*, 2012], will be able to investigate these processes directly in three dimensions.

In Figure 1a, an incident electron flux $\phi(x, t)$ transfers its power into a set of volume production rates $p(x, z, t)$ represented by the shaded contours, with z representing magnetic field line distance. The contours could represent production of excited atomic and molecular states, impact ionization rates, or heating rates. Figure 1b depicts, qualitatively, several representative response profiles arising as a consequence of p. This particular set of responses is consistent with what is observed within a poleward boundary intensification (e.g., Figure 6 at 1:05 UT) [*Semeter et al.*, 2005] or at the onset of a geomagnetic substorm [*Mende et al.*, 2003], where wave-particle coupling leads to a broad distribution of energies extending to several kiloelectron volts. For instance, for ~10 keV primaries, a peak in plasma density, N_e, appears at ~100 km altitude. Lower-energy electrons ionize F region altitudes.

The low-energy portion of the incident distribution heats the plasma in the topside ionosphere, enhancing T_e. This heating produces upward ambipolar diffusion and ion upwelling, represented by V_i. The plasma state parameters N_e, T_e, T_i, and V_i may be estimated through ISR ranging and spectral fitting [*Evans*, 1969].

The deposition of kinetic energy flux also excites a spectrum of optical emissions, exemplified in Figure 1b by the oxygen 630 nm profile (dashed line). Unlike plasma state parameters, photon volume production rates cannot be sensed directly. The directly measured quantity is a line-of-sight integral through the emitting region, several of which are depicted in Figure 1a. Access to the intrinsic volumetric quantity (photon volume emission rate) requires the use of tomographic techniques using multistation observations [*Frey et al.*, 1996; *Semeter et al.*, 1999]. More commonly, the integrated intensities are analyzed directly [*Rees and Luckey*, 1974; *Hecht et al.*, 1989; *Strickland et al.*, 1994; *Zettergren et al.*, 2007].

3. MATHEMATICAL DESCRIPTION OF THE FORWARD PROBLEM

The colocation of an ISR and passive optical diagnostics provides a unique collection of independent information about the ionospheric response to a particle magnetospheric event. The distribution of responses in altitude, wavelength, and time provides a "fingerprint" from which the causative electron energy spectrum may be determined. The mathematical treatment of this problem begins by computing the stationary flux ϕ (cm^{-2} s^{-1} eV^{-1} sr^{-1}) as a function of altitude, energy, and pitch angle produced by a differential number flux ϕ_{top} incident on the upper boundary of the atmosphere. This process is represented by the electron transport equation, which may be expressed as

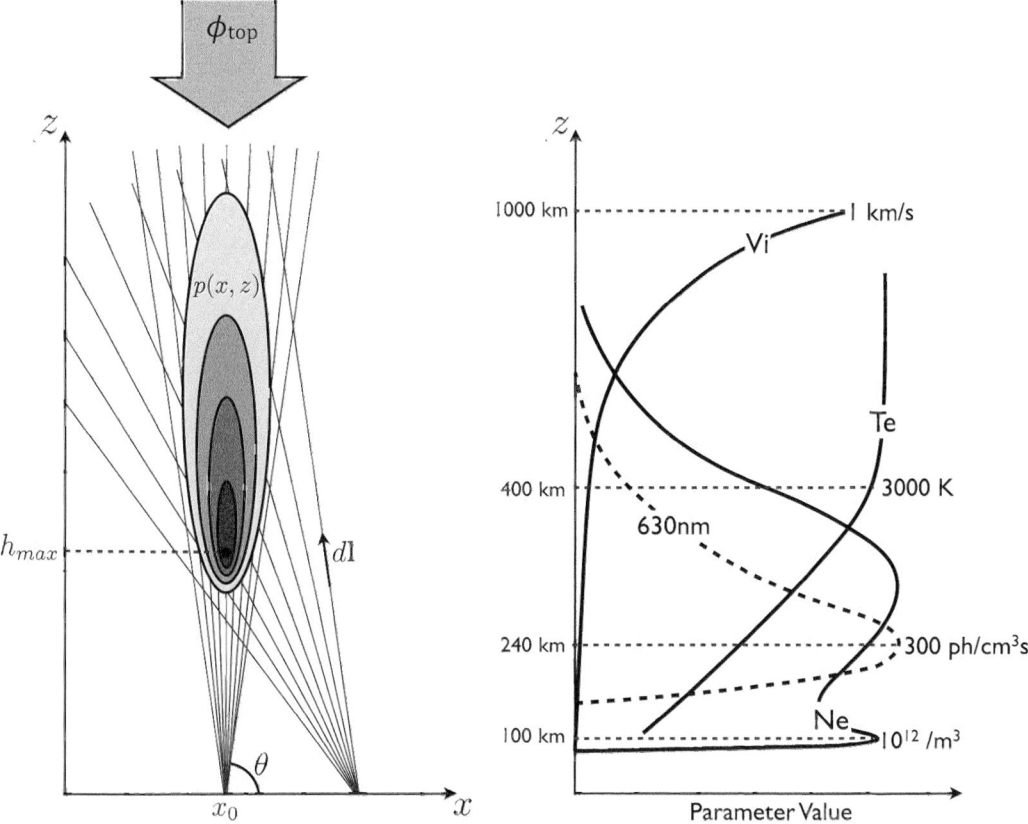

Figure 1. Schematic representation of the forward problem treated in this chapter. (a) A flux of incident electrons ϕ_{top} impinges the atmosphere, which heats, ionizes, and excites the neutral gases. These rates are collectively represented by $p(x, y)$. (b) Selected response profiles within an auroral arc. The solid lines represent parameters that can be directly resolved in range with an incoherent scatter radar (ISR). Plasma density N_e is enhanced in the E region. Electron temperature T_e is elevated in the F region due to the low-energy portion of the incident flux. Upward plasma flow V_i is produced via ambipolar diffusion and thermal ionospheric expansion. The dashed line shows a representative optical emission profile associated with the energy deposition process, which can only be directly resolved in a line-of-sight-integrated sense.

$$\mu \frac{\partial \phi(z, \mu, E)}{\partial z} = -\phi(z, \mu, E) + \phi_{top}(\mu, E)$$
$$+ n_e \frac{\partial (L(z, E)\phi(z, \mu, E))}{\partial E}$$
$$+ \int_{-1}^{1} d\mu' \int_{E}^{E_{max}} dE' R(E', \mu' \to E, \mu)\phi(z, \mu', E') \quad (1)$$

[*Lilensten and Blelly*, 2002], where μ, μ' are the cosines of scattered and incident electron pitch angles and E, E,' the energies of scattered and incident electrons, R the redistribution function describing the degradation from a state (E', μ') to state (E, μ). The loss function $L(E)$ appearing in the fourth term represents energy loss to the thermal electrons. For simplicity, equation (1) is for a single-species neutral atmosphere. See the works of *Strickland et al.* [1989] or *Lilensten and Blelly* [2002] for extension to the multispecies case.

Once $\phi(z, \mu, E)$ is established, the volume production rate for a given process p can be computed from a calculation of the form

$$p_{ij}(z) = \int n_i(z)\sigma_{ij}(E)\phi(z, E)dE \quad [\text{cm}^{-3}\ \text{s}^{-1}], \quad (2)$$

where n_i is the ith density of ground-state neutral species (N_2, O_2, or O), $\phi(z, E)$ is the pitch angle–integrated flux, and σ_{ij} is the electron impact cross section for the jth excitation process involving the ith species.

Equation (2) represents both excitation and ionization processes. Once the production rates are known, the resulting state densities may be computed from a 1-D continuity calculation,

$$\frac{\partial n_s}{\partial t} + \frac{1}{A} \frac{\partial An_s u_s}{\partial z} = p_s - l_s, \quad (3)$$

where n_s is the density associated with process p_s, A is the flux tube area, u_s is the diffusion velocity, and l_s is the loss rate.

The measurable responses by ISR or passive optical techniques (e.g., Figure 1b) fall into one of three categories. If process s corresponds to excitation of a prompt transition, then $p_s(z)$ represents a photon volume emission rate directly,

$$p_\lambda(z) = p_s(z), \qquad (4)$$

where λ is a wavelength designation for the transition. For optically thin emissions, the measured response is a line-of-sight integral through the emitting region (weighted by an appropriate extinction factor)

$$b = \int_0^\infty p_\lambda(\mathbf{l})\mathrm{d}l \quad [\mathrm{cm}^{-2}\ \mathrm{s}^{-1}], \qquad (5)$$

where $\mathbf{l}$ is the position vector defining the line of sight. If the emitting region is defined in 2-D Cartesian coordinates (x, z), and the observer's position is described by (x_0, z_0), then the observed brightness as a function of elevation angle θ can be expressed explicitly as

$$b_\lambda(\theta; x_0, z_0) = \int_0^\infty \int_0^\infty \delta[z - z_0 - (\tan\theta)(x - x_0)] p_\lambda(x, z) \mathrm{d}x \mathrm{d}y, \qquad (6)$$

where δ is the Dirac delta function, which is nonzero only when the argument lies along the line of sight. A tomographic inversion problem may be formulated using measurements of b_λ from different ground locations [*Semeter et al.*, 1999].

If process s corresponds to excitation of a metastable state with radiative lifetime $1/A_\lambda$, then one must consider diffusion and collisional loss in computing the measured emission rate. In cases where the source varies slowly compared to the radiative lifetime, we may ignore diffusion and express the volume emission rate as

$$p_\lambda(z) = \frac{A_\lambda}{\beta} p_s(z) \quad [\mathrm{cm}^{-3}\ \mathrm{s}^{-1}], \qquad (7)$$

where A_λ is the transition probability (first Einstein coefficient), and β is the collisional loss (or quenching) rate. The measured brightness b is then computed from equation (6).

Finally, if s corresponds to impact ionization processes, equation (3) is solved for each process and then summed to determine total density enhancement sensed by ISR,

$$n_e(z) = \sum_k n_k(z) \quad [\mathrm{cm}^{-3}]. \qquad (8)$$

Below 150 km, for sources ϕ_{top} that are stable for time scales longer than a few seconds, we typically assume $p_s = l_s$ and, again, avoid the complications of time history in our forward model [e.g., *Semeter and Kamalabadi*, 2005].

4. CHARACTERISTIC RESPONSES

Equations (1) to (8) constitute a model that maps an incident electron spectrum ϕ_{top} to a set of measurable ionospheric responses. The central question in this work is whether this model is invertible, i.e., given some set of measures of the ionospheric response, can we reliably estimate ϕ_{top}? If so, might we then be able to use this result to study other responses of the ionosphere-thermosphere system that are self-consistently calculated from this driver, but not directly observed, such as changes in ion composition, or species-specific ion upflow rates?

To consider invertibility, we first examine the sensitivity of measured responses to changes in the incident electron spectrum. We address this using the TRANSCAR model [*Blelly et al.*, 2005; *Lilensten and Blelly*, 2002]. TRANSCAR is a 1-D time-dependent flux tube model that solves for the density, drift velocity, temperature, and heat flux for seven different ion species (O^+, H^+, N^+, N_2^+, NO^+, O_2^+, and e^-). The model incorporates both a fluid module, which solves the eight-moment equations along a geomagnetic field line, and a kinetic module [*Lummerzheim and Lilensten*, 1994] that solves the transport equation for suprathermal electrons (equation (1)). The fluid and kinetic modules are dynamically coupled; at each time step, the fluid module provides thermal electron density and temperature to the kinetic module, and the kinetic module provides ionization, excitation, and heating rates to the fluid module. The background neutral thermospheric densities and temperatures for TRANSCAR are provided by the Mass Spectrometer Incoherent Scatter-90 (MSIS-90) model [*Hedin*, 1991].

Figure 2 shows characteristic responses of the ionosphere to an impinging electron beam, computed using the TRANSCAR model. In each panel, the columns represent either field-aligned profiles of volume emission rate (panels a–c) or plasma state parameters (panels f–i) computed for a discrete set of monoenergetic electron beams of increasing energy, with net energy flux held constant at 0.7 mW m^{-2}. In each case, the beam was applied for 5 min.

The Figure 2 encapsulates several important features about how the ionosphere responds to auroral precipitation. For incident energies <100 eV (approaching the average ionization potential of ~20 eV), a comparatively larger portion of the energy is deposited as heat (panel g). Heating drives an upward ambipolar ion diffusion (panel h) and a net outward ion flux (panel i). Hence, soft precipitation is associated with increased topside scale height and ion outflow to the magnetosphere [*Richards*, 1995; *Zettergren et al.*, 2008]. Between

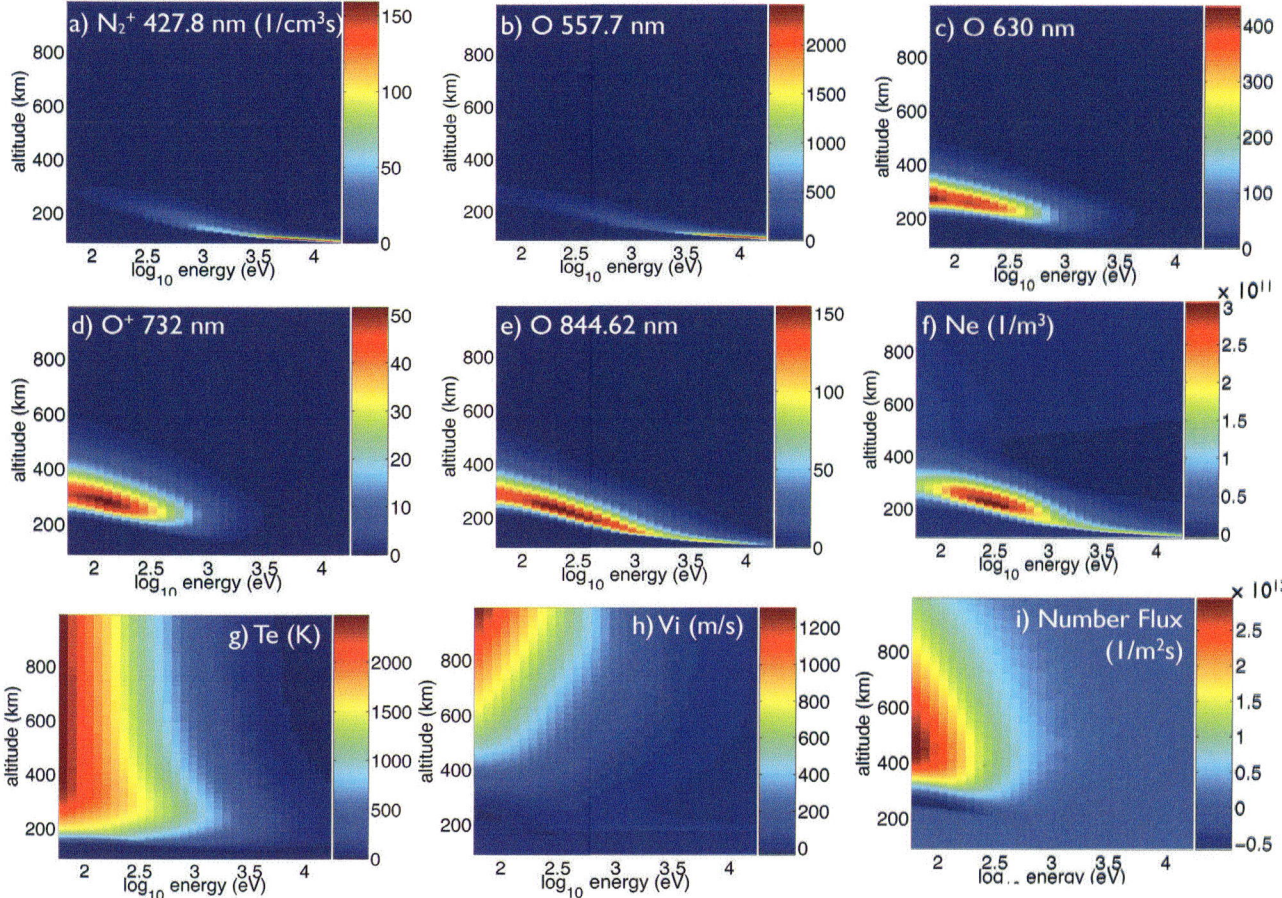

Figure 2. (a–e) Volume emission profiles for several prominent auroral features, computed for a series of monoenergetic beams of increasing energy. The net energy flux has been held constant at 0.7 mV m^{-2}. (f–i) Change in ionospheric state parameters computed over the same set of beams. These panels represent different representations of matrix **K** in equation (10).

100 eV and 1 keV, we find the greatest sensitivity between beam energy and altitude of peak ionization/emission. It is in this range that tomographic analysis of optical emissions is most effective for studying auroral energetics [*Semeter et al.*, 1999]. Above 1 keV, the oxygen 630 and 732 nm emissions become negligible due to quenching by neutral species, and the prompt 427.8 nm emission increases as the relative amount of N_2 increases, providing the basis for the efficacy of the "red-to-blue" ratio in determining beam energy [*Rees and Luckey*, 1974]. The oxygen 846 nm emission, however, remains significant in this energy range because it has multiple sources and is not quenched. This fact has been exploited to access neutral composition factors from auroral analysis [*Hecht et al.*, 1999]. Also, above 1 keV, the O 557.7 nm emissions become increasingly significant, although it has not been widely used in quantitative analyses due to modeling uncertainties [*Campbell and Brunger*, 2010].

None of the parameters plotted in Figure 2 are observed directly. Optical emissions (panels a–e) are observed in a line-of-sight-integrated sense through imaging spectroscopy. Plasma state parameters (panels f–i) may be accessed via fitting of the plasma autocorrelation function, which is estimated by ISR with resolution defined by the radar ambiguity function. We may define parallel paths from magnetospheric driver to measured quantity for both the optical and radar cases. This is illustrated schematically in Figure 3. For the case of discrete auroral drivers, the magnetospheric energy influx can be approximated as an admixture of Poynting flux $S(t)$ and particle kinetic energy flux $\phi_{top}(E, t)$ [*Thayer and Semeter*, 2004]. For the neutral atmosphere, we must rely on a model, such as MSIS which, when subject to a specified solar EUV source, produces the background ionosphere. Applying magnetospheric drivers to this system produces the set of plasma state responses and optical emission profiles exemplified in Figure 2.

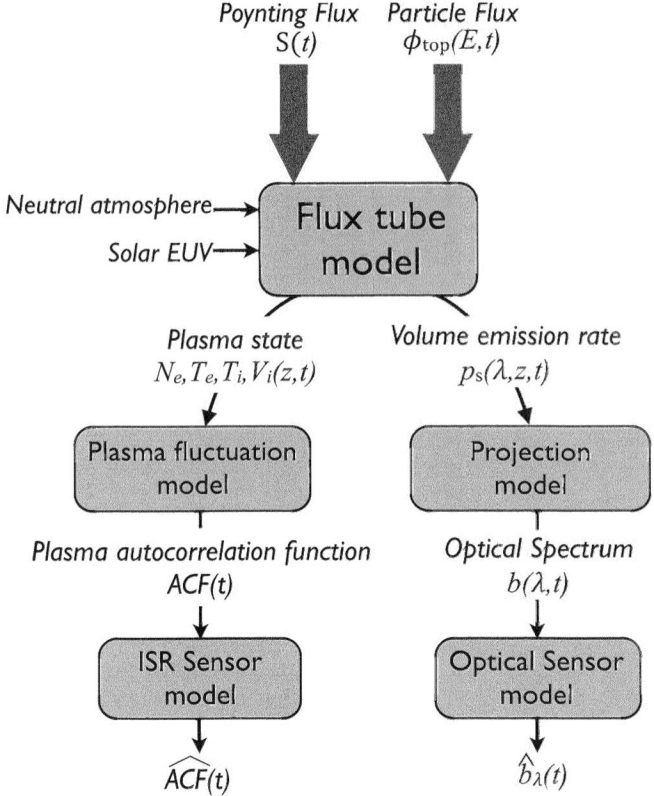

Figure 3. Flow chart showing the inputs and models used in our ISR and optical analysis. It is possible to formulate inverse problems along multiple paths in this diagram.

Following the left side of Figure 3, a given plasma state defines the spectrum of ion-acoustic fluctuations. An ISR produces an estimate of this function, denoted $\widehat{ACF}(t)$, with resolution and fidelity determined by the frequency, antenna pattern, pulse pattern, and integration time of the radar. Following the right side, the spatial distribution of volume emission rate p defines the wavelength-specific brightness b for a given position and look direction. A photometric sensor estimates this function, denoted $\hat{b}_\lambda(t)$. An inverse problem can be defined through either path.

5. FREDHOLM FORMULATION

Although the response functions of Figure 2 are the result of complicated calculations, they, nonetheless, exhibit smooth variations in altitude and intensity as beam energy is varied. We may thus be justified in representing this complex physical model by a simple linear integral equation of the form

$$g(s) = \int k(s,t) f(t) \, dt, \qquad (9)$$

where f represents the input function (electron beam), g represents the response, and k is the kernel, which maps input to response. Equations of this form (formally, a Fredholm equation of the first kind) arise naturally in a great many inverse problems. The general solution approach involves some form of discretization of both the measured response and the unknown input, resulting in a matrix equation

$$\mathbf{g} = \mathbf{K}\mathbf{f} + \mathbf{e}, \qquad (10)$$

where the vector $\mathbf{e}$ represents uncertainties in the measurement and, for Bayesian approaches, the model. The problem of estimating $\mathbf{f}$ becomes one of finding $\mathbf{K}^+$, the "pseudo-inverse" of $\mathbf{K}$. The designation pseudo-inverse indicates that $\mathbf{K}$ is not, in general, full rank, a subject that will be treated in section 7.

The panels in Figure 2 represent realizations of $\mathbf{K}$ for the case of mapping an energy distribution onto an altitude distribution. For instance, if the input is taken as the primary electron number flux spectrum $\phi_{top}(E)$, and the output is the resulting plasma density enhancement $n_e(z)$, then equation (9) takes the form

$$n_e(z) = \int k(z,E)\phi_{\text{top}}(E)dE, \quad (11)$$

with $k(z, E)$ represented in matrix form by Figure 2f. This equation represents a well-studied inverse problem in ISR studies of the aurora [e.g., *Osepian and Kirkwood*, 1996; *Semeter and Kamalabadi*, 2005]. In practice, the pitch angle distribution is prescribed a priori (e.g., downgoing for Alfvénic aurora, isotropic for inverted-V aurora) such that the estimated function is $\phi_{\text{top}}(E)$.

The output $g(s)$ may, similarly, be taken as a photon volume emission rate, $p_\lambda(z)$ (e.g., Figures 2a–2e). But unlike $n_e(z)$, the altitude-dependent emission rate is not accessed directly, requiring a tomographic analysis using multistation measurements [e.g., *Semeter and Mendillo*, 1997]. A more common approach is to work with line-of-sight-integrated emissions (i.e., brightnesses) directly. For the case of photometric observations in the magnetic zenith, we may combine equations (2) and (5) to obtain the Fredholm equation

$$b(\lambda) = \int k(\lambda, E)\phi_{\text{top}}(E)dE, \quad (12)$$

with matrix representation

$$\mathbf{b} = \mathbf{K}\boldsymbol{\phi} + \epsilon. \quad (13)$$

The independent information for estimating $\phi_{\text{top}}(E)$ is now embodied in the distribution of spectral intensities. Although we have formulated the problem in λ rather than z, the form of equations (11) and (12) are identical. In either case, $\phi_{\text{top}}(E)$ is determined via solution of a system of linear equations in the form of equation (10).

6. TIME HISTORY

An important consideration in formulating an auroral inverse problem is time history. Equation (11), for instance, is only valid for >1 keV electrons, which penetrate to altitudes <150 km. At these altitudes, the ion lifetime is <10 s, and so steady state may be reasonably assumed in equation (3) (i.e., $p = l$) [*Semeter and Kamalabadi*, 2005]. However, as mentioned previously, there is great interest in the 100 eV to 1 keV energy range, where precipitation associated with dynamic Alfvénic aurora [*Chaston et al.*, 2002; *Semeter et al.*, 2005] and cusp processes [*Strangeway et al.*, 2000] occurs. The standard emission used to study this energy range is oxygen red line. But interpretation of this emission is complicated [*Semeter et al.*, 1996].

To avoid the complications of time history for studies of soft but dynamic electron precipitation, *Zettergren et al.* [2007] analyzed emissions of atomic oxygen and molecular nitrogen in the NIR region of the auroral spectrum using equation (12). Figure 4 illustrates the sensitivity of spectral emissions associated with electron precipitation in this range. Plotted are normalized height-integrated emission rates as a function of beam energy, computed using the TRANSCAR model. Also plotted is the ion upflow rate, illustrating a strong negative correlation with beam energy.

The brightness curves in Figure 4, when stacked into a matrix form, constitute a discrete representation of $k(\lambda, E)$ in equation (12). When properly scaled, the elements of the vector ϕ in the discrete representation $\mathbf{b} = \mathbf{K}\boldsymbol{\phi} + \epsilon$ will then represent samples of the differential number flux spectrum ($\text{cm}^{-2}\,\text{s}^{-1}\,\text{eV}^{-1}\,\text{sr}^{-1}$) (see the works of *Semeter and Kamalabadi* [2005] and *Zettergren et al.* [2007] for implementation details).

Estimating parameters of the incident energy spectrum from auroral spectral brightnesses has also been well studied, but the problem has not traditionally been formulated using the linear system approach. *Hetch et al.* [1989] used measurements in four wavelength bands to estimate four parameters: the energy flux and characteristic energy of an assumed Maxwellian incident electron spectrum, and two atmospheric composition factors. The fit was done using an unconstrained least squares estimator. By contrast, both *Semeter and Kamalabadi* [2005] and *Zettergren et al.* [2007] formulated an underdetermined inverse problem, corresponding to a rank-deficient forward model $\mathbf{K}$. To find a unique solution, maximum entropy regularization was used. A benefit of this approach is that it places no a priori constraint on the shape of the ϕ_{top}, while still seeking the smoothest possible solution. This could be particularly important in the low-energy range of interest, where ϕ_{top} can have variable morphology (e.g., bump on tail, flat top, Maxwellian [*Lynch et al.*, 2007]).

7. INVERSION APPROACHES

Solving inverse problems in the form of equation (10) requires inversion of the matrix $\mathbf{K}$ [*Hansen*, 1992]. Two issues distinguish this problem a simple matrix inversion. The first is that the number of measurements and unknowns are, in general, not equal. So, we must seek a pseudo-inverse of our forward model, which we denote as $\mathbf{K}^+$. Having more measurements than unknowns does not guarantee an overdetermined problem. Frequently, the geometry or other feature of the inverse problem results in $\mathbf{K}$ being rank deficient (or ill conditioned), regardless of its dimensions. The second related issue is the presence of noise. Standard inversion techniques from linear algebra that may work on idealized data, fail in the presence of measurement uncertainties.

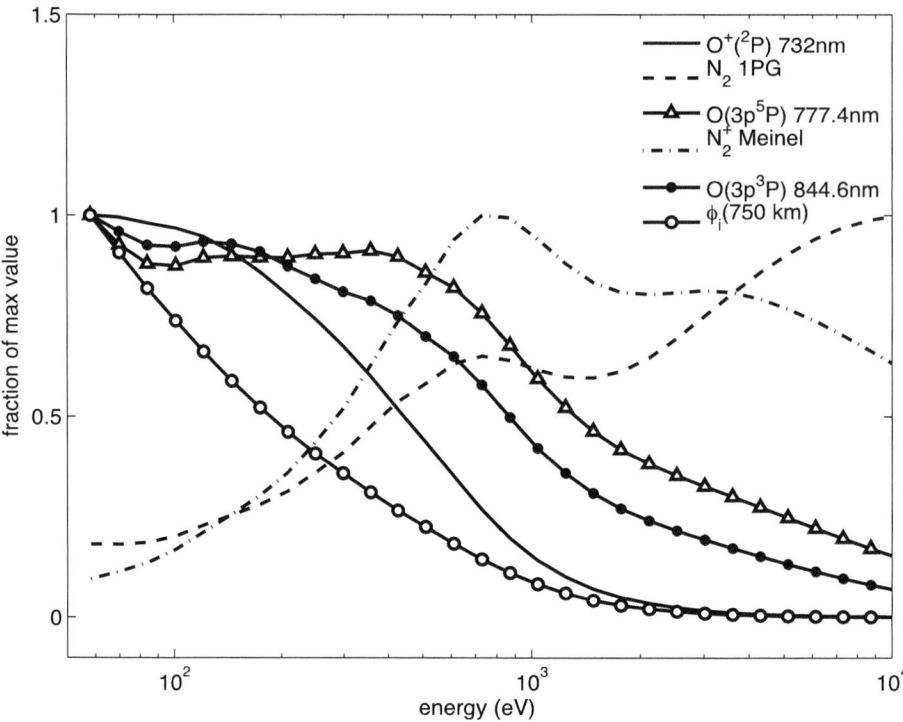

Figure 4. Dependence of selected near-infrared auroral emission features on incident beam energy, plotted as the fraction of maximum value after 5 min exposure to the monoenergetic beams used in Figures 2. The upward ion number flux ($\phi = n_i u_i$) is also plotted, extracted at 750 km.

Tikhonov regularization provides a framework for finding a unique solution to a rank-deficient inverse problem. The approach defines the solution as the one that minimizes a cost function of the form

$$\hat{\mathbf{f}} = \arg\min_{\mathbf{f}}\{\|\mathbf{Kf} - \mathbf{g}\|^2 + \alpha\|\mathbf{Lf}\|^2\}, \qquad (14)$$

where α balances the competing interests of data fidelity (expressed as distance between measured and modeled data) and a constraint on the solution behavior, and $\mathbf{L}$ represents an operator acting on the solution space $\mathbf{f}$. Many standard inversion approaches can be shown to be equivalent to equation (14) with appropriate selection of regularization term [*Demoment*, 1989]. For instance, the degenerate case of $\alpha = 0$ results in the least squares solution with minimum norm, given by

$$\hat{\mathbf{f}} = (\mathbf{K}^T\mathbf{K})^{-1}\mathbf{K}^T\mathbf{g}. \qquad (15)$$

For the rank-deficient problems of interest here, $\hat{\mathbf{f}}$ from equation (15) will be highly sensitive to noise in the measurements, including round-off error. One common approach to limiting solution variability and, hence, noise sensitivity, is a truncated singular value decomposition (TSVD) inversion,

$$\hat{\mathbf{f}} = \sum_{i=1}^{m} \frac{\mathbf{v_i}\langle\mathbf{u_i^T}, \mathbf{g}\rangle}{\sigma_i}, \qquad (16)$$

where m is the rank of $\mathbf{K}$, $\mathbf{v_i}$ and $\mathbf{u_i}$ are the column vectors of the unitary matrices $\mathbf{V}$ and $\mathbf{U}$, respectively, and σ_i are the singular values, or the diagonal elements of Σ in the singular value decomposition of $\mathbf{K}$: $\mathbf{K} = \mathbf{U\Sigma V}^T$. The reciprocal singular values act as gain terms. Limiting the number of singular values limits the dimensions of the solution, thus limiting high-frequency variations. This approach is essentially equivalent to equation (14) with $\mathbf{L} = \mathbf{I}$. The drawback of this approach is that the solution is entirely defined by the properties of the data and does not allow for inclusion of prior knowledge (statistical or physical) about the solution.

Bayesian approaches guide the solution by incorporating statistical prior information about *f*. For instance, for $\mathbf{L} = \mathbf{I}$, and α given by the ratio of data variance to model variance, the minimum mean square estimator (MMSE) results

$$\hat{\mathbf{f}} = \Sigma_m \mathbf{K}^T(\mathbf{K}\Sigma_v\mathbf{K}^T\mathbf{K} + \Sigma_e)^{-1}\mathbf{g}, \qquad (17)$$

where Σ_g and Σ_f are the error covariances for $\mathbf{g}$ and $\mathbf{f}$, respectively. This approach has been used by *Heinselman and Nicolls* [2008] in the estimation of ionospheric convective flows from monostatic line-of-sight projections resolved by ISR. Here the rank deficiency is handled by imposing confidence intervals on the data and model, rather than by imposing smoothness. These notions are similar, and the similar nature of the regularization parameter for MMSE and TSVD is not surprising.

Finally, if **L** corresponds to a measure of the entropy of **f**, then variants of the popular maximum entropy method (MEM) emerge. The maximum entropy solution may be viewed as the one that is maximally noncommittal with respect to the unavailable information [*Skilling*, 1990]. We favor this approach for the inverse problems of interest here. In particular, the solution, which maximizes the Berg entropy, $-\log \mathbf{f}$ [*De Pierro*, 1991], has been found to perform well for forward models such as those encapsulated in Figure 2, which contain both sharp (bottomside) and soft (topside) "edges." The noncommittal nature of MEM is typically able to find smooth bump-on-tail and Maxwellian-like solutions for $\phi_{top}(E)$ even for well-underdetermined problems in both optical [*Zettergren et al.*, 2007] and ISR [*Semeter and Kamalabadi*, 2005] inverse problems. The question of validation remains, and this will be addressed in section 8.

8. VALIDATION

For any linear inverse problem, regardless of condition number, it is always possible to impose some form of regularization that yields a unique solution. The question comes down to, what is the best way to deal with the rank deficiency in the forward model, and how reliable is the solution? Direct validation of $\hat{\phi}_{top}$ using in situ measurements is challenging due to the extreme requirements for ground-space conjunction [e.g., *Semeter et al.*, 2005]. Here we apply an internal self-consistency test between optical and ISR results. While not a direct validation, the approach lends confidence to the technique and suggests a powerful role for relatively inexpensive optical instrumentation for remote diagnosis of magnetosphere-ionosphere interactions.

We consider the inverse problem described by equation (12) which, in matrix form, becomes

$$\phi_{top} = \mathbf{K}^+ \mathbf{b}. \qquad (18)$$

Zettergreen et al. [2008] studied this problem using maximum entropy estimation applied to an eight-element vector b of spectral brightnesses in the NIR. Figure 5 summarizes

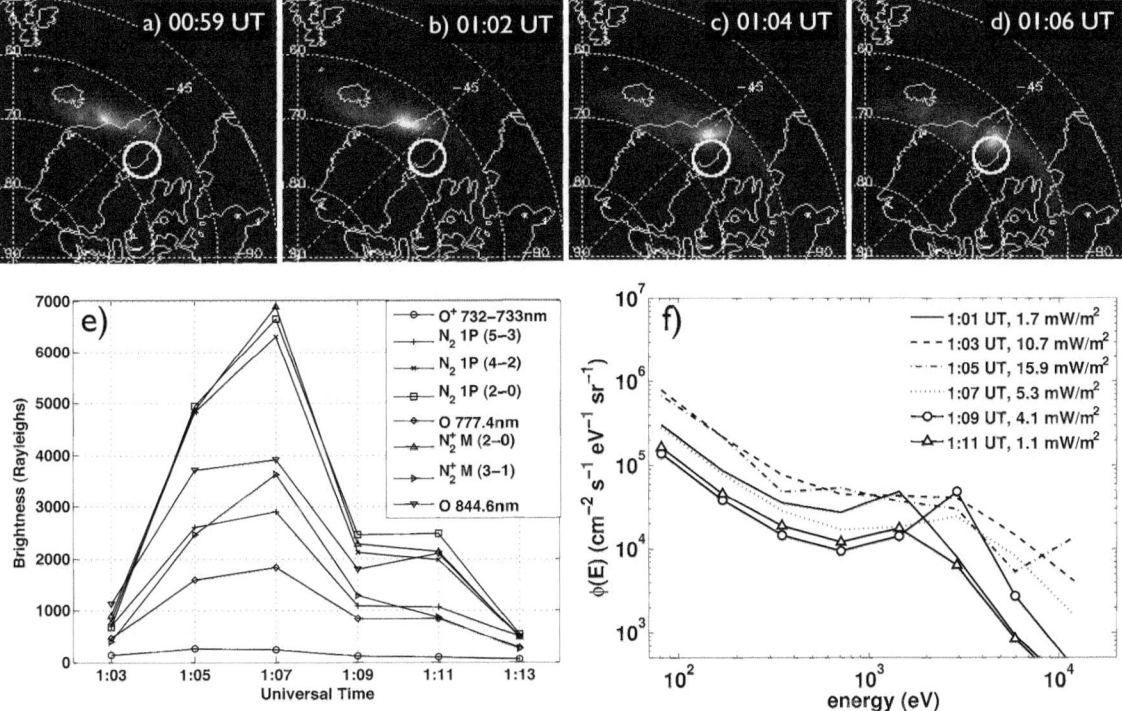

Figure 5. (a–d) Sequence of FUV images recorded from the NASA IMAGE satellite as a poleward boundary intensification (PBI) transited the zenith at Sondrestrom, indicated by a white circle. (e) Spectral brightnesses as a function of time observed during the event. (f) Corresponding differential number flux spectra estimated by applying maximum entropy inversion to equation (13).

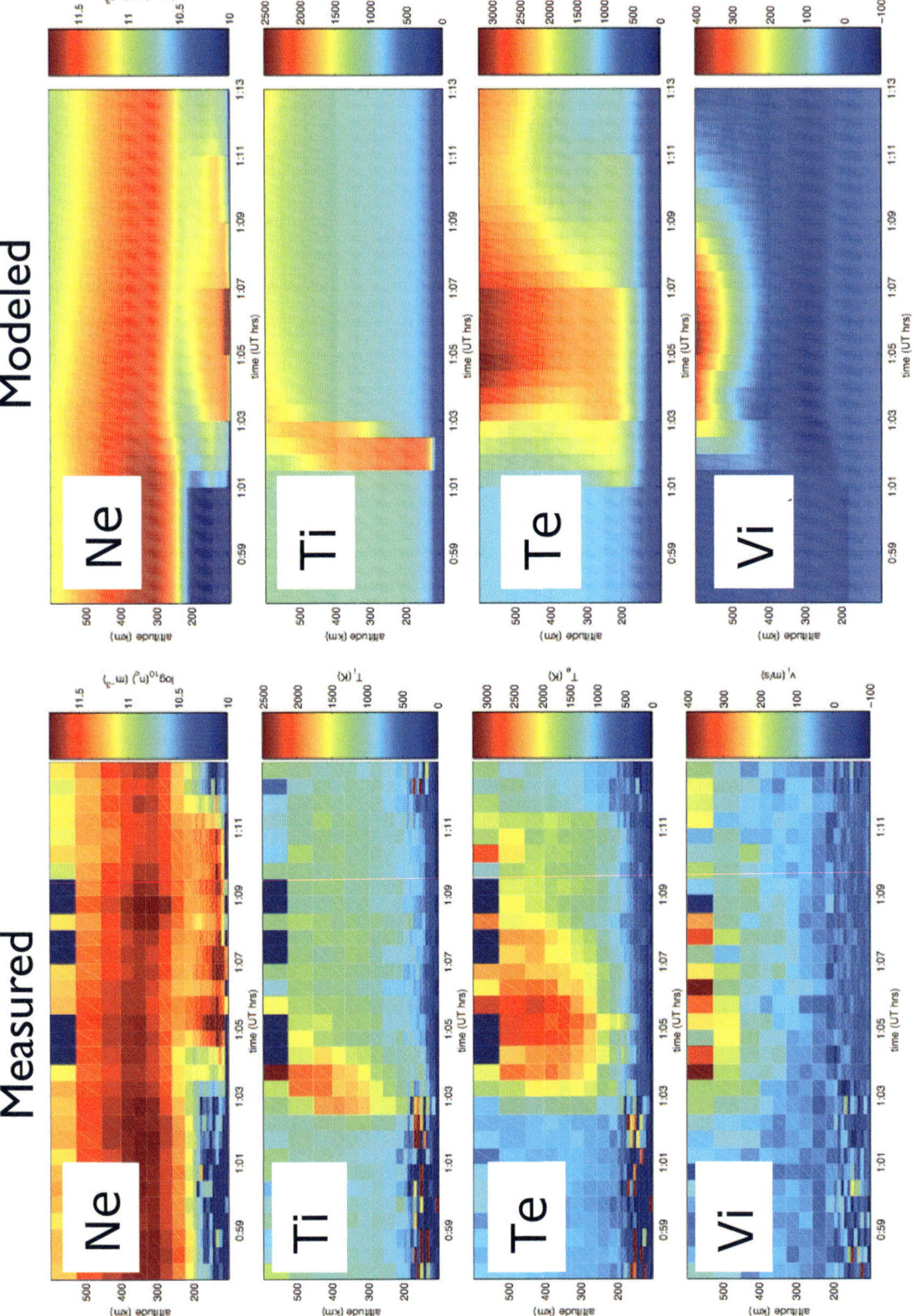

Figure 6. (left) Ionospheric state parameters measured in the magnetic zenith by the Sondrestrom ISR for the event of Figure 5. (right) Modeled state parameters produced by applying estimated electron spectra as a time-dependent boundary condition to the TRANSCAR model.

the results from this case study. Panels a–d show FUV images recorded by the IMAGE satellite as an intense poleward boundary auroral feature transited into the field of view of radar and optical instruments at Sondrestrom (white circle). Panel e shows the eight NIR spectral curves used in the inversion of this event and panel f the corresponding estimates of $\hat{\phi}_{top}(E)$. For the 10 min period analyzed, the suprathermal electron peak is seen to shift from ~1 to ~10 keV and back to ~1 keV. Evidence for an increase in the 0.1 to 1 keV range is also seen from 01:01 to 01:05 UT.

The colocation of an ISR and optical spectrometer enables cross-validation of results obtained solely by one instrument or the other. In Figure 6 (right), the recovered $\hat{\phi}_{top}(E,t)$ plotted in Figure 5f has been reapplied as a time-dependent upper boundary condition to TRANSCAR and the resulting time-dependent variation in the plasma state computed. Figure 6 (left) shows the corresponding parameters sensed directly by the Sondrestrom ISR during this interval. The comparison shows good agreement in all parameters. The agreement extends to certain subtle effects. The reduction in bottomside F region density, for instance, is a consequence of enhanced recombination caused by ion heating and the attendant conversion from atomic (O^+) to molecular (NO^+) ions [*Zettergreen et al.*, 2010, 2011].

The agreement between measured and modeled T_i is artificial, in the sense that T_i is controlled by the strength of the convective electric field [*Zettergreen et al.*, 2010], which is not reliably observed by optical means. In this case, the electric field was adjusted to 65 mV m^{-1} to obtain agreement with the observed T_i enhancement outside of the PBI event.

9. FUTURE DIRECTIONS: ACCESSING HIDDEN STATES OF THE I-T SYSTEM

Although Figure 6 does not constitute direct validation of the inverse method or the forward model, it provides a strong demonstration of internal consistency. The fact that ISR state parameters were successfully estimated through analysis of optical measurements alone is an important departure from a standard model-data comparison [e.g., *Hecht et al.*, 1999]. The suggestion that ϕ_{top} may be reliably estimated in the <1 keV electron energy range using NIR spectroscopy opens new possibilities for diagnosing the ionosphere-thermosphere system via optical means.

Furthermore, the good agreement between plasma state parameters derived from these results and direct ISR observables suggest an intriguing question: To what extent can we believe unobservable parameters in the forward model? For instance, Figure 7a shows species-specific ion upflow rates at 1000 km altitude, extracted from the model run used in Figure 6. Figure 7a shows expected temporal morphologies, such as the evacuation of lighter ions first. Figures 7b and 7c illustrate another parameter that is not readily observed by ISR, ion composition. The panels show, for instance, a nuanced change to the topside profiles preevent and postevent, with a bulk upward shift of all species present. As the model currently exists, the neutral atmosphere is a static background. This could have a significant effect on the reliability of such topside dynamics.

Finally, an auroral arc and its surrounding region of influence may be thought of as a cohesive system, with identifiable

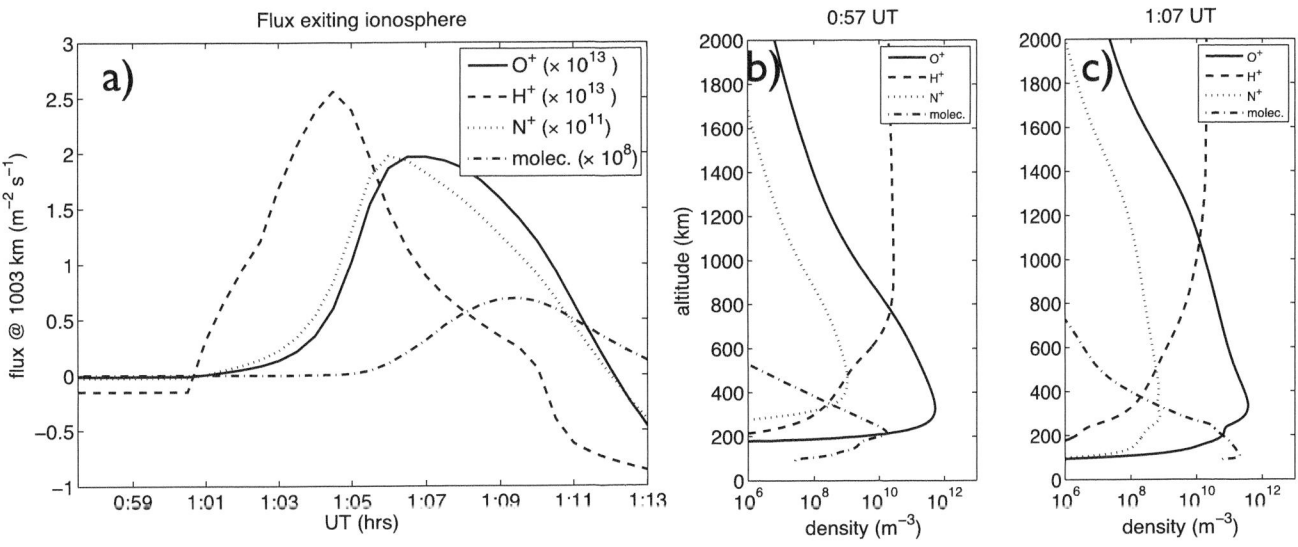

Figure 7. (a) Time-dependent upflow rates for three major ion species estimated using data from the event summarized in Figures 5 and 6. (b, c) Estimated field-aligned ion concentration profiles before (b) and during (c) the PBI transit.

inputs and responses. When driven by an admixture of kinetic and electromagnetic energy influx, the ionosphere-thermosphere system reconfigures in such a way as to dissipate this free energy [*Thayer and Semeter*, 2004]. Here we have considered only one component of that system response, the suprathermal electron flux. Regions adjacent to the optical aurora are known to experience enhanced convective flow and attendant increases in T_i [*Marklund*, 1984]. *Zettergren et al.* [2011] has developed an inverse problem that uses measurements of enhanced T_i in these regions (e.g., Figure 6) to estimate ion composition effects. *Semeter et al.* [2010] has developed composite imaging techniques able to study ionization, heating, and flows in three dimensions. A future challenge is to develop inverse strategies that consider the aggregate responses of the auroral arc system collectively in order to better understand time-dependent magnetosphere-ionosphere interactions.

Acknowledgments. This work was supported by the U.S. Air Force Office of Scientific Research under contract FA9550-12-1-018 and by the National Science Foundation under grants AGS-0960078 and AGS-1000302.

REFERENCES

Blelly, P.-L. (1996), 8-moment fluid models of the terrestrial high-latitude ionosphere between 100 and 3000 km, in *Solar-Terrestrial Energy Program: Handbook of Ionospheric Models*, edited by R. Schunk, pp. 53–72, Utah State Univ., Logan.

Blelly, P.-L., C. Lathuillère, B. Emery, J. Lilensten, J. Fontanari, and D. Alcaydé (2005), An extended TRANSCAR model including ionospheric convection: Simulation of EISCAT observations using inputs from AMIE, *Ann. Geophys.*, *23*, 419–431.

Campbell, L., and M. J. Brunger (2010), Modeling of kinetic, ionospheric and auroral contributions to the 557.7-nm nightglow, *Geophys. Res. Lett.*, *37*, L22104, doi:10.1029/2010GL045444.

Chaston, C. C., J. W. Bonnell, L. M. Peticolas, C. W. Carlson, J. P. McFadden, and R. E. Ergun (2002), Driven Alfven waves and electron acceleration: A FAST case study, *Geophys. Res. Lett.*, *29*(11), 1535, doi:10.1029/2001GL013842.

De Pierro, A. (1991), Multiplicative iterative methods in computed tomography, in *Mathematical Methods in Tomography*, edited by G. Herman, A. Louis and F. Natterer, pp. 1441–1450, Springer, Berlin.

Demoment, G. (1989), Image reconstruction and restoration: Overview of common estimation structures and problems, *IEEE Trans. Acoust. Speech Signal Process.*, *37*(12), 2024–2036.

Evans, J. (1969), Theory and practice of ionospheric study by Thomson scatter radar, *Proc. IEEE*, *57*, 496–530.

Frey, S., H. U. Frey, D. J. Carr, O. H. Bauer, and G. Haerendel (1996), Auroral emission profiles extracted from three-dimensionally reconstructed arcs, *J. Geophys. Res.*, *101*(A10), 21,731–21,741.

Hansen, P. (1992), Numerical tools for analysis and solution of Fredholm integral equations of the first kind, *Inverse Probl.*, *8*, 849–872.

Hecht, J. H., A. B. Christensen, D. J. Strickland, and R. R. Meier (1989), Deducing composition and incident electron spectra from ground-based auroral optical measurements: Variations in oxygen density, *J. Geophys. Res.*, *94*(A10), 13,553–13,563, doi:10.1029/JA094iA10p13553.

Hecht, J. H., A. B. Christensen, D. J. Strickland, T. Majeed, R. L. Gattinger, and A. V. Jones (1999), A comparison between auroral particle characteristics and atmospheric composition inferred from analyzing optical emission measurements alone and in combination with incoherent scatter radar measurements, *J. Geophys. Res.*, *104*(A1), 33–44, doi:10.1029/1998JA900009.

Hedin, A. E. (1991), Extension of the MSIS Thermosphere Model into the middle and lower atmosphere, *J. Geophys. Res.*, *96*(A2), 1159–1172.

Heinselman, C. J., and M. J. Nicolls (2008), A Bayesian approach to electric field and E-region neutral wind estimation with the Poker Flat Advanced Modular Incoherent Scatter Radar, *Radio Sci.*, *43*, RS5013, doi:10.1029/2007RS003805.

Lilensten, J., and P. L. Blelly (2002), The TEC and F2 parameters as tracers of the ionosphere and thermosphere, *J. Atmos. Sol. Terr. Phys.*, *64*, 775–793.

Lummerzheim, D., and J. Lilensten (1994), Electron transport and energy degradation in the ionosphere: Evaluation of the numerical solution, comparison with laboratory experiments and auroral observations, *Ann. Geophys.*, *12*, 1039–1051.

Lynch, K. A., J. L. Semeter, M. Zettergren, P. Kintner, R. Arnoldy, E. Klatt, J. LaBelle, R. G. Michell, E. A. MacDonald, and M. Samara (2007), Auroral ion outflow: Low altitude energization, *Ann. Geophys.*, *25*, 1967–1977.

Marklund, G. T. (1984), Auroral arc classification scheme based on the observed arc-associated electric field pattern, *Planet. Space Sci.*, *32*(2), 193–211.

Mende, S. B., C. W. Carlson, H. U. Frey, L. M. Peticolas, and N. Østgaard (2003), FAST and IMAGE-FUV observations of a substorm onset, *J. Geophys. Res.*, *108*(A9), 1344, doi:10.1029/2002JA009787.

Osepian, A., and S. Kirkwood (1996), High-energy electron fluxes derived from EISCAT electron density profiles, *J. Atmos. Terr. Phys.*, *58*, 479–487.

Rees, M. H., and D. Luckey (1974), Auroral electron energy derived from ratio of spectroscopic emissions 1. Model computations, *J. Geophys. Res.*, *79*(34), 5181–5186.

Richards, P. G. (1995), Effects of auroral electron precipitation on topside ion outflows, in *Cross-Scale Coupling in Space Plasmas*, *Geophys. Monogr. Ser.*, vol. 93, edited by L. Horwitz, N. Singh and L. Burch, pp. 121–126, AGU, Washington, D. C., doi:10.1029/GM093p0121.

Semeter, J., and F. Kamalabadi (2005), Determination of primary electron spectra from incoherent scatter radar measurements of the auroral E region, *Radio Sci.*, *40*, RS2006, doi:10.1029/2004RS003042.

Semeter, J., and M. Mendillo (1997), A nonlinear optimization technique for ground-based atmospheric emission tomography, *Trans. Geosci. Remote Sens.*, *35*(5), 1105–1116.

Semeter, J., M. Mendillo, J. Baumgardner, J. Holt, D. E. Hunton, and V. Eccles (1996), A study of oxygen 6300 Å airglow production through chemical modification of the nighttime ionosphere, *J. Geophys. Res.*, *101*(A9), 19,683–19,699.

Semeter, J., M. Mendillo, and J. Baumgardner (1999), Multispectral tomographic imaging of the midlatitude aurora, *J. Geophys. Res.*, *104*(A11), 24,565–24,585.

Semeter, J., C. J. Heinselman, G. G. Sivjee, H. U. Frey, and J. W. Bonnell (2005), Ionospheric response to wave-accelerated electrons at the poleward auroral boundary, *J. Geophys. Res.*, *110*, A11310, doi:10.1029/2005JA011226.

Semeter, J., T. Butler, C. Heinselman, M. Nicolls, J. Kelly, and D. Hampton (2009), Volumetric imaging of the auroral ionosphere: Initial results from PFISR, *J. Atmos. Sol. Terr. Phys.*, *71*, 738–743, doi:10.1016/j.jastp.2008.08.014.

Semeter, J., T. W. Butler, M. Zettergren, C. J. Heinselman, and M. J. Nicolls (2010), Composite imaging of auroral forms and convective flows during a substorm cycle, *J. Geophys. Res.*, *115*, A08308, doi:10.1029/2009JA014931.

Skilling, J. (1990), Quantified maximum entropy, in *Maximum Entropy and Bayesian Methods*, edited by P. Fougére, pp. 341–350, Kluwer Acad., Dordrecht, The Netherlands.

Strangeway, R. J., C. T. Russell, C. W. Carlson, J. P. McFadden, R. E. Ergun, M. Temerin, D. M. Klumpar, W. K. Peterson, and T. E. Moore (2000), Cusp field-aligned currents and ion outflows, *J. Geophys. Res.*, *105*(A9), 21,129–21,141.

Strickland, D. J., D. L. Book, T. P. Coffey, and J. A. Fedder (1976), Transport equation techniques for the deposition of auroral electrons, *J. Geophys. Res.*, *81*(16), 2755–2764.

Strickland, D. J., R. R. Meier, J. H. Hecht, and A. B. Christensen (1989), Deducing composition and incident electron spectra from ground-based auroral optical measurements: Theory and model results, *J. Geophys. Res.*, *94*(A10), 13,527–13,539.

Strickland, D. J., J. H. Hecht, A. B. Christensen, and J. Kelly (1994), Relationship between energy flux Q and mean energy $<E>$ of auroral electron spectra based on radar data from the 1987 CEDAR Campaign at Sondre Stromfjord, Greenland, *J. Geophys. Res.*, *99*(A10), 19,467–19,473.

Thayer, J. P., and J. Semeter (2004), The convergence of magnetospheric energy flux in the polar atmosphere, *J. Atmos. Sol. Terr. Phys.*, *66*, 807–824, doi:10.1016/j.jastp.2004.01.035.

Zettergren, M., and J. Semeter (2012), Ionospheric plasma transport and loss in auroral downward current regions, *J. Geophys. Res.*, *117*, A06306, doi:10.1029/2012JA017637.

Zettergren, M., J. Semeter, P.-L. Blelly, and M. Diaz (2007), Optical estimation of auroral ion upflow: Theory, *J. Geophys. Res.*, *112*, A12310, doi:10.1029/2007JA012691.

Zettergren, M., J. Semeter, P.-L. Blelly, G. Sivjee, I. Azeem, S. Mende, H. Gleisner, M. Diaz, and O. Witasse (2008), Optical estimation of auroral ion upflow: 2. A case study, *J. Geophys. Res.*, *113*, A07308, doi:10.1029/2008JA013135.

Zettergren, M., J. Semeter, B. Burnett, W. Oliver, C. Heinselman, P.-L. Blelly, and M. Diaz (2010), Dynamic variability in F-region ionospheric composition at auroral arc boundaries, *Ann. Geophys.*, *28*, 651–664, doi:10.5194/angeo-28-651-2010.

Zettergren, M., J. Semeter, C. Heinselman, and M. Diaz (2011), Incoherent scatter radar estimation of F region ionospheric composition during frictional heating events, *J. Geophys. Res.*, *116*, A01318, doi:10.1029/2010JA016035.

J. Semeter, Boston University, 8 St. Mary's St., Boston, MA 02215, USA. (jls@bu.edu)

M. Zettergren, Physical Sciences Department, Embry-Riddle Aeronautical University, Daytona Beach, FL, USA.

AGU Category Index

Auroral ionosphere, 309
Current systems, 57, 201
Cyberinfrastructure, 273
Data assimilation, 171, 273
Electric fields, 57, 107, 119, 133, 201
Equatorial ionosphere, 107, 259
General or miscellaneous, 133
Instruments and techniques, 241, 309
Ion chemistry and composition, 29, 161
Ion chemistry of the atmosphere, 3
Ionization processes, 29, 73
Ionosphere/atmosphere interactions, v, 1, 3, 13, 57, 85, 161, 181, 259, 273
Ionosphere/magnetosphere interactions, 201, 309
Ionospheric disturbances, 101, 133, 251, 299
Ionospheric dynamics, v, 1, 49, 73, 119, 145, 161, 201, 251
Ionospheric irregularities, 49, 107, 217, 241, 251, 299
Ionospheric physics, 217, 283
Ionospheric storms, 283

Middle atmosphere dynamics, 181
Midlatitude ionosphere, 29, 49, 101, 133
Model verification and validation, 145
Modeling and forecasting, v, 1, 3, 29, 49, 57, 107, 119, 145, 241, 251, 283, 299
Modeling, 73, 85, 273
Particle precipitation, 309
Plasma temperature and density, 161, 259
Plasma waves and instabilities, 217, 241
Polar cap ionosphere, 3
Pressure, density, and temperature, 39
Solar radiation and cosmic ray effects, 119
Stochastic phenomena, 217
Theoretical modeling, 13, 171
Thermosphere: composition and chemistry, 13
Thermosphere: energy deposition, 39, 85
Thermospheric dynamics, 13, 73, 85, 101, 145, 171, 181, 259
Tides and planetary waves, 101, 171, 181
Tomography and imaging, 283

Index

Note: Page numbers followed by f and t indicate figures and tables, respectively.

A

Accelerometers, for satellite guidance, 40
Acoustic waves, 101, 102
Advanced Research Project Agency (ARPA), 111, 112f
AFRL (Air Force Research Laboratory)
  low-latitude PBMOD for C/NOFS mission. *See* Low-latitude PBMOD
Air Force Research Laboratory (AFRL)
  low-latitude PBMOD for C/NOFS mission. *See* Low-latitude PBMOD
Alfvén wave mode, 205–206, 207
ALTAIR radar, 244–245
Altitudinal dependence, GITM and nonhydrostatic process, 91–96, 93f, 94f, 95f
AMIE (assimilative mapping of ionospheric electrodynamics) procedure, 74
ARPA (Advanced Research Project Agency), 111, 112f
Assimilative mapping of ionospheric electrodynamics (AMIE), 74, 289, 289f
Atmospheric gravity waves (AGW). *See* Gravity waves (GW)
Auroral processes, model-based inversion of. *See* Inversion, model-based

B

Barometric wind, 14–15
Bayesian analysis, EnKF and, 274, 275, 279
Biases, data assimilation and, 280
BiCGStab method, 246
Biconjugate gradient stabilized (BiCGStab) method, 242
Birkeland currents, in M-I coupling, 201, 202, 205, 206, 211, 213
Bottom boundary(ies), NOGAPS-ALPHA as
  NCAR/TIEGCM. *See* Navy operational global atmospheric prediction system-advanced level physics high altitude (NOGAPS-ALPHA) model
Bottom-type layers, irregularities in, 226–227, 227f, 230
Boundary conditions, 41–42, 46
  in TIE-GCM model, 75
Boundary value analysis, 217, 221, 227, 228
Bourret's integral equation, in Farley-Buneman turbulence, 224–225, 235
Buoyancy acceleration, GITM and nonhydrostatic process, 87–91, 88f, 89f, 90f
Burnside factor (BF), 134, 137

C

CAM (community atmospheric model), 183–184
CCMC (Community Coordinated Modeling Center), 146, 147, 149, 304
  NASA, NCAR TIE-GCM and. *See* Thermosphere-ionosphere-electrodynamics general circulation model (TIE-GCM)
CEDAR (coupling energetics and dynamics of atmospheric regions)
  ETI challenge (2009–2010)
    IT model, evaluation. *See* Ionosphere/thermosphere (IT) models, systematic evaluation
CESM (community earth system model), 183, 184
Challenge, ETI (2009-2010)
  CEDAR, IT model evaluation. *See* Ionosphere/thermosphere (IT) models, systematic evaluation
CHAMP satellite, 146, 147, 286–288, 287f
  density measurements by, 111, 113, 114f–115f
  electron density, predictions, 150, 151t, 152t
  mass density, 277–278
  neutral density, predictions, 153, 154t, 155t
  observations
    data, IT model evaluation, 146
    low-latitude ionosphere and thermosphere. *See* Low latitude CHAMP satellite observations
Chemistry of $O(^1D)$ in Nightglow, 245t
Christmas Island (CXI)
  AFRL SCINDA ground station at, 109, 110f–111f, 111
  R-T linear growth rate in, 109, 110f–111f
CIRs (corotating interaction regions), 121, 122
CLIMO drift model, as driver for PBMOD, 111, 113, 113f, 114f–115f
CMEs (coronal mass ejections), 300

326 INDEX

CMIT (coupled magnetosphere-ionosphere-thermosphere model)
   NCAR TIE-GCM. *See* Thermosphere-ionosphere-electrodynamics general circulation model (TIE-GCM)

C/NOFS. *See* Communication/Navigation Outage Forecasting System (C/NOFS)

Collisional drift wave instabilities, 230–231, 232f

Collisional interchange instability, 226, 227

Collisional shear instability (CSI), 227–228, 241–242

Collisionless plasma flow, 7–8, 8f. *See also* Ionosphere-thermosphere (I-T) system

Communication/Navigation Outage Forecasting System (C/NOFS)
   mission, low-latitude PBMOD for. *See* Low-latitude PBMOD
   satellite, in ESF irregularities, 218

Communication speed, in M-IT coupling along magnetic field line, 208, 209

Community atmospheric model (CAM), 183–184

Community Coordinated Modeling Center (CCMC), 146, 147, 149, 304
   NASA, NCAR TIE-GCM and. *See* Thermosphere-ionosphere-electrodynamics general circulation model (TIE-GCM)

Community earth system model (CESM), 183, 184

Community model, of coupled thermosphere/ionosphere system
   NCAR TIEGCM. *See* Thermosphere-ionosphere-electrodynamics general circulation model (TIE-GCM)

Conductances, magnetospheric, 60, 62f
   Hall conductance, 61, 62, 65
   zonal Pedersen conductance, 61, 62, 65

Conductivities, ionospheric, 61–62

Constellation Observing System for Meteorology, Ionosphere and Climate/Formosa Satellite 3 (COSMIC/FORMOSAT-3) mission, 277, 278

Constellation Observing System for Meteorology Ionosphere and Climate (COSMIC), 75, 79
   electron density, 278, 279f

Continuity equation, 41–42
   in Earth's thermosphere, 15
   in GITM model, 97

Cooling and heating, ions
   ESF plumes and, 254, 256, 256f

Coriolis effect, earth's thermosphere and, 15–16

Coronal mass ejections (CMEs), 300

Corotating interaction regions (CIRs), 121, 122

COSMIC (Constellation Observing System for Meteorology Ionosphere and Climate), 75, 79
   electron density, 278, 279f

COSMIC/FORMOSAT-3 (Constellation Observing System for Meteorology, Ionosphere and Climate/Formosa Satellite 3) mission, 277, 278

Coulomb collisions, 6

Coupled IT models, physics-based, 147t, 149, 153

Coupled magnetosphere-ionosphere-thermosphere model (CMIT)
   NCAR TIE-GCM. *See* Thermosphere-ionosphere-electrodynamics general circulation model (TIE-GCM)

Coupled thermosphere ionosphere model (CTIM), 19, 86

Coupled thermosphere ionosphere plasmasphere, 86

Coupled Thermosphere Ionosphere Plasmasphere Electrodynamics model (CTIPe), 19, 135, 182
   physics-based coupled IT models, 147, 147t, 149, 153, 158

Coupling, ionosphere and magnetosphere, 57, 60–61, 62f

Coupling energetics and dynamics of atmospheric regions (CEDAR)
   ETI challenge (2009–2010)
      IT model, evaluation. *See* Ionosphere/thermosphere (IT) models, systematic evaluation

Coupling processes, in IT, 161–168
   ion-electron collisions, 161–163, 162f
      in F region, 162–163, 163f, 164f
      overview, 161–162, 162f
   ion-neutral collisions
      in lower thermosphere, 165–167, 166f, 167f
      in upper thermosphere, 164, 165, 165f

Cross-field dynamics
   numerical methods in ionospheric modeling, 53–55
      Eulerian grid, 54–55
      Lagrangian grid, 54, 54f

CTIM (coupled thermosphere ionosphere model), 19, 86

CTIPe (coupled thermosphere ionosphere plasmasphere electrodynamics) model, 19, 182
   physics-based coupled IT models, 147, 147t, 149, 153, 158

Current system, ionospheric, 66–68

"Curvature" terms, 16

Customers, system impacts and, 300–302, 301t, 302t

CXI (Christmas Island)
   AFRL SCINDA ground station at, 109, 110f–111f, 111
   R-T linear growth rate in, 109, 110f–111f

D

Daily zonal wavenumber one (DW1), tide component, 175, 176

Data
   CHAMP mission and, 260
   and models for upper atmosphere, 40–41
   observational, CHAMP satellite for
      IT model evaluation, 146

Data assimilation, upper atmosphere, 273–281
   EnKF, 274–276
      application to TIEGCM, 275–276, 275t
      filter algorithms, 274–275
   error issues, model, 280
      biases, 280
      error growth, 280
   estimation of model parameters, 278, 279–280, 280f
   geophysical information content of different observations, 277–278
      CHAMP mass density, 277–278
      COSMIC electron density, 278, 279f
   overview, 273–274
   thermosphere and ionosphere
      effects of self-consistent treatment, 276–277, 276f
      EnKF application, 275–276, 275t
Data assimilation algorithm, 293, 296
Data assimilation models, of ionosphere
   physics-based, 147, 147t, 149
Data Assimilation Research Testbed (DART), 273, 274, 275
DE3. *See* Diurnal eastward propagating wavenumber 3 component (DE3)
Defense Meteorological Satellite Program (DMSP), 111
   density measurements by, 111, 113, 114f–115f
Density
   Earth's thermosphere and, 18–19, 20f, 21f
   and temperature structure
      of ESF plumes. *See* Equatorial spread *F* (ESF), plumes
DE (Dynamics Explorer) satellite
   ion-neutral collisions in upper thermosphere, 164, 165, 165f
Descending phase of solar cycle 23 (2002–2008), modeling, 128–129, 129f
Dispersion relation, for FBGD, 233–234
Diurnal eastward propagating wavenumber 3 component (DE3), 181, 182, 183, 185, 187, 189f, 193, 194f
Diurnal variation of vertical drift velocities, 141f
Diurnal westward propagating wave 1 (DW1), 181, 183, 185, 187, 188f, 191, 192, 192f
"Divergence wind," 15
DMSP (Defense Meteorological Satellite Program)
   density measurements by, 111, 113, 114f–115f
Doppler orbitography and radiopositioning integrated by satellite (DORIS), 285
Doppler spectra, *E* region irregularities and, 221, 222f, 223, 223f
Downward electron heat flow in polar cap, 5. *See also* Ionosphere-thermosphere (I-T) system
Downward heat conduction, 39, 43, 45f, 46
Downwelling, 13, 16, 18, 19
D region
   absorption product, 302, 303f
   of ionosphere
      HF radio transmissions, 299–300, 301

DW1 (daily zonal wavenumber one), tide component, 175, 176
Dynamics Explorer (DE) satellite
   ion-neutral collisions in upper thermosphere, 164, 165, 165f

E

Earth's thermosphere
   physical characteristics and modeling, 13–26
      continuity equation, 15
      Coriolis effect, 15–16
      energy equations, 17
      gas law and hydrostatic balance, 14–15
      global wind, 18–19, 20f, 21f
      horizontally stratified fluid, 15
      ion drag, 16
      Lagrangian *vs.* Eulerian frames of reference, 15
      momentum equations, 16
      neutral composition, 17–18
      neutral composition bulge, 21–22, 23f
      other common assumption, 15–16
      overview, 13–14
      temperature, density, and composition structure, 18–19, 20f, 21f
      thermal expansion, 19, 21, 22f
      viscous drag, 16
      winds and nonlinearities at high latitude, 22, 24–25, 24f, 25f
Eddy diffusion, in NCAR-TIEGCM, 42
EIA (equatorial ionization anomaly), 182, 251
   ionization crests within ESF plume and, 253
   low latitude CHAMP satellite observations, of ionosphere, 260, 262–263, 262f
Electric current, in M-IT coupling, 211
Electric field mapping, in M-IT coupling
   along field line, 211
Electric potential, in ionospheric electrodynamics, 65–66
   PDE for, 57, 59, 60, 61, 62f
Electrodynamics, ionospheric. *See* Ionospheric electrodynamics modeling
Electrodynamics thermosphere ionosphere (ETI) challenge, 2009–2010
   CEDAR, IT model evaluation. *See* Ionosphere/thermosphere (IT) models, systematic evaluation
Electron density, 7f, 9f, 284, 288, 288f
   analyses, RMSD of, 276
   COSMIC, 278, 279f
   predictions, at CHAMP track, 150, 151t, 152t
Electron heat flux, 5
Electron quenching, 33, 34
Electron temperature morning overshoot (MO), 263
Electrostatic coupling, of M-IT
   inductive-dynamic *vs.*, 203–213
      governing equations, differences in, 203–205

inductive-dynamic *vs.*, (*continued*)
   physical description of coupling processes, differences, 205–207, 206f, 207f
   physical quantities, differences. *See* Physical quantities, M-IT coupling and
   reflection, effect of, 208
EMA (equatorial mass density anomaly)
   ionospheric forcing and, 265–266
   lower atmospheric forcing and
   wave-4 structure, 267
Empirical models
   for ionosphere/thermosphere, 147t, 148, 148t, 153
   IRI model. *See* International reference ionosphere (IRI) models
   MSIS model. *See* Mass spectrometer and incoherent scatter (MSIS) models
   of thermosphere, 40, 41
Energetics, and composition in thermosphere. *See* Thermospheres, energetics and composition
Energy equations, 41–42
   Earth's thermosphere and, 17
   in GITM model, 97–98
Energy input and output, in TIE-GCM model, 74–75
Energy source, global, 10
Energy transfer, 6
EnKF. *See* Ensemble Kalman filtering (EnKF)
Ensemble Kalman filtering (EnKF), 284, 296f
   upper atmosphere data assimilation with. *See also* Data assimilation, upper atmosphere
     application to TIEGCM, 275–276, 275t
     filter algorithms, 274–275
Equatorial electron temperature, 263–264, 263f
Equatorial ionization anomaly (EIA), 4, 133–134, 182, 251
   ionization crests within ESF plume and, 253
   low latitude CHAMP satellite observations, of ionosphere, 260, 262–263, 262f
Equatorial mass density anomaly (EMA)
   ionospheric forcing and, 265–266
   lower atmospheric forcing and
   wave-4 structure, 267
Equatorial plasma irregularities, magnetic signatures of, 264
Equatorial spread *F* (ESF), 241
   irregularity, ionosphere variability and, 183
   low latitude CHAMP satellite obeservations, 264
   plasma irregularities and, 218
   plumes, density and temperature structure, 251–257
     ionization crests within, 252–254, 253f, 254f
     overview, 251
     SAMI3 simulations, 252
     thermodynamics, 254–256, 255f, 256f
     time evolution of, 256–257, 256f

Equatorial wind jet, ionospheric forcing and, 266–267, 266f
E region
   electron density, 32
   ionosphere, GPS/GNSS system and, 299, 300
   ionospheric electrodynamics modeling. *See* Ionospheric electrodynamics modeling
   $O^+(^4S)$ sources and sinks, 33
E region irregularities, 218–219, 220–226
   Doppler spectra and, 221, 222f, 223, 223f
   Farley-Buneman wave theory
     FBGD instability. *See* Farley-Buneman gradient drift (FBGD)
     turbulence. *See* Farley-Buneman turbulence
Error correlation, vertical, 287
Error issues, model
   for data assimilation, 280
     biases, 280
     error growth, 280
ESF. *See* Equatorial spread *F* (ESF)
Estimating Model Parameters From Ionospheric Reverse Engineering (EMPIRE), 283, 284, 290–293, 291f, 294f, 295f. *See also* Ionospheric Data Assimilation Four Dimensional (IDA4D)
Estimation method. *See also* Plasmasphere-ionosphere data assimilation model
   linear, 284–285, 285f
   nonlinear, 285–286
ETI (electrodynamics thermosphere ionosphere) challenge, 2009-2010
   CEDAR, IT model evaluation. *See* Ionosphere/thermosphere (IT) models, systematic evaluation
Eulerian frame of reference
   Lagrangian *vs.*, 15
Eulerian grid, numerical methods in ionospheric modeling, 54–55
EUVAC model, ionospheric conductivities and, 61, 62
EVE (Extreme Ultraviolet Experiment), 306
Exobase, 13, 14
Extreme Ultraviolet Experiment (EVE), 306

F

Fabry-Perot interferometers (FPI), 41, 42
   measurements, 87
FAC (field-aligned currents), net, 57, 60, 61, 66, 68
FAIM model, 109
FAIs (field-aligned irregularities), 219
Faraday's law, M-IT coupling and, 201, 202, 203, 213
Farley-Buneman gradient drift (FBGD)
   dispersion relation, 233–234
   instabilities, 219, 220–224, 222f, 223f

Farley-Buneman turbulence, 224–225
  Bourret's integral equation in, 224–225, 235
  stochastic differential equation formalism, 224–225, 235
Farley-Buneman wave theory, 218–219
Fast wave mode, 205, 206
FBGD. *See* Farley-Buneman gradient drift (FBGD)
Field-aligned current (FAC) flow, 248
Field-aligned currents (FAC), net, 57, 60, 61, 66, 68
Field-aligned dynamics
  numerical methods in ionospheric modeling, 50–53
    fully implicit algorithm, 50–51
    numerical comparison, 52–53, 53f
    semi-implicit algorithm, 51–52
Field-aligned irregularities (FAIs), 219
Field line interhemispheric plasma (FLIP) model, 29, 30, 32, 32f, 263
  $O^+(^4S)$ sources and sinks, 33, 33f
  5 September 2005
    solar cycle variation, model, 30–32, 32f
      modeled ion densities for, 32, 32f
      model neutral densities for, 30–32, 32f
Filter algorithms, EnKF, 274–275
FLIP (field line interhemispheric plasma) model, 29, 30, 32, 32f, 33, 33f, 263
Flux-corrected transport (FCT) technique, 102
Flux tube, auroral, 309, 310
  forward problem and, 310, 311f
  for optical emissions, 309, 310, 312, 313
Forcing parameter estimation, EnKF and, 278, 279–280, 280f
Forecasting
  D region enhancements, 301
  flares, 301–302, 306
  geomagnetic storm, 300, 301, 303, 305
  ionospheric conditions, 302, 303–306
  space environment. *See* National Oceanic and Atmospheric Administration (NOAA); Space Weather Prediction Center (SWPC)
Forecasting Ionospheric Real-Time Scintillation Tool, by SWPT, 306
Forward problem, invertibility of
  flux tubes in, 310, 311f
  mathematical description, 310, 311–312
  overview of, 310, 311f
"Fossil" plume state, 251, 256, 257
Four-dimensional variational assimilation (4DVAR), 284
FPI (Fabry-Perot interferometers), 41, 42
  measurements, 87
Fredholm formulation, model-based inversion and, 314–315
F region
  ion-electron collisions in, 162–163, 163f, 164f
  ionosphere, GPS/GNSS system and, 299, 300

ionospheric electrodynamics modeling in. *See* Ionospheric electrodynamics modeling
irregularities, 226–230, 227f, 229f
  bottom-type layers, 226–227, 227f, 230
  collisional interchange instability, 226, 227
  collisional shear instability, 227–228
  numerical simulation of plasma irregularities, 228, 229f, 230
  R-T instability and, 226–228, 230
  SLT in, 226
  Taylor-Goldstein equation, 228
Fully implicit algorithm
  numerical methods in ionospheric modeling, 50–51
  semi-implicit algorithm *vs.*, 52–53, 53f

G
GAIA (ground-to-topside model of atmosphere and ionosphere for aeronomy), 265–266, 267
GAIM (global assimilation of ionospheric measurements), 85–86
Gas law, 14–15
Gauss-Markov Kalman filter (GMKF) model, 149
General circulation model (GCM), 273, 274
"Generalized Rayleigh-Taylor" instability, 226, 227, 241–242
Geographic longitude (GLON)
  for driving global simulations, 111, 113, 114f–115f
Geomagnetic perturbations, 66–68, 67f
Geomagnetic storms, 42–46, 43f, 44f, 45f
  forecasting, 300, 301, 303, 305
Geophysical information content
  of different observations, 277–278
    CHAMP mass density, 277–278
    COSMIC electron density, 278, 279f
Geostrophic balance, 18
GFS (global forecast system) model, 305
GITM. *See* Global ionosphere-thermosphere model (GITM)
Global assimilation of ionospheric measurements (GAIM), 85–86
Global Assimilation of Ionospheric Measurements-Gauss Markov (GAIM-GM) model, 3–4
Global forecast system (GFS) model, 305
Global Ionosphere Plasmasphere model (GIP) model, 135
Global ionosphere-thermosphere model (GITM), 85–98, 147, 147t, 149, 150, 153
  equations in, 86, 96–98
    continuity equation, 97
    energy equation, 97–98
    for ions, 98
    momentum equations, 97
    Navier-Stokes equation, 86

Global ionosphere-thermosphere model (GITM) (*continued*)
 global MHD model and, coupling, 87
 IGRF magnetic field in, 86
 nonhydrostatic processes and, 87–96
  altitudinal dependence, 91–96, 93f, 94f, 95f
  buoyancy acceleration and vertical winds, 87–91, 88f, 89f, 90f
  Joule heating, enhancements. See Joule heating
 overview, 85–86
 simulations, with MSIS and IRI, 86
 variables in, 96–97
Global navigation satellite systems (GNSS), 299, 300
 customers and, 300–302, 301t, 302t
 SAMI3 comparisons, WHI 2008 and, 121–125, 122f, 123f, 124f
Global positioning system (GPS), 299, 300
 customers and, 300–302, 301t, 302t
 in SWPT, 306, 306f
Global-scale oscillations, 101
Global scale wave/wind model (GSWM), 41, 42, 173, 174, 175f, 176
Global thermosphere, 101
Global thermosphere model (GTM), 102. See also Traveling atmospheric disturbance (TAD)
Global wind, Earth's thermosphere and, 18–19, 20f, 21f
GLON (geographic longitude)
 for driving global simulations, 111, 113, 114f–115f
GMKF (Gauss-Markov Kalman filter) model, 149
GNSS (global navigation satellite systems), 299, 300
 customers and, 300–302, 301t, 302t
 SAMI3 comparisons, WHI 2008 and, 121–125, 122f, 123f, 124f
Governing equations, for M-IT coupling, 203–205
G/P (gravitational and pressure-gradient) forces, on plasma, 63–65, 64f, 65f
GPS (global positioning system), 299, 300
 customers and, 300–302, 301t, 302t
 in SWPT, 306, 306f
GRACE mission, 268
Gravitational and pressure-gradient (G/P) forces, on plasma, 63–65, 64f, 65f
Gravity wave (GW) coupling, 101, 102, 103f, 104. See also Traveling atmospheric disturbance (TAD)
Gravity waves (GW), 101
 in initiating ESF, 218
Ground-to-topside model of atmosphere and ionosphere for aeronomy (GAIA), 265–266, 267
GSWM (global scale wave/wind model), 41, 42, 173, 174, 175f, 176
Guass-Markov Kalman filter, 2, 284

H

Hall conductance, 61, 62, 65
Heat flux, at top of ionosphere, 163, 164f, 167
Heating
 and cooling, ions
  ESF plumes and, 254, 256, 256f
  rate, in M-IT coupling, 212–213
 in thermosphere, 40, 42–46, 43f, 44f, 45f
Heating function, 103f
HEUVAC (high-resolution version of solar EUV irradiance model for aeronomic calculations) model, 30
High-resolution version of solar EUV irradiance model for aeronomic calculations (HEUVAC) model, 30
High-speed streams (HSSs), 121, 122
Homopause (turbopause), 18
Horizontally stratified fluid, Earth's thermosphere and, 15
Horizontal Wind Model-07, 244
Horizontal wind model (HWM-93), 135
Horizontal wind models (HWMs), 149
HSSs (high-speed streams), 121, 122
HWM-07, empirical wind model, 109
HWMs (horizontal wind models), 149
Hydrostatic balance, Earth's thermosphere and, 14–15

I

IDEA (Integrated Dynamics in Earth's Atmosphere), 305, 306
IGRF (international geomagnetic reference field) model, 58, 66, 69, 86, 109
IGS (International Global Navigation Satellite Systems Service) TEC, SAMI3 comparisons
 WHI 2008 and, 121–125, 122f, 123f, 124f
Incoherent scatter radar (ISR), 285, 309, 313, 315
 optical analysis and, 309, 310, 312, 314, 314f, 315
 optical spectrometer and, 319
Incoherent Scatter Radar Model (MSISE-00), 135
Incoherent scatter returns
 on ARPA Long-range Tracking and Identification Radar on KWA, 111, 112f
Inductive-dynamic coupling, of M-IT, 203–213
 electrostatic *vs.*, 203–213
  governing equations, differences in, 203–205
  physical description of coupling processes, differences, 205–207, 206f, 207f
  physical quantities, differences. See Physical quantities, M-IT coupling and
  reflection, effect of, 208
Inertia
 electron, in collisional drift wave instabilities, 230
 inertial regime, F region irregularities and, 226, 233
 ion, for Farley-Buneman waves, 220, 233
 ion, in SAMI3 simulations, 252

Integrated Dynamics in Earth's Atmosphere (IDEA), 305, 306
Integrated Sun-Earth System for Operational Environment (ISES-OE) model, 119–120, 120f
International geomagnetic reference field (IGRF) model, 58, 66, 69, 86, 109, 135
International Global Navigation Satellite Systems Service (IGS)
  TEC, SAMI3 comparisons
    WHI 2008 and, 121–125, 122f, 123f, 124f
International Reference Ionosphere (IRI) model, 77, 85, 102, 147, 147t, 148, 153, 156, 302
  version 2007, 262
International space environment services, 302
Inversion, model-based
  of auroral processes, 309–320
    characteristic responses, 312–314, 313f, 314f
    forward problem. See Forward problem
    Fredholm formulation, 314–315
    future directions, 319–320, 319f
    inversion approaches, 315, 316–317
    overview, 309–310
    time history, 315, 316f
    validation, 317–319, 317f, 318f
Ion acoustic speed, Farley-Buneman waves and, 218, 221, 223
Ion drag, in Earth's thermosphere, 16
Ion-electron collisions
  coupling processes in IT, 161–163, 162f
    in F region, 162–163, 163f, 164f
    overview, 161–162, 162f
Ion frictional heating, 165, 166, 168
Ion inertia
  for Farley-Buneman waves, 220, 233
  in SAMI3 simulations, 252
Ionization, 102
  crests, within ESF plumes, 252–254, 253f, 254f
  photoelectron impact, 30
  photoionization. See Photoionization (hv)
  processes, 29, 30f
  reaction rate coefficients, 31t
Ion momentum equation, in GITM model, 98
Ion-neutral collisions
  in lower thermosphere, 165–167, 166f, 167f
  in upper thermosphere, 164, 165, 165f
Ionosphere model, time-dependent, 5
Ionosphere-plasmasphere coupling, 8, 9f. See also Ionosphere-thermosphere (I-T) system
Ionosphere plasmasphere electrodynamics (IPE) model, 305
Ionospheres, 107
  observations of, low latitude CHAMP satellite, 260–264, 261f. See also Low latitude CHAMP satellite observations

  EIA, 260, 262–263, 262f
  equatorial electron temperature, 263–264, 263f
  magnetic signatures of equatorial plasma irregularities, 264
  photochemistry, solar cycle changes in, 29–36
  products and services, customers and requirements for, 299–307
    operational models, 302, 303f
    overview, 299
    space weather impacts, 299–300
    SWPT. See Space Weather Prediction Testbed (SWPT)
    system impacts, 300–302, 301t, 302t
Ionosphere/thermosphere (IT) models
  coupling processes in. See Coupling processes, in IT
Ionosphere/thermosphere (IT) models, systematic evaluation, 145–158
  metrics, 148t, 149–150
  models, 146–149, 147t
    coupled IT models, physics-based, 147t, 149, 153
    data assimilation models, physics-based, 147, 147t, 149
    empirical, 147t, 148, 148t, 153
    ionospheric models, physics-based, 147t, 148–149
  observational data, 146
  overview, 145–146
  results, 150–156
    electron density, 150–153, 151t, 152t
    neutral density, 153, 154t, 155t
    vertical drifts at Jicamarca, 153, 156, 156t, 157t
Ionosphere-thermosphere (I-T) system
  characteristics, 3
  collisionless plasma flow, 7–8, 8f
  downward electron heat flow in polar cap, 5
  ionosphere-plasmasphere coupling, 8, 9f
  ion temperature anisotrophy, 5, 7f
  lower atmosphere wave effects, 10
  neutral rain on thermosphere, 9–10
  plasma and neutral density structures, 8–9, 10f, 11f
  polar wind and auroral ion outflow, 4, 5f
  subauroral red (SAR) arcs, 5–7, 7f
  thermoelectric heat flow in return current regions, 5, 6f
Ionosphere/thermosphere (I/T) system
  EnKF data assimilation and. See Data assimilation, upper atmosphere
  hidden states, 319–320, 319f
Ionospheric conductivities, 61–62
Ionospheric Data Assimilation Four Dimensional (IDA4D)
  CHAMP validation of, 286–288, 287f
  data sources, 286
  development of, 283, 284
  linear estimation method, 284–285, 285f
  nonlinear estimation method, 285–286
  plasmasphere-ionosphere data assimilation model, 293–294, 294f, 296f

Ionospheric Data Assimilation Four Dimensional (IDA4D) (*continued*)
  and scientific investigations
    EMPIRE results, 290–293, 291f, 294f, 295f
    IDA4D results, 288–290, 289f, 290f
Ionospheric densities, low-latitude
  plasma density. *See* Plasma density
  statistical study, 111, 113–115, 113f, 114f–115f
Ionospheric electrodynamics convection patterns, 289f
Ionospheric electrodynamics modeling, 57–69
  current system, 66–68
  equations, 58–60
  future directions, 68–69
  geomagnetic perturbations, 66–68, 67f
  ionosphere-magnetosphere coupling, 57, 60–61, 62f
  ionospheric conductivities, 61–62
  overview, 57–58
  plasma gravity and pressure-gradient currents, 63–65, 64f, 65f
  solving for electric potential, 65–66
  thermospheric winds, 62–63
Ionospheric forcing, thermospheric response, 265–267
  equatorial mass density anomaly, 265–266
  equatorial wind jet, 266–267, 266f
Ionospheric irregularities, 217–235
  collisional drift wave instabilities, 230–231, 232f
  E region. *See E* region irregularities
  F region. *See F* region, irregularities
  fundamentals, 219–220
  overview, 217–219
  plasma density, 217, 219–220
    in *F* region, 226, 227f, 228, 229f, 230
Ionospheric models
  low-latitude PBMOD for C/NOFS mission. *See* Low-latitude PBMOD
  numerical methods in. *See* Numerical methods, in ionospheric modeling
  physics-based, 147t, 148–149
Ionospheric simulations using SAMI3, long-term, 119–129
  descending phase of solar cycle 23 (2002–2008), modeling, 128–129, 129f
  model description, 121, 122f
  overview, 119–121, 120f
  SAMI3 comparisons with IGS TEC during WHI 2008, 121–125, 122f, 123f, 124f
  vertical drifts, modeling, 125–128, 126f, 127f, 128f
Ionospheric wind dynamo, 58, 59, 64
Ion temperature anisotrophy, 5, 7f. *See also* Ionosphere-thermosphere (I-T) system

Ion velocity drift meter (IVM), 108, 109, 111, 113f, 114f–115f, 115, 128
IPE (ionosphere plasmasphere electrodynamics) model, 305
IRI (international reference ionosphere) models, 77, 85, 147, 147t, 148, 153, 156, 302
  version 2007, 262
Irregularities, ionospheric. *See* Ionospheric irregularities
ISES-OE (Integrated Sun-Earth System for Operational Environment) model, 119–120, 120f
ISES-OE model is Another Model of Ionosphere (SAMI3) long-term simulations of ionosphere using. *See* Ionospheric simulations using SAMI3, long-term
ISR. *See* Incoherent scatter radar (ISR)
IT. *See* Ionosphere/thermosphere (IT) models
IVM (ion velocity drift meter), 108, 109, 111, 113f, 114f–115f, 115, 128

J

Jacchia-Bowman 2008 (JB2008), 147, 147t, 148, 153
Jacchia models, for upper atmosphere, 41
Jacobian matrix, 286
JB2008 (Jacchia-Bowman 2008), 147, 147t, 148, 153
Jet Propulsion Laboratory-Global Assimilative Ionospheric Model (JPL-GAIM), 147, 147t, 149, 150, 153
Jicamarca, vertical ion drifts
  predicting, 153, 156, 156t, 157t
Jicamarca latitudinal plane, 244f
Jicamarca Unattended Long-Term Studies of Ionosphere and Atmosphere (JULIA), 146
Joule heating, 13, 17, 284
  defined, 17, 212
  neutral composition bulge and, 21
Joule heating, enhancements
  in nonhydrostatic processes, 87
    altitudinal dependence, 91–96, 93f, 94f, 95f
    buoyancy acceleration and vertical wind, 87–91, 88f, 89f, 90f
JPL-GAIM (Jet Propulsion Laboratory-Global Assimilative Ionospheric Model), 147, 147t, 149, 150, 153
JULIA (Jicamarca Unattended Long-Term Studies of Ionosphere and Atmosphere), 146

K

KWA (Kwajalein Atoll)
  incoherent scatter returns on ARPA Long-range Tracking and Identification Radar on, 111, 112f
  low-latitude ionosphere stimulation in, 109, 110f–111f, 111
Kwajalein Atoll (KWA)
  incoherent scatter returns on ARPA Long-range Tracking and Identification Radar on, 111, 112f
  low-latitude ionosphere stimulation in, 109, 110f–111f, 111

L

Lagrangian frame of reference, Eulerian *vs.*, 15
Lagrangian grid, numerical methods in ionospheric modeling, 54, 54f
Large angle and spectrometric coronagraph (LASCO) model, 120
LASCO (large angle and spectrometric coronagraph) model, 120
LFM (Lyon-Fedder-Mobarry) magnetosphere model, 73
Local coupling, in M-IT
   via collisions, 201, 202
Long-range coupling, M-IT coupling. *See* Magnetosphere-ionosphere-thermosphere (M-IT) coupling
Long-range tracking and identification radar
   ARPA, incoherent scatter returns on, 111, 112f
Lower atmosphere, around solstice, 190–193, 190f, 191f, 192f, 193f, 194f, 195f
Lower atmosphere wave effects, 10. *See also* Ionosphere-thermosphere (I-T) system
Lower atmospheric forcing, thermospheric response, 267–269
   terminator wave structures, 267, 268f
   thermosphere and ionosphere effects during SSW, 267–269, 269f
   wave-4 structure of EMA and neutral wind, 267
Low latitude CHAMP satellite observations, 259–269
   CHAMP mission and data, 260
   ionosphere, observations of, 260–264, 261f
      EIA, 260, 262–263, 262f
      equatorial electron temperature, 263–264, 263f
      magnetic signatures of equatorial plasma irregularities, 264
   overview, 259–260
   thermosphere, observations of, 264–269
      ionospheric forcing, thermospheric response. *See* Ionospheric forcing
      lower atmospheric forcing, thermospheric response. *See* Lower atmospheric forcing
      solar forcing, thermospheric response, 265
Low-latitude PBMOD, for C/NOFS mission, 107–115
   ionospheric densities, statistical study, 111, 113–115, 113f, 114f–115f
   model description and output, 108–111, 108f, 110f–111f, 112f
   overview, 107–108
Low latitudes of upper thermosphere
   during geomagnetic storms, causes of temperature variations in, 42–46, 43f, 44f, 45f
Lyon-Fedder-Mobarry (LFM) magnetosphere model, 73

M

Magnetic Apex coordinate system, 135
Magnetic field line, in M-IT coupling
   communication speed along, 208, 209
   electric field mapping along, 211
Magnetic local time (MLT), 289
Magnetic perturbations, 66–68, 67f
   in M-IT coupling, 211
Magnetic signatures of equatorial plasma irregularities, 264
Magnetosphere-ionosphere-thermosphere (M-IT) coupling, 40, 201–213
   inductive-dynamic *vs.* electrostatic coupling, 203–213
      governing equations, differences in, 203–205
      physical description of coupling processes, differences, 205–207, 206f, 207f
      physical quantities, differences. *See* Physical quantities, M-IT coupling and
      reflection, effect of, 208
   overview, 201–203
Magnetospheres
   dynamics of, 58
   ionosphere and, coupling, 57, 60–61, 62f
   net FAC from, 57, 60, 61
Magnetospheric conductances, 60, 62f
   Hall conductance, 61
   Zonal Pedersen conductance, 61
Magnetospheric model
   CCMC and, 304
   requirements, 304
March equinox, 136f
Mass spectrometer, for satellite guidance, 40
Mass spectrometer and incoherent scatter (MSIS) models, 39, 40f, 41, 85, 109, 147, 147t, 148, 153, 154t, 157, 158
   MSIS-90, 312
Maximum entropy method (MEM), 317
Medium-scale traveling ionospheric disturbances (MSTIDS), 219
MEM (maximum entropy method), 317
Meridional and zonal direction, momentum equations for, 16
Mesopause, 39
Mesosphere and lower thermosphere (MLT) region, 75
Metrics, IT model performance, 148t, 149–150
Middle latitudes of upper thermosphere, 42–46, 43f, 44f, 45f
Millstone Hill ISR, 290, 291f
Minimum mean square estimator (MMSE), 316, 317
M-IT. *See* Magnetosphere-ionosphere-thermosphere (M-IT) coupling
MLT (mesosphere and lower thermosphere) region, 75
MMSE (minimum mean square estimator), 316, 317
MO (morning overshoot), electron temperature, 263
Model comparisons in american sector (75°W), 137–141
Model discrepancies, sources of, 141–142
Model error growth, 280
Model output, in TIE-GCM model, 75

Models, for CEDAR ETI Challenge, 146–149, 147t
 empirical models, 147t, 148, 148t, 153
 physics-based models
  coupled IT, 147t, 149, 153
  data assimilation models, physics-based, 147, 147t, 149
  ionospheric, 147t, 148–149
Models, for upper atmosphere, 40–41
 Jacchia models, 41
 semiempirical models, 41
 theoretical first principle models, 41
Modified magnetic apex coordinates, 59
Momentum equations, 8
 in GITM model, 97
 for meridional and zonal direction, 16
Monthly averaged zonal mean winds, from TIE-GCM and NOGAPS-ALPHA models, 174, 175f
Morning overshoot (MO), electron temperature, 263
MSIS (mass spectrometer and incoherent scatter) models, 39, 40f, 41, 85, 109, 147, 147t, 148, 153, 154t, 157, 158
 MSIS-90, 312
MSTIDS (medium-scale traveling ionospheric disturbances), 219

N

NASA (National Aeronautics and Space Administration), 304
 CCMC, NCAR TIE-GCM and. See Thermosphere-ionosphere-electrodynamics general circulation model (TIE-GCM)
National Aeronautics and Space Administration (NASA), 304
 CCMC, NCAR TIE-GCM and. See Thermosphere-ionosphere-electrodynamics general circulation model (TIE-GCM)
National Center for Atmospheric Research (NCAR), 3, 273
 TIE-GCM. See Thermosphere-ionosphere-electrodynamics general circulation model (TIE-GCM)
National Centers for Environmental Prediction, 305
National Oceanic and Atmospheric Administration (NOAA), 299, 302, 304, 306
National Research Laboratory-Mass Spectrometer Incoherent Scatter E00 model, 243
National Weather Service, 302, 305
Naval Research Laboratory (NRL), 39, 40f, 41
 ISES-OE program, 119–120
Naval Research Laboratory (NRL) Mass Spectrometer, 135
Naval Research Laboratory Mass Spectrometer and Incoherent Scatter Extended-00 (NRLMSISE-00), 147, 148
Navier-Stokes equations, GITM and, 86
Navy operational global atmospheric prediction system-advanced level physics high altitude (NOGAPS-ALPHA) model
 as bottom boundary for NCAR/TIEGCM, 171–179
  descriptions, 172–174, 173f
  monthly averaged zonal mean winds, 174, 175f
  overview, 171–172
  results, 174–178
  sudden stratospheric warming (January 2009), changes, 177, 178, 178f
  tides, 172, 175–177, 176f, 177f, 178f
 forecast/assimilation system, 80
NCAR (National Center for Atmospheric Research)
 TIE-GCM. See Thermosphere-ionosphere-electrodynamics general circulation model (TIE-GCM)
Neutral composition, Earth's thermosphere and, 17–18
 bulge, 21–22, 23f
Neutral density, predictions
 at CHAMP track, 153, 154t, 155t
Neutral density perturbation, 9
Neutral rain on thermosphere, 9–10. See also Ionosphere-thermosphere (I-T) system
Neutral temperature perturbations, 104, 104f
Neutral wind, lower atmospheric forcing and, 267
Nightglow, chemistry of $O(^1D)$ in, 245t
NOAA (National Oceanic and Atmospheric Administration), 299, 302, 304, 306
NOGAPS-ALPHA. See Navy operational global atmospheric prediction system-advanced level physics high altitude (NOGAPS-ALPHA) model
Nonhydrostatic equilibrium flows, 102
Nonhydrostatic processes, GITM and. See Global ionosphere-thermosphere model (GITM)
Nonlinearities, at high latitude, 22, 24–25, 24f, 25f
North Pole (NP), temperature at
 WACCM-X model variability on various time scales and, 185, 186f
NP (North Pole), temperature at
 WACCM-X model variability on various time scales and, 185, 186f
NRL (Naval Research Laboratory), 39, 40f, 41
 ISES-OE program, 119–120
NRLMSISE-00 (Naval Research Laboratory Mass Spectrometer and Incoherent Scatter Extended-00), 147, 148
NRLSSI (NRL's solar spectral irradiance) model, 120
NRL's solar spectral irradiance (NRLSSI) model, 120
NRLWang-Sheeley model, 120
N^+ sources and sinks, 35–36, 36f
"Nudging," TIE-GCM winds, 173–174, 176
Numerical methods, in ionospheric modeling, 49–55
 cross-field dynamics, 53–55
  Eulerian grid, 54–55
  Lagrangian grid, 54, 54f

field-aligned dynamics, 50–53
  fully implicit algorithm, 50–51
  numerical comparison, 52–53, 53f
  semi-implicit algorithm, 51–52
 general equations, 49–50
 overview, 49
Numerical simulations of equatorial spread F, 242–248
 airglow imagery simulation, 245–246
 coherent/incoherent scatter radar simulation, 244–245
 conservation laws, 242
 empirical model drivers, 243–244
 numerical scheme, 242–243
 simulation setup, 244
 in situ satellite probe simulations, 246–248

O

Observational data, CHAMP satellite for
 IT model evaluation, 146
Observing system simulation experiments (OSSEs), 274, 276, 278
$O^+(^2D)$ sources and sinks, 33–34, 34f
Ohm's law
 in M-IT coupling, 201, 203, 204, 205
  for electric field in inductive-dynamic coupling, 211
Operational models, 302, 303f
 D Region absorption product, 302, 303f
 IRI model, 302
 USTEC model, 302, 303f
$O^+(^2P)$ sources and sinks, 34, 34f
Optical emissions
 auroral flux tube for, 309, 310, 312, 313
 ISR and, 309, 310, 312, 314, 314f, 315
Orographic forcing, 101
Oscillations, global-scale, 101
O^+2 sources and sinks, 35, 35f
OSSEs (observing system simulation experiments), 274, 276, 278
$O^+(^4S)$ sources and sinks, 32–33, 33f
Ovation Prime model, 304, 305, 305f

P

Parameterized ionospheric model (PIM), 243–244
Partial differential equation (PDE)
 for electric potential, 57, 59, 60, 61, 62f
 net FAC and, 57, 61
Participating models in equatorial-primo, 137
PBMOD (Physics-based ionospheric model)
 low-latitude, for C/NOFS mission. See Low-latitude PBMOD
PDE. See Partial differential equation (PDE)
PE (prediction efficiency), model performance and, 146, 148t, 149, 150, 151t, 153, 154t, 156, 156t

"Penetration electric field," 212, 213
Photochemistry, of ionosphere and thermosphere
 solar cycle changes, 29–36
Photoelectron fluxes, HEUVAC irradiances for, 29, 30
Photoelectron impact ionization, 29–30
 NO^+ sources and sinks, 35
 N^+ sources and sinks, 36, 36f
 $O^+(^2D)$ sources and sinks, 34f
 $O^+(^2P)$ sources and sinks, 34f
 $O^+(^4S)$ sources and sinks, 33, 33f
Photoionization (hv), 29
 NO^+ sources and sinks, 35
 N^+ sources and sinks, 36, 36f
 N^+2 sources and sinks, 34
 $O^+(^2D)$ sources and sinks, 33
 $O^+(^2P)$ sources and sinks, 34
 O^+2 sources and sinks, 35
 $O^+(^4S)$ sources and sinks, 33, 33f
Physical quantities, M-IT coupling and
 differences in, 208–213, 209t
  communication speed along magnetic field line, 208, 209
  electric current, 211
  electric field mapping along field line, 211
  heating rate, 212–213
  magnetic perturbation, 211
  Poynting vector, 211–212
  transient time, 209–210
  velocity mapping in weakly collisional regions, 211
Physics-based ionospheric model (PBMOD)
 low-latitude, for C/NOFS mission. See Low-latitude PBMOD
Physics-based models
 coupled IT, 147t, 149, 153
 data assimilation, 147, 147t, 149
 ionospheric, 147t, 148–149
Planar Langmuir probe (PLP), 260
 electron density, 146
Planetary and tidal wave variability
 WACCM-X simulation, in upper atmosphere. See Whole atmosphere community climate model with thermosphere extension (WACCM-X)
Planetary waves, 101, 102
Plasma and neutral density structures, 8–9, 10f, 11f. See also Ionosphere-thermosphere (I-T) system
Plasma blobs, 251, 252, 257
Plasma density, 111, 115
 ionospheric, irregularities. See Ionospheric irregularities
Plasma drift velocity, 135
Plasma G/P forces, 63–63, 64f, 65f
Plasma irregularities
 equatorial, magnetic signatures of, 264

Plasma Languin Probe (PLP), 286
Plasma outflow, 4
Plasmasphere, 8
Plasmasphere-ionosphere data assimilation model, 293–294, 294f, 296f. *See also* Ionospheric Data Assimilation Four Dimensional (IDA4D)
Plasma steepening, 233
PLP (planar Langmuir probe), 260
   electron density, 146
Polar cap potential, 103, 103f
Polar wind and auroral ion outflow, 4, 5f. *See also* Ionosphere-thermosphere (I-T) system
Poynting vector, in M-IT coupling, 211–212
PRE (prereversal enhancement), 125, 128, 262
Prediction efficiency (PE), model performance and, 146, 148t, 149, 150, 151t, 153, 154t, 156, 156t
Prereversal enhancement (PRE), 125, 128, 138–141, 262
PRIMO (Problems Related to Ionospheric Models and Observations) project, 262
PRIMO (Problems Related to Ionospheric Models and Observations) study, 4
Problems Related to Ionospheric Models and Observation at Equatorial Region (Equatorial-PRIMO), 134
Problems Related to Ionospheric Models and Observations (PRIMO) project, 262
Pseudo-inverse, of forward model, 314, 315
Pulsating geomagnetic storms, 102, 103f

Q

QBO (quasibiennial oscillation), 184
   modulation, DW1 and, 185, 187
QSPWs (quasistationary planetary waves), 182, 183, 185, 187, 190, 191
Quasibiennial oscillation (QBO), 184
   modulation, DW1 and, 185, 187
Quasi-hydrostatic balance, Earth's thermosphere and, 14–15
Quasistationary planetary waves (QSPWs), 182, 183, 185, 187, 190, 191
Quenching, electron, 33, 34

R

Radio communication, HF
   customer needs and requirements, 300–302, 301t, 302t
   during disasters/crisis, 301
   space weather and ionospheric variability on, 299–300, 301
Range-time-intensity (RTI) maps, 245
Rank-deficient inverse problem, 315, 316, 317
Ratio of (maximum (max)-minimum (min)), model performance and, 146, 147t, 148t, 149, 150, 152t, 153, 155t, 156

Ratio of maximum amplitude, model performance and, 146, 148t, 149, 150, 152t, 153, 155t, 156, 157t
Rayleigh-Taylor (R-T) instability
   F region irregularities and, 226–228, 230
Rayleigh-Taylor (R-T) linear growth rate
   in CXI, 109, 110f–111f
Reflection, effect
   in M-IT coupling, 208
RMSD (root-mean squared difference)
   of electron density analyses, 276
   of neutral mass density analysis, 277f, 278, 279f, 280f
   pre-assimilation and postassimilation analysis, 278, 279f
RMS error, model performance and, 146, 148t, 149, 150, 151t, 153, 154t, 156, 156t
Root-mean squared difference (RMSD)
   of electron density analyses, 276
   of neutral mass density analysis, 277f, 278, 279f, 280f
   pre-assimilation and postassimilation analysis, 278, 279f
Rossby number, in nonlinear system, 22
R-T (Rayleigh-Taylor) linear growth rate
   in CXI, 109, 110f–111f

S

SAID (subauroral ion drift)
   in topside ionosphere, 166, 166f, 167
SAMI3 (ISES-OE model is Another Model of Ionosphere)
   long-term simulations of ionosphere using. *See* Ionospheric simulations using SAMI3, long-term
1_SAMI3_HWM07 model, 147, 147t, 149, 156
1_SAMI3_HWM93 model, 147, 147t, 149, 150, 153, 156
Sami is Another Model of Ionosphere (SAMI2) model, 162, 163f
SAMI2 (Sami is Another Model of Ionosphere) model, 162, 163f
SAMI3 simulations, in ESF, 252
SAR (synthetic aperture radar), ESF irregularities, 218
Scherliess-Fejer (SF) empirical model
   in PBMOD simulations, 111, 113f, 115
SCINDA (Scintillation Network Decision Aid)
   ground station at CXI, 109, 110f–111f, 111
Scintillation Network Decision Aid (SCINDA)
   ground station at CXI, 109, 110f–111f, 111
SDO (Solar Dynamics Explorer), 306
Self-consistent models, 135t, 138t
Self-consistent treatment
   of thermosphere-ionosphere coupling, 276–277, 276f
Semidiurnal westward propagating wave 2 (SW2), 181, 183, 185, 187, 188f, 189f, 191, 192, 192f, 193, 193f, 194f, 195f
Semidiurnal zonal wavenumber two (SW2), tide component, 175, 176
Semiempirical models, for upper atmosphere, 41

INDEX 337

Semi-implicit algorithm
    numerical methods in ionospheric modeling, 51–52
        fully implicit algorithm vs., 52–53, 53f
17 March 1990
    solar cycle variation, model, 30–32, 32f
        modeled ion densities for, 32, 32f
        model neutral densities for, 30–32, 32f
        NO^+ sources and sinks, 35, 35f
        N^+ sources and sinks, 35–36, 36f
        N^+_2 sources and sinks, 34–35, 35f
        $O^+(^2D)$ sources and sinks, 33–34, 34f
        $O^+(^2P)$ sources and sinks, 34, 34f
        O^+_2 sources and sinks, 35, 35f
        $O^+(^4S)$ sources and sinks, 32–33, 33f
SF (Scherliess-Fejer) empirical model
    in PBMOD simulations, 111, 113f, 115
Shock formation, 102
Simulations
    of ionosphere using SAMI3, long-term. See Ionospheric simulations using SAMI3, long-term
    in ionospheric irregularities, 218
Sinks, of ions scale with solar activity
    N^+, 35–36, 36f
    N^+_2, 34–35, 35f
    NO^+, 35, 35f
    O^+_2, 35, 35f
    $O^+(^2D)$, 33–34, 34f
    $O^+(^2P)$, 34, 34f
    $O^+(^4S)$, 32–33, 33f
6 September 2005
    solar cycle variation, model
        NO^+ sources and sinks, 35, 35f
        N^+ sources and sinks, 35–36, 36f
        N^+_2 sources and sinks, 34–35, 35f
        $O^+(^2D)$ sources and sinks, 33–34, 34f
        $O^+(^2P)$ sources and sinks, 34, 34f
        O^+_2 sources and sinks, 35, 35f
        $O^+(^4S)$ sources and sinks, 32–33, 33f
Slow wave mode, 205, 206
SLT (solar local time)
    in F region irregularities, 226
Solar cycle 23
    declining phase of, 76, 77
    descending phase (2002-2008)
        modeling of, 128–129, 129f
Solar cycle changes, in photochemistry of ionosphere and thermosphere, 29–36
    5 September 2005, 30–32, 32f
    overview, 29–30, 30f
    results, 30–36
        NO^+ sources and sinks, 35, 35f

N^+ sources and sinks, 35–36, 36f
N^+_2 sources and sinks, 34–35, 35f
$O^+(^2D)$ sources and sinks, 33–34, 34f
$O^+(^2P)$ sources and sinks, 34, 34f
O^+_2 sources and sinks, 35, 35f
$O^+(^4S)$ sources and sinks, 32–33, 33f
        seventeenth17 March 1990, 30–32, 32f, 33, 33f
Solar Dynamics Explorer (SDO), 306
Solar flare, 299–300, 301–302
Solar forcing, thermospheric response, 265
Solar local time (SLT)
    in F region irregularities, 226
Solar maximum, 29, 30, 32, 108, 111
    atmospheric structure for, 39, 40f
Solar minimum, 29, 30, 32, 108, 111, 115
Solar terminator (ST)
    lower atmospheric forcing and, 267, 268f
Solar terrestrial relations observatory (STEREO) model, 120
Solstice, lower and upper atmosphere around, 190–193, 190f, 191f, 192f, 193f, 194f, 195f
Sounding of Atmosphere with Broadband Emission Radiometry (SABER), 171
Sound waves, 101, 102
Sources, of ions scale with solar activity
    N^+, 35–36, 36f
    N^+_2, 34–35, 35f
    NO^+, 35, 35f
    O^+_2, 35, 35f
    $O^+(^2D)$, 33–34, 34f
    $O^+(^2P)$, 34, 34f
    $O^+(^4S)$, 32–33, 33f
Sources of model discrepancies, 141–142
South Pole (SP), temperature at
    WACCM-X model variability on various time scales and, 185, 187f
SP (South Pole), temperature at
    WACCM-X model variability on various time scales and, 185, 187f
Space weather
    on ionosphere-thermosphere (I/T) system, 299–300
    types, 299–300
Space Weather Forecast Office, 302
Space Weather Prediction Center (SWPC), 299, 302, 306
    SWPT and, 302, 303–306
Space Weather Prediction Testbed (SWPT), 299, 302, 303–306, 306–307
    Forecasting Ionospheric Real-Time Scintillation Tool, 306
    GPS availability, 306
    for magnetospheric model, 304–305
    NASA CCMC, 304
    Ovation Prime model, 304, 305, 305f

Space Weather Prediction Testbed (SWPT) (continued)
  WAM model, 305–306
  WSA-Enlil model, 303, 304f, 305
Spectra, Doppler
  E region irregularities and, 221, 222f, 223, 223f
Spectral analysis, 220, 233
  E region irregularities and, 221, 222f, 223, 223f
SSW (stratosphere sudden warming) events, 182, 183, 190, 192
  January 2009, changes in, 177, 178, 178f
  lower atmospheric forcing and, 267–269, 269f
ST (solar terminator)
  lower atmospheric forcing and, 267, 268f
Steeping, 228, 233
STEREO (Solar Terrestrial Relations Observatory) model, 120
Stochastic differential equation formalism
  Farley-Buneman turbulence and, 224–225, 235
Storm-enhanced density (SED), 284
Stratosphere sudden warming (SSW) events, 182, 183, 190, 192
  January 2009, changes in, 177, 178, 178f
  lower atmospheric forcing and, 267–269, 269f
Subauroral ion drift (SAID)
  in topside ionosphere, 166, 166f, 167
Subauroral red (SAR) arcs, 5–7, 7f. See also Ionosphere-thermosphere (I-T) system
"Super fountain," 251, 254
Superrotation, defined, 266
Supersonic neutral winds, 102
SW2. See Semidiurnal westward propagating wave 2 (SW2)
SW2 (semidiurnal zonal wavenumber two), tide component, 175, 176
SWPC (Space Weather Prediction Center), 299, 302, 306
  SWPT and, 302, 303–306
SWPT. See Space Weather Prediction Testbed (SWPT)
Synthetic aperture radar (SAR), ESF irregularities, 218
Systematic quantitative assessment, of IT models. See Ionosphere/thermosphere (IT) models, systematic evaluation
System impacts, customers and
  GPS/GNSS system, 299, 300–302, 301t, 302t
  HF radio communication system. See Radio communication, HF

T
TAD. See Traveling atmospheric disturbance (TAD)
Taylor-Goldstein equation, 228
TEC (total electron content), 300
  IGS, SAMI3 comparisons
    WHI 2008 and, 121–125, 122f, 123f, 124f
  variation spectrum, 109
Temperatures
  in Earth's thermosphere, 18–19, 20f, 21f
  high, in upper atmosphere. See Thermosphere
ion
  from DMSP, 163, 164f
  and electron, at top of ionosphere, 162, 163f
  measurements from space, 41
  variations in low and middle latitudes of upper thermosphere during geomagnetic storms, causes, 42–46, 43f, 44f, 45f
Terdiurnal zonal wavenumber three (TW3), tide component, 175, 176
Terminator wave structures, lower atmospheric forcing and, 267, 268f
TGCM (thermosphere general circulation model), 19
Theoretical first principle models, for upper atmosphere, 41
Thermal expansion, Earth's thermosphere and, 19, 21, 22f
Thermodynamic equation, 41–42
Thermodynamics, of ESF plumes, 254–256, 255f, 256f. See also Equatorial spread F (ESF), plumes
Thermoelectric coefficient, 5
Thermoelectric heat flow in return current regions, 5, 6f. See also Ionosphere-thermosphere (I-T) system
Thermosphere, Ionosphere, Mesosphere Energetics and Dynamics (TIMED) measurement, 120
Thermosphere, neutral rain on, 9–10
Thermosphere general circulation model (TGCM), 19
Thermosphere-ionosphere-electrodynamics general circulation model (TIE-GCM), 19, 86
  NCAR, 58, 61, 62, 63, 64, 64f, 65
  community model of coupled thermosphere/ionosphere system, 73–80
    development history, 73–74
    energy input and output, 74–75
    equations, boundary conditions, and model output, 75, 76t
    future improvement and development plans, 79–80
    model description, 74–75
    overview, 73
    validation examples, 75, 76–79, 77f, 77t, 78f, 79f
  EnKF data assimilation. See Data assimilation, upper atmosphere
  model descriptions, 173–174
  NOGAPS-ALPHA as bottom boundary for. See Navy operational global atmospheric prediction system-advanced level physics high altitude (NOGAPS-ALPHA) model
  physics-based coupled IT model, 147, 147t, 149, 150, 153, 156
Thermosphere ionosphere general circulation model (TIGCM), 19
Thermosphere-ionosphere-mesosphere-electrodynamics general circulation model (TIME-GCM), 64, 65f, 73
  NCAR, 171, 172

Thermosphere-ionosphere nested grid (TING) model, 3
Thermosphere/ionosphere system
  NCAR TIE-GCM, community model of coupled. *See* Thermosphere-ionosphere-electrodynamics general circulation model (TIE-GCM)
Thermospheres
  defined, 39
  of earth, physical characteristics and modeling. *See* Earth's thermosphere, physical characteristics and modeling
  energetics and composition, 39–46
    boundary conditions, 41–42
    causes of temperature variations in low and middle latitudes of upper thermosphere during geomagnetic storms, 42–46, 43f, 44f, 45f
    continuity equation, 41–42
    data and models for, 40–41
    energy equation, 41–42
    neutral species in, 39–40
    overview, 39–40, 40f
    thermal and compositional structure, 40f
  lower, ion-neutral collisions in, 165–167, 166f, 167f
  observations of, low latitude CHAMP satellite, 264–269. *See also* Low latitude CHAMP satellite observations
  ionospheric forcing, thermospheric response. *See* Ionospheric forcing
  lower atmospheric forcing, thermospheric response. *See* Lower atmospheric forcing
  solar forcing, thermospheric response, 265
  photochemistry, solar cycle changes in, 29–36
  upper, ion-neutral collisions in, 164, 165, 165f
Thermospheric responses
  ionospheric forcing. *See* Ionospheric forcing
  lower atmospheric forcing. *See* Lower atmospheric forcing
  solar forcing, 265
"Thermospheric spoon," 18
Thermospheric winds, modeling, 62–63
Tides, 101, 102
  and planetary wave variability, WACCM-X simulation, in upper atmosphere. *See* Whole atmosphere community climate model with thermosphere extension (WACCM-X)
  TIE-GCM and NOGAPS-ALPHA, 172, 175–177, 176f, 177f, 178f
TIE-GCM. *See* Thermosphere-ionosphere-electrodynamics general circulation model (TIE-GCM)
TIGCM (Thermosphere ionosphere general circulation model), 19
Tikhonov regularization, 316
TIMED (Thermosphere, Ionosphere, Mesosphere Energetics and Dynamics) measurement, 120
Time evolution, of ESF plumes, 256–257, 256f

TIME-GCM (thermosphere-ionosphere-mesosphere-electrodynamics general circulation model), 64, 65f, 73
  NCAR, 171, 172
Time history, in auroral inverse problem, 315, 316f
Time scales, WACCM-X model variability on various, 184–187, 186f, 187f, 188f, 189f
Total electron content (TEC), 300
  IGS, SAMI3 comparisons
    WHI 2008 and, 121–125, 122f, 123f, 124f
  variation spectrum, 109
Total electron content (TEC), 285
Total electron content unit (TECU), 287
Total variation diminishing (TVD) condition, 228
TRANSCAR model, 312, 313f, 315, 319
Transient time, in M-IT coupling, 209–210
Transport processes in ionosphere, 134–137
Traveling atmospheric disturbance (TAD)
  global thermosphere-ionosphere model, 102
  overview of, 101–102
  simulation results, 104–105, 104f
  simulation setup, 102–104, 103f
Troposphere, 8
Truncated singular value decomposition (TSVD) inversion, 316, 317
TSVD (truncated singular value decomposition) inversion, 316, 317
Turbopause (homopause), 18
Turbulent mixing, in thermosphere, 39
TVD (total variation diminishing) condition, 228
TW3 (terdiurnal zonal wavenumber three), tide component, 175, 176

U

United States Total Electron Content (USTEC) model, 302, 303f
Upper atmosphere
  around solstice, 190–193, 190f, 191f, 192f, 193f, 194f, 195f
  change in dynamics in, 39
  data and models for, 40–41
  electron densities in, 40
  thermal and compositional structure of, 39, 40f
  WACCM-X simulation of tidal and planetary wave variability. *See* Whole Atmosphere Community Climate Model with thermosphere extension (WACCM-X)
Upward heat conduction, 43, 46
Upwelling, 13, 17, 18, 19, 21
USTEC (United States Total Electron Content) model, 302, 303f
USU-GAIM (Utah State University-Global Assimilative Ionospheric Model), 147, 147t, 149, 150

USU-IFM (Utah State University-Ionosphere Forecast Model), 147, 147t, 148, 149, 150, 153
Utah State University-Global Assimilative Ionospheric Model (USU-GAIM), 147, 147t, 149, 150
Utah State University-Ionosphere Forecast Model (USU-IFM), 147, 147t, 148, 149, 150, 153

V

Validation
  auroral inverse problem and, 317–319, 317f, 318f
  TIEGCM model, examples, 75, 76–79, 77f, 77t, 78f, 79f
Vector electric field instrument (VEFI), 108, 109, 111, 113f, 114f–115f, 115, 128
VEFI (vector electric field instrument), 108, 109, 111, 113f, 114f–115f, 115, 128
Velocity mapping, in M-IT coupling
  in weakly collisional regions, 211
Vertical drifts, modeling
  simulations of ionosphere, SAMI3 and, 125–128, 126f, 127f, 128f
Vertical drift velocities, diurnal variation of, 141f
Vertical ion drifts, predicting
  at Jicamarca, 153, 156, 156t, 157t
Vertical winds
  in Earth's upper atmosphere, 15
  GITM and nonhydrostatic process, 87–91, 88f, 89f, 90f
Vibrational excitation, of nitrogen, 29, 32, 33, 35, 36
Viscosity, in Earth's thermosphere, 16

W

WACCM (whole-atmosphere community climate model), 80
WAM (whole atmosphere model), 182, 190, 305–306
Wang-Sheely-Arge (WSA)-Enlil model, 303, 304f, 305
WATS (wind and temperature spectrometer), 39, 41, 42
Wave effects, lower atmosphere, 10
Wave heating, Farley-Buneman wave theory and, 218, 223, 224
Wave-4 structure, of EMA
  lower atmospheric forcing and, 267
Wave variability, tidal and planetary
  WACCM-X simulation, in upper atmosphere. *See* Whole Atmosphere Community Climate Model with thermosphere extension (WACCM-X)
Weather research and forecasting model, GITM and, 87
WHI 2008 (whole heliosphere interval 2008)
  SAMI3 comparisons with IGS TEC during, 121–125, 122f, 123f, 124f
Whole-atmosphere community climate model (WACCM), 80
Whole atmosphere community climate model with thermosphere extension (WACCM-X)
  overview, 183–184
  simulation of tidal and planetary wave variability in upper atmosphere, 181–195
    lower and upper atmosphere around solstice, 190–193, 190f, 191f, 192f, 193f, 194f, 195f
    overview, 181–183
    results, 184–193, 186f, 187f, 188f, 189f, 190f, 191f, 192f, 193f, 194f, 195f
    variability on various time scales, 184–187, 186f, 187f, 188f, 189f
Whole atmosphere model (WAM), 182, 190, 305–306
Whole heliosphere interval 2008 (WHI 2008)
  SAMI3 comparisons with IGS TEC during, 121–125, 122f, 123f, 124f
Wind and temperature spectrometer (WATS), 39, 41, 42
Wind dynamo, 58, 59, 64
WINDII (wind imaging interferometer), 41, 42
Wind imaging interferometer (WINDII), 41, 42
Wind-induced diffusion, 18
Winds
  at high latitude, 22, 24–25, 24f, 25f
  thermospheric, modeling, 62–63
World Meteorological Organization, 302
WSA (Wang-Sheely-Arge)-Enlil model, 303, 304f, 305

Z

Zonal and meridional direction, momentum equations for, 16
Zonal Pedersen conductance, 61